Tree Physiology

Volume 7

Series editors
Frederick C. Meinzer, Corvallis, USA
Ülo Niinemets, Tartu, Estonia

More information about this series at http://www.springer.com/series/6644

Eustaquio Gil-Pelegrín · José Javier Peguero-Pina
Domingo Sancho-Knapik
Editors

Oaks Physiological Ecology. Exploring the Functional Diversity of Genus *Quercus* L.

Editors
Eustaquio Gil-Pelegrín
Unit of Forest Resources
Agrifood Research and Technology Centre
Aragón, Zaragoza
Spain

Domingo Sancho-Knapik
Unit of Forest Resources
Agrifood Research and Technology Centre
Aragón, Zaragoza
Spain

José Javier Peguero-Pina
Unit of Forest Resources
Agrifood Research and Technology Centre
Aragón, Zaragoza
Spain

ISSN 1568-2544
Tree Physiology
ISBN 978-3-319-88713-5   ISBN 978-3-319-69099-5 (eBook)
https://doi.org/10.1007/978-3-319-69099-5

© Springer International Publishing AG 2017
Softcover re-print of the Hardcover 1st edition 2017
This work is subject to copyright. All rights are reserved by the Publisher, whether the whole or part of the material is concerned, specifically the rights of translation, reprinting, reuse of illustrations, recitation, broadcasting, reproduction on microfilms or in any other physical way, and transmission or information storage and retrieval, electronic adaptation, computer software, or by similar or dissimilar methodology now known or hereafter developed.
The use of general descriptive names, registered names, trademarks, service marks, etc. in this publication does not imply, even in the absence of a specific statement, that such names are exempt from the relevant protective laws and regulations and therefore free for general use.
The publisher, the authors and the editors are safe to assume that the advice and information in this book are believed to be true and accurate at the date of publication. Neither the publisher nor the authors or the editors give a warranty, express or implied, with respect to the material contained herein or for any errors or omissions that may have been made. The publisher remains neutral with regard to jurisdictional claims in published maps and institutional affiliations.

Printed on acid-free paper

This Springer imprint is published by Springer Nature
The registered company is Springer International Publishing AG
The registered company address is: Gewerbestrasse 11, 6330 Cham, Switzerland

Contents

1   **Oaks and People: A Long Journey Together** 1
    Eustaquio Gil-Pelegrín, José Javier Peguero-Pina
    and Domingo Sancho-Knapik

2   **An Updated Infrageneric Classification of the Oaks:
Review of Previous Taxonomic Schemes and Synthesis
of Evolutionary Patterns** 13
    Thomas Denk, Guido W. Grimm, Paul S. Manos,
    Min Deng and Andrew L. Hipp

3   **The Fossil History of *Quercus*** 39
    Eduardo Barrón, Anna Averyanova, Zlatko Kvaček,
    Arata Momohara, Kathleen B. Pigg, Svetlana Popova,
    José María Postigo-Mijarra, Bruce H. Tiffney,
    Torsten Utescher and Zhe Kun Zhou

4   **Physiological Evidence from Common Garden Experiments for
Local Adaptation and Adaptive Plasticity to Climate in American
Live Oaks (*Quercus* Section *Virentes*): Implications for
Conservation Under Global Change** 107
    Jeannine Cavender-Bares and José Alberto Ramírez-Valiente

5   **Oaks Under Mediterranean-Type Climates:
Functional Response to Summer Aridity** 137
    Eustaquio Gil-Pelegrín, Miguel Ángel Saz, Jose María Cuadrat,
    José Javier Peguero-Pina and Domingo Sancho-Knapik

6   **Coexistence of Deciduous and Evergreen Oak Species
in Mediterranean Environments: Costs Associated
with the Leaf and Root Traits of Both Habits** 195
    Alfonso Escudero, Sonia Mediavilla, Manuel Olmo,
    Rafael Villar and José Merino

7	The Role of Hybridization on the Adaptive Potential of Mediterranean Sclerophyllous Oaks: The Case of the *Quercus ilex* x *Q. suber* Complex 239
	Unai López de Heredia, Francisco María Vázquez and Álvaro Soto
8	The Anatomy and Functioning of the Xylem in Oaks 261
	Elisabeth M. R. Robert, Maurizio Mencuccini and Jordi Martínez-Vilalta
9	The Role of Mesophyll Conductance in Oak Photosynthesis: Among- and Within-Species Variability 303
	José Javier Peguero-Pina, Ismael Aranda, Francisco Javier Cano, Jeroni Galmés, Eustaquio Gil-Pelegrín, Ülo Niinemets, Domingo Sancho-Knapik and Jaume Flexas
10	Carbon Losses from Respiration and Emission of Volatile Organic Compounds—The Overlooked Side of Tree Carbon Budgets 327
	Roberto L. Salomón, Jesús Rodríguez-Calcerrada and Michael Staudt
11	Photoprotective Mechanisms in the Genus *Quercus* in Response to Winter Cold and Summer Drought 361
	José Ignacio García-Plazaola, Antonio Hernández, Beatriz Fernández-Marín, Raquel Esteban, José Javier Peguero-Pina, Amy Verhoeven and Jeannine Cavender-Bares
12	Growth and Growth-Related Traits for a Range of *Quercus* Species Grown as Seedlings Under Controlled Conditions and for Adult Plants from the Field 393
	Rafael Villar, Paloma Ruiz-Benito, Enrique G. de la Riva, Hendrik Poorter, Johannes H. C. Cornelissen and José Luis Quero
13	Drought-Induced Oak Decline—Factors Involved, Physiological Dysfunctions, and Potential Attenuation by Forestry Practices 419
	Jesús Rodríguez-Calcerrada, Domingo Sancho-Knapik, Nicolas K. Martin-StPaul, Jean-Marc Limousin, Nathan G. McDowell and Eustaquio Gil-Pelegrín
14	Physiological Keys for Natural and Artificial Regeneration of Oaks 453
	Jesús Pemán, Esteban Chirino, Josep María Espelta, Douglass Frederick Jacobs, Paula Martín-Gómez, Rafael Navarro-Cerrillo, Juan A. Oliet, Alberto Vilagrosa, Pedro Villar-Salvador and Eustaquio Gil-Pelegrín

15 Competition Drives Oak Species Distribution and Functioning in Europe: Implications Under Global Change 513
Jaime Madrigal-González, Paloma Ruiz-Benito, Sophia Ratcliffe, Andreas Rigling, Christian Wirth, Niklaus E. Zimmermann, Roman Zweifel and Miguel A. Zavala

Index 539

Chapter 1
Oaks and People: A Long Journey Together

Eustaquio Gil-Pelegrín, José Javier Peguero-Pina and Domingo Sancho-Knapik

Abstract Genus *Quercus* L. has been closely associated to humans throughout the history, with empirical evidences of such relationship before the appearance of *Homo sapiens* strictly speaking. Since then, mankind has obtained different basic resources from oaks, from acorns as food, charcoal for metal melting or wood as key material for different works. Such relation has been especially strong in some areas where oaks are considered as "tree of life" or "people's species". Moreover, the interest of scientists in the study of this genus has provided a lot of new discovers in different areas of the socalled plant sciences. Genus *Quercus*, comprising more than 400 species found throughout the Northern Hemisphere in a lot of contrasted habitats, have been the case study in many papers about taxonomy, palaeobotany, plant physiology or basic and applied ecology. This fact is summarized in this chapter, serving as a preface to this book.

Mankind has established a close relationship with oaks, which are deeply rooted in the folklore, mythology or even religion of many human cultures (Ciesla 2002; Chassé 2016; Out 2017). In fact, Goren-Inbar et al. (2000) recovered *Quercus* sp. rests among the "edible species" found in the Acheulean (Middle Pleistocene, 780,000 years ago) archaeological site of Gesher Benot Ya'aqov (Israel). In a later study, Goren-Inbar et al. (2002) went beyond and suggested that hominins population that occupied this site during this period consumed acorns from *Q. calliprinos* and *Q. ithaburensis*. So, this evidence dates such relationship between oaks and "human" transcends our existence as *Homo sapiens*.

The practise of acorn eating by human hunter-gatherer cultures has been well documented by archaeologists since the Palaeolithic (Cacho 1986; Chassé 2016). Many other archaeological evidences seem to indicate the importance of acorns eating for the survival of pre-Neolithic cultures of the eastern Mediterranean Basin

---

E. Gil-Pelegrín (✉) · J. J. Peguero-Pina · D. Sancho-Knapik
Unidad de Recursos Forestales, Centro de Investigación y Tecnología
Agroalimentaria de Aragón, Gobierno de Aragón, Avda, Montañana
930, 50059 Saragossa, Spain
e-mail: egilp@cita-aragon.es

© Springer International Publishing AG 2017
E. Gil-Pelegrín et al. (eds.), *Oaks Physiological Ecology. Exploring the Functional Diversity of Genus* Quercus *L.*, Tree Physiology 7,
https://doi.org/10.1007/978-3-319-69099-5_1

(Natufian), living under a Mediterranean-type climate with associated oak woodlands (e.g. McCorriston 1994; Olszewski 2004). This dietary resource remained very important for sedentary cultures, as evidenced by many archaeological sites corresponding to Pre-Roman cultures of the Iberian Peninsula (Pereira-Sieso and García-Gómez 2002). Moreover, the consumption of acorns by people during famine moments, associated to poor harvests, is reported until the 18th century (García-Gómez et al. 2002).

A well-known example of ancestral dependence on the resources offered by oaks is given also by the native cultures of southeastern USA (Fagan 2004). Anderson (2007) used the expression "bread of life" for acorns, due to their paramount importance for the indigenous diet of this region. However, this author went further when defined such relationship, since he also proposed the expression "tree of life" for oaks in their relation with these cultures due to the many other benefits they obtained from these trees. In the same direction, Long et al. (2016) proposed to consider *Q. kelloggii* as a "cultural keystone" species for the indigenous cultures of California and Oregon, in the sense that Garibaldi and Turner (2004) gave to this concept. These indigenous managed traditionally oak forests (Anderson 2005), even with the use of fire to control the dominance of conifers (Ciesla 2002; Anderson 2007), yielding a mutual benefit for humans and oaks, as part of their "traditional ecological knowledge" (Long et al. 2016).

The use nowadays of different resources offered by *Quercus* species to the inhabitants in the "Middle Hills" of Central Himalaya (Shrestra et al. 2013) is another example of the complex dependence between oak woodlands and humans. The many resources that people obtains from these forest communities are so extent that Singh and Singh (1986) used the term "people's species" for oaks living in this region, with a very special mention to the banj oak (*Q. leucotrichophora*). As is indicated by these authors, the oak woodlands in Central Himalaya offer different benefits for humans, such as forage for cattle, firewood, or compost from leaves to manure the crop fields. Moreover, at a landscape scale, the existence of oaks is clearly related to the amount and quality of spring water, besides the critical influence on soil conservation in a territory of high slopes. However, this complex relationship is fragile and the sustainability is dependent on the exploitation intensity. Shrestra et al. (2013) analysed the coexistence of oaks and humans in Nepal, with especial reference to the situation of *Q. semecarpifolia* stands. They indicated that, while the presence of humans in these "Middle Hills" of the Central Himalaya goes back several millennia, a severe increase in the Nepal population (due to growth and migration) during the last century has dealt to a sobreexplotation of these habitats, with negative effects on forest structure, regeneration and species diversity associated to these ecosystems (Christensen and Heilmann-Clausen 2009). The existence of a high disturbance due to human pressure on these forests has been also pointed out in other areas of the Central Himalaya, severely reducing the viability and the area of banj oak forests (Singh et al. 2014).

In the Iberian Peninsula, a particular exploitation regime of the mediterranean woodlands is the development of cleared oak forests or savannah-like woodlands of *Quercus ilex* subsp. *rotundifolia* or *Q. suber*, constituting the so-called "dehesas" in

Spain or "montados" in Portugal (Rodriguez-Estevez et al. 2012). This agrosilvopastoral system, as considered in Olea and San Miguel-Ayanz (2006), allows humans to obtain different resources from this anthropogenic habitat since the Neolithic (López-Sáez et al. 2007). The maintenance of a traditional extensive pig farming based on an autochthonous porcine breed foraging acorns during the "montanera" is one of the most important benefits obtained from these woodlands (Rodríguez-Estévez et al. 2009). The consumption of acorns has a positive influence on the lipid profile of the carcasses (Cava et al. 1997), which confers a high quality and value-added to those pork products obtained and, hence, to the "dehesa" (Gaspar et al. 2007). However, this agrosilvopastoral system is threatened by severe processes of oak decline, severely affecting the oak stands since the beginning of the 1980s (Gil-Pelegrín et al. 2008). This particular oak decline process, frequently called "seca", has a proven biological component, with the fungal species *Phytophthora cinnamomi* Rands. as the recognised agent since the very first studies (Brasier et al. 1993; Tuset et al. 1996). To date the concern is maintained, as death of holm and cork oaks persists until now (Avila et al. 2016).

Obviously, oaks have been a source of wood for millennia, and evidences of that are found in different archaeological sites (e.g. De'Athe et al. 2013; Out 2017; Ruiz-Alonso et al. 2017). In some areas, as northern Spain, different studies indicate a continuous use of deciduous *Quercus* species as firewood during several millennia, from the early Neolithic to the early Bronze Age (Ruiz-Alonso et al. 2017). Other archaeological sites allow interpreting the wood of *Quercus* species in a burial context, giving a symbolic value to this genus in many areas of Europe during the Neolithic and subsequent ages (Out 2017). The study of Iron-Age sites in the United Kingdom suggested that humans established a management of oak stands to obtain a regular source of woodland for metalworking (De'Athe et al. 2013). This management is based on the high capability of *Quercus* species for resprouting after cutting (Giovannini et al. 1992), with the production of many small stems in short rotation cycles, which length depends on the species and environmental conditions (Corcuera et al. 2006). This practise, known as coppicing, has been regularly used over time (Barberó et al. 1990), to the extent that this coppice stands are the most common structure in oak woodlands of southern Europe (Serrada et al. 1992; Amorini et al. 1996; Montes et al. 2004). As stated by Cañellas et al. (1994), the traditional practice of coppicing has been reduced since the middle of the 20th century in many areas of Spain, due to important sociological changes that implied a drastic reduction in the demand of firewood or charcoal. This fact leads to the existence of many overaged coppice stands of different *Quercus* species with a reduced growth (Cañellas et al. 1996). Moreover, Corcuera et al. (2006) found that this reduced growth has a negative effect on the hydraulic conductivity of the stems together with a higher vulnerability to water stress during the summer drought period. A progressive transformation of these oak coppices into high forest by thinning may have positive effects on tree growth and soil water availability (Fedorová et al. 2016) or even a greater tolerance to droughts (Rodríguez-Calcerrada et al. 2011).

Oak wood have been also a basic resource through the history as key material for naval construction (e.g. Giachi et al. 2017). It has been documented a massive demand for timber as raw material for the respective fleets during the Modern Age (mainly during the 15th and 16th centuries) in Portugal (Reboredo and Pais 2014) and Spain (Wing 2012), which gave a strategic importance to the oak woodlands there. Furthermore, the use of oak barrel in winemaking is a crucial practice to ensure a high quality of the final product, as oak wood adds different compounds that contribute to improve the wine flavour and colour (e.g. Chira and Teissedre 2013). In this sense, the so-called "American oak" (*Quercus alba*), the "French oak" (*Q. robur* and *Q. petraea*) and *Q. pyrenaica* (Jordão et al. 2006) are the most common species for barrel cooperage, with a differential influence on wine characteristics (see Chira and Teissedre 2015). Additionally, a bottle of good wine is corked with a very particular product obtained also from the bark of a very concrete *Quercus* species. Effectively, the properly called cork oak (*Q. suber*) has traditionally provided the key material for that purpose, indicating that oaks play an outstanding role in the process of wine-making.

Another important contribution of oaks to the economy in many areas of southern Europe (Italy, France and Spain) and Australia are the production of edible fungi of very high added value and internationally related to the haute cuisine (Reyna and García-Barreda 2014) We are particularly referring to the hypogeous fruiting body of different species of the genus *Tuber*, the so-called truffles, in mycorrhizal symbiosis with different *Quercus* species, with *Q. ilex* as the most common host. As stated by Aumeeruddy-Thomas et al. (2012), gathering these mushrooms is a recognised and documented activity in France since the Middle Age. Concerning *Tuber melanosporum*, one of the "quintessential truffle" (Reyna and García-Barreda 2014), it can be artificially produced in planted truffle orchards since the 19th century (Olivier et al. 1996) in many areas under Mediterranean-type climates. In Spain, truffle production has been a new opportunity in areas suffering from severe problems of depopulation (e.g. the province of Teruel), which adds a social value to truffle and to the *Quercus* species, without which this cultivation would not be possible.

Thus, genus *Quercus* has been culturally and economically linked to humans since millennia. But, otherwise, oaks have been food for thought in plant sciences as this book reflects. *Quercus* L. (*Fagaceae*) has an outstanding role in the vegetation of the Northern Hemisphere and can be considered the most diverse Northern Temperate tree genus. It comprises ca. 400 tree and shrub species distributed among contrasting phytoclimates, from temperate and subtropical forests to mediterranean evergreen woodlands (Manos et al. 1999; Kremer et al. 2012). More specifically, oaks live in a great variety of environments, from subalpine forests (e.g. the Alborz Mountains in northern Iran) to semiarid forests (Afghanistan, the Mediterranean region and western North America) and riparian and swamp forests in different parts of the world (for example wetlands of Florida, Alabama, or the riparian forests of the Danube river), even touching the Tropics in SE Asia and Central South America. Exploring this complexity is an opportunity and also a challenge for naturalists.

This wide geographical range and phytoclimatic diversity has a direct expression in terms of taxonomic complexity. A considerable number of studies about phylogeny of oaks have been carried out in order to get an overview about this large modern diversity of oaks (see Chap. 2 for a comprehensive review). Up to now, all of the successive infrageneric classifications of *Quercus* have recognized the same major groups (see Denk and Grimm 2010 and references therein). In Chap. 2 of this book, Denk et al. propose a revised classification of *Quercus* based on pollen morphology that includes two subgenera, Quercus and Cerris. On the one hand, subgenus Quercus comprises 5 sections: section Ponticae, section Virentes, section Protobalanus, section Quercus, and section Lobatae. On the other hand, subgenus Cerris comprises 3 sections: section Cerris, section Ilex, and section Cyclobalanopsis. Subgenus Cerris is confined to the Old World, while subgenus Quercus is distributed throughout Northern Hemisphere.

Genus *Quercus* is an ancient lineage of *Fagaceae* whose first records are pollen grains of the late Paleocene age, as described in Chap. 3 by Barrón et al. Since the Eocene, oaks diversified and spread throughout the Northern Hemisphere. As today, they inhabited very different environments, both temperate and cold-temperate regions (suggesting an unequivocal Arctotertiary origin) or tropical and subtropical realms (indicating a Palaeotropical origin). However, one particular group of fossil *Quercus* species, those belonging to section Cyclobalanopsis (sensu Denk et al. in Chap. 2 of this book and formerly subgenus Cyclobalanopsis), always shows a tropical-subtropical distribution, being common on the Paleotropical Tethyan shores of North America and Eurasia during great part of the Cenozoic. The analysis of the information given by the palaeontological records in contrast with the present distribution and ecology of closely related species may serve for a good reconstruction of the palaeoecology of the Northern Hemisphere since the Cenozoic.

The ecological importance and functional diversity of genus *Quercus* has been also addressed in many ecophysiological studies concerning the response of different oak species to several abiotic stress factors. Chapter 4, by Cavender-Bares and Ramírez-Valiente, explores the adaptative response of an interesting study case in a particular lineage of American oaks, namely the live oaks (*Quercus* section *Virentes* Nixon), with species such as *Quercus virginiana*, *Q. geminata*, *Q. fusiformis* or *Q. oleoides*. Several of these live oaks span the Tropical-Temperate divide (Koehler et al. 2012), which implies the existence of different conditions of water availability besides different temperature registers during winter, including freezing values. These authors offer new insights about the existence of local adaptative responses within species in terms of withstanding both drought and freezing. Moreover, they also report interspecific differences in this set of closely phylogenetically related species with similar physiognomic features.

The existence of a drought period during summer, due to the combination of a temperature monthly maximum and a precipitation minimum is the most distinctive characteristic of the Mediterranean-type climates. Chapter 5, by Gil-Pelegrín et al., delves into the ecophysiological features of the many *Quercus* species inhabiting areas under this particular climate, as compared with oaks from areas with

Temperate, wet Tropical and dry Tropical climates. In this chapter, two very different mediterranean oaks are recognised and compared: (i) evergreen and sclerophyllous species (e.g. *Quercus ilex*) and (ii) malacophyllous or semi-sclerophyllous winter deciduous oaks (e.g. *Quercus faginea*). The coexistence of these two leaf habits in mediterranean oaks seems to be the consequence of different paleogeographical origins, with winter-deciduous (from an Arctotertiary geoflora) and evergreen (from the Palaeotropical geoflora) co-occurring in a complex patchwork. In most mediterranean areas, the balance has tipped in favor of evergreen species (such as *Quercus ilex* subsp. *rotundifolia*) through the ancestral alteration of the soil by humans, as these species with a longer leaf life span seem to better response to the splitted vegetative period induced by summer drought and winter cold. The leaf life span in *Quercus* species under mediterranean-type climates is the central aim of Chap. 6, where Escudero et al. state that the most striking difference among *Quercus* species inhabiting these areas is the dichotomy represented by the deciduous and evergreen habits, which major implications in terms of leaf anatomy, carbon gain, cost construction and maintenance.

Quercus L. has been proposed as an outstanding genus to understand how hybridation and introgression influence the evolution of plants. Thus, in Chap. 7, López de Heredia et al. revised several evidences of ancient introgressions between two mediterranean evergreen oaks, namely *Q. ilex* and *Q. suber*, and update estimations of present hybridation rates. These authors concluded that these processes seem to be a very relevant mechanism explaining some distribution and ecological patterns of these species, especially during glaciations.

Concerning the hydraulic conductivity of the xylem, genus *Quercus* has been the object of many studies since the very first steps in the study of this key topic in tree functioning, both from anatomical and biophysical or physiological points of view. Some of the seminal ideas proposed by Zimmermann (1983), such as the architecture of the water-conduction pathway of a tree or the segmentation hypothesis were early explored in species of *Quercus* (e.g. Cochard and Tyree 1990; Lo Gullo and Salleo 1993). In Chap. 8, Robert et al. explore the bibliography concerning the xylem anatomy in oak species, focussing on the overall variation in the xylem structural and functional features, with special incidence on the different performance of the ring-porous (with few wide vessels) and diffuse-porous wood (with numerous narrow vessels) within this genus.

Oaks have also been subject of study in several processes related to CO_2 assimilation (mesophyll conductance and photosynthesis in Chap. 9) and loss (respiration and volatile organic compounds emission in Chap. 10). Moreover, it should be noted that oaks are main targets in seminal papers concerning the implementation of new methodologies and techniques for the estimation of mesophyll conductance. Thus, the method based on the simultaneous measurement of gas exchange and chlorophyll fluorescence parameters was firstly used in *Q. ilex* (Di Marco et al. 1990) and *Q. rubra* (Harley et al. 1992). This method was firstly compared with the stable carbon isotope fractionation technique (Evans et al. 1986) for both species by Loreto et al. (1992). More recently, the validation of mesophyll conductance modelled on the basis of anatomical characteristics has been carried out in several deciduous and evergreen oak species (Tomás et al. 2013; Peguero-Pina et al. 2016, 2017). These studies in different *Quercus* species have reinforced the prevailing role of leaf anatomy in mesophyll conductance and net CO_2 assimilation, as stated by Peguero-Pina et al. in Chap. 9. Oaks also constitute an excellent taxonomic group to study the variability in carbon losses from respiration and the emission of volatile organic compounds among different plant functional types and environmental conditions (see Chap. 10 by Salomón et al.). These authors conclude that, besides carbon assimilation, it would be necessary a comprehensive understanding of carbon loss in oaks to accurately assess carbon cycling in current and future scenarios of climate change. In other cases, some physiological mechanisms have also been firstly showed in *Quercus* species, such as some photo protective mechanisms stated in Chap. 11 by García-Plazaola et al. Thus, a xanthophyll cycle involving the so-called "lutein epoxide cycle" was described for the first time in non-parasitic woody plants by García-Plazaola et al. (2002) in eight oak species. In line with this, Peguero-Pina et al. (2009) found that *Q. ilex* and *Q. coccifera* showed a drought-mediated chronic photoinhibition and an overnight retention of de-epoxidated forms of xantophyll cycle (i.e. anteraxanthin and zeaxanthin).

Villar et al., in Chap. 12, show how long *Quercus* provides an interesting study case for analysing the traits involved in growth processes, due to the high variability of functional traits there found, following the postulates of the leaf economics spectrum. In this chapter, it is evidenced that seedlings of *Quercus* species were characterized, among other traits, by a low relative growth rate (RGR) and a high root mass ratio (RMR), while leaf dry mass per area ratio (LMA) explains most of the differences in RGR among oak species. Moreover, the proportion of biomass in leaves and roots decreased with tree size, by contrast increasing the biomass in stems. According to this, bigger trees grow more slowly. In spite of this, the authors conclude that seedling RGR under controlled conditions is positively related with that of adult trees in the field. Such conclusion offers a way for comparative studies at a wide scale.

Chapter 13, by Rodríguez-Calcerrada et al., evidences how much oaks have been threatened in many areas of the world, trough massive oak decline processes which have affected species belonging to clearly separate taxonomic groups, with different leaf habit and/or physiological performance. This fact has promoted the concern of foresters and scientific community about the different oak decline processes reported in Central Europe and Northeastern USA since the 18th and 19th centuries (Millers et al. 1989; Thomas 2008). Since then and up to now, more episodes of oak decline were reported over a wide range of sites in most forested places of the northern hemisphere, including deciduous and evergreen species, as described in Chap. 13 by Rodríguez-Calcerrada et al. These authors note that most oak decline episodes have been observed after extreme climatic events (severe droughts, waterlogging or after consecutive events of winter freezing), but they have also been associated to different pathogens and site conditions. In most cases, the interaction of at least two stress agents, where one of them is often an extreme climatic event, has triggered important outbreaks of decline.

In order to preserve the oak woodlands from vanishing, if these massive decline processes become more frequent, the improvement of new seedling recruitment may be a challenge to be solved. In genus *Quercus*, the inherent biological limitations to the natural regeneration of oaks, and especially in a degraded landscape by human intervention, makes their natural expansion quite difficult. For this reason, it is necessary the implementation of techniques to facilitate this process in an artificial way (as is reviewed in Chap. 14 by Pemán et al.), from cultivation methods in the nursery phase to the final installation in the field.

This book also highlights the overall importance of oaks from an ecological viewpoint (see Chap. 15 by Madrigal-González et al.). Madrigal-González et al. state that oak forests are highly valued ecosystems from the viewpoint of human economical and cultural interests, and their distribution and physiognomy has been greatly modulated by humans since the Neolithic (Barbero et al. 1990). Madrigal-González et al. in Chap. 15 also conclude that, not only climatic fluctuations, but also agricultural intensification and, more recently, widespread agricultural land abandonment associated with human migration from rural to urban areas are recognized as major forces leading to recent oak encroachment, expansion or decline in different European regions.

Acknowledgements We thank Elena Martí Beltrán, Rut Cuevas Calvo, Oscar Mendoza Herrer, José Sánchez Mesones and Francisco Garín García for their disinterested and kind collaboration in different tasks during the realization of this book.

References

Amorini E, Biocca M, Manetti MC, Motta E (1996) A dendroecological study in a declining oak coppice stand. Ann Sci For 53:731–742

Anderson EN (2005) Tending the wild: native american knowledge and the management of california's natural resources. University of California Press, Berkeley

Anderson MK (2007) Indigenous uses, management, and restoration of Oaks of the far western United States. Technical Note No 2. NRCS, National Plant Data Center, Washington DC

Aumeeruddy-Thomas Y, Therville C, Lemarchand C, Lauriac A, Richard F (2012) Resilience of sweet chestnut and truffle holm-oak rural forests in Languedoc-Roussillon, France: roles of social ecological legacies, domestication, and innovations. Ecol Soc 17:12

Avila JM, Gallardo A, Ibáñez B, Gómez-Aparicio L (2016) *Quercus suber* dieback alters soil respiration and nutrient availability in Mediterranean forests. J Ecol 104:1441–1452

Barberó M, Bonin G, Loisel R, Quézel P (1990) Changes and disturbances of forest ecosystems caused by human activities in the western part of the Mediterranean basin. Vegetatio 87:151–173

Brasier CM, Robredo F, Ferraz JFP (1993) Evidence for *Phytophthora cinnamomi* involvement in Iberian oak decline. Plant Pathol 42:140–145

Cacho C (1986) Nuevos datos sobre la transición del magdaleniense al epipaleolítico en el Pais Valenciano: el Tossal de la Roca. Boletín del Museo Arqueológico Nacional IV(2):117–129

Cañellas I, Montero G, San Miguel A, Montoto JL, Bachiller A (1994) Transformation of rebollo oak coppice (*Quercus pyrenaica* Willd.) into open woodlands by thinning at different intensities. Inv Agric Sist Rec For 3:71–78

Cañellas I, Montero G, Bachiller A (1996) Transformation of quejigo oak (*Quercus pyrenaica* Lam.) coppice forest into high forest by thinning. Ann Inst Sper Selvic 27:143–147

Cava R, Ruiz J, López-Bote C, Martín L, García C, Ventanas J, Antequera T (1997) Influence of finishing diet on fatty acid profiles of intramuscular lipids, triglycerides and phospholipids in muscles of the Iberian pig. Meat Sci 45:263–270

Chassé B (2016) Eating acorns: what story do the distant, far and near past tell us, and why. Int Oaks 27:107–135

Chira K, Teissedre PL (2013) Extraction of oak volatiles and ellagitannins compounds and sensory profile of wine aged with French winewoods subjected to different toasting methods: Behaviour during storage. Food Chem 40:168–177

Chira K, Teissedre PL (2015) Chemical and sensory evaluation of wine matured in oak barrel: effect of oak species involved and toasting process. Eur Food Res Technol 240:533–547

Christensen M, Heilmann-Clausen J (2009) Forest biodiversity gradients and the human impact in Annapurna Conservation Area, Nepal. Biodivers Conserv 18:2205–2221

Ciesla WM (2002) Non-wood forest products from temperate broad-leaved trees. FAO Technical Papers 15, FAO, Rome

Cochard H, Tyree MT (1990) Xylem dysfunction in *Quercus*: vessel sizes, tyloses, cavitation and seasonal changes in embolism. Tree Physiol 6(4):393–407

Corcuera L, Camarero JJ, Sisó S, Gil-Pelegrín E (2006) Radial-growth and wood-anatomical changes in overaged *Quercus pyrenaica* coppice stands: functional responses in a new Mediterranean landscape. Trees 20:91–98

De'Athe R, Barnett C, Cooke NN, Grimm JM, Jones GP, Leivers M, McKinley JI, Pelling R, Schuster J, Stevens CJ, Foster W, James SE (2013) Early iron age metalworking and Iron Age/early Romano-British settlement evidence along the Barton Stacey to Lockerley gas pipeline. Proc Hampshire Field Club Archaeol Soc 68:29–63

Denk T, Grimm GW (2010) The oaks of western Eurasia: traditional classifications and evidence from two nuclear markers. Taxon 59:351–366

Di Marco G, Manes F, Tricoli D, Vitale E (1990) Fluorescence parameters measured concurrently with net photosynthesis to investigate chloroplastic CO_2 concentration in leaves of *Quercus ilex* L. J Plant Physiol 136:538–543

Evans J, Sharkey T, Berry J, Farquhar G (1986) Carbon isotope discrimination measured concurrently with gas exchange to investigate CO_2 diffusion in leaves of higher plants. Funct Plant Biol 13:281–292

Fagan BM (2004) Before California: an archaeologist looks at our earliest inhabitants. Roman Altamira, Lanham MD

Fedorová B, Kadavý J, Adamec Z, Kneifl M, Knott R (2016) Response of diameter and height increment to thinning in oak-hornbeam coppice in the southeastern part of the Czech Republic. J For Sci 62:229–235

García-Gómez E, Pereira-Sieso J, Ruiz-Taboada A (2002) Aportaciones al uso de la bellota como recurso alimenticio por las comunidades campesinas. Cuad Soc Esp Cien For 14:65–70

García-Plazaola JI, Errasti E, Hernández A, Becerril JM (2002) Occurrence and operation of the lutein epoxide cycle in *Quercus* species. Funct Plant Biol 29:1075–1080

Garibaldi A, Turner N (2004) Cultural keystone species: implications for ecological conservation and restoration. Ecol Soc 9:1

Gaspar P, Mesías FJ, Escribano M, Rodriguez de Ledesma A, Pulido F (2007) Economic and management characterization of dehesa farms: implications for their sustainability. Agrofor Syst 71:151–162

Giachi G, Capretti C, Lazzeri S, Sozzi L, Paci S, Lippi MM, Macchioni N (2017) Identification of wood from Roman ships found in the docking site of Pisa (Italy). J Cult Herit 23:176–184

Gil-Pelegrín E, Peguero-Pina JJ, Camarero JJ, Fernández-Cancio A, Navarro-Cerrillo R (2008) Drought and forest decline in the Iberian Peninsula: a simple explanation for a complex phenomenom? In: Sánchez JM (ed) Droughts: causes, effects and predictions. Nova Science Publishers Inc, New York, pp 27–68

Giovannini G, Perulli D, Piussi P, Salbitano F (1992) Ecology of vegetative regeneration after coppicing in macchia stand in central Italy. Vegetatio 99–100:331–343

Goren-Inbar N, Feibel CS, Verosub KL, Melamed Y, Kislev ME, Tchernov E, Saragusti I (2000) Pleistocene Milestones on the Out-of-Africa Corridor at Gesher Benot Ya'aqov, Israel. Science 289(5481):944–947

Goren-Inbar N, Sharon G, Melamed Y, Kislev M (2002) Nuts, nut cracking, and pitted stones at Gesher Benot Ya'aqov, Israel. PNAS 99(4):2455–2460

Harley PC, Loreto F, Di Marco G, Sharkey TD (1992) Theoretical considerations when estimating the mesophyll conductance to CO_2 flux by analysis of the response of photosynthesis to CO_2. Plant Physiol 98:1429–1436

Jordão AM, Ricardo-da-Silva JM, Laureano O, Adams A, Demyttenaere J, Verhé R, De Kimpe N (2006) Volatile composition analysis by solid-phase microextraction applied to oak wood used in cooperage (*Quercus pyrenaica* and *Quercus petraea*): effect of botanical species and toasting process. J Wood Sci 52:514–521

Koehler K, Center A, Cavender-Bares J (2012) Evidence for a freezing tolerance-growth rate trade-off in the live oaks (*Quercus* Series *Virentes*) across the tropical-temperate divide. New Phytol 193:730–744

Kremer A, Abbott AG, Carlson JE, Manos PS, Plomion C, Sisco P, Staton ME, Ueno S, Vendramin GG (2012) Genomics of Fagaceae. Tree Genet Genomes 8:583–610

Lo Gullo MA, Salleo S (1993) Different vulnerabilities of *Quercus ilex* L. to freeze- and summer drought-induced xylem embolism: an ecological interpretation. Plant Cell Environ 16(5):511–519

Long JW, Goode RW, Gutteriez RJ, Lackey JJ, Anderson MK (2016) Managing California black oak for tribal ecocultural restoration. J Forest. doi:10.5849/jof.16-033

López-Sáez JA, López-García P, López-Merino L, Cerrillo-Cuenca E, Gonzalez-Cordero A, Prada-gallardo A (2007) Origen prehistórico de la dehesa en Extremadura: una perspectiva paleoambiental. Revista de Estudios Extremeños 63:493–510

Loreto F, Harley PC, Di Marco G, Sharkey TD (1992) Estimation of mesophyll conductance to CO_2 flux by three different methods. Plant Physiol 98:1437–1443

Manos PS, Doyle JJ, Nixon KC (1999) Phylogeny, biogeography, and processes of molecular differentiation in *Quercus* subgenus *Quercus* (Fagaceae). Mol Phylogenet Evol 12:333–349

McCorriston J (1994) Acorn eating and agricultural origins: California ethnographies as analogies for that Ancient Near East. Antiquity 68:97–107

Millers I, Shriner DS, Rizzo D (1989) History of hardwood decline in the Eastern United States. Gen. Tech. Rep. NE-126. Broomall, PA: U. S. Department of Agriculture, Forest Service, Northeastern Forest Experiment Station, 75 p

Montes F, Cañellas I, Del Río M, Calama R, Montero G (2004) The effects of thinning on the structural diversity of coppice forests. Ann For Sci 61:771–779

Olea L, San Miguel-Ayanz A (2006) The Spanish dehesa. A traditional Mediterranean silvopastoral system linking production and nature conservation. Opening Paper. 21st General Meeting of the European Grassland Federation, Badajoz, Spain

Olivier JM, Savignac JC, Sourzat P (1996) Truffe et trufficulture. Fanlac, Perigueux

Olszewski DL (2004) Plant food subsistence issues and scientific inquiry in the early Natufian. In: Delage C (ed) The Last Hunter-Gatherer Societies in the Near East. John and Erica Hedges (BAR International Series 1320), Oxford, pp 189–209

Out WA (2017) Wood usage at Dutch Neolithic wetland sites. Quatern Int 436:64–82

Peguero-Pina JJ, Sancho-Knapik D, Morales F, Flexas J, Gil-Pelegrín E (2009) Differential photosynthetic performance and photoprotection mechanisms of three Mediterranean evergreen oaks under severe drought stress. Funct Plant Biol 36:453–462

Peguero-Pina JJ, Sisó S, Sancho-Knapik D, Díaz-Espejo A, Flexas J, Galmés J, Gil-Pelegrín E (2016) Leaf morphological and physiological adaptations of a deciduous oak (*Quercus faginea* Lam.) to the Mediterranean climate: a comparison with a closely related temperate species (*Quercus robur* L.). Tree Physiol 36:287–299

Peguero-Pina JJ, Sisó S, Flexas J, Galmés J, García-Nogales A, Niinemets Ü, Sancho-Knapik D, Saz MÁ, Gil-Pelegrín E (2017) Cell-level anatomical characteristics explain high mesophyll conductance and photosynthetic capacity in sclerophyllous Mediterranean oaks. New Phytol 214:585–596

Pereira-Sieso J, García-Gómez E (2002) Bellotas, el alimento de la Edad de Oro. Arqueoweb 4. Departamento de Prehistoria UCM

Reboredo F, Pais J (2014) Evolution of forest cover in Portugal: A review of the 12th–20th centuries. J For Res 25:249–256

Reyna S, García-Barreda S (2014) Black truffle cultivation: a global reality. For Syst 23:317–328

Rodríguez-Calcerrada J, Pérez-Ramos IM, Ourcival JM, Limousin JM, Joffre R, Rambal S (2011) Is selective thinning an adequate practice for adapting Quercus ilex coppices to climate change? Ann For Sci 68:575–585

Rodriguez-Estevez V, Sanchez-Rodriguez M, Arce C, García AR, Perea JM, Gomez-Castro AG (2012) Consumption of acorns by finishing iberian pigs and their function in the conservation of the Dehesa Agroecosystem. In: Kaonga M (ed) Agroforestry for biodiversity and ecosystem services—science and practice. InTech, Rijeka, pp 1–22

Rodríguez-Estévez V, García A, Peña F, Gómez AG (2009) Foraging of Iberian fattening pigs grazing natural pasture in the Dehesa. Livest Sci 120:135–143

Ruiz-Alonso M, Zapata L, Pérez-Díaz S, López-Sáez JA, Fernández-Eraso J (2017) Selection of firewood in northern Iberia: Archaeobotanical data from three archaeological sites. Quatern Int 431:61–72

Serrada R, Allué M, San Miguel A (1992) The coppice system in Spain. Current situation, state of art and major areas to be investigated. Ann Inst Sper Selvic 23:266–275

Shrestra KB, Måren IE, Arneberg E, Sah JP, Vétaas OR (2013) Effect of anthropogenenic disturbance on plant species diversity in oak forests in Nepal, Central Himalaya. Int J Biodivers Sci Ecosyst Serv Manag 9:21–29

Singh SP, Singh JS (1986) Structure and function of the Central Himalayan oak forests. Proc Indian Acad Sci (Plant Sci) 96:159–189

Singh N, Tamta K, Tewari A, Ram J (2014) Studies on vegetational analysis and regeneration status of *Pinus roxburghii*, Roxb. and *Quercus leucotrichophora* forests of Nainital Forest Division. Glob J Sci Front Res C Biol Sci, 14(3) (Version 1.0)

Thomas FM (2008) Recent advances in cause-effect research on oak decline in Europe. CAB Rev Pers Agric Vet Sci Nutr Nat Resour 3(037):1–12

Tomás M, Flexas J, Copolovici L, Galmés J, Hallik L, Medrano H, Ribas-Carbó M, Tosens T, Vislap V, Niinemets Ü (2013) Importance of leaf anatomy in determining mesophyll diffusion conductance to CO_2 across species: quantitative limitations and scaling up by models. J Exp Bot 64:2269–2281

Tuset JJ, Hinarejos C, Mira JL, Cobos JM (1996) Implicación de *Phytophthora cinnamomi* Rands en la enfermedad de la «seca» de encinas y alcornoques. Bol San Veg Plagas 22:491–499

Wing JT (2012) Keeping Spain afloat: state forestry and imperial defense in the sixteenth century. Environ Hist 17:116–145

Zimmermann MH (1983) Xylem structure and the ascent of sap. Springer-Verlag, Berlin

Chapter 2
An Updated Infrageneric Classification of the Oaks: Review of Previous Taxonomic Schemes and Synthesis of Evolutionary Patterns

Thomas Denk, Guido W. Grimm, Paul S. Manos, Min Deng and Andrew L. Hipp

Abstract In this chapter, we review major classification schemes proposed for oaks by John Claudius Loudon, Anders Sandøe Ørsted, William Trelease, Otto Karl Anton Schwarz, Aimée Antoinette Camus, Yuri Leonárdovich Menitsky, and Kevin C. Nixon. Classifications of oaks (Fig. 2.1) have thus far been based entirely on morphological characters. They differed profoundly from each other because each taxonomist gave a different weight to distinguishing characters; often characters that are homoplastic in oaks. With the advent of molecular phylogenetics our view has considerably changed. One of the most profound changes has been the realisation that the traditional split between the East Asian subtropical to tropical subgenus *Cyclobalanopsis* and the subgenus *Quercus* that includes all other oaks is artificial. The traditional concept has been replaced by that of two major clades, each comprising three infrageneric groups: a Palearctic-Indomalayan clade including Group Ilex (Ilex oaks), Group Cerris (Cerris oaks) and Group Cyclobalanopsis (cycle-cup oaks), and a predominantly Nearctic clade including Group Protobalanus (intermediate or golden cup oaks), Group Lobatae (red oaks) and Group Quercus (white oaks,

---

T. Denk (✉)
Swedish Museum of Natural History, Box 50007, 10405 Stockholm, Sweden
e-mail: thomas.denk@nrm.se

G. W. Grimm
45100 Orléans, France

P. S. Manos
Duke University, Durham, NC 27708, USA

M. Deng
Shanghai Chenshan Plant Science Research Center, Chinese Academy of Sciences, 201602 Shanghai, China

A. L. Hipp
The Morton Arboretum, Lisle, IL 60532-1293, USA

A. L. Hipp
The Field Museum, Chicago, IL 60605, USA

Fig. 2.1 Classification schemes for *Quercus* from Loudon to Nixon. Colour coding denotes the actual systematic affiliation of species included in each taxon: of the 'Old World' or 'mid-latitude clade' section *Cyclobalanopsis* (cycle-cup oaks, yellow), section *Cerris* (Cerris oaks; orange), and section *Ilex* (Ilex oaks; green); and of the 'New World' or 'high-latitude clade' section *Quercus* (white oaks s.str.; blue), sections *Virentes* (cyan) and *Ponticae* (dark blue), section *Protobalanus* (intermediate oaks; purple), and section *Lobatae* (red oaks; red). Colour gradients are proportional, i.e. reflect the proportion of species with different systematic affiliation included in each taxon. Names in bold were treated as genera. Note: Menitsky (1984) and Trelease (1924) only treated the Eurasian and American oaks, respectively, and provided classifications in (nearly) full agreement with current phylogenies

with most species in America and some 30 species in Eurasia). In addition, recent phylogenetic studies identified two distinct clades within a wider group of white oaks: the Virentes oaks of North America and a clade with two disjunct endemic species in western Eurasia and western North America, *Quercus pontica* and *Q. sadleriana*. The main morphological feature characterising these phylogenetic lineages is pollen morphology, a character overlooked in traditional classifications. This realisation, along with the now available (molecular-)phylogenetic framework, opens new avenues for biogeographic, ecological and evolutionary studies and a re-appraisal of the fossil record. We provide an overview about recent advances in these fields and outline how the results of these studies contribute to the establishment of a unifying systematic scheme of oaks. Ultimately, we propose an updated classification of *Quercus* recognising two subgenera with eight sections. This classification considers morphological traits, molecular-phylogenetic relationships, and the evolutionary history of one of the most important temperate woody plant genera.

2.1 History of Classifications of Oaks

In his original work, Carl von Linné listed 14 species of oaks from Europe and North America: the white oaks *Q. alba*, *Q. æsculus* (= *Q. petraea* (Matt.) Liebl.), *Q. robur*, and *Q. prinus* (status unresolved); the red oaks *Q. rubra*, *Q. nigra*, and *Q. phellos*; the Cerris oaks *Q. cerris*, *Q. ægilops* (= *Q. macrolepis* Kotschy), *Q. suber*; and the Ilex oaks *Q. ilex*, *Q. coccifera*, *Q. gramuntia* (= *Q. ilex*), and *Q. smilax* (= *Q. ilex*) (Linné 1753). This number had increased to 150 species when Loudon (1838, 1839) provided the first infrageneric classification of oaks recognising ten sections based on reproductive and leaf characters. Eight of Loudon's sections (*Albæ*, *Prinus*, *Robur*; *Nigræ*, *Phellos*, *Rubræ*; *Cerris*; *Ilex*) were based on species described by Linné (Fig. 2.1). New additions were the (fully) evergreen south-eastern North American "Live Oaks", sect. *Virentes*; and the "Woolly-leaved Oaks", sect. *Lanatæ*, of Nepal (including an Ilex oak and a species that was later recognised as a cycle-cup oak). Loudon's classification is remarkable in one aspect: he established the fundamental subdivision of European oaks (his sections *Cerris*, *Ilex*, and *Robur*). This subdivision, although modified, occurs in nearly all later classifications and corresponds to clades in most recent molecular-phylogenetic

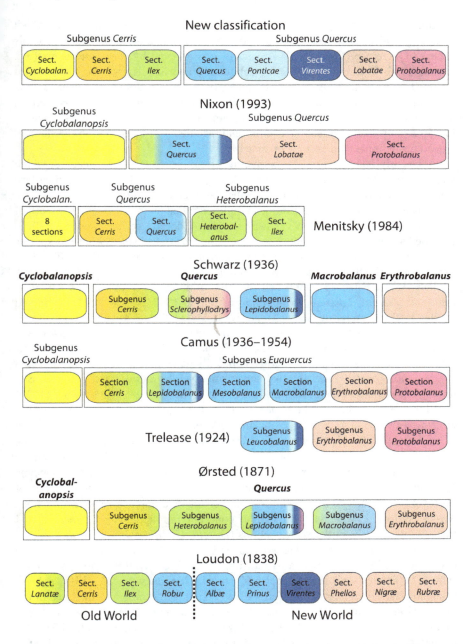

trees (cerroid, ilicoid, and roburoid oaks; cf. Denk and Grimm 2010; A. Hipp and co-workers, work in progress).

Ørsted (1871) can be credited for recognising an important Asian group of oaks hardly known at the time of Loudon and originally associated with *Cyclobalanus*

(= *Lithocarpus*): the cycle-cup oaks of subtropical and tropical East Asia, which Ørsted considered distinct from *Quercus* as genus *Cyclobalanopsis*, within his subtribe Quercinae (Fig. 2.1). This concept was adopted by later researchers (e.g. Camus 1936–1938; Nixon 1993; as subgenera) and is still used for the Flora of China (Huang et al. 1999; Flora of China 2016). Within the second genus of the Quercinae, *Quercus*, Ørsted recognised five subgenera with a total of 16 sections and about 184 species. His work is the first to treat oaks in a global context; Loudon, and later Camus, Trelease, and Menitsky, treated the Nearctic and Palearctic-Indomalayan taxa independently.

In the early 20th century, two competing classification concepts emerged, which were henceforth used by researchers (partly until today). The central/eastern European tradition followed in principle the classification system of Schwarz (1936), whereas the western/southern European tradition relied on the monographic work of Camus (1936–1938, 1938–1939, 1952–1954). A decade earlier, Trelease (1924) provided a comprehensive treatment of the American oaks listing about 371 species (nearly half of them new) in 138 series and three subgenera/sections (Fig. 2.1): *Leucobalanus* (white oaks), *Erythrobalanus* (red oaks), and *Protobalanus* (intermediate oaks). Thus, he established the tripartition of the genus in the Americas (sections *Quercus*, *Lobatae*, *Protobalanus*; (Jensen 1997; Manos 1997; Nixon and Muller 1997). Camus and Schwarz (partly) followed Trelease regarding the classification of the American oaks, but disagreed with respect to the oaks of Eurasia and North Africa, specifically on how to classify the American oaks in relation to their Eurasian counterparts. Camus followed Ørsted's general scheme, but recognised a single genus *Quercus* with the two subgenera *Cyclobalanopsis* and *Quercus*. She downgraded Ørsted's subgenera in *Quercus* to sections (Fig. 2.1). Schwarz (1936) also followed in principle the concepts of Ørsted, but raised Ørsted's categories, erecting a two tribe system (Cyclobalanopsideae, Querceae) with two genera each (*Cyclobalanopsis* + *Erythrobalanus*, *Macrobalanus* + *Quercus*). A novelty in the system of Schwarz was the subgenus *Sclerophyllodrys* (Fig. 2.1), in which he accommodated many sclerophyllous oaks of Eurasia, Trelease's subgenus *Protobalanus* (including an Asian series *Spathulatae*), and six evergreen series of Trelease's subgenus *Erythrobalanus*. Another major difference relative to Camus was that Schwarz adopted Ørsted's global concept by grouping North American and Eurasian white oaks in the same sections (*Dascia, Gallifera, Prinus, Roburoides*).

The most recent monographic work towards a new classification of oaks was the one of Menitsky (1984, translated into English in 2005) dealing with Asian oaks (Fig. 2.1). Except for a single species (*Q. suber*), Menitsky placed all Ilex oaks in subgenus *Heterobalanus*, while Cerris oaks (except for *Q. suber*) formed one of the two sections in subgenus *Quercus* (the other section included the white oaks). Menitsky's account is the only morphology-based system that correctly identified the natural groups of Eurasian oaks confirmed later by palynological and molecular data. In the same way, Trelease's sections of American oaks also have been confirmed as natural groups.

The latest and currently most widely used (e.g. Govaerts and Frodin 1998; see also www.wikipedia.org and www.internationaloaksociety.org) classification is by

Nixon (1993), published as a review. Nixon adopted the concept of Camus but merged her sections *Cerris*, which comprised Cerris and Ilex oaks, and *Euquercus*, comprising the remaining Ilex oaks and the white oaks, into a single section *Quercus*. According to this latest modification of Ørsted's more than 150 years old scheme, the genus *Quercus* is divided into two subgenera, the cycle-cup oaks (*Cyclobalanopsis*) and all remaining oaks (*Quercus*). Subgenus *Quercus* includes two natural sections, one comprising the red oaks (sect. *Lobatae*) and one comprising the intermediate oaks (sect. *Protobalanus*), and a heterogeneous, artificial, northern hemispheric section *Quercus* including all white oaks, Cerris and Ilex oaks (Fig. 2.1).

2.2 Change in Criteria for Classification

There are two major causes for the differences in the traditional, morphology-based classifications of oaks: (1) the weighing of morphological characters, (2) the geographic regions considered. Convergent morphological evolution is a common phenomenon in the genus *Quercus* and the Fagaceae in general (Oh and Manos 2008; Kremer et al. 2012). For instance, Loudon's (1838) descriptions for the distantly related sections *Ilex* (Eurasian *Q. ilex* and relatives) and *Virentes* (North American *Q. virens* Ait. [= *Q. virginiana* Miller], a white oak relative) are essentially identical. For similar reasons, Ørsted (1871) included a section *Ilex* in his subgenus *Lepidobalanus* (white oaks in a broad sense), while expanding this section to include evergreen North American white oaks (the sect./subsect. *Virentes* of Loudon, Trelease, Camus, etc.) On the other hand, the Himalayan Ilex oak *Q. lanata* was included in Ørsted's section *Prinus* of North American white oaks. The assumption that leaf texture can be used to assign species to higher taxonomic groups on a global scale supports Schwarz' largely artificial subgenera (and genera to some degree). Using the descriptions by Trelease, the Eurasian Ilex oaks would still fall in his subgenus *Protobalanus*, and the same is true for the descriptions in Nixon (1993) and the Flora of North America (Manos 1997).

Nixon's concept of a section *Quercus* including all white, Cerris and Ilex oaks primarily relies on the basal position of aborted ovules in these groups. Much earlier, de Candolle (1862b) noted this feature as being variable in different oak species, and Camus (1936–1938, p. 40f) emphasised that this trait is stable not only within a species, but also characterises groups of species (but see general descriptions in Menitsky 1984). Nixon also adopted Camus' concept of subgenus *Cyclobalanopsis* (aborted ovules always apical; but see general description provided by Huang et al. 1999). Apical abortive ovules are also found in most but not all subsections of sect. *Erythrobalanus* (the red oaks) and in the castanoid genera. Therefore, Nixon suggested that basal abortive ovules are a synapomorphy of his sect. *Quercus*. Subsequent work has shown that the position of aborted ovules in the mature seeds of *Quercus* is the result of different developmental processes and less

Table 2.1 Different contributions of placenta and funiculus to the position of aborted ovules in mature seeds of *Quercus*

Section	*Quercus Ponticae Virentes*	*Lobatae*	*Protobalanus*	*Cyclobalanopsis*	*Cerris*	*Ilex*
Position of aborted ovules	Basal	Apical Type I	Apical, basal, or lateral	Apical Types I, III	Apical, basal or lateral Type II	Basal or lateral Types II, III
Placenta	Sessile	Elongated	?	Elongated	Sessile (compressed)	Sessile or elongated
Funiculus	Sessile	Sessile	?	Sessile or elongated	Sessile or elongated	Sessile or elongated

Type I: apical/lateral aborted ovules by elongated placenta, **Type II**: by elongated funiculus, **Type III**: both elongated placenta and funiculus. Other Fagaceae (*Castanea, Castanopsis, Lithocarpus, Trigonobalanus*) have Types I & III aborted ovules. All other Fagaceae have apical aborted ovules. Information compiled from Borgardt and Pigg (1999), Borgardt and Nixon (2003), Deng (2007), Deng et al. (2008), and Min Deng, unpublished data

stable than originally assumed (Borgardt and Pigg 1999; Borgardt and Nixon 2003; Deng et al. 2008) (Table 2.1).

The only two classification schemes that recognised the same groups later recovered in molecular studies are those by Trelease (1924) and Menitsky (1984). Notably, these monographs were restricted to American and Eurasian oaks, respectively. Therefore, they did not run the risk of creating artificial groups including morphologically similar but unrelated Old World and New World species.

2.3 Changing from Morphology to Molecules

The first molecular phylogeny of *Quercus* including a comprehensive oak sample is the one of Manos et al. (2001) based on sequences of the nuclear ITS region and plastid RFLP data. While Manos et al.'s molecular phylogeny included only a limited sample of Old World species, it challenged the traditional views of Ørsted until Nixon. Instead, the intermediate and white oaks grouped with the red oaks, forming the 'New World Clade', but not with the Cerris and Ilex oaks. The latter formed an 'Old World Clade' that later would be shown to include the cycle-cup oaks (Manos et al. 2008). While the red oaks and cycle-cup oaks were resolved in well-supported and distinct clades within their respective subtrees, the situation appeared more complex for Camus' section *Cerris* (including a few Ilex oaks) and the white oaks (Manos et al. 2001). The lack of unambiguous support may be one reason, why morphologists and oak systematists did not readily implement the new evidence (e.g. Borgardt and Nixon 2003; le Hardÿ de Beaulieu and Lamant 2010; see also www.internationaloaksociety.org). The other reason is probably that the two new clades lacked compelling, unifying morphological traits.

Plastid gene regions commonly used in plant phylogenetics turned out to be less useful for inferring infrageneric and inter- to intraspecific relationships in oaks. This is mainly because the plastid genealogy is largely decoupled from taxonomy and substantially affected by geography (e.g. Neophytou et al. 2010, 2011; Simeone et al. 2016; Pham et al. 2017). Using genus- to family-level plastid data sets, even when combined with nuclear data, oaks are consistently recognised as a diphyletic group. This is best illustrated in Manos et al. (2008): one moderately supported main clade comprises the 'New World Clade' of oaks and *Notholithocarpus*, a monotypic Fagaceae genus of western North America; the other major clade comprises the Eurasian Fagaceae *Castanea* and *Castanopsis*, and the 'Old World Clade' of *Quercus*. The phenomenon is also seen in broadly sampled plastid data sets and can produce highly artificial molecular phylogenies (e.g. Xiang et al. 2014; Xing et al. 2014) as discussed in Grímsson et al. (2016). Nevertheless, all currently available plastid data reject the traditional subdivision into two subgenera *Cyclobalanopsis* and *Quercus*: the overall signal (e.g. Manos et al. 2008) is in line with the 'New World/Old World Clade' concept introduced by Oh and Manos (2008).

In view of the problems encountered with plastid sequence data, oak molecular phylogenetics concentrated on nuclear-encoded sequence regions. Nine years after the study by Manos et al. (1999), the first ITS phylogeny was confirmed and supplemented by data from a single-copy nuclear gene region, the *Crabs Claw* (*CRC*) gene (Oh and Manos 2008). Denk and Grimm (2010) provided an updated Fagaceae ITS tree including more than 900 individual sequences of oaks (including c. 600 newly generated for western Eurasian species taking into account substantial intra-individual variation). Their data on the 5S intergenic spacer (over 900 sequences), a multicopy nuclear rDNA gene region not linked with the ITS region, supported three groups of western Eurasian oaks as originally conceived by Menitsky (1984). Hubert et al. (2014) compiled new data from six single-copy nuclear gene regions and combined the new data with ITS consensus sequences (based on Denk and Grimm 2010) and *CRC* sequence data (Oh and Manos 2008). Most recently, Hipp et al. (2015) showed a tree based on a large, nuclear reduced representation next-generation sequencing (RADseq) data set. All these data sets and analyses support the recognition of two, reciprocally monophyletic groups of oaks (Fig. 2.2) that can be formalised as two subgenera with eight phylogenetic lineages (Hubert et al. 2014; Hipp et al. 2015), accepted here as sections that match the morphological groups originally perceived by Trelease (1924) and Menitsky (1984):

- **Subgenus *Quercus***, the 'New World clade' (Manos et al. 2001) or 'high-latitude clade' (Grímsson et al. 2015; Simeone et al. 2016), including
  - the North American intermediate oaks, **section *Protobalanus*** (= Trelease's subgenus of the same name);
  - the western Eurasian-western North American disjunct **section *Ponticae***;
  - the American "southern live oaks", **section *Virentes***;

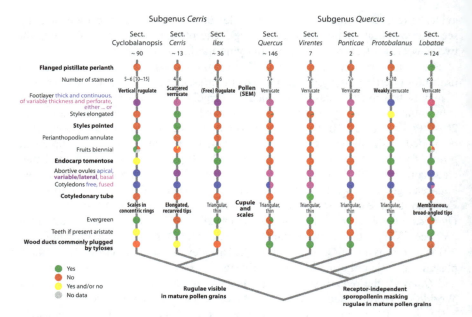

Fig. 2.2 Revised sectional classification of oaks and diagnostic characters of lineages. The basic phylogenetic relationships of the six infrageneric groups of oaks are shown, formalised here as sections in two monophyletic subgenera, subgenus *Cerris* ('Old World' or 'mid-latitude clade') and subgenus *Quercus* ('New World' or 'high-latitude clade'). **Section-specific traits in bold**; subgenus-diagnostic traits indicated at the respective branches of the schematic phylogenetic tree (Hubert et al. 2014; Hipp et al. 2015). Most traits are shared by more than one section of oaks including non-sister-lineages (normal font); they evolved convergently or are potentially plesiomorphic traits. Some are variable within a section as indicated by (semi-)proportional pie charts. Nonetheless, each section can be diagnosed by unique, unambiguous character suites. Note: 'yes' (green) and 'no' (red) refers to whether the mentioned trait is observed or not in members of the section, but should not be generally viewed as derived or ancestral

- all white oaks from North America (= Trelease's subgenus *Leucobalanus*) and Eurasia (= Menitsky's section *Quercus*), **section *Quercus***; and
- the North American red oaks, **section *Lobatae*** (= Trelease's subgenus *Erythrobalanus*).

- **Subgenus *Cerris***, the exclusively Eurasian 'Old World clade' (or 'mid-latitude clade'), including

  - the cycle-cup oaks of East Asia (including Malesia), **section *Cyclobalanopsis*** (former [sub]genus *Cyclobalanopsis* of Ørsted, Camus, Schwarz, Menitsky, and Nixon);
  - the Ilex oaks, **section *Ilex*** (= Menitsky's subgenus *Heterobalanus* minus *Q. suber*); and
  - the Cerris oaks, **section *Cerris*** (= Menitsky's section *Cerris* plus *Q. suber*).

2.4 Revised Subgeneric and Sectional Classification of Oaks

The following information for diagnostic morphological characters for the recognised groups of oaks is based mostly on information provided in Trelease (1924), Camus (1936–1938, 1938–1939, 1952–1954), Schwarz (1936, 1937), Menitsky (1984), le Hardÿ de Beaulieu and Lamant (2010), and the Floras of China (Huang et al. 1999) and North America (Flora of North America Editorial Commitee 1997). Information on pollen morphology is from Rowley et al. (1979), Solomon (1983a, b), Rowley and Claugher (1991), Rowley (1996), Rowley and Gabarayeva (2004), Denk and Grimm (2009), Makino et al. (2009), and Denk and Tekleva (2014). Updated information on the position of aborted ovules and the relative contributions of placenta and funiculus to it is from Borgardt and Nixon (2003), Deng et al. (2008), and Min Deng (unpublished data).

If no reference is provided, most monographers (Trelease 1924; Schwarz 1936; Camus 1936–1938; Schwarz 1937; Camus 1938–1939, 1952–1954; Menitsky 1984) agreed on a particular character. Trelease (1924) emphasised the importance of wood characters for delimitation of major groups of American oaks. According to Trelease (1924), Menitsky (1984), and Akkemik and Yaman (2012) the type of wood porosity and presence or absence of tyloses plugging vessels of early-wood are clade-specific to some degree.

Group-specific traits are highlighted by italics (see also Fig. 2.2).

2.4.1 *Genus* Quercus

1753, Sp. Pl., 1: 994.

Lectotype: *Quercus robur* L. (selected by Britton and Brown, Ill. Fl. N. U.S. ed. 2. 1: 616, 7 Jun 1913; confirmed by Green, in Sprague, Nom. Prop. Brit. Bot.: 189, Aug 1929)

Trees 20–30(–55) m high, or **shrubs**; monoecious, evergreen or deciduous; **propagating** from seeds (saplings) or, occasionally, vegetative propagation (ramets); **bark** smooth or deeply furrowed or scaly or papery, corky in some species; **wood** ring-porous or (semi) diffuse-porous, tyloses common in vessels of early-wood or rarely present; **terminal buds** spherical to ovoid, terete or angled, all scales imbricate; **leaves** spirally arranged, stipules deciduous and inconspicuous or sometimes retained until the end of the vegetative period; **lamina** chartaceous or coriaceous, lobed or unlobed, margin entire, dentate or dentate with bristle-like extensions; **primary venation** pinnate; **secondary venation** eucamptodromous, brochidodromous, craspedodromous, semicraspedodromous, or mixed; intersecondary veins present or absent; **inflorescences** unisexual in axils of leaves or bud scales, usually clustered at base of new growth; **staminate inflorescences** lax, racemose to spicate; **pistillate inflorescence** usually stiff, a simple spike, with terminal cupule and sometimes one to several sessile, lateral cupules; **staminate flowers** subsessile, in dichasial clusters of 1–3(–7) (section *Cyclobalanopsis*) or solitary; subtending bracts persistent or caducous, commonly longer than the perianth, sepals connate to varying degrees forming a shallowly or deeply lobed perianth, stamens (2–)6(–15), anthers short or long, apically notched or apiculate to mucronate, pistillodes reduced and replaced by a tuft of silky hairs; **pollen** monad, medium-sized or small (size categories according Hesse et al. 2009), 3-colp(or)ate, shape prolate, outline in polar view trilobate or rounded, in equatorial view elliptic to oval, tectate, columellate; pollen ornamentation (micro) rugulate, (micro) rugulate-perforate, or (micro) verrucate, (micro)verrucate-perforate; foot layer discontinuous or continuous, of even or uneven thickness; **pistillate flower** one per cupule, with 1–2 subtending bracts, sepals connate, (3–)6(–9) lobed, either situated directly on the tip of the ovary or on the perianthopodium (stylopodium); carpels and styles 3–6, occasionally with staminodes, styles with a broad stigmatic surface on adaxial suture of style (less prominent in section *Cyclobalanopsis*); **ovules** pendent, anatropous or semi-anatropous; **position of aborted ovules** apical, basal, or lateral depending on whether or not the placenta and/or funiculus are secondarily elongated; **fruit** a one-seeded nut (acorn) with a proximal scar, fruit maturation annual or biennial, nut one per cup, round in cross-section, not winged, cotyledons free or fused; **endocarp** glabrous or tomentose; **cup** covering at least base of nut,

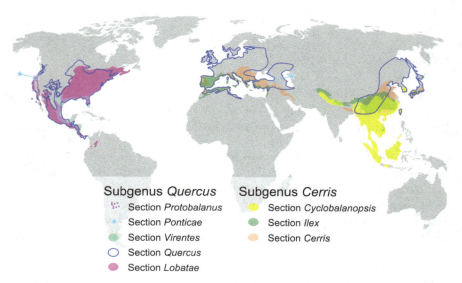

Fig. 2.3 Geographic distribution of the eight sections of *Quercus*. Distribution data from Browicz and Zieliński (1982), Menitsky (1984), Costa Tenorio et al. (2001), Deng (2007), Fang et al. (2009), and Manos (2016)

with lamellate rings or scaly; **scales** imbricate and flattened or tuberculate, not or weakly to markedly reflexed; **chromosome number X** = 12. Around 400 species mostly in the Northern Hemisphere (Fig. 2.3).

2.4.2 Subgenus Quercus

Receptor-independent sporopollenin masking rugulae in mature pollen grains (Rowley and Claugher 1991; Rowley 1996).

2.4.2.1 Section *Protobalanus* (Intermediate Oaks)

Quercus section *Protobalanus* (Trelease) Schwarz, Notizbl. Bot. Gart. Berlin-Dahlem, 13/116: 21 (1936)
  Quercus subgenus *Protobalanus* Trelease, in Standley, Contr. US Natl. Herb. 23:176 (1922).—*Quercus* section *Protobalanus* (Trelease) Camus, Les Chênes, 1: 157 (1938).—*Quercus* section *Protobalanus* (Trelease) Schwarz, Notizbl. Bot. Gart. Berlin-Dahlem, 13/116: 21 (1936) p.p.
  Type: *Quercus chrysolepis* Liebm. (Trelease, Proc. Natl. Acad. Sci. 2: 627, 1916; confirmed by Nixon, Ann. Sci. For. 50, suppl. 1: 32s, 1993)
  Stamens 8–10, with apiculate apices (Trelease 1924); *pollen ornamentation weakly verrucate*, perforate (Denk and Grimm 2009); footlayer thick and continuous (Denk and Tekleva 2014); styles short to long, elliptic in cross-section; stigmata abruptly dilated; stigmatic surface extending adaxially along stylar suture (Trelease 1924; Manos 1997); fruit maturation biennial (Trelease 1924; Camus 1952–1954; Manos 1997); endocarp tomentose (Trelease 1924; Camus 1952–1954; Manos 1997); position of abortive ovules basal, lateral or apical, can be variable within a single plant (Manos 1997); cup scales triangular and fused at the base, thickened and compressed into rings, often tuberculate and obscured by glandular trichomes, with sharp angled tips; leaf dentitions spinose; wood diffuse porous, tyloses rarely present in vessels of early-wood (Trelease 1924).
  Five species in southwestern North America and northwestern Mexico (Manos 1997).

2.4.2.2 Section *Ponticae*

Quercus section *Ponticae* Stefanoff., Ann. Univ. Sofia, ser. 5, 8: 53 (1930)
  Quercus ser. *Sadlerianae* Trelease, Oaks of America: 111 (1924).—*Quercus* subsect. *Ponticae* Menitsky (Stefanoff) A.Camus, Bull. Soc. Bot. Fr., 81: 815 (1934).—*Quercus* ser. *Ponticae* Schwarz, Notizbl. Bot. Gart. Berlin-Dahlem, 13/116: 11 (1936).
  Lectotype (here designated): *Quercus pontica* K.Koch

Shrubs or small trees, rhizomatous; number of stamens mostly 6 (Trelease 1924; Camus 1952–1954); pollen ornamentation verrucate (Denk and Grimm 2009); footlayer of variable thickness and perforate (Denk and Tekleva 2014); staminate catkins up to 10 cm long; styles short, fused or free, elliptic in cross-section; stigmata abruptly or gradually dilated (Schwarz 1936); fruit maturation annual; endocarp glabrous; position of abortive ovules basal; cotyledons free; cup scales slightly tuberculate with sharp angled apices, occasionally with attenuated tips (Trelease 1924; Gagnidze et al. 2014); leaves evergreen or deciduous, chestnut-like, stipules large, persistent or early shed, number of secondary veins 10–15(–25), dentate, teeth simple or compound (in *Q. pontica*), sharply mucronate or with thread-like, curved upwards extension; leaf buds large, *bud scales loosely attached* (Trelease 1924; Schwarz 1936; Menitsky 1984); wood ring porous or diffuse porous, large vessels commonly plugged by tyloses.

Two species in mountainous areas of north-eastern Turkey and western Georgia (Transcaucasia) and in western North America (northern-most California, southern-most Oregon; Trelease 1924; Menitsky 1984; Gagnidze et al. 2014) (Fig. 2.3).

2.4.2.3 Section *Virentes*

Quercus section *Virentes* Loudon, Arbor. Frut. Brit., 3: 1730, 1918 (1838).
  Quercus ser. *Virentes* Trelease, Oaks of America: 112 (1924).
  Type: *Quercus virens* Aiton (= *Q. virginiana* Mill.)
  Trees or rhizomatous shrubs; pollen ornamentation verrucate (Denk and Grimm 2009); footlayer of variable thickness and perforate (Denk and Tekleva 2014); styles short, fused or free, elliptic in cross-section; stigmata abruptly or gradually dilated (Schwarz 1936); fruit maturation annual; cup scales narrowly triangular, free or fused at the base, thinly keeled and barely tuberculate with sharp angled apices; leaves evergreen or subevergreen (Trelease 1924; Nixon and Muller 1997); wood diffuse porous, tyloses abundant in large vessels (Trelease 1924); *cotyledons fused* (de Candolle 1862a; Engelmann 1880); *germinating seed with elongated radicle/ epicotyl forming a tube* (Nixon 2009); hypocotyl region produces a tuberous fusiform structure.

Seven species in south-eastern North America, Mexico, the West Indies (Cuba), and Central America (Muller 1961; Cavender-Bares et al. 2015) (Fig. 2.3).

2.4.2.4 Section *Quercus* (White Oaks)

Quercus section *Albae* Loudon, Arbor. Frut. Brit., 3: 1730 (1838).—*Quercus* section *Prinus* Loudon, Arbor. Frut. Brit., 3: 1730 (1838).—*Quercus* section *Robur* Loudon, Arbor. Frut. Brit., 3: 1731 (1838).—*Quercus* section *Gallifera* Spach, Hist. Nat. Veg., 11:170 (1842).—*Quercus* section *Eulepidobalanus* Oerst., Vidensk. Meddel. Naturhist. Foren. Kjøbenhavn 1866, 28: 65 (1866–1867) p.p.—

Quercus section *Macrocarpae* Oerst., Vidensk. Meddel. Naturhist. Foren. Kjøbenhavn 1866, 28: 68 (1866–1867).—*Quercus* section *Diversipilosae* C.K. Schneid., Handb. Laubholzk., 1: 208 (1906).—*Quercus* section *Dentatae* C.K. Schneid., Handb. Laubholzk., 1: 209 (1906).—*Quercus* section *Mesobalanus* A. Camus, Bull. Soc. Bot. Fr., 81: 815 (1934).—*Quercus* section *Roburoides* Schwarz, Notizbl. Bot. Gart. Berlin-Dahlem 13: 10 (1936).—*Quercus* section *Robur* Schwarz, Notizbl. Bot. Gart. Berlin-Dahlem 13: 12 (1936).—*Quercus* section *Dascia* (Kotschy) Schwarz, Notizbl. Bot. Gart. Berlin-Dahlem 13: 14 (1936).

Stamens $\geq$ 7 (Trelease 1924; Camus 1936–1938, 1938–1939); pollen ornamentation verrucate (Denk and Grimm 2009); footlayer of variable thickness and perforate (Denk and Tekleva 2014); styles short, fused or free, elliptic in cross-section; stigmata abruptly or gradually dilated; stigmatic surface extending adaxially along stylar suture (all authors; best illustrated in Schwarz 1936); fruit maturation annual; endocarp glabrous or nearly so; cotyledons free or fused; position of abortive ovules basal (de Candolle 1862b; confirmed/accepted by later authors), placenta and funiculus sessile; cup scales triangular, free or fused at the base, thickened, keeled and often tuberculate with sharp angled apices, occasionally with attenuated tips; leaf dentitions typically without bristle-like, aristate tips; wood ring porous, large vessels in (early-)wood commonly plugged by tyloses (Trelease 1924; see Akkemik and Yaman 2012).

Ca. 146 species in North America, Mexico, Central America, western Eurasia, East Asia, and North Africa (Nixon and Muller 1997).

2.4.2.5 Section *Lobatae* (Red Oaks)

Quercus section *Lobatae* Loudon, Hort. Brit., 385 (1830).

Quercus section *Integrifoliae* Loudon, Hort. Brit., 384 (1830) p.p.—*Quercus* section *Mucronatae* Loudon, Hort. Brit., 385 (1830) p.p.—*Quercus* section *Rubrae* Loudon, Arbor. Frut. Brit., 3: 1877 (1838; see also Loudon 1839).—*Quercus* section *Nigrae* Loudon, Arbor. Frut. Brit., 3: 1890 (1838; see also Loudon 1839).—*Quercus* section *Phellos* Loudon, Arbor. Frut. Brit., 3: 1894 (1838; see also Loudon 1839).—*Quercus* section *Erythrobalanus* Spach, Hist. veg. Phan., 11:160 (1842). —*Quercus* subgenus *Erythrobalanus* (Spach) Oerst., Vidensk. Meddel. Naturhist. Foren. Kjøbenhavn, 28: 70 (1866–1867).—*Erythrobalanus* (Spach) O.Schwarz (as genus), Notizbl. Bot. Gart. Berlin-Dahlem 13: 8 (1936).

Lectotype: *Quercus aquatica* (Lam.) Walter (= *Q. nigra* L.) (Nixon, Ann. Sci. For. 50, suppl.1: 30s, 1993).

Pistillate perianth forming a characteristic flange (Schwarz 1936, Fig. 2.1; Nixon 1993; Jensen 1997); number of stamens $\leq$ 6 (Trelease 1924; Camus 1952–1954); pollen ornamentation verrucate (Denk and Grimm 2009); footlayer of variable thickness and perforate (Denk and Tekleva 2014); styles elongated, linear, outcurved, elliptic in cross-section; stigmata slightly dilated, spatulate to oblong;

stigmatic surface extending adaxially along stylar suture (Trelease 1924); perianthopodium conical, often annulate (Trelease 1924); fruit maturation biennial, rarely annual (le Hardÿ de Beaulieu and Lamant 2010); endocarp tomentose; cotyledons free or sometimes basally fused; position of abortive ovules apical or rarely lateral to basal (de Candolle 1862b; Trelease 1924), placenta sessile or elongated, funiculus sessile; *cupule fused with peduncle forming a 'connective piece'* (compare Denk and Meller 2001) for *Fagus*), connective piece covered with small scales similar to those on the cupule; *cup scales* triangular and free, mostly thin, membranous and smooth *with broadly angled tips*; leaf teeth and lobes typically with bristle-like extensions, teeth reduced to bristles in entire or nearly entire leaves; wood ring-porous or semi ring-porous, late-wood markedly porous, tyloses in vessels of early-wood rarely present (Trelease 1924).

Ca. 124 species in North America, Mexico, Central America, and Colombia in South America (Jensen 1997).

2.4.3 Subgenus Cerris

Quercus subgenus *Cerris* Oerst., Vidensk. Meddel. Naturhist. Foren. Kjøbenhavn 1866, 28: 77 (1866–1867).

Rugulae visible in mature pollen grains or weakly masked (Solomon 1983a, b; Denk and Grimm 2009; Makino et al. 2009; Denk and Tekleva 2014).

2.4.3.1 Section *Cyclobalanopsis*

Quercus sect. *Cyclobalanopsis* (Oerst.) Benth. & Hook. f., Gen. Plant. 3, 408 (1880).

Cyclobalanopsis Oerst. (as genus), Vidensk. Meddel. Naturhist. Foren. Kjøbenhavn 1866, 28: 77 (1866–1867), nom. conserv.—*Quercus* sect. *Cyclobalanopsis* (Oerst.) Benth. & Hook. f., Gen. Plant. 3, 408 (1880).—*Quercus* subgenus *Cyclobalanopsis* (Oerst.) Schneider, Ill. Handb. Laubholzk. 1, 210 (1906).

Type: *Quercus velutina* Lindl. ex Wall., non Lam. (vide Farr and Zijlstra 2017).

Staminate flowers in groups of 1–3(–7) along inflorescence axis (Menitsky 1984; Nixon 1993); stamens 5–6 (Huang et al. 1999) to 10–15 (Ohwi 1965); *pollen ornamentation vertical-rugulate* (Denk and Grimm 2009); footlayer thick and continuous or of variable thickness and perforate (Denk and Tekleva 2014); styles short to very short (<3 to <1 mm), elliptic in cross-section; stigmata dilated, subcapitate; stigmatic surface not forming a prominent stigmatic groove (Camus 1936–1938; Menitsky 1984; Nixon 1993; Huang et al. 1999); *perianthopodium annulate with 3–5 distinct rings* (Schwarz 1936); fruit maturation annual or biennial (Camus 1936–1938; le Hardÿ de Beaulieu and Lamant 2010); endocarp tomentose or rarely glabrous (Camus 1936–1938); cotyledons free; position of abortive ovules apical (Camus 1936–1938; Menitsky 1984) [note: according to Huang et al. (1999) the position is variable, but no details are provided in the species descriptions], placenta elongated reaching the apical part of the seed, where vascular bundles enter the seed and the aborted ovules, funiculus sessile or with short petiole; *cupule with concentric lamellae*; leaves evergreen; leaf dentitions with bristle-like extensions or not; wood diffuse porous, tyloses very rarely present in vessels of early-wood (Menitsky 1984).

Ca. 90 species in tropical and subtropical Asia including the southern Himalayas (Huang et al. 1999).

2.4.3.2 Section *Ilex*

Quercus section *Ilex* Loudon, Arbor. Frut. Brit., 3: 1730, 1899 (1838).

Quercus subgenus *Heterobalanus* Oerst., Vidensk. Meddel. Naturhist. Foren. Kjøbenhavn 1866, 28: 69 (1866–1867).—*Quercus* subgenus *Heterobalanus* (Oerst.) Menitsky, Duby Azii, 89 (1984) [Oaks of Asia, 133, (2005)].—*Quercus* section *Heterobalanus* (Oerst.) Menitsky, Duby Azii, 89 (1984) [Oaks of Asia, 134, (2005)].—*Quercus* subsection *Ilex* (Loudon) Guerke sensu Menitsky, Duby Azii, 97, (1984) [Oaks of Asia, 151, (2005)].

Type: *Quercus ilex* L.

Stamens 4–6 (Schwarz 1937); *pollen ornamentation rugulate* (Denk and Grimm 2009); footlayer thick and continuous or of variable thickness and perforate (Denk and Tekleva 2014); styles medium-long, apically gradually dilated, recurved, v-shaped in diameter; stigmata slightly subulate; stigmatic surface extending adaxially along stylar suture (Schwarz 1937; Menitsky 1984); fruit maturation annual or biennial (Camus 1936–1938, 1938–1939; Menitsky 1984; le Hardÿ de Beaulieu and Lamant 2010) [note that observations by Menitsky partly differ from those of Camus and le Hardÿ de Beaulieu and Lamant]; endocarp tomentose (Schwarz 1936; Camus 1936–1938, 1938–1939; Schwarz 1937; Menitsky 1984); cotyledons free; position of the abortive ovules basal or lateral, placenta and funiculus sessile or elongated; cup scales triangular, free or fused at the base, mostly thin, membranous, often keeled and tuberculate with sharp angled apices, occasionally with slightly raised tips or narrowly triangular, well-articulated, thickened with elongated recurved tips (as in *Q. alnifolia*, *Q. baronii*, *Q. coccifera*, *Q. dolicholepis*); leaves evergreen, dentitions spinose or with bristle-like extensions; wood diffuse porous, tyloses rarely present in vessels of early-wood (Menitsky 1984).

Ca. 36 species in Eurasia and North Africa (Menitsky 1984; Denk and Grimm 2010; Deng et al. 2017).

2.4.3.3 Section *Cerris*

Quercus section *Cerris* Dumort., Florula Belgica: 15 (1829).

Quercus section *Cerris* Loudon, Arbor. Frut. Brit., 3: 1730 (1838).—*Quercus* section *Eucerris* Oerst., Vidensk. Meddel. Naturhist. Foren. Kjøbenhavn 1866, 28: 75, nom. illeg. (1867).—*Quercus* section *Erythrobalanopsis* Oerst., Vidensk. Meddel. Naturhist. Foren. Kjøbenhavn 1866, 28: 76 (1867).—*Quercus* section *Castaneifolia* O.Schwarz, Feddes Repert., 33: 322 (1934).—*Quercus* section *Vallonea* O.Schwarz, Feddes Repert., 33: 322 (1934).—*Quercus* section *Aegilops* (Reichenb.) O.Schwarz, Notizbl. Bot. Gart. Berlin-Dahlem, 13/116: 19 (1936).

Type: *Quercus cerris* L.

Number of stamens 4–6; *pollen ornamentation scattered verrucate* (Denk and Grimm 2009); footlayer of variable thickness and perforate (Denk and Tekleva 2014); *styles* elongated, outcurved, *pointed*, v-shaped in diameter; stigmatic area linear; stigmatic surface extending adaxially along stylar suture; fruit maturation biennial, variable only in *Q. suber* (Camus 1936–1938; le Hardÿ de Beaulieu and Lamant 2010); endocarp tomentose (Camus 1936–1938); cotyledons free; position of abortive ovules basal, lateral or apical (de Candolle 1862b; Camus 1936–1938; Schwarz 1937; Menitsky 1984), placenta sessile, funiculus sessile or elongated; *cup scales* narrowly triangular, well-articulated, thickened and keeled *with elongated, well-developed recurved tips*; leaf dentitions typically with bristle-like extensions; wood (semi-)ring-porous, tyloses in vessels of early-wood present but not common (Trelease 1924; Akkemik and Yaman 2012).

Ca. 13 species in Eurasia and North Africa (Menitsky 1984).

2.5 Infrasectional Classification: The Big Challenge

The main challenge for oak systematics in the coming years will be a meaningful classification below the sectional rank. Nuclear-phylogenomic data (Hipp et al. 2015) within *Quercus* recover subclades that occur in well-defined biogeographic regions (e.g. western North America, eastern North America, western Eurasia, East Asia etc.), a sorting not so clear from traditional sequence data. Only two New World-Old World disjuncts are recognised (e.g. section *Ponticae*, section *Quercus*; McVay et al. 2017). Ongoing phylogenomic work (Hipp et al. 2014, 2015, 2017; McVay et al. 2017) is beginning to reveal structure within the sections *Lobatae* and *Quercus* that corresponds to regional diversity within the Americas. Preliminary phylogenetic analyses suggested that the early evolutionary branches of *Lobatae* include many of the lobed-leaf species groups of North America (Hipp et al. 2015). The first branch, however, comprises the seven Californian taxa (*Agrifoliae* sensu Trelease), followed by various groups containing mostly temperate species that sort out into well-defined subclades. For section *Quercus*, analyses suggest some uncertainty at the base of the clade, specifically regarding the position of the Eurasian subclade. Previous morphology-based treatments of the Eurasian white oaks ('roburoids') suggested close affinities to certain eastern North American species, like *Q. montana* Willd. (series *Prinoideae* of Trelease), based on a similar (e.g. 'prinoid') leaf morphology (Axelrod 1983). In the most recent time calibrated tree (Hipp et al. 2017; McVay et al. 2017) the roburoids are nested within the white oak (s.str.) clade and diverged at around 25–30 Ma (Oligocene) from a North American, fully temperate clade including the type species of the section *Quercus* (*Q. alba*) and morphologically similar species such as *Q. montana* (see also Pearse and Hipp 2009). Dispersed pollen provide evidence for the presence of the white oak s.l. lineage (pollen types found today in sections *Quercus* and *Ponticae* but not *Virentes*) in the middle Eocene of western Greenland and the Baltic amber region of northern Europe (Grímsson et al. 2015, 2016). Hence, the early radiation of this lineage involved the North Atlantic land bridge. Pollen and leaf fossil evidence from Eocene and Oligocene strata in East Asia further indicate migration from North America via the Bering land bridge. The expansion of this East Asian branch of white oaks (early roburoids) gave rise to the (modern) western Eurasian roburoids. This is in some agreement with the latest dated tree proposing a crown age of ca. 15–20 Ma for the roburoids. By that time, the final radiation within the North American white oaks (s.str.; section *Quercus*) might have been completed (Hipp et al. 2017).

Lack of resolution using traditional sequence data, e.g. identical ITS variants found in North American and Eurasian white oaks, and the relative young inferred root (stem) and crown ages of the roburoids can be explained by recent episodic migration from North America to Europe across the North Atlantic land bridge in addition to probably (very) large population sizes of temperate (white) oaks. Ancient hybridisation between roburoids and section *Ponticae* has been identified as one possible source of potentially misleading phylogenetic data (McVay et al. 2017).

However, the current, partly preliminary results also indicate that (sub)sections/series recognised in the monographs of Trelease, Camus, and Menitsky do not always correspond to groups identified when using molecular sequence data. In some cases, the molecular-defined groups may seem counterintuitive. For example, the western Eurasian Ilex oaks *Quercus alnifolia*, *Q. aucheri*, *Q. coccifera* and *Q. ilex* are resolved as a monophyletic group (Denk and Grimm 2010; Hipp et al. 2015), but were placed into two subgenera and three sections by Schwarz (1936), and two sections and three subsections by Camus (1936–1938, 1938–1939) and Menitsky (1984) due to conspicuous differences in indumentum, leaf margin, and cup scales. Similar mismatches between traditional classification and DNA-based groups are encountered for all large infrageneric groups and will pose a major challenge when searching for morphological criteria to subdivide sections within oaks. For example, while it has long been noticed that characters of the indumentum of the abaxial leaf surface provide valuable information for species delimitation (Manos 1993; Nixon 2002; Tschan and Denk 2012; Deng et al. 2015, 2017), these characters appear to have evolved convergently in related and unrelated groups (Tschan and Denk 2012; Deng et al. 2017).

2.6 Fossil Record

Section *Protobalanus*—In addition to pollen of section *Lobatae*, Grímsson et al. (2015) found dispersed pollen similar to pollen of section *Protobalanus* in middle Eocene deposits of western Greenland. Unambiguous leaf fossils and dispersed pollen of section *Protobalanus* are known from the latest Eocene-earliest Oligocene Florissant Formation (ca. 34 Ma; Bouchal et al. 2014). The section had a western and northern North American distribution during the Paleogene and became restricted to western North America during the Neogene.

Section *Ponticae*—Miocene and Pliocene leaf fossils assigned to *Quercus pontica miocenica* Kubát have traditionally been compared with the extant *Q. pontica* (e.g. Andreánszky 1959; Gagnidze et al. 2014). Presently, these fossils are included within the fossil-species *Q. gigas* Göppert, a fossil representative of section *Cerris* (Walther and Zastawniak 1991; Kvaček et al. 2002). These leaf remains and early Cenozoic fossils from Arctic regions (see Grímsson et al. 2016) are superficially similar to leaves of section *Ponticae* but lack the characteristic dentition of *Q. pontica*. Hence, there is currently no reliable fossil record of this section because chestnut-like foliage might have evolved in parallel in various modern and extinct lineages of Fagaceae or represents the ancestral state within genus *Quercus* or subgenus *Quercus*.

Section *Virentes*—For similar reasons as outlined above for section *Ponticae* there is no reliable fossil record of this modern section.

Section *Quercus*—Leaf fossils from middle Eocene deposits of Axel-Heiberg Island, Canadian Arctic (ca. 45 Ma; McIver and Basinger 1999) most likely belong to section *Quercus*. The leaf fossils are strikingly similar to the extant East Asian *Quercus aliena* Blume var. *acutiserrata* Maxim. ex Wenzig. From the roughly coeval Baltic amber of northern Europe, Crepet (1989) described in situ pollen of male flowers, which may represent section *Quercus*. Fagaceous remains from the Baltic amber are currently revised by Eva-Maria Sadowski, Göttingen, and may represent more than one section of *Quercus*. The leaf fossil-taxon *Quercus kraskinensis* Pavlyutkin from the early Oligocene of the Primorskii Region (Pavlyutkin 2015) is similar to the material from Axel-Heiberg Island and most likely belongs to section *Quercus*. The Paleogene radiation of the white oak lineage might have involved both the Bering and the North Atlantic land bridges. Lobed oaks of section *Quercus* are also known from the Oligocene of Central Asia and Northeast Asia (e.g. Krishtofovich et al. 1956; Tanai and Uemura 1994), and North America (Bouchal et al. 2014). Hence, the section had a scattered distribution during the Paleogene that included low, mid-, and high latitudes. In the Neogene, section *Quercus* was widespread across the entire Northern Hemisphere (Borgardt and Pigg 1999).

Section *Lobatae*—Oldest fossils that can securely be assigned to section *Lobatae* are dispersed pollen from the middle Eocene of western Greenland (latest Lutetian to earliest Bartonian, 42–40 Ma). The pollen shows a sculpturing found today only in members of section *Lobatae* (Grímsson et al. 2015). A further record of entire-margined, lanceolate foliage with preserved epidermal structures from middle Eocene (48–38 Ma) deposits of Central Europe—described as *Quercus subhercynica* (Kvaček and Walther 1989) and originally assigned to section *Lobatae*—was subsequently transferred to the extinct genus *Castaneophyllum*, a fossil-genus for which morphological affinities to certain castaneoid genera have been established using leaf epidermal characteristics (Kvaček and Walther 2010 [2012]). For North America, Daghlian and Crepet (1983) described cups and acorns associated with lobate leaves from the Oligocene (Rupelian, ca. 30 Ma) Catahoula Formation in Texas (*Quercus oligocenensis*). Markedly similar leaf records from the early Oligocene of northeast Asia, *Quercus sichotensis* Ablaev et Gorovoj, *Q. ussuriensis* Krysh., *Q. arsenjevii* Ablaev et Gorovoj, and *Q. kodairae* Huzioka assigned to section *Cerris* (Tanai and Uemura 1994; Pavlyutkin 2015) clearly also belong to section *Lobatae*. Hence, the section had a mid- to high latitude northern hemispheric distribution during the Paleogene; during the Neogene it was present in Europe (Jähnichen 1966; Kovar-Eder and Meller 2003) and North America.

Section *Cerris*—The earliest unambiguous record of section *Cerris*, *Quercus gracilis* (Pavlyutkin) Pavlyutkin, comes from early Oligocene leaf fossils from the Russian Far East (Pavlyutkin et al. 2014). In western Eurasia, earliest evidence of section *Cerris* comes from dispersed pollen from late Oligocene/ early Miocene (ca. 23 Ma) deposits of Central Europe (Kmenta 2011). The section has a rich Neogene fossil record in Eurasia (e.g. Mai 1995; Song et al. 2000; Yabe 2008).

Section *Ilex*—The Paleogene record of Ilex oaks is so far limited to dispersed pollen from East Asia (Hainan Island, China, Changchang Formation, Lutetian-Bartonian, ca. 40 Ma; Hofmann 2010; Spicer et al. 2014) and Central Europe (Germany, Rupelian, ca. 33 Ma; Denk et al. 2012). From the Changchang Formation, Spicer et al. (2014) also reported leaf morphotypes (OTUs 68 and 71) that most likely belong to section *Ilex*. There is also leaf fossil evidence of section *Ilex* from 26 Ma strata in Tibet (Zhou Zhekun, personal communication). Section *Ilex* has a rich fossil record in Neogene deposits across Eurasia (e.g. Denk et al. 2017).

Section *Cyclobalanopsis*—*Quercus paleocarpa* (Manchester 1994) cupules and nuts from the Eocene (Lutetian, ca. 48 Ma) of western North America are possibly the oldest fossils belonging to section *Cyclobalanopsis*, but without preserved stigmas the assignment of these fruits remains ambiguous. Additionally, fossilised fruits of *Cyclobalanopsis nathoi* from the middle Eocene of Japan (Huzioka and Takahashi 1973) may belong to section *Cyclobalanopsis* based on the shape of nuts. Hofmann (2010) reported dispersed pollen grains from the middle Eocene of Hainan Island, China (Changchang Formation, Lutetian-Bartonian, 48–38 Ma; Spicer et al. 2014). From the Changchang Formation, Spicer et al. (2014) also reported leaf morphotypes (OTUs 61, 62 [partly], 63, 67) that most likely belong to section *Cyclobalanopsis*. From the Oligocene of south-western China, several fossil-species based on leaf impressions have been assigned to cycle cup oaks (Writing Group of Cenozoic Plants of China [WGCPC] 1978). Evidence for assignment to section *Cyclobalanopsis* is based on the number, arrangement, and course of secondary veins, the dentition, and the attenuate leaf apex (e.g. *Q.*

parachampionii Cheu and Liu; *Q. paraschottkyana* Wang and Liu). This section has a Paleogene record in mid to low latitude East Asia and western North America, while it is restricted to Asia during the Neogene (e.g. Jia et al. 2015). No reliable records are known from Europe.

2.7 Conclusion and Outlook

Recent molecular phylogenetic studies consistently suggest two major clades within oaks, one comprising three Old World groups (sections *Cyclobalanopsis*, *Ilex*, and *Cerris*), the other comprising three New World groups (sections *Protobalanus*, *Virentes*, and *Lobatae*,) and two northern hemispheric groups (sections *Ponticae* and *Quercus*). This is in contrast to the established view that *Cyclobalanopsis* oaks are sister to the remainder of the genus *Quercus*. The reason for this conflict is that morphological characters evolved convergently in all major groups of oaks and even outside oaks in other Fagaceae (e.g. concentric cupula rings). Important conserved morphological and diagnostic characters are pollen sculpturing and ultrastructure (Fig. 2.2).

Based on the new molecular and morphological evidence the infrageneric classification of *Quercus* is revised. A major challenge for future studies will be the molecular and morphological circumscription of infrasectional groups and their biogeographic and ecological characterisation. In this context, comparative morphological investigations of the seed ontogeny will be important to document the distribution of type I, II and III developmental pathways of aborted ovule positions (Table 2.1) across all sections. Some characters that have been described mainly on the basis of herbarium material, such as the annual or biennial mode of maturation, need to be reinvestigated in the field. In (fully) evergreen species of sections *Ilex*, *Protobalanus*, and *Cyclobalanopsis*, the fruiting twigs do not produce new growth in the second year after pollination (pseudo-annual maturation sensu Nixon 1997) and therefore may erroneously be interpreted as annual. In a number of high-mountain species of section *Ilex*, maturation may take much longer than previously assumed, with time periods of up to three to four years between pollination and mature seeds (Min Deng, unpublished data). Considering the oak fossil record, it is noteworthy that Paleogene plant-bearing deposits from East Asia and the Far East have so far been understudied. Once recovered, this fossil record should contribute to a better understanding of the emergence of major groups within oaks.

Acknowledgements We thank John McNeill for valuable comments. This work was supported by the Swedish Research Council (VR, grant to TD). GWG acknowledges financial support by the AMS Wien.

Appendix 2.1

At https://doi.org/10.6084/m9.figshare.5547622.v1, we provide an electronic appendix including the following information (which may be subject to future updates): *(i)* an overview of earlier systematic schemes for oaks (genera, subgenera, sections) in comparison to the new classification; *(ii)* diagnostic morphological traits reported by earlier taxonomists extracted from the original literature; *(iii)* a comprehensive list of formerly and currently accepted species of oaks, compiled from the cited oak monographs and complemented by further data sources.

References

Akkemik Ü, Yaman B (2012) Wood anatomy of Eastern Mediterranean species. Verlag Kessel, Remagen
Andreánszky G (1959) Die Flora der Sarmatischen Stufe in Ungarn. Akadémiai Kiadó, Budapest
Axelrod DI (1983) Biogeography of oaks in the Arcto-Tertiary Province. Ann Missouri Bot Gard 70:629–657
Bentham G, Hooker JD (1880) Genera plantarum, vol 3. L. Reeve & Co., Williams & Norgate, London
Borgardt SJ, Nixon KC (2003) A comparative flower and fruit anatomical study of *Quercus acutissima*, a biennial-fruiting oak from the *Cerris* group (Fagaceae). Am J Bot 90:1567–1584
Borgardt SJ, Pigg KB (1999) Anatomical and developmental study of petrified *Quercus* (Fagaceae) fruits from the Middle Miocene, Yakima Canyon, Washington, USA. Am J Bot 86:307–325
Bouchal J, Zetter R, Grímsson F, Denk T (2014) Evolutionary trends and ecological differentiation in early Cenozoic Fagaceae of western North America. Am J Bot 101:1–18
Britton NL, Brown A (1913) An illustrated flora of northern United States, Canada and British possessions. Scribner & Sons, New York
Browicz K, Zieliński J (1982) Chorology of trees and shrubs in South-West Asia and adjacent regions. Polish Scientific Publishers, Warsaw, Poznan
Camus A (1936–1938) Les Chênes. Monographie du genre *Quercus*. Tome I. Genre *Quercus*, sous-genre *Cyclobalanopsis*, sous-genre *Euquercus* (sections *Cerris* et *Mesobalanus*). Texte. Paul Lechevalier, Paris
Camus A (1938–1939) Les Chênes. Monographie du genre *Quercus*. Tome II. Genre *Quercus*, sous-genre *Euquercus* (sections *Lepidobalanus* et *Macrobalanus*). Texte. Paul Lechevalier, Paris
Camus A (1952–1954) Les Chênes: Monographie du genre *Quercus*. Tome III. Genre *Quercus*: sous-genre *Euquercus* (sections *Protobalanus* et *Erythrobalanus*) et genre *Lithocarpus*. Texte. Paul Lechevalier, Paris
Cavender-Bares J, Gonzalez-Rodriguez A, Eaton DAR, Hipp AL, Beulke A, Manos PS (2015) Phylogeny and biogeography of the American live oaks (*Quercus* subsection *Virentes*): a genomic and population genetics approach. Mol Ecol 24:3668–3687
Costa Tenorio M, Morla Juarista C, Sáinz Ollero H (2001) Los bosques Ibéricos. Una interpretación geobotánica. Planeta, Barcelona
Crepet WL (1989) History and implications of the early North American fossil record of Fagaceae. In: Crane PR, Blackmore S (eds) Evolution, systematics, and fossil history of the Hamamelidae Vol 2: 'Higher' Hamamelidae. Systematic Association Special Vol 40B. Clarendon, Oxford, pp 45–66

Daghlian CP, Crepet WL (1983) Oak catkins, leaves and fruits from the Oligocene Catahoula Formation and their evolutionary significance. Am J Bot 70:639–649
de Candolle A (1862a) Étude sur l'espèce á l'occasion d'une révision de la famille des Cupulifères. Arch Sci Phys Nat II 15:211–237, 326–365
de Candolle A (1862b) Note sur un charactère observé dans les fruits des chênes. Ann sci nat ser 4 (18):49–58
Deng M (2007) Anatomy, taxonomy, distribution and phylogeny of *Quercus* Subg. *Cyclobalanopsis* (Oersted) Schneid. (Fagaceae). Ph.D. thesis, Kunming Institute of Botany, Chinese Academy of Sciences and Graduate School of Chinese Academy of Sciences, Beijing
Deng M, Zhou Z-K, Chen Y-Q, Sun W-B (2008) Systematic significance of the development and anatomy of flowers and fruit of *Quercus schottkyana* (subgenus *Cyclobalanopsis*: Fagaceae). Int J Plant Sci 169:1261–1277
Deng M, Hipp A, Song Y-G, Li Q-S, Coombes A, Cotton A (2015) Leaf epidermal features of *Quercus* subgenus *Cyclobalanopsis* (Fagaceae) and their systematic significance. Bot J Linn Soc 176:224–259
Deng M, Jiang X-L, Song Y-G, Coombes A, Yang X-R, Xiong Y-S, Li Q-S (2017) Leaf epidermal features of *Quercus* Group Ilex (Fagaceae) and their application to species identification. Rev Palaeobot Palynol 237:10–36
Denk T, Grimm GW (2009) Significance of pollen characteristics for infrageneric classification and phylogeny in *Quercus* (Fagaceae). Int J Plant Sci 170:926–940
Denk T, Grimm GW (2010) The oaks of western Eurasia: traditional classifications and evidence from two nuclear markers. Taxon 59:351–366
Denk T, Meller B (2001) Systematic significance of the cupule/nut complex in living and fossil *Fagus*. Int J Plant Sci 162:869–897
Denk T, Tekleva MV (2014) Pollen morphology and ultrastructure of *Quercus* with focus on Group Ilex (= *Quercus* Subgenus *Heterobalanus* (Oerst.) Menitsky): implications for oak systematics and evolution. Grana 53:255–282
Denk T, Grímsson F, Zetter R (2012) Fagaceae from the early Oligocene of Central Europe: persisting New World and emerging Old World biogeographic links. Rev Palaeobot Palynol 169:7–20
Denk T, Velitzelos D, Güner HT, Bouchal JM, Grímsson F, Grimm GW (2017) Taxonomy and palaeoecology of two widespread western Eurasian Neogene sclerophyllous oak species: *Quercus drymeja* Unger and *Q. mediterranea* Unger. Rev Palaeobot Palynol 241:98–128
Dumortier B-C (1829) Florula Belgica. J. Casterman, Tournai
Engelmann G (1880) The acorns and their germination. Trans Acad Sci St Louis 4:190–192
Fang J, Wang Z, Tang Z (2009) Atlas of woody plants in China. Volumes 1–3 and index. Higher Education Press, Beijing
Farr ER, Zijlstra G (2017) Index Nominum Genericorum (Plantarum). 1996+. http://botany.si.edu/ing/. Last accessed 25 June 2017
Flora of China (2016) eFloras: Flora of China. http://www.efloras.org/flora_page.aspx?flora_id=2. Last accessed 2 Nov 2016
Flora of North America Editorial Commitee (1997) Flora of North America north of Mexico, vol 3. Oxford University Press, New York
Gagnidze R, Urushadze T, Pietzarka U (2014) *Quercus pontica* Enzyklopädie der Holzgewächse: Handbuch und Atlas der Dendrologie 63. Erg. Lfg. 04/13. Wiley-VCH, Weinheim, pp 1–8
Govaerts R, Frodin DG (1998) World checklist and bibliography of Fagales (Betulaceae, Corylaceae, Fagaceae and Ticodendraceae). Royal Botanic Gardens, Kew
Grímsson F, Zetter R, Grimm GW, Krarup Pedersen G, Pedersen AK, Denk T (2015) Fagaceae pollen from the early Cenozoic of West Greenland: revisiting Engler's and Chaney's Arcto-Tertiary hypotheses. Plant Syst Evol 301:809–832
Grímsson F, Grimm GW, Zetter R, Denk T (2016) Cretaceous and Paleogene Fagaceae from North America and Greenland: evidence for a Late Cretaceous split between *Fagus* and the remaining Fagaceae. Acta Palaeobot 56:247–305

Hesse M, Halbritter H, Zetter R, Weber M, Buchner R, Frosch-Radivo A, Ulrich S (2009) Pollen terminology—an illustrated handbook. Springer, Wien, New York

Hipp AL, Eaton DAR, Cavender-Bares J, Fitzek E, Nipper R, Manos PS (2014) A framework phylogeny of the American oak clade based on sequenced RAD data. PLoS ONE 9:e93975

Hipp AL, Manos P, McVay JD, Cavender-Bares J, González-Rodriguez A, Romero-Severson J, Hahn M, Brown BH, Budaitis B, Deng M, Grimm G, Fitzek E, Cronn R, Jennings TL, Avishai M, Simeone MC (2015) A phylogeny of the world's oaks. Botany 2015, Edmonton. Available at http://2015.botanyconference.org/engine/search/index.php?func=detail&aid=1305

Hipp AL, Manos PS, González-Rodríguez A, Hahn M, Kaproth M, McVay JD, Valencia Avalos S, Cavender-Bares J (2017) Sympatric parallel diversification of major oak clades in the Americas and the origins of Mexican species diversity. New Phytol. doi:10.1111/nph.14773

Hofmann C-C (2010) Microstructure of Fagaceae pollen from Austria (Paleocene/Eocene boundary) and Hainan Island (?middle Eocene). 8th European Palaeobotany-Palynology Conference. Hungarian Natural History Museum, Budapest, p 119

Huang C, Zhang Y, Bartholomew B (1999) Fagaceae. In: Wu Z-Y, Raven PH (eds) Flora of China 4. Cycadaceae through Fagaceae. Science Press and Missouri Botanical Garden Press, Beijing, St. Louis, pp 314–400

Hubert F, Grimm GW, Jousselin E, Berry V, Franc A, Kremer A (2014) Multiple nuclear genes stabilize the phylogenetic backbone of the genus *Quercus*. Syst Biodivers 12:405–423

Huzioka K, Takahashi E (1973) The Miocene flora of Shimonoseki, Southwest Honshu, Japan. Bull Nat Sci Mus 16:115–148

Jähnichen H (1966) Morphologisch-anatomische Studien über strukturbietende, ganzrandige Eichenblätter des Subgenus *Euquercus—Quercus lusatica* n. sp.—im Tertiär Mitteleuropas. Monatsber Dtsch Akad Wiss Berlin 8:477–512

Jensen RJ (1997) *Quercus* Sect. *Lobatae* G. Don in J. C. Loudon, Hort. Brit. 385. 1830. In: Flora of North America Editorial Committee (ed) Flora of North America North of Mexico Vol 3. Missouri Botanical Garden Press, St. Louis, pp 447–468

Jia H, Jin P, Wu J, Wang Z, Sun B (2015) *Quercus* (subg. *Cyclobalanopsis*) leaf and cupule species in the late Miocene of eastern China and their paleoclimatic significance. Rev Palaeobot Palynol 219:132–146

Kmenta M (2011) Die Mikroflora der untermiozänen Fundstelle Altmittweida, Deutschland. M.Sc. thesis, University of Vienna, Vienna, Austria. http://othes.univie.ac.at/15964/

Kovar-Eder J, Meller B (2003) The plant assemblages from the main seam parting of the western sub-basin of Oberdorf, N Voitsberg, Styria, Austria (Early Miocene). Cour Forschungsinst Senckenberg 241:281–311

Kremer A, Abbott AG, Carlson JE, Manos PS, Plomion C, Sisco P, Staton ME, Ueno S, Vendramin GV (2012) Genomics of Fagaceae. Tree Genet Genomes 8:583–610

Krishtofovich AN, Palibin IV, Shaparenko KK, Yarmolenko AV, Baykovskaya TN, Grubov VI, Iljinskaya IA (1956) Oligotsenovaya flora gory Ashutas v Kazakhstane (Oligocene flora of Ashutas Mount in Kazakhstan). Komarov Bot Inst Acad Sci SSSR Publ 145 Ser 8 Palaeobot 1:1–241 (in Russian)

Kvaček Z, Walther H (1989) Palaeobotanical studies in Fagaceae of the European tertiary. Plant Syst Evol 162:213–229

Kvaček Z, Walther H (2010 [2012]) European Tertiary Fagaceae with chinquapin-like foliage and leaf epidermal characteristics. Feddes Repert 121:248–267

Kvaček Z, Velitzelos D, Velitzelos E (2002) Late Miocene Flora of Vegora, Macedonia, N. Greece. Korali Publications, Athens

le Hardÿ de Beaulieu A, Lamant T (2010) Guide illustré des Chênes. 2 vols. Edilens, Geer

Linné C (1753) Species Plantarum. Vol 2. Laurentii Salvii, Stockholm

Loudon JC (1830) Loudon's Hortus Brittanicus. A. & R. Spottiswoode, London

Loudon JC (1838) Arboretum et Fruticetum Brittanicum, vol III. Printed for the author by A. Spottiswoode, London

Loudon JC (1839) Part II. The Jussieuean arrangement. In: Loudon JC (ed) Loudon's Hortus Brittanicus A new edition. A. Spottiswode, London, pp 491–704

Mai DH (1995) Tertiäre Vegetationsgeschichte Europas. Gustav Fischer Verlag, Jena, Stuttgart, New York
Makino M, Hayashi R, Takahara H (2009) Pollen morphology of the genus *Quercus* by scanning electron microscope. Sci Rep Kyoto Prefect Univ Life Environ Sci 61:53–81
Manchester SR (1994) Fruits and seeds of the Middle Eocene nut beds flora, Clarno Formation, Oregon. Palaeontogr Am 58:1–205
Manos PS (1993) Foliar trichome variation in *Quercus* section *Protobalanus* (Fagaceae). SIDA Contr Bot 15:391–403
Manos PS (1997) *Quercus* Sect. *Protobalanus* (Trelease) A.Camus. In: Flora of North America Editorial Committee (ed) Flora of North America North of Mexico, vol 3. Missouri Botanical Garden Press, St. Louis, p 468ff
Manos PS (2016) Systematics and biogeography of the American oaks. Int Oak J 27:23–36
Manos PS, Doyle JJ, Nixon KC (1999) Phylogeny, biogeography, and processes of molecular differentiation in Quercus subgenus Quercus (Fagaceae). Mol Phyl Evol 12:333–349
Manos PS, Zhou ZK, Cannon CH (2001) Systematics of Fagaceae: Phylogenetic tests of reproductive trait evolution. Int J Plant Sci 162:1361–1379
Manos PS, Cannon CH, Oh S-H (2008) Phylogenetic relationships and taxonomic status of the paleoendemic Fagaceae of Western North America: recognition of a new genus, *Notholithocarpus*. Madroño 55:181–190
McIver EE, Basinger JF (1999) Early Tertiary floral evolution in the Canadian High Arctic. Ann Missouri Bot Gard 86:523–545
McVay JD, Hipp AL, Manos PS (2017) A genetic legacy of introgression confounds phylogeny and biogeography in oaks. Proc R Soc B 284:20170300
Menitsky YL (1984) Duby Azii. Nauka, Leningrad [St. Petersburg]
Menitsky YL (2005) Oaks of Asia [translated from the Russian original of 1984]. Science Publishers, Enfield, NH
Muller CH (1961) The live oaks of the series *Virentes*. Am Midland Nat 65:17–39
Neophytou C, Dounavi A, Fink S, Aravanopoulos FA (2010) Interfertile oaks in an island environment: I. High nuclear genetic differentiation and high degree of chloroplast DNA sharing between *Q. alnifolia* and *Q. coccifera* in Cyprus. A multipopulation study. Eur J For Res 130:543–555
Neophytou C, Aravanopoulos FA, Fink S, Dounavi A (2011) Interfertile oaks in an island environment: II. Limited hybridization *Quercus alnifolia* Poech and *Q. coccifera* L. in a mixed stand. Eur J For Res 130:623–635
Nixon KC (1993) Infrageneric classification of *Quercus* (Fagaceae) and typification of sectional names. Ann Sci For 50:25s–34s
Nixon KC (1997) Fagaceae. In: Flora of North America Editorial Committee (ed) Flora of North America North of Mexico. Oxford University Press, New York, p. 436–537
Nixon KC (2002) The oak biodiversity of California and adjacent regions. In: Standiford RB, McCreary D, Purcell KL (eds) Proceedings of the 5th Symposium on Oak Woodlands: Oaks in California's Changing Landscape. USDA Forest Service General Technical Report PSW-GTR-184. Pacific Southwest Research Station, San Diego
Nixon KC, Muller CH (1997) *Quercus* Sect. *Quercus* Linneus. In: Flora of North America Editorial Committee (ed) Flora of North America North of Mexico Vol 3. Missouri Botanical Garden Press, St. Louis
Oh S-H, Manos PS (2008) Molecular phylogenetics and cupule evolution in Fagaceae as inferred from nuclear CRABS CLAW sequences. Taxon 57:434–451
Ohwi J (1965) Flora of Japan (English ed., edited by F. G. Meyer and E. H. Walker). Smithsonian Institution, Washington, DC
Ørsted AS (1866–1867) Bidrag til egeslægtens systematik. Vidensk Medd naturhist Foren Kjöbenhavn 28:11–88
Ørsted AS (1871) Bidrag til Kundskab om Egefamilien. Kongl Danske Vidensk Selsk Biol Skr 5 naturvidensk math Afd 6:331–538

Pavlyutkin BI (2015) The genus *Quercus* (Fagaceae) in the early Oligocene flora of Kraskino, Primorskii Region. Paleontol J 49:668–676

Pavlyutkin BI, Chekryzhov IU, Petrenko TI (2014) Geology and floras of lower Oligocene in the Primorye. Dalnauka, Vladivostok

Pearse IS, Hipp AL (2009) Phylogenetic and trait similarity to a native species predict herbivory on non-native oaks. Proc Natl Acad Sci 106:18097–18102

Pham KK, Hipp AL, Manos PS, Cronn RC (2017) A time and a place for everything: phylogenetic history and geography as joint predictors of oak plastome phylogeny. Genome. doi:10.1139/gen-2016-0191

Rowley JR (1996) Exine origin, development and structure in pteridophytes, gymnosperms and angiosperms. In: Jansonius J, McGregor DC (eds) Palynology, Principles and Applications. American Association of Stratigraphic Palynologists Foundation, Dallas, pp 443–462

Rowley JR, Claugher D (1991) Receptor-independent sporopollenin. Bot Acta 104:316–323

Rowley JR, Gabarayeva NI (2004) Microspore development in *Quercus robur* (Fagaceae). Rev Palaeobot Palynol 132:115–132

Rowley JR, Skvarla JJ, Ferguson IK, El-Gazhaly G (1979) Pollen wall fibrils lacking primary receptors for sporopollenin. In: Bailey GW (ed) In: Proceedings of the 37th Annual Meeting Electron Microscopy Society of America, San Antonio, TX, August 13–17, 1979. Claitor, Baton Rouge, pp 340–341

Schneider CK (1906) Illustriertes Handbuch der Laubholzkunde, vol 1. Gustav Fischer, Jena

Schwarz O (1934) In: Krause, K.: Beiträge zur Flora Kleinasiens, IV. Feddes Repert 33: 321–328

Schwarz O (1936) Entwurf zu einem natürlichen System der Cupuliferen und der Gattung *Quercus* L. Notizbl Bot Gart Mus Berlin-Dahlem Bd. 13 Nr. 116: 1–22

Schwarz O (1937) Monographie der Eichen Europas und des Mittelmeergebietes. Repertorium specierum nov. regni vegetabilis, Sonderbeihefte D. Selbstverlag Friedrich Fedde, Dahlem-Berlin

Simeone MC, Grimm GW, Papini A, Vessella F, Cardoni S, Tordoni E, Piredda R, Franc A, Denk T (2016) Plastome data reveal multiple geographic origins of *Quercus* Group Ilex. PeerJ 4:e1897

Solomon AM (1983a) Pollen morphology and plant taxonomy of red oaks in eastern North America. Am J Bot 70:495–507

Solomon AM (1983b) Pollen morphology and plant taxonomy of white oaks in eastern North America. Am J Bot 70:481–492

Song SY, Krajewska K, Wang YF (2000) The first occurrence of the *Quercus* section *Cerris* Spach fruits in the Miocene of China. Acta Palaeobot 40:153–163

Spach E (1842) Histoire naturelle des végétaux. Phanerogames, vol 11. Schneider & Langrand, Paris

Spicer RA, Herman AB, Liao W, Spicer TEV, Kodrul TM, Yang J, Jin J (2014) Cool tropics in the Middle Eocene: evidence from the Changchang Flora, Hainan Island, China. Palaeogeogr Palaeoclimatol Palaeoecol 412:1–16

Sprague TA (1929) International Botanical Congress Cambridge (England), 1930. Nomenclature proposals by British botanists. Wyman & Sons, London

Standley PC (1922) Trees and shrubs of Mexico. Contr US Natl Herb 23. Government Printing Office, Washington

Tanai T, Uemura K (1994) Lobed oak leaves from the Tertiary of East Asia with reference to the oak phytogeography of the northern hemisphere. Trans Proc Palaeontol Soc Japan 173: 343–365

Trelease W (1916) The oaks of America. Proc Natl Acad Sci 2:626–629

Trelease W (1924) The American Oaks. Mem Natl Acad Sci 20. Washington Government Printing Office, Washington, DC

Tschan G, Denk T (2012) Trichome types, foliar indumentum and epicuticular wax in the Mediterranean gall oaks, *Quercus* subsection *Galliferae* (Fagaceae): implications for taxonomy, ecology and evolution. Bot J Linn Soc 139:611–644

Walther H, Zastawniak E (1991) Fagaceae from Sosnica and Malczyce (near Wrocław, Poland). A revision of original materials by Goeppert 1852 and 1855 and a study of new collections. Acta Palaeobot 31:153–199

Writing Group of Cenozoic Plants of China (WGCPC) (1978) Cenozoic plants from China. Fossil Plants of China 3. Science Press, Beijing (in Chinese)

Xiang X-G, Wang W, Li R-Q, Lin L, Liu Y, Zhou Z-K, Li Z-Y, Chen Z-D (2014) Large-scale phylogenetic analyses reveal fagalean diversification promoted by the interplay of diaspores and environments in the Paleogene. Perspect Plant Ecol Syst 16:101–110

Xing Y, Onstein RE, Carter RJ, Stadler T, Linder HP (2014) Fossils and large molecular phylogeny show that the evolution of species richness, generic diversity, and turnover rates are disconnected. Evolution 68:2821–2832

Yabe A (2008) Plant megafossil assemblage from the lower Miocene Ito-o Formation, Fukui Prefecture, Central Japan. Mem Fukui Prefect Dinosaur Mus 7:1–24

Chapter 3
The Fossil History of *Quercus*

Eduardo Barrón, Anna Averyanova, Zlatko Kvaček,
Arata Momohara, Kathleen B. Pigg, Svetlana Popova,
José María Postigo-Mijarra, Bruce H. Tiffney, Torsten Utescher
and Zhe Kun Zhou

Abstract The evolution of plant ecosystems during the Cenophytic was complex and influenced by both abiotic and biotic factors. Among abiotic forces were tectonics, the distribution of continents and seas, climate, and fires; of biotic factors were herbivores, pests, and intra- and interspecific competition. The genus *Quercus* L. (Quercoideae, Fagaceae) evolved in this context to become an established member of the plant communities of the Northern Hemisphere, commencing in the Paleogene and spreading to a diverse range of environments in the later Cenozoic. Its palaeontological record, dominated by leaves and pollen, but also including wood, fruits and flowers, is widespread in Eurasia and North America. Consequently, a great number of species have been described, from the 19th century to the present day. Although *Quercus* is currently an ecologically and economically important component of the forests in many places of the Northern

---

E. Barrón (✉)
Museo Geominero, Instituto Geológico y Minero de España—IGME,
Ríos Rosas 23, 28003 Madrid, Spain
e-mail: e.barron@igme.es

A. Averyanova · S. Popova
Laboratory of Palaeobotany, Komarov Botanical Institute,
2 Professor Popov Street, 197376 St. Petersburg, Russia

Z. Kvaček
Faculty of Science, Institute of Geology and Palaeontology, Charles University,
Albertov 6, 128 43 Praha 2, Czech Republic

A. Momohara
Graduate School of Horticulture, Chiba University, 648 Matsudo,
Chiba 271-8510, Japan

K. B. Pigg
School of Life Sciences and Biodiversity Knowledge Integration Center (BioKIC),
Arizona State University, PO Box 874501, Tempe, AZ 85287-4501, USA

J. M. Postigo-Mijarra
Departamento de Sistemas y Recursos Naturales, Escuela Técnica
de Ingeniería de Montes, Forestal y del Medio Natural, Universidad
Politécnica de Madrid, Ciudad Universitaria s/n, 28040 Madrid, Spain

---

© Springer International Publishing AG 2017
E. Gil-Pelegrín et al. (eds.), *Oaks Physiological Ecology. Exploring the Functional Diversity of Genus* Quercus *L.*, Tree Physiology 7,
https://doi.org/10.1007/978-3-319-69099-5_3

Hemisphere and Southeastern Asia, no comprehensive summary of its fossil record exists. The present work, written by an international team of palaeobotanists, provides the first synthesis of the fossil history of the oaks from their appearance in the early Paleogene to the Quaternary.

3.1 Introduction

Genus *Quercus* Linnaeus 1753

1753—*Quercus*, Linné. Species Plantarum Vol. 2, p. 995

1870–1872—*Quercus*, Schimper. Traité de Paléontologie végétale Vol. 2, p. 616

Diagnosis (sensu Schimper): *Arbores, rarius frutices, pro more sylvas vastas efformantes. Folia alternantia, petiolata, caduca (in regione frigidiore) vel persistentia (in regione calidiore), membranacea, subcoriacea, et coriacea, quam maxime variablia, margine simpliciter vel repetito-lobata, lobulata, crenata, dentata vel subspinosa, rarius integra, sæpius anguste incrassato-marginata, laevia vel pubescentia, pinnatinervia; nervo medio plus minus valido ad apicem producto, nervis secundariis craspedodromis, camptodromis in foliis integris, mixtis in foliis pro parte integris and pro parte dentatis vel crenatis. Flores masculi in amento gracili solitarii, rarius ternati, perigonio regulariter vel irregulariter 4–7—lobato; flores feminei gemmacei, axillares, in rachi communi sessiles, bracteis et squamulis multiseriatis imbricatis, in cupulam (involucrum) floris basin recipientem connatis. Fructus e cupula solida squamosa vel zonata nunquam spinosa, et e glande plus minus emersa vel subinclusa constans.*

Oaks (*Quercus* L., Fagaceae) are woody angiosperms known to humans from prehistoric times (Pereira Sieso and García Gómez 2002). Currently, they are one of the most significant and diverse elements in Northern Hemisphere forest ecosystems (Fig. 3.1), comprising ca. 400–500 species of trees and shrubs in North and Central America, Colombia, Eurasia and northern Africa (Nixon 1997; Govaerts and Frodin 1998). Their highest diversity occurs in Central America and Southeast

---

B. H. Tiffney
Department of Earth Science and College of Creative Studies,
University of California, Santa Barbara, CA 93106-6110, USA

T. Utescher
Senckenberg Research Institute, Frankfurt am Main, Germany

T. Utescher
Steinmann Institute, University of Bonn, Nussalle 8, 531 15 Bonn, Germany

Z. K. Zhou
Xishuangbanna Tropical Botanical Garden, Chinese Academy of Sciences,
Kunming 650244, China

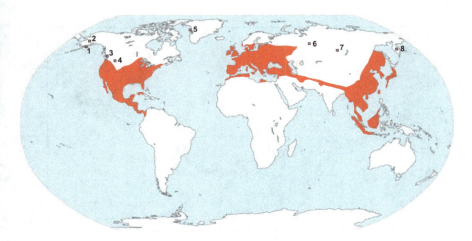

Fig. 3.1 Modern distribution of the genus *Quercus* modified from Camus (1936–1954) and Manos et al. (1999). Selected northern localities with the presence of fossils belonging to oaks: 1. Early/middle Miocene, Seldovia Point, Alaska (Wolfe 1980), 2. Early Miocene, Sanctuary Formation, Central Alaska (Leopold and Liu 1994), 3. Early Eocene, Princeton Chert, British Columbia, Canada (Grímsson et al. 2016), 4. Oligocene, Ruby Basin, Montana, USA (Lielke et al. 2012), 5. Middle Eocene, Qeqertarsuatsiaat Island, Greenland (based on pollen, Grímsson et al. 2016), 6. Oligocene, Dunaevskiy Yar, Siberia, Russia (Iljinskaja 1982), 7. Miocene, Kozhevnikovo, Siberia, Russia (Iljinskaja 1982), 8. Late Eocene, Duktylikich River, Kamchatka, Russia (Budantsev 1997)

Asia, with a lower number of species in western America, western Eurasia and the Mediterranean (Nixon 2006; Simeone et al. 2016).

Oaks are common or even dominant species in a wide variety of habitats, including temperate deciduous forests, subtropical and tropical savannas, cloud forests, tropical montane forests and Mediterranean vegetation (Nixon 2006). *Quercus* is also economically significant in temperate and (sub-) tropical areas of the Old and the New Worlds (Camus 1936–1954). Their fruits are widely exploited for food by both people and livestock and their woods are used for construction and fuel (Nixon 2006). These are the reasons for the early botanical (i.e. de Candolle 1862), and palaeobiological interest in the genus (Schimper 1870–1872; de Saporta 1888; Zittel 1891).

Currently, *Quercus* is placed in the subfamily Quercoideae of the Fagaceae, together with *Castanea* Mill., *Castanopsis* (D. Don) Spach, *Chrysolepis* Hjelmq., *Lithocarpus* Blume, and *Trigonobalanus* Forman. Its infrageneric classification has been a matter of controversy (e.g. Camus 1936–1954; Schwarz 1936; Menitsky 1984; Nixon 1993). Traditionally, oaks were classified in two subgenera: *Cyclobalanopsis* (Oersted) Schneider (cycle-cup oaks) and *Quercus* L. (scale-cup oaks). Subgenus *Cyclobalanopsis* is restricted largely to subtropical and tropical regions in Southeast Asia (Fig. 3.2) and can be distinguished by fruit morphology and DNA-based evidence (Manos et al. 1999). Subgenus *Quercus* is divided in

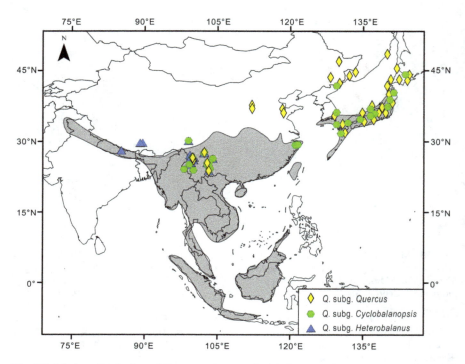

Fig. 3.2 Modern distribution of *Quercus* subgenus *Cyclobalanopsis* (in grey; modified from Xu et al. 2016) and presence of fossil oaks in East Asia (Section *Quercus* = yellow diamond, Subgenus *Cyclobalanopsis* = green hexagon, Section *Heterobalanus* = blue triangle)

three sections: *Lobatae* Loudon (red oaks; North and South America; Fig. 3.3), *Protobalanus* (Trelease) Schwarz (intermediate oaks from western North America; Fig. 3.3) and *Quercus* L. (white oaks: Eastern and Western Hemispheres) (Nixon 1993; Manos et al. 1999).

According to Nixon (1993) two groups of white oaks (groups *Ilex* and *Cerris*), both included currently in section *Quercus*, need more analysis to confirm their final taxonomic status. This classification has been widely accepted and used over the last twenty-five years. However, Denk and Grimm (2009, 2010) recently presented a new classification in which six informal infrageneric units are recognized: a first subgroup includes the *Quercus* group *Quercus* (white oaks), group *Lobatae* (red oaks) and group *Protobalanus* (golden cup oaks) and a second subgroup is formed by the group *Ilex* (Ilex oaks), group *Cerris* (Cerris oaks) and group *Cyclobalanopsis* (cycle cup oaks).

Tracing the evolution of the genus *Quercus* in the fossil record is a complex matter. Although the genus possesses a consistent set of reproductive features, it also exhibits marked vegetative variation (Tucker 1974; Manos et al. 1999). Intraspecific morphological variation within living *Quercus* is not uncommon and hybridization may play a significant role in these patterns (Manos et al. 1999;

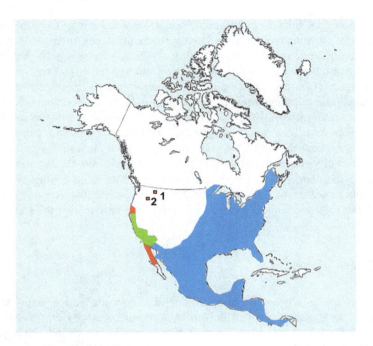

Fig. 3.3 Modern distribution of the American endemic sections *Lobatae* (in blue) and *Protobalanus* (in red; the green colour corresponds to the area where the species of the two sections are both present), modified from Camus (1936–1954) and Manos et al. (1999). Species of the section *Lobatae* are also present in Colombia. During the Cenozoic, section *Lobatae* was widespread in Europe and Western Asia (see the text). Section *Protobalanus* was always restricted to western North America with early occurrences in 1. Montana during the Eocene–Oligocene border and Oligocene (Becker 1969) and 2. Idaho during the Oligocene and late Miocene (Axelrod 1998; Buechler et al. 2007)

Nixon 2006). For this reason, phylogenetic relationships based only on leaf morphology are not completely reliable. Some authors believe that pollen ornamentation is the only morphological character that unambiguously can be used to distinguish extant members of most groups, except for the *Quercus* and *Lobatae* groups that share the same pattern (Denk and Grimm 2010).

The first attribution of fossil to *Quercus* was made by Schimper (1870–1872), who focused on leaf features and included the genera *Trigonobalanus* and *Lithocarpus* in his concept of *Quercus*. In Late Cretaceous and Paleocene floras, different leaves with a general fagaceous morphology have been assigned to extinct genera such as *Dryophyllum* Debey ex Saporta and *Quercophyllum* Fontaine (Mai 1995). However, their systematic affinities are in need of thorough re-examination.

Fossil leaves, with and without cuticle, wood, pollen, fruits and flowers belonging to *Quercus* are common in Oligocene through Quaternary floras of the Northern Hemisphere. Given the abundance of leaves in the fossil record of oaks, it would be remiss not to note the difficulty of distinguishing species of closely related

Quercus based on foliage, both among extant oaks and particularly in the fossil record where identifications are frequently based upon one or a few specimens. Clearly, given the morphological variability in living oaks, it is quite possible that many of the established fossil taxa do not represent true biological species and may be misassigned. Consequently, the number of fossil species of oaks is most likely exaggerated, and in many cases, these fossil species need a serious revision.

To date, no synthesis of *Quercus* exists that integrates palaeobotanical and palynological data with palaeogeographic, palaeoecological and palaeoclimatic changes throughout the Cenozoic and Quaternary. The aims of the present work are: (1) to clarify the true first occurrences of the genus on all continents, (2) to summarize the main species of *Quercus* considered to be of ecological or taxonomic importance in the Paleogene and Neogene, and (3) to reconstruct the main traits of the palaeoecological role and evolution of *Quercus* over the Cenozoic.

3.2 Methodology

Fossil oaks are usually identified by means of leaves and pollen grains and less commonly, by woods, fruits and flowers. Systematists working with extant plants can study hundreds of herbarium specimens and sample populations. Consequently, they can circumscribe both parent species and the morphological intermediates among them that form a hybrid complex and use morphometrics to establish the boundaries of the species within the complex (e.g., Jensen et al. 1993) to create hypotheses that can be tested with molecular studies. However, this is not possible in the fossil record. Thus one can but sympathize with Chaney (1944, p. 342) in his reflections upon of *Quercus winstanleyi* Chaney from the Troutdale Flora: "*There are times in the life of every paleontologist when difficulties involved in the sound treatment of his material become insurmountable. At such times it is necessary to make decisions which are not wholly consistent with all the known facts, and which may be at variance with the best taxonomic procedure. Such a compromise seems inescapable in the case of the oak leaves of the Troutdale flora, which we are describing as a new species in spite of close resemblances to previously described Tertiary oaks.*"

We recognize two major limitations of the present synthesis that may be confusing to readers. Firstly, in some cases we have accepted the literature as it stands without attempting to verify the identifications of individual fossils. Further, many species are based on very brief diagnoses and limited material. For instance, Axelrod (1998) names six new species using very short paragraphs of description, based upon one or a very few, often incomplete, specimens. Consequently, we have not included many references in our summation, especially those from older literature, where fragments of leaves were often used for describing new species that currently need revision. This is a conservative, but probably a more accurate approach.

Secondly, we have not addressed the larger question of whether certain existing fossil species can continue to stand alone, or should be synonymized, as suggested by previous authors. The species synonymies are well exemplified in the case of the famous Miocene flora from Oehningen (Switzerland). At this site, Heer (1853, 1856, 1859) described around forty species of oaks from leaf remains. After an intensive revision made by Hantke (1965), the number of the species were reduced only to five. One need only look at the tremendous thought and time that Chaney and Axelrod (1959) spent in re-assigning previously described specimens of *Quercus* to species (pp. 164–172), and the associated systematic revisions (pp. 219–222) to appreciate the enormity of such a task.

Even following such work, uncertainties still exist. Leckey and Smith (2015), in their review of the geologic history of gall wasps in western North America, specifically note that (pp. 237–238): "*Following the opinions of previous workers, Q. hannibali and Q. dayana are considered to be synonyms of Q. pollardiana* (Axelrod 1983; Fields 1996), *although other authors continue to use the name Q. hannibali* (Buechler et al. 2007)". Further, Fields (1996) has suggested in his Ph.D. thesis that *Q. consimilis* Newberry be subsumed in *Q. simulata* Knowlton, and Buechler et al. (2007) placed *Q. winstanleyi* Chaney as a junior synonym of *Q. columbiana* Chaney. As if such uncertainties at the species level were not confusing enough, there is the question of whether the widespread western North American species *Q. simulata* should be assigned to *Lithocarpus* or *Castanopsis* (Fields 1996).

From these observations, it is clear that particularly the foliage of fossil oaks is ripe for an in-depth revision, although this could be an immense task. It may be aided in the present day by more detailed and innovative approaches to the identification of foliar remains (Huff et al. 2003; Ellis et al. 2009) such as the epidermal studies (see e.g., Kvaček and Walther 1989; Walther and Zastawniak 1991). It is also possible that insights from palaeoentomology may provide some guidance, as cynipoid gall wasps have tight host relationships with *Quercus* (Diéguez et al. 1996; Stone et al. 2009) and have a distinct fossil record (Erwin and Schick 2007; Liu et al. 2007b; Holden et al. 2015; Leckey and Smith 2015). Thus, while we provide a summation of the known record, we recognize it is fraught with errors that will be corrected by future research.

In most cases, the description of fossil material includes a comparison with one or more living species, allowing provisional assignment of the fossil to a section of the genus. In cases where conflicts have occurred in the literature, we have not ascribed the fossil to a section. Clearly, given the morphological variability in living oaks, it is quite possible that many of the established fossil taxa do not represent true biological species and may be mis-allied to section. Indeed, Bouchal et al. (2014, p. 1339) note that some leaf species that earlier authors assigned to section *Quercus* (*Q. dumosoides* MacGinitie, *Q. mohavensis* Axelrod) can in fact not "*unambiguously be referred*" to either section *Quercus* or *Lobatae*. We therefore emphasize that, in the following summation, assignments to section based upon leaves are based upon several levels of assumption, and are open to revision with further study.

The fossil wood record is usually defined to genus, but some wood anatomical features can delimit intrageneric types (see e.g., Pl. 3.8-10). Wiemann et al. (1998) were able to group 30 species of extant temperate oaks from North America into three wood types: the live oaks (e.g., *Q. virginiana* L.), the white oaks, both of section *Quercus*, and the red oaks, of section *Lobatae*. Wood of the live oaks is diffuse porous or semi-ring-porous (and difficult to distinguish from that of the fagaceous genus *Lithocarpus* Blume), while the red and white oak types are ring porous, but can be separated on the basis of vessel and ray features.

Most pollen records are not offered in sufficient detail to permit sectional assignment and are thus treated as a record of the genus. Recent work by Bouchal et al. (2014) building on advances in the study of oak pollen by Denk and Grimm (2009) provides taxonomically useful resolution of fagaceous pollen in western North America and Europe at the sectional level through a detailed, same grain, LM/SEM analysis of micromorphological features. This approach leads the way to future re-analysis of other reports of *Quercus* pollen allowing sub-generic resolution.

Palaeoclimates were determined using the coexistence approach (CA) method (Mosbrugger and Utescher 1997; Utescher et al.2014), employing ClimStat software and the Palaeoflora database (Utescher and Mosbrugger 2015). The latter contains climate information for more than 1800 extant plant taxa at the global scale.

Specimens were photographed from the following institutions: University of Washington, Burke Museum of Natural History and Culture, USA; Florida Museum of Natural History, Gainesville, Florida, USA; University of South Alabama, Mobile, Alabama, USA; University of California at Santa Barbara, California, USA; Arizona State University, Tempe, Arizona, USA; Museo Geominero (IGME), Museu de Geologia de Barcelona and Museu de Geologia del Seminari de Barcelona, Spain; Museum für Naturkude, Stuttgart, Germany; National Museum, Prague, Czech Republic; Hungarian Natural History Museum, Budapest, Hungary; Komarov Botanical Institute, St. Petersburg, Russia; National Museum of Nature and Science, Kyoto, Japan; Laboratory of Palaeoecology, Xishuangbanna, Tropical Botanical Garden, Chinese Academy of Sciences, China.

3.3 Results

The oldest mesofossil of fagaceous affinity is *Archaefagaecea* Takahashi, Friis, Herendeen and Crane from the early Coniacian of Japan (Takahashi et al. 2008). Friis et al. (2011), also report the presence of fossil flowers with distinctive fagaceous characters from Santonian strata, indicating that the Fagaceae was present by the mid–Late Cretaceous.

3.3.1 The Cretaceous and Paleocene record

3.3.1.1 North America

Santonian and Campanian (Late Cretaceous) mesofossils of fagaceous affinities have been described from eastern North America. These include the Santonian *Antiquacupula sulcana* Sims et al. (1998) and *Protofagacea allonensis* Herendeen et al. (1995), staminate flowers, associated fruits and cupules from the Campanian in central Georgia. Additional fruits and flowers are known from the Campanian of Massachusetts (Taylor et al. 2012). The most common Cretaceous fossils attributed to *Quercus* in North America are leaves, particularly in the older literature, e.g., *Q. wardiana* Lesquereux from the Cenomanian Dakota Formation of Kansas (Lesquereux 1892) and Black Hills, South Dakota (Ward 1899). None of Lesquereux's species occur in the Cenozoic (LaMotte 1952).

Wang (2002) reviewed the angiosperm leaf record of the Dakota Formation, but offered few synonymies and updates of Lesquereux's material. The only *Quercus*-related occurrence he discussed is that of the leaf type *Quercophyllum tenuinerve* Fontaine. *Quercophyllum* is a form genus created by Fontaine (1889) to include leaf remains similar to those of *Quercus*. Species of this genus have been identified in Late Cretaceous floras from North America and Europe (e.g., Dorf 1942; Němejc and Kvaček 1975; Váchová and Kvaček 2009).

In more recent literature, Bell (1957) described *Quercus richardsonii* from the Turonian–Santonian Comox Coal Field in the Nanaimo Group of Vancouver Island, British Columbia, Canada. However, recent accounts of megafloral assemblages of this region do not mention this or any other species of *Quercus* (e.g., Jonsson and Hebda 2015). In the Late Cretaceous Eagle Formation of Montana, Van Boskirk (1998) cites *Quercus* leaves but these have not been confirmed. In his review of the Cretaceous floras of the Rocky Mountains, Crabtree (1987) recognizes forms assignable to Fagales by the early Campanian but does not specify any assignments to *Quercus*. *Quercus viburnifolia*? cited by Dorf (1942) from the Late Maastrichtian Lance Formation is listed by LaMotte (1952), as also occurring in the Paleocene Denver Formation of Golden, Colorado. However, recent treatments of the Denver Formation lack *Quercus* (Barclay et al. 2003), raising the likelihood this report is erroneous. Lozinsky et al. (1984) refer to the occurrence of cf. *Q. viburnifolia* and cf. *Dryophyllum* leaves in the upper Coniacian–Santonian Crevasse Canyon Formation in association with dinosaurs from the overlying McRae Formation of south central New Mexico. They cite two unpublished master's theses (Lozinsky 1982; Wallin 1983) without further substantiation. We could find no further reference to *Quercus* in this area.

Recently, dispersed pollen attributed to Fagaceae has been described in the Campanian of Elk Basin (Wyoming) as *Paraquercus campania* by Grímsson et al. (2016). This pollen shows similarities with both *Eotrigonobalanus* and *Quercus*.

Paleocene reports of *Quercus* leaves in North America are known largely from the work of Roland Brown from localities in Wyoming, North Dakota, Montana and Colorado and include five species: *Q. sullyi* Newberry, *Q. groenlandica* Heer, *Q. macneili* R.W. Brown, *Q. yulensis* R.W. Brown and *Q. asymmetrica* Trelease (Brown 1962; Budantsev and Golovneva 2009). Manchester (2014) reviewed these occurrences. He united *Q. sullyei* with similar forms described as *Dicotylophyllum flexuosum* (Newberry) Wolfe, *Meliosma longifolia* (Heer) Hickey, and *Dyrana* (Newberry) Golovneva, in the form taxon *Dyrana flexuosa* and suggested them to be of possible platanaceous affinity (Wolfe 1966; Hickey 1977; Golovneva 2000; Manchester 2014). Manchester (1999) reassigned *Q. groenlandica* to *Fagopsiphyllum groenlandicum* (Heer) Manchester for fagaceous leaves that cannot be assigned to genus. This "*Quercus*" taxon has also been reported from the Bighorn Basin of Wyoming (Wing et al. 1995). The last three species Manchester recognized informally as "*Quercus*" *macneili*, "*Quercus*" *yulensis* and "*Quercus*" *asymmetrica* with the provision that the assignment of these fossils, even at the level of family, is suspect in the absence of the "*unequivocal cupulate fruits*" distinctive of the family in the associated sediments. Another Paleocene report of *Quercus* (Benammi et al. 2005) is based upon pollen from an unpublished Master's thesis (Altamira-Areyán 2002) and requires further validation.

3.3.1.2 Europe and Western Asia

To date, no Cretaceous remains are attributable to the *Quercus* in Europe. The earliest evidence of fossils related to this genus indicates a late Paleocene age. However, some of these occurrences are suspect. One of the few late Paleocene sites that can be verified is that of Ménat, France (originally thought to be Eocene, the Ménat site is now dated by K-Ar to ca. 56 Ma; Vincent et al. 1977; Michon and Merle 2001). From this locality, leaves have been assigned to *Q. lonchitis* Unger, *Q. parceserrata* Saporta and Marion or *Q. provectifolia* Saporta. Some macroremains from the Heer collection from Ménat were also attributed to catkins and acorns of *Quercus* (Laurent 1912). Within these fossils, two leaves in excellent state of preservation have been identified as *Quercus subfalcata* Friedrich, which occurred with *Platanus schimperi* Saporta and Marion and other taxa such as *Palaeocarpinus-Craspedodromophyllum*, *Casholdia* and a *Sassafras*-like morphotype (Piton 1940; Kvaček 2010). *Q. subfalcata* could be the earliest occurrence of the genus in Europe if its taxonomic status can be confirmed.

Quercus is clearly present in Europe at the Paleocene–Eocene transition (ca. 55 Ma) at St. Pankraz (Austria), which was placed in the northwestern Tethyan realm (Hofmann et al. 2011). These authors identify *Quercoidites*, pollen related to *Quercus* infrageneric group *Ilex* sensu Denk and Grimm (2009). *Quercus* formed part of communities of megathermal (e.g. *Lannea*, Arecaceae, Chloranthaceae, Icacinaceae) and mesothermal taxa (e.g. *Eotrigonobalanus*, *Ilex*, *Parthenocissus*). The authors state that the palynoflora represents a "*warm temperate evergreen-deciduous forest*", which can be classified as a kind of subtropical flora with temperate elements, typical of a warm and wet, but not tropical climate. Although the St. Pankraz pollen shares a number of similarities with pollen of extant members of group *Ilex*, Denk et al. (2012) note that this pollen cannot with certainty be assigned to this group.

3.3.1.3 Eastern Asia

Quercus cretaceoxylon Suzuki and Ohba is the first fossil record of *Quercus*-type wood from East Asia, which was described from the Upper Cretaceous of the Upper Yezo Group in Mikasa City, Hokkaido, Japan (Suzuki and Ohba 1991). This taxon, described from a piece of silicified wood, exhibits typical red-oak type wood with distinct ring porosity and radial arrangement of medium to small, round, thick-walled latewood pores. It also presents transition from early to latewood gradual, abundant tracheids and wood parenchyma.

In northern Asia the first occurrence of "*Quercus*" is *Q. tsagajanica* Pojark. from the lower Paleocene of Bureinskyi Tzagayan (Far Eastern Russia). This species shows rounded or oval leaves with small, sharp teeth (Iljinskaja 1982). Likewise, isolated "*Quercus*" fossils have been cited from the Upper Cretaceous of Kazakhstan as *Q. buroinensis* Shilin (1983). According to Iljinskaja (1982), the generic identification is doubtful because of its poor preservation. Moreover, in several Late Cretaceous localities of Kazakhstan, two pollen species have been attributed to oaks, *Q. aurita* Bolkhovitina and *Q. sparsa* (Mart.) Samoil (Boitzova and Panova 1966). However, these pollen grains need evaluation using modern techniques.

Possible early evidence for oaks in Asia comes from leaf materials recovered in the early Paleocene of the Russian Far East (Primorje) being assigned to the genus only (localities Sobolevka, Augustovska; Akhmetiev 1988; Kodrul 1999). The same strata provided also pollen materials (*Quercus* sp.). The palynological record of oaks became more diverse in this region from the Ypresian, including various morphotypes (*Quercus conferta* Kit. [=*Q. frainetto* Ten.], *Q. forestdalensis* Trav., *Q. graciliformis* Boitz., *Q. gracilis* Korth.; Pavlyutkin and Petrenko 2010) that are, however, not referable to any extant taxonomic group.

In light of our survey, we agree with authors such as Jones (1986), Zhou (1993) and Xing et al. (2013) that pre-Paleogene, and perhaps pre-Eocene occurrences of *Quercus* macroremains are generally represented by poorly preserved fossils that lack critical features needed for certain identification and need to be treated with caution.

3.3.2 *North America: Eocene through Pliocene*

Currently, North America hosts three sections of the genus *Quercus*: (i) Section *Lobatae* (the "red oaks"; Fig. 3.3), represented by 35 species dominantly of eastern North America with six species in North America west of the Rocky Mountains; (ii) Section *Protobalanus* (the "intermediate oaks"), represented by 4 species, all western North American (Fig. 3.3), and (iii) Section *Quercus* (the "white oaks") represented by 51 species, 22 of which occur west of the Rocky Mountains (Flora of North America 1997). These three sections are represented in the North American fossil record, as well as the endemic Asian section *Cyclobalanopsis* (the "ring-cupped oaks"; Manchester 1994) and possibly the Eurasian group *Cerris* (Becker 1969).

In general terms, the Paleogene and Neogene fossil record of *Quercus* is dominated by leaf specimens in western North America, particularly in the Oligocene through Pliocene, and generally by pollen records in eastern North America (cf. Ochoa et al. 2012; Baumgartner 2014), reflecting the relative paucity of Cenozoic deposits yielding megafossils in the East. Petrified woods are known from several Eocene and Neogene sites in western North America (Wheeler and Manchester 2002; InsideWood 2004 onwards; Wheeler and Dillhoff 2009). Wood is even rarer in the East, except for the Miocene of Vermont (Spackman 1949) and an informal report from the Oligocene of east Texas (Singleton 2001).

Records of fruits are uncommon, ranging from the middle Eocene acorn from Clarno, Oregon (Plate 3.1 2; Manchester 1994) and the late Eocene of Oregon (Manchester and McIntosh 2007) and the LaPorte Flora of California (Plate 3.1 5; Tiffney, unpublished) to Oligocene forms in Texas (Daghlian and Crepet 1983; Crepet 1989) and Oregon (Bridge Creek flora; Meyer and Manchester 1997), Miocene specimens from Washington (Borgardt and Pigg 1999; Plate 3.1 4, 6), Vermont (Plate 3.1 3; Tiffney 1994) and the Miocene/Pliocene of Tennessee (Liu 2011).

3.3.2.1 The Eocene Records

The Early to Middle Eocene

Unequivocal evidence of *Quercus* in the fossil record lies with acorns. The oldest North American acorn known to date is *Quercus paleocarpa* Manchester, from the middle Eocene Clarno Formation of Oregon (~ 44 Ma; Manchester 1994, Plate 3.1 2). The Clarno acorns have involucral scales arranged in concentric rings, a feature that occurs in both the Asian *Quercus* subgenus *Cyclobalanopsis*, as well as in the genus *Lithocarpus*. However, Manchester (1994) suggests their assignment to *Cyclobalanopsis* based on the presence of woody, rather than papery involucral scales. Leaves (Manchester 1981) and wood (Scott and Wheeler 1982; Wheeler and Manchester 2002) consistent with *Quercus* are also known from Clarno.

The Clarno flora is often compared with coeval European floras, including the Eocene London Clay (Reid and Chandler 1933; Collinson 1984) and the middle Eocene flora of Messel, Germany (Collinson et al. 2012). One might also include the permineralized middle Eocene Princeton chert of British Columbia and the Appian Way flora of Vancouver Island (Mindell et al. 2007, 2009; Pigg and DeVore 2016). It is interesting that of these floras, Fagaceae appears only at Clarno and Appian Way, in contrast to the other families shared by these floras (Pigg and DeVore 2016).

Additional reproductive structures are known from the middle Eocene in the Southeast. *Quercus oligocenensis* Daghlian and Crepet is an incomplete staminate catkin from the middle Eocene Claiborne Formation of Tennessee (Wang et al. 2013). This taxon was first described from the Oligocene Catahoula Formation of Texas, which shares many floral elements with the Claiborne Formation (DeVore et al. 2014). The Claiborne Formation also provides the oldest record of *Quercus* pollen in the Southeast (Frederiksen 1981, 1988; Graham 1999a; Burnham and Graham 1999). In the Northwest, the microthermal assemblages of Republic, Washington and other Okanogan Highlands localities, now recognized as latest early Eocene, contain fagaceous leaves (Greenwood et al. 2016). However, acorns have not been recovered from this area and the assignment of the leaves to particular genera within the Fagaceae is *"not unequivocal"* (Gandolfo 1996). No evidence exists at these localities for acorns or leaves that could be assigned conclusively to *Quercus* (KB Pigg, Personal observation). MacGinitie (1969), in describing the early to middle Eocene Green River flora, dismissed earlier reports of *Q. castaneopsis* Lesquereux and *Q. drymeja* Unger as unsupported, but recognized two new species, *Q. cuneatus* MacGinitie and *Q. petros* MacGinitie, comparing these to modern species in sections *Quercus* and *Lobatae* respectively.

Pollen assigned to *Quercus* has been listed as occurring in the Republic, Princeton ("Allenby"), McAbee, Hat Creek and Horsefly floras by Moss et al. (2005). It has also been reported and illustrated from McAbee, British Columbia, based on light microscopy (Dillhoff et al. 2005) and from the Princeton Chert, British Columbia using SEM single-grain studies (Grímsson et al. 2016). No corresponding, unequivocal megafossil remains are known at these sites. At higher latitudes in North America, the pollen type *Quercoidites* is reported from the Alaskan Eocene (Frederiksen et al. 2002; Bouchal et al. 2014). The presence of *Quercus* or a *Quercus*-like plant at high latitudes is further supported by foliage (Basinger 1991; McIver and Basinger 1999) and pollen (McIntyre 1991) from the middle Eocene Buchanan Lake Formation of Axel Heiberg Island, Northwest Territories and pollen from the middle Eocene Margaret Formation of Ellesmere Island (Jahren 2007; Eberle and Greenwood 2012), and west Greenland (Grímsson et al. 2015).

The middle Eocene leaf record of *Quercus* is scant in part because of the difficulty in distinguishing leaves of *Quercus* from those of other fagaceous genera. Fagaceous leaves with features of both *Quercus* and the modern chestnut *Castanea* Mill. have been difficult to distinguish from one another, although *Quercus* leaves have a fimbrial vein that is lacking in *Castanea* (Manchester 1999). This situation has led to the need for caution in assigning leaves to these taxa, especially early in their evolutionary history. An example is provided by MacGinitie (1941), who identified *Quercus nevadensis* MacGinitie leaves from the Chalk Bluffs and Buckeye Flat sites in the Central Sierra Nevada, California. He compared this species to *Q. glauca* Thunberg and *Q. hiananensis* Merrill, but also noted its similarity to *Castanopsis* and suggested that it "*...may possibly represent fossils of that genus*". In Chalk Bluffs he also lists *Q. distincta* Lesq. and *Q. eoxalapensis* MacGinitie. In both cases their modern relative species are in the section *Lobatae*.

Thus it appears that section *Lobatae* is present in the middle Eocene in southeastern (Claiborne Formation; Wang et al. 2013) and western (Chalk Bluffs; MacGinitie 1941) North America and subgenus *Cyclobalanopsis* (Manchester 1994) in western North America.

Fossil oak wood from the middle Eocene is known from Clarno and Post (Oregon) and was assigned to *Quercinium crystallifera* Scott and Wheeler, a species with features found within Fagaceae and with noted similarities to extant evergreen oaks "*of the* Quercus/Lithocarpus *wood type*" (InsideWood 2004 onwards; Wheeler et al. 2006). Additionally, an acorn with apparently whorled scales is reported from Post (Manchester and McIntosh 2007).

The middle Eocene Yellowstone Fossil Forest yields both early reports of leaves (Knowlton 1899) and of permineralized woods (Wheeler et al. 1978; Pl. 3.2 8–10). Wood at Yellowstone is identified as *Quercinium* Unger emend. Brett. *Quercinium* was considered originally to represent ring porous woods with two types of rays (Unger 1842). Brett (1960) recognized that fossil woods assigned to this genus are similar to both those of modern evergreen species of *Quercus* and those of *Lithocarpus*. He emended the genus to include fossil woods with the anatomical features that characterize these two wood types.

Wheeler et al. (1978) placed one Yellowstone species in *Quercinium amethystianum* Wheeler, Scott and Barghoorn, and reexamined material of a second species, *Q. lamarense* Knowlton emend. Wheeler, Scott and Barghoorn. Both species are considered to be "evergreen oaks" and "indistinguishable from *Lithocarpus*". Wheeler et al. (1978) also summarize earlier species described from Yellowstone, including *Q. knowltonii* Felix (Felix 1896) and *Quercus rubida* Beyer (Beyer 1954). *Q. knowltonii* is inadequately known for comparison with other species, whereas *Q. rubida* is ring-porous and thus unlike the other forms known from Yellowstone. The affinity of this last taxon with Fagaceae is not clear.

The Late Eocene

The late Eocene Florissant Fossil Beds of central Colorado (34 Ma) contain a well-studied megafossil (MacGinitie 1953; Manchester 2001) and palynomorph flora (Bouchal et al. 2014), both of which document the occurrence of *Quercus*. MacGinitie (1953) identified the presence of nine separate species from a variety of Florissant localities, all based upon leaves. In his review of these, Manchester (2001) recognized leaves only to the level of *Quercus* spp. Bouchal et al. (2014) reviewed these leaves and concluded that, while three of these species were likely not *Quercus*, the six others included representatives of sections *Quercus*, *Lobatae*, *Protobalanus*, with two species resembling extant species in both *Quercus* and *Lobatae*.

Pollen from Florissant was initially studied by Leopold and Clay-Poole (2001) who found two forms, one "typical" *Quercus* type, and a second, prolate form they referred to "Quercoid, long axial pollen" based on light microscopy. They suggested that the "quercoid" pollen might be an extinct genus. Pollen was examined using SEM as well as light microscopy by Bouchal et al. (2014). Taken in parallel with the leaf record, these authors suggested the presence of sections *Quercus*, *Lobatae*, *Protobalanus* and *Quercus/Lobatae*. Additionally, they confirmed the presence of the subgenus *Cyclobalanopsis*, potentially supporting Manchester's (1994) report from the Clarno Formation that indicated the presence of this section in the Paleogene of North America.

Platen (1908) described three species of wood of the evergreen oak type from the late Eocene of California: *Quercinium anomalum, Q. solderderi,* and *Q. wardii. Q. anomalum* was later reassigned to *Quercoxylon anomalum* (Platen) Mädel-Angeliewa (Mädel-Angeliewa 1968), while *Quercinium wardii* was reassigned to *Lithocarpoxylon wardii* (Platen) Suzuki and Ohba (InsideWood 2004 and onward). The morphogenus *Quercoxylon* was described by Kräusel (1939) including fossil wood resembling *Quercus/Lithocarpus*. InsideWood accepts the two transfers to *Quercoxylon* but refers to *Lithocarpoxylon* as a synonym.

3.3.2.2 The Oligocene

A great number of oak species have been described in Oligocene North American floras, mainly from leaf remains. Becker monographed several floras in closely adjacent Oligocene basins in southwestern Montana, reporting 9 taxa (Becker 1961, 1969, 1973). More recent work (Lielke et al. 2012) confirms that these floras fall at the Eocene–Oligocene boundary or the early Oligocene. From the Ruby Basin, Becker (1961) recognized *Q. brooksi* Becker, *Q. consimilis* Newberry, *Q. convexa* Lesq., and *Q. mohavensis* Axelrod. From the Beaverhead Basins and the York Ranch Flora (Becker 1969, 1973), five species that are widespread and are also common in the Neogene have been described: *Q. eoprinus* Smith, *Q. prelobata* Condit, *Q. pseudolyatra* Lesq., *Q. winstaleyi* Chaney (=*Q. columbiana* Chaney per Buechler, et al. 2007), and *Q. dispersa* (Lesq.) Axelrod.

The posited modern relatives of these species include members of sections *Lobatae, Protobalanus* and *Quercus*. Interestingly, *Q. prevariabilis* Becker that was described in the Beaverhead Basins, was compared with five living species, four of which (*Q. acutissima* Carruthers, *Q. chinensis* Bunge, *Q. variabilis* Blume, *Q. bungeana* F. B. Forbes [a synonym of the preceding] and *Q. chenii* Nakai), are all of group *Cerris*. One living relative species (*Q. serrata* Thunb.) was in section *Quercus*. If the dominant group of modern relatives is correct, this is the only record of group *Cerris* we are currently aware of in North America.

From the Early Oligocene Bridge Creek flora of Oregon, Meyer and Manchester (1997) recognized leaves of *Quercus berryi* Trelease at three localities, and *Q. consimilis* Newberry at seven localities. Additionally they recognized acorns and cupules at 5 localities as *Quercus* spp., and suggested their affinities with either sections *Lobatae* or *Quercus*.

In his study of the middle Oligocene Lower Haynes Creek Florule (Idaho), Axelrod (1998) recognized six new species: *Q. bilobata* Axelrod, *Q. castormontis* Axelrod, *Q. haynesii* Axelrod, *Q. lemhiensis* Axelrod, *Q. moyei* Axelrod, and

Q. snookensis Axelrod, all based on leaves. Of these, two are allied with section *Protobalanus*, one with *Quercus* and three left with unclear affinities. Three of these are based upon a single leaf and a fourth upon two specimens. Additionally, Axelrod reports *Q. predayana* MacGinitie of section *Protobalanus*.

Likewise, trigonobalanoid, castaneoid and fagaceous remains occur together in the Catahoula Formation near Huntsville, east Texas (Daghlian and Crepet 1983; Crepet 1989; DeVore et al. 2014). There, staminate catkins, leaves, acorns and woods related to *Quercus* were collected. *Q. oligocenensis* was described from staminate catkins. It possesses *in situ* pollen and exhibits perianth with features of extant members of section *Lobatae* (red oaks). Leaves, placed in *Q. catahoulaensis* Daghlian and Crepet, have the aristate tips on their lobes that are also characteristic of red oaks. In contrast, the acorns found at the Huntsville site, designated *Q. huntsvillensis* Daghlian and Crepet, have cupulate scales more like those of white oaks, suggesting both sections (*Lobatae* and *Quercus*) were present in the Oligocene in eastern Texas (Daghlian and Crepet 1983; Crepet 1989; DeVore and Pigg 2010; DeVore et al. 2014). The Catahoula Formation also has yielded petrified wood compared to that of extant *Q. virginiana* ("live oak") from Jasper County, east Texas (Singleton 2001). However, confirmation of the details of this material is needed.

3.3.2.3 The Neogene

Western North America

Based on primarily on leaves, the fossil record of *Quercus* indicates it was thoroughly established in the central Rocky Mountains in the later Paleogene with sections *Lobatae*, *Protobalanus*, *Quercus* and, possibly, group *Cerris* present (Becker 1961, 1969, 1973; Bouchal et al. 2014). The Neogene record is dominated by leaf floras preserved in ash fall and aqueous settings from the Pacific Northwest to southern California, but it is notable that these Neogene records generally cease at the western margin of the Rocky Mountains, although rare records of *Quercus* pollen occur in the Neogene of Colorado and Wyoming (Leopold and MacGinitie 1972).

Acorns occur as compressions in Neogene floras (e.g., Condit 1944; Smiley and Rember 1985) but are often not well enough preserved to permit identification to section or comparison with living species. Axelrod assigned several acorns from the Miocene of Nevada to *Q. hannibali* Dorf of section *Protobalanus*, although these exhibit varying degrees of detail (see Axelrod 1956 [pl. 28, Fig. 6], 1985 [pl. 25, Fig. 7], 1991 [pl. 12, Figs. 4–5]). Others offer better preservation, e.g. acorns in their cups attached to a twig from the Miocene Succor Creek Flora illustrated in Taylor and Taylor (1993, Fig. 22.89; Plate 3.1 1).

Borgardt and Pigg (1999) examined over 120 specimens of acorns from the middle Miocene of Washington. These are preserved in exquisite detail, including a range of developmental stages from very small, presumably young (or possibly aborted) fruits to mature ones containing embryos. Of these, 55 specimens were serially sectioned (Plate 3.1 4, 6). Twenty-six were assigned to *Q. hiholensis* Borgardt and Pigg, of section *Quercus*, based on the superior position of the surviving and developed ovule, position of the stylopodium, and the shape of the involucre scales. Within the white oaks, *Q. hiholensis* is interpreted as an annual fruiting oak on the basis of its reduced perianth parts, in contrast to the larger, interlocking perianth segments of biennial fruiting oaks (Borgardt and Pigg 1999). The remaining specimens were designated as *Quercus* sp.

The Neogene wood record for western North American *Quercus* includes several Miocene and Pliocene localities. Platen (1908) reported *Q. lesquereuxii* from the Pliocene of California; this was later reassigned to *Quercoxylon lesquereuxii* (Platen) Mädel-Angeliewa (Mädel-Angeliewa 1968). Webber (1933) described *Quercus ricardensis* Webber from the Pliocene Ricardo Formation of southern California and compared it with the wood of living *Q. agrifolia* Née (section *Lobatae*). Boeshore and Jump (1938) described *Quercinium album* from the late Miocene Payette Formation of Idaho allying it with Section *Quercus*. Prakash identified three types of oak wood from the middle Miocene Vantage floras of Washington. The first (Prakash and Barghoorn 1961a) was *Quercus leuca* Prakash and Barghoorn, compared with *Q. alba* L. of section *Quercus*. The second (Prakash and Barghoorn 1961b) was *Q. sahnii* Prakash and Barghoorn, allied with section *Lobatae*, and the third (Prakash 1968) as *Quercoxylon compactum* Prakash.

In their subsequent revision of the Vantage woods, Wheeler and Dillhoff (2009) suggested these three species were variations of a single type that they referred to *Q. leuca* Prakash and Barghoorn of section *Quercus*. They also remarked that other informal reports of red oak woods from the Columbia Plateau basalts needed to be verified by further investigation. Other reports of fossil *Quercus* wood in western North America exist (e.g., Call and Tidwell 1988; mid–late Miocene of Nevada).

The two most commonly reported leaf species in the Neogene (*Quercus hannibali* Dorf and *Q. simulata* Knowlton) both first appear in at the Eocene–Oligocene boundary in Montana (Becker 1969), but are widespread in the Miocene. *Q. hannibali* occurs in a minimum of 17 Neogene floras including seven floras in Nevada (Plate 3.2 3), five in California, two in Idaho and three in Oregon, and survives to 3.5 Ma in California. Its suggested affinity is with section *Protobalanus*. *Q. simulata* occurs in a minimum of 22 Neogene floras including eleven in Nevada, four in California, four in Oregon, two in Idaho and one in Washington and lasts until 7 Ma in Nevada. However, while *Q. simulata* and its likely synonym *Q. consimilis* Newberry (Becker 1969; Fields 1996) are widely attributed to *Quercus* in the older literature, there is much uncertainty about the placement of these species. Several authors have allied *Q. simulata* (Axelrod 1956; Graham 1965; Becker 1969) and *Q. consimilis* (Becker 1961) with *Q. myrsinaefolia* Blume and *Q. salicina* Blume (=*Q. stenophylla* Makino) of subgenus *Cyclobalanopsis*. However, Axelrod (1985) rejected comparison to members of *Cyclobalanopsis* and instead compared some examples of *Q. simulata* to *Q. chrysolepis* Liebmann in section *Protobalanus*, while reassigning others to *Lithocarpus* Blume. Rember (1991) proposed the new combination *Lithocarpus simulata* (Knowlton) Rember. In an extensive review of the

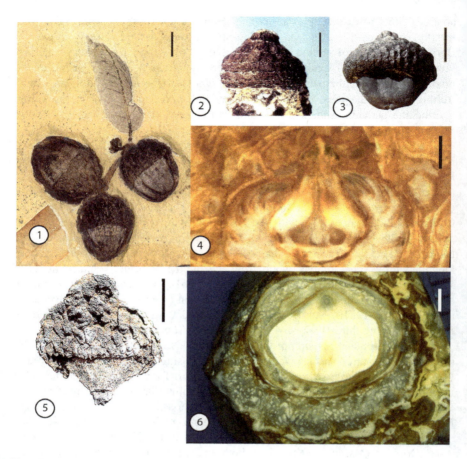

Plate 3.1 1 *Quercus hannibali* Dorf. Acorns in attachment to small twig. Middle–late Miocene Succor Creek Flora of Idaho and Oregon, USA. Courtesy of Patrick F. Fields and Darlene and Howard Emry, collectors. Photo courtesy of Steven R. Manchester. Scale: 10 mm. **2** *Quercus paleocarpa* Manchester. Middle Eocene Clarno Formation, Oregon, USA (see Manchester 1994). Photo courtesy of Steven R. Manchester. Scale: 5 mm. **3** *Quercus* sp. Middle Miocene Brandon Lignite, Vermont, USA (see Tiffney 1994). Photo courtesy of Bruce H. Tiffney. Scale: 5 mm. **4** *Quercus hiholensis* Borgardt and Pigg (Paratype), Middle Miocene Yakima Canyon, Washington, USA. Longitudinal section of anatomically preserved young acorn (see Borgardt and Pigg 1999 for details) Specimen UWBM B4101/93-1. University of Washington, Burke Museum of Natural History and Culture. Photo courtesy of Kathleen B. Pigg. Scale: 1 mm. **5** *Quercus* sp. Late (?) Eocene La Porte Flora, California, USA. Photo courtesy of Bruce H. Tiffney. Scale: 2.5 mm. **6** *Quercus hiholensis* Borgardt and Pigg (Holotype), Middle Miocene Yakima Canyon, Washington, USA. Longitudinal section of mature acorn, showing internal anatomy (see Borgardt and Pigg 1999). Specimen UWBM B4101/55126. University of Washington, Burke Museum of Natural History and Culture. Photo courtesy of Kathleen B. Pigg. Scale: 2.5 mm

species, Fields (1996) suggested that it is not clear if the affinity of *Q. simulata/ consimilis* lies with *Quercus, Castanopsis* or *Lithocarpus*. Because of this uncertainty, we have not included occurrences of *Q. simulata* or *Q. consimilis* in recounting the diversity of species and sections in the following paragraph.

Of the species of Neogene *Quercus* that we have tallied (and we emphasize that this summation is not complete), 13 occur only at a single locality. Five of these species occur in floras in California. Idaho, Nevada, Oregon and Washington each have two floras hosting a unique species. Another 17 Neogene species occur in two to seven floras: twelve in California, four in Nevada, three in Oregon and two in Idaho and one in Alaska. It is of note that, of the 30 species that first appear in the Neogene, about half (16) appear between 18 and 15 Ma, another ten between 13.5 Ma and ∼11 Ma, three at ∼9 Ma and one at 6 Ma.

Of the species that appear between 18 and 15 Ma, six are ascribed to section *Quercus*, six to section *Lobatae* and four to section *Protobalanus*. Of these early species in section *Protobalanus*, three first appear in California and one in Nevada. Six species in section *Lobatae* first appeared between 18 and 15 Ma in California, Nevada and Oregon, while two species in section *Quercus* first appeared in California and two in Oregon, and single species first appearing in Idaho and Alaska. Of the ten species that appeared between ∼12 and ∼9 Ma, seven first appeared in California, six of which were in section *Quercus*. From this limited data set, it would not appear possible to infer any patterns of changing sectional diversity through the Neogene or across geography.

This pattern of diversification of new species in the Miocene of western North America could be quite reasonably interpreted as reflecting an evolutionary radiation of oaks involving all three sections. However, it also possible that this apparent diversification could result from a taphonomic bias that creates a false timing signal. This was a time of active volcanism and tectonic changes in western North America creating ash falls and depositional basins. Consequently, it is possible that the upsurge in new species reflects the greater availability of preservational environments, and in fact, some or many of the species might have evolved earlier but not been preserved. Certainly, Oligocene and earliest Miocene floras are not as numerous west of the Rocky Mountains as are middle and later Miocene floras.

Eastern and Central North America

The Neogene record is represented largely by pollen in eastern North America. Further, there is a lack of fossil Neogene floras in the central portion of the continent. Consequently, it is difficult to track the dynamics of the changing ranges of the various sections of the genus on a continental basis. Pollen of *Quercus* is widespread in the Neogene of eastern North America from the Gulf Coast along the Atlantic coast to New England (see Hall et al. 1980; Ochoa et al. 2012, Appendix F; Baumgartner 2014). While widespread, the current data lack the precision to identify the sections present.

Fruits are reported in three cases. One morphology of acorn, accompanied by three types of leaves, occurs in the late Miocene Brandywine flora of Maryland (McCartan et al. 1990), although neither fruit nor foliage are attributed to section. Similarly, two apparent acorn morphologies are present in the late early Miocene Brandon Lignite of Vermont, both tentatively attributable to section *Lobatae* (Tiffney 1977), however these have not been studied in detail (Plate 3.1 3). These are accompanied by wood (Spackman 1949) and pollen (Traverse 1955). *Quercus virginiana* Mill. leaves and associated acorns are also noted from the Miocene/Pliocene Gray Site flora of eastern Tennessee (Liu 2011; Ochoa et al. 2012; Baumgartner 2014). These occurrences of *Q. virginiana* post-date the postulated

◀**Plate 3.2** **1** *Quercus falcata* Michaux leaf. Pliocene Citronelle Formation, Alabama, USA. Photo courtesy of Brian Axsmith. Scale: 10 mm. **2** *Quercus virginiana* Miller leaf. Pliocene Citronelle Formation, Alabama, USA. Photo courtesy of Brian Axsmith. Scale: 10 mm. **3** *Quercus hannibali* Dorf leaf. Early Miocene Buffalo Canyon Flora, Nevada, USA (see Axelrod 1991). Photo courtesy of Bruce H. Tiffney. Scale: 5 mm. **4** Pollen of *Quercopollenites* sp. SEM detail of the scabrate to verrucate ornamentation of the same pollen grain shown in Fig. 7. Late Miocene La Cerdanya Basin, Eastern Pyrenees, Spain (see Barrón 1996). Photo courtesy of E. Barrón. Scale: 2 µm. **5** Light microscope photomicrograph of an individual pollen grain of *Quercopollenites granulatus* Nagy. Late Miocene La Cerdanya Basin, Eastern Pyrenees, Spain. Photo courtesy of E. Barrón. Scale: 10 µm. **6** Light microscope photomicrograph of an individual pollen grain of *Quercoidites microhenrici* (Potonié) Potonié, Thomson and Thiergart ex Potonié. Late Oligocene As Pontes Basin, NW Spain (see Casas-Gallego 2017). Photo courtesy of M. Casas-Gallego. Scale: 5 µm. **7** SEM photomicrograph of an individual pollen grain of *Quercopollenites* sp. Late Miocene La Cerdanya Basin, Eastern Pyrenees, Spain (see Barrón 1996). Photo courtesy of E. Barrón. Scale: 10 µm. **8** *Quercus amethystianum* Wheeler, Scott and Barghoorn. Tangential section through wood showing large and small vessels, wide and narrow rays and fibers. Eocene Amethyst Mountain, Yellowstone National Park, USA, Photo from InsideWood, Fagaceae, 2054A. Courtesy of Elisabeth A. Wheeler. Scale: 20 µm. **9** *Quercus amethystianum* Wheeler, Scott and Barghoorn. Radial section through wood showing ray (at left), vessel elements (at right) and crossfield pitting. Eocene Amethyst Mountain, Yellowstone National Park, USA, Photo from InsideWood, Fagaceae, 2054A. Courtesy of Elisabeth A. Wheeler. Scale: 20 µm. **10** *Quercus amethystianum* Wheeler, Scott and Barghoorn. Tangential section through wood showing large ray. Eocene Amethyst Mountain, Yellowstone National Park, USA, Photo from InsideWood, Fagaceae, 2054A. Courtesy of Elisabeth A. Wheeler. Scale: 20 µm

divergence date for *Quercus* section *Quercus*, subsection *virentes*, suggested by Cavender-Bares et al. (2015).

Berry (1909, 1916) noted three species of leaves of *Quercus* from the Miocene Calvert Cliffs Formation of Virginia and Washington D.C., allying two with section *Quercus* and the third, *Q. lehmanni* Hollick, with section *Lobatae*, previously reported from the Calvert formation of Maryland (Hollick 1904). Stults et al. (2016) report leaves of sections *Quercus* and *Lobatae* from the Miocene Hattiesburg Formation of Mississippi and Stults and Axsmith (2015, Table 4) reported leaves of *Q. virginiana* (section *Quercus*), and *Q.* cf. *nigra* L. and *Q. falcata* Michaux (both section *Lobatae*) from the mid-Pliocene Citronelle Formation of Alabama (see Plate 3.2 1–2). Berry (1952) summarizes the Pleistocene occurrences of 14 species of *Quercus* from the coastal plain of eastern North America. Of the 11 living species, all are native to eastern North America today.

Central North America has only a few, rare Neogene plant sites, the best known of which is the middle Miocene Kilgore Flora of Nebraska (MacGinitie 1962). This flora hosts four species of *Quercus*, three in section *Quercus* and one in section *Lobatae*. Of these four, two are unique to the Kilgore Flora, one, *Q. remingtoni* Condit also occurs in the Miocene (Condit 1944) and Pliocene (Axelrod 1980) of California, and the last one, *Q. argentum* Knowlton (which has been synonymized with *Q. turneri*) also occurs in the late Miocene of Nevada (Axelrod 1940). *Quercus* is conspicuously absent from the Mio-Pliocene floras of the Ogallala Formation of the High Plains (Chaney and Elias 1936; Thomasson 1987; Gabel

et al. 1998), and the early Pliocene Pipe Creek Flora of Indiana (Farlow et al. 2001; Shunk et al. 2008; Ochoa et al. 2016).

Central America

Currently *Quercus* occurs in upland portions of Central America, south to the northern Andes. The history of this spread is currently recorded by a sparse pollen record. The oldest, though possibly suspect, record is from the early middle Miocene Mendez Flora of northern Chiapas, Mexico (Graham 1999b). This is followed by Miocene-Pliocene records from the Padre Miguel Group of Guatemala

◀**Plate 3.3 1** *Quercus alexeevii* Pojarkova. Late Oligocene of Ashutas, Kazakhstan. Komarov Botanical Institute, St. Petersburg, Russia: coll. 2113, sample 1335. Scale: 5 mm. **2** *Quercus palaeoserrata* Iljinskaja. Early-middle Miocene of Kiin-Kerish, Zaisan Depression, East Kazakhstan. Komarov Botanical Institute, St. Petersburg, Russia: coll. 4337, sample 206. Scale: 5 mm. **3** *Quercus protopontica* Iljinskaja. Early-Middle Miocene of Kiin-Kerish, Zaisan Depression, East Kazakhstan. Komarov Botanical Institute, St. Petersburg, Russia: coll. 4337, sample 124. Scale: 5 mm. **4** *Quercus kiinkerishica* Iljinskaja. Early-Middle Miocene of Kiin-Kerish, Zaisan Depression, East Kazakhstan. Komarov Botanical Institute, St. Petersburg, Russia: coll. 4337, sample 138. Scale: 5 mm. **5** Cupule of *Quercus sibirica* Dorof. Oligocene of Dunaevskiy Yar, Western Siberia, Russia. Komarov Botanical Institute, St. Petersburg, Russia: sample 47/4. Scale: 2.5 mm. **6** Acorn of *Quercus sibirica* Dorof. Oligocene of Dunaevskiy Yar, Western Siberia, Russia. Komarov Botanical Institute, St. Petersburg, Russia: sample 47/3. Scale: 2.5 mm. **7** *Quercus pseudocastanea* Göppert emend. Walther and Zastawniak. Early-Middle Miocene of Kiin-Kerish, Zaisan Depression, East Kazakhstan. Komarov Botanical Institute, St. Petersburg, Russia: coll. 2113, sample 1394. Scale: 5 mm. **8** *Quercus sosnowskyi* Kolakovskii. Pliocene, of Meore-Atara, Abkhaziya. Komarov Botanical Institute, St. Petersburg, Russia: coll. 441, sample 184. Scale: 5 mm

and the Mio-Pliocene Gatun Formation of Panama (Graham 1999b). In the northernmost Andes (near Bogota), the first appearance of the genus is at 478,000 years ago, becoming established by 330,000 years ago (Van't Veer and Hooghiemstra 2000).

3.3.3 Europe and Western Asia: Eocene through Pliocene

In the European continent, the genus *Quercus* is represented currently by 25–29 species grouped in the subgenus *Quercus* in groups *Ilex* (=*Sclerophyllodrys*), *Cerris* and *Quercus* (Schwartz 1964). The latter comprises the highest number of taxa (ca. 18 species) and it is widely widespread in Europe, including emblematic species in Europe such as *Q. robur*. Regarding the Ex-USSR territory, its modern flora has only 18 native species, all of them belonging to subgenus *Quercus*. Groups *Ilex* and *Cerris* are mainly distributed in the Mediterranean region and include a few evergreen taxa (e.g., *Q. ilex* and *Q. suber*) (Schwartz 1964).

3.3.3.1 The Eocene

The fossils assigned to *Quercus* for the Eocene show that the genus was well diversified and widespread during this period in Europe. The earliest occurrences come from the Arctic regions of Greenland.

The first occurrence of *Quercus* in this area is indicated by McIntire (1991), who mentions the occurrence of pollen of the section *Lobatae/Quercus* in the Axel Heiberg Island, ca. 45 Ma. From the same sediments, McIver and Basinger (1999) identified as "?*Trigonobalanus*", a set of leaves and cupules that could be also related to section *Lobatae/Quercus* as well as cupules with attached nuts in association with *Castanea, Fagus* and *Trigonobalanopsis* (Grímsson et al. 2015).

In the western Greenland, in the Aamaruutissaa Member (Lutetian–Bartonian, ca. 42–40 Ma), seven different pollen types belonging to *Quercus* have been identified, which show a clear and notable diversity of the genus since the middle Eocene (Grímsson et al. 2015). This palynological record has been related with four different "groups" into the genus: *Quercus, Lobatae, Protobalanus* and *Ilex*. According to Grímsson et al. (2015), the presence of the group *Ilex* would be related to an old lineage of this group on the Island.

In central Europe first evidence for pollen related to *Quercus* (*Quercoidites microhenrici*; extinct group of Quercoideae, often related to the genus *Quercus*), comes from the late Paleocene of the Polish Lowlands (Grabowska 1996) and from early Eocene marine strata (Ypresian, NP 12, ca. 51 Ma) of the Wursterheide Well (N Germany) (Meyer 1989). *Quercopollenites asper* (Thomson & Pflug) Kohlman-Adamska and Ziembinska-Tworzydlo, possibly related to the section *Lobatae*, was reported from the Messel oil shale, S Germany (ca. 47 Ma; Lenz et al. 2011), and middle Eocene lignite seams of the Helmstedt open cast mine (Lenz 2000).

The first possible European occurrence of *Quercus* on the basis of cuticular studies is of middle Eocene age. *Quercus subhercynica* Walther & Kvaček is described from a lauroid leaf fragment from sediments from the Königsaue mine of Germany (Kvaček and Walther 1989). However, this record is regarded as ambiguous in light of re-examination (Kvaček, personal observation). *Quercus haraldii* Knobloch and Kvaček has been described from the Eocene of the Staré Sedlo Formation in Central Europe. However, the precise diagnosis of these leaves (Knobloch and Konzalová 1998) also remains uncertain.

Quercus is first recognized in the Iberian Peninsula from pollen grains of the middle Bartonian Collbàs Formation (Ebro Basin) (Cavagnetto and Anadón 1996). In the late Eocene (Priabonian), pollen of *Quercus* has also been identified in the Ebro Basin (Cavagnetto and Anadón 1996) as well as in a dry area of Central Europe (Krutzsch et al. 1992). Dry phases of the Bartonian–Priabonian transition could have been influential in the spread of the xerophilous lineages of oaks (Cavagnetto and Anadón 1996).

Woods related to *Quercus* occur beginning in the Eocene (Gregory et al. 2009), and are included mostly in the fossil genus *Quercoxylon* (e.g., Privé 1975). The earliest occurrence in Europe of *Quercoxylon* (*Q. sempervirens* Gottwald) is from the Eocene of Germany (Gottwald 1966). Eocene fossil wood assigned to *Quercinium* has also been reported from Great Britain, for instance *Quercinium porosum* Brett, a silicified wood from the sands of the Woolwich and Reading Series, and *Q. pasanioides* Brett, a calcified wood from the London Clay (Brett 1960).

Staminate catkins and stellate trichomes with clear affinities with *Quercus* are preserved in Baltic ambers (middle Eocene, ca. 44 Ma sensu Kosmowska-Ceranowicz 1987) of northern Europe and Russia. Staminate flowers of *Quercus meyeriana* (Göppert and Berendt) Unger and *Q. taeniato-pilosa*

Conwentz were described and figured by Conwentz (1886), Friis and Crepet (1987) and Sadowsky et al. (2015). However, according to Nixon (1993), these fossils need further investigation since they occur in association with fruits of trigonobalanoid aspect.

A diverse array of leaf species of *Quercus* commonly occurs in the Eocene floras of Eurasia. Borsuk (1956) indicates the existence of two species, *Q. rectinervis* Borsuk and *Q. olafsenii* Heer, in the early Eocene of Sakhalin. Beginning in the mid-late Eocene, records of the genus are widespread in Eurasia and locally played an important role in the ecosystems. One of the most commonly distributed oaks from the late Eocene of the European part of Russia was *Q. pseudoneriifolia* Vikulin. This common species is related to the section *Lobatae*, and has linear lanceolate leaves with an entire margin (Vikulin 2011). The stratigraphic range of this species extends into the late Miocene in European Russia, northern of Kazakhstan, Ukraine, Georgia and Azerbaijan (Avakov 1979; Iljinskaja 1982).

In the Eocene of northern Kazakhstan, several common species with sclerophyllous leaves were described, including *Q. korniloviae* Makul. from the middle Eocene of Karasor and *Q. takyrsoriana* Makul. from the late Eocene of Takuirsor (Iljinskaja 1982). However, oaks are not present at the end of Eocene in the eastern part of Kazakhstan where floras were dominated by *Dryophyllum*. This genus completely disappeared at the turn of the Eocene–Oligocene and was replaced in the early Oligocene by a wide variety of Fagaceae including numerous oaks such as *Q. palaeoserrata* Iljinskaja (Plate 3.3 2), *Q. protopontica* Iljinskaja (Plate 3.3 3), *Q. kiinkerishica* Iljinskaja (Plate 3.3 4) and *Q. zaisanica* Iljinskaja (Iljinskaja 1991). These species were characterized by elongated to ovate, toothed leaves similar to those of the recent East Asian *Q. serrata* Thunb. and *Q. mongolica* C. Koch.

Pollen grains of *Quercus* are common in several localities of the late Eocene of Kazakhstan, including *Q. conferta* Boitzova, *Q. gracilis* Boitzova, *Q. graciliformis* Boitzova and *Q. sparsa* Boitzova (Boitzova and Panova 1966). Likewise, oak pollen has been cited in the late Eocene of the Ob-Irtysh interfluve in the West Siberian Plain (Kondinskaya and Yudina 1989). *Q. graciliformis* and *Q. gracilis* are characteristic of many other late Eocene localities of Ukraine, Caucasus, Kazakhstan, Siberia and the Far East. These two species replaced more ancient pollen complexes in the Priabonian and disappeared in the early Oligocene. This change may reflect the climatic cooling at the Eocene–Oligocene border.

3.3.3.2 The Oligocene

Quercus has been identified in a significant number of Oligocene European sites. It has been reported in Bouches du Rhône, Provence (France), but without further taxonomic detail or precise age (Chateneuf and Nury 1995), occurring with mesophilous taxa such as *Acer, Ulmus, Carpinus,* Juglandaceae, *Liquidambar, Fagus, Cornus, Castanea,* etc. Gastaldo et al. (1998) indicated the occurrence of pollen and leaves of *Quercus* in the Oligocene materials of the Thierbach member of the Weißelster Basin (Germany) in association with riparian and mesophilous

elements. Pollen belonging to groups *Cerris* and *Ilex* has been found in fossil floras from Altmittweida (Saxony, Germany) forming part of late Oligocene–early Miocene riparian and swamp forest environments around lakes and marshes which surrounded mesophytic forests (Kmenta and Zetter 2013). In northwest Germany (Lower Rhine Basin) *Quercoidites henrici/microhenrici* and *Quercopollenites asper* are regularly present in continental to shallow marine Chattian strata of the

◄**Plate 3.4** **1** *Quercus mediterranea* Unger. Late Miocene of Erdőbénye-Barnamáj, Hungary. Hungarian Natural History Museum, Budapest, Hungary: specimen BP 54.83. Photo courtesy of Boglárka Erdei. Scale: 10 mm. **2** *Quercus hispanica* Rérolle emend. Barrón, Postigo-Mijarra and Diéguez. Late Miocene La Cerdanya Basin, Eastern Pyrenees, Spain. Museo Geominero, Madrid, Spain: specimen MGM 1046 M. Scale: 10 mm. **3** *Quercus faginea* Lam. Pleistocene, Tubilla del Agua outcrop, Burgos, Spain. Collection of R. Iglesias-González: specimen TAG-TP-85. Photo courtesy of Raúl Iglesias-González. Scale: 10 mm. **4** *Quercus drymeja* Unger. Late Miocene La Cerdanya Basin, Eastern Pyrenees, Spain. Museu de Geologia de Barcelona, Barcelona: specimen MGB V9522. Scale: 10 mm. **5** *Quercus drymeja* Unger. Late Miocene La Cerdanya Basin, Eastern Pyrenees, Spain. Museo Geominero, Madrid, Spain: specimen MGM 1064 M. Scale: 10 mm. **6** Impression of a cupule of *Quercus* sp. occurring with leaves of *Quercus rhenana*. Early Miocene Most Basin, Vršovice, Louny Distric, North Bohemia, Czech Republic (see Kvaček and Hurník 2000). National Museum, Prague, Czech Republic: specimen NM G 1937. Photo courtesy Zlatko Kvaček. Scale: 3 mm. **7** *Quercus rhenana* (Kräusel and Weyland) Knobloch and Kvaček. Early Miocene Bílina mine, Most Basin, North Bohemia, Czech Republic. National Museum, Prague, Czech Republic. Photo courtesy of Zlatko Kvaček. Scale: 10 mm. **8** *Quercus neriifolia* A. Braun ex Unger. Late Miocene La Cerdanya Basin, Eastern Pyrenees, Spain. Museu de Geologia del Seminari de Barcelona, Barcelona: specimen MGSB 69428. Photo courtesy of Evaristo Aguilar. Scale: 10 mm

so-called Lower Seam Group, Köln Fm (Von der Brelie et al. 1981). In Denmark, Oligocene *Quercus* pollen was found in association with *Fagus* and *Betula* pollen (Larsson et al. 2010). Pollen grains referrable to the *Quercus* (*robur* type) and group *Cerris* (*cerris*/*crenata* type) are reported from the Enspel Maar Lake deposits (W Germany, late Oligocene, MP 28) (Hermann 2007). Pollen grains found in the opencast mine at Cospuden (Germany) indicate the presence of two lineages in *Quercus*: *Ilex* and *Quercus/Lobatae* (Denk and Grimm 2009, 2010) that is in agreement with the evidence of the macrofossil record (Denk et al. 2012). The pollen from Cospuden assigned to the group *Ilex* would be the earliest evidence of this group for Europe.

For southwestern Europe, pollen attributed to *Quercus* is represented by *Quercoidites microhenrici* (Potonié) Potonié, Thomson and Thiergart ex Potonié (Plate 3.2 6), *Quercopollenites granulatus* Nagy, *Quercopollenites rubroides* Kohlman-Adamska and Ziembińska-Tworzydło, *Verrutricolporites irregularis* Roche and Schuler, and *V. theacoides* Roche and Schuler in the Rupelian of the northwestern Iberia (Casas-Gallego 2017). Likewise, a palynological record assigned to the *Q. ilex-coccifera* type (Sarral Formation) has been cited for the Ebro basin (Cavagnetto and Anadón 1996).

Ovate-lanceolate leaves with an entire margin have been related to a large number of oak species (e.g., de Saporta 1865, 1867; Sanz de Siria 1992). For example, leaves of this type are found at the early Oligocene Cervera site (Ebro Basin, Spain) where an oak-laurel forest vegetation developed (Sanz de Siria 1992; Barrón et al. 2010). Leaf remains attributed to *Quercus* were assigned to seven different species, but require a comprehensive revision that assesses cuticular features.

Cuticular studies reveal that many of the supposed laurel-like oak species cannot be confidently attributed to *Quercus*. According to Kvaček and Walther (1989),

Quercus lyellii Heer, from the European Oligocene, is now included into the morphogenus *Dryophyllum* Debey ex Saporta as *D. furcinerve* (Rossmässler) Schmalhausen forma *lyellii* (Heer) Kvaček and Walther (i.e., *Eotrigonobalanus furcinervis*). The fossil genus *Dryophyllum* includes leaves that were believed initially to be fagaceous, but are now partially considered within the Juglandaceae (Jones and Dilcher 1988; Jones et al. 1988).

◀**Plate 3.5 1** Cuticle of *Quercus roburoides* Gaudin showing anomocytic stomata and hair basis. Pliocene of Frankfurt, Germany. Forschungsinstitut Seckenberg, Frankfurt am Main, Germany. Specimen: SM B 11805.2. Photo courtesy Zlatko Kvaček. Scale: 50 μm. **2** Cuticle of the same specimen (SM B 11805.2) of *Quercus roburoides* showing stellate hairs. Photo courtesy of Zlatko Kvaček. Scale: 50 μm. **3** Mummified leaf of *Quercus roburoides* Gaudin. Pliocene of Willershausen, Germany. Forschungsinstitut Seckenberg, Frankfurt am Main, Germany. Specimen: SM B 11831. Photo courtesy of Zlatko Kvaček. Scale: 10 mm. **4** *Quercus praeerucifolia* Straus. Pliocene of Willershausen, Germany. Staatliches Museum für Naturkude in Stuttgart, Germany. Specimen: SM B 15142. Photo courtesy of Zlatko Kvaček. Scale: 10 mm. **5** *Quercus roburoides* showing detail of stellate hairs seen in Fig. 3.2. Photo courtesy of Zlatko Kvaček. Scale: 50 μm. **6** *Quercus praeerucifolia* Straus. Pliocene of Willershausen, Germany. Staatliches Museum für Naturkude in Stuttgart, Germany. Specimen: SM B 151413. Photo courtesy of Zlatko Kvaček. Scale: 10 mm. **7** Mummified leaf of *Quercus praecastaneifolia* Knobloch. Pliocene of Willershausen, Germany. Staatliches Museum für Naturkude in Stuttgart, Germany. Specimen: SM B 11850. Photo courtesy of Zlatko Kvaček. Scale: 10 mm. **8** *Quercus pseudorobur* Kováts (Holotype). Late Miocene of Erdőbénye-Barnamáj, Hungary. Hungarian Natural History Museum, Budapest, Hungary: specimen BP 62.21.1. Photo courtesy of Boglárka Erdei. Scale: 10 mm. **9** *Quercus kubinyii* (Kováts ex Ettingshausen) Berger (Syntype). Late Miocene of Erdőbénye-Barnamáj, Hungary. Hungarian Natural History Museum, Budapest, Hungary: specimen BP 64.95.1. Photo courtesy of Boglárka Erdei. Scale: 10 mm

Quercus rhenana (Kräusel and Weyland) Knobloch and Kvaček is a well-known evergreen oak with laurel-like leaves (Plate 3.4 7), cyclocytic stomata and massive hair-bases. According to Knobloch and Kvaček (1976) and Kvaček and Walther (1989), it belongs most probably to the section *Lobatae* and was distributed in Central Europe from the Oligocene to the middle Miocene. This species had importance in the formation of Miocene lignites (Kovar-Eder et al. 1998). It inhabited river-banks and swampy areas, sometimes as the dominant species (Kvaček 1998; Kovar-Eder et al. 2001). The presumed ancestor of this species was the Oligocene *Q. praerhenana* Walther and Kvaček, which differs by the architecture of its leaves and the lack of trichomes on its cuticle (Walther 1999).

European Oligocene oaks with toothed leaf margins are mainly referred to *Q. lonchitis* Unger and *Q. praekubinyii* Walther and Kvaček, both of which have been compared to the group *Cerris* (Mai and Walther 1991; Walther 1999). These species inhabited mixed mesophytic forests with more than 50% evergreen thermophilous elements. Similarly, *Q. cerverensis* Sanz de Siria was an element of the Iberian evergreen sclerophyllous-laurophyllous formations together with Lauraceae and Fabaceae during the early Oligocene (Sanz de Siria 1992). The Oligocene *Q. pseudoalexeevii* Vikulin, from European Russia, possessed sharp teeth (Vikulin 1987). This species was replaced in the mid-Oligocene–Miocene of the east Russia and Kazakhstan by *Q. alexeevii* Pojarkova (Plate 3.3 1) that showed rather similar leaves (Iljinskaja 1982). The record of *Q. furuhjelmii*? Heer from the late Oligocene of Kazakhstan and Western Siberia seems to be the first one for oaks with *robur*-like leaves. According to Menitsky (1969), it could be the possible ancestor of the modern section *Quercus*. Cupules of this section attributed to the species *Q. parazaisanica* Iljinskaja and perhaps to *Q. sibirica* Dorof (Plate 3.3 5–6) have been

Plate 3.6 **1** *Quercus kobatakei* Tanai and Yokoyama. Eocene-Oligocene, Kobe, Hyogo, Japan. National Museum of Nature and Science, Kyoto, Japan. Specimen: NSM PP-26969. Photo courtesy of Dr. Atsushi Yabe. Scale: 10 mm. **2** *Quercus sichotensis* Ablaev and Gorovoi. Eocene-Oligocene, Kobe, Hyogo, Japan. National Museum of Nature and Science, Kyoto, Japan. Specimen: NSM PP- 16357. Scale: 10 mm. **3** *Quercus ussuriensis* Kryshtofovich. Early Oligocene, Wakamatsuzawa, Hokkaido, Japan. National Museum of Nature and Science, Kyoto, Japan. Specimen: NSM PP- 16353. Scale: 10 mm. **4** *Quercus ishikariensis* Tanai. Late middle Eocene, Yubari, Hokkaido, Japan. National Museum of Nature and Science, Kyoto, Japan. Specimen: NSM PP- 10585. Scale: 10 mm. **5** *Quercus kitamiana* Tanai. Early Oligocene, Wakamatsuzawa, Hokkaido, Japan. National Museum of Nature and Science, Kyoto, Japan. Specimen: NSM PP- 10635. Scale: 10 mm. **6** *Cyclobalanopsis ezoana* Tanai. Early Oligocene, Wakamatsuzawa, Hokkaido, Japan. National Museum of Nature and Science, Kyoto, Japan. Specimen: NSM PP- 10634. Scale: 10 mm. **7** *Cyclobalanopsis nagatoensis* Tanai and Uemura. Late Oligocene, Noda, Yamaguchi, Japan. National Museum of Nature and Science, Kyoto, Japan. Specimen: NSM PP- 10374. Scale: 10 mm

found in the Oligocene of Kazakhstan and Western Siberia, respectively (Iljinskaja 1982, 1991).

European Oligocene woods assigned to *Quercoxylon* also have been identified. Examples include *Q. intermedium* Petrescu and Velitzelos from Romania (Iamandei et al. 2012) and *Q. bavaricum* Selmeier and *Q. lecointrei* Gazeau & Koeniguer from France (Limagne at Bussières, Puy-de-Dôme) (Privé-Gill et al. 2008). The earliest woods assigned to *Quercus* from North Africa (Sahara and Egypt) are of Oligocene age (Biondi et al. 1985), namely *Quercoxylon retzianum* Kräusel, which is known only from the Petrified Forest of El Cairo (Kräusel 1939; El-Saadawi et al. 2011). This taxon formed part of a plant community in which taxa currently widespread in savanna environments of Africa were common (e.g. *Bombacoxylon, Terminalioxylon, Dalbergioxylon* or *Detarioxylon*). Another post-Eocene species also described for the North Africa (Algeria) is *Q. gevinii* Boureau (Dupéron-Laudoueneix and Dupéron 1995). *Quercinium* has been recorded in the Oligocene of several localities of Azerbaijan as *Q. uniradiatum* (J. Felix) Jarm., but cannot be attributed to a particular section of *Quercus* (Iljinskaja 1982).

3.3.3.3 The Neogene

Evergreen oaks with laurel-like leaves are prominent in the early Miocene. For example, *Q. neriifolia* A. Braun ex Unger is a poorly studied species related to the section *Lobatae* (Plate 3.4 8), which was found at several locations of southern Europe (Brambilla and Penati 1987; Barrón et al. 2014). In the Pannonian of Eastern Styria (Austria), Kovar-Eder and Hably (2006) described the species "*Quercus*" *rhenanasimilis* which is similar to *Q. rhenana* but with different cuticular features. Likewise, in the Piazencian of Meximieux (eastern France), de Saporta and Marion (1876) described the species *Q. praecursor* that they related to the holm oak (*Q. ilex*, group *Ilex*). *Quercus praecursor* has sclerophyllous laminae and is one of the last oaks with laurel-like leaves in the Neogene of Europe. Unfortunately, the lack of epidermal studies prevents us from being able to relate *Q. praecursor* to a particular section of extant oaks.

The occurrence of a ring-cupped oak, *Cyclobalanopsis stojanovii* Palamarev and Kitanov, in the early Pliocene of the Beli Brjag coal Basin (Bulgaria) is remarkable (Palamarev and Kitanov 1988). This species has elliptical leaves with a serrate margin and craspedodromous venation. It is considered as an endemic of the mesophytic Pliocene East European forests, which were distinguished by numerous species of Fagaceae (Palamarev and Ivanov 2003).

The oaks with toothed or lobed leaf-margins of the group *Cerris* such as *Q. kubinyii* (Kovats ex Ettingshausen) Berger (Plate 3.5 9), inhabited Europe from the early Miocene (Burdigalian) (Teodoridis and Kvaček 2006; Mai 2007) to the Piazencian (Roiron 1992), and are an important element of the Neogene European floras. This species has been also found in the early Miocene of Western Siberia, the middle Miocene of Ukraina and the late Miocene of Abkhasia (Iljinskaja 1982; Shvaryova and Mamchur 2003). It inhabited both riparian and mesophytic environments, as well as forests with a mixture of deciduous and palaeotropical broad-leaved elements (Kovar-Eder et al. 2001). According to Kvaček et al. (2011), the living relatives of *Q. kubinyii* are *Q. libani* Oliv. from Western Asia, and *Q. variabilis* Blume and *Q. acutissima* Carruth. from China.

Oaks of the group *Cerris* diversified in Europe through the Neogene, resulting in a range of toothed or lobed leaf species (Knobloch and Velitzelos 1986; Stephyrtza 1990; Striegler 1992). *Q. gigas* Göppert emend. Walther and Zastawniak (=*Q. czeczottiae* Hummel, *Q. pontica miocenica* Kubát) and *Q. pseudocastanea* Göppert emend. Walther and Zastawniak (Plate 3.3 7) were probably the most relevant oaks in the leaf assemblages. Both species lived in Europe and Ukraina from the middle Miocene to the early Pliocene (Iljinskaja 1982; Hummel 1983; Walther and Zastawniak 1991; Worobiec and Lesiak 1998). *Quercus pseudocastanea* was also widespread throughout Russia, northern Caucasus, Armenia, Georgia, Kazakhstan, Abkhazia and western Siberia (Iljinskaja 1982; Shvaryova and Mamchur 2003). The first record of *Q. pseudocastanea* comes from the late Oligocene of Kazakhstan and Bashkiria (Russia) (Iljinskaja 1982). In addition, *Q.* cf. *pseudocastanea* has been found in the late Pliocene of the Czech Republic (Bůžek et al. 1985). Sometimes *Q. kubinyii*, *Q. pseudocastanea* and *Q. gigas* occur together in the same late Miocene localities of Central and southern Europe (Knobloch 1969, 1988; Meller 1989; Martinetto et al. 2007).

Quercus cruciata Al. Braun in Stitzenberger and *Q. buchii* Weber from the Oligocene and Miocene of Europe originally considered members of sections *Quercus* or *Lobatae* (Hantke 1965) have been excluded from the genus *Quercus* and transferred on account of differences in the epidermal structure to the fossil genus *Pungiphyllum* of uncertain affinities (Frankenhäuser and Wilde 1995).

One of the most widespread pre-Mediterranean elements in Europe was *Q. mediterranea* Unger (Plate 3.4 1). It first appears in the upper Oligocene (Palamarev 1989) and is a typical floral component during the Neogene in all of Europe. This species shows a set of leaf features that remain fairly stable throughout the Neogene of western Eurasia (Denk et al. 2017). It was a sclerophyllous, evergreen tree (Kvaček et al. 2002) whose epidermal structure and leaf morphology suggest affinities to the recent *Q. coccifera* L. of the group *Ilex* (Kvaček and Walther 1989). In Southern Europe, *Q. mediterranea* usually co-occurs with *Q. drymeja* Unger (Kvaček et al. 1993, 2002; Barrón et al. 2016) from the early/middle Miocene boundary (Kovar-Eder et al. 2004).

Quercus drymeja was also a sclerophyllous tree showing very polymorphic leaves (Plate 3.4 4–5; Barrón 1998). The early Miocene specimens attributed to *Q. drymeja* are very similar to those of *Q. lonchitis* Unger (e.g. Knobloch and Kvaček 1981), which is a species retained for European late Paleogene oaks (Kvaček et al. 1993). The stratigraphic distribution of *Q. drymeja* is almost identical to that of *Q. mediterranea*. Living relatives of both species are within the group *Ilex* (Denk and Grimm 2010). According to Denk et al. (2017), *Quercus drymeja* and *Q. mediterranea* should be considered as part of a morphotype complex that formed forests in fully humid or summer-wet climates. In several places, these two species occur during the middle Miocene together with other sclerophyllous species of oaks

such as *Q. zoroastri* Unger (e.g. Kovar-Eder et al. 2004; Kvaček et al. 2011). However, *Q. zoroastri* is now included into the morphotype complex of *Q. drymeja* (as morphotype *Q. drymeja zoroastri*, see Denk et al. 2017).

It is also interesting to note that *Q. drymeja* and *Q. mediterranea* were associated with pre-mediterranean oaks in the late Miocene. On the one hand, they occur

◄**Plate 3.7 1** *Quercus tibetensis* Xu, Su and Zhou. Late Miocene of Tibet, China. Xishuangbanna, Tropical Botanical Garden, Chinese Academy of Sciences, China. Specimen: 2014009. Scale: 10 mm. **2** *Quercus praedelavayi* Xing and Zhou (Holotype). Late Miocene of Xianfeng flora, central Yunnan province, southwestern China (see Xing et al. 2013). Xishuangbanna, Tropical Botanical Garden, Chinese Academy of Sciences, China. Specimen: HLT 450A. Scale: 10 mm. **3** *Quercus sinomiocenica* Hu and Chaney, Late Miocene Xiaolongtan Formation of Yunnan, China. Nanjing Institute of Geology and Palaentology, Chinese Academy of Sciences, China. Specimen: nj198503. Scale: 10 mm. **4** *Quercus tenuipilosa* Hu and Zhou. Late Pliocene Ciying Formation in Kunming, Yunnan province, southwestern China. Kunming Institute of Botany, Chinese Academy of Sciences, China. Specimen: HST254 HLT. Scale: 10 mm. **5** *Quercus tenuipilosa* Hu and Zhou. Late Pliocene Ciying Formation in Kunming, Yunnan province, southwestern China. Kunming Institute of Botany, Chinese Academy of Sciences, China. Specimen: HST 751. Scale: 10 mm. **6** *Quercus preguyavaefolia* Tao. Late Pliocene, Yunnan, China. Paleoecology Laboratory, Xishuangbanna, Tropical Botanical Garden, Chinese Academy of Sciences, China. Specimen: YP 109. Scale: 10 mm. **7** *Quercus preguyavaefolia* Tao. Late Pliocene, Yunnan, China. Paleoecology Laboratory, Xishuangbanna, Tropical Botanical Garden, Chinese Academy of Sciences, China. Specimen: YP109 YP 10901. Scale: 10 mm. **8** *Quercus praedelavayi* Xing and Zhou (Holotype, counterpart). Late Miocene of Xianfeng flora, central Yunnan, China. Xishuangbanna, Tropical Botanical Garden, Chinese Academy of Sciences, China. Specimen: HLT 450B. Scale: 10 mm

together with *Q. hispanica* Rérolle emend. Barrón, Postigo-Mijarra and Diéguez in Western Europe (Plate 3.4 2; Grangeon 1953, 1958; Barrón 1998; Barrón et al. 2016). *Q. hispanica* is similar to several southern European extant species belonging to the section *Quercus*, especially *Q. humilis* Mill., *Q. faginea* Lam. spp. *faginea* and *Q. lusitanica* Lam (Barrón et al. 2014). On the other hand, *Q. drymeja* and *Q. mediterranea* frequently appear together with *Q. sosnowskyi* Kolakovskii in the late Miocene of the Eastern Mediterranean area (Kvaček et al. 2002; Palamarev and Tsenov 2004; Velitzelos et al. 2014). *Quercus sosnowskyi* has been related with the cork oak (*Q. suber* L., group *Cerris*) by virtue of its epidermal features, leaf shape and cupules. *Q. suber* is an emblematic Mediterranean evergreen, sclerophyllous tree which now inhabits southwestern Europe and the North of Africa (Magri et al. 2007). However, *Q. sosnowskyi* is also a characteristic Neogene element of the Balkan-soutwest Asian area (Plate 3.3 8; Palamarev and Ivanov 2003). Kolakovskii (1964) considers this extremely polymorphic species as a possible link between the Chinese and Mediterranean oaks since it dominated a special type of sclerophyllous forest in the late Miocene of Abkhazia.

From the late Miocene, deciduous red oak (roburoid) species of the section *Quercus* inhabited mesophytic forests. The most frequent species in Europe were *Q. pseudorobur* Kováts (Plate 3.5 8) and *Q. roburoides* Gaudin (Plate 3.5 1–3, 5) while *Q. kodorica* Kolakovskii was common in Abkhazia (Iljinskaja 1982; Van der Burgh 1993; Hably and Kvaček 1998; Knobloch 1998; Kvaček et al. 2008; Teodoridis et al. 2015). The two first species differ in cuticular aspects and can be related to the extant *Q. petraea* (Mattuschka) Lieblein and *Q. hartwissiana* Steven. During the Pliocene, roburoid species are usually associated with oaks of the group

Cerris such as *Q. kubinyii*, *Q. praecastaneaefolia* Knobloch (Plate 3.5 7) or *Q. praeerucifolia* Straus (Plate 3.5 4, 6).

At the end of the Pliocene, leaves similar to recent species such as *Q. coccifera*, *Q. ilex*, *Q. cerris* L, *Q. canariensis* Willd., *Q. robur* L. and *Q. suber* have been described by several authors (Depape 1912; Iljinskaja 1982; Roiron 1992). The lack of epidermal studies prevents us from confirming these identifications with certainty. However, the presence of red oaks in the Pleistocene of Europe and North Africa is accepted by a number of palaeobotanists (see e.g., Arambourg et al. 1953; Follieri 1979; Roiron 1983; Martinetto et al. 2014).

Quercus is identified from pollen grains throughout the Neogene (Plate 3.2 4–7). The pollen species *Quercoidites henrici* (Potonié) Potonié, Thomson and Thiegart ex Potonié and *Q. microhenrici* (Potonié) Potonié, Thomson and Thiegart ex Potonié, (Plate 3.2 6) usually recorded from the Paleogene to Pliocene (e.g., Benda 1971; Chateauneuf 1972; Sittler and Schuler 1974; Solé de Porta and de Porta 1977; Valle and Civis 1978; Pais 1979; Sittler 1984; Kohlman-Adamska 1993; Alcalá et al. 1996; Alcalá 1997; Barrón et al. 2006; Stuchlik et al. 2014), may represent ancient thermophilous oak types of unknown affinity (Doláková 2004; Stuchlik et al. 2014).

Pollen of section *Quercus* was recorded in the early/middle Miocene transition of Poland (Stuchlik et al. 2014). *Quercopollenites porasper* (Pflug) Kohlman-Adamska & Ziembińska-Tworzydło was described in the locality of Chłapowo. This form was similar to the pollen of *Q. robur* L. while *Quercopollenites sculptus* Kohlman-Adamska & Ziembińska-Tworzydło and *Q. granulatus* (Plate 3.5), may be related to *Q. petraea* and *Q. frainetto* Ten. From the middle Miocene, pollen of the section *Lobatae* (*Quercopollenites asper* [Pflug and Thomson in Thomson and Pflug] Kohlman-Adamska & Ziembińska-Tworzydło and *Q. rubroides* Kohlman-Adamska & Ziembińska-Tworzydło) is recorded in central, eastern and southern Europe (Alcalá et al. 1996; Ashraf et al. 1996; Slodkowska 2004; Ivanov et al. 2007; Stuchlik et al. 2014). In central European early to middle Miocene lignite deposits such as the Rhenish Main Seam, Lower Rhine Basin (Germany), both morphotypes may attain very high proportions of over 50% of non-bisaccate pollen (Von der Brelie 1968), and have a clear affinity to wet forest swamp vegetation including as well *Taxodium* (Huhn et al. 1997).

The pollen type *Quercus ilex-coccifera* also has been identified in Neogene sediments (e.g., Bessedik 1984, 1985; Zheng 1990; Jiménez-Moreno et al. 2005; Jiménez-Moreno and Suc 2007). In the Langhian of Northern Africa (Northeastern Tunisia) pollen of *Q. ilex-coccifera* type and *Q. suber* is very common, suggesting the source plants were important components of the sclerophyllous forest (Moktar and Mannaï-Tayech 2016). However, the lack of illustrations of this pollen type in the most consulted works prevents us to relate it to the group *Ilex*. Recently, pollen of the group *Ilex* was identified in the Portuguese Piacenzian (Vieira et al. 2011) with Scanning Electron Microscopy (SEM).

Generally, the pollen of *Quercus* becomes common in palynological assemblages in the middle Miocene. It can comprise up to 15% of assemblages in the Langhian (Bessedik 1984; Kohlman-Adamska 1993; Gardère and Pais 2007;

Jiménez-Moreno 2006), and can exceed this value from the Serravalian (Jiménez-Moreno 2006; Barrón et al. 2010). In the Piacenzian, the Mediterranean seasonality was established in South Europe (Suc and Cravatte 1982; Bessais and Cravatte 1988). Late Pliocene pollen spectra thus present conspicuous percentages of evergreen *Quercus* that sometimes surpass the values of deciduous oaks in several sites of the Mediterranean region (Suc 1980; Barrón et al. 2010; Jiménez-Moreno et al. 2013).

Oak fruits (acorns) are common in the European late Neogene (e.g., Plate 3.4 6). Different types of acorns have been described in older works (see e.g. Göppert 1855; Massalongo and Scarabelli 1859; de Saporta 1888; Grangeon 1958). However, most of the fruits studied correspond to the group *Cerris* and the section *Quercus* (Günther and Gregor 1997). Frequently, acorns are represented by fragments of cupules whose identification is problematic. Generally, they are not attached to leafy twigs, and are usually detached from their cupules.

Several types of cupules have been found that have conspicuous scales that relate them to the group *Cerris* (Iljinskaja 1982; Hummel 1983; Walther and Zastawniak 1991). These include *Q. kustanaica* Kornilova and *Q. popovii* Kornilova from the early Miocene of Kazakhstan; *Q. cerrisaecarpa* Kolakovskii and *Q. microcerrisaecarpa* Kolakovskii from the late Miocene of Abkhazia; and *Q. sapperi* (Menzel) Mai ex Hummel, *Q. variabilis* Hummel and *Q. microcerrisaecarpa* from the late Miocene and early Pliocene of Poland. The Polish species have also been found associated with *Q. kubinyii* in the Pliocene of Hungary (Hably and Kvaček 1997). Roburoid oaks of the section *Quercus* are well represented by cupules of *Q. robur* and *Q. petraea* types in late Miocene and Pliocene of Central and South Europe (Van der Burgh 1997; Martinetto 2015).

The morphogenus *Quercoxylon* has been cited widely in the Neogene (e.g. Böhme et al. 2007; Iamandei et al. 2011). For example, in the late Miocene of Goznica (SW Poland), Dyjor et al. (1992) described a specimen of *Quercoxylon* sp. that may be related to *Q. cerris*. A large number of European Miocene fossil woods related to *Quercus* were described as *Quercoxylon bavaricum* Selmeier, a form that has been related to the extant *Q. robur* (Iamandei et al. 2011). In the Tortonian of Portugal and the Pliocene of the Northwestern Spain, woods attributed to the cork oak (*Q. suber*) were identified in older studies (de Carvalho 1958; Losa-Quintana 1978). However, these woods should be revised in comparison with other species of the group *Cerris*. Occurrences of Neogene woods in the territory of ex-USSR are not numerous. They are limited to two Ukrainian species of *Quercinium*, *Q. rossicum* Merckl. and *Q. montanum* (Merckl.) J. Felix (Iljinskaja 1982).

3.3.4 East Asia: Eocene through Pliocene

The Flora of China considers *Quercus* and *Cyclobalanopsis* as separate genera (Huang et al. 1999) based on whether the cupule is imbricate-scaled or lamellate, respectively (Camus 1936–1954). However, other sources often consider *Cyclobalanopsis* a subgenus of *Quercus*, as we do here. Today China is the current center of diversity for Subgenus *Cyclobalanopsis* with 69 species in subtropical and tropical areas (Fig. 3.2), of which 43 are endemic. In addition, there are 35 species of the subgenus *Quercus* in China, with 15 of them endemic (Huang et al. 1999). Fifteen oak species now inhabit Japan. Seven of these belong to the subgenus *Quercus* and 8 to the subgenus *Cyclobalanopsis*, with two species of *Cyclobalanopsis* endemic to Japan.

3.3.4.1 The Eocene

In Japan the earliest, unequivocal occurrence of *Quercus* is *Cyclobalanopsis naitoi* Huzioka from the late middle Eocene (early Bartonian) Ube flora in western Japan, represented by leaves and compressed acorns (Huzioka and Takahashi 1970; Tanai 1995). The second oldest record is *Q. ishikariensis* Tanai (leaves, Plate 3.6 4) and *Q. sp.* (cupule) from the Ikushunbetsu Formation (middle Bartonian) in Ishikari, Hokkaido (Tanai 1995). Tanai (1995) assigned *Q. ishikariensis* to the group *Cerris* based on its areoles formed by thick veins, although veinlets with branching were not preserved. A specimen formerly described as *Q. kushiroensis* Tanai from the late Bartonian Shakubetsu Formation in Kushiro (Hokkaido), was reidentified to *Q. ishikariensis* (Tanai 1970, 1995). Another *Quercus* record from the Ikushunbetsu Formation displays a bowl-shaped cupule with an outer surface that has no sign of concentric rings and appears to be covered by short appressed scales; this was referred to the subgenus *Quercus*, excluding the group *Cerris*, by Tanai (1995).

Quercus pollen is also well represented in Eocene sediments from Hokkaido with its abundance clearly influenced by the Terminal Eocene Climate Cooling (Sato 1994). Oak pollen decreased while *Tsuga* pollen increased, from the middle Eocene Ishikari Group to the late Eocene Poronai Group.

Quercus occurs in China from the early Eocene as the palynological record of *Quercoidites* indicates (Quan et al. 2012a). From the middle Eocene, oak pollen became abundant in the region (Zhang 1995). However, macroremains are rare. Two fossil leaves from the middle Eocene Jijuntun Formation were identified as *Quercus* sp. in Funshun and Huadian in Northeast China (Writing Group of Cenozoic Plants of China 1978; Manchester et al. 2005; Quan et al. 2012b), and appear most similar to the group *Cerris*. *Quercus* leaf fossils from the Huadian formation (Manchester et al. 2005; Quan et al. 2012a) were identified as cf. *Quercus berryi* Trelease by Manchester et al. (2005). At present, there are no other Eocene *Quercus* records from other regions of East Asia.

3.3.4.2 The Oligocene

The latest Eocene–earliest Oligocene fossil floras of the Kobe Group, in western Japan, includes diverse *Quercus* morphotypes that have close affinity with modern species found in East Asia and beyond. *Quercus miovariabilis* Hu and Chaney has lanceolate leaves with many aristate-tipped teeth, traits representative of modern East Asian species of the group *Cerris*, including *Q. variabilis* Blume, *Q. acutissima* Carruth., and *Q. chenii* Nakai. This species occasionally is associated with fossil cupules with slender scales projecting outward (Hori 1976, 1987). *Quercus miovariabilis* has a long stratigraphic distribution, being well represented in fossil assemblages between the Oligocene and Pliocene.

Another species belonging to group *Cerris* is *Q. ussuriensis* Kryshtofovich (Plate 3.6 3), which has a wide, oblong lamina with an inequilateral base, deltoid teeth with short, aristate tips, and irregular branching veinlets. These characteristics are similar to modern *Q. macrolepis* Kotschy and *Q. pyrami* Kotschy (=*Quercus ithaburensis* Decne subsp. *macrolepis* [Kotschy] Hedge and Yalt) of southeastern Europe and Western Asia (Tanai and Uemura 1994). *Q. ussuriensis* occurs in Oligocene localities in North Korea, southern Primorye, and Hokkaido (early Oligocene Wakamatsuzawa locality).

Lobed-leafed species of *Quercus* with a close affinity with sections *Prinus* (=group *Quercus* sensu Denk and Grimm 2009) and *Quercus* of East Asia, characterize the Kobe flora and other Oligocene floras in Northeastern Asia. *Q. sichotensis* Ablaev and Gorovoi, reported from Kobe, Hokkaido (Wakamatsuzawa) and southern Primorye (Plate 3.6 2), has leaves with four to six pairs of deeply compound lobes with narrow sinuses extending close to the midvein and subsidiary teeth. Its leaf shape and venation patterns are similar to North American section *Prinus* (*Q. alba* L., *Q. macrocarpa* Michx. and *Q. garryana* Douglas ex Hook) and south European section *Quercus* (*Q. pyrenaica* Willd. and *Q. frainetto* Ten.) (Tanai and Uemura 1994). *Quercus kobatakei* Tanai and Yokoyama described from the Kobe Group (Plate 3.6 1; Tanai and Yokoyama 1975) also has deeply lobed leaves but more broadly opened sinuses with rounded bases, similar to modern section *Prinus* such as *Q. alba* and *Q. lyrata* Walter of eastern North America. Tanai and Uemura (1994) included *Q. pseudolyrata* reported from the Oligocene in southern Primorye (Klimova 1976) into *Q. kobatakei*. *Quercus kodairae* Huzioka has dentate margins with large, lobe-like teeth with acute tips, similar to *Q. petraea* of the section *Quercus* in Europe (Tanai and Uemura 1994), and was widely distributed during the Oligocene in North Korea and Primorye. In addition, a fossil wood reported from the Kobe Group exhibits white oak (section *Prinus*) type (Terada and Handa 2009).

Taxa belonging to section *Prinus* that resemble East Asian species are also reported from the Oligocene. *Quercus protoserrata* Tanai and Onoe, which is similar with modern *Q. serrata* Murray, is widely distributed in East Asia and is reported from the Kobe Group (Hori 1976, 1987). *Quercus kitamiana* Tanai (Plate 3.6 5) described from the early Oligocene in Wakamatsuzawa (Hokkaido), has a narrow, obovate leaf with an acute base and dentate margin, except for part of the lower one-thirds or one-fourths of the blade, which is entire; this species is similar to extant *Q. griffithii* Hook and Thomson ex Miquel and *Q. aliena* Blume of East Asia and modern *Q. prinus* Willd. of eastern North America (Tanai 1995).

Four species of subgenus *Cyclobalanopsis* recorded from the Oligocene in Japan include taxa that also occur in the Neogene and are referable to modern East Asian species: (i) *C. mandraliscae* (Gaudin) Tanai (Hori 1976, 1987), reported for the Kobe Group, has a lanceolate lamina with serrated margins, similar to modern *C. longinus* (Hayata) Schottky of Taiwan; (ii) *C. ezoana* Tanai (Plate 3.6 6), described from the Wakamatsuzawa flora (late Rupelian) in Hokkaido, has an elliptic lamina with more than 15 thick secondary veins and small, sharp teeth and possible tomentose undersurface, resembling *Quercus oxyodon* Miq. and *Q. lamellosa* Sm. of central and southwestern China (Tanai 1995); (iii) *C. nagatoensis* Tanai and Uemura (Plate 3.6 7) described from the Noda flora (earliest Chattian) in western Japan is characterized by an entire margin and reticulate tertiary venation, and it is closely related to the modern *Quercus hui* Chun of S. China (Tanai and Uemura 1991). On the other hand, (iv) *C. protoacuta* (K. Suzuki) Huzioka and Uemura has an attenuate apex, entire margin, and highly developed fine veins, and it is closely related to the modern *Quercus acuta* Thunb. from Japan (Uemura et al. 1999).

The fossil record of *Quercus* in the Oligocene from other Asian countries is limited. Three species have been recorded in the early Oligocene strata of Kazakhstan: *Quercus alexeevii* Pojarkova, *Q. drymeja* Unger and *Q.* cf. *cerris* L. (Zhilin 1989). *Q. alexeevii* and *Q. drymeja* were found in the Ascheayrykian flora, whereas the three species together occurred in the Murunchink flora (Zhilin 1989). Two *Quercus* fossil species, *Q. kodairae* Huzioka and *Quercus* sp. were reported from Sanhe flora of northeastern China (Guo and Zhang 2002). *Q. kodairae* was widely distributed in North Korea and Japan in Oligocene as well (Guo and Zhang 2002).

3.3.4.3 The Neogene

In early–middle Miocene times, the Russian Far East was characterized by oaks with roburoid and castanoid leaves. *Quercus ussuriensis* is related to the group *Cerris* (Iljinskaja 1982). *Quercus bersenevii* Ablaev and Iljinskaja, *Q. sinomiocenica* Hu and Chaney (Plate 3.7 3) and *Q. praemongolica* Ablaev and Iljinskaja, are all related to the section *Quercus* (Iljinskaja 1982; Ablaev and Vassiliev 1998). However, Iljinskaja (1982) had misgivings about the taxonomic relationship of *Q. praemongolica* since it was found in association with cupules that are not typical for section *Quercus*. During the middle Miocene of this region, section *Quercus* was represented by the castanoid-like leaves species *Q. protoserrata* Tanai and Onoe, which was also widespread in the Neogene of Japan (Pavlyutkin 2005).

Quercus sichotensis, which had *robur*-like leaves, was probably one of the species that made up the first stage of differentiation of the section *Prinus* (now integrated into section *Quercus*; Nixon 1993) during the middle–late Miocene of Rettichovka and Kraskino (Far East) (Menitsky 1984). At the same time, the castanoid-like taxon *Q. miocrispula* Huzioka was recorded for Chiornaya Rechka (Iljinskaja 1982). To date, there is no evidence for the genus in High Arctic Cenozoic sites of the Russian Far East.

Throughout China, at least 64 species have been recorded based on macrofossils. Among these, three groups are recognized based on leaf characteristics. All species of the first group, *Quercus* subgenus *Quercus*, have toothed margins, with either dentate teeth (teeth with their axes perpendicular to the axis of the leaf margin, e.g., *Q. mongolica* Fisch. ex Ledeb., *Q. dentata* Thunb.), or serrated teeth (teeth with their axes inclined toward the leaf margin, e.g., *Q. miovariabilis*), or with long spines (margin with long sharp teeth formed by the secondary vein extending into the tooth and fusing with it). The tertiary veins may be slightly convex or straight percurrent or forked occasionally. Most of this group was distributed throughout the North of China, similar to the current distribution of deciduous *Quercus* species in this territory.

In the fossil floras of China, the second type of *Quercus* represented by leaves is the group *Cerris*. *Cerris*, which has been represented so far only by leaf impressions (e.g., Hu and Chaney 1940; Zhou 1993). The first occurrence of a fruit of this group is the compression of an acorn with imbricate scales from the middle Miocene of Shandong province, which was referred to *Q.* cf. *cerrisaecarpa* (Song et al. 2000). This specimen can be compared with the extant *Q. acutissima* Carruth., *Q. chenii* Nakai and *Q. variabilis* Blume.

The third group of *Quercus* recognized in China is section *Heterobalanus* (=group *Ilex* sensu Denk and Grimm 2009) of the subgenus *Quercus* that includes oaks with sclerophyllous and obovate leaves. Leaf apices are rounded and the leaf margins are mainly entire or with a few spinose teeth in some species. Their midveins generally display a zigzag pattern and are branched at the apex, and secondary veins are branched near the margin; the third order veins are weakly percurrent and predominantly alternate (Zhou et al. 1995). These characteristics make this section easily distinguishable from other Chinese species. Several species such as *Q. preguyavaefolia* Tao (Plate 3.7 6–7; Tao et al. 2000) of this section exhibit leaves similar to some Mediterranean sclerophyllous evergreen oaks, such as *Q. ilex* and *Q. suber*. Section *Heterobalanus* has currently 9–11 species with a core distribution in the Himalayas-Hengduan Mountains where they have left an abundant Neogene fossil record and they are still the dominant species (Zhou et al. 2007). In this area, the oldest records of section *Heterobalanus* are *Q. namlingensis* Li and Guo and *Q. wulongensis* Li and Guo (Zhou 1992). Both are from middle Miocene formations (15 Ma) of Namling, Xizang (Li and Guo 1976). Fossils related to this section have also been reported from the late Miocene Xiaolongtan and Xiangfeng floras in Yunnan (SW China) (Xia et al. 2010; Xing 2010).

Pollen also indicates that oaks were widespread in China. For example, three species, one of them related to the subgenus *Cyclobalanopsis*, were identified in the late early to early middle Miocene of Shandong province (Liu and Leopold 1992); four species (*Quercopollenites asper*, *Quercoidites henrici*, *Q. microhenrici* and *Q. minor*) were recognized in the sediments of Yunnan (Yao et al. 2011); and three morphotaxa of deciduous origin, two of evergreen habits and three oaks whose habit is uncertain were found in the Zhejiang province (Liu et al. 2007a). In pollen spectra from lower and middle Miocene sites of central Japan, evergreen *Quercus* occasionally forms the dominant element, together with *Carya* and *Liquidambar*, and exhibits a higher percentage than deciduous *Quercus* (Yamanoi 1989). The diversity of *Quercus* pollen grains would suggest that there were numerous oak species growing in China during Miocene times (Wang 1994).

In the late Pliocene, oaks of the section *Heterobalanus* became the main components of the floras of Yingping, Eryuan, and Lanping, all of which are located in the Hengduan Mountains, on the eastern margin of the Himalayas. Based on the fossil history and anatomical and physiological characteristics of section *Heterobalanus*, Zhang et al. (2005) hypothesized how this group of *Quercus* came to dominate forest ecosystems in the Hengduan Mountains. They proposed that the section *Heterobalanus* originated in subtropical, broadleaf forests. As the climate cooled with the uplift of the Hengduan Mountains (Zhou 1993), it produced an unfavorable environment for most broadleaved evergreen trees. However, these oaks had xerophytic leaf characteristics, such as dense hairs, thick cuticles, lignified epidermal cell walls and cuticles, and a low stomatal density (Zhou et al. 2003). Thus, they may have been well adapted for the environmental change and become a dominant species in the forested regions (Zhou and Coombes 2001). Although the physiological and ecological adaptation of evergreen sclerophyllous *Quercus* has long interested ecologists (He et al. 1994), the ecophysiology of montane plants in the Hengduan Mountains has not been well studied (Terashima et al. 1993). Recent molecular research (Du et al. 2016; Meng et al. 2017) supports the hypothesis of Zhou et al. (2007), that when the mountains uplifted, this oak group was able to colonize the mountain niches.

In modern oaks, cupule scale characteristics are a key feature used to distinguish the two oak subgenera, however fossils of cupules are rare in this area. Ring-cupped oaks can also be distinguished from other oaks based on their leaf characteristics. Most ring-cup oaks have an elliptical leaf shape and are in the mesophyll size range. The margins may be entire or serrated or with short spines. The secondary veins are uniform and parallel, and the tertiary veins are strongly percurrent.

In some cases, the trichomes are useful to distinguish species; e.g., *Q. delavayi* Franch., *Q. glauca* Thunb., and *Q. schottkyana* Rehder and E.H. Wilson are very similar in leaf shape, size, and vein pattern, but have different trichome characteristics. Leaves of *Q. delavayi* have dense, yellow, fascicular trichomes, which can be distinguished from those of *Q. glauca* and *Q. schottkyana*, which have only single hairs. Fortunately, the bases of trichomes can be preserved in the cuticle of some fossils, and they can help identify fossil *Quercus* and recognize their nearest living relatives.

Xing et al. (2013) found fascicular trichome bases in late Miocene fossils and they were able to identify as *Q. praedelavayi* Xing and Zhou (Plate 3.7 2, 8). Hu et al. (2014), in contrast, described similar late Pliocene fossils with unicellular and multicellular trichome bases that they included in the species *Q. tenuipilosa* Hu and Zhou (Plate 3.7 4–5). The nearest living relative of these two Neogene species is *Q. delavayi*.

Six *Quercus* species records belonging to subgenus *Cyclobalanopsis* were reported from early Miocene (21–16.5 Ma) Jinggu flora (Writing Group of Cenozoic Plants of China 1978), Yunnan, southwest China. All of them, including

Q. decora Tao, *Q. lahtenoisii* Colani (=*Dryophyllum relongtanense* Colani), *Q. parachampionii* Chen and Liu, *Q. parahelferiana* Chen and Tao, *Q. parschottkyana* Wang and Liu, and *Quercus* sp. were identified from leaf impressions (Writing Group of Cenozoic Plants of China 1978). These appear to have occupied subtropical evergreen broad-leaf to temperate forests (Mehrotra et al. 2005).

A ring-cupped oak, *Quercus tibetensis* Xu, Su and Zhou was reported from Tibet (Plate 3.7 1) from strata that were considered to be of late Miocene age. The occurrence of this species seems indicate the presence of subtropical forests in the core area of the Qinghai–Tibetan Plateau Tibet in the late Miocene (Xu et al. 2016). Today, most species of ring-cupped oaks live in forests in tropical or subtropical climates, although a few are temperate. Most species are distributed in deciduous and evergreen broad-leaf forests and deciduous broadleaf mixed forests near the northern border of the subtropical zone in China.

Neogene *Quercus* specimens found commonly in Japan became increasingly similar to extant taxa of the same area. The percentage of species with entire-margined leaves in the early Miocene floras persisted through temperature fluctuations between warmer (ca. 21–19 and 17–15 Ma) and cooler (23–21 and 19–17 Ma) stages (Tanai 1991; Momohara in press). Species of subgenus *Cyclobalanopsis* were major components of fossil floras alongside other evergreen broad-leaved trees in Honshu during warmer stages (Tanai 1961). In addition to *C. protoacuta* and *C. mandraliscae* already noted from the Oligocene, *C. nathorsti* Kryshtofovich, *C. protosalicina* Suzuki, and *C. praegilva* Kryshtofovich are relatives of modern *C. glauca*, *C. salicina*, and *C. gilva* Blume, respectively, These species were common in early and middle Miocene floras and are found occasionally in late Miocene and Pliocene floras. Deciduous oaks, such as *Q. miovariabilis* and *Q. protoserrata,* were more dominant in warmer stage assemblages than cooler ones.

Morphotypes assignable to group *Ilex* are recorded from the early Miocene Yoshioka flora in southern Hokkaido. *Q. elliptica* Tanai et Suzuki has coriaceous, elliptic leaves with entire margins or sometimes small teeth near the apex, similar to *Q. phylliraeoides* A. Gray in east Asia and *Q. chrysolepis* Liebm. in western North America (Tanai and Suzuki 1963). *Q. koraika* Tanai, with ovate and aristate-serrated leaves, is similar to modern *Q. tarokoensis* Hayata in Taiwan, is included in this flora. This species also has been described from the middle Miocene in Korea (Tanai and Suzuki 1963).

Pollen shows that evergreen oaks were significant in the late middle Miocene to the earliest late Miocene of the Himi area of Toyama Prefecture in Central Japan (Wang et al. 2001). Deciduous *Quercus* that dominates current temperate forests in Japan increased from the late middle to late Miocene. In addition with *Q. protoserrata*, species include *Q. protoaliena* Ozaki, related to modern *Q. aliena* Blume, *Q. miocrispula* Huzioka, related to modern *Q. crispula* Blume (=*Q. mongolica* Fisch. ex Ledeb.), and *Q. protodentata* Tanai et Onoe, related to modern *Q. dentata* Thunb (Uemura 1988). During the late Miocene, a conspicuous decrease of both deciduous and evergreen oak pollen types were recorded in Central Japan (Wang et al. 2001).

3.4 Biogeographical and Palaeoclimatic Implications

Today, subgenus *Quercus* occurs in the Northern Hemisphere and has its core distribution in cool to warm temperate climates. At its southern limit, the subgenus occurs at higher elevations. Subgenus *Cyclobalanopsis*, which had a much wider distribution throughout the Cenozoic, is now restricted to eastern and southeastern Asia where it extends to the lowlands and thus touches the tropical climate realm (Fig. 3.2).

The northern border of the distribution of *Quercus* generally follows a zonal orientation roughly coinciding with the 2 °C mean annual temperature isotherm (Global Biodiversity Information Facility; http://www.gbif.org). In Europe, the effect of the Gulf Stream Current permits the genus to extend further north—up to 63°N in Scandinavia (cf. Atlas Florae Europaeae). In contrast, on the Pacific coast of Eurasia, 48°N is not exceeded. Modern oaks tend to avoid extremely continental climate conditions of the continental interior with summer drought and very cold winters (with mean temperature of the coldest month—CMT <20 °C).

Oak species may tolerate a considerable variety of habitats and climates. Group *Cerris* includes species adapted to seasonal drought and hence are native to regions with Mediterranean type climates. In the Koeppen-Geiger climate system (Peel et al. 2007), subgenus *Quercus* has its main distribution in cool to warm temperate climates, partly fully humid (indicated as: Dfa, Dfb, Cfa, Cfb), or seasonally dry as Sub-Mediterranean and Mediterranean (Csb, Csa) and the snow climates with hot summers of Asia Minor and the Caucasus (Dsa, Dsb). In Central America, oak species may exist under equatorial winter dry climates (Aw) but mainly at higher altitudes.

In parallel with the current day, oaks were often an important component of the Cenozoic plant communities, dominating leaf assemblages of western North America and Europe. For example, oak specimens comprised 42.4 and 65.1% of the leaves collected from the Miocene Sucker Creek and Trout Creek floras of Oregon, respectively (Graham 1965), and over 44% of the Eastgate and 85% Middlegate floras of the Miocene Nevada (Axelrod 1985). In Europe, *Q. rhenana* and *Q. sosnowskyi* predominate in several paleofloras (Kovar-Eder et al. 2001; Kvaček et al. 2002).

In addition, numerous oak species can coexist in a given flora. Seven species were recorded in the Miocene Tehachapi flora from California (Axelrod 1939) including four species of the section *Quercus* and three of the *Protobalanus*; six species co-occur in the Miocene Mascall flora of Oregon (Chaney 1959; Chaney and Axelrod 1959), including 2 species each representing sections *Quercus*, *Protobalanus* and *Lobatae*, and also six in the Pliocene Sonoma flora of California (Dorf 1930; Axelrod 1944) including 3 species of section *Quercus*, 2 of section *Protobalanus* and one of section *Lobatae*; five to six species have been identified in the Spanish Miocene of La Cerdanya Basin (Barrón et al. 2014); and five species coexist in the Miocene floras from Temblor in California (Renney 1972), the Pickett Creek in Idaho (Buechler et al. 2007) and Yatağan Basin in Turkey (Güner et al.

2017), and also five in the Pliocene Mount Eden and Broken Hill floras of California (Axelrod 1950, 1980).

In western Eurasia, conditions close to tropical (paratropical) existed at times in the earlier Paleogene. In these settings, oaks were rare and their assignment to modern sections difficult. For example, *Quercoxylon sempervirens* Gottwald is described from the Lutetian Upper Seam of Helmstedt, Germany (Wilde 1989), and *Quercus* sp. from the Barthonian Grés à Palmiers in southern France (Vaudois-Miéja 1985). Using the Koeppen-Geiger climate system (Peel et al. 2007), the Coexistence Approach (CA)-based climate reconstructions reveals the cool end of tropical (Af) to the warm end of warm temperate perhumid (Cfa) Koeppen-Geiger climate for both sites (mean annual temperature [MAT]: 22–25 °C; cold month mean [CMT]: Lutetian 17–23 °C, Barthonian: 15–17 °C; mean precipitation of the warmest month [MPwarm]: 120–190 mm). Vegetation reconstruction at the level of Plant Functional Types (PFTs) reveals subtropical rain and laurel forest for the Helmstedt flora and warm, mixed evergreen-deciduous forest for the Grés à Palmiers palaeoflora (Utescher and Mosbrugger 2007), coinciding with the CA-based climatic values. The presence of oaks under the very warm conditions existing in western Eurasia in the earlier Paleogene is noteworthy when discussing the origin of the genus, however the implications for the subsequent diversification of the genus are unclear.

Oligocene evergreen, laurophyllous oak species belonging to section *Lobatae* (e.g., *Q. neriifolia*, *Q. lyellii*, Heer, *Q. armata* Saporta, *Q. oligodonta* Saporta) together *Q. praerhenana*, recovered from various late Oligocene to Burdigalian sites of Central Europe (including Lusatian Basins), existed under warm temperate conditions (MAT 15–17 °C; CMT 5–10 °C) (see Walther 1999; Grein et al. 2013; Table 3.1).

Although the group *Ilex* includes species that are today restricted to summer-dry climates, the morphology of *Q. drymeja* of the Oligocene and Miocene more closely resembles East Asian species and hence is not considered to represent an indicator of Mediterranean type climate (Velitzelos et al. 2014). *Q. neriifolia* is either interpreted as being part of the zonal vegetation (e.g., Chattian of Bobov Dol, Bulgaria; see Bozukov et al. 2009), or as swamp forest element (Pangaion Range, Miocene, Greece; see Velitzelos et al. 2014), comparable to the ecology of various modern red oaks (e.g., *Q. laurifolia* Michx, *Q. nigra* L., *Q. phellos* L.). CA-based climate estimates for the Chattian Bobov Dol site, located in the eastern part of the Tethyan Archipelago and including *Q. drymeja* and *Q. neriifolia*, exemplify humid, warm temperate Cfa climate conditions (MAT ca. 16–21 °C; CMT ca. 8–14 °C; mean annual precipitation [MAP] ca. 1200–1300 mm; MPwarm ca. 90–170 mm; Table 3.1).

Oaks underwent a substantial radiation in the Neogene of western Eurasia (Fig. 3.4). Among the species emerging in the early Miocene, *Q. rhenana* might have had an affinity to thermophilous wetland vegetation existing in central and Eastern Europe and the Central Paratethys, where the taxon was part of thermophilous, dominantly broadleaf evergreen vegetation, the so-called "Mastixioid Floras" (Mai 1995). Most of the early Miocene species are referred to the groups *Cerris* and *Ilex*, e.g., *Q. ilicoides*, *Q. kubinyii* and *Q. mediterranea*. Among these thermophilous and

Table 3.1 Important European Cenozoic oak taxa and their palaeoclimatic conditions

Species	MAT min	MAT max	CMT min	CMT max	MAP min	MAP max	MMP drymin	MMP drymax
Q. petraea	9.3	20.5	−2.7	13.5	826	1187	24	63
Q. pubescens	13.3	13.9	2.2	3.8	979	998	–	–
Q. roburoides	11.2	16.5	−1.3	8.7	897	1151	24	56
Q. ilicoides	13.3	21.1	0	13.3	823	1355	42	61
Q. drymeja	12.2	20.5	−0.5	13.3	735	1437	19	66
Q. kubinyii	12.2	20.8	−0.1	14.8	759	1613	8	67
Q. pseudocastanea	11.6	21.1	−0.5	13.3	735	1355	11	59
Q. mediterranea	12.2	19.5	0.4	13.3	735	1356	21	62
Quercus ex. gr. Ilex	13.3	17.4	2.2	8.3	823	1206	23	24
Q. gigas	13.3	21.1	−0.1	13.3	867	1355	32	43
Q. trojana foss.	13.3	16.5	−0.1	5.8	867	1231	24	56
Q. suber foss.	13.7	18.3	7.4	10.9	735	1333	11	59
Q. cerris foss.	13.3	16.5	0.1	5.8	867	1179	32	51
Q. neriifolia	13.4	21.8	−0.5	13.3	700	1613	11	64
Q. lyellii	15.6	21.1	1.8	12.6	439	1360	5	43
Q. rhenana	15.6	20.5	2.7	14.8	810	1362	25	51

Palaeoclimatic ranges considering their lowest (min) and highest (max) limits for mean annual temperature (MAT), mean temperature of the coldest month (CMT), mean annual precipitation (MAP), and mean precipitation of the driest month (MMPdry). These are based on palaeoclimate reconstructions using the Coexistence Approach (CA) for a total of 178 sites (mainly NECLIME datasets published in PANGAEA (http://www.pangaea.de)

sclerophyllous oaks, *Q. ilicoides*, whose nearest living relative is in the group *Ilex*, may tolerate Mediterranean type of climate; but with an MP warm range of 0–116 mm (Palaeoflora database), these taxa are not necessarily indicative of significant seasonal drought. Thus, their occurrence in the palaeobotanical record does not contradict the suggested overall persistence of permanently humid (Cfa) climates in western Eurasia during the early Miocene (Bruch et al. 2011). One of the more important middle Miocene species of the group *Cerris* was *Q. pseudocastanea*. It was common in the mixed mesophytic forest vegetation of Central Europe, the Central Paratethys realm and the northern Mediterranean (Mai 1995) but may have had also an affinity to deciduous riverside forests (Belz 1992; Walther and Eichler 2010). A Messinian site in the northern Mediterranean Integrated Plant Record (IPR) vegetation analysis (Kovar-Eder and Kvaček 2007; Kovar-Eder et al. 2008) including *Q. pseudocastanea* reveals a transitional vegetation type between "Broad-leaved Evergreen Forest" and "Mixed Mesophytic Forest" (abbreviated: BLEF/MMF) (Monte Tondo, Tossignano; Teodoridis et al. 2015). Another common species of the group *Cerris*, *Q. gigas*, was part of the deciduous lowland vegetation widespread in the Pannonian realm in cooler climatic phases (Utescher et al. 2017).

Late Miocene roburoid oaks of the section *Quercus* such as *Q. roburoides* commonly occur in Europe in the Messinian to Pliocene deciduous hardwood riverside forests of the Lower Rhine Basin, Germany (Belz 1992; Van der Burgh 1993).

Regarding most recent exchange between Old and New World lineages of oaks, Hubert et al. (2014) suggested the occurrence of *"phases of unhindered gene flow via the North-Atlantic Land Bridge in Group Quercus until at least 8 Ma* (Denk et al. 2010) *and Beringia until the latest Pliocene or Pleistocene interglacials"*. The record of *Quercus* in eastern North America is consonant with possible geographic exchange across the North Atlantic through the late Miocene. However, the Alaskan palynological record, while recording *Quercoidites* in the Eocene (Frederiksen et al. 2002), suggests that the genus disappeared from Alaska by about 12 Ma, last appearing in the Betulaceae zone, Polygonaceae subzone, of White et al. (1999). The current megafossil record of *Quercus* in the Alaskan Neogene consists of *Q. furuhjelmi* Heer of the early to early middle Miocene (Wolfe 1980; Wolfe in Lathram et al. 1965; Wolfe in Leopold and Liu 1994), a species compared with living *Q. sadleriana* R. Brown, *Q. prinoides* Willd. and *Q. mongolica* Fisch. ex Ledeb, all of section *Quercus*. This suggests that the last potential Beringian exchange would have occurred earlier in the Miocene.

Subsequent reports from central and eastern North America (e.g., Berry 1952; MacGinitie 1962; Tiffney 1977; Liu 2011; Stults and Axsmith 2015; Stults et al. 2016) indicate the continued presence of both sections *Quercus* and *Lobatae* through the rest of the Tertiary in eastern North America. In addition, sections *Lobatae* and *Quercus* similarly continue to be present through the rest of the Neogene in western North America, but in parallel with other mesic elements, are increasingly confined to more westerly occurrences, responding to the cooling and drying trends of the later Tertiary (Axelrod 1973; Schierenbeck 2014; Bouchal et al. 2014). It is possible that these climatic changes also influenced three other oak taxa. It appears that species of section *Protobalanus*, if anything, adapted to these changes, remaining restricted to southwestern North America (Fig. 3.3), where warmer and dryer climates are the norm. By contrast, it is possible that the representatives of subgenus *Cyclobalanopsis* (present in the mid-late Eocene) and group *Cerris* (possibly present at the Eocene-Oligocene border) were confined to the west at the outset, and were sufficiently moisture-loving and thermophilic that they, like many other taxa present in the Paleogene of the west coast of North America, went regionally extinct at the end of the Paleogene in response to climatic change (Wolfe 1978, 1985). However, we note that Buechler et al. (2007) suggested that *Q. oberlii* Buechler, Dunn and Rember of the Miocene of Idaho (which they compared to species in section *Lobatae*) was also compared with *Q. saliciana* Blume of East Asia, a member of section *Cyclobalanopsis*. In Western Eurasia, the macrofossil record of *Cyclobalanopsis* is characterized by low diversity (Fig. 3.4). A single species probably related to this subgenus, namely *Cyclobalanopsis stojanovii* Palamarev and Kitanov, is intermittently reported in eastern Paratethys realm since the early Oligocene (Palamarev and Ivanov 2003). The possibly lastest record of this subgenus in western Eurasia dates to the early Pliocene (Bozukov et al. 2011).

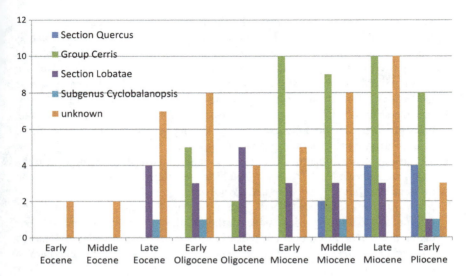

Fig. 3.4 Species diversity of Cenozoic oaks of Western Eurasia based on floral lists of a total of 237 fossil sites (most of them NECLIME data sets published in PANGAEA; http://www.pangaea.de)

During the Quaternary, oaks were affected by successive glaciations, resulting in the appearance of recent species (e.g., Plate 3.4 3). Today, northern latitudes are inhabited by deciduous species, while evergreen taxa appear in tropical-subtropical and Mediterrranean areas. This most recent history is largely dominated by palynological records, as macrofossils are uncommonly reported (e.g., Berry 1952; Postigo-Mijarra et al. 2007; Iglesias-González 2015). The recent anthrophic disturbance of terrestrial ecosystems, especially in the Northern Hemisphere, has led to the disappearance of some populations of *Quercus* and a reduction of the range of others. Consequently, some populations of *Quercus* have become extinct or are seriously endangered. One example of this was the communities of *Quercus-Carpinus* from the Canary Islands that became extinct when humans arrived on the islands around 3000 years BP (De Nascimento et al. 2009).

3.5 Conclusions

1. Although *Quercus* has been reported from the Cretaceous (e.g. *Quercophyllum* or *Quercus cretaceoxylon*), the earliest possible occurrence of *Quercus* in the fossil record is in the Paleocene. The Paleocene European macrofossil *Quercus subfalcata* Friedrich, which still need a confirmation of its taxonomic status, could be the earliest occurrence of the genus. Currently, *Quercoidites* pollen grains in the Late Paleocene (ca. 55 Ma) at the St Pankraz site (Austria) is the unequivocally earliest evidence of the genus.

2. The earliest evidence of *Quercus* in North America is *Quercus paleocarpa*, from the middle Eocene Clarno formation of Oregon (ca. 44 Ma), which is related to the subgenus *Cyclobalanopsis*. Likewise, the first record from East Asia corresponds to the middle Eocene *Cyclobalanopsis naitoi* from western Japan, which is represented by leaves and acorns. The middle Eocene *Quercus ishikariensis* Tanai from Japan may be the first occurrence of the group *Cerris*.
3. In general terms, the Palaeogene leaf fossil record of *Quercus* is scant, a fact probably related to the difficulty in distinguishing leaves of *Quercus* from those of other fagaceous genera, especially *Castanea*.
4. The paleobotanical record shows that the genus was well diversified since the Late Paleogene in North America including the subgenus *Cyclobalanopsis* and the sections *Quercus*, *Lobatae* and *Protobalanus*.
5. Many of the species described from leaves in Paleogene sediments (e.g., Europe and North America) need further studies in order to confirm their taxonomic status. In Europe, the species *Q. rhenana* (laurel-like leaves) and *Q. lonchitis*, *Q. praekubinyii* and *Q. pseudoalexeevii* (toothed leaf margins) are of particular interest.
6. Since the mid-late Eocene, oak records occur frequently in the territory of the former USSR, and *Quercus* clearly played locally an important role in these ecosystems (e.g., *Q. pseudoneriifolia*).
7. In late Eocene, sections *Quercus*, *Lobatae* and *Protobalanus* as well as the ring-cupped oaks, all were present in North America whereas sections *Lobatae* and the subgenus *Cyclobalanopsis* occurred in Europe.
8. From the early Oligocene, group *Cerris* diversified in Eurasia. However, it became extinct in North America by the Neogene.
9. Section *Protobalanus* appeared in the Eocene of North America and seems to be always limited to this continent.
10. Deciduous oaks were common in northern temperate regions in the Paleogene. The southern part of the North Hemisphere was inhabited by evergreen oaks in warmer climates. Palynological data indicate that sections *Quercus*, *Lobatae* and *Protobalanus* as well as the group *Ilex* were already present in middle Eocene Arctic localities.
11. The Neogene fossil record of *Quercus* in western North America is primarily dominated by leaves. The two most common leaf species were *Q. hannibali* and *Q. simulata*. The affinities of *Q. simulata* have been debated and it may represent another genus of Fagaceae.
12. Sclerophyllous evergreen oaks of the groups *Cerris* and *Ilex* (=*Heteroblanus*) were important elements of Pre-Mediterranean vegetation of central and south Europe during the Neogene. Some significant species for this period were *Q. gigas*, *Q. pseudocastanea*, *Q. kubinyii*, *Q. mediterranea* and *Q. drymeja*. The subgenus *Cyclobalanopsis* is also present in Europe (as *C. stojanovii*) in the Pliocene.
13. Also in the Neogene, *Heterobalanus* species colonize the Chinese Himalayas and Hengduan Mountains where cold and xeric environmental conditions developed.

14. The late Oligocene records of *Quercus furuhjelmii?* Heer from Kazakhstan and Western Siberia seem to be the first one for oaks with deciduous, roburoid leaves. Due to the progressive Neogene climate cooling, these trees, which mainly belonged to section *Quercus*, colonized southern areas, replacing subtropical evergreen oak species. Oaks with *robur*-like leaves became common in Europe from the late Miocene.
15. Representatives of the subgenus *Cyclobalanopsis* disappeared from North America probably by the late Paleogene, and Europe at the end of the Neogene. Now, this subgenus is present only in East Asia.
16. The genus *Quercus* was a common element of the Cenozoic ecosystems inhabiting temperate and tropical/subtropical environments of the North Hemisphere. It attained an almost modern distribution in the Neogene.
17. An occurrence of "phases of unhindered gene flow" via the North Atlantic Land Bridge in section *Quercus* until at least 8 Ma and Beringia until the latest Pliocene of even Pleistocene interglacials has been suggested. The fossil record does not contradict the North Atlantic connection, but suggests that the Beringian route was closed to *Quercus* by 12 Ma.
18. Fossil wood attributed to *Quercinium* and *Quercoxylon* is recorded since the Eocene. Staminate catkins and stellate trichomes are much more scant. In both cases, paleoecological or taxonomic conclusions from these macroremains are very difficult to establish.
19. Carpological, cuticular and new SEM studies are necessary to trace the evolution of this genus over the Cenozoic.

Acknowledgements This work was performed as part of the research projects: Progress Q45 (Charles University, Prague), CGL2015-68604-P and CSO2015-65216-C2-2-P (MINECO, Spain), and is a contribution to NECLIME (Neogene Climate Evolution in Eurasia, www.neclime.de). The study of fossil oaks from Russia and Central Asia was provided by the Russian grant RFBR N 16-04-00946-a. The study of the Chinese fossil Quercus was supported by the NSFC (U1502231) grant. We thank Eustaquio Gil-Pelegrín, José Javier Peguero-Pina and Domingo Sancho-Knapik (CITA, Zaragoza, Spain); Isabel Rábano, Ana Rodrigo, Enrique Peñalver and Silvia Menéndez (Museo Geominero, IGME, Madrid, Spain); Celia Santos and Ana Bravo (Museo Nacional de Ciencias Naturales—CSIC, Madrid, Spain); Raul Iglesias-González (E.T.S. de Ingeniería de Montes, Forestal y del Medio Natural, Universidad Politécnica de Madrid, Spain); Jaume Gallemí and Vicent Vicedo (Museu de Geologia de Barcelona, Spain); Sebastián Calzada and Evaristo Aguilar (Museu de Geologia del Seminari de Barcelona, Spain); Manuel Casas-Gallego (CGG Services, Wales, UK); Boglárka Erdei and Lilla Hably (Hungarian Natural History Museum, Budapest); Johanna Kovar-Eder (Staatliches Museum für Naturkunde Stuttgart, Germany); Volker Wilde, Martin Müller and Karin Schmidt (Senckenberg Forschungsinstitut, Frankfurt/M., Germany); Jiří Kvaček (National Museum of Prague, Czech Republic); Edoardo Martinetto, University of Turin, Italia; Patrick Fields and Darlene and Howard Emry (USA); Brian Axsmith, University of South Alabama, Mobile (USA); Elisabeth A. Wheeler, North Carolina State University and InsideWood (USA); Alexey Hvalj (Laboratory of Palaeobotany, Komarov Botanical Institute, St. Petersburg, Russia); Atsushi Yabe (National Museum of Nature and Science, Tokyo, Japan). The authors thank Steven R. Manchester (Florida Museum of Natural History, University of Florida, Gainesville, Florida, USA) and Melanie L. DeVore (Georgia College and State University, Milledgeville, Georgia, USA) for reading a portion of the manuscript and offering criticisms and suggestions.

References

Ablaev AG, Vassiliev IV (1998) The Miocene Kraskinskaja flora of Primorye. Rossiĭskaia akad nauk. Dal'nevostochnoe otdelenie. Tikhookeanskiĭ okeanologicheskiĭ institut. Vladivostok (in Russian)

Akhmetiev MA (1988) Cenozoic floras of the Sikhote-Alin. Geol Inst Acad Sci 1988:1–48 (in Russian)

Alcalá B (1997) Prospección palinológica en el Neógeno de Teruel. Teruel 85(1):9–20

Alcalá B, Benda L, Ivanovic-Calzaga Y (1996) Erste palynologische Untersuchungen zur Altersstellung des Neogen-Beckens von Xinzo de Limia (Prov. Orense, Spanien). Newsl Stratigr 34(1):31–38

Altamira-Areyán A (2002) Las litofacies y sus implicaciones de la cuenca sedimentaria Cutzamala-Tiquicheo, Estado de Guerrero y Michoacán, México. M.Sc. thesis, Universidad Nacional Autónoma de México, México DF

Arambourg C, Arènes J, Depape G (1953) Contribution à l'étude des flores fossils quaternaires de l'Afrique du Nord. Arch Mus Natl Hist Nat Paris 2:1–85

Ashraf AR, Mosbrugger V, Hebbeker U (1996) Palynologie und Palynostratigraphie in der Niederrheinischen Bucht. Teil 2: Pollen. Palaeontographica Abt B 241:1–98

Avakov GS (1979) Miocene flora of Medjuda. Tbilisi (in Russian)

Axelrod DI (1939) A Miocene flora from the western border of the Mohave Desert. Carnegie Inst Wash Publ 516:1–144

Axelrod DI (1940) The Mint Canyon Flora of southern California: a preliminary statement. Am J Sci 238:577–585

Axelrod DI (1944) The Sonoma Flora. In: Chaney RW (ed) Pliocene Floras of California and Oregon. Carnegie Inst Wash Publ 553:167–206

Axelrod DI (1950) Studies in late Tertiary paleobotany. Carnegie Inst Wash Publ 590:1–323

Axelrod DI (1956) Mio-Pliocene floras from west-central Nevada. Univ Calif Publ Geol Sci 33:1–322

Axelrod DI (1973) History of the Mediterranean ecosystem in California. In: di Castri F, Mooney HA (eds) Ecological studies. Analysis and synthesis, vol 7, pp 225–277

Axelrod DI (1980) Contributions to the Neogene paleobotany of California. Univ Calif Publ Geol Sci 121:1–212

Axelrod DI (1983) Biogeography of oaks in the Arcto-Tertiary province. Ann Missouri Bot Gard 70:629–657

Axelrod DI (1985) Miocene floras from the Middlegate Basin, west-central Nevada. Univ Calif Publ Geol Sci 129:1–280

Axelrod DI (1991) The Miocene Buffalo Canyon flora, western Nevada. Univ Calif Publ Geol Sci 135:1–180

Axelrod DI (1998) The Oligocene Haynes Creek Flora of eastern Idaho. Univ Calif Publ Geol Sci 143:1–99

Barclay RS, Johnson KR, Betterton WJ, Dilcher DL (2003) Stratigraphy and megaflora of a K-T boundary section in the eastern Denver Basin, Colorado. Rocky Mt Geol 38:45–71

Barrón, E (1996) Estudio tafonómico y análisis paleoecológico de la macro y microflora miocena de la Cuenca de la Cerdaña. Ph.D. thesis, Universidad Complutense de Madrid, Madrid

Barrón E (1998) Presencia del género *Quercus* Linné (Magnoliophyta) en el Vallesiense (Neógeno) de la Cerdaña (Lérida, España). Bol Geol Min 109(2):121–150

Barrón E, Lassaletta L, Alcalde Olivares C (2006) Changes in the early Miocene palynoflora and vegetation in the east of the Rubielos de Mora Basin (SE Iberian Ranges, Spain). N Jb Geol Paläontol Abh 242(2/3):171–204

Barrón E, Rivas-Carballo MR, Postigo-Mijarra JM, Alcalde-Olivares C, Vieira M, Castro L, Pais J, Valle-Hernández M (2010) The Cenozoic vegetation of the Iberian Peninsula: a synthesis. Rev Palaeobot Palynol 162:382–402

Barrón E, Postigo-Mijarra JM, Diéguez C (2014) The late Miocene macroflora of the La Cerdanya Basin (Eastern Pyrenees, Spain): towards a synthesis. Palaeontographica Abt B 291:85–129

Barrón E, Postigo-Mijarra JM, Casas-Gallego M (2016) Late Miocene vegetation and climate of the La Cerdanya Basin (eastern Pyrenees, Spain). Rev Palaeobot Palynol 235:99–119

Basinger JF (1991) The fossil forests of the Buchanan Lake Formation (early tertiary), Axel Heiberg Island, Canadian High Arctic: preliminary floristics and paleoclimate. Geol Surv Can Bull 403:39–65

Baumgartner KA (2014) Neogene climate change in eastern North America: a quantitative reconstruction. Electronic theses and dissertations paper 2348 http://dc.etsu.edu/etd/2348

Becker HF (1961) Oligocene plants from the upper Ruby River Basin, southwest Montana. Geol Soc Am Mem 82:1–127

Becker HF (1969) Fossil plants of the Tertiary Beaverhead Basins in southwestern Montana. Paleontographica Abt B 127:1–142

Becker HF (1973) The York Ranch Flora of the Upper Ruby River Basin, southwestern Montana. Palaeontographica Abt B 143:18–93

Bell W (1957) Flora of the Upper Cretaceous Nanaimo group of Vancouver Island, British Columbia. Geol Surv Canada Mem 293:1–84

Belz G (1992) Systematisch-palaeooekologische und palaeoklimatologische Analyse von Blattfloren im Mio/Pliozaen der Niederrheinischen Bucht (NW-Deutschland). Dissertation, Univ. Tuebingen, Tuebingen

Benammi M, Centeo-García E, Martínez-Hernández E, Morales-Gámez M, Tolson G, Urrutia-Fucugauchi J (2005) Presencia de dinosaurios en la Barranca Los Bonetes en el sur de México (Región de Tiquicheo, Estado de Michoacán) y sus implicaciones cronoestratigráficas. Rev Mex Cien Geol 22:429–435

Benda L (1971) Grundzüge einer pollenanalytischen Gliederung des türkischen Jungtertiärs (Känozoikum und Braunkohlen der Türkei. 4). Beih Geol Jahrb 113:3–45

Berry EW (1909) A Miocene flora from the Virginia coastal plain. J Geol 17:19–30

Berry EW (1916) The physical conditions indicated by the flora of the Calvert formation. US Geol Surv Prof Pap 98:61–73

Berry EW (1952) The Pleistocene plant remains of the coastal plain of eastern North America. Palaeobotanist 1:79–98

Bessais E, Cravatte J (1988) Les écosystèmes végétaux pliocènes de Catalogne Méridionale. Variations latitudinales dans le domaine Nord-Ouest Méditerranéen. Geobios 21(1):49–63

Bessedik M (1984) The early Aquitanian and upper Langhian-lower Serravallian environments in the Northwestern Mediterranean Region. Paléobiol Cont 14(2):153–179

Bessedik M (1985) Réconstitution des environnements miocènes des regions nord-ouest mediterranéennes a partir de la Palynologie. Ph.D. thesis, Univ Montpellier II, Montpellier

Beyer AF (1954) Some petrified woods from the Specimen Ridge area of Yellowstone National Park. Am Midland Naturalist 51:553–567

Biondi E, Koeniguer JC, Privé-Gill C (1985) Bois fossiles et vegetations arborescentes des régions méditerranéenes durant le Tertiare. G Bot Ital 119:167–196

Boeshore I, Jump JT (1938) A new fossil oak from Idaho. Am J Bot 25:307–311

Böhme M, Bruch A, Selmeier A (2007) The reconstruction of early and middle Miocene climate and vegetation in Southern Germany as determined from the fossil wood flora. Palaeogeogr Palaeoclimatol Palaeoecol 253:91–114

Boitzova EP, Panova LA (1966) Fagaceae. In: Paleopalynology, vol 1. Leningrad, Nedra, p 352 (in Russian)

Borgardt SJ, Pigg KB (1999) Anatomical and developmental study of petrified *Quercus* (Fagaceae) fruits from the Middle Miocene, Yakima Canyon, Washington, USA. Am J Bot 86:307–325

Borsuk MO (1956) Paleogene flora of Sakhalin. Trab All-Union Prosp Invest Geol Inst 12:1–131 (in Russian)

Bouchal J, Zetter R, Grimsson F, Denk T (2014) Evolutionary trends and ecological differentiation in early Cenozoic Fagaceae of western North America. Am J Bot 101:1332–1349

Bozukov V, Utescher T, Ivanov D (2009) Late Eocene to early Miocene climate and vegetation of Bulgaria. Rev Palaeobot Palynol 153:360–374

Bozukov VS, Utescher T, Ivanov D, Tsenov B, Ashraf AR, Mosbrugger V (2011) New results for the fossil macroflora of the Beli Breg Linite Basin. West Bulgaria. Phytol Balcanica 17(1):3–19

Brambilla G, Penati F (1987) Le filliti mioceniche del Colle dela Badia di Brescia. Osservazioni sistematiche, cronologiche e ambientali. Ann Mus Civico Sci Nat Brescia 23:79–102

Brett D (1960) Fossil oak wood from the British Eocene. Palaeontology 3(1):86–92

Brown RW (1962) Paleocene Flora of the Rocky Mountains and Great Plains. US Geol Surv Prof Pap 375:1–119

Bruch AA, Utescher T, Mosbrugger V, NECLIME members (2011) Precipitation patterns in the Miocene of Central Europe and the development of continentality. Palaeogeogr Palaeoclimatol Palaeoecol 304:202–211

Budantsev LY (1997) Late Eocene flora of the Western Kamchatka. Nauka, St. Petersburg (in Russian)

Budantsev LY, Golovneva LB (2009) Fossil Flora of Arctic, II. Palaeogene Flora of Spitsbergen. Russian Academy of Sciences, St. Petersburg

Buechler WK, Dunn MT, Rember WC (2007) Late Miocene Pickett Creek Flora of Owyhee County, Idaho. Contrib Mus Paleontol Univ Michigan 31:305–362

Burnham RJ, Graham A (1999) The history of Neotropical vegetation: new developments and status. Ann Missouri Bot Gard 86:546–589

Bůžek C, Kvaček Z, Holý F (1985) Late Pliocene palaeoenvironment and correlation of the Vildštejn floristic complex within Central Europe. Rozpr Cesk Akad Ved Rada Mat Prírod Ved 95(7):1–72

Call V, Tidwell WD (1988) Fossil wood from the Miocene of Wall Canyon Creek, northwestern Nevada. Am J Bot 75(2):104–105

Camus A (1936–1954) Les chênes. Monographie du genre *Quercus* et *Lithocarpus*. Encyclopedie économique de sylviculture, Paul Lechevalier éditeur, Paris

Casas-Gallego M (2017) Estudio palinológico del Oligoceno-Mioceno Inferior de la cuenca de As Pontes (Galicia, España). Ph.D. thesis, Universidad Autónoma de Madrid, Madrid

Cavagnetto C, Anadón P (1996) Preliminary palynological data on floristic and climatic changes during the Middle Eocene-Early Oligocene of eastern Ebro Basin, northeast Spain. Rev Palaeobot Palynol 92:281–305

Cavender-Bares J, Gonzalez-Rodriguez A, Eaton DAR, Hipp AAL, Beulke A, Manos PS (2015) Phylogeny and biogeography of the American live oaks (*Quercus* subsection *Virentes*): a genomic and population genetics approach. Mol Ecol 24:3668–3687

Chaney RW (1944) The Troutdale Flora. In: Chaney RW (ed) Pliocene Floras of California and Oregon. Carnegie Inst Wash Publ 553:323–352

Chaney RW (1959) Miocene Floras of the Columbia Plateau. Part 1. Composition and interpretation. Carnegie Inst Wash Publ 617:1–134

Chaney RW, Axelrod DI (1959) Miocene Floras of the Columbia Plateau. Part 2. Systematic considerations. Carnegie Inst Wash Publ 617:135–237

Chaney RW, Elias MK (1936) Late Tertiary floras from the High Plains. Carnegie Inst Wash Publ 476:1–46

Chateauneuf JJ (1972) Étude palynologique de l'Aquitanien. Bull Bur Rech Géol Min 2(14):59–65

Chateauneuf JJ, Nury D (1995) La flore de l'Oligocène de Provence méridionale: implications stratigraphiques, environmentales et climatiques. Geol Fr 2:43–45

Collinson ME (1984) Fossil plants of the London Clay. Field guide to fossils, 1. The Palaeontological Association, London

Collinson ME, Manchester SR, Wilde V (2012) Fossil fruits and seeds of the middle Eocene Messel biota, Germany. Abh Senckenb Ges Naturforsch 570:1–251

Condit C (1944) The Remington Hill Flora. In: Chaney RW (ed) Pliocene Floras of California and Oregon. Carnegie Inst Wash Publ, vol 553, pp 21–56

Conwentz H (1886) Die flora des Bernsteins, Zweiter Band; Die Angiospermen des Bernsteins. Engelmann, Danzig

Crabtree DR (1987) Angiosperms of the northern Rocky Mountains: Albian to Campanian (Cretaceous) megafossil floras. Rev Palaeobot Palynol 74:707–747

Crepet WL (1989) History and implications of the early North American fossil record of Fagaceae In: Crane PR, Blackmore S (eds) Evolution, systematics and fossil history of the Hamamelidae. Vol. 2 "Higher" Hamamelidae. Oxford University Press, Oxford

Daghlian CP, Crepet WL (1983) Oak catkins, leaves and fruits from the Oligocene Catahoula Formation and their evolutionary significance. Am J Bot 70:639–649

de Candolle A (1862) Étude sur l'espèce à l'occasion d'une révision de la famille des cupulifères. Ann Sci Nat Bot 4ème Ser. 18:59–110

de Carvalho A (1958) Identificação de un possível fóssil de sobreiro (*Quercus suber*) proveniente de solos do Mioceno lacustre do Alentejo. Bol Soc Broteriana 32:75–79

De Nascimento L, Willis KJ, Fernández-Palacios JM, Criado C, Whittaker RJ (2009) The long-term ecology of the lost forests of La Laguna, Tenerife (Canary Islands). J Biogeogr 36 (3):499–514

de Saporta G (1865) Études sur la végétation du sud-est de la France à l'époque tertiaire. Deuxième partie. Flore d'Armissan et de Peryac, dans le Bassin de Narbonne (Aude). Ann Sci Nat Bot 5ème Ser 4:5–264

de Saporta G(1867) Études sur la végétation du Sud-Est de la France à l'époque tertiaire. Ann Sci Nat Bot 5ème Ser 8:5–136

de Saporta G (1888) Origine paléontologique des arbres cultivés o utilisés par l'homme. Académie de Médecine, Paris

de Saporta G, Marion A-F (1876) Recherches sur les végétaux fossiles de Meximieux. Arch Mus Hist Nat Lyon 1:131–335

Denk T, Grimm GW (2009) Significance of pollen characteristics for infrageneric classification and phylogeny in *Quercus* (Fagaceae). Int J Plant Sci 170:926–940

Denk T, Grimm GW (2010) The oaks of western Eurasia: traditional classifications and evidence from two nuclear markers. Taxon 59:351–366

Denk T, Grímsson F, Zetter R (2010) Episodic migration of oaks to Iceland: evidence for a North Atlantic 'land bridge' in the latest Miocene. Am J Bot 97:276–287

Denk T, Grímsson F, Zetter R (2012) Fagaceae from the early Oligocene of Central Europe: persisting new world and emerging old world biogeographic links. Rev Palaeobot Palynol 169:7–20

Denk T, Velitzelos D, Güner TH, Bouchal JM, Grímsson F, Grimm G (2017) Taxonomy and palaeoecology of two widespread western Eurasian Neogene sclerophyllous oak species: *Quercus drymeja* Unger and *Q. mediterranea* Unger. Rev Palaeobot Palynol 241:98–128

Depape G (1912) Note sur quelques chênes miocènes et pliocènes. Rev Gén Bot 24:355–372

DeVore ML, Pigg KB (2010) Floristic composition and comparison of Middle Eocene to Late Eocene and Oligocene floras in North America. Bull Geosci 85:111–134

DeVore ML, Pigg KB, Dilcher DL, Freile D (2014) *Catahoulea grahamii* gen. et sp. nov.: fagaceous involucres from the Oligocene Catahoula Formation, central Texas and the middle Eocene Claiborne Formation of Kentucky and Tennessee, USA. In: Stevens WD, Montiel OM, Raven P (eds) Paleobotany and biogeography A Festschrift for Alan Graham in his 80th year. Missouri Botanical Garden Press, St. Louis (Chapter 3)

Diéguez C, Nieves-Aldrei JL, Barrón E (1996) Fossil galls (zoocecids) from the Upper Miocene of La Cerdaña (Lérida, Spain). Rev Palaeobot Palynol 94:329–343

Dillhoff RM, Leopold EB, Manchester SR (2005) The McAbee flora of British Columbia and its relation to the early-middle Eocene Okanagan Highlands flora of the Pacific Northwest. Can J Earth Sci 42:151–166

Doláková N (2004) Discussion of some thermophile palynomorphs from the Miocene sediments in the Carpathian Foredeep (Czech Republic) and Modrý Kamen basin (Slovakia). Acta Palaeobot 44(1):79–85

Dorf E (1930) Pliocene Floras of California. Carnegie Inst Wash Publ 412:1–108

Dorf E (1942) Upper Cretaceous Floras of the Rocky Mountain region: Flora of the Lance Formation at its type locality, Niobrara County, Wyoming. Carnegie Inst Wash Publ 508:83–168

Du FK, Hou M, Wang W, Mao K, Hampe A (2016) Phylogeography of *Quercus aquifolioides* provides novel insights into the Neogene history of a major global hotspot of plant diversity in south-west China. J Biogeogr 44:294–307

Dupéron-Laudoueneix M, Dupéron J (1995) Inventory of Mesozoic and Cenozoic woods from Equatorial and North Equatorial Africa. Rev Palaeobot Palynol 84:439–480

Dyjor S, Kvaček Z, Łańcucka-Środoniowa M, Pyszyński W, Sadowska A, Zastawniak E (1992) The younger tertiary deposits in the Gozdnica region (SW Poland) in the light of recent palaeobotanical research. Polish Bot Stud 3:3–129

Eberle JJ, Greenwood DR (2012) Life at the top of the greenhouse Eocene world—a review of the Eocene flora and vertebrate fauna from Canada's High Arctic. Geol Soc Am Bull 124:3–23

Ellis B, Daley DC, Hickey LJ, Johnson KR, Mitchell JD, Wilf P, Wing SL (2009) Manual of leaf architecture. Cornell University Press, Ithaca

El-Saadawi W, Kamal-El-Din MM, Attia Y, El-Faramawi MW (2011) The wood flora of the Cairo Petrified Forest, with five Paleogene new legume records for Egypt. Rev Palaeobot Palynol 167:184–195

Erwin DM, Schick KN (2007) New Miocene oak galls (Cynipini) and their bearing on the history of cynipid wasps in western North America. J Paleontol 81:568–580

Farlow JO, Sunderman JA, Havens JJ, Swinehart AL, Holman JA, Richards RL, Miller NG, Martin RA, Hunt RM Jr, Storrs GW, Curry BB, Fluegeman RH, Dawson MR, Flint MET (2001) The Pipe Creek Sinkhole biota, a diverse Late Tertiary continental fossil assemblage from Grant County, Indiana. Am Midland Natur 145:367–378

Felix J (1896) Untersuchungen über fossile Hölzer. Verh Z Deutsch Geol Ges 48:249–260

Fields PF (1996) The Succor Creek Flora of the Middle Miocene Sucker Creek Formation, southwestern Idaho and eastern Oregon: systematics and paleoecology. Ph.D. thesis, Michigan State University, Michigan

Flora of North America editorial committee (1997) Magnoliophyta: Magnoliidae and Hamamelidae, vol 3. Oxford, New York

Follieri M (1979) Ricerche paleobotaniche sulla serie di Torre in Pietra (Roma). Quaternaria 21:73–86

Fontaine WM (1889) The Potomac or younger Mesozoic flora. US Geol Surv Monogr 15:1–377

Frankenhäuser H, Wilde V (1995) Stachelspitzige Blätter aus dem Mitteleozän von Eckfeld (Eifel). Abh Staatl Mus Mineral Geol Dresden 4:97–115

Frederiksen NO (1981) Middle Eocene to early Oligocene plant communities of the Gulf Coast US. In: Gray J, Boucot WBN (eds) Communities of the Past. Hutchinson Ross, Stroudsbury

Frederiksen NO (1988) Sporomorph biostratigraphy, floral changes and paleoclimatology. Eocene and earliest Oligocene of the eastern Gulf Coast. US Geol Surv Prof Pap 1448:1–68

Frederiksen NO, Edwards LE, Ager TA, Sheehan TP (2002) Palynology of Eocene strata in the Sagavanirktok and Canning Formations on the North slope of Alaska. Palynology 26:59–93

Friis EM, Crepet WL (1987) Time of appearance of floral features. In: Friis EM, Chaloner R, Crepet WL (eds) The origins of angiosperms and their biological consequences. Cambridge University Press, Cambridge

Friis EM, Crane PR, Pedersen KR (2011) Early flowers and angiosperm evolution. Cambridge University Press, Cambridge

Gabel ML, Backlund DC, Haffner J (1998) The Miocene macroflora of the northern Ogallala Group, northern Nebraska and southern South Dakota. J Paleontol 72:388–397

Gandolfo MA (1996) The presence of Fagaceae (oak family) in sediments of the Klondike Mountain Formation (Middle Eocene) Republic, Washington. Wash Geol 24:20–21

Gardère P, Pais J (2007) Palynologic data from Aquitaine (SW France) Middle Miocene Sables Fauves Formation. Climatic evolution. Ciênc Terra (UNL) 15:151–161

Gastaldo RA, Riegel W, Püttmann W, Linnemann UG, Zetter R (1998) A multidisciplinary approach to reconstruct the Late Oligocene vegetation in Central Europe. Rev Palaeobot Palynol 101:71–94

Golovneva LB (2000) Early Palaeogene floras of Spitsbergen and North Atlantic floristic exchange. Acta Univ Carol Geol 44:39–50
Göppert HR (1855) Die Tertiäre Flora von Schossnitz in Schlesien. Heyn'sche Buchhandlung E, Remer, Görlitz
Gottwald H (1966) Eozäne Holzer aus Bruankohle von Helmstedt. Palaeontographica B 119:76–93
Govaerts R, Frodin DG (1998) World checklist and bibliography of Fagales (Betulaceae, Corylaceae, Fagaceae and Ticodendraceae). Royal Botanic Gardens, Kew
Grabowska I (1996) Flora sporowo-pyłkowa. In: Malinowska L, Piwocki M (eds) Budowa Geologiczna Polski. Vol. 3. Atlas skamieniałości przewodnich i charakterystycznych. 3a. Kenozoik. Trzeciorzęd. 1. Paleogen. Polska Agencja Ekologiczna, Warszawa
Graham A (1965) The Sucker Creek and Trout Creek Miocene floras of southeastern Oregon. Kent State Univ Bull 53:1–147
Graham A (1999a) Late Cretaceous and Cenozoic history of North American vegetation. Oxford University Press, Oxford
Graham A (1999b) The Tertiary history of the northern temperate element in the northern Latin American biota. Am J Bot 86:32–38
Grangeon P (1953) La flore pontienne de Gourgouras (Ardèche). Bull Soc Géol Fr 6$^{\grave{e}me}$ Ser 3:303–320
Grangeon P (1958) Contribution à l'étude de la paléontologie végétale du Massif du Coiron (Sud-Est du Massif Central Français). Ph.D. thesis, Univ. Clermont, Clermont-Ferrand
Greenwood DR, Pigg KB, Basinger JF, DeVore ML (2016) A review of paleobotanical studies of the Early Eocene Okanagan (Okanogan) Highlands floras of British Columbia, Canada, and Washington, USA. Can J Earth Sci 53:548–564
Gregory M, Poole I, Wheeler EA (2009) Fossil dicot wood names an annotated list with full bibliography. IAWA J, Suppl 6:1–220
Grein M, Oehm C, Konrad W, Utescher T, Kunzmann L, Roth-Nebelsick A (2013) Atmospheric CO_2 from the late Oligocene to early Miocene based on photosynthesis data and fossil leaf characteristics. Palaeogeogr Palaeoclimat Palaeoecol 374:41–51
Grímsson F, Zetter R, Grimm GW, Pedersen GK, Pedersen AK, Denk T (2015) Fagaceae pollen from the early Cenozoic of West Greenland: revisiting Engler's and Chaney's arctotertiary hypothesis. Plant Syst Evol 301:809–832
Grímsson F, Grimm GW, Zetter R, Denk T (2016) Cretaceous and Paleogene Fagaceae from North America and Greenland: evidence for a Late Cretacous split between *Fagus* and the remaining Fagaceae. Acta Palaeobot 56:247–305
Güner TH, Bouchal JM, Köse N, Göktaş F, Mayda S, Denk T (2017) Landscape heterogeneity in the Yatağan Basin (southwestern Turkey) during the middle Miocene inferred from plant macro fossils. Paleontographica B (in press)
Günther T, Gregor HJ (1997) Computeranalyse neogener Frucht- und Samenfloren europas. Band 5: Artennachweise und stratigraphische Problematik. Doc Nat 50(5)1–150
Guo SX, Zhang GF (2002) Oligocene Sanhe flora in Longjing county of Jilin, Northeast China. Acta Palaeontol Sin 41(2):193–210
Hably L, Kvaček Z (1997) Early Pliocene plant megafossils from the volcanic area in West Hungary. In: Hably L (ed) Early Pliocene volcanic environment, flora and fauna from Transdanubia, West Hungary. Hungarian Research Fund (OTKA), Budapest, pp 5–151
Hably L, Kvaček Z (1998) Pliocene mesophytic forests surrounding crater lakes in western Hungary. Rev Palaeobot Palynol 101:257–269
Hall RE, Poppe LJ, Ferrebee WM (1980) A stratigraphic test well, Martha's Vineyard, Massachusetts. Description of Pleistocene to Upper Cretaceous sediments recovered from 262 meters of test coring. US Geol Surv Bull 1488:1–19
Hantke R (1965) Die fossilen Eichen und Ahorne aus den Molasse der Schweiz und von Oehningen (Süd-Baden). Eine Revision der von Oswald Heer diesem Gattungen zugeordneten Reste. Neuj natur Gesel Zürich 167:1–140
He JS, Chen WL, Wang XL (1994) Morphological and anatomical features of *Quercus* section *suber* and its adaptation to the ecological environment. Acta Phytoecol Sin 18:219–227

Heer O (1853) Übersicht der Tertiärflore der Schweiz. Mitt naturf Ges Zürich 7:88–153

Heer O (1856) Flora tertiaria Helvetiae. Die Tertiäre Flora der Schweiz. Vol 2. Die apetalen Dicotyledonen. Wurster and Company, Winterthur

Heer O (1859) Flora tertiaria Helvetiae. Die Tertiäre Flora der Schweiz. Vol 3. Die gamopetalen und polypetalen Dicotyledonen. Anhang. Allgemeiner Theil. Wurster and Company, Winterthur

Herendeen PS, Crane PR, Drinnan AN (1995) Fagaceous flowers, fruits, and cupules from the Campanian (Late Cretaceous) of central Georgia, USA. Int J Plant Sci 156:93–116

Hermann M (2007) Eine palynologische Analyse der Bohrung Enspel—Rekonstruktion der Klima- und Vegetationsgeschichte im Oberoligozän. Ph.D. thesis, University of Tübingen, Germany

Hickey LJ (1977) Stratigraphy and paleobotany of the Golden Valley Formation (early tertiary) of western North Dakota. Geol Soc Am Mem 150:1–181

Hofmann CC, Mohamed O, Egger H (2011) A new terrestrial palynoflora from the Palaeocene/Eocene boundary in the northwestern Tethyan realm (St. Pankraz, Austria). Rev Palaeobot Palynol 166:295–310

Holden AR, Erwin DM, Schick KN, Gross J (2015) Late Pleistocene galls from the La Brea Tar Pits and their implications for cynipine wasp and native plant distribution in southern California. Quaternary Res 84:358–367

Hollick AW (1904) Systematic Paleontology; Plantae Angiospermae. In: Maryland Geologic Survey Miocene. Johns Hopkins University Press, Baltimore

Hori J (1976) On the study of the Kobe flora from the Kobe Group (Late Miocene age), Rokko highland. Nihon Chibaku-kaikan, Kyoto (in Japanese)

Hori J (1987) Plant fossils from the Miocene Kobe flora. Hyogo Biology Society, Fukusaki (in Japanese)

Hu HH, Chaney RW (1940) A Miocene flora from Shantung province, China. Carnegie Inst Wash Publ 507(1–82):1–147

Hu Q, Xing Y, Hu JJ, Huang Y, Ma H, Zhou Z-K (2014) Evolution of stomata and trichome density of the *Quercus delavayi* complex since the late Miocene. Chin Sci Bull 59(3):310–319

Huang CC, Chang YT, Bartholomew B (1999) Fagaceae. In: Wu CY, Raven PH (eds) Flora of China. Science Press, Beijing

Hubert FGW, Jousslin E, Berry B, Franc A, Kremer A (2014) Multiple nuclear genes stabilize the phylogenetic backbone of the genus *Quercus*. Syst Biodiver 12:405–423

Huff PM, Wilf P, Azumah EJ (2003) Digital future for paleoclimate estimation from fossil leaves? Preliminary results. Palaios 18:266–274

Huhn B, Utescher T, Ashraf AR, Mosbrugger V (1997) The peat-forming vegetation in the Middle Miocene of the Lower rhine Embayment, an analysis based on palynological data. Med Nederlands Inst Toegepaste Geow 58:211–218

Hummel A (1983) The Pliocene leaf flora from Ruszów near Żary in Lower Silesia, SW Poland. Pr Muz Ziemi 36:9–104

Huzioka K, Takahashi E (1970) The Eocene flora of the Ube coal-field, southwest Honshu, Japan. J Min Coll Akita Univ A 4:1–88

Iamandei E, Iamandei S, Diaconu F (2011) Fossil woods in the collections of Drobeta-Tr Severin Museum. Acta Paleontol Romaniae 5(7):199–218

Iamandei S, Iamandei E, Bozukov V, Tsenov B (2012) Oligocene fossil woods from Rhodopes. Bulgaria. Acta Paleontol Romaniae 9(2):15–25

Iglesias-González R (2015) Evolución de la vegetación en el sector biogeográfico Castellano Cantábrico en el Cuaternario final, a través del registro tobáceo. Ph.D. thesis, Universidad Politécnica de Madrid, Madrid

Iljinskaja IA (1982) *Quercus* L. In: Takhtajan AL (ed) Iskopaemye tsvetkovye rasteniya SSSR, vol 2. Fossil Flowering Plants of the USSR. Ulmaceae–Betulaceae, Nauka, Leningrad, pp 89–114 (in Russian)

Iljinskaja IA (1991) Relations of the Early Oligocene flora of Kiin-Kerish Mountain with the modern flora. In: Zhilin SG (ed), Development of the flora in Kazakhstan and Russian Plain from the Eocene to the Miocene. Lectures in memory of A.N. Kryshtofovich. Ser. vol 2, pp 22–28 (in Russian)

InsideWood (2004 onwards) InsideWood: a web resource for hardwood anatomy. http://insidewood/lib.ncsu.edu

Ivanov D, Ashraf AR, Mosbrugger V (2007) Late Oligocene and Miocene climate and vegetation in the Eastern Paratethys area (northeast Bulgaria), based on pollen data. Palaeogeogr Palaeoclimatol Palaeoecol 255:342–360

Jahren AH (2007) The Arctic forest of the middle Eocene. Ann Rev Earth Planetary Sci 35:509–540

Jensen R, Hokanson S, Isebrands J, Hancock J (1993) Morphometric variation in oaks of the Apostle Islands in Wisconsin: evidence of hybridization between *Quercus rubra* and *Q. ellipsoidalis* (Fagaceae). Am J Bot 80:1358–1366

Jiménez-Moreno G (2006) Progressive substitution of a subtropical forest for a temperate one during the middle Miocene climate cooling in Central Europe according to palynological data from cores Tengelic-2 and Hidas-53 (Pannonian Basin, Hungary). Rev Palaeobot Palynol 142:1–14

Jiménez-Moreno G, Suc JP (2007) Middle Miocene latitudinal climatic gradient in Western Europe: Evidence from pollen records. Palaeogeogr Palaeoclimatol Palaeoecol 253:208–225

Jiménez-Moreno G, Rodríguez-Tovar FJ, Pardo-Igúzquiza E, Fauquette S, Suc JP, Müller P (2005) High-resolution palynological analysis in late early-middle Miocene core from the Pannonian Basin, Hungary: climatic changes, astronomical forcing and eustatic fluctuations in the Central Paratethys. Palaeogeogr Palaeoclimatol Palaeoecol 216:73–97

Jiménez-Moreno G, Burjachs F, Expósito I, Oms O, Carrancho A, Villalaín JJ, Agustí J, Campeny G, Gómez de Soler B, Van der Made J (2013) Late Pliocene vegetation and orbital-scale climate changes from the western Mediterranean area. Global Planet Change 108:15–28

Jones JH (1986) Evolution of the Fagaceae. The implications of foliar features. Ann Missouri Bot Gard 73:228–275

Jones JH, Dilcher DL (1988) A study of the *"Dryophyllum"* leaf forms from the Paleogene of southeastern North America. Palaeontographica Abt B 208:53–80

Jones JH, Manchester SR, Dilcher DL (1988) *Dryophyllum* Debey ex Saporta, Juglandaceous not Fagaceous. Rev Palaeobot Palynol 56:205–211

Jonsson CHW, Hebda RJ (2015) Macroflora, paleogeography, and paleoecology of the Upper Cretaceous (Turonian?–Santonian) Saanich Member of the Comox Formation, Saanich Peninsula, British Columbia, Canada. Canadian J Earth Sci 52:519–536

Klimova RS (1976) Fagaceae from Miocene flora of western Primorye. Palaeontol Z 1976(1):104–109 (in Russian)

Kmenta M, Zetter R (2013) Combined LM and SEM study of the upper Oligocene/lower Miocene palynoflora from Altmittweida (Saxony): Providing new insights into Cenozoic vegetation evolution of Central Europe. Rev Palaeobot Palynol 195:1–18

Knobloch E (1969) Tertiäre Floren von Mähren. Moravské Museum, Brno

Knobloch E (1988) Neue Ergebnisse zur Flora aus der Oberen Sübwassermolasse von Aubenham bei Ampfing (Krs. Muhldorf am Inn). Doc Nat 42:1–27

Knobloch E (1998) Der pliozäne Laubwald von Willershausen am Harz. Doc Nat 120:1–306

Knobloch E, Konzalová M (1998) Comparison of the Eocene plant assemblages of Bohemia (Czech Republic) and Saxony (Germany). Rev Palaeobot Palynol 101:29–41

Knobloch E, Kvaček Z (1976) Miozäne Blätterfloren vom Westrand der Böhmischen Masse. Rozpr Ústred ústav Geol Praha 42:1–129

Knobloch E, Kvaček Z (1981) Miozäne Pflanzenreste aus der Umgebung von Tamsweg (Niedere Tauern). Acta Univ Carol Geol 2:95–120

Knobloch E, Velitzelos E (1986) Die obermiozäne Flora von Likudi bei Elassona/Thessalien, Griechenland. Doc Nat 29:5–20

Knowlton FH (1899) Fossil flora of the Yellowstone National Park. US Geol Surv Monogr Ser 32(2):651–791

Kodrul T (1999) Paleogene phytostratigraphy of the South Sakhalin. Geol Inst RAS Transactions 1999(519):1–146 (in Russian)

Kohlman-Adamska A (1993) Pollen analysis of the Neogene deposits from the Wyrzysk Region, North-Western Poland. Acta Palaeobot 31(3):91–297

Kolakovskii AA (1964) The Pliocene flora of Kodor. Sukhumskiy Bot Sad Monogri 1:1–200 (in Russian)

Kondinskaya L, Yudina E (1989) Palynocomplexes of Paleogene and Neogene of Ob-Irtysh interfluve of the West Siberian Plain and their stratigraphic significance. In: Cenozoic of Siberia and North-east of USSR, Nauka. Sib. Otd., Novosibirsk, 192 pp, pp 5–59 (in Russian)

Kosmowska-Ceranowicz B (1987) Charakterystuka mineralogiczno-petrograficzna bursztynonosnych osadów Eocenu w okolicach Chlapowo oraz osadów Paleogenu Polnocnej Polski. Biul Inst Geol 356:29–50

Kovar-Eder J, Hably L (2006) The flora of Mataschen—a unique plant assemblage from the Late Miocene of eastern Styria (Austria). Acta Palaeobot 46(2):157–233

Kovar-Eder J, Kvaček Z (2007) The integrated plant record (IPR) to reconstruct Neogene vegetation: the IPR-vegetation analysis. Acta Palaeobot 47(2):391–418

Kovar-Eder J, Meller B, Zetter R (1998) Comparative investigations on the basal fossiliferous layer at the opencast mine Oberdorf (Köflach-Voitsberg lignite deposit, Styria, Austria; Early Miocene). Rev Palaeobot Palynol 101:125–145

Kovar-Eder J, Kvaček Z, Meller B (2001) Comparing Early to Middle Miocene floras and probable vegetation types of Oberdorf N Voitsberg (Austria), Bohemia (Czech Republic), and Wackersdorf (Germany). Rev Palaeobot Palynol 114:83–125

Kovar-Eder J, Kvaček Z, Ströbitzer-Hermann M (2004) The Miocene flora of Parschlug (Styria, Austria)—Revision and synthesis. Ann Naturhist Mus Wien 105A:45–159

Kovar-Eder J, Jechorek H, Kvaček Z, Parasiv V (2008) The integrated plant record: an essential tool for reconstructing Neogene zonal vegetation in Europe. Palaios 23:97–111

Kräusel R (1939) Ergebnisse der Forschungsreisen Prof. E. Stromers in den Wüsten Ägyptens, IV. Die fossilen Floren Ägyptens 3. Die fossilen Pflanzen Ägyptens, E-1. Abh Bayer Akad Wiss München NF 47:1–140

Krutzsch W, Blumenstengel H, Kiesel Y, Rüffle L (1992) Paläobotanische Klimagliederung des Alttertiärs (Mitteleozän bis Oberoligozän) in Mitteldeutschland und das Problem der Verknupfung mariner und kontinentaler Geliederungen (klassische Biostratigraphien–paläobotanisch–ökologische Klimastratigraphie–Evolutions–Stratigraphie der Vertebraten). N Jb Geol Paläontol Abh 186:137–253

Kvaček Z (1998) Bílina: a window on Early Miocene marshland environments. Rev Palaeobot Palynol 101:111–123

Kvaček Z (2010) Forest flora and vegetation of the European early Palaeogene—a review. Bull Geosci 85:3–16

Kvaček Z, Hurník S (2000) Revision of early Miocene plants preserved in baked rocks in the North bohemian Tertiary. Acta Mus Natl Pragae Ser B Hist Nat 56(1–2):1–48

Kvaček Z, Walther H (1989) Paleobotanical studies in Fagaceae of the European tertiary. Plant Syst Evol 162:213–229

Kvaček Z, Mihajlovic D, Mihajlovic D, Vrabac S (1993) Early Miocene flora of Miljevina (Eastern Bosnia). Acta Palaeobot 33(1):53–89

Kvaček Z, Velitzelos D, Velitzelos E (2002) Late Miocene flora of Vegora Macedonia N. Univ Athens, Athens, Greece

Kvaček Z, Teodoridis V, Gregor HJ (2008) The Pliocene leaf flora of Auenheim, Northern Alsace (France). Doc Nat 155(10):1–108

Kvaček Z, Teodoridis V, Roiron P (2011) A forgotten Miocene mastixioid flora of Arjuzanx (Landes, SW France). Palaeontographica Abt B 285:3–111

LaMotte RS (1952) Catalog of the Cenozoic Plants of North America through 1950. Geol Soc Am Mem 51:1–381

Larsson LM, Vajda V, Dybkjær K (2010) Vegetation and climate in the latest Oligocene–earliest Miocene in Jylland, Denmark. Rev Palaeobot Palynol 159:166–176

Lathram EH, Pomeroy JS, Berg H, Loney RA (1965) Reconnaissance Geology of Admiralty Island, Alaska. US Geol Surv Bull 1181-R:R1–R48
Laurent L (1912) Flore fossile des schistes de Menat (Puy-de-Dôme). Ann Mus Hist Nat Marseille 14:1–246
Leckey EH, Smith DM (2015) Host fidelity over geologic time: restricted use of oaks by oak gallwasps. J Paleontol 89:236–244
Lenz OK (2000) Paläoökologie eines Küstenmoores aus dem Eozän Mitteleuropas am Beispiel der Wulfersdorfer Flöze und deren Begleitschichten (Helmstedter Oberflözgruppe, Tagebau Helmstedt). Ph.D. thesis, University of Göttingen, Germany
Lenz OK, Wilde V, Riegel W (2011) Short-term fluctuations in vegetation and phytoplankton during the Middle Eocene greenhouse climate: a 640-kyr record from the Messel oil shale (Germany). Int J Earth Sci (Geol Rundsch) 100:1851–1874
Leopold EB, Clay-Poole ST (2001) Florissant leaf and pollen floras of Colorado compared: climatic implications. In: Evanoff E, Gregory-Wodzicki KM, Johnson KR (eds) Fossil flora and stratigraphy of the Florissant Formation, Colorado. Proc Denver Mus Nat Sci Ser 4:17–69
Leopold EB, Liu G (1994) A long pollen sequence of Neogene age. Alaska Range. Quatern Int 22 (23):103–140
Leopold EB, MacGinitie HD (1972) Development and affinities of Tertiary floras in the Rocky Mountains. In: Graham A (ed) Floristics and Paleofloristics of Asia and Eastern North America. Elsevier Amsterdam
Lesquereux L (1892) The flora of the Dakota Group. US Geol Surv Monogr 17:1–400
Li HM, Guo SX (1976) The Miocene flora of Nanmuling of Xizang. Acta Paleontol Sin 15 (4):598–609 (in Chinese)
Lielke K, Manchester SR, Meyer HW (2012) Reconstructing the environment of the northern Rocky Mountains during the Eocene/Oligocene transition: constraints from the palaeobotany and geology of south-western Montana, USA. Acta Palaeobot 52:317–358
Liu, YS(C) (2011) Why are the oak acorns from the late Neogene Gray Site so small? Mid Continental Paleobotanical Colloquium vol 28, Raleigh, North Carolina, p 9
Liu G, Leopold EB (1992) Paleoecology of a Miocene flora from the Shanwang Formations, Shandong Province, Northern East China. Palynology 16:187–212
Liu, YS(C), Zetter R, Ferguson DK, Mohr B (2007a) Discriminating fossil evergreen and deciduous *Quercus* pollen: A case study from the Miocene of eastern China. Rev Palaeobot Palynol 145:289–303
Liu Z, Engel MS, Grimaldi DA (2007b) Phylogeny and geological history of the cynipoid wasps (Hymenoptera: Cynipoidea). Am Mus Novit 3583:1–48
Losa-Quintana JM (1978) Estudio mineralógico y estructural de un fósil de Galicia. An Inst Bot Cavanilles 35:235–243
Lozinsky RP (1982) Geology and late Cenozoic history of the Elephant Butte area, Sierra County, New Mexico. M.Sc. thesis, University of New Mexico, New Mexico
Lozinsky RP, Hunt AP, Wolberg DL, Lucas SG (1984) Late Cretaceous (Lancian) dinosaurs from the McRae Formation, Sierra County, New Mexico. New Mexico Geol 6:72–77
MacGinitie HD (1941) A middle Eocene flora from the central Sierra Nevada. Carnegie Inst Wash Publ 534:1–178
MacGinitie HD (1953) Fossil plants of the Florissant Beds, Colorado. Carnegie Inst Wash Publ 599:1–198
MacGinitie HD (1962) The Kilgore flora. Univ Calif Publ Geol Sci 35:67–158
MacGinitie HD (1969) The Eocene Green River flora of northwestern Colorado and northeastern Utah. Univ Calif Publ Geol Sci 83:1–203
Mädel-Angeliewa E (1968) Eichen- und Pappelholz aus der pliozänen Kohle im Gebiet von Baccinello (Toskana, Italien). Geol Jb 86:433–447
Magri D, Fineschi S, Bellarosa R, Buonamici A, Sebastiani F, Schirone B, Simeone MC, Vendramin GG (2007) The distribution of *Quercus suber* chloroplast haplotypes matches the palaeogeographical history of the western Mediterranean. Mol Ecol 16:5259–5266
Mai DH (1995) TertiäreVegetationsgeschichte Europas. Gustav Fisher Verlag, Stuttgart

Mai DH (2007) The floral change in the Tertiary of Rhön mountains (Germany). Acta Palaeobot 47:135–143
Mai DH, Walther H (1991) Die oligozänen und untermiozänen Floren Nordwest-Sachsens und des Bitterfelder Raumes. Abh Staatl Mus Min Geol Dresden 38:1–230
Manchester SR (1981) Fossil plants of the Eocene Clarno Nut Beds. Oregon Geol 43:75–81
Manchester SR (1994) Fruits and Seeds of the Middle Eocene Nut Beds Flora, Clarno Formation, Oregon. Palaeontographica Am 58:1–205
Manchester SR (1999) Biogeographical relationships of North American Tertiary floras. Ann Missouri Bot Gard 86:472–522
Manchester SR (2001) Update on the megafossil flora of Florissant, Colorado, USA. In: Evanoff E, Gregory-Wodzicki KM, Johnson KR (eds) Fossil flora and stratigraphy of the Florissant Formation, Colorado. Proc Denver Mus Nat Sci Ser 4(1):137–161
Manchester SR (2014) Revisions to Roland Brown's North American Paleocene flora. Sb Nár Muz Praze Ser B Hist Nat 70:153–210
Manchester SR, McIntosh WC (2007) Late Eocene silicified fruits and seeds from the John Day Formation near Post, Oregon. PaleoBios 27:7–17
Manchester SR, Chen ZD, Geng BY, Tao JR (2005) Middle Eocene flora of Huadian, Jilin province, Northeastern China. Acta Palaeobot 45:3–26
Manos PS, Doyle JJ, Nixon KC (1999) Phylogeny, biogeography, and processes of molecular differentiation in *Quercus* subgenus *Quercus* (Fagaceae). Mol Phylogen Evol 12:333–349
Martinetto E (2015) Monographing the Pliocene and the early Pleistocene carpofloras of Italy: methodological challenges and current progress. Palaeontographica Abt B 293:57–99
Martinetto E, Uhl D, Tarabra E (2007) Leaf physiognomic indications for a moist warm-temperate climate in NW Italy during the Messinian (Late Miocene). Palaeogeogr Palaeoclimatol Palaeoecol 253:41–55
Martinetto E, Bertini A, Basilici G, Baldanza A, Bizarri R, Cherin M, Gentili S, Pontini MR (2014) The plant record of the Dunarobba and Pietrafitta sites in the Plio-Pleistocene palaeoenvironmental context of Central Italy. Alpine Mediterr Quatern 27(1):29–72
Massalongo A, Scarabelli G (1859) Studii sulla flora fossile e Geologia Stratigrafica del Senigalliese. Ignazio Galeati e figlio, Imola
Mehrotra RC, Liua X-Q, Lia Ch-S, T, Wanga Y-F, Chauhan MS (2005) Comparison of the tertiary flora of southwest China and northeast India and its significance in the antiquity of the modern Himalayan flora. Rev Palaeobot Palynol 135:145–163
McCartan L, Tiffney BH, Wolfe JA, Ager TA, Wing SL, Sirkin LA, Ward LW, Brooks J (1990) Late tertiary floral assemblage from upland gravel deposits of the southern Maryland Coastal Plain. Geology 18:311–314
McIntyre DJ (1991) Pollen and spore flora of an Eocene forest, eastern Axel Heiberg Island, N.W.T. Geol Surv Can Bull 403:83–98
McIver EE, Basinger JF (1999) Early Tertiary floral evolution in the Canadian high Arctic. Ann Missouri Bot Gard 86:523–545
Meller B (1989) Eine Blatt-Flora aus den Obermiozänen Dinotherien-Sanden (Vallesium) von Sprendlingen (Rheinhessen). Doc Nat 54:1–101
Meng H-H, Su T, Gao X-Y, Li J, Jiang X-L, Zhou Z-K (2017) Warm-cold colonization: Response to the rising of Himalayan-Tibetan Plateau. Mol Ecol. doi:10.1111/mec.14092
Menitsky GL (1969) The ancestors and the evolution of Caucasian and Southwest Asian oaks belonging to the subsection *Quercus*. Bot J 54(11):1675–1688 (in Russian)
Menitsky GL (1984) The oaks of Asia. Nauka, Leningrad (in Russian)
Meyer KJ (1989) Forschungsbohrung Wursterheide. Pollen und Sporen des Tertiaer aus der Forschungsbohrung Wursterheide, Nordwest-Deutschland. Geol Jahrb Reihe A: Allgemeine und regionale Geologie BRD und Nachbargebiete, Tektonik, Stratigraphie, Palaeontologie 111:523–539
Meyer HW, Manchester SR (1997) The Oligocene Bridge Creek Flora of the John Day Formation, Oregon. Univ Calif Publ Geol Sci 141:1–195

Michon L, Merle O (2001) The evolution of the Massif Central rift, spatio-temporal distribution of the volcanism. Bull Soc Géol Fr 172:201–211

Mindell RA, Stockey RA, Beard G (2007) *Cascadiacarpa spinosa* gen. et sp. nov. (Fagaceae), castaneoid fruits from the Eocene of Vancouver Island. Canada. Am J Bot 94:351–361

Mindell RA, Stockey RA, Beard G (2009) Permineralized *Fagus* nuts from the Eocene of Vancouver Island, Canada. Int J Plant Sci 170:551–560

Moktar NB, Mannaï-Tayech B (2016) Climatic control on vegetation and sedimentary dynamics during the Miocene (Burdigalian to Langhian) in northeastern Tunisia. Rev Micropaléontol 59:1–17

Momohara A (in press) Influence of mountain formation on floral diversification in Japan based on macrofossil evidence. In: Hoorn C, Antonelli A (eds) Mountain, climate and biodiversity. Wiley

Mosbrugger V, Utescher T (1997) The coexistence approach—a method for quantitative reconstructions of tertiary terrestrial paleoclimate data using plant fossils. Palaeogeogr Palaeoclimatol Palaeoecol 134:61–86

Moss PT, Greenwood DR, Archibald SB (2005) Regional and local vegetation community dynamics of the Eocene Okanagan Highlands (British Columbia—Washington State). Can J Earth Sci 42:187–204

Němejc F, Kvaček Z (1975) Senonian plant macrofossils from the region of Zliv and Hluboká (near České Budějovice) in South Bohemia. Universita Karlova, Praha

Nixon KC (1993) Infrageneric classification of *Quercus* (Fagaceae) and typification of section names. Ann Sci For 50(Suppl. 1):25s–34s

Nixon KC (1997) *Quercus*. In: Flora of North America Editorial Committee (ed), Flora of North America, North of Mexico, vol 3. Oxford University Press, New York, pp 445–447

Nixon KC (2006) Global and neotropical distribution and diversity of oak (Genus *Quercus*) and oak forests. In: Kappelle M (ed) Ecology and Conservation of Neotropical Montane Oak Forests. Ecol Stud 185:3–12

Ochoa D, Whitelaw M, Liu Y-S, Zavada M (2012) Palynology of Neogene sediments at the Gray Fossil Site, Tennessee, USA: Floristic implications. Rev Palaeobot Palynol 184:36–48

Ochoa D, Zavada MS, Liu Y, Farlow JO (2016) Floristic implications of two contemporaneous inland upper Neogene sites in the eastern US: Pipe Creek Sinkhole, Indiana, and the Gray Fossil Site, Tennessee (USA). Palaeobiodiv Palaeoenviron 96:239–254

Pais J (1979) La végétation de la basse vallée du Tage (Portugal) au Miocène. Ann Géol Pays Hellén Hors Ser Fasc 2:933–942

Palamarev E (1989) Paleobotanical evidences of the Tertiary history and origin of the Mediterranean sclerophyll dendroflora. Plant Syst Evol 162:93–107

Palamarev E, Ivanov D (2003) A contribution to the Neogene history of Fagaceae in the Central Balkan area. Acta Palaeobot 43(1):51–59

Palamarev E, Kitanov G (1988) Fossil macroflora of the Beli Brjag Coal Basin. In: Velchev V, Markova M, Palamarev E, Vanev S (eds) 100th Anniversary of the National Academy A. Stojanov, Bulgarian Academy of Sciences, Sofia, pp 183–206 (in Bulgarian)

Palamarev E, Tsenov B (2004) Genus *Quercus* in the Late Miocene flora of Baldevo Formation (Southwest Bulgaria): taxonomical composition and palaeoecology. Phytol Balcanica 10(2–3):147–156

Pavlyutkin BI (2005) The Mid-Miocene Khanka Flora of the Primorye. Vladivostok, Dalnauka (in Russian)

Pavlyutkin BI, Petrenko TI (2010) Stratigrafiya paleogen-neogenovykh otlozhenii Primor'ya. Dal'nauka, Vladivostok (in Russian)

Peel MC, Finlayson BL, McMahon TA (2007) Updated world map of the Köppen–Geiger climate classification. Hydrology and earth system sciences discussions. Eur Geosci Union 11(5):1633–1644

Pereira Sieso J, García Gómez E (2002) Bellotas, el alimento de la edad de oro. Arqueoweb 4(2):1–29

Pigg KB, DeVore ML (2016) A review of the plants of the Princeton chert (Eocene, British Columbia, Canada). Botany 94:661–681

Piton L (1940) Paléontologie du gisement éocène de Menat (Puy-de-Dôme), flore et faune. Mém Soc Hist Nat Auvergne 1:1–303

Platen P (1908) Untersuchungen fossiler Hölzer aus dem Westen der Vereinigen Staaten von Nordamerika. Naturforsh Ges Leipzig 34(1–155):161–164

Postigo-Mijarra JM, Burjachs F, Manzaneque FG, Morla C (2007) A palaeoecological interpretation of the lower–middle Pleistocene Cal Guardiola site (Terrassa, Barcelona, NE Spain) from the comparative study of wood and pollen samples. Rev Palaeobot Palynol 146:247–264

Prakash U (1968) Miocene fossil woods from the Columbia basalts of central Washington III. Palaeontographica Abt B 122:183–200

Prakash U, Barghoorn ES (1961a) Miocene fossil woods from the Columbia Basalts of central Washington. J Arnold Arbor 42:165–203

Prakash U, Barghoorn ES (1961b) Miocene fossil woods from the Columbia Basalts of central Washington, II. J Arnold Arbor 42:347–362

Privé C (1975) Étude de quelques bois de chênes tertiaries du Massif Central, France. Palaeontographica Abt B 153:119–140

Privé-Gill C, Cao N, Legrand P (2008) Fossil wood from alluvial deposits (reworked Oligocene) of Limagne at Bussières (Puy-de-Dôme, France). Rev Palaeobot Palynol 149:73–84

Quan C, Liu Y-S(C), Utescher T (2012a) Paleogene temperature gradient, seasonal variation and climate evolution of northeast China. Palaeogeogr Palaeoclimatol Palaeoecol 313–314:150–161

Quan C, Liu Y-S(C), Utescher T (2012b) Eocene monsoon prevalence over China: a paleobotanical perspective. Palaeogeogr Palaeoclimatol Palaeoecol 365–366:302–311

Reid EM, Chandler ME (1933) Flora of the London Clay. British Museum of Natural History, London

Rember CW (1991) Stratigraphy and paleobotany of Miocene Lake sediments near Clarkia, Oregon. Ph.D. thesis. Univ Idaho, Moscow

Renney KM (1972) The Miocene Temblor Flora of west central California. M.Sc. thesis. Univ California, Davis

Roiron P (1983) Nouvelle étude de la macroflore plio-pléistocene de Crespià (Catalogne, Espagne). Geobios 16(6):687–715

Roiron P (1992) Flores, végétation et climats du Neogène Mediterranéen: apports de macroflores du Sud de la France et du Nord-Est de l'Espagne. Ph.D. thesis. Univ Montpellier II, Montpellier

Sadowski EM, Seyfullah LJ, Sadowski F, Fleischmann A, Hermann Behling H, Schmidt AR (2015) Carnivorous leaves from Baltic amber. PNAS 112(1):190–195

Sanz de Siria A (1992) Estudio de la macroflora oligocena de las cercanías de Cervera (Colección Martí Madern del Museo de Geología de Barcelona). Treb Mus Geol Barcelona 2:269–379

Sato S (1994) On the palynoflora in the Paleogene in the Ishikari Coal Field, Hokkaido, Japan. J Fac Sci Hokkaido Univ Ser 4. Geol Min 23:555–559

Schierenbeck KA (2014) Phylogeography of California, an introduction. Univ Calif Press, Berkeley

Schimper NP (1870–1872) Traité de Paléontologie végétale. Ou, la flore du monde primitif dans ses raports avec les formations géologiques et la flore du monde actuel, vol. 2. Baillière et fils, Paris

Schwarz O (1936) Entwurf zu einem natürlichen System der Cupuliferen und der Gattung *Quercus* L. Notizbl Bot Gart Berlin-Dahlem 13(116):1–22

Schwarz O (1964) *Quercus* L. In: Tutin TG, Heywood VH, Burges NA, Valentine DH, Walters SM, Webb DA (eds) Flora Europaea, vol. 1: Lycopodiaceae to Platanaceae. Cambridge University Press, Cambridge

Scott RA, Wheeler EA (1982) Fossil woods from the Eocene Clarno Formation of Oregon. Bull Int Ass Wood Anatom (NS) 3:135–154

Shilin PV (1983) Upper Cretaceous floras of the Lower Syrdarya elevation. Paleontol J 2:105–112 (in Russian)
Shunk AJ, Driese SG, Farlow JO, Zavada MS, Zobaa MK (2008) Late Neogene paleoclimate and paleoenvironment reconstructions from the Pipe Creek Sinkhole, Indiana, USA. Palaeogeogr Palaeoclimatol Palaeoecol 274:173–184
Shvaryova N, Mamchur A (2003) The Miocene flora of the Velykaya Ugol'ka (Transkarpathians). Lviv, p 144 (in Russian)
Simeone MC, Grimm GW, Papini A, Vessella F, Cardoni S, Tordoni E, Piredda R, Franc A, Denk T (2016) Plastome data reveal multiple geographic origins of *Quercus* Group *Ilex*. PeerJ 4:e1897. doi:10.7717/peerj.1897
Sims HJ, Herendeen PS, Lupia R, Christopher RA, Crane PR (1998) Fossil flowers with Normapolles pollen from the Late Cretaceous of southeastern North America. Rev Palaeobot Palynol 106:131–151
Singleton S (2001) Fossil wood of the Oligocene Catahoula Formation, Jasper County, Texas http://www.hgms.org/StartPageHTMLFiles/PaleoPetrifiedWoodArticles/2001-8JasperWoodArticle.html
Sittler C (1984) Essai de zonation palynologique des dépôts paléogènes des bassins tributaires de la vallée du Rhône et du Midi Méditerranéen. Géol Fr 1–2:85–90
Sittler C, Schuler M (1974) Comparison palynologique de dépôts oligo-miocènes des Limagnes (Massif Central) et de Manosque (Alpes de Provence). Mém Bur Rech Géol Min 78(2):571–578
Slodkowska B (2004) Palynological studies of the Paleogene and Neogene deposits from the Pomeranian Lakeland area (NW Poland). Polish Geol Instit Spec Pap 14:1–116
Smiley CJ, Rember WC (1985) Composition of the Miocene Clarkia Flora. In: Smiley CJ (ed) Late Cenozoic history of the Pacific Northwest. American Association for the Advancement of Science, Pacific Division, San Francisco
Solé de Porta N, de Porta J (1977) Primeros datos palinológicos del Messiniense (=Turoliense) de Arenas del Rey (Provincia de Granada). Stvd Geol Salmant 13:67–88
Song S, Krajewska K, Wang Y (2000) The first occurrence of the *Quercus* section *Cerris* Spach fruits in the Miocene of China. Acta Palaeobot 40(2):153–163
Spackman W (1949) The flora of the Brandon Lignite. Geological aspects and a comparison of the flora with its modern equivalents. Ph.D. thesis, Harvard University, Cambridge, Massachusetts
Stephyrtza AG (1990) The families Fagaceae and Aquifoliaceae in the Early Sarmatian flora of Bursuk (Moldavia). Bot J 75(10):1442–1449 (in Russian)
Stone GN, Hernández-López A, Nicholls JA, Di Pierro E, Pujade-Villar J, Melika G, Cook JM (2009) Extreme host plant conservatism during at least 20 million years of host plant pursuit by oak gallwasps. Evolution 63:854–869
Striegler U (1992) Bemerkungen zu den Eichenblättern des Blättertons von Wischgrund (Miozän, Niederlausitz)—Vorläufige Mitteilung. Doc Nat 70:54–61
Stuchlik L, Ziembińska-Tworzydło M, Kohlman-Adamska A, Grabowska I, Słodkowska B, Worobiec E, Durska E (2014) Atlas of the pollen and spores of the Polish Neogene. Volume 4 —Angiosperms (2). Polish Academy of Sciences, Kraków
Stults DZ, Axsmith B (2015) New plant fossil records and paleoclimate analyses of the late Pliocene Citronelle Formation flora, U.S. Gulf Coast. Palaeontol Electronica 18.3.47A.http://palaeo-electronica.org/content/2015/1318-citronelle-flora-climate
Stults DZ, Axsmith B, McNair D, Alford M (2016) Preliminary investigation of a diverse Miocene megaflora from the Hattiesburg Formation, Mississippi. Bot Soc Am, Ann Bot Conference, Savannah Georgia, p 138
Suc JP (1980) Contribution à la connaissance du Pliocène et du Pleistocène inférieur des régions mediterranéennes d'Europe Occidentale par l'analyse palynologique des dépots du Languedoc-Roussillon (Sud de la France) et de la Catalogne (Nord-Est de l'Espagne). Ph.D. thesis, Univ Montpellier II, Montpellier
Suc JP, Cravatte J (1982) Étude palynologique du Pliocène de Catalogne (Nord-Est de l'Espagne). Apports à la connaissance de l'histoire climatique de la Méditerranée occidentale et implications chronostratigraphiques. Paléobiol Cont 13(1):1–31

Suzuki M, Ohba H (1991) A revision of fossil woods of *Quercus* and its allies in Japan. J Japan Bot 66(5):255–274

Takahashi M, Friis EM, Herendeen PS, Crane PR (2008) Fossil flowers of Fagales from the Kamikitaba locality (early Coniacian); Late Cretaceous of northwestern Japan. Int J Plant Sci 169:899–907

Tanai T (1961) Neogene floral change in Japan. J Fac Sci Hokkaido Univ Geol Min 4(11):119–398

Tanai T (1970) The Oligocene floras from the Kushiro coal field, Hokkaido, Japan. J Fac Sci Hokkaido Univ Geol Min 4(14):383–514

Tanai T (1991) Tertiary climate and vegetation changes in the Northern Hemisphere. J Geogr 100:951–966 (in Japanese, with English abstract)

Tanai T (1995) Fagaceous leaves from the Paleogene of Hokkaido, Japan. Bull Natl Sci Mus Tokyo C 21:71–101

Tanai T, Suzuki N (1963) Tertiary Floras of Japan (I), Miocene Floras. The Collaborating Assoc Commemorate 80th Anniversary Geol Surv Japan, vol 1, pp 9–149

Tanai T, Uemura K (1991) The Oligocene Noda flora from the Yuya-wan area of the western end of Honshu, Japan. Part 1. Bull Natl Sci Mus Tokyo C 17:57–80

Tanai T, Uemura K (1994) Lobed oak leaves from the Tertiary of East Asia with reference to the oak phytogeography of the Northern Hemisphere. Trans Proc Palaeontol Soc Japan New Ser 173:343–365

Tanai T, Yokoyama A (1975) On the lobed oak leaves from the Miocene Kobe Group, western Honshu, Japan. J Fac Sci Hokkaido Univ Geol Min 4(17):129–141

Tao JR, Zhou Z-K, Liu YS (2000) The evolution of the Late Cretaceous-Cenozoic flora in China. Science Press, Beijing (in Chinese with English abstract)

Taylor TN, Taylor EL (1993) The biology and evolution of fossil plants. Prentice Hall, Englewood Cliffs

Taylor DW, HU S, Tiffney BH (2012) Fossil floral and fruit evidence for the evolution of unusual developmental characters in Fagales. Bot J Linn Soc 168:353–376

Teodoridis V, Kvaček Z (2006) Palaeobotanical research of the Early Miocene deposits overlying the main coal seam (Libkovice and lom Members) in the Most Basin (Czech Republic). Bull Geosci 81(2):93–113

Teodoridis V, Kvaček Z, Sami M, Utescher T, Martinetto E (2015) Palaeoenvironmental analysis of the Messinian macrofossil floras of Tossigniano and Monte Tondo (Vena del Gesso Basin, Romagna Apennines, Northern Italy). Acta Mus Nat Pragae B Hist Nat 71(3–4):249–292

Terada K, Handa K (2009) Fossil woods from the Paleogene Kobe Group in Hyogo Prefecture, Japan (preliminary report). Mem Fukui Prefectural Dinosaur Mus 8:17–29

Terashima I, Massuzawa T, Ohba H (1993) Photosynthetic characteristic of a giant alpine plant. *Rheum nobile* Hook F. Thoms. and of some other alpine species measured to 4300 m, in the eastern Himalaya. Nepal Oecol 95:194–201

Thomasson JR (1987) Late Miocene plants from northeastern Nebraska. J Paleontol 61:1065–1079

Tiffney BH (1977) Contributions to a monograph of the fruit and seed flora of the Brandon Lignite. Ph.D. thesis, Harvard University, Cambridge, Massachusetts

Tiffney BH (1994) Re-evaluation of the age of the Brandon Lignite (Vermont USA) based on plant megafossils. Rev Palaeobot Palynol 82:299–315

Traverse A (1955) Pollen analysis of the Brandon Lignite of Vermont. US Bur Min Rep Invest 5151:1–107

Tucker JM (1974) Patterns of parallel evolution of leaf form in New World oaks. Taxon 23:129–154

Uemura K (1988) Late Miocene floras in Northeast Honshu, Japan. Natural Science Museum, Tokyo

Uemura K, Doi E, Takahashi F (1999) Plant megafossil assemblage from the Kiwado Formation (Oligocene) from Ouchiyama-kami in Yamaguchi Pref., western Honshu, Japan. Bull Mine City Mus 15:1–59 (in Japanese)

Unger F (1842) Synopsis lignorum fossilium plantarum acramphibryarum. In: Endlichen S (ed) Genera plantarum secundum ordines naturales disposita, Supplement 2, Appendix. Apud Fridericum Beck, Vienna

Utescher T, Bruch A A, Erdei B, François L, Ivanov D, Jacques F M B, Kern A K, Liu Y-S, Mosbrugger V, Spicer R A (2014) The Coexistence Approach—Theoretical background and practical considerations of using plant fossils for climate quantification. Palaeogeogr Palaeoclimatol Palaeoecol 410:58–73

Utescher T, Erdei B, Hably L, Mosbrugger V (2017) Late Miocene vegetation of the Pannonian Basin. Palaeogeogr Palaeoclimatol Palaeoecol 467:131–148

Utescher T, Mosbrugger V (2007) Eocene vegetation patterns reconstructed from plant diversity - A global perspective. Palaeogeogr Palaeoclimatol Palaeoecol 247:243–271

Utescher T, Mosbrugger V (2015) The Palaeoflora database. http://www.geologie.unibonn.de/palaeoflora

Váchová Z, Kvaček J (2009) Palaeoclimate analysis of the flora of the Klikov Formation, Upper Cretaceous, Czech Republic. Bull Geosci 84(2):257–268

Valle MF, Civis J (1978) Investigaciones palinológicas en el Plioceno inferior de Can Albareda (Barcelona). Palinología 1(extraord):463–468

Van Boskirk K (1998) The flora of the Eagle Formation and its significance for Late Cretaceous floristic evolution. Ph.D. thesis, Yale University, New Haven, Connecticut

Van der Burgh J (1993) Oaks related to *Quercus petraea* from the upper Tertiary of the Lower Rhenish Basin. Palaeontographica Abt B 230:195–201

Van der Burgh J (1997) Miocene floras in the Lower Rhenish Basin and their ecological interpretation. Rev Palaeobot Palynol 52:299–366

Van't Veer R, Hooghiemstra H (2000). Montane forest evolution during the last 650,000 years in Colombia: a multivariate approach based on pollen record Funza-1. J Quatern Sci 15:329–346

Vaudois-Miéja N (1985) La flore des grès à palmiers de l'Ouest de la France. Bull Sect Sci 8:259–273

Velitzelos D, Bouchal JM, Denk T (2014) Review of the Cenozoic floras and vegetation of Greece. Rev Palaeobot Palynol 204:56–117

Vieira M, Poças E, Pais J, Pereira D (2011) Pliocene flora from S. Pedro da Torre deposits (Minho, NW Portugal). Geodiversitas 33(1):71–85

Vikulin SV (1987) A new oak species of subgenus Erythrobalanus (Fagaceae) in early Oligocene flora from Pasekovo (south of the Middle Russian upland). Bot J 72(4):518–522 (in Russian)

Vikulin SV (2011) Thermophilic Fagaceae: *Quercus, Lithocarpus, Castanopsis* from the Late Eocene of the Southern European Russia. Lectures in memory of A.N. Kryshtofovich. Ser. 7:128–147 (in Russian)

Vincent PM, Aubert M, Boivin P, Cantagrel JM, Lenat JF (1977) Découverte d'un volcanisme paléocène en Auvergne, les maars de Ménat et leursannexes; étude géologique et géophysique. Bull Soc Géol Fr 19:1057–1070

Von der Brelie G (1968) Zur mikrofloristischen Schichtengliedrung im Rheinischen Braunkohlerevier. Fortschr Geol Rheinl Westfalen 16:85–102

Von der Brelie G, Hager H, Weiler H (1981) Pollenflora und Phytoplankton in den Kölner Schichten sowie deren Lithostratigraphie im Siegburger Graben. Fortschr Geol Rheinld u Westf 29:21–48

von Linné C (1753) Species Plantarum, vol II. Laurentii Salvii, Stockholm

Wallin ET (1983) Stratigraphy and paleoenvironments of the Engle coal field, Sierra County, New Mexico. M.Sc. thesis, New Mexico Institute of Mining and Technology, New Mexico

Walther H (1999) Die Tertiärflora von Kleinsaubernitz bei Bauztze. Palaeontographica Abt B 249:63–174

Walther H, Eichler B (2010) Die Neogene Flora von Ottendorf-Okrilla bei Dresden. Geol Saxonica 56(2):193–234

Walther H, Zastawniak E (1991) Fagaceae from Sosnica and Malczyce (near Wroclaw, Poland). A revision of original materials by Goeppert 1852 and 1855 and a study of new collections. Acta Palaeobot 31:153–199

Wang WM (1994) Palaeofloristic and palaeoclimatic implications of Neogene Palynoflora in China. Rev Palaeobot Palynol 82:239–250

Wang H (2002) Diversity of Angiosperm leaf megafossils from the Dakota Formation (Cenomanian, Cretaceous), North Western Interior, USA. Ph.D. thesis, University of Florida, Gainesville, Florida

Wang W-M, Saito T, Nakagawa T (2001) Palynostratigraphy and climatic implications of Neogene deposits in the Himi area of Toyama Prefecture, Central Japan. Rev Palaeobot Palynol 117:281–295

Wang H, Blanchard J, Dilcher DL (2013) Fruits, seeds, and flowers from the Warman clay pit (Middle Eocene Claiborne Group), western Tennessee, USA. Palaeo-Electronica 16(3)31A:1–73. palaeo-electronica.org/content/2013/545-eocene-plants-from-tennessee

Ward LF (1899) The Cretaceous Formation of the Black Hills as indicated by the fossil plants. Annu Rep US Geol Surv 19:523–712

Webber IE (1933) Woods from the Ricardo Pliocene of Last Chance Gulch, California. Carnegie Inst Wash Publ 412:113–134

Wheeler EA, Dillhoff TA (2009) The middle Miocene wood flora of Vantage, Washington, USA. Int Assoc Wood Anat J Suppl 7:1–101

Wheeler EA, Manchester SR (2002) Woods of the Eocene Nut Beds Flora. Int Assoc Wood Anat J Suppl 3:1–188

Wheeler EA, Scott RA, Barghoorn ES (1978) Fossil dicotyledonous woods from Yellowstone National Park II. J Arnold Arbor 59:1–31

Wheeler EA, Manchester SR, Wiemann M (2006) Eocene woods of central Oregon. PaleoBios 26:1–6

White JM, Ager TA, Adam DP, Leopold EB, Liu G, Jetté H, Schweger CE (1999) Neogene and Quaternary quantitative palynostratigraphy and paleoclimatology from sections in Yukon and adjacent Northwest Territories and Alaska. Geol Surv Can Bull 543:1–30

Wiemann MC, Wheeler EA, Manchester SR, Portier KM (1998) Dicotyledonous wood anatomical characters as predictors of climate. Palaeogeogr Palaeoclimatol Palaeoecol 139:83–100

Wilde V (1989) Untersuchungen zur Systematik der Blattreste aus dem Mitteleozan der Grube Messel bei Darmstadt (Hessen, Bundesrepublik Deutschland). Cour Forsch-Inst Senckenberg 115:1–213

Wing SL, Alroy J, Hickey LJ (1995) Plant and Mammal diversity in the Paleocene to early Eocene of the Bighorn Basin. Palaeogeogr Palaeoclimatol Palaeoecol 115:117–155

Wolfe JA (1966) Tertiary Plants from the Cook Inlet region, Alaska. US Geol Surv Prof Pap 398B:1–32

Wolfe JA (1978) A paleobotanical interpretation of Tertiary climates in the Northern Hemisphere. Am Sci 66:694–703

Wolfe JA (1980) The Miocene Seldovia Point Flora from the Kenai Group, Alaska. US Geol Surv Prof Pap 1105:1–52

Wolfe JA (1985) Distribution of major vegetational types during the Tertiary. Am Geophys Union Publ Monogr 32:357–375

Worobiec G, Lesiak MA (1998) Plant megafossils from the Neogene deposits of Stawek-1A (Belchatów, Middle Poland). Rev Palaeobot Palynol 101:179–208

Writing Group of Cenozic Plant of China (WGCPC) (1978) Fossil Plants from China, within Fossil Plants of China, vol III. Science Press, Beijing (in Chinese)

Xia K, Su T, Liu YS(Christopher), Xing YW, Jacques FMB, Zhou ZK (2010) Quantitative climate reconstructions of the late Miocene Xiaolongtan megaflora from Yunnan, southwest China. Palaeogeogr Palaeoclimatol Palaeoecol 276:80–86

Xing Y (2010) The Late Miocene Xianfeng flora, Yunnan, Southwest China and its quantitative palaeoclimatic reconstructions. Ph.D. thesis, Kunming Institute of Botany, Chinese Academy of Sciences, Kunming, China (in Chinese, with English abstract)

Xing Y, Hu J, Jacques FMB, Wang L, Su T, Huang Y, Liu Y-U(C), Zhou Z-K (2013) A new *Quercus* species from the Upper Miocene of southwestern China and its ecological significance. Rev Palaeobot Palynol 193:99–109

Xu H, Su T, Zhang S-T, Deng M, Zhou Z-K (2016) The first fossil record of ring-cupped oak (*Quercus* L. subgenus *Cyclobalanopsis* (Oersted) Schneider) in Tibet and its paleoenvironmental implications. Palaeogeogr Palaeoclimatol Palaeoecol 442:61–71

Yamanoi T (1989) Palynoflora of the Middle Miocene sediments in Noto Peninsula, Central Japan. Professor Hidekuni Matsuo Memorial Vol, pp 5–13 (in Japanese with English abstract)

Yao Y-F, Bruch AA, Mosbrugger V, Li C-S (2011) Quantitative reconstruction of Miocene climate patterns and evolution in Southern China based on plant fossils. Palaeogeogr Palaeoclimatol Palaeoecol 304:291–307

Zhang YY (1995) Outline of Palaeogene palynofloras of China. Acta Palaeontol Sin 34(2):212–227

Zhang SB, Zhou Z-K, Hu H (2005) Photosynthetic performances of *Quercus pannosa* vary with altitude in Hengduan Mountains, southwest China. Forest Ecol Manag 212:291–301

Zheng Z (1990) Végétations et climats néogenes des Alpes maritimes Franco-Italiennes d'après les donées de l'analyse palynologique. Paléobiol Cont 17:217–244

Zhilin SG (1989) History of the development of the temperate forest flora in Kazakhstan, U.S.S.R. from the Oligocene to the Early Miocene. Bot Rev 55(4):205–330 (in Russian)

Zhou Z-K (1992) A taxonomical revision of fossil evergreen sclerophyllous oaks from China. Acta Bot Sin 34(12):954–961 (in Chinese)

Zhou Z-K (1993) The fossil history of *Quercus*. Acta Bot Yunnanica 15:21–33 (in Chinese, with English abstract)

Zhou Z-K, Coombes A (2001) *Quercus* sect. *Heterobalanus*—an interesting group of evergreen oaks. Int Dendrology Soc Yearbook 2001:18–24

Zhou Z-K, Wilkinson H, Wu ZY (1995) Taxonomical and evolutionary implications of the leaf anatomy and architecture of *Quercus* L. subgen. *Quercus* from China. Cathaya 7:1–34

Zhou Z-K, Pu CX, Chen WY (2003) Relationship between the distributions of *Quercus* sect. *Heterobalanus* (Fagaceae) and uplift of Himalayas. Adv Earth Sci 18(6):884–889 (in Chinese)

Zhou Z-K, Yang Q-S, Xia K (2007) Fossil of *Quercus* sect. *Heterobalanus* can help explain the uplift of the Himalayas. Chin Sci Bull 52(2):238–247

Zittel KA (1891) Traité de Paléontologie, partie 2. Paléophytologie 1. Octave Doin Éditeurs, Paris

Chapter 4
Physiological Evidence from Common Garden Experiments for Local Adaptation and Adaptive Plasticity to Climate in American Live Oaks (*Quercus* Section *Virentes*): Implications for Conservation Under Global Change

Jeannine Cavender-Bares and José Alberto Ramírez-Valiente

Abstract Climate is known to be a critical factor controlling the broad-scale distribution of plants but often the physiological basis for species distribution limits is not well understood, nor is the extent to which populations within species are locally adapted to climate. Reciprocal transplant experiments designed to test for local adaptation are difficult to conduct and interpret in long-lived species, like oaks. Linking the physiological tolerances of species to their climatic distributions is an alternative approach to understanding adaptation to climate, and is important in predicting future distributions of species under changing climatic conditions. Here we synthesize a series of studies in a single lineage of American oaks that span the temperate tropical divide and encompass a range of precipitation and edaphic regimes, to determine (1) the physiological basis for adaptation to seasonal winter and seasonal drought and (2) the variation among populations that associated with climate variation and can be interpreted as local adaptation. We focus primarily on a series of common gardens that allow us to determine the genetically based differences in functional and physiological traits as well as the genetically based responses to contrasting temperature or precipitation regimes. We show that variation in freezing tolerance among closely related species is greater than variation among populations within species. Nevertheless, freezing tolerance varies predictably with climate of origin and is negatively associated with growth rate. In contrast, drought tolerance mechanisms vary more among populations within a

---

J. Cavender-Bares (✉)
Department of Ecology, Evolution and Behavior, University of Minnesota, Minneapolis, MN, USA
e-mail: cavender@umn.edu

J. A. Ramírez-Valiente (✉)
Department of Ecology and Genetics, INIA—Forest Research Centre (CIFOR), Madrid, Spain
e-mail: josealberto.ramirezvaliente@gmail.com

© Springer International Publishing AG 2017
E. Gil-Pelegrín et al. (eds.), *Oaks Physiological Ecology. Exploring the Functional Diversity of Genus* Quercus *L.*, Tree Physiology 7,
https://doi.org/10.1007/978-3-319-69099-5_4

single species, at least for the most widely distributed species, *Quercus oleoides,* than between species. Within this species, climate of origin predicts a suite of leaf physiological traits, and there is evidence for evolutionary trade-off between desiccation avoidance and desiccation resistance. Combined, these results show evidence for local adaptation to both freezing and drought stress within species, as well as adaptive differentiation between closely related species, despite phylogenetic conservatism in functional traits and highly similar physiognomy across the American live oak clade. The results inform conservation efforts aimed at preventing extinction of tree species in the face of global change.

4.1 Introduction

A central biological question under rapidly changing global climatic conditions is the extent to which species are locally adapted to climate. Climate is a driving force in evolution (Etterson 2004a; Jump et al. 2006; Ramírez-Valiente et al. 2010; Shaw and Etterson 2012), even though niche conservatism is widespread (Crisp et al. 2009b; Wiens et al. 2010). Physiological traits that are linked to tolerance of seasonal temperature variation and water availability are known to vary considerably among woody taxa and are thought to delimit species distributions across climatic gradients in both temperate and tropical biomes (Larcher 1960, 2000; Sakai and Larcher 1987; Koerner and Larcher 1988; Engelbrecht and Kursar 2003; Tyree et al. 2003; Brodribb and Holbrook 2006; Engelbrecht et al. 2006). Such traits are likely to be under strong selection in relation to climate. We have used a small "model clade" of live oaks (*Quercus* section *Virentes* Nixon) that span a range of climates from the temperate zone to the tropics to examine evidence for adaptive evolution in response to seasonal winter and seasonal drought. The evolutionary history and ecological distribution of the live oaks are well understood, providing a platform for investigation of adaptive change. More broadly, the oaks of the Americas contribute a large fraction of the total forest biomass and woody diversity of North America in both the US and Mexico (Cavender-Bares 2016). Climate change scenarios predict warmer climates in southern and southeastern regions of eastern North America and drier climates in most regions of Mexico Central America by 2100 (IPCC 2007), but the seasonal timing of decreases in rainfall are uncertain (Karmalkar et al. 2008). We used the live oaks as a system to address the question of whether variation between populations and between species in both sensitivity to chilling and freezing stress and resistance to drought corresponds to the climate of origin. We compare our studies of the American live oaks, trees adapted to wet summers and either dry or cold winters, to parallel work on old world Mediterranean oak species, which also experience seasonality in precipitation and temperature but are adapted to cold temperatures that are seasonally offset from low rainfall. These oaks are similar in appearance and many functional attributes but display important physiological and phenological differences as a consequence of the contrasting patterns in seasonality.

Our goal was to examine evidence for local adaptation to climate using a series of common garden experiments, where environmental variation was limited and individual plants were randomized across this variation. Distinguishing between plasticity and genetically-based variation has important implications for understanding range limits and how plants respond to changing environments. As part of these investigations, we tested for environmentally-induced changes in resistance to cold and to drought in plants grown under contrasting and experimentally manipulated climate regimes (temperature or precipitation). These experiments allowed us to examine plasticity in response to contrasting climatic regimes and to evaluate evidence in support of adaptive plasticity. We examined variation within and among species in response to freezing stress in five of the live oaks but focused on the two most widespread species, *Quercus virginiana* (temperate biome) and *Quercus oleoides* (tropical biome). In examining adaptation to drought and seasonality in precipitation, we focused primarily on *Quercus oleoides*, which spans a range of precipitation regimes, all of which fall within the classification of seasonally-dry tropics. We planted common gardens in the field and also reciprocally transplanted populations to test directly for local adaptation and supplemented water at two times during the year to decipher consequences of water limitation. We review these common garden studies to synthesize what we have learned about genetically based variation within and among closely related species of this lineage in response to the range of climatic variation that it encompasses. We suggest future steps needed to (1) better understand the evolution of climate responses in oak ecosystems and the evolutionary potential of oaks to adapt to increasing drought severity expected in future decades and (2) provide guidance on conservation to prevent extinction of threatened oaks.

4.2 The Live Oaks, *Quercus* Section *Virentes*, as a Study System

The live oaks consist of seven species of interfertile brevideciduous or semievergreen oaks that span the tropical temperate divide in southern USA, Mexico and Central America and the Caribbean (Muller 1961; Nixon 1985; Nixon and Muller 1997; Cavender-Bares et al. 2015) (Fig. 4.1).

Q. virginiana and *Q. oleoides* are the two most broadly distributed members of the live oaks. *Q. virginiana* extends from the outer banks of southern Virginia and North Carolina in the U.S. into northern Mexico, and *Q. oleoides* extends from northern Mexico to northwestern Costa Rica. We hypothesized that the wide range of climatic variation encountered by the two species throughout their ranges has led to both interspecific and intraspecific variation as a consequence of adaptation to contrasting climates. The majority of the work we present on population-level differentiation is within and between these two species.

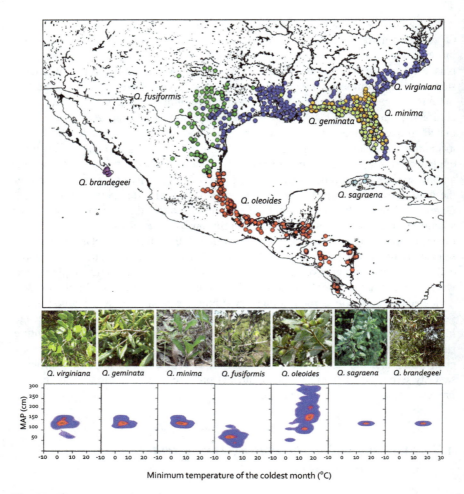

Fig. 4.1 The live oaks, *Virentes*, are a monophyletic lineage of interfertile American oaks that span the southeastern US, Mexico, Central America and the Caribbean and occur across a range of climates that vary in both minimum temperature and precipitation. The live oaks fall within the white oak group (Nixon and Muller 1997; Cavender-Bares et al. 2015). Shown are the geographic distributions of the seven species in the lineage, their phenotypes, and their climatic distribution in terms of mean annual precipitation (cm) and minimum temperature of the coldest month (°C). Modified from Cavender-Bares et al. (2015)

4.3 Adaptation to Winter Stress: Species and Population Responses to Freezing Under Different Climate Regimes

Freezing is considered a major barrier to migration and a strong selective force (Sakai and Weiser 1973; Larcher 2000; Cavender-Bares 2005; Zanne et al. 2014). Freezing temperatures can cause lethal injuries in living plant tissues, and the ability of different species to avoid or tolerate freezing stress through various mechanisms can go a long way in explaining their geographic distributions (Parker 1963; Burke et al. 1976). Live oaks are likely strongly limited by freezing, given that they are not deciduous. Deciduousness is a very common adaptation to freezing winters, and has evolved repeatedly in oaks (Hipp et al. 2017). However, the live oaks are restricted to mild climates and are considered evergreen or brevideciduous, but not deciduous, in taxonomic treatments (Miller and Lamb 1985; Nixon 1985; Nixon and Muller 1997). While they occur in the temperate zone, they are found only in climates where winters are fairly mild but where subzero temperatures are nevertheless frequent. Tree species, such as the live oaks, that remain active during winter are subject to freezing of living and nonliving tissues. Freezing can cause intracellular ice formation, which can kill the cells, or extra-cellular ice formation, which may lead to cellular dehydration and cell membranes damage (Fujikawa and Kuroda 2000).

We tested the hypothesis that populations at different latitudes within species are differentially adapted to cold and freezing stress. However, given the tendency for phylogenetic conservatism in traits (Ackerly and Reich 1999; Wiens et al. 2010), we hypothesized that, alternatively, populations within species could be equally tolerant of cold and freezing, due to ancestral acquisition and conservatism of such traits, even though only some populations currently experience these stresses. The ability to cold acclimate is itself an evolved trait, and the capacity to undergo morphological shifts that protect against freezing damage is characteristic of temperate species. We therefore anticipated that the growth climate would influence the freezing response, and exposure to winter temperatures would lead to cold acclimation in populations adapted to cold such they would show less damage in response to freezing stress.

In an initial controlled environment experiment with two populations of both *Q. virginiana* (temperate species) and *Q. oleoides* (tropical species), Cavender-Bares (2007) found ecotypic differentiation in cold and freezing sensitivity between populations within species and between species across a latitudinal gradient, when grown under either tropical or temperate growth conditions. In response to short term freezing, both the North Carolina and Florida populations from the temperate *Q. virginiana* showed small losses of photosynthetic function (dark acclimated quantum yield, assessed as variable to maximum chlorophyll fluorescence, F_v/F_m), 24 h after freezing at −10, −5, −2 °C compared to before freezing. In contrast, Belize and Costa Rica populations from the tropical *Q. oleoides* showed very large losses in photosynthetic function after freezing at −10 °C. However, the extent of damage to the photosynthetic apparatus was a consequence of growing conditions, and populations within *Q. virginiana* showed different degrees of damage, depending on climate of origin. Plasticity in the responses of populations to freezing stress is an important adaptation. Plasticity is adaptive if it results in a higher fitness across environments (van Kleunen and Fisher 2005). The case for claiming adaptive plasticity for increased freezing tolerance in response to cold exposure is not disputed. The widely observed changes in cell wall properties and other functions that reduce intracellular freezing and frost damage among most freezing tolerant plants, called "cold acclimation," increases plant survival (Steponkus 1984; Wisniewski and Ashworth 1985; Huner et al. 1993; Cavender-Bares 2005). Cold acclimation encompasses the range of physiological and morphological changes that occur in response to chilling and prepare a plant to encounter freezing stress. Overall, *Q. virginiana* plants showed an ability to cold acclimate, while *Q. oleoides* populations did not. When grown in a tropical treatment exposed to consistently warm growth conditions both *Q. virginiana* populations showed greater loss of photosynthetic function when exposed to decreasing minimum temperatures, as indicated by a decline in F_v/F_m, than when acclimated to three months of winter chilling (Fig. 4.2a, b). Across populations and species, the decline in F_v/F_m under both climate conditions corresponded to climate of origin. Under the tropical treatment (Fig. 4.2a), while both populations in both species suffered declines in F_v/F_m after freezing at −10 °C, the northern most population from North Carolina showed only a minimal decline in F_v/F_m and plants in the Florida population of *Q. virginiana* showed an intermediate response at −10 °C. The ability of *Q. virginiana* populations grown in the temperate treatment to cold

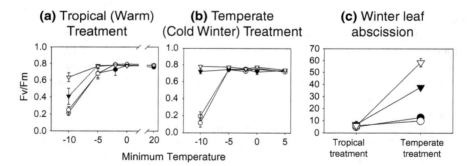

Fig. 4.2 Effects of minimum temperature (0, −2, −5, and −10 °C) on recovery of photochemical efficiency in dark leaves **a** in plants grown under warm conditions all year (tropical treatment), and **b** in plants acclimated to cold temperatures (temperate treatment). Dark F_v/F_m was measured on leaves of branches with a stem submersed in water before and 24 h after a dark freezing cycle. Immediately after the freezing cycle, branches were placed in a dark cabinet at room temperature. Plants were measured in February. **c** Percentage of leaf loss during the interval between November and February in the tropical and temperate growth treatments. *Q. virginiana* is represented by triangles (open = North Carolina; closed = Florida), and *Q. oleoides* is represented by circles (open = Belize; closed = Costa Rica). Redrawn from Cavender-Bares (2007)

acclimate was apparent from the lack of decline in F_v/F_m, even at -10 °C. In contrast, the two *Q. oleoides* populations showing very large declines under both treatments with values of F_v/F_m that dropped below 0.2 at -10 °C.

We also observed a striking difference in leaf abscission in responses to cold exposure both between populations within *Q. virginiana* and between the two species that corresponded to latitude (Fig. 4.2c), which again demonstrates adaptive plasticity. Abscission in response to cold provides evidence for evolution towards deciduousness. The northern population of *Q. virginiana* from NC lost nearly 60% of its leaves in response to prolonged cold exposure that reached 4 °C at night, while the Florida population abscised less than 40% compared to only 6% for all populations in the tropical treatment. Meanwhile, the tropical *Q. oleoides* showed only approximately 10% abscission in response to the same cold exposure, and not significantly different from the abscission rate in the tropical treatment.

In a second study (Cavender-Bares et al. 2011), we again found genetically based differences in freezing sensitivity between *Q. virginiana* and *Q. oleoides*; although variation within species was not significant in this case. As before, F_v/F_m was measured in a replicated controlled environment experiment under tropical and winter-treated temperate conditions, before freezing and 12 h after freezing at -5, -7 and -10 °C. Critical freezing temperatures (the freezing temperature at which $F_v/F_m = 0.4$) were more negative for *Q. virginiana* than *Q. oleoides*, indicating higher freezing tolerance in the temperate species. However, populations within species did not differ significantly (Fig. 4.3c). In the temperate treatment, where plants were acclimated to cold temperatures for three months prior to freezing, all *Q. virginiana* populations showed an increase in freezing tolerance relative to when they were grown in consistently warm conditions. *Q. oleoides* populations showed no ability to increase freezing tolerance when exposed to cold.

In the most extensive study of the series examining freezing tolerance, Koehler et al. (2012), tested maternal families from multiple populations within five species of the *Virentes* for freezing tolerance and response to cold exposure (Fig. 4.4). In this study we used the electrolyte leakage method to examine intracellular cell death of stems in response to freezing, as well as loss of chlorophyll function in leaves, as before. Minimum temperatures in the climate of origin of maternal families across all species strongly predicted freezing tolerance, cold acclimation ability, and growth rates. Maternal families from climates with colder winters had slower growth rates and greater freezing tolerance than those from milder climates. As a consequence, we found evidence for an evolved trade-off between freezing tolerance and growth rate, such that the maternal families from warmer latitudes within and across species showed faster growth rates but lower freezing tolerance than the maternal families from colder latitudes. Live oaks from lower latitudes had much high freezing tolerance and ability to acclimate to freezing, but lower growth rates in the absence of cold stress.

In a study on the same species, populations and maternal families, Ramírez-Valiente et al. (2015) found significant variation among populations and species, as well as increasing anthocyanin content with minimum temperature in the climate of origin. The maternal families with higher freezing tolerance (Fig. 4.4)

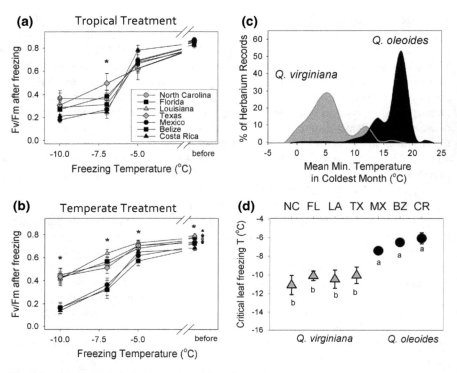

Fig. 4.3 Genetically based differences in freezing sensitivity between *Q. virginiana* and *Q. oleoides*. The dark-acclimated quantum yield of photosynthesis (F_v/F_m) was measured in a common garden experiment under **a** tropical conditions and **b** winter-treated temperate conditions, before freezing and 12 h after freezing at −5, −7 and −10 °C. Asterisks indicate temperatures and treatments for which the two species were significantly different ($\alpha = 0.05$). Small symbols to the right in **b** indicate dark-acclimated F_v/F_m values of leaves warmed at 22 °C for 48 h. **c** Climatic distributions showing the percentage of herbarium record occurrences at each temperature for the mean minimum temperature in the coldest month. These occurrence localities were used in the Maxent model to predict climatic distributions for *Q. oleoides* and *Q. virginiana*, which showed that minimum temperature was an effective climate variable for predicting species occurrence and was significantly different between the species ($P < 0.0001$). **d** A ∼3 °C difference in critical freezing temperatures between the two species was apparent. Quadratic curves fitted to F_v/F_m responses of individuals to three freezing temperatures (not shown) allowed the prediction of critical freezing temperatures (the freezing temperature at which $F_v/F_m = 0.4$) for each population. Redrawn from Cavender-Bares et al. (2011)

had lower anthocyanin accumulation (Fig. 4.5). Our interpretation is that maternal families with lower freezing tolerance use anthocyanins as a general protective mechanism in response to cold, by attenuating light and/or neutralizing reactive oxygen species to diminish the risk of photodamage under low temperatures (Pietrini et al. 2002; Gould 2004; Hughes et al. 2012).

All of these studies provide clear evidence for adaptive divergence in freezing tolerance and the ability to acclimate to cold winters among species with contrasting climates of origin. Within *Q. virginiana*, we also found adaptive differentiation among populations in cold acclimation ability and freezing tolerance. In other words, families and populations within the species have different levels of adaptive plasticity in response to cold, depending on climate of origin. The ability to increase freezing tolerance after cold exposure is entirely absent in the tropical species, *Q. oleoides,* and remains untested in *Q. brandegeei* and *Q. sagraena* in Cuba. Lacking freezing and cold tolerance, which is hypothesized to be maintained at significant metabolic cost (Burke et al. 1976; Guy 2003; Savage and Cavender-Bares 2013). This lack of cold acclimation ability and freezing tolerance, more generally, helps explain why *Q. oleoides* appears to use anthocyanins as a general mechanisms to reduce photoprotective stress under cold conditions, particularly in young leaves (Ramírez-Valiente et al. 2015). Our current working hypothesis is that the live oaks lost freezing tolerance after radiation into Mexico and Central America, given that the oaks colonized the temperate zone first (Hipp et al. 2017). The alternative possibility is that the live oaks originated in tropical climates and gained freezing tolerance. Regardless, these studies demonstrate population level local adaptation in freezing tolerance in the widely distributed temperate *Q. virginiana*, and either adaptive loss of freezing tolerance and cold acclimation ability in the tropical *Q. oleoides,* or adaptive acquisition of these attributes in *Q. virginiana*. Molecular evidence from candidate genes lends support to the hypothesis that the live oaks lost freezing tolerance, given strong conservatism and purifying selection in a core gene responsible for cold acclimation ability (Meireles et al. 2017). In these same populations of *Q. virginiana* and *Q. oleoides*, we studied two cold response candidate genes ICE1, a key gene in the cold acclimation pathway, and HOS1, which modulates cold response by negatively regulating ICE1. Meireles et al. (2017) found that that HOS1 experienced recent balancing selection. This finding indicates that evolution has favored diversity in cold tolerance modulation through balancing selection in HOS1, perhaps due to the range of climatic environments the species experience across their ranges. At a deeper evolutionary scale, a codon based model of evolution revealed the signature of negative (or purifying) selection in ICE1. In the same analysis, three positively selected codons were identified in HOS1, possibly a signature of the diversification of *Virentes* into warmer climates from a freezing adapted lineage of oaks. It thus appears that, while evolution has favored diversity in cold tolerance modulation through balancing selection in HOS1, it has maintained core cold acclimation ability, given evidence for purifying selection in ICE1.

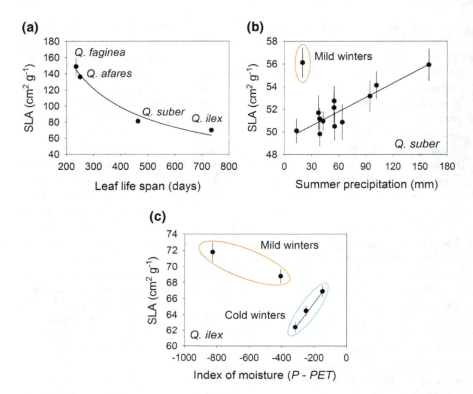

Fig. 4.4 Minimum temperature of the coldest month in the climate of origin predicts leaf **a** and stem **b** freezing tolerance in maternal families grown under nonstressed tropical (gray) and after exposure to cold temperatures in temperate conditions (black) based on leaf decline in F_v/F_m after freezing at −10 °C and stem index of injury after freezing at −15 °C. Minimum temperature of the coldest month further predicts leaf cold acclimation ability ((tropical-temperate)/tropical for decline in F_v/F_m after freezing at −10 °C) **c** and stem cold acclimation ability ((tropical-temperate)/tropical for index of injury after freezing at −15 °C) **d**. **e** and **f** show the trade-offs between growth rate (absolute growth rate, AGR) in tropical and temperate conditions and freezing tolerance across maternal families from four live oak species. Leaf freezing tolerance and stem freezing tolerance are both higher in maternal families within and among species with lower growth rates. Species: *Q. virginiana*, squares; *Q. geminata*, crosses; *Q. fusiformis*, circles; *Q. oleoides*, triangles. Redrawn from Koehler et al. (2012). Regressions are shown as least squares fitted lines

4.4 Species and Population Responses to Water Availability Under Different Climate Regimes

Water limitation is a second major barrier in the ability of plants to occupy a given biome and accounts for major shifts in the Earth's species composition (Pennington 2006; Crisp et al. 2009a; Anderegg et al. 2016). Within the same climatic zone,

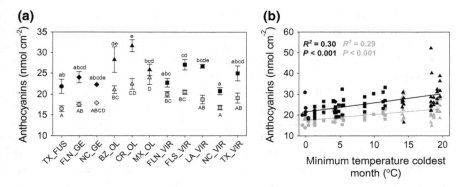

Fig. 4.5 a Means for populations (within species) under temperate (*black*) and tropical (*gray*) treatments for anthocyanin concentration measured in the reddest leaf of the plants. Bars indicate standard errors. Different letters indicate significant differentiation ($P < 0.05$) within the temperate and tropical treatments. Populations: *Q. fusiformis* (*circles*), TX_FUS—Texas; *Q. geminata* (*diamonds*), FLN_GE—Northern Florida; NC_GE—North Carolina; *Q. oleoides* (*triangles*), CR_OL—Costa Rica; BZ_OL—Belize; MX_OL—Mexico; *Q. virginiana* (*squares*), FLS_VIR—Southern Florida; FLN_VI—northern Florida; LA_VIR—Louisiana; TX_VIR—Texas; NC_VIR—North Carolina. **b** Relationship between anthocyanins and minimum temperatures of the coldest month in the source of origin of the maternal families under temperate (*black*) and tropical (*gray*) treatments. *Q. fusiformis* (*circles*), *Q. geminata* (*diamonds*), *Q. oleoides* (*triangles*) and *Q. virginiana* (*squares*). Points represent mean maternal-family values

water limitation can be caused by topographic variation and soil type. In these cases, water availability and soil fertility often covary (Cavender-Bares et al. 2004). *Quercus virgininiana* and *Quercus geminata,* two co-occurring temperate live oak species, known to be sister species, occur in contrasting soils and hydraulic regimes in the southeastern US (Cavender-Bares and Pahlich 2009). They provide a good test of sympatric divergence in function in relation to microhabitat water availability. *Quercus virginiana* occurs on moister and richer soils than *Q. geminata* based on ecological studies in Florida (Myers 1990; Cavender-Bares et al. 2004) as well as taxonomic treatments of the species (Kurz and Godfrey 1962; Nixon and Muller 1997). Cavender-Bares et al. (2004) found significant niche differentiation across soil types between the two species based on soil moisture and soil fertility (pH, calcium content, exchangeable NH_4 and NO_3, and exchangeable P). *Q. virginiana* occurrs on moister, more nutrient rich, and higher pH sites than *Q. geminata*. *Quercus virginiana* also has a broader distribution across the range of variation in all of the edaphic factors relative to *Q. geminata* (Cavender-Bares and Pahlich 2009). When sympatric Florida populations of each species that occur in different microhabitats are grown in a common environment, *Q. virginiana* has faster growth and higher photosynthetic rates per gram of leaf tissue than *Q. geminata,* corresponding to its more resource rich native habitat. The resource allocation patterns of *Q. virginiana* support its faster growth strategy; it has thinner leaves (higher specific leaf area, SLA) and allocates more to leaf area relative to total plant biomass, thus maximizing light capture and total plant photosynthetic capacity. *Quercus virginiana* also allocates less to root mass than shoot mass. In contrast, the slower growth strategy of *Q. geminata* is accompanied by a greater investment in roots relative to shoots and lower allocation to leaf area per unit biomass. Lower evaporative surface area and greater proportional belowground biomass permits *Q. geminata* to conserve water. This conservative strategy matches the lower water availability in their native habitat. Based on measurements from stems naturally grown in the field, *Q. virginiana* has higher stem specific hydraulic conductance than *Q. geminata* and lower Huber values than *Q. geminata* (Cavender-Bares and Holbrook 2001). The significant functional differentiation between the species observed both in common gardens and in naturally occurring populations corresponds to habitat differentiation and provides evidence for adaptive divergence between these sympatric sister species. Adaptive differentiation must have either occurred in sympatry, a possibility given contrasting phenology and flowering times (Cavender-Bares and Pahlich 2009) or in allopatry prior to secondary contact. In a subsequent series of studies, we found similar kinds of adaptive differentiation within a single species that spans a range of climates and soils as we explain in Sect. 4.5.

4.5 Intraspecific Variation in Seasonally-Dry Tropical Climates in the Widely Distributed Tropical Live Oak, *Quercus Oleoides*

Across latitudes, variation in the timing and amount of precipitation establishes contrasting selection pressures that may be anticipated to lead to adaptive differentiation in populations and local adaptation. Yet it is not well understood the extent to which local adaptation occurs in long-lived tree species. Maintenance of high genetic variation and plastic responses to the environment are other important means of persisting in variable environments, particularly when generation times are long and an individual tree may experience a range of environments throughout the course of its lifespan (Shaw and Etterson 2012; Meireles et al. 2017). *Quercus oleoides* is a long-lived species widely distributed in seasonally dry tropical forests (SDTFs) of Central America. This species usually forms mono-dominant stands and influences local hydrologic budgets and soil conservation (Boucher 1981; Klemens et al. 2011). It is considered to have a evergreen or brevi-deciduous leaf habit depending on the population of origin (Muller 1942). This species is a useful study system to explore evolution of drought resistance strategies in SDTFs because it spans a large gradient of dry-season aridity and wet season rainfall in the region. Here we review and synthesize the main findings from recent studies relative to the drought response exhibited by *Q. oleoides* to seasonal water variation.

Vast areas in tropical latitudes are characterized by seasonally dry climates, which are particularly abundant in Mesoamerica. Seasonally dry tropical forests (SDTFs) in this region exhibit nearly constant temperatures throughout the year but have marked variation in precipitation. Usually, rainfall exhibits a bimodal distribution with maximums in June and October and minimums between March and April. The length of the dry season might vary between two and seven months, and its severity is variable across the region. Xeric environments can also experience drought events during the wet season (Ananthakrishnan and Soman 1989; Nicholls and Wong 1990).

In an initial study of local adaptation to contrasting precipitation regimes using a reciprocal transplant experiment within Costa Rica, Deacon and Cavender-Bares (2015) found that upland and lowland populations of *Q. oleoides* both had higher fitness, in terms of both growth and survival, in upland environments, where precipitation was higher and water limitation less severe during the dry season. The results clarified that water was more limiting to fitness in the lowland environment than in the upland. A later field common garden study in the same lowland region again showed that water limitation during the dry season reduced seedling fitness from both the upland and lowland populations by decreasing survival. Furthermore, water supplementation at the low elevation site during the dry season resulted in an increase in emergence of seedlings and subsequent fitness from seeds produced late in the season (Center et al. 2016). The upland and lowland Costa Rican populations originate from environments that span the full range of precipitation variation across the entire species range. However, in two separate transplant experiments, we found no evidence of local adaptation of these two populations within Costa Rica through reciprocal transplanting, despite barriers to gene flow that could have permitted it (Center 2015; Deacon and Cavender-Bares 2015). In the latter study Center (2015), reciprocal transplanting included populations and sites in the upland and lowland regions in Costa Rica as well as in a very xeric region in southern Honduras. However, even including this broader span of populations we found no evidence for local adaptation, although biotic factors may have interfered (Center 2015). Detecting local adaptation under complicated field conditions in long-lived species is difficult, however, and sometimes better evidence can be obtained for adaptive differentiation in physiological function in controlled environments.

Making inferences about local adaptation based on functional differentiation in traits requires a clear understanding of the expectations for how traits should vary with climatic and soil conditions. Seasonally dry tropical forests in Central MesoAmerica are typically dominated by trees with very different leaf life spans, including drought deciduous and evergreen species (Borchert et al. 2002; Givnish 2002; Bowman and Prior 2005; Klemens et al. 2011; Vico et al. 2015). Deciduous species usually have thin leaves with high specific leaf area, short life spans and high investment in photosynthetic tissues per leaf mass. For this reason, they are thought to sustain high photosynthetic rates in the wet season under high soil water potentials but to abscise their leaves in the drought season to reduce water loss via transpiration (Reich and Borchert 1984; Eamus and Prichard 1998). Theoretical models predict that this acquisitive resource-use strategy is beneficial for carbon, nutrients and water balances in SDTFs when the dry season is longer or more severe because it maximizes carbon uptake and nutrient use when water is not limiting and minimizes water loss during the long dry season

(Cornelissen et al. 1996; Givnish 2002; Bowman and Prior 2005; Poorter and Markesteijn 2008). In contrast, as the dry season becomes shorter or less severe, a conservative resource use strategy with increased drought tolerance is hypothesized to be beneficial for species that inhabit SDTFs because it allows carbon assimilation throughout the entire year including during the dry season (Oertli et al. 1990; Niinemets 2001; Read and Sanson 2003; Wright et al. 2005; Bowman and Prior 2005; Poorter et al. 2009; Markesteijn et al. 2011). In general, a conservative resource use strategy is associated with leaves with greater investment in structural components with low SLA, high leaf thickness and high lignin concentration (Reich 2014). These leaves have lower photosynthetic rates but can maintain function much longer (Parkhurst and Loucks 1972; Fetcher 1981; Niinemets 2001; Read and Sanson 2003; Markesteijn et al. 2011). Species that use resources conservatively also tend to possess adaptations traits that allow them to be functionally active at low soil water potentials such as adaptations that reduce xylem cavitation (e.g., narrow vessels with resistant pit membranes, high stem wood density) and traits that maintain leaf turgor (Brodribb et al. 2003). Photoprotection is also expected to vary with leaf lifespan and exposure to dry season drought (Demmig-Adams and Adams 2006; Savage et al. 2009). To the extent that leaves remain functional during the dry season, they would be expected to increase xanthophyll pigments that aid in energy dissipation when water becomes limiting, stomatal conductance declines and photosynthesis quenches less of the incoming absorbed solar radiation.

In several common garden studies testing for ecophysiological differentiation among populations, Ramírez-Valiente et al. (2015), Ramírez-Valiente and Cavender-Bares (2017) and Ramírez-Valiente et al. (2017) demonstrated evolutionary divergence in leaf functional traits among populations of *Q. oleoides* from contrasting precipitation regimes that vary in dry season length and severity. They found, somewhat counterintuitively, that more mesic populations tend to face greater water stress because they maintain functional leaves for longer during the dry season. Ramírez-Valiente et al. (2015) observed that in response to drought, *Q. oleoides* populations originating from mesic areas increased the de-epoxidation rates of the xanthophyll cycle more than xeric populations (Fig. 4.6). They showed that differences in physiological mechanisms, particularly the activation of the xanthophyll cycle, were much higher among populations within species than among different species. The nature of variation within and among species thus contrasts that observed for freezing tolerance, which showed greater differentiation between species than within them. Differences in SLA among *Q. oleoides* populations were also higher than differences observed among live oak species. This study showed for the first time that populations from more mesic areas tended to have more sclerophyllous leaves with higher capacity for photoprotection in this tropical oak (Ramírez-Valiente et al. 2015). The interpretation is that the mesic populations maintain leaves for longer during the dry season and need to continue to maintain function with increasing water stress.

Consistent patterns of variation in leaf thickness and specific leaf area with the index of moisture in the location of the source populations were found in multiple common garden studies in both the greenhouse and in the field (Ramírez-Valiente

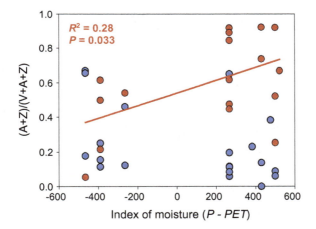

Fig. 4.6 Relationship between the de-epoxidation state of the xanthophyll cycle measured in *Quercus oleoides* seedlings established in a greenhouse experiment and the index of moisture (precipitation—potential evapotranspiration) of families in their climate of origin. Red: dry treatment, blue: well-watered treatment. Redrawn from Ramírez-Valiente et al. (2015). Circles represent family means

and Cavender-Bares 2017; Ramírez-Valiente et al. 2017). Regardless of the location of the experiment, watering treatment and sampled populations, a consistently negative association between specific leaf area and the index of moisture of the source has been observed (Fig. 4.7). Differences in SLA among populations were mainly due to leaf thickness and to a lesser extent to leaf density. In fact, a positive association between leaf thickness and the index of moisture of the source were also observed across studies (Ramírez-Valiente and Cavender-Bares 2017; Ramírez-Valiente et al. 2017).

These studies further revealed that *Q. oleoides* populations had similar values of stomatal conductance and water use efficiency (WUE) (Ramírez-Valiente and Cavender-Bares 2017; Ramírez-Valiente et al. 2017) but significantly differed in water potential at the turgor loss point and leaf abscission in response to drought. Differences in these two traits were again associated with the climate of origin, consistent with variation in leaf morphology. Overall, our findings reveal that populations from more mesic sites have smaller sclerophyllous leaves (lower SLA and higher thickness) and greater drought tolerance (lower π_{tlp}) (Fig. 4.8a) than populations from more xeric sites, which have larger mesophyllous leaves (higher SLA and lower thickness) and increase leaf abscission in response to drought (Fig. 4.8b). Since populations had similar stomatal conductance and WUE, leaf senescence in response to drought may have been favored in populations from more xeric soil conditions as a means of reducing water loss (Jonasson et al. 1997; Condit et al. 2000; Franklin 2005; Stevens et al. 2016). The increased drought tolerance and more durable leaves observed in the most mesic areas within the distribution range of *Q. oleoides* would allow maintaining photosynthetic activity under lower

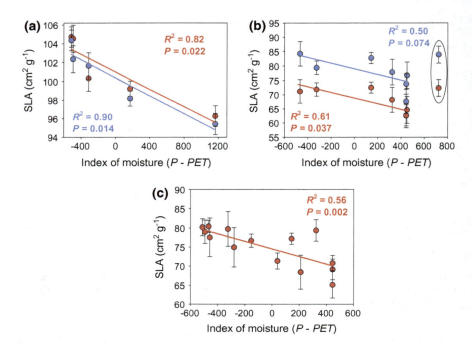

Fig. 4.7 Relationship between specific leaf area and the index of moisture (Precipitation − potential evapotranspiration) in *Quercus oleoides* seedlings established in three common garden experiments: **a** Greenhouse experiment (red: dry treatment, blue: well-watered treatment) (Ramírez-Valiente and Cavender-Bares 2017), **b** Field experiment **a**, established in Honduras in 2011 with eight populations (red: dry season, blue: wet season) (Ramírez-Valiente et al. 2017) and **c** Field experiment **b**, established in Honduras in 2012 with fourteen populations (measured only in the dry season) (Ramírez-Valiente and Cavender-Bares unpublished). Circles represent population means. Bars indicate standard errors. Index of moisture values are population means in the climate of origin

water potentials. Thus, the association between drought resistance strategies, leaf morphology and climate of populations for *Q. oleoides* agrees with the postulates by the resource-use hypothesis in SDTFs (Borchert 1994; Medina 1995; Condit 1998; Givnish 2002; Bowman and Prior 2005; Choat et al. 2007; Tomlinson et al. 2013; Vico et al. 2015). Our findings for *Q. oleoides* are consistent with temporal studies, which show that decreasing rainfall in the dry season enhances the relative abundance of deciduous species in tropical dry forests over time (Enquist and Enquist 2011) and by spatial analyses at small scales, which show that dry deciduous species preferentially occupy drier microhabitats than evergreen species (Comita and Engelbrecht 2009). In sum, these findings suggest that water availability is a key factor driving the spatial and temporal dynamics of functional strategies in the tropics and provide experimental evidence that selection has favored an increase in deciduousness with increasing dry season severity in *Quercus oleoides*.

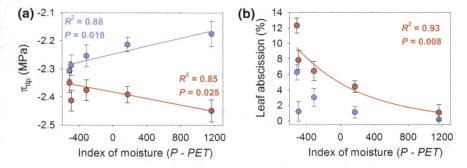

Fig. 4.8 a Relationship between the index of moisture (Precipitation − potential evapotranspiration) and water potential at the turgor loss point (π_{tlp}) in one-year old *Quercus oleoides* seedlings established in a greenhouse experiment (red: dry treatment, blue: wall-watered treatment) (Ramírez-Valiente and Cavender-Bares 2017). Circles represent population means. Bars indicate standard errors. **b** Relationship between the index of moisture (Precipitation − potential evapotranspiration) and leaf abscission in one-year old *Quercus oleoides* seedlings established in a greenhouse experiment (red: dry treatment, blue: wall-watered treatment). Redrawn from Ramírez-Valiente and Cavender-Bares (2017). Circles represent population means. Bars indicate standard errors

4.6 Comparison of the Seasonally Dry Tropical *Q. Oleoides* to Mediterranean Oaks

It is interesting to note that the trends observed in *Q. oleoides* are opposite to those reported by Barlett et al. (2012) at global scale, who found a positive relationship between water availability and π_{tlp} in a metanalysis. This inconsistency across studies probably reflects contrasting patterns in the evolution of drought tolerance among- and within-biomes. Differences in π_{tlp} among functional types within biomes are broadly documented (Bartlett et al. 2012). In fact, species adapted to avoid or escape from dry seasons usually exhibit lower drought tolerance (increased osmotic potentials and turgor loss points) in dry tropical and temperate ecosystems (Medina 1995).

Our results also contrast with those found in intraspecific studies on evergreen oak species from seasonally-dry temperate zones (i.e. Mediterranean-type ecosystems), which show that populations from xeric climates with long dry seasons have a conservative resource-use strategy (Gratani et al. 2003; Ramírez-Valiente et al. 2010, 2014; Niinemets 2015). We speculate that the differences in the patterns of variation of functional traits between tropical and Mediterranean oaks are probably related to temperatures in the wet season. In Mediterranean-type ecosystems carbon assimilation is limited by both water deficit in summer and low temperatures in winter (Larcher 2000; Nardini et al. 2000; Cavender-Bares 2005; Flexas et al. 2014; Granda et al. 2014; Niinemets 2016). In contrast to Mediterranean ecosystems, which have cold winters, in seasonally-dry tropical ecosystems, temperature is not a limiting factor for photosynthesis in the wet season. Mediterranean species face a

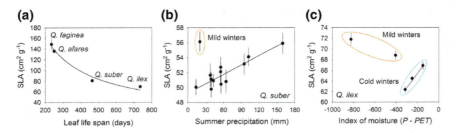

Fig. 4.9 **a** Specific leaf area (SLA, cm^2 g^{-1}) in relation to leaf lifespan for four Mediterranean species from northern Africa and southern Europe, grown and measured in a common garden in France, showing species means and standard errors from sampled individuals. Modified from Cavender-Bares et al. (2005). **b** Relationship between summer precipitation and SLA in seven-year-old saplings of *Quercus suber* established in a field common garden experiment (Ramírez-Valiente et al. 2010). Circles represent population means. Bars indicate standard errors. **c** Relationship between the index of moisture and specific leaf area (SLA) in seedlings of five populations of *Quercus ilex* from contrasting climates (data from García-Nogales et al. 2016). Circles represent population means. Bars indicate standard errors

range of severity in drought stress in summer and mild to more significant freezing stress in winter. Across species of Mediterranean oaks, generally, those with longer leaf lifespans, including the evergreen *Q. ilex* (holm oak), which maintains its leaves for well over two years, and *Q. suber* (cork oak), which maintains leaves for well over one full year, have much lower SLA than species with shorter leaf lifespans, including the deciduous species such as *Q. afares* and *Q. faginea*, from northern Africa, (Fig. 4.9a). Species with lower SLA also have lower leaf nitrogen concentration and are much more resistant to freezing (Cavender-Bares et al. 2005). This direction of variation in these traits is consistent with leaf economic spectrum. Within the two evergreen species, *Q. ilex* and *Q. suber*, the nature and severity of seasonal stress drives leaf variation either in the same direction as the LES or in the opposite direction, similar to *Q. oleoides*. We hypothesize that a conservative resource-use strategy with long leaf life span, thick leaves, high density tissues and high water use efficiency is beneficial in terms of carbon, nutrient and water balance for species inhabiting areas with long dry seasons and cold winters. Thick leaves with higher investment in structural tissues are more resistant to water stress but also to freezing temperatures (Cavender-Bares 2005; Granda et al. 2014). They have higher construction costs but might be offset by a longer payback interval (Williams et al. 1989; Eamus and Prichard 1998).

The patterns of response to cold and drought found in oak species that inhabit seasonally-dry areas as well as those that span the temperate-tropical divide provide key results to understand patterns of resource-use strategies in oaks. For example, in the Mediterranean evergreen oak species, *Q. suber*, one of the studied populations exhibited higher SLA than expectations based on its low precipitation in summer. That "outlier" population was located in the southernmost area of their distribution ranges in the Iberian Peninsula (Cadiz province, Spain), characterized by mild temperatures in winter. *Q. suber* from this location was also found to be highly sensitive to low temperatures in winter (Aranda et al. 2005). Similar results have been reported for the evergreen oaks *Q. ilex*. In a recent study with five holm oak populations, García-Nogales et al. (2016) found that populations from the Iberian Peninsula characterized by cold winters tended to have a positive association between SLA and precipitation, indicating that mesic populations had higher SLA as observed in *Q. suber* and *Q. faginea*. In contrast, southern populations from North Africa with significantly lower index of moisture had higher SLA than Iberian populations, following the pattern shown by *Q. oleoides* in the seasonally dry tropical forest, where winter is absent.

4.7 Population Differentiation in Growth and Photosynthesis

Despite the strong population patterns observed for leaf morphology and drought resistance strategies, our results for growth and photosynthetic rates were not consistent across studies. In analyses performed with data from four common garden trials established in Honduras, we found population-level differentiation in height growth (Fig. 4.10). Populations from areas with longer or more severe dry seasons had higher growth rates in height than mesic populations, which agrees with the resource-use hypothesis for SDTFs. However, the reverse pattern of variation was also observed. Specifically, the results derived from a greenhouse experiment with five *Q. oleoides* populations revealed a positive association between growth rates and the index of moisture (Ramírez-Valiente and Cavender-Bares 2017), contrary to expectations based on the resource-use hypothesis. One possibility for this unexpected positive association between the index of moisture and growth could be the influence of the Rincón population,

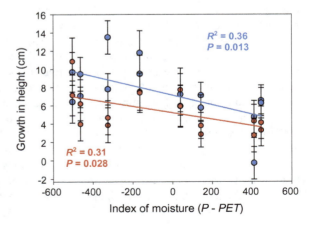

Fig. 4.10 Relationship between growth in height (cm) and index of moisture (Precipitation—potential evapotranspiration) measured in four common garden experiments established in Honduras with natural and manipulated precipitation regimes (blue: water supply in the dry season, red: no water supply in the dry season). Redrawn from Ramírez-Valiente et al. (2017)

a mesic isolated population from Costa Rica. This population showed an outstanding growth rate in this experiment. In fact, once it was removed from the analysis, the relationship between the index of moisture and growth rate became no longer significant ($R = 0.330$, $P = 0.001$ including Rincón vs. $R = -0.066$, $P = 0.578$ excluding Rincón). Rincón de la Vieja is a high elevation population from Costa Rica that exhibits a marked neutral genetic differentiation relative to lowland populations (Deacon and Cavender-Bares 2015). This isolation could have leaded the evolution of particular traits conferring high relative growth rates in this population, even though we detected no fitness advantage in this population.

Patterns of population-level variation for photosynthetic rates were not consistent with the resource-use hypothesis. Populations from xeric areas did not show higher photosynthetic rates as expected. Furthermore, SLA was negatively associated with A_{mass} or relative growth rate (RGR) under well-watered conditions in different studies (Ramírez-Valiente et al. 2017). Xeric populations undergo a short period of water deficit (July–August) within the wet season during which, precipitation is lower than potential evapotranspiration. This "little dry season" has an impact on physiology of species similar to the actual dry season. It is possible that this 'little dry season' may have constrained the evolution of increased A_{mass} and RGR under favorable conditions of water and have promoted some leaf drought resistance during the wet season in xeric populations of *Q. oleoides* that experience this unpredictable water shortage (Choat et al. 2007). Consistent with this idea, we found a positive relationship between turgor loss point (π_{tlp}) and index of moisture when *Q. oleoides* grew under well-watered conditions and the lack of plasticity in π_{tlp} in response to water availability for xeric populations (see also next section).

4.8 Plasticity in Response to Drought

In long-lived species, plasticity is a critical means of surviving spatial and temporal environmental variation and may be more important than local adaptation in tolerating seasonal stress. Studies on *Q. oleoides* have showed a high phenotypic plasticity to water availability in growth rates, gas exchange, leaf morphology and photochemistry (Ramírez-Valiente et al. 2015, 2017; Ramírez-Valiente and Cavender-Bares 2017). Unlike in the case of cold acclimation discussed earlier, whether the plasticity we observed in this suite of traits is adaptive and therefore able to evolve in response to natural selection does not have an easy answer. Several studies on annual or short-lived species have shown an association between plasticity in response to water availability (or other resource) and fitness (Sultan 1995; Dudley 1996; Sultan 1996; Donohue et al. 2000). Plasticity is also considered adaptive if genotypes of a given species differ in phenotypic plasticity and the direction of the response is consistent with expectations based on the environment. Studies on *Q. oleoides* found population-level differentiation in plasticity of three functional traits: photoprotective pigments, water potential at the turgor loss point and leaf abscission that was associated with the index of moisture of the

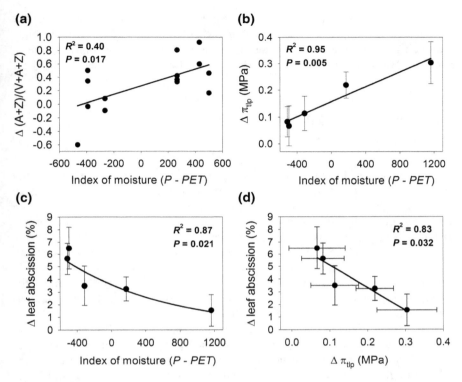

Fig. 4.11 a Plasticity (Δ) in the de-epoxidation state of the xanthophyll cycle (Δ = dry treatment − well-watered treatment) in relation to the index of moisture. Points represent family means means. Redrawn from Ramírez-Valiente et al. (2015). Circles represent family means. Plasticity (Δ) in turgor loss point **b** and leaf abscission **c** (Δ = dry treatment − well-watered treatment) in relation to the index of moisture of the population. Points represent population means. Bars indicate standard errors. **d** Relationship between plasticity in water potential at turgor loss point (π_{tlp}) and plasticity in leaf abscission (Δ = dry treatment − well-watered treatment). Points represent population means. Bars indicate standard errors. **b–d** are redrawn from Ramírez-Valiente and Cavender-Bares (2017)

populations. Specifically, we found that populations originating from mesic areas tended to have higher plasticity in the de-epoxidation rates of the xanthophyll cycle (Fig. 4.11a) and higher capacity of osmoregulation (i.e. high plasticity in the water potential at the turgor loss point) (Fig. 4.11b) whereas populations from more xeric populations had higher plasticity in leaf abscission (which is probably associated with leaf life span) (Fig. 4.11c). These observations together with the fact that the direction of the phenotypic change is consistent with expectations based on the response to drought suggest that phenotypic plasticity is adaptive for these traits and could be subjected to natural selection. A trade off between plasticity in drought avoidance via leaf abscission with plasticity in drought tolerance via osmotic adjustment, suggests that plants have evolved flexibility in one kind of response or the other but not both (Fig. 4.11d). It is important to point out, however, that the variation in the response patterns within each of the populations is quite high, reinforcing other work in this system showing high diversity within populations (Center 2015; Deacon and Cavender-Bares 2015; Cavender-Bares et al. 2011).

4.9 Response of Oaks to Past Climate Change Provides Lessons for the Future

Understanding the physiological limits of species, and the nature of variation within and among species, is critical to understanding species responses to climate change and in providing guidelines for conservation. To the extent to which species are composed of locally adapted populations with narrow climatic tolerances, on one extreme, or of populations that have little variation among them but broad climatic tolerances, at the other extreme, different conservation strategies are required. There are real limits to adaptive potential in long-lived organisms relative to the rate of climate change. Even during past climatic changes that occurred during the expansion of the desert and Mediterranean biomes some 5-million years ago, evidence is mounting that a live oak species, *Quercus brandegeei* (Fig. 4.12c, d), currently found only in the Cape region of southern Baja California, underwent severe range retraction in response to increased drought (Cavender-Bares et al. 2015). Coalescence models using molecular data indicate that the species once occupied a much more extensive range and had a population size >100-fold larger than its current population size. The species underwent range contraction as the Earth continued to cool and dry forming both Mediterranean and desert ecosystems globally. Rather than adapting to the novel climatic regime, now inhabited largely by desert and chaparral species, *Q. brandegeei* became a relictual population restricted to the edges of the ephemeral river beds in the Cape of Baja California. It is now an IUCN red-listed species.

4.10 Evolutionary Potential of Oak Populations

Future work in this system is aimed at addressing the adaptive potential of the live oaks to respond through genetic change to future climate change, a topic of increasingly highlighted importance (Etterson 2004a, b, 2008; Davis et al. 2005; Shaw and Etterson 2012). The genetic architecture of traits determines the potential rate of adaptive evolution in response to a changing environment. Response to natural selection requires genetic variation for traits (Falconer and Mackay 1996), although genetic correlations of various kinds can enhance or impede the rate of evolutionary change (Etterson and Shaw 2001). In particular, voluntary change can be enhanced if the sign of a genetic correlation follows the direction of selection

Fig. 4.12 a Photographs of the common garden experiment in Honduras, **b** the greenhouse experiment at the University of Minnesota, and **c, d** of the endangered *Quercus brandegeei* in the Cape region of Southern Baja California, Mexico

but impeded if these are antagonistic. Facilitating or antagonistic relationships between the genetic architecture of traits and the direction of selection can occur between different life history stages (Schluter et al. 1991), between pairs of traits in a single life history stage (Conner and Via 1992; Caruso 2004) and between trait expression in different environments (Dickerson 1955; Via 1993; Etterson 2004b). Future work elucidating the underlying genetic architecture of physiological traits at different life history stages and as expressed in different environments will help predict the potential for evolutionary responses of these important long-lived species to future climates. Parallel to these efforts, advances in deciphering the genes and gene expression patterns associated with adaptations to climate is critical; important progress in oak systems has been made recently (Gugger et al. 2016, 2017). Understanding the nature and distribution of physiological and genome-wide variation within species facilitates conservation efforts. To that end, we are working with botanical gardens to link the knowledge we have gained to help them with in situ and ex-situ conservation of threatened oak species.

4.11 Conclusions

We synthesized evidence from a series of common garden studies for adaptive differentiation in physiological function both within and among closely related species in response to low temperature and drought. While direct experimental evidence for local adaptation to climatic variation associated with precipitation is lacking, the pattern of trait variation as well as the direction of plasticity are consistent with local adaptation to climate and adaptive plasticity. That nature of the variation in function that exists within and among populations has conservation implications. From these studies we have learned that the variation represented in the species cannot be captured within a single location. At the same time, variation in freezing tolerance is greatest between species, despite important evidence for population differentiation within *Q. virginiana*. In contrast, variation in drought tolerance, in some cases, is greater among populations within a species than between species of live oaks. The patterns may depend on the degree and nature of climatic variation that the populations within a species encounter. And despite clear evidence for evolution in response to climate, the case of *Q. brandegeei* demonstrates that even after 5 million years of exposure to drought conditions, the species maintains its conserved niche in well-drained soils with seasonal water availability, unable to migrate into the surrounding desert. As a consequence, it is a relictual species that will very likely perish without human intervention. The work synthesized here is fundamental to understanding and protecting the oaks, a critical group of species that contributes much to human well-being.

Acknowledgements The studies represented here were funded by grants from the University of Minnesota and the National Science Foundation (IOS: 0843665) to J.C-B. and a fellowship from the Severo Ochoa excellence program to J.A.R-V.

References

Ackerly DD, Reich PB (1999) Convergence and correlations among leaf size and function in seed plants: a comparative test using independent contrasts. Am J Botany 86:1272–1281

Ananthakrishnan R, Soman MK (1989) Statistical distribution of daily rainfall and its association with the coefficient of variation of rainfall series. Int J Climatol 9:485–500

Anderegg WRL, Klein T, Bartlett M, Sack L, Pellegrini AFA, Choat B, Jansen S (2016) Meta-analysis reveals that hydraulic traits explain cross-species patterns of drought-induced tree mortality across the globe. Proc Nat Acad Sci 113:5024–5029

Aranda I, Castro L, Alia R, Pardos JA, Gil L (2005) Low temperature during winter elicits differential responses among populations of the Mediterranean evergreen cork oak (*Quercus suber*). Tree Physiol 25:1085–1090

Bartlett MK, Scoffoni C, Sack L (2012) The determinants of leaf turgor loss point and prediction of drought tolerance of species and biomes: a global meta-analysis. Ecol Lett 5:393–405

Borchert R (1994) Soil and stem water storage determine phenology and distribution of tropical dry forest trees. Ecology 75:1437–1449

Boucher DH (1981) Seed predation by mammals and forest dominance by *Quercus oleoides*, a tropical lowland oak. Oecologia 49:409–414

Bowman D, Prior L (2005) Why do evergreen trees dominate the Australian seasonal tropics? Austr J Bot 53:379–399

Brodribb TJ, Holbrook NM (2006) Declining hydraulic efficiency as transpiring leaves desiccate: two types of response. Plant Cell Environ 29:2205–2215

Brodribb TJ, Holbrook NM, Edwards EJ, Gutierrez MV (2003) Relations between stomatal closure, leaf turgor and xylem vulnerability in eight tropical dry forest trees. Plant Cell Environ 26:443–450

Burke MJ, Gusta LV, Quamme HA, Weiser CJ, Li PH (1976) Freezing injury in plants. Ann Rev of Plant Physiol 27:507–528

Caruso CM (2004) The quantitative genetics of floral trait variation in *Lobelia*: potential constraints on adaptive evolution. Evolution 58:732–740

Cavender-Bares J (2005) Impacts of freezing on long-distance transport in woody plants. In: Holbrook MN, Zwieniecki M (eds) Vascular transport in plants. Elsevier Inc., Oxford, pp 401–424

Cavender-Bares J (2007) Chilling and freezing stress in live oaks (*Quercus* section *Virentes*): intra- and interspecific variation in PS II sensitivity corresponds to latitude of origin. Photosynth Res 94:437–453

Cavender-Bares J (2016) Diversity, distribution and ecosystem services of the North American Oaks. Int Oaks 27:37–48

Cavender-Bares J, Holbrook NM (2001) Hydraulic properties and freezing-induced cavitation in sympatric evergreen and deciduous oaks with, contrasting habitats. Plant Cell Environ 24:1243–1256

Cavender-Bares J, Pahlich A (2009) Molecular, morphological, and ecological niche differentiation of sympatric sister oak species, *Quercus virginiana* and *Q. geminata* (Fagaceae). Am J Bot 96:1690–1702

Cavender-Bares J, Kitajima K, Bazzaz F (2004) Multiple trait associations in relation to habitat differentiation among 17 Floridian oak species. Ecol Monogr 74:635–662

Cavender-Bares J, Cortes P, Rambal S, Joffre R, Miles B, Rocheteau A (2005) Summer and winter sensitivity of leaves and xylem to minimum freezing temperatures: a comparison of cooccurring Mediterranean oaks that differ in leaf lifespan. New Phytol 168:597–612

Cavender-Bares J, Gonzalez-Rodriguez A, Pahlich A, Koehler K, Deacon N (2011) Phylogeography and climatic niche evolution in live oaks (*Quercus* series *Virentes*) from the tropics to the temperate zone. J Biogeogr 38:962–981

Cavender-Bares J, Gonzalez-Rodriguez A, Eaton DAR, Hipp AAL, Beulke A, Manos PS (2015) Phylogeny and biogeography of the American live oaks (*Quercus* subsection *Virentes*): a genomic and population genetics approach. Mol Ecol 24:3668–3687

Center A (2015) Physiological and fitness consequences of seasonal rainfall variation in neotropical live oak seedlings (*Quercus oleoides*): implications for global change. University of Minnesota, Saint Paul

Center A, Etterson JR, Deacon NJ, Cavender-Bares J (2016) Seed production timing influences seedling fitness in the tropical live oak *Quercus oleoides* of Costa Rican dry forests. A J Bot 103:1407–1419

Choat B, Sack L, Holbrook N (2007) Diversity of hydraulic traits in nine *Cordia* species growing in tropical forests with contrasting precipitation. New Phytol 175:686–698

Comita L, Engelbrecht B (2009) Seasonal and spatial variation in water availability drive habitat associations in a tropical forest. Ecology 90:2755–2765

Condit R (1998) Ecological implications of changes in drought patterns: shifts in forest composition in Panama. Clim Change 39:413–427

Condit R, Watts K, Bohlman S, Pérez R, Foster R, Hubbell S (2000) Quantifying the deciduousness of tropical forest canopies under varying climates. J Veg Sci 11:649–658

Conner J, Via S (1992) Natural selection on body size in Tribolium: possible genetic constraints on adaptive evolution. Heredity 69:73–83

Cornelissen JHC, Diez PC, Hunt R (1996) Seedling growth, allocation and leaf attributes in a wide range of woody plant species and types. J Ecol 84:755–765
Crisp M, Arroyo M, Cook L, Gandolfo M, Jordan G (2009a) Phylogenetic biome conservatism on a global scale. Nature 458:754–756
Crisp MD, Arroyo MTK, Cook LG, Gandolfo MA, Jordan GJ (2009b) Phylogenetic biome conservatism on a global scale. Nature 458:754–756
Davis MB, Shaw RG, Etterson JR (2005) Evolutionary responses to changing climate. Ecology 86:1704–1714
Deacon NJ, Cavender-Bares J (2015) Limited pollen dispersal contributes to population genetic structure but not local adaptation in *Quercus oleoides* forests of Costa Rica. PLoS ONE 10: e0138783
Demmig-Adams B, Adams WW (2006) Tansley review: photoprotection in an ecological context: the remarkable complexity of thermal energy dissipation. New Phytol 172:11–21
Dickerson G (1955) Genetic slippage in response to selection for multiple objectives. Cold Spr Harb Symp Quant Biol 20:213–224
Donohue K, Messiqua D, Pyle EH, Heschel MS, Schmitt J (2000) Evidence of adaptive divergence in plasticity: density- and site-dependent selection on shade-avoidance responses in Impatiens capensis. Evolution 54:1956–1968
Dudley SA (1996) The response to differing selection on plant physiological traits: evidence for local adaptation. Evolution 50:103–110
Eamus D, Prichard H (1998) A cost-benefit analysis of leaves of four Australian savanna species. Tree Physiol 18:537–545
Engelbrecht BMJ, Kursar TA (2003) Comparative drought-resistance of seedlings of 28 species of co-occurring tropical woody plants. Oecologia 136:383–393
Engelbrecht BMJ, Dalling JW, Pearson TRH, Wolf RL, Galvez DA, Koehler T, Tyree MT, Kursar TA (2006) Short dry spells in the wet season increase mortality of tropical pioneer seedlings. Oecologia 148:258–269
Enquist B, Enquist C (2011) Long-term change within a Neotropical forest: assessing differential functional and floristic responses to disturbance and drought. Glob Chang Biol 17:1408–1424
Etterson JR (2004a) Evolutionary potential of *Chamaecrista fasciculata* in relation to climate change I. Clinal patterns of selection along an environmental gradient in the Great Plains. Evolution 58:1446–1458
Etterson JR (2004b) Evolutionary potential of *Chamaecrista fasciculata* in relation to climate change: II. Genetic architecture of three populations reciprocally planted along an environmental gradient in the Great Plains. Evolution 58:1459–1471
Etterson JR (2008) Evolution in response to climate change. In: Carroll S, Fox C (eds) Conservation biology: evolution in action. Oxford University Press, Oxford, p 145
Etterson J, Shaw R (2001) Constraint to adaptive evolution in response to global warming. Science 294:151–154
Falconer DS, Mackay TFC (1996) Introduction to quantitative genetics. Prentice Hall, New York
Fetcher N (1981) Leaf Size and Leaf Temperature in Tropical Vines. Am Nat 117:1011–1014
Flexas J, Diaz-Espejo A, Gago J, Gallé A, Galmés J, Gulías J, Medrano H (2014) Photosynthetic limitations in Mediterranean plants: a review. Environl Exp Bot 103:12–23
Franklin D (2005) Vegetative phenology and growth of a facultatively deciduous bamboo in a monsoonal climate. Biotropica 37:343–350
Fujikawa S, Kuroda K (2000) Cryo-scanning electron microscopic study on freezing behavior of xylem ray parenchyma cells in hardwood species. Micron 31:669–686
Givnish TJ (2002) Adaptive significance of evergreen vs. deciduous leaves: solving the triple paradox. Silva Fenn 36:703–743
Gould K (2004) Nature's Swiss army knife: the diverse protective roles of anthocyanins in leaves. J Biomed Biotechnol 2004:314–320
Granda E, Scoffoni C, Rubio-Casal A, Sack L, Valladares F (2014) Leaf and stem physiological responses to summer and winter extremes of woody species across temperate ecosystems. Oikos 123:1281–1290

Gratani L, Meneghini M, Pesoli P, Crescente M (2003) Structural and functional plasticity of *Quercus ilex* seedlings of different provenances in Italy. Trees 17:515–521

Gugger PF, Cokus SJ, Sork VL (2016) Association of transcriptome-wide sequence variation with climate gradients in valley oak (*Quercus lobata*). Tree Genet Genom 12:15

Gugger PF, Peñaloza-Ramírez JM, Wright JW, Sork VL (2017) Whole-transcriptome response to water stress in a California endemic oak, *Quercus lobata*. Tree Physiol 37:632–644

Guy CL (2003) Freezing tolerance of plants: current understanding and selected emerging concepts. Can J Bot 81:1216–1223

Hipp AL, Manos PS, González-Rodríguez A, Hahn M, Kaproth M, McVay JD, Avalos SV, Cavender-Bares J (2017) Sympatric parallel diversification of major oak clades in the Americas and the origins of Mexican species diversity. New Phytol. doi:10.1111/nph.14773

Hughes NM, Burkey KO, Cavender-Bares J, Smith WK (2012) Xanthophyll cycle pigment and antioxicant profiles of winter-red (anthocyanic) and winter-green (acyanic) angiosperm evergreen species. J Exp Bot 63:1895–1905

Huner NPA, Oquist G, Hurry VM, Krol M, Falk S, Griffith M (1993) Photosynthesis, photoinhibition and low-temperature acclimation in cold tolerant plants. Photosynth Res 37:19–39

IPCC (2007) Climate Change 2007: the physical science basis. contribution of working group i to the fourth assessment report of the intergovernmental panel on climate change. Cambridge University Press, New York

Jonasson S, Medrano H, Flexas J (1997) Variation in leaf longevity of *Pistacia lentiscus* and its relationship to sex and drought stress inferred from leaf $\delta^{13}C$. Funct Ecol 11:282–289

Jump A, Hunt J, Peñuelas J (2006) Rapid climate change-related growth decline at the southern range edge of *Fagus sylvatica*. Glob Change Biol 12:2163–2174

Karmalkar AV, Bradley RS, Diaz HF (2008) Climate change scenario for Costa Rican montane forests. Geophys Res Lett 35. doi:10.1029/2008GL033940

Klemens JA, Deacon NJ, Cavender-Bares J (2011) Pasture recolonization by a tropical oak and the regeneration ecology of seasonally dry tropical forests. In: Dirzo R, Young HS, Mooney HA, Ceballos G (ed) Seasonally Dry Tropical Forests. Island Press/Center for Resource Economics, pp 221–237

Koehler K, Center A, Cavender-Bares J (2012) Evidence for a freezing tolerance—growth rate trade-off in the live oaks (*Quercus* series *Virentes*) across the tropical-temperate divide. New Phytol 193:730–744

Koerner C, Larcher W (1988) Plant life in cold climates. In: Long SP, Wodward FI (eds) Plants and temperature. Society of Experimental Biology, Cambridge, pp 25–57

Kurz H, Godfrey RK (1962) Trees of Northern Florida. University of Florida, Gainesville

Larcher W (1960) Transpiration and photosynthesis of detached leaves and shoots of *Quercus pubescens* and *Q. ilex* during desiccation under standard conditions. Bull Res Counc Isr 8:213–224

Larcher W (2000) Temperature stress and survival ability of Mediterranean sclerophyllous plants. Plant Biosyst 134:279–295

Markesteijn L, Poorter L, Paz H, Sack L, Bongers F (2011) Ecological differentiation in xylem cavitation resistance is associated with stem and leaf structural traits. Plant, Cell Environ 34:137–148

Medina E (1995) Diversity of life forms of higher plants in neotropical dry forests. In: Bullock S, Mooney H, Medina E (eds) Seasonally dry tropical forests. Cambridge University Press, Cambridge, pp 221–242

Meireles JE, Beulke A, Borkowski D, Romero-Severson J, Cavender-Bares J (2017) Balancing selection maintains diversity in a cold tolerance gene in broadly distributed live oaks. Genome in press

Miller HA, Lamb SH (1985) Oaks of North America. Naturegraph Publishers Inc, Happy Camp, California

Muller CH (ed) (1942) The central American species of *Quercus*. United States Department of Agriculture, Washington, DC

Muller SC (1961) The origin of *Quercus fusiformis* small. J Linn Soc 58:186–192
Myers RL (1990) Scrub and High Pine. In: Myers RL, Ewel JJ (eds) Ecosystems of Florida. University of Central Florida Press, Orlando, pp 150–193
Nardini A, Salleo S, Gullo MAL, Pitt F (2000) Different responses to drought and freeze stress of *Quercus ilex* L. growing along a latitudinal gradient. Plant Ecol 148:139–147
Nicholls N, Wong KK (1990) Dependence of rainfall variability on mean rainfall, latitude, and the Southern Oscillation. J Clim 3:163–170
Niinemets Ü (2001) Global-scale climatic controls of leaf dry mass per area, density, and thickness in trees and shrubs. Ecology 82:453–469
Niinemets Ü (2015) Is there a species spectrum within the world-wide leaf economics spectrum? Major variations in leaf functional traits in the Mediterranean sclerophyll *Quercus ilex*. New Phytol 205:79–96
Niinemets Ü (2016) Does the touch of cold make evergreen leaves tougher? Tree Physiol 36:267–272
Nixon KC (1985) A Biosystematic Study of *Quercus* Series *Virentes* (the live oaks) with Phylogenetic Analyses of Fagales, Fagaceae and *Quercus*, Ph.D. Thesis. University of Texas, Austin
Nixon KC, Muller CH (1997) *Quercus* Linnaeus sect. *Quercus* White oaks. In: Flora of North America Committee (ed) Flora of North America, North of Mexico. Oxford University Press, New York, pp 436–506
Oertli JJ, Lips SH, Agami M (1990) The strength of sclerophyllous cells to resist collapse due to negative turgor pressure. Acta Oecologica 11:281–289
Parkhurst DF, Loucks OL (1972) Optimal leaf size in relation to environment. J Ecol 60:505–537
Parker J (1963) Cold resistance in woody plants. Bot Rev 29:123–201
Pennington RT (2006) Neotropical Savannas and seasonally dry forests: plant diversity, biogeography, and conservation. CRC Press, Taylor & Francis Group, New York
Pietrini F, Iannelli M, Massacci A (2002) Anthocyanin accumulation in the illuminated surface of maize leaves enhances protection from photo-inhibitory risks at low temperature, without further limitation to photosynthesis. Plant, Cell Environ 25:1251–1259
Poorter H, Niinemets Ü, Poorter L, Wright IJ, Villar R (2009) Causes and consequences of variation in leaf mass per area (LMA): a meta-analysis. New Phytol 182:565–588
Poorter L, Markesteijn L (2008) Seedling traits determine drought tolerance of tropical tree species. Biotropica 40:321–331
Ramírez-Valiente JA Cavender-Bares J (2017) Evolutionary trade-offs between drought resistance mechanisms across a precipitation gradient in a seasonally dry tropical oak (*Quercus oleoides*). Tree Physiol 1–13. doi:10.1093/treephys/tpx1040
Ramírez-Valiente JA, Sánchez-Gómez D, Aranda I, Valladares F (2010) Phenotypic plasticity and local adaptation in leaf ecophysiological traits of 13 contrasting cork oak populations under different water availabilities. Tree Physiol 30:618–627
Ramírez-Valiente J, Valladares F, Sánchez-Gómez D, Delgado A, Aranda I (2014) Population variation and natural selection on leaf traits in cork oak throughout its distribution range. Acta Oecol 58:49–56
Ramírez-Valiente JA, Koehler K, Cavender-Bares J (2015) Climatic origins predict variation in photoprotective leaf pigments in response to drought and low temperatures in live oaks (*Quercus* series *Virentes*). Tree Physiol 35:521–534
Ramírez-Valiente JA, Center A, Sparks SP, Sparks KL, Etterson JR, Longwell T, Pilz G, Cavender-Bares J (2017) Population-level differentiation in growth rates and leaf traits in seedlings of the neotropical live oak *Quercus oleoides* grown under natural and manipulated precipitation regimes. Front Plant Sci 8:585
Read J, Sanson GD (2003) Characterizing sclerophylly: the mechanical properties of a diverse range of leaf types. New Phytol 160:81–99
Reich PB (2014) The world-wide 'fast–slow' plant economics spectrum: a traits manifesto. J Ecol 102:275–301

Reich PB, Borchert R (1984) Water stress and tree phenology in a tropical dry forest in the lowlands of Costa Rica. J Ecol 61–74
Sakai A, Larcher W (1987) Frost survival of plants: responses and adaptations to freezing stress. Springer-Verlag, Berlin
Sakai A, Weiser CJ (1973) Freezing resistance of trees in North America with reference to tree regions. Ecology 54:118–126
Savage JA, Cavender-Bares J (2013) Phenological cues drive an apparent trade-off between freezing tolerance and growth in the family *Salicaceae*. Ecology 94:1708–1717
Savage J, Cavender-Bares J, Verhoeven A (2009) Habitat generalists and wetland specialists in the genus *Salix* vary in their photoprotective responses to drought. Funct Plant Biol 36:300–309
Schluter D, Price TD, Rowe L (1991) Conflicting selection pressures and life history trade-offs. Proc Roy Soc B 246:11–17
Shaw RG, Etterson JR (2012) Rapid climate change and the rate of adaptation: insight from experimental quantitative genetics. New Phytol 195:752–765
Steponkus PL (1984) Role of the plasma membrane in freezing injury and cold acclimation. Annu Rev Plant Physiol 35:543–584
Stevens N, Archibald S, Nickless A, Swemmer A, Scholes R (2016) Evidence for facultative deciduousness in *Colophospermum mopane* in semi-arid African savannas. Austr Ecol 41:87–96
Sultan SE (1995) Phenotypic plasticity and plant adaptation. Acta Bot Neerl 44:363–383
Sultan SE (1996) Phenotypic plasticity for offspring traits in *Polygonum persicaria*. Ecology 77:1791–1807
Tomlinson K, Poorter L, Sterck F, Borghetti F, Ward D, Bie S, Langevelde F (2013) Leaf adaptations of evergreen and deciduous trees of semi-arid and humid savannas on three continents. J Ecol 101:430–440
Tyree MT, Engelbrecht BMJ, Vargas G, Kursar TA (2003) Desiccation tolerance of five tropical seedlings in Panama. Relationship to a field assessment of drought performance. Plant Physiol 132:1439–1447
Van Kleunen M, Fischer M (2005) Constraints on the evolution of adaptive phenotypic plasticity in plants. New Phytol 166:49–60
Via S (1993) Adaptive phenotypic plasticity: target of by-product of selection in a variable environment? Am Nat 142:352–365
Vico G, Thompson S, Manzoni S, Molini A, Albertson J, Almeida-Cortez J, Fay P, Feng X, Guswa A, Liu H, Wilson T, Porporato A (2015) Climatic, ecophysiological, and phenological controls on plant ecohydrological strategies in seasonally dry ecosystems. Ecohydrology 8:660–681
Wiens JJ, Ackerly DD, Allen AP, Anacker BL, Buckley LB, Cornell HV, Damschen EI, Davies TJ, Grytnes JA, Harrison SP, Hawkins BA, Holt RD, McCain CM, Stephens PR (2010) Niche conservatism as an emerging principle in ecology and conservation biology. Ecol Lett 13:1310–1324
Williams K, Field CB, Mooney HA (1989) Relationships among leaf construction cost, leaf longevity, and light environment in rain-forest plants of the genus Piper. Am Nat 133:198–211
Wisniewski ME, Ashworth EN (1985) Changes in the ultrastructure of xylem parenchyma cells of peach (*Prunus persica*) and red oak (*Quercus rubra*) in response to a freezing stress. Am J Bot 72:1364–1376
Wright IJ, Reich PB, Cornelissen JH, Falster DS, Groom PK, Hikosaka K, Lee W, Lusk CH, Niinemets U, Oleksyn J, Osada N, Poorter H, Warton DI, Westoby M (2005) Modulation of leaf economic traits and trait relationships by climate. Glob Ecol Biogeogr 14:411–421
Zanne AE, Tank DC, Cornwell WK, Eastman JM, Smith SA, FitzJohn RG, McGlinn DJ, O'Meara BC, Moles AT, Reich PB, Royer DL, Soltis DE, Stevens PF, Westoby M, Wright IJ, Aarssen L, Bertin RI, Calaminus A, Govaerts R, Hemmings F, Leishman MR, Oleksyn J, Soltis PS, Swenson NG, Warman L, Beaulieu JM (2014) Three keys to the radiation of angiosperms into freezing environments. Nature 506:89–92

Chapter 5
Oaks Under Mediterranean-Type Climates: Functional Response to Summer Aridity

Eustaquio Gil-Pelegrín, Miguel Ángel Saz, Jose María Cuadrat, José Javier Peguero-Pina and Domingo Sancho-Knapik

Abstract Mediterranean-type climates are characterized by warm or hot summers, mild or cold winters and, especially, by the existence of a summer drought period driven by the low or even nule precipitation during this season. Mediterranean-type climates are represented in different areas of the world, both in the Northern and the Southern Hemisphere. Specifically, regarding the existence of *Quercus* under these climatic conditions, two main geographical areas should be considered, namely the Mediterranean Basin in the Palearctic and California (USA) and Baja California (Mexico) in the Nearctic. Despite the relatively low geographical extension of the areas occupied by oaks under this type of climate, it has deserved its own phytoclimatical entity since the first geobotanical synthesis at a global scale. Although evergreen and sclerophyllous oak species are widely assumed as a prototype of mediterranean oaks, both palaeoecological evidences and present biogeographical analysis confirm the co-existence of this oak type with winter-deciduous species of the same genus. In this chapter, the different advantages and disadvantages of both phenological patterns (evergreeness and winter-deciduousness) are presented. Moreover, the strategies for saving water through the overall leaf size reduction, the stomatal control of water losses or some xeromorphic traits for a further reduction of transpiration are also shown. Finally, the development of a high resistance to drought-induced cavitation, as a way for coping with low water potential during dry periods, is discussed.

E. Gil-Pelegrín (✉) · J. J. Peguero-Pina · D. Sancho-Knapik
Centro de Investigación y Tecnología Agroalimentaria de Aragón, Gobierno de Aragón, Unidad de Recursos Forestales, Avda. Montañana 930, 50059 Saragossa, Spain
e-mail: egilp@cita-aragon.es

M. Á. Saz · J. M. Cuadrat
Departamento de Geografía y Ordenación del Territorio, Universidad de Zaragoza, 50009 Saragossa, Spain

© Springer International Publishing AG 2017
E. Gil-Pelegrín et al. (eds.), *Oaks Physiological Ecology. Exploring the Functional Diversity of Genus* Quercus *L.*, Tree Physiology 7,
https://doi.org/10.1007/978-3-319-69099-5_5

5.1 Key Features of Mediterranean-Type Climates Worldwide

The Mediterranean-type climates include a set of sub-varieties with common characteristics: presence of warm or hot summers, mild or cold winters and a rainfall regime characterized by a severe summer drought (Lionello et al. 2006). Total rainfall exceeds 300 mm on average (Grove et al. 1977), although it is not unusual to find locations with values above 2000 mm due to the influence of orography (Cuadrat et al. 2007). Less than 20% of the annual precipitation occurs during summer, while more than 50% of it falls during the cold season (Deitch et al. 2017).

The typical monthly temperature and precipitation regime of the Mediterranean-type climates of the Northern Hemisphere is shown in Fig. 5.1, with an aridity period during summer that can be also extended from late spring to early autumn, and a high concentration of precipitation during winter. The combination of a maximum of temperature and a minimum of precipitation induces the summer aridity period that better charactherizes the Mediterranean-type climates (Walter 1985; Breckle 2002). The aridity period is defined as the time-span where the temperature values are above the precipitation values in a Gaussen-type ombrothermic diagram as shown in Fig. 5.1 (P = 2T). Köppen classified these climates as Cs (Köppen 1936), although subdivided into Csa and Csb according to the temperature of the warmer month (above or below 22 °C respectively).

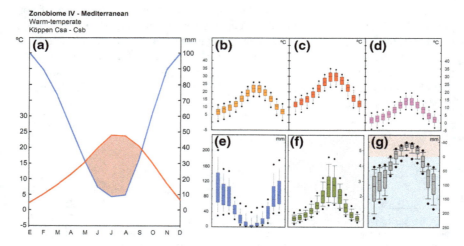

Fig. 5.1 Typical Mediterranean-type climate (Northern Hemisphere) with warm and dry summers and mild and rainy winters (**a**). **b**, **c** and **d** show the mean, maximum and minimum monthly temperature values respectively extracted from a set of more than 2000 points with Cs climate from those in Appendix 5.1. **e** and **f** show the monthly precipitation and vapour pressure deficit (VPD) respectively from the same set of points. **g** shows the difference between 2P and T, indicating aridity (pink area) according to Gaussen (see text for details)

Beyond this common feature of the Mediterranean-type climates, with the existence of a dry period during summer, a high variability is registered in the precipitation or temperature values among localities. Box plots of Fig. 5.1b–e have been calculated taking into account the mean temperature and precipitation data of the WorldClim V2.0 database (Fick and Hijmans 2017), extracted from a set of more than 2000 points with a Cs climate (Peel et al. 2007) and confirmed presence of *Quercus* species (data set in Appendix 5.1). The climatic variability, although also observed in the thermal regime (Fig. 5.1b–d), is especially extreme when precipitation is considered (Fig. 5.1d). Regarding aridity, Fig. 5.1g, which shows the distribution of arid months according to the criterium above described, evidences that summer months are expected to be arid in most if not all the analyzed points, even extending this aridity period far from the early autumn in some locations. The variability in the atmospheric dryness (estimated as vapour pressure deficit, thereafter VPD, kPa) is quite high among sites under Mediterranean-type climates (Fig. 5.1f). Maximum values above 4 kPa during mid summer are found in some locations, while VPD remains below 2 kPa in others during the same period. The outstanding importance of this parameter will be considered below in this chapter in terms of water losses by transpiration and the mechanisms developed by different *Quercus* species to cope with such high atmospheric demand during the hottest days of the summer.

The regions under Mediterranean-type climates represent a relatively small proportion of the continental areas of the world, if compared with other type of climates. Moreover, they appear fragmented in different territories of the Northern and Southern Hemispheres. Despite these facts, it must be highlighted that this climate has deserved its own phytoclimatic consideration. In this sense, Schimper (1903) dedicated a chapter of his "Plant Geography upon a Physiological Basis", a very early phytogeographical synthesis of the vegetation of the earth, to the so-called "District of the warm temperate belts with moist winters", where the author included those areas under Mediterranean-type climate (namely Mediterranean Basin, Cape Region, South and West Australia, California and Chile). Later, Walter (1985) proposed the Zonobiome IV, "of sclerophyllic woodlands" or "Zonobiome of the arido-humid winter rain region"(Breckle 2002), while Rivas-Martínez et al. (2011) proposed the consideration of Macrobioclimate for this climatic type.

As mentioned before, there are several regions in the world that show these climatic characteristics, always located on the western face of the continents in both hemispheres between 30° and 40° of latitude (Lionello et al. 2006). Specifically, Mediterranean-type climates can be found in (i) regions of Europe, Africa and Asia surrounding the Mediterranean Basin (excluding Egypt, Libya and most part of Tunisia) and extending to the south of Turkey and northern Syria; (ii) the Pacific shore of North America, extending through the state of California in USA and New California in Mexico; (iii) the central coast of Chile; (iv) a large area of western and southern Australia; and (v) the Cape region of South Africa.

These regions are located near the latitudinal limit between temperate and tropical latitudes, in areas where the shift of the subtropical high pressure cells to higher latitudes during summer causes atmospheric stability and, as a consequence, the absence of rainfall. In addition to the typical dry summer, a high frequency of meteorological droughts (García-Ruiz et al. 2011) associated to high atmospheric pressures is frequent during spring and autumn. Besides, the influence of high pressures during the winter associated to the strong cooling of the earth surface in the Eurasian and American continents can also induce dry periods during winter. At the same time, the presence of cold air masses in the middle troposphere can favour atmospheric instability and, as a consequence, events of intense rainfall (Serrano-Muela et al. 2013). In addition, cold and heat waves are not rare in areas under Mediterranean-type climates, associated to the latitudinal displacement of air masses with very contrasted temperatures. The consequences on natural systems, anthropic activities and human health of such cold or heat waves have been widely recognised (Trigo et al. 2005).

Despite the apparent severity of this climate for life, these regions are especially rich in terms of biological diversity (Cowling et al 1996). Variations in (i) the time distribution of precipitation, (ii) the length of the aridity period, (iii) the annual thermal amplitude or (iv) the frequency of extreme events, produce a complex mosaic of environments. Such diversity of climatic variations in space favours the existence of different species of animals and plants that have been able to develop physiological, morphological or behavioral adaptative responses. However, the current physiognomy and floristic composition of the vegetation in areas under Mediterranean-type climate cannot be fully understood without considering the human influence. The five regions under Mediterranean-type climate above indicated are characterized by an intense human occupation that radically transformed the primitive landscape., Some of the most advanced societies in history have been developed in the Mediterranean Basin (Büntgen et al. 2011), with an enormous capacity for influencing and modifying the natural environment. Moreover, such changes in soil occupation, through its effect on the albedo, could have altered the regional atmospheric circulation in the Mediterranean Sea and influenced some aspects of the climate, at least since 4000 BP (Reale and Dirmeyer 2000). Nevertheless, the human land use has also contributed to enhance the diversity of ecological situations and, as a consequence, the current interest in these landscapes (Lasanta et al. 2016). In this context, the coexistence of mediterranean-type oaks, both evergreen and winter-deciduous, will be discussed below.

5.2 Mediterranean-Type Climates and Oaks: The Areas of Northern Hemisphere Under This Climate

Although five regions are recognised as representative of Mediterranean-type climate (see above), the aim of this chapter is to focus on the response of oak species to the environmental constrains imposed by this climatic type. The areas with presence of *Quercus* spp. under Mediterranean-type climate are those of the

Northern Hemisphere, namely Mediterranean Basin and the pacific shores in California (USA) and Baja California (Mexico).

5.2.1 Mediterranean Basin

Mediterranean Basin is the region with the most heterogeneous climatic regime among the areas under such climatic type, with an outstanding influence of the orography and the thermal and dynamic behavior of the sea masses (Bethoux et al. 1999). The annual precipitation ranges from 400 to 800 mm, with a mean annual temperature ranging from 16 to 19 °C (García Ruiz et al. 2013). The relief of the Mediterranean Basin is associated with the Alpine Orogeny, especially active in the areas between the African and Eurasian plates. There are several geographical factors related to the latitudinal situation of the Mediterranean Basin that are superimposed to the atmospheric patterns: (i) differences in altitude between mountains and valleys, (ii) the proximity to the sea masses and (iii) the exposure to westerlies, All these factors induce the existance of a complex climatic mosaic and extreme climatic gradients (Castro-Díez et al. 1997).

Although the existence of dry summers and rainy winters is a common feature of the climate in the Mediterranean Basin, the genesis of thermal anticyclones inside the continents during winter (Wallen 1970) can drastically reduce the precipitation registered during the colder months of the year, producing a second precipitation minimum (Cuadrat et al. 2007). This situation can be persistent if the winter high pressures cell located in Siberia comes in contact with the Azores high tropical cell. In fact, the lack of winter precipitation during two consecutive years has been related to episodes of massive oak decline in the Iberian Peninsula (Corcuera et al 2004b). The influence of high pressures disappears during the spring, with the arrival of fronts of humid air from the Atlantic. Thus, this season is the wettest period of the year in most areas of the Mediterranean Basin, especially in the Iberian Peninsula. Apart from that, the strong cyclonic activity of the Mediterranean sea and the high temperatures of the water mass after the summer heating provoke abundant precipitation in other areas of the Mediterranean Basin, especially in the coastal areas of the eastern Iberian peninsula, Italy and Greece.

The climate in the Mediterranean Basin is affected by atmospheric patterns at a global scale. There are a great amount of studies demonstrating the influence of NAO (North Atlantic Oscillation) on winter precipitation in the western Mediterranean Basin (Hurrell et al. 2004) as well as the EA (East Atlantic pattern) on precipitation anomalies in the eastern areas (Krichak and Alpert 2005). The influence of ENSO (*El Niño*-Southern Oscillation) seems also to be important on winter precipitation in the eastern Mediterranean Basin (Pozo-Vázquez et al. 2001), although some evidences of positive influence on autumn precipitation have been also found in some western territories, such as Spain and Morocco (Mariotti et al. 2002). The influence of other teleconnection patterns have also been proposed, namely the "Mediterranean Sea-Caspian Pattern", the "Southern Europe-North Atlantic Pattern" and the

"Western Mediterranean Oscillation" (Lionello et al. 2006), but further research is needed to confirm their impact on the climate in this area.

5.2.2 Pacific Shore of USA and Mexico: California and Baja California

The Mediterranean climate in California (USA) is located from the Pacific coastal regions to the piedmont of Sierra Nevada, including a large fragment of the Central Valley, where climate becomes warmer and drier. There are also areas under Mediterranean-type climate in northern Baja California (Mexico), with warm and dry summers, mild winters, high interannual variability of precipitation and high frequency of severe droughts. Climatic gradients in this region are related to changes in latitude, proximity to the Pacific Ocean, altitude and continentality. Mediterranean-type climates are also largely influenced by the pressure systems of the Pacific Ocean and the high and low pressure cells of the interior of the American continent. The location of the "Pacific High", an extensive area of high pressures related to its subtropical latitudinal position, is responsible of the summer drought and the north-south gradient of precipitation. Its gradual shift in winter allows the arrival of humid fronts associated with the atmospheric dynamic of the temperate zone. The rainfall is higher when the fronts reach the foothills of the coastal mountain ranges and Sierra Nevada (Bryson and Hare 1974).

There are also other patterns of general circulation at a cyclonic scale that affect the Mediterranean-type climates of this region, such as ENSO, Pacific Decadal Oscillation (PDO) and Pacific North American Pattern (PNA). During positive ENSO phases, dry and warm conditions are expected in the north of California and Oregon while the north-south dipole causes wet anomalies in Southern California (Trouet et al. 2009). During negative ENSO phases, the opposite climatic conditions are registered (Trouet et al. 2009). PDO seems to have a greater influence on climatic variability at temporal decadal scales (Mantua et al. 1997). Finally, PNA, that determines the circulation in the middle layers of the troposphere, causes a movement of the cyclonic systems towards the north or the south of the analyzed area modifying the annual patterns of precipitation (Wallace and Gutzler 1981).

5.3 Climatic Transitions in Mediterranean Areas

As stated before, despite the common feature of a summer drought period, Mediterranean-type climates are quite diverse, with a marked influence on the physiognomy of the vegetation. From a phytoclimatic perpective, and taking in mind the presence of different oak species, the genuine Mediterranean-type climate can evolve towards (i) warmer and drier sub-climates, in transition to arid ones (Fig. 5.2a),

(ii) milder and more humid ones, in transition to temperate climates (Fig. 5.2b) or (iii) drier and cooler variants, closer to the typical steppe climates (Fig. 5.2c).

These climatic transitions can be expressed in terms of Walter's Zonobiomes (Walter 1985; Breckle 2002) (Fig. 5.3). According to this classification, the Zonobiome IV or "Zonobiome of the Sclerophyllic Woodlands" is considered the genuine Mediterranean-type climate, and *Q. ilex* the representative oak species. When aridity increases, in a progressive transition to Zonobiome III or "Zonobiome of Hot Deserts", *Q. ilex* would be substituted by *Q. coccifera*, which is present in areas of semi-arid climates (Vilagrosa et al. 2003a, b, 2010). The transition to Zonobiome VI or "Zonobiome of Deciduous Forests" would imply a shortening of the summer aridity period, a cooler winter and an overall higher precipitation. Most of the Iberian Peninsula can be considered to be under this transitional climate, with small leaved, winter-deciduous oaks, such as *Q. faginea* (Sánchez de Dios et al. 2009). The transition to dry but cooler conditions, towards the Zonobiome VII or "Zonobiome of Steppes and Cold Deserts" should imply the dominance of some oak species that are able to withstand such cold climatic subtypes, such as *Q. ilex* subsp. *rotundifolia* in the western Mediterranean Basin and *Q. baloot* in southwestern Asia.

5.4 The History of Oaks Under Mediterranean-Type Climates. Is There a Single Prototype of Mediterranean *Quercus*?

Areas under Mediterranean-type climates sustain a significant part of the world's terrestrial biomass, net primary productivity and biodiversity with almost the 20% of the known vascular plant species diversity on Earth (Atjay et al. 1979;

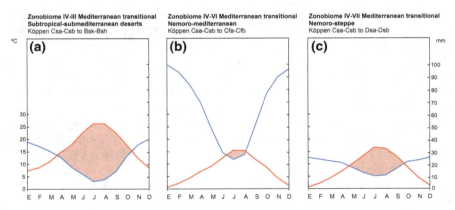

Fig. 5.2 Climograms of transition zones within the mediterranean climate to arid (**a**), temperate (**b**) and cold steppe (**c**). Zonobiomes according to Walter's classification (Breckle 2002)

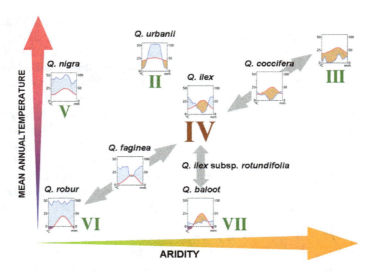

Fig. 5.3 Transition from Zonobiome IV to more arid (III), humid-colder (VI) and colder (VII) conditions and species of *Quercus* characteristic of these transitions. Zonobiomes according to Walter's classification (Breckle 2002). See text for details

Valiente-Banuet et al. 2006), conferring to these areas a current great importance. However, from a geological time scale, Mediterranean-type climates could be considered nearly negligible, as the paleontological evidences suggest a relatively recent development (Ackerly 2009; Rundel et al. 2016). Most authors considered that summer drought that characterized this type of climate was not established on the Northern Hemisphere until the end of the Tertiary—beginning of the Quaternary (between 7 and 2.8 Million year depending on the region) (Suc 1984; Verdú et al. 2003; Jiménez-Moreno et al. 2010; Millar 2012).

Long before the development of the Mediterranean-type climates, at the beginning of the Tertiary, the current mediterranean regions like in many other parts of the world had a warm and wet climate characterized by a precipitation regime distributed throughout the year (Verdú et al. 2003; Valiente-Banuet et al. 2006; Ackerly 2009; Millar 2012). Under these humid climatic conditions, the vegetation in these areas of the Northern Hemisphere was formed by rich oak-laurel-madrone evergreen woodlands (Valiente-Banuet et al. 2006) where sclerophyllous taxa occurred as minor elements (Axelrod 1975, 1989), constituting in most cases the understory of these woodlands (Valiente-Banuet et al. 2006). During the Early Tertiary, arid climates and dry environments started to develop (Millar 2012) and laurophyllous forests had to adapt to these new conditions (Axelrod 1977). While a lot of plant species became extinct, evergreen sclerophyllous shrubs present in the understory expanded geographically (Valiente-Banuet et al. 2006). The resulting type of flora can be compared with the current evergreen sclerophyllous broad-leaved forest in China (Suc 1984; Rundel et al. 2016) or with the rain forest in Southern Mexico (Millar 2012).

Arid climates and dry environments continued to develop during the Middle-Late Tertiary, modifying the composition of the flora (Millar 2012). In this time, some authors already stated the common presence of sclerophyllous leaves corresponding to *"ilex"* or *"suber"* oaks type in Southern Europe (Kovar-Eder 2003) and to *"chrysolepis"* type in Western USA (Retallack 2004). Afterwards, at the end of the Tertiary, with this general palaeoclimatic trend towards greater aridity, summer precipitation started to decrease in the current mediterranean areas (Axelrod 1973; Suc 1984). The change to Mediterranean-type climates led surviving taxonomic components of the ancient flora, including some of the most abundant modern genera, such as *Quercus* L. (Valiente-Banuet et al. 2006), to seek refuge in today's current mediterranean areas (Axelrod 1975). Furthermore, current xerophitic taxa present in southern Europe (*Quercus ilex*-type, *Phillyrea*, *Olea*, *Cistus*, *Pistacia*) rose by degrees to higher frequencies and appear to have been the most resistant to the new climatic conditions (Suc 1984). A similar situation occurred in California, when the intensification of Mediterranean-type climates derivated in a dominance of evergreen *Quercus* species (Millar 2012). That is, the archetypal evergreen sclerophyllous oak that lives under arid conditions in contemporary communities of mediterranean regions might have its origin in pre-mediterranean lineages that existed during the Tertiary in a belt around North America and Eurasia, where the climate was warm and wet (Herrera 1992; Verdú et al. 2003; Valiente-Banuet et al. 2006).

Not only evergreen sclerophyllous *Quercus* species like *Q. ilex* or *Q. chrysolepis* inhabit the current northern hemisphere mediterranean regions. Deciduous oaks like *Q. faginea* or *Q. lobata* also coexist under a Mediterranean-type climate regime. According to Mai (1991), there was also a temperate broad-leaved deciduous vegetation (the so-called Arctotertiary Flora) that occupied high northern latitudes both in North America and in Eurasia in the Early Tertiary. Ancestors of the current white and black oaks (*Quercus* and *Lobatae* groups respectively, according to Denk and Grimm 2009) seem to have been part of this flora (Nixon 2002; Kovar-Eder 2003). During the Middle and Late Tertiary, a global trend in the decrease of temperatures caused a shift of this deciduous vegetation towards southern latitudes (Mai 1991; Kovar-Eder 2003). This implied the appearance of arctotertiary elements in the mediterranean areas (Mediterranean Basin and California) prior to the onset of Mediterranean-type climates (Axelrod 1973; Kovar-Eder 2003; Ackerly 2009). The current existence of deciduous oak types living under Mediterranean-type climates might have implied a subsequent adaptation of the arctotertiary oak ancestors to the spreading dry Mediterranean regime, in opposition to the pre-adapted evergreen sclerophyllous (Suc 1984; Blumler 1991; Nixon 2002; Peguero-Pina et al. 2016a).

5.5 Co-occurrence of Evergreen and Winter-Deciduous Oaks Under Mediterranean-Type Climates

Nowadays there are two types of oaks cohabiting in the current mediterranean regions of the Northern Hemisphere: the more acknowledged broadleaved evergreen sclerophyllous type and the scarcely recognized winter-deciduous malacophyllous type (Barbero et al. 1992; Castro-Díez and Montserrat-Martí 1998; Damesin et al. 1998; Mediavilla and Escudero 2004). The former has been commonly associated to the archetype of the Mediterranean-type climate (Walter 1985) and has served to develop the concept of "convergent evolution" between mediterranean regions (Mooney and Dunn 1970; Cody and Mooney 1978; Shmida 1981; Shmida and Whittaker 1984). Nevertheless, evergreen sclerophyllous oaks can also be found in non Mediterranean-type climates (Poudyal et al. 2004; Zhang et al. 2005; Singh et al. 2006), being important components of the vegetation in several regions (such as Arizona, Mexico, Northern Pakistan, India, Nepal and Afghanistan) that receive their precipitation maximum in summer (Blumler 2005 and references therein). The winter-deciduous type, in spite of being less recognized, can also be considered an adaptation-type to extreme summer drought as effective as the evergreen type (Blumler 1991; Barbero et al. 1992; Scarascia-Mugnozza et al. 2000). In fact, winter-deciduous oaks are often dominant in many areas of California (Barbour 1988; Griffin 1971, 1973, 1977; Vankat 1982) and the Mediterranean Basin, especially in the eastern areas (Barbero et al. 1992; Radoglou 1996). Moreover, some species such as *Q. ithaburensis*, can be found in some of the most arid Mediterranean-subtype climates (Dufour-Dror and Ertas 2004). The coexistence of both types of oaks under mediterranean conditions was already reported by Schimper (1903) and further evidenced by other authors (Tognetti et al. 1998; Nardini et al. 1999; Manes et al. 2006; Montserrat-Martí et al. 2009). Appendix 5.1 shows how both types of *Quercus* species can be found in locations under genuine Mediterranean-type climates (e.g. Csa, Csb), to such an extent that is hard to assume a single prototype of "mediterranean oak". Furthermore, this co-occurrence has been pointed out to be one of the distinctive features of the Mediterranean Biome (Givnish 2002; Baldocchi et al. 2010; Noce et al. 2016).

Although a general explanation for the coexistance of both mediterranean oaks - evergreen and winter deciduous—deserves more research, it can be broadly assumed that such co-occurrence is a matter of local variations in climate or in soil characteristics. Thus, mediterranean deciduous oaks would be able to dominate those sites characterized by relatively higher precipitation and lower winter temperature (Chabot and Hicks 1982) and/or those zones with heavier and fertile soils that can maintain high levels of water content (Salleo et al. 2002; Moreno et al. 2011). On the contrary, evergreen oaks may be restricted by competition or predominate in the dryest places and/or coarse, rocky, or infertile substrates that promote low soil water contents (Blumler 1991; Gasith and Resh 1999; Manes et al 2006; Di Paola et al. 2017). In this sense, *Q. ilex*, as a genuine evergreen oak, which can grow on a variety of climates and substrata in its western distribution area in the Mediterranean Basin (Barbero et al. 1992), does not succesfully compete with co-occurring winter deciduous *Quercus* species, such as *Q. faginea*, except under conditions of poor soil development (Peguero-Pina et al. 2015). According to this idea, the coexistance of both types of mediterranen oaks would be possible in a heterogeneous environment, while homogenity should imply the dominance of a single type in the landscape.

As a consequence of the paramount importance of soil water in the occurrence of evergreen or winter deciduous oaks in areas under Mediterranean-type climates, the human action along the centuries constitutes a crucial factor explaining most present distribution patterns (Dufour-Dror and Ertas 2004). Records of human influence on *Quercus* species distribution are already found since 6000 BP, where forests of deciduous oaks started to be replaced by evergreens in the western Mediterranean Basin (Follieri et al. 1988; Reille and Pons 1992; Riera-Mora and Esteban-Amat 1994). These replacements may be due to erosion and degradation of the substrate that harms deciduous species for the benefit of evergreens. An example of this replacement was reported by Peguero-Pina et al. (2015), who found that stands of deciduous *Quercus subpyrenaica* living in scarce soils suffered premature withering, while evergreen *Q. ilex* stands remained intact. In a different way, the distribution of an oak species can also be favoured by human activity if the species is usable by humans. This is the case of *Q. suber*, which niche in the western Mediterranean Basin has been enlarged at the expense of *Q. canariensis* (Urbieta et al. 2008). That is, human activity can influence the distribution pattern at local or regional scale. Moreover, Sánchez de Dios et al. (2009) predicted that climatic and environmental changes, accelerated by human activity, could reduce dramatically the submediterranean territories of the Iberian Peninsula and, therefore, the extension of the submediterranean oaks *Q. pubescens*, *Q. pyrenaica* and *Q. faginea*. These authors stated, with appropiate prudence, that the reduction could have a serious impact on biodiversity.

The co-occurrence of evergreen and winter-deciduous *Quercus* species offers the opportunity for studying the role of these two phenological patterns from a functional perspective (Kikuzawa 1991), especially under the environmental constrains imposed by the Mediterranean-type climate (Montserrat-Martí et al. 2009). Effectively, winter-deciduousness is the prevailing leaf habit in areas under Temperate climate with cold winters (Walter 1985; Breckle 2002), as leaf shedding before the winter colds has been long recognized as a mechanism for evading the physiological drought imposed by low temperatures (Schimper 1903). The absence of leaves during the winter months restricts the photosynthetic activity to the warm period, which is expected to be favourable enough for a winter-deciduous oak in terms of carbon gain. Peguero-Pina et al. (2016a) described the typical environmental conditions imposed by the Temperate climate using *Q. robur* as a reference. Under the summer conditions that characterize the genuine habitat for this species —mild temperatures, high water availability and low VPD_{max}—oaks can support a high leaf area, both in terms of leaf size and number of leaves per shoot. This fact enables this species to achieve a very favourable carbon gain during the vegetative period. As soil and atmospheric drought increase in the transition to the areas under Mediterranean-type climate at lower latitudes, these genuine temperate species are substituted by the so-called "submediterranean" oaks (Himrane et al. 2004; Sánchez de Dios et al. 2009). In the Iberian Peninsula, the existance of certain summer aridity (as the balance between precipitation and evapotranspiration) and VPD_{max} close to 3 kPa are considered of paramount importance for explaining the substitution of *Q. robur* by *Q. faginea*, with much lower values of leaf area (Peguero-Pina et al. 2016a). However, in spite of the outstanding benefits for reducing transpiration in drier climates, a reduced leaf area also diminishes the photosynthetic potential of this species. Effectively, the severe reduction in transpiring leaf area, together with the possible existance of water strees during the vegetative period in the submediterranean areas (Corcuera et al. 2004a, b; Pasho et al. 2011; Peguero-Pina et al. 2015), causes an overall decrease of the net assimilation during the vegetative period (Peguero-Pina et al. 2016a). Thus, the vegetative period may be reduced to a few weeks in spring and autumn (Montserrat-Martí et al. 2009), when the environmental conditions do not limit the photosynthetic activity (Abadía et al. 1996; Mediavilla and Escudero 2003). The reduction in photosynthesis during the driest days of the summer has been reported in several winter deciduous mediterranean *Quercus* species of Europe or North America, or "submediterranean" oaks in the term here used (Damesin and Rambal 1995; Xu and Baldocchi 2003; Poyatos et al. 2008; Siam et al. 2009). In fact, Montserrat-Martí et al. (2009) suggested that a phenological advance in bud bursts and a delayed leaf shedding in mediterranean winter deciduous oaks would allow a prolongation of the leaf life-span, while increasing the chance of suffering early frosts. This idea, which may be of paramount relevance to understand the ecology of transitional areas under Mediterranean-type climates, deserves further empirical evidences taken into account a wider spectrum of winter deciduous mediterranean oaks.

The access to deep water in well-developed soils has been proposed as the ecological mechanism explaining the survival of winter-deciduous oaks in areas with certain summer aridity (Esteso-Martínez et al. 2006; Urbieta et al. 2008; Moreno et al. 2011). A further increase in the summer aridity due to climatic factors (González-Rebollar et al. 1995; del Río and Penas 2006) or to an exacerbated water scarcity due to soil degradation (Blondel and Aronson 1995; Salleo et al. 2002; Corcuera et al. 2005a; Peguero-Pina et al. 2014) is widely associated to a substitution of winter-deciduous oaks by evergreen species in areas under Mediterranean-type climates.

However, under such circumstances, even evergreen oaks are not always able to keep a maximum photosynthetic activity throughout the vegetative period. A significant reduction in carbon gain during the driest period, as compared to the maximum values recorded in unstressed specimens, has been widely reported in these species (Faria et al. 1998; Limousin et al. 2010a; Vaz et al. 2010). The shortening of the growing season due to summer drought is accentuated by the low temperatures during the winter, which severely impaired the photosynthetic activity in mediterranean evergreen oaks (García-Plazaola et al. 1999; Corcuera et al. 2005a, b). In fact, such combination of two stress periods (summer drought and winter frost) and two favourable seasons (spring and autumn) can be considered the real essence of the Mediterranean-type climate concerning the ecophysiology of the woody vegetation (Mitrakos 1980; Castro-Díez and Montserrat-Martí 1998; Montserrat-Martí et al. 2009). The need for splitting the vegetative activity into these two favourable periods (in the sense given by Kikuzawa 1991) may confer a functional advantage to evergreen species such as *Q. ilex*, as leaves from the previous year can start to contribute to the overall plant carbon gain as soon as the temperature rises in the spring (Corcuera et al. 2005b), well before the average bud burst date reported in this species (Castro-Díez and Montserrat-Martí 1998; Ogaya and Peñuelas 2003). Under such circumstances, 1-year old leaves in *Q. ilex* (i.e. those produced in the previous spring) positively contribute to the whole canopy carbon assimilation (Escudero and Mediavilla 2003), even though different ageing processes would reduce the photosynthetic rate (Niinemets et al. 2005) as compared to that recorded for current-year leaves (Mediavilla and Escudero 2003). As a possible consequence of their long lifespan, it has been suggested that evergreen leaves need to be structurally reinforced to withstand biotic or abiotic damages (Turner 1994; Poudyal et al. 2004), which has been proposed as the reason explaining their higher LMA (Gonzalez-Zurdo et al. 2016). As further described, this idea could establish a link between the condition of sclerophyllous and evergreen in these mediterranean oaks.

5.6 Sclerophylly in Mediterranean Oaks. Functional Explanations

Evergreen sclerophyllous woody plants have long been regarded as one of the most typical components of the mediterranean-type vegetation (Walter 1985). Among them, different oak species with sclerophyllous leaves are principal constituent of such woodlands and/or shrubsland in many areas of the Mediterranean Basin (Barbero et al. 1992, Bozzano and Turok 2003), Southern Asia (Meher-Homji 1973) or California (Goulden 1996). This convergence in foliar attributes among different areas under Mediterranean-type climates around the world (Barbour and Minnich 1990; Cowling and Witkowski 1994; Cowling et al. 1996) has induced the search for a common factor explaining such response in woody plants and their dominance under the ecological conditions imposed by this climate. However, the functional association between sclerophylly and the Mediterranean-type climate has been controversial since the first geobotanical synthesis of the distribution of vegetation types over the earth's surface.

In this sense, Schimper (1903) wrote one of the earliest physiological approximations to the "botanical geography" at a global scale. This author, when described the so-called "mild temperate districts with winter rain and prolonged summer-drought" (areas under Mediterranean-type climate, as we consider in this chapter), stated that they were inhabited by "evergreen xerophylous" trees and shrubs with stiff, thick and leathery leaves (sclerophyllous). This seminal proposal, that established a link between sclerophylly (condition of having hard leaves) and xeromorphism (morphological or anatomical responses to live in a dry or physiologically dry habitat), has been subjected to a continuous discussion (e.g. Loveless 1962; Beadle 1966; Seddon 1974; Oertli et al. 1990; Turner 1994; Salleo and LoGullo 1990; Nardini et al. 1996; Salleo et al. 1997). In fact, the controversy about the adaptive role of sclerophylly continues in the present century (Lamont et al. 2002; Read et al. 2006; Verdú et al. 2007; Rubio de Casas et al. 2009; Read et al. 2016).

The existence of a summer drought period has long been recognised as a severe ecological constrain for plant life in Mediterranean-type climates (Peñuelas et al. 1998; Joffre et al. 2007; Medrano et al. 2008; Nardini et al. 2014; Niinemets and Keenan 2014). This evidence has induced to consider that hard leaves of the mediterranean woody plants can be a functional adaptation to withstand water stress during the drought period (Mooney and Dunn 1970; Levitt 1980; Mooney 1982; Savé et al. 1999; Sardans and Peñuelas 2013) in a climate with a severe seasonality, in accordance with Schimper's interpretation of sclerophylly (Lamont et al. 2002).

In order to perform a deep analysis of sclerophylly, it should be taken into account that this is a term that refers to a textural feature of the leaves. So, sclerophyllous leaves would be harder, stronger, tougher or stiffer than "soft" or malacophyllous leaves (Read and Sanson 2003). To obtain a quantitative value for such leaf trait, different mechanical analysis through fracture tests must be done (Lucas and Pereira 1990; Choong et al. 1992; Aranwela et al. 1999; Westbrook et al. 2011; Onoda et al. 2011). However, one of the more widespread index of sclerophylly used in ecophysiological studies is the ratio of leaf dry weight to leaf area (Cowling and Campbell 1983; Witkowski and Lamont 1991; Cowling and Witkowski 1994; Salleo and LoGullo 1990; Salleo et al. 1997; Groom and Lamont 1999), or Leaf Mass per Area (hereafter LMA) as the most common designation. Although this ratio does not directly reflect mechanical properties but mass allocation and related processes (Onoda et al. 2011; Read et al. 2016), it may be used as a good proxy of them, as different studies have revealed (Choong et al. 1992; Edwards et al. 2000; Read and Sanson 2003; Westbrook et al. 2011). So, most of the studies concerning sclerophylly in mediterranean plants do really deal with differences in LMA.

Although some authors have given a threshold value for sclerophylly in terms of LMA values (Salleo and LoGullo 1990; Flexas et al. 2014), it must be better considered as a relative index for comparative studies. In this sense, Corcuera et al. (2002) compared different *Quercus* species grown under the same environmental conditions, in terms of leaf morphology (leaf area and LMA) and parameters derived from pressure-volume curves. In this study, the different species were a priori grouped in three different categories according to their phytoclimate sensu Walter (1985), namely: (i) mediterranean oaks (*Q. agrifolia, Q. chrysolepis, Q. coccifera, Q. ilex* ssp. *ilex, Q. ilex* ssp. *ballota* = *Q.ilex* ssp. *rotundifolia, Q. suber*); (ii) transitional nemoro-mediterranean oaks (*Q. cerris, Q. faginea Q. frainetto, Q. pyrenaica*); and (iii) nemoral oaks (*Q. alba, Q. laurifolia, Q. nigra, Q. petraea, Q. robur, Q. rubra, Q. velutina*). The average LMA value for mediterranean oaks in this study reached 150 g m^{-2}, which is higher than 120 g m^{-2}, the threshold value suggested by Flexas et al. (2014) for true sclerophytes. Mean LMA value of transitional nemoro-mediterranean oaks was close to 80 g m^{-2}, while the nemoral oaks showed a mean value lower than 70 g m^{-2}. Villar and Merino (2001) also reported values for different *Quercus* species from 3 different habitats, with a strong correspondence with the classification by Corcuera et al. (2002). Thus, the values for evergreen oaks from "xeric" or "mesic" mediterranean forests (*Q. agrifolia*, *Q. coccifera*, *Q. rotundifolia* = *Q. ilex* ssp. *rotundifolia*, *Q. suber*) were higher—mean ca. 160 gm^{-2}—than those for deciduous oaks from "mesic" mediterranean forests (*Q. douglasii*, *Q. faginea*, *Q. keloggii*, *Q. lobata*, *Q. pyrenaica*)—mean ca. 112 g m^{-2}—and from temperate forests (*Q. rubra*), with a LMA around 83 g m^{-2}. Furthermore, the mean LMA value for the whole species of the "xeric mediterranean forest" in the study of Villar and Merino (2001) was the higher among the different habitats compared, besides the leaf construction cost when expressed on an area basis (g glucose m^{-2}).

In fact, values close or higher than 200 g m^{-2} have been reported for *Q. ilex* (Bussotti et al. 2002; Villar-Salvador et al. 2004; Serrano et al. 2005) or *Q. coccifera* (Castro-Díez et al. 1997; Vilagrosa et al. 2003a; Rubio de Casas et al. 2009). At the other end of the scale, values close or lower than 50 g m^{-2} have been obtained for European temperate oak species, such as *Q. petraea* and *Q. robur* (Burghardt and Riederer 2003, Withington et al. 2006; Giertych et al. 2015). It can be accepted therefore that evergreen oaks living in areas under Mediterranean-type climate have sclerophyllous leaves, assuming LMA and sclerophylly as closely linked, especially when compared with winter deciduous from Temperate climates where water availability during the vegetative period is not limited. Besides the rise of LMA in oaks from climates with water scarcity during the vegetative period, other studies have evidenced a within-species increase of LMA for *Q. ilex* provenances in the drier sites of their natural range (Castro-Díez et al. 1997; Bussotti et al. 2002; Ogaya and Peñuelas 2007; Niinemets 2015), which may support the Schimper's proposal. Do these empirical evidences confirm the seminal idea proposed by Schimper (1903), implying the association between sclerophylly and xeromorphism in these evergreen *Quercus* species? This question, which has been recognised as one of the most ancient controversies in ecology (Groom and Lamont 1997; Lamont et al. 2002), is far from being answered with a common consensus, as the functional association between sclerophylly and xeromorphism in the mediterranean oaks has not been perfectly established up until now.

A widely accepted alternative hypothesis to explain sclerophylly comes from some classic studies of the flora and vegetation of the Australian Continent (Beadle 1953; 1954) where the existence of soils with low phosphorus content are common (Kooyman et al. 2017). Since the first studies by Beadle (1953; 1954), the identification between xeromorphism and sclerophylly in the vegetation growing in low fertile soils has been a matter of discussion. Thus, Beadle (1966) used the term "low fertility xeromorphs" for these plant species with sclerophyllous leaves of the Australian Flora. This author considered that these Australian sclerophylls were not xerophytes, as the sclerophylly in these plants was not a response to water scarcity but an accentuation of some leaf features through a reduction in leaf area. Loveless (1961) defined sclerophyllous leaves according to their relative proportion of fibre content to protein content. This ratio, the so called "Loveless sclerophylly index", gives an estimation of the ratio of cell wall to cell content and were found to be higher under limited phosphate uptake. Furthermore, Loveless (1962) offered a functional interpretation to the link between phosphorus availability and sclerophylly. So, sclerophylly could be interpreted as the consequence (or the "expression", in the words of the author) of a plant metabolism adapted to low phosphate uptake due to the importance of this macronutrient for protein synthesis. Since this interpretation, other papers have supported the idea that sclerophylly is a response to oligotrophic soils (Sobrado and Medina 1980; Gonçalves-Alvim et al. 2006). However, this proposal fails when explaining the existence of sclerophyllous vegetation in some areas under Mediterranean-type climate with a relatively high nutrient content (cf. Verheye and de la Rosa 2005; Sardans and Peñuelas 2013) or the prevalence of mesophytic (low LMA) vegetation in different forest ecosystems where a limited phosphorus supply has been found to reduce productivity (Knox et al. 1995; Sardans et al. 2004; Zhu et al. 2013). Furthermore, Cavender-Bares et al. (2004), when studied the habitat differentiation among co-occurring *Quercus* species growing in North Central Florida (USA), did not find a significative correlation between soil phosphorus content and LMA in these oaks with contrasting leaf traits.

Besides summer drought or the presence of poor soils, winter temperature is another outstanding factor of paramount importance for the physiognomy of the Mediterranean Flora, as firstly reported by Mitrakos (1980, 1982). Effectively, previous phytoclimatical classification considered Mediterranean-type climates as affected by dry summer but mild winters (cf. Schimper 1903), disregarding the existence of many continental areas where the vegetative activity of genuine Mediterranean evergreen plants, such as *Q. ilex*, is limited by low temperatures during winter (Corcuera et al. 2005b). One direct effect of low temperatures could be a rise in LMA in such evergreen oaks, e.g. *Q. ilex* and *Q. suber*, as suggested by Oliveira and Peñuelas (2002), Ogaya and Peñuelas (2007) and González-Zurdo et al. (2016). In our opinion, the proposal that LMA can be explained by low winter temperatures needs to be supported by physiological studies offering a mechanistic explanation for this empirical evidence.

Niinemets (2016) explored some functional reasons that could explain a higher LMA in plants under cold climates, considering the effect of whole leaf thickening and/or cell wall thickening, both factors positively modifying LMA (Westbrook et al. 2011). The higher amount of water content per area in thicker leaves would affect the freezing rate of the leaf tissues or the incidence of ice formation during the freeze-thaw cycles in the stability of cell membranes. Moreover, thicker cell walls could constitute a mechanical advantage during the process of cell desiccation while freezing. Effectively, evergreen mediterranean oaks show values of cell wall thickness higher than those reported in winter deciduous *Quercus* species (Peguero-Pina et al. 2016a, 2017a, b). Furthermore, both data in Fig. 5.4 and in Table 5.1 reflect that leaf thickness is also especially high in the Mediterranean evergreen oaks when a large set of *Quercus* species from different habitats but growing under the same environmental condition (common garden) are compared. However, those *Quercus* species grouped as "evergreen arid" in Fig. 5.4, such as

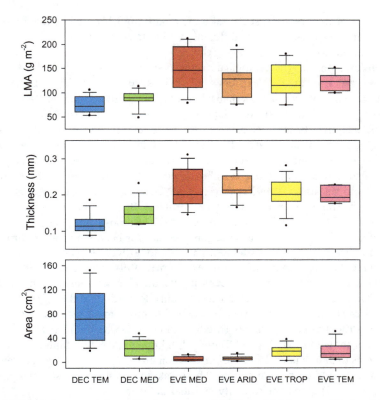

Fig. 5.4 Box-plot representation of leaf mass area (LMA), leaf thickness and leaf area from a survey of 76 oak species living in a common garden under a humid temperate climate (Jardín Botánico de Iturrarán 43° 13′N, 02 °01′W, 70 m a.s.l., Gipuzkoa, Spain). Species were classified according with their leaf habit and climate origin in one of the following groups: deciduous temperate (DEC TEM), deciduous mediterranean (DEC MED), evergreen mediterranean (EVE MED), evergreen arid (EVE ARID), evergreen tropical (EVE TROP) and evergreen temperate (EVE TEM) (See Apendix 5.1 for species classification)

Q. miquihuaensis or *Q. hypoxantha*, also showed a high LMA in spite of not being affected by winter frost, neither in their natural habitats nor in the common garden where they are grown. In the same way, the variation range in LMA found by Cavender-Bares et al. (2004) in 17 co-ocurring *Quercus* species of North Central Florida cannot be considered a differential response to low winter temperatures.

Cell wall thickness and whole leaf thickness can also be considered a physiological response to cope with the conditions imposed by the summer in Mediterranean-type climates (Peguero-Pina et al. 2016b, 2017a), as will be discussed below. In fact, Ogaya and Peñuelas (2007), when compared a large set of *Q. ilex* population of Catalonia (Northeastern Spain), concluded that LMA was higher "in the drier sites and especially in the colder sites". Assuming that the same leaf attributes can guarantee a good performance both during the summer drought and the winter frost periods, such leaf traits modifying LMA can be interpreted as a concomitant response to the different stress factors affecting the evergreen mediterranean flora.

Therefore, one single ecological factor does not seem to provide an unambiguous response to the existence of oak species with high LMA under Mediterranean-type climate, perhaps because sclerophylly may be a syndrome (Fonseca et al. 2000; Read and Sanson 2003; Read et al. 2016) in response to different adaptive factors (Nardini 1996; Read et al. 2006). In this sense, an interesting proposal interpreted the sclerophylly as a sum of anatomical features aimed to increase the leaf mechanical resistance to withstand damages (Turner 1994; Read and Sanson 2003; Westbrook et al. 2011; Read et al. 2016), due to abiotic factors (Niklas 1999; Poudyal et al. 2004) and/or biotic factors, with a particular reference to insect herbivory (Grubb 1986; Wright and Vincent 1996; Ribeiro and Basset 2007; Peeters et al. 2007; Barbosa and Fernandes 2014). According to this proposal, the selection pressure for becoming tough, through an increased sclerophylly, should be higher in long-lived leaves and in habitats imposing a limited supply of resources (Turner 1994). Effectively, it has been reported a LMA value for evergreens higher than those measured in deciduous species (John et al. 2017 and references therein). Regarding to this, a shift in LMA has been explained through an increased leaf reinforcement by a higher accumulation of structural carbohydrates (Mediavilla et al. 2008). From our own results, and as it is shown in Fig. 5.4 and Table 5.1, a higher LMA was found in evergreen oak species than in winter deciduous ones when growing in a common garden, with independence of their natural habitat. These data may confirm the seminal proposal of Turner (1994) about sclerophylly and leaf defence.

This explanation about sclerophylly, when applied to the mediterranean oaks, would be able of integrate, in fact, most of the circumstances affecting the leaves of such species: (i) a limited access to water and nutrients, (ii) a short vegetative period imposed by cold winters and dry summers and (iii) the need for extending the leaf life-span to exploit the most favourable climatic period through the year (Corcuera et al. 2005b). However, this interpretation of sclerophylly, in spite of its apparent robustness, faces with the finding that the defence against phytophagus insects can

Table 5.1 Mean values ± SE of leaf mass area (LMA), leaf thickness and leaf area from a survey of 76 oak species living in a common garden under a humid temperate climate

	LMA	Thickness (mm)	Area (cm^2)
DEC TEM	77 ± 5 a	0.119 ± 0.007 a	73.8 ± 12.2 a
DEC MED	89 ± 4 a	0.149 ± 0.09 a	23.2 ± 3.6 b
EVE MED	148 ± 12 b	0.218 ± 0.015 b	6.0 ± 1.0 c
EVE ARID	123 ± 11 b	0.225 ± 0.010 b	6.7 ± 1.1 c
EVE TROP	122 ± 10 b	0.203 ± 0.011 b	17.8 ± 2.9 b
EVE TEM	122 ± 5 b	0.199 ± 0.006 b	18.4 ± 3.9 b

DEC deciduous, *EVE* evergreen, *TEM* temperate, *MED* Mediterranean, *ARID* arid, *TROP* tropical. Letters indicate statistically significant differences (Tukey test, $P < 0.05$)

also be also achieved by the accumulation of deterrent substances in the leaf tissues (Feeny 1970; Levin 1976; Bernays 1981; Gonçalves-Alvim et al. 2006; Read et al. 2009), independently of leaf toughness. Even more, some authors have suggested that such chemical defence should be more significant for the leaf defence than the mechanical reinforcement (Kouki and Manetas 2002; Onoda et al. 2011).

From a global perspective, sclerophylly might be better considered as a non-specific response affecting different vegetation types under particular constraining factors, which may help to understand the complexity of this long-standing debate. It may worth focussing again the debate to the sclerophyllous oak species under Mediterranean climate, revisiting the Schimper (1903) proposal from a physiological more than a phytogeographical perspective. With this aim, a detailed examination of LMA as a proxy of sclerophylly must be done.

A high LMA may be the consequence of a high leaf thickness, a high accumulation of dense tissues or both (Onoda et al. 2011). Figure 5.6 evidences how leaf thickness increases with LMA in the whole set of oak species, with evergreen oaks showing the higher values of LMA and leaf thickness (Table 5.1; Fig. 5.5). A thicker leaf might have some functional advantages, especially under the conditions imposed by the Mediterranean-type climate. In this sense, Peguero-Pina et al. (2016b) found (i) a lower maximum stomatal conductance ($g_{s,max}$), (ii) a higher leaf thickness but (iii) an equal net photosynthesis per area (A_N) in *Q. coccifera* plants living under mediterranean conditions when compared with the same species grown under a humid temperate climate. The reduced $g_{s,max}$ is a consequence of a partial closure of the stomatal pore with epicuticular waxes (Roth-Nebelsick et al. 2013), and allows this species to significantly reduce the water losses under the high VPD experienced during summer in a Mediterranean-type climate. The goal of keeping a constant A_N values in spite of a drastic reduction in the CO_2 uptake via the stomata is achieved through a higher

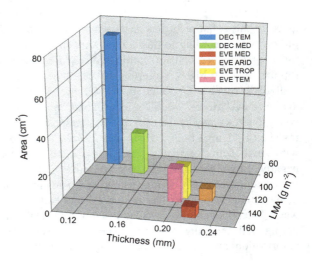

Fig. 5.5 Representation of the mean values of leaf mass area (LMA), leaf thickness and leaf area of the data from Fig. 5.4 in a 3D diagram. Note that evergreen (EVE) oak groups are clearly segregated from deciduous (DEC) groups; mediterraneans (MED) seems to be quite segregated from temperate (TEMP) and tropical (TROP). Climatic classification and species involved in this figure are shown in Apendix 5.1

amount of protein content per area in a thicker mesophyll. In addition, a study comparing different *Q. ilex* provenances (Peguero-Pina et al. 2017b) showed that A_N is positively correlated with leaf thickness, as a greater leaf volume in thicker leaves were associated with a greater leaf nitrogen content and a higher maximum velocity of carboxylation ($V_{c,max}$). This idea is in accordance with Niinemets (2015), who also reported increases in leaf nitrogen content per area with leaf thickness and LMA in *Q. ilex*. Moreover, Peguero-Pina et al. (2017a) have found that the leaf thickening of evergreen mediterranean oaks could be a way for achieving high A_N values through the improvement of the mesophyll conductance to CO_2 by different anatomical modifications.

Thus, leaf thickening in mediterranean evergreen oaks would yield adaptive advantages in terms of CO_2 uptake and water losses under dry atmospheres, although implying a concomitant rise in LMA, and besides the above mentioned effect on frost resistance in the cold locations. Anyway, evergreen arid, evergreen tropical and evergreen temperate oak species do not differ statistically in LMA with evergreen mediterranean oaks in a common garden survey (Table 5.1; Fig. 5.4). How long the arguments given above for the evergreen mediterranean species can be extrapolated to the other groups or not deserves further investigation.

The shape of the fitting curve in Fig. 5.6 suggests a saturation response of leaf thickness as LMA increases, which could indicate a further rise in density by dry mass accumulation per area once the highest thickness values are raised. It should be noted that the highest individual values recorded for LMA are those found in mediterranean evergreen species (Fig. 5.6). Moreover, the small plot in the upper left side of Fig. 5.6 shows that the mediterranean evergreen group has the highest LMA mean value (in spite of the lack of statistical significance) among the evergreen oaks from all habitats considered. This fact may be interpreted as a higher densification of leaf tissues in mediterranean evergreen oaks. Which are these other ways of densification of leaf tissues and how can be interpreted as a way for coping with the conditions imposed by Mediterranean-type climates?

Oertli (1986) gave a functional interpretation of sclerophylly in terms of resistance to the collapse of leaf cells when subjected to dehydration far below their turgor loss. According to this author, smaller cells or cells with thicker walls (both inducing a higher density) could withstand substantial negative turgor pressure (higher than 1.6 MPa), buffering the negative effects of such collapse of the cell wall (cytorrhysis). Oertli et al. (1990) proposed that the mechanism based on avoiding cytorrhysis would be considered "an adaptation to suboptimal moisture conditions", as may be the case of Mediterranean plants. This idea has been recently revisited by Ding et al. (2014), who insisted on the benefits of developing tissues with small cell size under water stress conditions. According to these authors, negative turgor pressure may be considered as a desiccation avoidance mechanism in plants of arid zones besides osmotic adjustment, without the need for investment of metabolic resources. The negative impact of leaf cell collapse under negative water potential may be associated to the physical damage of the cell membrane during cell buckling and polyelectrolyte leakage (Farrant 2000). Such strong deformation in the cell shape during dehydration was evidenced by cryo-SEM

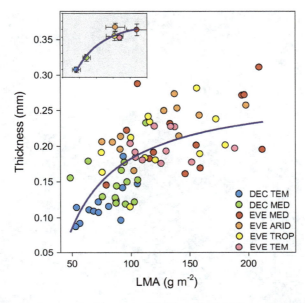

Fig. 5.6 Relationship between leaf mass area (LMA) and leaf thickness from a survey of 76 oak species living in a common garden under a humid temperate climate (Jardín Botánico de Iturrarán 43° 13′N, 02° 01′W, 70 m a.s.l., Gipuzkoa, Spain). Equation: Polynomial Inverse First Order, $y = 0.28 - (9.7/x)$ $R^2 = 0.42$ $P < 0.0001$. *DEC* deciduous, *EVE* evergreen, *TEM* temperate, *MED* mediterranean, *TROP* tropical. Mean values ± SE of these parameters are plotted in the upper left side of the figure. Note that most of the evergreen mediterranean oaks increase LMA without an increase in thickness. Species involved in this figure are shown in Appendix 5.1

imaging of cell shape changes during dehydration in *Q. muehlenbergii* (Sancho-Knapik et al. 2011; Zhang et al. 2016) in *Q. rubra*, two deciduous oaks from Temperate climates with low LMA values. Figure 5.7 shows a group of mesophyll cells of *Q. muehlenbergii* below the turgor loss point, with evident signals of cell buckling and tension of the plasmodesmata. The specific resistance to such deformation in the mesophyll cells could explain the differences in the leakage of polyelectrolytes with leaf desiccation between temperate (Epron and Dreyer 1992) and mediterranean *Quercus* species (Vilagrosa et al. 2010).

Another possible link between leaf sclerophylly, by increasing leaf density, and the ability for coping with drought may be established through the functional importance of the bulk elastic modulus at full turgor (ε_{max}) in the plant water relationships (Niinemets 2001). A higher ε_{max} would also allow a higher drop in leaf water potential for a given symplasmic water loss, with evident benefits in water limited conditions (Corcuera et al. 2002; Singh et al. 2006). Salleo and LoGullo (1990) and Burghardt and Riederer (2003) reported a high positive correlation between LMA and ε_{max}, which should indicate that denser tissues could also be more rigid in terms of Pressure-Volume relationships. This idea fails to fully explain all the empirical evidences found in the Mediterranean woody flora, with

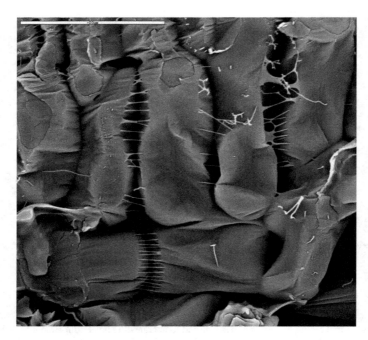

Fig. 5.7 Detailed Cryo-SEM micrograph of the spongy mesophyll in a leaf of *Q. muehlenbergii* below turgor loss point. Note the evidenced cell buckling and the tension in plasmodesmata. Scale bar 20 μm

some findings that suggest a lack of correlation between LMA and ε_{max} (LoGullo et al. 1986; Nardini et al. 1996). Nevertheless, two circumstances should taken into consideration: (i) to the extent of our present knowledge, ε_{max} depends on the rigidity of mesophyll cell walls (Nardini et al. 1996 and references therein) and (ii) LMA may also be dependent on other anatomical features and not only the consequence of a cell wall thickening (inducing a higher rigidity) in the living tissues of the mesophyll. So, in order to further study this relationship, LMA and ε_{max} data obtained from the literature for different *Quercus* species were plotted (Fig. 5.8), yielding a non-linear relationship between both parameters. In this plot, the species are grouped into temperate and mediterranean, and these last grouped again into evergreen and winter deciduous. In this relation, the low LMA values of deciduous temperate oaks are associated with low values of ε_{max}. The other extreme is occupied by the evergreen mediterranean oaks, which show the highest values for these parameters. As LMA reaches higher values, further increases in this parameter does not imply a proportional increase in ε_{max}, suggesting the existence of other factors that may influence LMA without further effects on the PV relationships.

In order to determine a mechanistic trade-off between LMA and ε_{max}, a leaf feature that simultaneously affects both parameter should be explored. Such feature could be the thickness of the mesophyll cell wall, which reaches a higher value

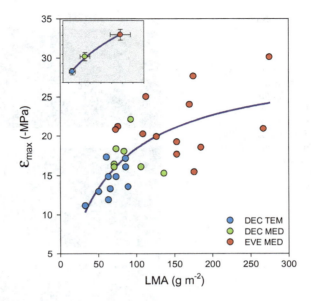

Fig. 5.8 Relationship between leaf mass area (LMA) and cell wall maximum bulk modulus of elasticity (ε_{max}) for oak species belonged to one of the following groups: deciduous temperate (DEC TEM), deciduous mediterranean (DEC MED) and evergreen mediterranean (EVE MED). Equation: Hyperbola, $y = 29.6 * x/(61.0 + x)$ $R^2 = 0.50$ $P < 0.0001$. Mean values ± SE of these parameters are plotted in the upper left side of the figure. Data obtained from Salleo and LoGullo (1990), Nardini et al (1996), Salleo and Nardini (2000), Corcuera et al (2002), Burghardt and Riederer (2003), Villar-Salvador et al. (2004), Peguero-Pina et al (2016a). The species classification into groups according to Appendix 5.1

in evergreen mediterranean oaks than in winter deciduous oaks (Peguero-Pina et al. 2016a, 2017a). When the cell wall thickness of mesophyll cells was plotted against ε_{max} (Peguero-Pina et al. 2017b), a strong positive correlation between both parameters was found, which open the possibility for further studies to confirm this relationship and its possible role in the ecophysiological response of plants to Mediterranean habitats.

Some studies have analysed the functional role of other dense tissues that contribute to increase LMA. Thus, Salleo et al. (1997) studied the possible role of dense tissues as an apoplastic water source that could "migrate" to the symplast in diurnal cycles. In fact, the role of these dense tissues (sclerenchyma or colenchyma) in terms of their contribution to the inner water pathway among leaf tissues (Heide-Jorgensen 1990) should be considered, as they are especially developed in leaves of evergreen mediterranean oaks (Fig. 5.9).

Finally, a high vein density would also increase the LMA due to the anatomical features of the bundle and its associated tissues. Scoffoni et al. (2011) found that leaves with a higher major vein density (the sum of the density of the first, second and third order veins) showed higher resistance to drought-induced cavitation,

Fig. 5.9 Transverse section of the mesophyll leaf of the deciduous temperate *Quercus robur* (**a**), deciduous mediterranean *Q. broteroi* (**b**), evergreen mediterranean *Q. ilex* subsp. *rotundifolia* (**c**), evergreen arid *Q. hintoniorum* (**d**) and evergreen tropical *Q. rugosa* (**e**). Bright areas mainly coincide with thick cell walls in epidermal structures, vascular bundles and bundle sheath extensions

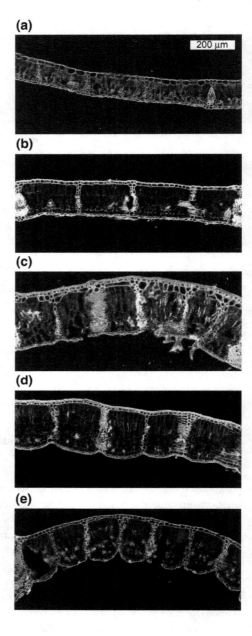

which was confirmed by Nardini et al. (2012a) in different *Quercus* and *Acer* species. In the same direction, Nardini et al. (2012b) found that a higher LMA paralleled a higher investment of dry matter to transport ability in sun leaves with respect to shade leaves of *Q. ilex*. A higher major vein density in the populations of

Q. ilex from xeric areas within its distribution range was evidenced by Peguero-Pina et al. (2014). The transversal sections in Fig. 5.9 also indicate a lower vein density in *Q. robur*, as a representative temperate oak, than in the evergreen congeneric species, especially when compared with *Q. ilex*. This fact implicitly may relate a higher LMA derived from a higher vein density with a better tolerance to the water stress experienced under the arid conditions of the mediterranean summer.

In essence, the sclerophyllous leaves of mediterranean oaks can be regarded as a "non-specific" response to many ecological factors inducing such response, in spite of the existence of several links among LMA and critical traits for plants living in dry climates (such as leaf thickness, wall thickness of the mesophyll cells, ε_{max} or major veins density). This fact may ultimately confirm or support the ancient interpretation of the sclerophylly in mediterranean areas given by Schimper (1903).

5.7 Leaf Size Reduction in Mediterranean Oaks

From a global perspective, it has been long considered that an increase of dryness, in the sense of its effect on plant physiology, promotes the reduction in leaf size (Schimper 1903). Therefore, an association between small leaves and dry habitats is usually reported (Dolph and Dilcher 1980; Givnish 1987; Sultan and Bazzaz 1993; Gibson 1998; McDonald et al. 2003; Ackerly 2004; Peguero-Pina et al. 2014). In this way, plant species should have the greatest leaf size in the tropics, with a decrease towards the subtropics, an increase towards warm temperate forests, and a decrease towards the poles (Givnish 1976). In accordance with the leaf size pattern of this author, a study of leaf morphology in Europe vegetation revealed that small leaves appeared mostly in warm climates of South Europe while large leaves were most abundant in cool climates of the North (Traiser et al. 2005). Focusing on *Quercus*, Corcuera et al. (2002) found the same trend: oak species from Mediterranean-type climates had smaller leaf areas than oaks from Nemoral or Temperate climates. When comparing a broader set of oaks and climatic types (Figs. 5.4 and 5.5), the evergreen mediterranean species and those from arid climates show the smallest leaf size, supporting the global trend of leaf reduction under drought conditions. In fact, leaf reduction in mediterranean oaks has been proposed as one of the key traits that allow these species to withstand water deficit (Baldocchi and Xu 2007; Peguero-Pina et al. 2014).

A reduction in leaf size is associated with higher major vein densities (MVD) (Fig. 5.10) and higher leaf resistance to hydraulic conductivity losses (Scoffoni et al. 2011). These authors stated that, for all else being equal, a higher major vein density would allow a more safety water transport due a higher redundancy. Thus, when the linear relationship between leaf area and leaf water potential at 80% loss of conductivity (PLC80) provided by Scoffoni et al. (2011) is applied to the mean leaf area values of Fig. 5.5, temperate oaks would have a leaf PLC80 value of −0.7 MPa, deciduous mediterranean and tropical oaks a value around −2 MPa, whereas the PLC80 values for evergreen mediterranean and arid

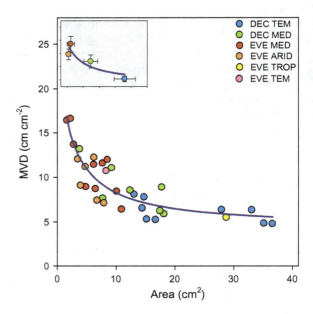

Fig. 5.10 Relationship between leaf area and mayor vein density (MVD). Equation: Polynomial Inverse Third Order, $y = 4.38 + (45.4/x) - (72.6/x^2) + (52.6/x^3)$ $R^2 = 0.76$ $P < 0.0001$. *DEC* deciduous, *EVE* evergreen, *TEM* temperate, *MED* mediterranean, *TROP* tropical. Mean values ± SE of these parameters are plotted in the upper left side of the figure. Note that evergreen mediterranean oaks are slightly on top of the evergreen arid oaks. Species involved in this figure are shown in Appendix 5.1. Data obtained from Sack et al. (2012) and Peguero-Pina et al. (2014, 2016a)

oaks would be −6 MPa. The empirical results in Peguero-Pina et al. (2015), when compared cavitation in leaves of *Q. ilex* and *Q. subpyrenaica* (evergreen and winter deciduous respectively), seems to confirm this idea. Thus, the larger leaves from deciduous temperate group would have less resistance to drought-induced cavitation than the other four groups of oaks studied. Evidently, further studies are needed to empirically confirm this hypothesis.

The development of small leaves either by reducing leaf width and length or by increasing lobation (McDonald et al. 2003) has been also related with high levels of radiation and high temperatures (Fonseca et al. 2000; Ackerly et al. 2002). A small leaf has a thinner leaf boundary layer that facilitates a sensible heat loss enabling a more rapid convective cooling in warm climates (Baldocchi and Xu 2007; Nicotra et al. 2008; Vogel 2009; Yates et al. 2010). This phenomenon has been study by Baldocchi and Xu (2007) by comparing the temperate *Quercus alba* and the Mediterranean *Q. douglasii*. These authors showed that the smaller leaves of *Q. douglasii* remain up to 2 °C cooler than the larger leaves of *Q. alba* under high light conditions. In this chapter we compare *Q. robur*, a deciduous temperate oak against *Q. faginea* and *Q. pyrenaica*, two deciduous mediterranean oaks (Fig. 5.11). The computation of the leaf energy balance in the three species (www.landflux.org/resources/Ecofiz_Tleaf_K2_v3.xls, Kevin Tu, U.C. Berkeley), for a windspeed of 1 m s^{-1} and 700 W m^{-2} of short-wave radiation, yielded a certain leaf temperature (T_{Leaf}) for a given leaf length (*d*) and environmental conditions (air temperature and air relative humidity). This computation show that an increase in *d* implies a concomitant increase in T_{Leaf} under the environmental conditions

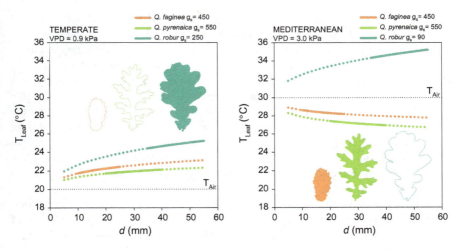

Fig. 5.11 Relationship between leaf temperature (T_{Leaf}) and leaf length in the direction of the wind (d) for three *Quercus* species (*Q. faginea*, *Q. pyrenaica*, *Q. robur*) and two climatic conditions: temperate (T = 20 °C RH = 60%) and mediterranean (T = 30 °C RH = 30%). Continuous line indicates the observed d range of the species. Filled leaf images indicate the real presence of the species in that climate. Values of stomatal conductance (g_s, mmol H_2O m^{-2} s^{-1}) for *Q. pyrenaica* were obtained from Mediavilla and Escudero (2004); for *Q. robur* from Grassi and Magnani (2005) and Peguero-Pina et al. (2016a); and for *Q. faginea* from Acherar and Rambal (1992), Mediavilla and Escudero (2004) and Peguero-Pina et al. (2016a). VPD: vapour pressure deficit (kPa). T_{Air}: air temperature (°C). T_{Leaf} obtained applying the Leaf Energy Balance Program by Kevin Tu, U.C. Berkeley (www.landflux.org/resources/Ecofiz_Tleaf_K2_v3.xls) for windspeed = 1 m s^{-1}, short-wave radiation = 700 W m^{-2}

typical of a Temperate climate in the three species under consideration. According to Dreyer et al. (2001), this fact may benefit the leaf functioning in terms of leaf photosynthetic capacity by reaching a better thermal regime for some processes related to carbon gain. In a mediterranean environment, *Q. robur* would follow the same trend, which may cause a serious overheating problem if T_{Leaf} achieve 40 °C, value considered to be near the upper limit of viable temperatures (Baldocchi and Xu 2007; and references therein). This problem could be mitigated by reducing drastically its leaf size. On contrary, *Q. faginea* and *Q. pyrenaica* show a phenomenon of leaf cooling which is higher as d increases (Fig. 5.11). This cooling effect is due to the combination of i) the high stomatal conductance (g_s) of these species and ii) the high VPD that allows a high water evapotranspiration. This effect on leaf temperature should persist as soil moisture kept available and VPD did not reach very high values. As we assume that water vapour concentration inside the leaf is temperature dependent (Nobel 1991), a lower leaf temperature associated to a possible leaf cooling should reduce the driving force for transpirational water losses and contribute to water saving.

If soil deficit (Peguero-Pina et al. 2015) or high VPD values (Mediavilla and Escudero 2004) induce stomatal closure, the water losses by transpiration would not be high enough to keep this effect of leaf cooling, and a leaf overheating could be developed (Gibson 1998). In fact, the results from the model evidence a small overheating ($\Delta T = + 1.2$ °C) in the small-leaved evergreen mediterranean *Q. ilex* ($d = 10$ mm) under "mediterranean conditions" (VPD = 3 kPa, $T_{air} = 30$ °C) due to the low g_s (160 mmol H_2O m^{-2} s^{-1}). Such overheating is much lower than the one calculated for *Q. robur* ($\Delta T = + 4.1$ °C), partly due to the higher value of g_s in *Q. ilex* (160 mmol H_2O m^{-2} s^{-1}) than in *Q. robur* (90 mmol H_2O m^{-2} s^{-1} at VPD = 3 kPa and $T_{air} = 30$ °C, or "mediterranean conditions") and also due to the smaller leaf size in the evergreen oak ($d = 40$ mm in *Q. robur* and $d = 10$ mm in *Q. ilex*).

Finally, the reduction in leaf size also decreases the total leaf area per shoot, which should induce a shift in the leaf specific hydraulic conductivity (LSC, ratio of stem conductivity to leaf area, kg m^{-1} s^{-1} MPa^{-1}), increasing the ability for supplying water to the transpiring leaves in dry atmospheres (Martínez-Vilalta et al. 2009; Peguero-Pina et al. 2011). This is the case of *Q. ilex* subsp. *rotundifolia*, which lives in more xeric habitats than *Q. ilex* subsp. *ilex*. A higher LSC was found in *Q. ilex* subsp. *rotundifolia* (Peguero-Pina et al. 2014), mainly due to the adjustment in single leaf area, more than by reducing the number of leaves per shoot, shoot sapwood area or the specific conductivity of the xylem (Ks, kg m^{-1} s^{-1} MPa^{-1}). Something similar was found when the Temperate *Q. robur* was compared with the Mediterranean *Q. faginea* (Peguero-Pina et al. 2016a). In spite of the large difference found in hydraulic conductivity (K_h, kg m s^{-1} MPa^{-1}) between both species (ca. seven times higher in *Q. robur*), *Q faginea* reached similar values of LSC by reducing fourfold its leaf size. That is, an adjustment between hydraulic conductivity and whole-shoot leaf area contributes to withstand Mediterranean conditions, especially high VPD (see Fig. 5.1f).

5.8 Leaf Stomatal Conductance. Water Saving and Water Spending Strategies

The stomatal conductance (g_s, mmol H_2O m^{-2} s^{-1}) of a plant estimates the rate of transpiration or water loss through the leaf stomata of that plant per unit leaf area. This parameter is a function of the stomatal density, stomatal size and stomatal aperture (Franks and Beerling 2009), with maximum values when stomata are fully open ($g_{s,max}$). Considering that $g_{s,max}$ occurs when a particular plant is living under non-stressed conditions (well watered soil, low VPD and saturating irradiance) (Epron and Dreyer 1993), we have compiled $g_{s,max}$ values from the literature for different *Quercus* species. When comparing these $g_{s,max}$ values (Fig. 5.12), we observe that the mean value found in the deciduous mediterranean group (470 ± 35 mmol H_2O m^{-2} s^{-1}) is statistically higher ($p < 0.05$) than those found

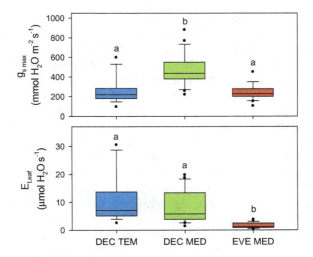

Fig. 5.12 Box-plot representation of maximum stomatal conductance ($g_{s,max}$) and transpiration of the whole leaf (E_{Leaf}) at VPD = 0.9 kPa (T = 20 °C RH = 60%) of deciduous temperate (DEC TEM), deciduous mediterranean (DEC MED) and evergreen mediterranean (EVE MED) oaks. Letters indicate statistically significant differences in the Mean (Tukey test, $P < 0.05$). Species involved in this figure are shown in Apendix 5.1. Data obtained from Acherar and Rambal (1992), Epron and Dreyer (1993), Rico et al. (1996), Filella et al. (1998), Morecroft and Roberts (1999), Nardini et al. (1999), Fotelli et al. (2000), Xu and Baldocchi (2003), Mediavilla and Escudero (2003, 2004), Grassi and Magnani (2005), Tognetti et al. (2007), Johnson et al. (2009), Maire et al. (2015), Peguero-Pina et al. (2015, 2016a, b, 2017a, b)

in the other two groups analyzed (266 ± 30 mmol H_2O m^{-2} s^{-1} for deciduous temperate and 243 ± 14 mmol H_2O m^{-2} s^{-1} for evergreen mediterranean). The high value found in deciduous mediterranean oaks can be considered counterintuitive, assuming that these oak species are able to survive in dry climates. However, when the whole leaf transpiration (mmol H_2O s^{-1} per leaf) is considered, the evident reduction in leaf size in these deciduous mediterranean species compensate for the higher $g_{s,max}$, yielding a similar value of whole leaf tranpiration than the one calculated for temperate oaks, although still higher than that for evergreen mediterranean species (Fig. 5.12). According to Mediavilla and Escudero (2003), these different strategies can be interpreted in terms of the concepts of water saver and water spender (Levitt 1980).

On the one hand, the deciduous mediterranean oaks would be considered water spender as they have higher values of $g_{s,max}$, higher transpiration rates (Fig. 5.12) and, furthermore, a lower stomatal sensitivity to atmospheric dryness (Mediavilla and Escudero 2003). In fact, while *Q. faginea* starts to close stomata above 3 kPa, *Q. ilex* subsp. *rotundifolia* starts the stomatal closure at 2 kPa (Fig. 5.13). A more efficient conductive system that allows a faster water supply to the leaves (Nardini et al. 1999; Sisó et al. 2001; Lo Gullo et al. 2005; Kröber et al. 2014) is needed to maintain the high transpiration in atmospheres with moderate or high VPD

(Nardini et al. 1999). The high transpiration rates may be coupled with a high photosynthetic rate and growing capacity under Mediterranean conditions (Levitt 1980; Blumler 1991; Mediavilla and Escudero 2003; Flexas et al. 2014) in spite of their lower leaf longevity. Nevertheless, being a water spender could lead to a more rapid consumption of soil water reserves (Mediavilla and Escudero 2004). The depletion of the soil water content should imply the development of more negative soil water potential that will induce the reduction of $g_{s,max}$ through the effect on the leaf water potential at predawn (Fig. 5.13; Acherar and Rambal 1992), with the consequent reduction in photosynthesis. For this reason, this strategy may require the existance of a large enough source of soil water during the vegetative period associated to deep and well-developed soils (Peguero-Pina et al. 2015).

On the other hand, the evergreen mediterranean oaks would be comparatively considered as water savers, due to the more conservative water-use characteristics: lower $g_{s,max}$, lower transpiration rates and higher stomatal sensitivity to VPD (Figs. 5.12 and 5.13). This ability for maintaining a better control of the water losses, besides their higher tolerance to low soil water availability (Acherar and Rambal 1992; Blumler 1991), would be the in the base of the present predominance of evergreen mediterranean oaks in many areas under Mediterranean-type climates (Archibold 1995). However, the handicap of this group would be their lower photosynthetic rates (Levitt 1980; Mediavilla and Escudero 2003; Flexas et al. 2014), which make them less competitive under non-stress conditions. Nevertheless, these rates might be slightly compensated with the longer leaf life span that allows these species to start the photosynthetic activity as soon as

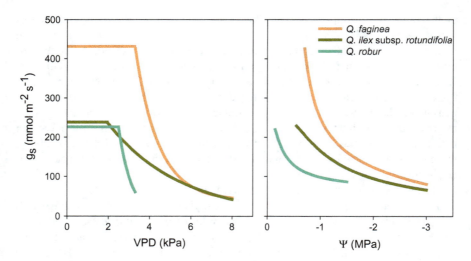

Fig. 5.13 Maximum stomatal conductance at a given level of vapour pressure deficit (VPD) and soil water potential (Ψ) for *Q. faginea*, *Q. ilex* subsp. *rotundifolia* and *Q. robur*. g_s: stomatal conductance. Data obtained from Mediavilla and Escudero (2004) and Grassi and Magnani (2005)

conditions are favourable. Together with these evergreen mediterranean oaks, other evergreen species living in xeric climates might share the same water saving behaviour. This is the case, for example, of *Q. turbinella*, an evergreen species living in semiarid non-Mediterranean environments of south-western USA that also shows low leaf conductances (Ehleringer and Smedly 1988; Ehleringer and Phillips 1996).

5.9 Mechanism for a Further Reduction of Stomatal Conductance: Waxes and Trichomes

Some studies have revealed that the water saving strategy in evergreen mediterranean oaks goes further the common mechanisms to reduce the stomatal conductance. Some species, such as *Quercus coccifera*, which is able to survive in the most arid Mediterranean-subtype climates (see Fig. 5.3), can develop leaf epidermal structures to reduce $g_{s,max}$ (Roth-Nebelsick et al. 2013). These structures are accumulations of epicuticular waxes in the protruding surface of the guard cells (Fig. 5.14). Waxes encrypt stomata, simulating a roof, and drastically reduce the effective pore dimensions. Consequently, this reduction in the pore area (from 30 to 5 μm^2) causes a permanent strong decrease in $g_{s,max}$ which leads to a high decrease in transpirational water loss. Furthermore, Peguero-Pina et al. (2016b) showed that *Q. coccifera* only develops these structures when habits in xeric areas (with high VPD). Under a Temperate climate the waxes produced by the leaf do not reduce the pore dimension as much as in Mediterranean-type climates, suggesting a plasticity of stomatal protection in relation to contrasting plant growth climatic conditions. Epicuticular waxes covering the stomatal rim have also been reported in other *Quercus* species, such as *Q. arizonica* (Scareli-Santos et al. 2013) or in *Q. infectoria* (Panahi et al. 2012a) and different taxa of the *Q. brantii* complex (Panahi et al. 2012b). Although the role of such epicuticular waxes it is not a matter of discusion in these studies, the habitat of these species may suggest a similar role than in *Q. coccifera*.

Other leaf epidermal structures commonly found in many species of the genus *Quercus* are trichomes (Hardin 1979; Morales et al. 2002; Panahi et al. 2012a, b; Scareli-Santos et al. 2013; He et al. 2014). These structures can be found scarcely distributed along the ribs, on the axil tufts, throughout the leaf blade or creating a continuous and dense layer both on the adaxial (upper) and abaxial (lower) side of the leaves in many oak species (Fig. 5.14). Trichomes are described in a widely set of plants with several and different functions, such as plant defence against herbivores and pathogens, mechanical protection to abrasion, or absorption of water and nutrients (Wagner et al 2004; Bickford 2016; and references therein). Among these functions, the role of a dense layer of trichomes in the abaxial surface (where the stomata are present) has been considered in this genus a way for reducing the water flow to the atmosphere, and so, a xeromorphic adaptation to dry environments (He et al. 2014). Hardin (1979) indicated that individuals of the same oak

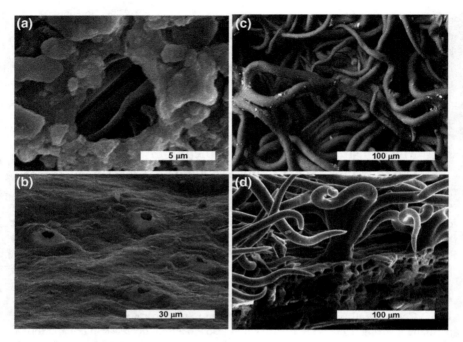

Fig. 5.14 Stomatal encryption with epicuticular waxes in *Quercus coccifera* (**a, b**). Details of the leaf abaxial trichome layer in *Q. ilex* supsp. *rotundifolia* (**c, d**). Upper (**a, c**) and lateral (**b, d**) views

species showed a denser leaf trichome layer when living in drier habitats. Moreover, from the comparison of the leaf abaxial pubescence in a wide set of oaks from different climate-types (Fig. 5.15), it can be concluded that: (i) oak species from habitats with some kind of aridity (during summer or winter) seem to be frequently pubescent beneath and (ii) the frequency of oak species showing a dense layer of trichomes in the abaxial side is higher in mediterranean and arid evergreen oaks. Could this result support the idea of a dense abaxial leaf pubescence as a xeromorphic trait in *Quercus*?

To the extent of our knowledge, and according to previous studies concerning this topic, the possible role of trichomes in water saving are at present merely speculative (Wagner et al. 2004). The trichome covering has proposed to modify the leaf boundary layer, which could increase the leaf diffusion resistance to water loss in the abaxial side (Ripley et al. 1999; Benz and Martin 2006). However, other authors rejected the effect of these structures in the boundary layer of leaves (Johnson 1975) or, though recognising a possible influence, suggested a minor effect on whole leaf transpiration (Ehleringer and Mooney 1978). More recently, it has been shown that the increment in the boundary layer resistance seems to be negligible compared with the resistance imposed by other components of the diffusion pathway (mesophyll, stomata, cuticle). In fact, Roth-Nebelsick et al. (2009) concluded that trichomes inside the stomatal crypt of *Banksia ilicifolia* had no

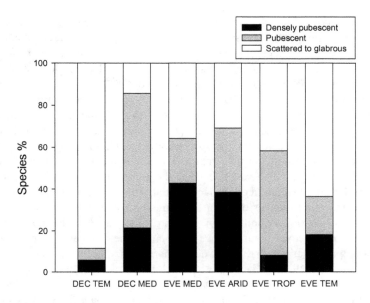

Fig. 5.15 Percentage of species in each oak group with the abaxial leaf surface densely pubescent (black), pubescent (grey) or scattered to glabrous (white). *DEC TEM* deciduous temperate, *DEC MED* deciduous mediterranean, *EVE MED* evergreen mediterranean, *EVE ARID*: evergreen arid, *EVE TROP* evergreen tropical and *EVE TEM* evergreen temperate. Species involved in this figure are shown in Appendix 5.1. Data was obtained from literature taxonomical descriptions (Negi and Naithani 1995; Felger et al. 2001; Fralish and Franklin 2002; Arizaga et al. 2009), personal observations and web pages (http://oaks.of.the.world.free.fr/ http://www.aiapagoeta.com/ http://forest.jrc.ec.europa.eu/ http://www.efloras.org)

significant influence on transpiration. Moreover, Schreuder et al. (2001) indicated that the trichome layer could stimulate the transition from a laminar to a turbulent flow implying an increase in the water transport conductance. In addition, a study on *Proteaceae* did not find the same relation between the presence of trichomes covering the stomata and aridity, suggesting that a dense hair layer covering the areas with stomata are not clearly associated to dry climates (Jordan et al. 2008). With these scarce and contradictory studies, more research is needed (e.g. taking into account the trichomes density) in order to clarify if the abaxial trichome layer (i) decreases the water conductance (helping the evergreen mediterranean oaks as water savers), (ii) is totally insignificant, or (iii) increases the water conductance (favouring the water-spender role of deciduous mediterranean).

However, some positive effects of leaf trichomes for plants living under dry and sunny climates have been proposed. The role of trichomes in the optical properties of the leaf has received an especial attention (Johnson 1975). In this way, leaf pubescence, which is able to reduce leaf absorptance of solar radiation by increasing leaf reflectance (Ehleringer 1984), causes a reduction of leaf temperatures and water losses (Ehleringer et al. 1976, 1981; Ehleringer and Björkman 1978; Ehleringer and Mooney 1978; Ehleringer 1981; Pérez-Estrada et al. 2000). This effect would make sense when trichomes densely cover the adaxial face of the leaf, due to the predominance of the solar incidence in the upper leaf side, but also in the abaxial face to avoid excessive absorption of ground radiation (Jonhson 1975). Trichomes located adaxially have also been considered as protective structures against excessive UV-B radiation (Grammatikopoulos et al. 1994; Karabourniotis and Bornman 1999; Savé et al. 2000; Manetas 2003) and excessive photosynthetic active radiation (Morales et al. 2002). These authors found that the presence of trichomes in the adaxial leaf surface of *Q. ilex* subsp. *ballota* was an important structural mechanism to reduce the susceptibility to photodamage preserving the photochemistry apparatus, as also reported in other mediterranean plant species (Galmés et al. 2007). This feature, besides the contribution of trichomes in water uptake through the leaf surface (Fernández et al. 2014), make adaxial trichomes of *Q. ilex* subsp. *ballota* leaves efficient structures under mediterranean xeric conditions. That is, adaxial trichomes can give protection against the excess of light energy due to high levels of irradiance and help in the foliar water uptake in sites with scarce precipitations that might not recharge the soil during summer.

Brewer and Smith (1997) suggested that repelling moisture away from the epidermis and, consequently, from stomatal pores, would reduce the interference of a layer of a liquid water with CO_2 uptake through the stomata. This implies that the presence of an abaxial trichome layer in oak species such as *Quercus ilex* subsp. *rotundifolia*, which seems to be highly hydrophobic (Fernández et al. 2014), may serve as a way to ensure the free movement of CO_2 inside the leaf in habitats prone to induce accumulation of water, such as foggy mountain areas. This may also explain the existence nowadays of such features in oaks that live under rainy habitats. Taking this into account and knowing that evergreen oaks existed before mediterranean xeric conditions, the hypothetic abaxial trichomes pre-adaptation to moist environments could serve afterwards in mediterranean *Quercus* history as a xeromorphic feature to cope with dry atmospheres. This is a matter that should be tested in further research.

5.10 Minimum Conductance. The Role of Cuticle and Leaf Size

The cuticle is a thin continuous membrane that terrestrial plants have developed to act as a barrier against uncontrolled water loss (Riederer and Schreiber 2001). This membrane consists of a polymer matrix (cutin), polysaccharides and associated solvent-soluble lipids (cuticular waxes) (Holloway 1982; Jeffree 1996), which make the cuticle efficient for water saving, translucent for photosynthetic radiation absorption, flexible and self-healing. Despite the cuticle role as an important water barrier, there is still a movement of water through the cuticle between the outer cell wall of the epidermis and the atmosphere adjacent to the plant that gives to the cuticle certain water permeability (Riederer and Schreiber 2001). This movement is based in a simple diffusion process along a gradient of the chemical potential of water. Individual water molecules are sorbed at one interface, follow a random path across the cuticle in a mostly lipophilic chemical environment and are desorved at the other interface (Frisch 1991; Riederer and Schreiber 2001; Kerstiens 2006). In this sense, the degree to which cuticles transmit water is called "cuticular permeance" (P). When the cuticle surface has stomata, as happens in leaves, the term often used is "minimum conductance" (g_{min}, mol m^{-2} s^{-1}) (Kerstiens 1996), which could be defined as the conductance of the leaf surface when stomata are completely closed.

According to Riederer and Schreiber (2001), xeromorphic plants growing in Mediterranean-type climates have lower values of cuticular permeance than deciduous plant species from Temperate climates. Focusing on *Quercus*, Fig. 5.16 reveals only a slight difference of g_{min} values between evergreen mediterranean and deciduous temperate oaks. Among deciduous oaks, no difference was noticed between mediterranean and temperate species. However, when the total leaf area per shoot is considered in order to calculate the amount of water transmitted through the cuticle at shoot level ($E_{min\ shoot}$, mmol s^{-1}), the differences between groups are magnified (Fig. 5.16). $E_{min\ shoot}$ values of evergreen mediterranean oaks are much lower than deciduous oaks values; and even, the values of the deciduous mediterranean oaks are lower than those deciduous oaks from Temperate climates. In this sense, although g_{min} in mediterranean oaks is only slightly lower than in temperate oaks, the lower total leaf area per shoot of mediterranean oaks induces a lower amount of water loss per shoot, which may benefit these species for withstanding water deficits.

5.11 Resistance to Cavitation

The constrain for plant growth or even survival imposed by summer drought in Mediterranean-type climates has promoted the search for common xeromorphic features in the flora under this climate (Thompson 2005; Medrano et al. 2008; Nardini et al. 2014). However, the existence of multiple strategies to withstand the summer drought among mediterranean woody plants, which therefore implies different functional traits, has been revealed in many papers (e.g. LoGullo and Salleo 1988; Vertovec et al. 2001; Ackerley 2004; Iovi et al. 2009; Bussotti et al. 2014), even when mediterranean *Quercus* species were compared (Salleo and LoGullo 1990; Salleo et al. 2002). The xylem vulnerability to drought-induced embolism (Sperry and Tyree 1988) can be explored in order to search for a single, common trait, that should confer a best performance under drought conditions in woody plant species (e.g. Bhaskar and Ackerley 2006; Pratt et al. 2007; Blackman et al. 2010). Maherali et al. (2004) compared the value of the xylem tension (MPa) inducing a loss of 50% of hydraulic conductivity (the socalled Ψ_{50}, P_{50} or even PLC$_{50}$) in 167 woody from different "vegetation types", namely "Mediterranean", "Desert", "Temperate Forest and Woodland", "Tropical Dry

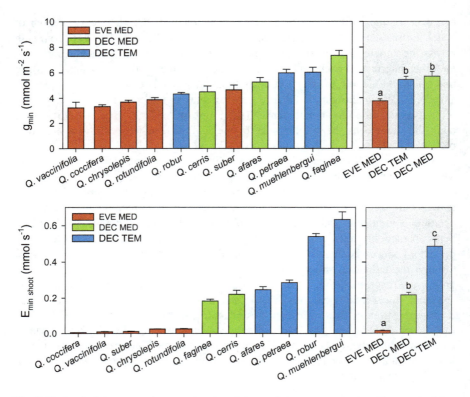

Fig. 5.16 Leaf minimum conductance (g_{min}) and shoot minimum transpiration ($E_{min\ shoot}$) of five evergreen mediterranean oaks (EVE MED), three deciduous mediterranean (DEC MED) and three deciduous temperate oaks (DEC TEM) living in a common garden under mediterranean climate conditions (CITA de Aragón 41° 39′N, 0° 52′W, 200 m a.s.l., Zaragoza, Spain). Mean values ± SE of these parameters are plotted in the right side of the figure. Letters indicate statistically significant differences (Tukey test, $P < 0.05$)

Forest" and "Tropical Rain Forest". According to these authors, plants from the "Mediterranean" vegetation type showed the lower median Ψ_{50} (ca. −5.5 MPa), followed by the group constituted by the "Desert" plants (ca. −4.5 MPa). This fact can be regarded as an indication of the noteworthy importance of this parameter in habitats affected by drought periods, such as those under Mediterranean-type climates. A wider global analysis allowed Choat et al. (2012) to find a significant relationship between the species Ψ_{50} and the mean annual precipitation in their respective origins, with a lower resistance in those woody plants from areas with higher rainfall. However, Maherali et al. (2004) reported a very high variability in Ψ_{50} within groups, including the Mediterranean group, where species with values close to −2 MPa besides other with values lower than −8 MPa are shown.

In the same way, Jacobsen et al. (2007a, b) reported a variation range of 10 MPa in Ψ_{50} among co-occurring mediterranean shrubs of the California chaparral and

Jacobsen et al. (2009) among shrubs from different Mediterranean-type regions of the earth. Nardini et al. (2014) plotted the Ψ_{50} of different Mediterranean woody species contained in the dataset of Choat el al (2012). In a similar way, besides the finding of a convergence in the mean values between geographical origins (namely the Mediterranean Basin and the California Chaparral), Nardini et al. (2014) also reported a range of variation around 8 MPa in Ψ_{50} which evidences the lack of a common response among all the mediterranean plant species in terms of this physiological threshold.

Concerning *Quercus* species, Vilagrosa et al. (2012) showed a good correlation between Ψ_{50} and the length of the aridity period according to the Gaussen-type ombrotermic diagram (see Fig. 5.1), a good proxy of the intensity of the climatic restriction to plant growth during the vegetative period (e.g. Peguero-Pina et al. 2016a). In fact, those species from the most xeric areas in the Mediterranean Basin showed a lower (more negative) Ψ_{50}, such as *Q. coccifera* (Ψ_{50} value close to −7 MPa, Vilagrosa et al. 2003b), which is able to inhabit semi-arid areas of southeastern Iberian Peninsula (up to 7 months of aridity). Pinto et al. (2012) also suggested that evergreen oaks, *Q. suber* and *Q. ilex* in their study, registered lower (more negative) Ψ_{50} values than temperate oaks. A meta-analysis using published data of Ψ_{50} for different *Quercus* species of the world revealed that Ψ_{50} in temperate species is higher (less negative) than in evergreen mediterranean (Figs. 5.17 and 5.18). However, the boxplots of the values clearly indicate a higher dispersion in these last species, evidenced by the width of the interquartile and interdecil ranges and the outliers at both extremes. This fact confirms the results by Maherali et al. (2004) and Nardini et al. (2014), and induces to consider that a lower Ψ_{50} is not a common trait in mediterranean oaks. From these data, it can be assumed the existence of evergreen mediterranean oaks with Ψ_{50} as high as or even higher than temperate species. On the contrary, it is also true that the most negative values of Ψ_{50} so far published in genus *Quercus* correspond to evergreen species inhabiting Mediterranean areas (Vilagrosa et al. 2003b; Peguero-Pina et al. 2014). The high resistance to drought-induced embolism in many of these evergreen oaks, as suggested by their Ψ_{50} values, can be regarded as an adaptation for withstanding severe droughts in the more xeric areas under Mediterranean-type climate, due to climatic but also to edaphic factors (Gil-Pelegrín et al. 2008; Peguero-Pina et al. 2015).

The high dispersion in Ψ_{50} among the mediterranean evergreen oaks (ca. 6 MPa) can also be found at within-species level. Thus, in *Q. berberidifolia*, the value of Ψ_{50} ranges from −0.7 MPa (Jacobsen et al. 2014) and −1.5 MPa (Jacobsen et al. 2007a) to −2.6 MPa (Bhaskar et al. 2007) and even −5.1 MPa (Jacobsen et al. 2007b). A wide variation range has been also published for *Q. ilex*, from ca. −2 MPa (Martínez-Vilalta et al. 2002) to −7.1 MPa in *Q. ilex* subsp. *rotundifolia* (Peguero-Pina et al. 2014). Nardini et al. (1996) considered *Q. ilex* to be "very vulnerable" to cavitation, suffering this phenomenon through the year in northeastern Italy. The high intraspecific variability in Ψ_{50} in *Q. ilex* has not a simple explanation. Pinto et al. (2012) analysed the possible influence of (i) the methodology used to obtain the vulnerability curve, (ii) the age of the plant and the growth conditions or the (iii) existence of a genetic variability in this species, as suggested by Corcuera et al. (2004a, b). The information published so far suggests that *Q. ilex* can be hardly considered a homogeneous taxon, both from an ecophysiological (Niinemets 2015; Peguero-Pina et al. 2014) or genetic perspective (Michaud et al. 1995; Lumaret et al. 2002), even within any of the presumed subspecies (Leiva and Fernández-Alés 1998; Pesoli et al. 2003; Valero-Galván et al. 2011, 2013). In fact, this evergreen oak is sorted in the lower extreme of drought tolerance, together with *Q. pedunculiflora*, among the oaks form Greece (Radoglou 1996; and references therein). The conception of this species as "very vulnerable" contrast with that concerning the most western populations of *Q. ilex* (David et al. 2007; Limousin et al. 2010b; Vaz et al. 2010; Andivia et al. 2012; Peguero-Pina et al. 2015), where this species is assumed to occupy dry areas in the Mediterranean region. A certain correspondence can be found with the phylogeography of *Q. ilex* (Lumaret et al. 2002), with a predominance of the subspecies *rotundifolia* in the western area (Iberian Peninsula, northwestern Africa), and the published data or references to the vulnerability to drought-induced embolism. So, most this variation found in this taxa may be attributable to the existence of two different ecological performance within the whole *Q. ilex* distribution area (Corcuera et al. 2004a, b; Gil-Pelegrín et al. 2008), which have adapted to contrasting habitats during their evolution.

In spite of the importance of Ψ_{50} as a proxy of the tolerance to drought-induced embolism, Urli et al. (2013) found that the water potential inducing a critical and non recuperable hydraulic failure in angiosperms was more negative than Ψ_{50} values, being close to the water potential which induces the loss of 88% of the stem hydraulic conductivity (the socalled Ψ_{88} or P_{88}). When this parameter is plotted, comparing temperate and evergreen mediterranean oaks (Fig. 5.17), it can be seen how much higher is the difference between these two groups as compared with that situation observed when Ψ_{50} is considered. The distribution of the Ψ_{88} values (Fig. 5.18) allows visualizing the difference between the two groups. In this case, and unlike the situation observed in the plots showing Ψ_{50} values (Figs. 5.17 and 5.18), the interquartile and interdecil ranges do not overlap and only two lower outliers of evergreen Mediterranean overlap with the interquartile range of temperate species. These two low values for Ψ_{88} were also reported for *Q. ilex*, the taxon with the higher variability in terms of hydraulic vulnerability performance, as above discussed. The very high (more negative) values for Ψ_{88} of mediterranean oaks, and according to Urli et al. (2013), clearly indicate an overall response of this group of *Quercus* species to withstand severe drought in their natural habitats. In fact, besides the values for Mediterranean evergreen species, Bhaskar et al. (2007) also reported a Ψ_{88} value of -10.5 MPa in *Q. sebifera*, an oak species living in dry areas of Eastern Sierra Madre (Mexico).

The published data for deciduous mediterranean oaks (namely *Q. faginea, Q. frainetto, Q. pubescens, Q. pyrenaica* and *Q. subpyrenaica*) point out that the mean value ($\pm$ SE) for Ψ_{50} and Ψ_{88} was found to be -3.4 (0.012) and -5.9 (0.062) MPa, respectively. These data would allow considering these species able to withstand a higher level of drought than the most genuine winter deciduous temperate ones, which is in accordance to their habitat climatic conditions. However, a

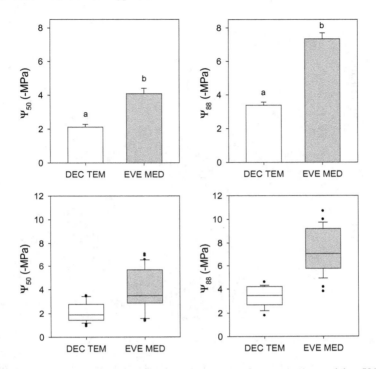

Fig. 5.17 Mean values ± SE (upper) and box-plot (lower) of stem water potential at 50% (Ψ_{50}) and 88% (Ψ_{88}) loss of conductivity for oak species belonged to the deciduous temperate (DEC TEM) and evergreen mediterranean (EVE MED) groups. Letters indicate statistically significant differences (Tukey test, $P < 0.05$). Climatic classification of the species involved in this figure is shown in Appendix 5.1. Data from Cochard and Tyree (1990), Cochard et al. (1992), Lo Gullo and Salleo (1993), Tyree and Cochard (1996), (1999), Martínez-Vilalta et al. (2002), Vilagrosa et al. (2003a), Corcuera et al. (2004a), Brodribb et al. (2003), Esteso-Martínez et al. (2006), Hacke et al. (2006), Maherali et al. (2006), Bhaskar et al. (2007), Jacobsen et al. (2007a, b), Li et al. (2008), Iovi et al. (2009), Limousin et al. (2010b), Christman et al. (2012), Nardini et al. (2012), Pinto et al. (2012), Sperry et al. (2012), Vaz et al. (2012), Paddock et al. (2013), Tobin et al. (2013), Urli et al. (2013), Jacobsen et al. (2014), Peguero-Pina et al. (2014), (2015), Venturas et al. (2016a, b). When values are not reported by authors, they were estimated from the graphical reinterpretation of the published vulnerability curves

high variability is found within this group, especially when Ψ_{88} is considered. The lowest values for this parameter have been published for *Q. pubescens* (−3.2 MPa, Tognetti et al. 1999) and *Q. pyrenaica* (−4 MPa, Corcuera et al. 2006). However, Tognetti et al. (1998) and Choat et al. (2012) reported a value much more negative for *Q. pubescens* (ca. −5.5 MPa), and Esteso-Martínez et al. (2006) found a Ψ_{88} value of −5.65 MPa for *Q. faginea*. Finally, it should be noted that the lower (more negative) values have been reported for *Q. frainetto* (−8.23 MPa, Iovi et al. 2009) and *Q. subpyrenaica* (−8.7 MPa, Peguero-Pina et al. 2015). The values reported in these two last species are very similar to the highest values found in evergreen

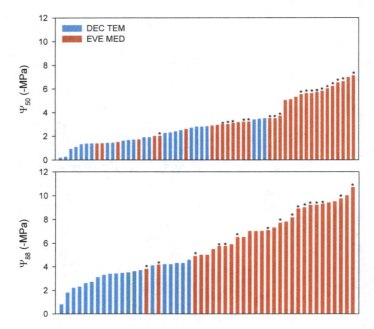

Fig. 5.18 Individual values of stem water potential at 50% (Ψ_{50}) and 88% (Ψ_{88}) loss of conductivity for deciduous temperate (DEC TEM, blue) and evergreen mediterranean (EVE MED, red) oak species. Data as in Fig. 5.17. Note the great variability in the individual values found for *Quercus ilex* (indicated in the figure by an asterisk). Climatic classification of the species involved in this figure is shown in Appendix 5.1

mediterranean oaks and would indicate an extreme resistance to suffer an irreversible hydraulic failure. The condition of "transitional species", as suggested in Corcuera et al. (2002) from the parameters derived from PV curves, can be reinforced by their performance in terms of vulnerability to drought-induced embolism.

Acknowledgements We thank Elena Martí Beltrán for her valuable help in the search of the species distribution. We thank also Francisco Garin García for his meaningful support in the recollection of oak leaves and identification of the species. We thank *Jardín Botánico de Iturrarán* for allowing us the recollection of the oak species used in the analysis on this chapter. Work of D S-K is supported by a DOC INIA contract co-funded by INIA and ESF.

Appendix 5.1

List of *Quercus* species (scientific name, infrageneric group according to Denk and Grimm (2009) and distribution) used in the figures of the chapter. Species are classified into six groups according to: (i) their leaf habit (evergreen or winter deciduous) and (ii) their climatic distribution conditions (Temperate, Tropical,

Mediterranean or Arid). For this purpose, geographical distribution coordinates for each species were obtained from herbarian data (Appendix 5.2) and overlapped on the climatic Köppen map. Köppen categories and zonobiomes sensu Walter (in brackets) are classified in four main groups: (i) *temperate*, without dry season (green); (ii) *tropical*, dry winter (blue); (iii) *mediterranean*, dry summer (red); and (iv) *arid*, arid (orange). For each group, represented by a particular colour, Köppen categories are listed according to their respective relevance.

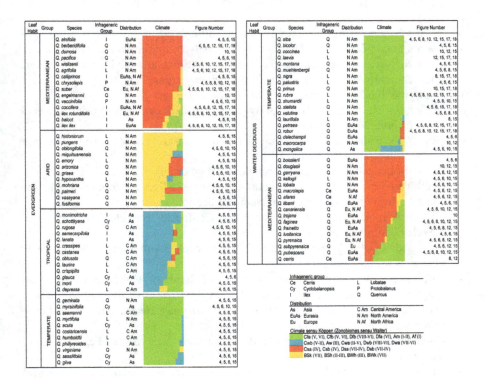

Appendix 5.2

List of data sources for species distribution.

- Archbold Biological Station (www.archbold-station.org)
- California State Parks (https://www.parks.ca.gov)
- CNPS Inventory Database (www.rareplants.cnps.org/simple.html)
- Delta State University (www.deltastate.edu)- Harvard University Herbaria (https://huh.harvard.edu/pages/digital-resources,kiki.huh.harvard.edu/databases)

- Desert Botanical Garden Herbarium Collection (https://www.dbg.org/desert-plant-research-center)
- Eastern Michigan University Herbarium (www.emich.edu/biology/facilities/herbarium.php)
- Fairchild Tropical Botanic Garden Virtual Herbarium (www.virtualherbarium.org)
- Global Biodiversity Information Facility (GBIF) (http://www.gbif.org)
- Herbarium WU Institute of Botany, University of Vienna (http://herbarium.univie.ac.at)
- Illinois Natural History Survey (www.inhs.illinois.edu)
- Intermountain Herbarium Utah State University (http://intermountainbiota.org)
- Instituto Nacional de Tecnología Agraria y Alimentaria. INIA (http://wwwx.inia.es/herbario/herbarioweb/default.asp?tabla=quercus)
- Kathryn Kalmbach Herbarium (www.gbif.org/dataset)
- Mountains Restoration Trust (www.mountainstrust.org-New York Botanical Garden (https://www.nybg.org)
- New York Botanical Garden Vascular Plant Database (http://sciweb.nybg.org/science2/hcol/allvasc/index.asp)
- Royal Botanical Garden Edinburgh (www.rbge.org.uk)
- Southwest Environmental Information Network-SEINet Arizona-New Mexico Chapter (http://swbiodiversity.org/seinet/collections)
- The Jepson Herbarium, University of California Berkeley (http://ucjeps.berkeley.edu/interchange)
- Plant Atlas. USF Water Institute, University of South Florida (http://www.plantatlas.usf.edu/)
- Tall Timbers Research Station University of Southern Mississippi Herbarium (http://www.herbarium.bio.fsu.edu)—UCLA Herbarium (https://sites.lifesci.ucla.edu/eeb-herbarium/)—University of Arizona Herbarium (https://cals.arizona.edu/herbarium/content/specimens)
- University of British Columbia Herbarium (www.biodiversity.ubc.ca/museum/herbarium)
- University of Florida Herbarium (https://www.floridamuseum.ufl.edu/herbarium)

References

Abadía A, Gil E, Morales F, Montañés L, Montserrat G, Abadía J (1996) Marcescence and senescence in a submediterranean oak (*Quercus subpyrenaica* E.H. del Villar): photosynthetic characteristics and nutrient composition. Plant, Cell Environ 19:685–694

Acherar M, Rambal S (1992) Comparative relations of four Mediterranean oak species. Vegetatio 99–100:177–184

Ackerly DD, Knight CA, Weiss SB, Barton K, Starmer KP (2002) Leaf size, specific leaf area and microhabitat distribution of woody plants in a California chaparral: contrasting patterns in species level and community level analyses. Oecologia 130:449–457

Ackerly DD (2004) Adaptation, niche conservatism, and convergence: comparative studies of leaf evolution in the California chaparral. Am Nat 163:654–671

Ackerly DD (2009) Evolution, origin and age of lineages in the Californian and Mediterranean floras. J Biogeogr 36:1221–1233

Andivia E, Carevic F, Fernández M, Alejano R, Vázquez-Piqué J, Tapias R (2012) Seasonal evolution of water status after outplanting of two provenances of Holm oak nursery seedlings. New For 43:815–824

Aranwela N, Sanson G, Read J (1999) Methods of assessing leaf-fracture properties. New Phytol 144:369–393

Archibold OW (1995) Ecology of world vegetation. Chapman & Hall, London

Arizaga S, Martínez-Cruz J Salcedo-Cabrales M, Bello-González MA (2009) Manual de la biodiversidad de encinos michoacanos. Semarnat, INE, Mexico

Atjay GL, Ketner P, Duvigneaud P (1979) Terrestrial primary production and phytomass. In: Bolin B, Degens ET, Kempe S, Ketner P (eds) The global carbon cycle, SCOPE Report 13. Wiley, UK, pp 129–181

Axelrod DI (1973) History of the Mediterranean ecosystem in California. In: Di Castri F, Mooney HA (eds) Mediterranean type ecosystems. Origin and structure. Springer, Berlin, pp 225–277

Axelrod DI (1975) Evolution and biogeography of Madrean-Tethyan sclerophyll vegetation. Ann Mo Bot Gar 62:280–334

Axelrod DI (1977) Outline history of California vegetation. In: Barbour MG, Major J (eds) Terrestrial vegetation of California. John Wiley, New York, pp 139–193

Axelrod DI (1989) Age and origin of Chaparral. The California Chaparral: paradigms revisited (ed. SC Keeley) Natural History Museum of Los Angeles County, Los Angeles, pp 7–19

Baldocchi DD, Xu L (2007) What limits evaporation from Mediterranean oak woodlands—the supply of moisture in the soil, physiological control by plants or the demand by the atmosphere? Adv Water Resour 30:2113–2122

Baldocchi DD, Ma S, Rambal S, Misson L, Ourcival JM, Limousin JM, Papale D (2010) On the differential advantages of evergreenness and deciduousness in Mediterranean oak woodlands: a flux perspective. Ecol Appl 20:1583–1597

Barbero M, Loisel R, Quèzel P (1992) Biogeography, ecology and history of Mediterranean *Quercus ilex* ecosystems. Vegetatio 99–100:19–34

Barbosa M, Fernandes GW (2014) Bottom-up effects on gall distribution. In: Fernandes GW, Santos JC (eds) Neotropical insect galls. Springer, Dordrecht, pp 99–113

Barbour MG (1988) Californian upland forests and woodlands. In: Barbour MG, Billings WD (eds) North American terrestrial vegetation. Cambridge University Press, Cambridge, pp 131–164

Barbour MG, Minnich RH (1990) The myth of chaparral convergence. Israel J Bot 39:453–463

Beadle NCW (1953) The edaphic factor in plant ecology with a special note on soil phosphates. Ecology 34:426–428

Beadle NCW (1954) Soil phosphate and the delimitation of plant communities in eastern Australia. Ecology 35:370–375

Beadle NCW (1966) Soil phosphate and its role in molding segments of the Australian flora and vegetation, with special reference to xeromorphy and sclerophylly. Ecology 47:992–1007

Benz BW, Martin CE (2006) Foliar trichomes, boundary layers, and gas exchange in 12 species of epiphytic *Tillandsia* (*Bromeliaceae*). J Plant Physiol 163:648–656

Bernays EA (1981) Plant tannins and insect herbivores: an appraisal. Ecol Entomol 6:353–360

Bethoux JP, Gentili B, Morin P, Nicolas E, Pierre C, Ruiz-Pino D (1999) The Mediterranean Sea: a miniature ocean for climatic and environmental studies and a key for climatic functioning of the North Atlantic. Prog Oceanogr 44:131–146

Bhaskar R, Ackerley DD (2006) Ecological relevance of minimum seasonal water potentials. Physiol Plant 127:353–359

Bhaskar R, Valiente-Banuet A, Ackerly DD (2007) Evolution of hydraulic traits in closely related species pairs from mediterranean and non-mediterranean environments of North America. New Phytol 176:718–726

Bickford CP (2016) Ecophysiology of leaf trichomes. Funct Plant Biol 43:807–814
Blackman CJ, Brodribb TJ, Jordan GJ (2010) Leaf hydraulic vulnerability is related to conduit dimensions and drought resistance across a diverse range of woody angiosperms. New Phytol 188:1113–1123
Blondel J, Aronson J (1995) Biodiversity and ecosystem function in the Mediterranean Basin: human and nonhuman determinants. In: Davis GW, Richardson DM (eds) Mediterranean-type ecosystems: the function of biodiversity. Springer, Berlin, pp 43–119
Blumler MA (1991) Winter-deciduous versus evergreen habit in Mediterranean regions: a model. USDA Forest Service Gen Tech Rep PSW-126
Blumler MA (2005) Three conflated definitions of Mediterranean climates. Middle States Geographer 38:52–60
Bozzano M, Turok J (2003) Mediterranean Oaks network, report of the second meeting, 2–4 may 2002-Gozo. Malta, International Plant Genetics Resources Institute, Rome, Italy
Breckle SW (2002) Walter's vegetation of the earth. The ecological systems of the geo-biosphere. 4th edn. Springer, Berlin
Brewer CA, Smith WK (1997) Patterns of leaf surface wetness for montane and subalpine plants. Plant, Cell Environ 20:1–11
Brodribb TJ, Holbrook NM, Edwards EJ, Gutierrez MV (2003) Relations between stomatal closure, leaf turgor and xylem vulnerability in eight tropical dry forest trees. Plant, Cell Environ 26:443–450
Bryson RA, Hare FK (1974) Climates of North America. World Survey of Climatology, vol 11. Elsevier Scientific Publishing Co., Amsterdam
Büntgen U, Tegel W, Nicolussi K, McCormick M, Frank D, Trouet V, Kaplan JO, Franz Herzig F, Heussner KU, Wanner H, Luterbacher J, Esper J (2011) 2500 years of European climate variability and human susceptibility. Science 33:578–582
Burghardt M, Riederer M (2003) Ecophysiological relevance of cuticular transpiration of deciduous and evergreen plants in relation to stomatal closure and leaf water potential. J Exp Bot 54:1941–1949
Bussotti F, Bettini D, Grossoni P, Mansuino S, Nibbi R, Soda C, Tani C (2002) Structural and functional traits of *Quercus ilex* in response to water availability. Environ Exp Bot 47:11–23
Bussotti F, Ferrini F, Pollastrini M, Alessio Fini A (2014) The challenge of Mediterranean sclerophyllous vegetation underclimate change: from acclimation to adaptation. Environ Exp Bot 103:80–98
Castro-Díez P, Pedro Villar-Salvador P, Pérez-Rontomé C, Maestro-Martínez M Montserrat-Martí G (1997) Leaf morphology and leaf chemical composition in three *Quercus* (Fagaceae) species along a rainfall gradient in NE Spain. Trees 11:127–134
Castro-Díez P, Montserrat-Martí G (1998) Phenological pattern of fifteen Mediterranean phanaerophytes from *Quercus ilex* communities of NE-Spain. Plant Ecol 139:103–112
Cavender-Bares J, Kitajima K, Bazzaz FA (2004) Multiple trait associations in relation to habitat differentiation among 17 Floridian oak species. Ecol Monogr 74:635–662
Chabot BF, Hicks DJ (1982) The ecology of leaf life spans. Annu Rev Ecol Syst 13:229–259
Christman MA, Sperry JS, Smith DD (2012) Rare pits, large vessels and extreme vulnerability to cavitation in a ring-porous tree species. New Phytol 193:713–720
Choat B, Jansen S, Brodribb TJ, Cochard H, Delzon S, Bhaskar R, Bucci SJ, Feild TS, Gleason SM, Hacke UG, Jacobsen AL, Lens F, Maherali H, Martínez-Vilalta J, Mayr S, Mencuccini M, Mitchell PJ, Nardini A, Pittermann J, Pratt RB, Sperry JS, Westoby M, Wright IJ, Zanne AE (2012) Global convergence in the vulnerability of forests to drought. Nature 491:752–755
Choong MF, Lucas PW, Ong JSY, Pereira B, Tan HTW, Turner IM (1992) Leaf fracture toughness and sclerophylly: their correlations and ecological implications. New Phytol 121:597–610
Cody ML, Mooney HA (1978) Convergence versus nonconvergence in Mediterranean-Climate ecosystems. Annu Rev Ecol Syst 9:265–321
Cochard H, Breda N, Granier A, Aussenac G (1992) Vulnerability to air-embolism of 3 European oak species (*Quercus petraea* (Matt) Liebl, *Quercus pubescens* Willd, L). Ann Sci For 49:225–233

Cochard H, Tyree MT (1990) Xylem dysfunction in *Quercus*: vessel sizes, tyloses, cavitation and seasonal changes in embolism. Tree Physiol 6:393–407
Corcuera L, Camarero JJ, Gil-Pelegrín E (2002) Functional groups in *Quercus* species derived from the analysis of pressure-volume curves. Trees 16:465–472
Corcuera L, Camarero JJ, Gil-Pelegrín E (2004a) Effects of a severe drought on *Quercus ilex* radial growth and xylem anatomy. Trees 18:83–92
Corcuera L, Camarero JJ, Gil-Pelegrín E (2004b) Effects of a severe drought on growth and wood anatomical properties of *Quercus faginea*. IAWA J 25:185–204
Corcuera L, Morales F, Abadía A, Gil-Pelegrín E (2005a) The effect of low temperatures on the photosynthetic apparatus of *Quercus ilex* subsp. *ballota* at its lower and upper altitudinal limits in the Iberian peninsula and during a single freezing-thawing cycle. Trees 19:99–108
Corcuera L, Morales F, Abadía A, Gil-Pelegrín E (2005b) Seasonal changes in photosynthesis and photoprotection in a *Quercus ilex* subsp. *ballota* woodland located in its upper altitudinal extreme in the Iberian Peninsula. Tree Physiol 25:599–608
Corcuera L, Camarero JJ, Sisó S, Gil-Pelegrín E (2006) Radial-growth and wood-anatomical changes in overaged *Quercus pyrenaica* coppice stands: functional responses in a new Mediterranean landscape. Trees 20:91–98
Cowling RM, Campbell BM (1983) The definition of leaf consistence categories in the fynbos biome and their distribution along an altitudinal gradient in the south eastern Cape. J S Afr Bot 49:87–101
Cowling RM, Witkowski ETF (1994) Convergence and non-convergence of plant traits in climatically and edaphically matched sites in Mediterranean Australia and South Africa. Aust J Ecol 19:220–232
Cowling RM, Rundel PW, Lamont BB, Arroyo MK, Arianoutsou M (1996) Plant diversity in mediterranean-climate regions. Trends Ecol Evol 11:362–366
Cuadrat JM, Saz MA, Vicente-Serrano S, González-Hidalgo JC (2007) Water resources and precipitation trends in Aragon (Spain). Int J Water Resour D 23:107–124
Damesin C, Rambal S (1995) Field study of leaf photosynthetic performance by a Mediterranean deciduous oak tree *(Quercus pubescens)* during a severe summer drought. New Phytol 131:159–167
Damesin C, Rambal S, Joffre R (1998) Co-occurrence of trees with different leaf habit: a functional approach on Mediterranean oaks. Acta Oecol 19:195–204
David TS, Henriques MO, Kurz-Besson C, Nunes J, Valente F, Vaz M, Pereira JS, Siegwolf R, Chaves MM, Gazarini LC, David JS (2007) Water-use strategies in two co-occurring Mediterranean evergreen oaks: surviving the summer drought. Tree Physiol 27:793–803
Deitch MJ, Sapundjieff MJ, Feirer ST (2017) Characterizing precipitation variability and trends in the world's Mediterranean-Climate areas. Water 9:259
del Río S, Penas A (2006) Potential distribution of semi-deciduous forests in Castile and Leon (Spain) in relation to climatic variations. Plant Ecol 185:269–282
Denk T, Grimm GW (2009) Significance of pollen characteristics for infrageneric classification and phyllogeny in *Quercus (Fagaceae)*. Int J Plant Sci 170(7):926–940
Ding Y, Zhang Y, Zheng QS, Tyree MT (2014) Pressure-volume curves: revisiting the impact of negative turgor during cell collapse by literature review and simulations of cell micromechanics. New Phytol 203:378–387
Di Paola A, Paquette A, Trabucco A, Mereu S, Valentini R, Paparella F (2017) Coexistence trend contingent to Mediterranean oaks with different leaf habits. Ecol Evol 7:3006–3015
Dolph GE, Dilcher DL (1980) Variation in leaf size with respect to climate in the tropics of the Western-hemisphere. Bull Torrey Bot Club 107:154–162
Dreyer E, Le Roux X, Montpied P, Duadet FA, Masson F (2001) Temperature response of leaf photosynthetic capacity in seedlings from seven temperate tree species. Tree Physiol 21:223–232
Dufour-Dror JM, Ertas (2004) A Bioclimatic perspectives in the distribution of *Quercus ithaburensis* Decne. Subspecies in Turkey and in the Levant. J Biogeogr 31:461–474
Edwards C, Read J, Sanson G (2000) Characterising sclerophylly: some mechanical properties of leaves from heath and forest. Oecologia 123:158–167

Ehleringer J (1981) Leaf absorptances of Mohave and Sonoran desert plants. Oecologia 49:366–370
Ehleringer J, Björkman O, Mooney HA (1976) Leaf pubescence: effects on absorptance and photosynthesis in a desert shrub. Science 192:376–377
Ehleringer J, Björkman O (1978) Pubescence and leaf spectral characteristics in a desert shrub, *Encelia farinosa*. Oecologia 36:151–162
Ehleringer J, Mooney HA, Gulmon SL, Rundel PW (1981) Parallel evolution of leaf pubescence in *Encelia* in coastal deserts of north and south America. Oecologia 49:38–41
Ehleringer J (1984) Ecology and ecophysiology of leaf pubescence in North American desert plants. In: Rodriguez E, Healey PL, Mehta I (eds) Biology and chemistry of plant trichomes. Plenum, New York, pp 113–132
Ehleringer J, Mooney HA (1978) Leaf hairs: effects on physiological activity and adaptive value to a desert shrub. Oecologia 37:183–200
Ehleringer JR, Smedly MP (1988) Stomatal sensitivity and water-use efficiency in oaks and their hybrids. In: Wallace A, McArthur ED, Haferkamp MR (eds) Symposium on shrub ecophysiology and biotechnology. USDA Forest Service Tech Rep INT-256, Ogden, UT, pp 98–102
Ehleringer JR, Phillips SL (1996) Ecophysiological factors contributing to the distributions of several *Quercus* species in the Intermountain West. Ann Sci Forest 53:291–302
Epron D, Dreyer E (1992) Effects of severe dehydration on leaf photosynthesis in *Quercus petraea* (Matt.) Liebl.: photosystem II efficiency, photochemical and non-photochemical fluorescence quenching and electrolyte leakage. Tree Physiol 10:273–284
Epron D, Dreyer E (1993) Long-term effects of drought on photosynthesis of adult oak trees (*Quercus petraea* Matt.) Liebl. and *Quercus robur* L.) in a natural stand. New Phytol 125:381–389
Escudero A, Mediavilla S (2003) Decline in photosynthetic nitrogen use efficiency with leaf age and nitrogen resorption as determinants of leaf life span. J Ecol 91:880–889
Esteso-Martínez J, Camarero JJ, Gil-Pelegrín E (2006) Competitive effects of herbs on *Quercus faginea* seedlings inferred from vulnerability curves and spatial-pattern analyses in a Mediterranean stand (Iberian System, northeastern Spain). Ecoscience 13(3):378–387
Faria T, Silvério D, Breia E, Cabral R, Abadia A, Abadia J, Pereira JS, Chaves MM (1998) Differences in the response of carbon assimilation to summer stress (water deficits, high light and temperature) in four Mediterranean tree species. Physiol Plant 102:419–428
Farrant JM (2000) A comparison of mechanisms of desiccation tolerance among three angiosperm resurrection plant species. Plant Ecol 151:29–39
Feeny P (1970) Seasonal changes in oak leaf tannins and nutrients as a cause of spring feeding by winter moth caterpillars. Ecology 51:565–581
Felger RS, Johnson MB, Wilson MF (2001) The trees of Sonora. Oxford University Press, Oxford, Mexico
Fernández V, Sancho-Knapik D, Guzmán P, Peguero-Pina JJ, Gil L, Karabourniotis G, Khayet M, Fasseas C, Heredia-Guerrero JA, Heredia A, Gil-Pelegrín E (2014) Wettability, polarity, and water absorption of holm oak leaves: effect of leaf side and age. Plant Physiol 166:168–180
Fick SE, Hijmans RJ (2017). WorldClim 2: new 1-km spatial resolution climate surfaces for global land areas. Int J Climatol doi:10.1002/joc.5086
Filella I, Llusià J, Piñol J, Peñuelas J (1998) Leaf gas exchange and fluorescence of *Phillyrea latifolia*, *Pistacia lentiscus* and *Quercus ilex* saplings in severe drought and high temperature conditions. Environ Exp Bot 39:213–220
Flexas J, Diaz-Espejo A, Gago J, Gallé A, Galmés J, Gulías J, Medrano H (2014) Photosynthetic limitations in Mediterranean plants: A review. Environ Exp Bot 103:12–23
Follieri M, Magri D, Sadori L (1988) 250,000-year pollen record from Valle di Castigliore (Roma). Pollen Spores 30:329–356
Fonseca CR, Overton JM, Collins B, Westoby M (2000) Shifts in trait-combinations along rainfall and phosphorus gradients. J Ecol 88:964–977
Fotelli MN, Radoglou M, Constantinidou H-IA (2000) Water stress responses of seedlings of four Mediterranean oak species. Tree Physiol 20:1065–1075

Fralish JS, Franklin SB (2002) Taxonomy and ecology of woody plants in North American forests (excluding Mexico and Subtropical Florida). Wiley, New York

Franks PJ, Beerling DJ (2009) Maximun leaf conductance driven by CO_2 effects on stomatal size and density over geologic time. PNAS 106:10343–10347

Frisch HL (1991) Fundamentals of membrane transport. Polym J 23:445–456

Galmés J, Medrano H, Flexas J (2007) Photosynthesis and photoinhibition in response to drought in a pubescent (var. minor) and a glabrous (var. palaui) variety of *Digitalis minor*. Environ Exp Bot 60:105–111

García-Ruiz JM, López-Moreno JI, Vicente-Serrano SM, Lasanta T, Beguería S (2011) Mediterranean water resources in a global change scenario. Earth Sci Rev 105:121–139

García-Ruiz JM, Nadal-Romero E, Lana-Renault N, Beguería S (2013) Erosion in Mediterranean landscapes: changes and future challenges. Geomorphology 198:20–36

García-Plazaola JI, Artetxe U, Duñabeitia MK, Becerril JM (1999) Role of photoprotective systems of Holm-Oak (*Quercus ilex*) in the adaptation to winter conditions. J Physiol 155:25–630

Gasith A, Resh VH (1999) Streams in Mediterranean climate regions: abiotic influences and biotic responses to predictable seasonal events. Annu Rev Ecol Syst 30:51–81

Gibson AC (1998) Photosynthetic organs of desert plants. Bioscience 48:911–920

Giertych MJ, Karolewski P, Oleksyn J (2015) Carbon allocation in seedlings of deciduous tree species depends on their shade tolerance. Acta Physiol Plant 37:216

Gil-Pelegrín E, Peguero-Pina JJ, Camarero JJ, Fernández-Cancio A, Navarro-Cerrillo R (2008) Drought and forest decline in the Iberian Peninsula: a simple explanation for a complex phenomenom? In: Sánchez JM (ed) Droughts: causes, effects and predictions. Nova Science Publishers Inc., New York, pp 27–68

Givnish T (1976) Leaf form in relation to environment: a theoretical study. PhD Thesis, Princeton University

Givnish TJ (1987) Comparative studies of leaf form: assessing the relative roles of selective pressures and phylogenetic constraints. New Phytol 106:131–160

Givnish TJ (2002) Adaptive significance of evergreen versus deciduous leaves: solving the triple paradox. Silva Fenn 36:703–743

Gonçalves-Alvim SJ, Korndorf G, Fernandes GW (2006) Sclerophylly in *Qualea parviflora* (*Vochysiaceae*): influence of herbivory, mineral nutrients, and water status. Plant Ecol 187:153–162

González-Zurdo P, Escudero A, Babiano J, García-Ciudad A, Mediavilla S (2016) Costs of leaf reinforcement in response to winter cold in evergreen species. Tree Physiol 36:273–286

Goulden ML (1996) Carbon assimilation and water-use efficiency by neighboring Mediterranean-climate oaks that differ in water access. Tree Physiol 16:417–424

González-Rebollar JL, García-Álvarez A, Ibáñez JJ (1995) A mathematical model for predicting the impact of climate changes on mediterranean plant landscapes. In: Zewer S, van Rompaey RSAR, Kok MTJ, Berk MM (eds) Climate change research: evaluation and policy implications. Elsevier, Amsterdam, pp 757–762

Grammatikopoulos G, Karabourniotis G, Kyparissis A, Petropoulou Y, Manetas Y (1994) Leaf hairs of olive (*Olea europaea*) prevent stomatal closure by ultraviolet-B radiation. Aust J Plant Physiol 21:293–301

Grassi G, Magnani F (2005) Stomatal, mesophyll conductance and biochemical limitations to photosynthesis as affected by drought and leaf ontogeny in ash and oak trees. Plant, Cell Environ 28:834–849

Griffin JR (1971) Oak regeneration in the upper Carmel Valley, California. Ecology 52:862–868

Griffin JR (1973) Xylem sap tension in three woodland oaks of central California. Ecology 54:152–159

Griffin JR (1977) Oak woodland. In: Barbour MG, Major J (eds) Terrestrial vegetation of California. Wiley, New York, pp 383–415

Groom P, Lamont BB (1997) Xerophytic implications of increased sclerophylly: interactions with water and light in *Hakea psilorrhyncha* seedlings. New Phytol 136:23–231

Groom PK, Lamont BB (1999) Which commn indices of sclerophylly best reflect differences in leaf structure? Ecoscience 6:471–474

Grove AT, Miles MR, Worthington EB, Doggett H, Dasgupta B, Farmer BH (1977) The geography of semi-arid lands [and discussion]. Philos Trans R Soc Lond B Biol Sci 278:457–475

Grubb PJ (1986) Sclerophylls, pachyphylls and pycnophylls: the nature and significance of hard leaf surfaces. In: Juniper B, Southwood R (eds) Insects and the plant surface. Edward Arnold, London, UK, pp 137–150

Hacke UG, Sperry JS, Wheeler JK, Castro L (2006) Scaling of angiosperm xylem structure with safety and efficiency. Tree Physiol 26:689–701

Hardin JW (1979) Patterns of variation in foliar trichomes of eastern North American *Quercus*. Am J Bot 66:576–585

He Y, Li N, Wang Z, Wang H, Yang G, Xiao L, Wu J, Sun B (2014) *Quercus yangyiensis* sp. nov. from the late Pliocene of Baoshan, Yunnan and its paleoclimatic significance. Acta Geol Sin 88:738–747

Heide-Jorgensen HS (1990) Xeromorphic leaves of *Hakea suaveolens* R. Br. IV. Ontogeny, structure and function of the sclereids. Aust J Bot 38:25–43

Herrera CM (1992) Historical effects and sorting processes as explanations for contemporary ecological patterns: character syndromes in Mediterranean Woody plants. Am Nat 140:421–446

Himrane H, Camarero JJ, Gil-Pelegrín E (2004) Morphological and ecophysiological variation of the hybrid oak *Quercus subpyrenaica* (*Q. faginea* × *Q. pubescens*). Trees 18:566–575

Holloway PJ (1982) Structure and histochemistry of plant cuticular membranes: an overview. In: Cutler DF, Alvin KL, Price CE (eds) The plant cuticle. Academic Press, London, pp 1–32

Hurrell JW, Hoerling MP, Phillips AS, Xu T (2004) Twentieth century North Atlantic climate change. Part I: assessing determinism. Clim Dyn 23:371–389

Iovi K, Kolovou C, Kyparissis A (2009) An ecophysiological approach of hydraulic performance for nine Mediterranean species. Tree Physiol 29:889–900

Jacobsen AL, Pratt RB, Davis SD, Ewers FW (2007a) Cavitation resistance and seasonal hydraulics differ among three arid Californian plant communities. Plant Cell Envir 30:1599–1609

Jacobsen AL, Pratt RB, Ewers FW, Davis SD (2007b) Cavitation resistance among 26 chaparral species of Southern California. Ecol Monogr 77:99–115

Jacobsen AL, Esler KJ, Pratt RB, Ewers FW (2009) Water stress tolerance of shrubs in mediterranean type climate regions: convergence of fynbos and succulent karoo communities with California shrub communities. Am J Bot 96:1445–1453

Jacobsen AL, Pratt RB, Davis SD, Tobin MF (2014) Geographic and seasonal variation in chaparral vulnerability to cavitation. Madroño 61:317–327

Jeffree CE (1996) Structure and ontogeny of plant cuticles. In: Kerstiens G (ed) Plant cuticles: an integrated functional approach. Bios Scientific Publishers, Oxford, pp 33–82

Jiménez-Moreno G, Fauquette S, Suc JP (2010) Miocene to Pliocene vegetation reconstruction and climate estimates in the Iberian Peninsula from pollen data. Rev Palaeobot Palyno 162:403–415

Joffre R, Rambal S, Damesin C (2007) Functional attributes in Mediterranean-type ecosystems. In: Valladares F, Pugnaire FI (eds) Functional plant ecology, 2nd edn. CRC Press, Boca raton, Fl, USA, pp 285–312

John GP, Scoffoni C, Buckley TN, Villar R, Poorter H, Sack L (2017) The anatomical and compositional basis of leaf mass per area. Ecol Lett 20(4):412–425

Johnson HB (1975) Plant pubescence: an ecological perspective. Bot Rev 41:233–258

Johnson DM, Woodruff DR, McCullo KA, Meinzer FC (2009) Leaf hydraulic conductance, measured in situ, declines and recovers daily: leaf hydraulics, water potential and stomatal conductance in four temperate and three tropical tree species. Tree Physiol 29:879–887

Jordan GJ, Weston PH, Carpenter RJ, Dillon RA, Brodribb TJ (2008) The evolutionary relations of sunken, covered and encrypted stomata to dry habitats in Proteaceae. Am J Bot 95:521–530

Karabourniotis G, Bornman JF (1999) Penetration of UV-A, UV-B and blue light through the leaf trichome layers of two xeromorphic plants, olive and oak, measured by optical fibre microprobes. Physiol Plant 105:655–661

Kerstiens G (1996) Cuticular water permeability and its physiological significance. J Exp Bot 47:1813–1832
Kerstiens G (2006) Water transport in plant cuticles: an update. J Exp Bot 57:2493–2499
Kikuzawa K (1991) A cost-benefit analysis of leaf habit and leaf longevity of trees and their geographical pattern. Am Nat 138:1250–1263
Knox RG, Harcombe PA, Elsik IS (1995) Contrasting patterns of resource limitation in tree seedlings across a gradient in soil texture. Can J Forest Res 25:1583–1594
Kooyman RM, Laffan SW, Westoby M (2017) The incidence of low phosphorus soils in Australia. Plant Soil 412:143–150
Köppen W (1936) Das geographische system der klimate. In: Köppen W, Geiger R (eds) Handbuch der Klimatologie 3. Gebrueder Borntraeger, Berlin
Kouki M, Manetas Y (2002) Toughness is less important than chemical composition of *Arbutus* leaves in food selection by *Poecilimon* species. New Phytol 154:399–407
Kovar-Eder J (2003) Vegetation dynamics in Europe during the Neogene. In: Reumer JWF, Wessels W (eds) Distribution and migration of tertiary mamals in Eurasia. A volume in honour of Hans de Bruijn. Deinsea 10:373–392
Krichak SO, Alpert P (2005) Decadal trends in the East Atlantic/West Russiapattern and the Mediterranean precipitation. Int J Climatol 25:183–192
Kröber W, Zhang S, Ehmig M, Bruelheide H (2014) Linking xylem hydraulic conductivity and vulnerability to the leaf economics spectrum-A cross-species study of 39 evergreen and deciduous broadleaved subtropical tree species. PLoS ONE 9(11):e109211
Lamont BB, Groom PK, Cowling RM (2002) High leaf mass per area of related species assemblages may reflect low rainfall and carbon isotope discrimination rather than low phosphorus and nitrogen concentrations. Funct Ecol 16:403–412
Lasanta T, Nadal-Romero E, Errea P, Arnáez J (2016) The effect of landscape conservation measures in changing landscape patterns: a case study in mediterranean mountains. Land Degrad Dev 27(2):373–386
Leiva MJ, Fernández-Alés (1998) Variability in seedling water status during drought within a *Quercus ilex* subsp. *ballota* population, and its relation to seedling morphology. For Ecol Manag 111:147–156
Levin DA (1976) The chemical defenses of plants to pathogens and herbivores. Annu Rev Ecol Syst 7:121–159
Levitt J (1980) Responses of plants to environmental stresses. Volume 2, Water, radiation, salt and other stresses, 2nd edn. Academic Press, New York
Li Y, Sperry JS, Taneda H, Bush SE, Hacke UG (2008) Evaluation of centrifugal methods for measuring xylem cavitation in conifers, diffuse- and ring-porous angiosperms. New Phytol 177:558–568
Limousin JM, Misson L, Lavoir AV, Martin NK, Rambal S (2010a) Do photosynthetic limitations of evergreen *Quercus ilex* leaves change with long-term increased drought severity? Plant, Cell Environ 33:863–875
Limousin JM, Longepierre D, Huc R, Rambal S (2010b) Change in hydraulic traits of Mediterranean *Quercus ilex* subjected to long-term throughfall exclusion. Tree Physiol 30:1026–1036
Lionello, P, Malanotte-Rizzoli, P, Boscolo, R, et al (2006) The Mediterranean climate: an overview of the main characteristics and issues. In: Lionello P, Malanotte-Rizzoli P, R Boscolo R (eds). Mediterranean Climate Variability. Elsevier, Amsterdam, p 1–26
Lo Gullo MA, Salleo S, Rosso (1986) Drought avoidance strategy in *Ceratonia Siliqua* L., a mesomorphic-leaved tree in the Xeric Mediterranean area. Ann Bot 58:745–756
Lo Gullo MA, Salleo S (1988) Different strategies of drought resistance in three Mediterranean sclerophyllous trees growing in the same environmental conditions. New Phytol 108:267–276
Lo Gullo MA, Salleo S (1993) Different vulnerabilities of *Quercus ilex* L. to freezeand summer drought-induced xylem embolism: an ecological interpretation. Plant, Cell Environ 16:511–519
Lo Gullo M, Nardini A, Trifilò P, Salleo S (2005) Diurnal and seasonal variations in leaf hydraulic conductance in evergreen and deciduous trees. Tree Physiol 25:505–512

Loveless AR (1961) A nutritional interpretation of sclerophylly based on differences in the chemical composition of sclerophyllous and mesophytic leaves. Ann Bot 25:168–184
Loveless AR (1962) Further evidence to support a nutritional interpretation of sclerophylly. Ann Bot 26:551–561
Lucas PW, Pereira B (1990) Estimation of the fracture toughness of leaves. Funct Ecol 4:819–822
Lumaret R, Mir C, Michaud H, Raynal V (2002) Phylogeographical variation of chloroplast DNA in holm oak (*Quercus ilex* L.). Mol Ecol 11:2327–2336
Maherali H, Pockman WT Jackson RB (2004) Adaptive variation in the vulnerability of woody plants to xylem cavitation. Ecology 85:2184–2199
Maherali H, Moura CF, Caldeira MC, Willson CJ, Jackson RB (2006) Functional coordination between leaf gas exchange and vulnerability to xylem cavitation in temperate forest trees. Plant, Cell Environ 29:571–583
Mai DH (1991) Palaeofloristic changes in Europe and the confirmation of the arctotertiary-palaeotropical geoflora concept. Rev Palaeobot Palyno 68:29–36
Maire V, Wright IJ, Prentice IC, Batjes NH, Bhaskar R, van Bodegom PM, Cornwell WK, Ellsworth D, Niinemets U, Ordonez A, Reich PB, Santiago LS (2015) Global effects of soil and climate on leaf photosynthetic traits and rates. Global Ecol Biogeogr 24:706–717
Manes F, Vitale M, Donato E, Giannini M, Puppi G (2006) Different ability of three Mediterranean oak species to tolerate progressive water stress. Photosynthetica 44(3):387–393
Manetas Y (2003) The importance of being hairy: the adverse effects of hair removal on stem photosynthesis of *Verbascum speciosum* are due to solar UV-B radiation. New Phytol 158:503–508
Mantua NJ, Hare SR, Zhang Y, Wallace JM, Francis RC (1997) A pacific interdecadal climate oscillation with impacts on salmon production. B Am Meteorol Soc 78:1069–1079
Mariotti A, Zeng N, Lau KM (2002) Euro-Mediterranean rainfall and ENSO a seasonally varying relationship. Geophys Res Lett 29:1621
Martínez-Vilalta J, Prat E, Oliveras I, Piñol J (2002) Xylem hydraulic properties of roots and stems of nine Mediterranean woody species. Oecologia 133:19–29
Martínez-Vilalta J, Cochard H, Mencuccini M, Sterck F, Herrero A, Korhonen JFJ, Llorens P, Nikinmaa E, Nolè A, Poyatos R, Ripullone F, Sass-Klaassen U, Zweifel R (2009) Hydraulic adjustment of Scots pine across Europe. New Phytol 184:353–364
McDonald PG, Fonseca CR, Overton JM, Westoby M (2003) Leaf-size divergence along rainfall and soil-nutrient gradients: is the method of size reduction common among clades? Funct Ecol 17:50–57
Mediavilla S, Escudero A (2003) Stomatal responses to drought at a Mediterranean site: a comparative study of co-occurring woody species differing in leaf longevity. Tree Physiol 23:987–996
Mediavilla S, Escudero A (2004) Stomatal responses to drought of mature trees and seedlings of two co-occurring Mediterranean oaks. For Ecol Manag 187:281–294
Mediavilla S, García-Ciudad A, García-Criado B, Escudero A (2008) Testing the correlations between leaf life span and leaf structural reinforcement in 13 species of European Mediterranean woody plants. Funct Ecol 22:787–793
Medrano H, Flexas J, Galmés (2008) Variability in water use efficiency at the leaf level among Mediterranean plants with different growth forms. Plant Soil 317:17–29
Meher-Homji (1973) A phytoclimatic approach to the problem of Mediterraneity in the Indo-Pakistan sub-continent. Feddes-Repertorium 83:757–788
Michaud H, Toumi HL, Lumaret R, Li TX, Romane F, Di Giusto F (1995) Effect of geographical discontinuity on genetic variation in *Quercus ilex* L. (holm oak). Evidence from enzyme polymorphism. Heredity 74:590–606
Millar CI (2012) Geologic, climatic, and vegetation history of California. In: Baldwin BG, Goldman DH, Keil DJ, Patterson R, Rosatti TJ, Wilken DH (eds) The Jepson Manual: Vascular Plants of California, 2nd edn. University of California Press, pp 49–67
Mitrakos KA (1980) A theory for Mediterranean plant life. Acta Oecol 1:245–252

Mitrakos K (1982) Winter low temperatures in mediterranean-type ecosystems. Ecol Mediterr 8:95–102
Montserrat-Martí G, Camarero JJ, Palacio S, Pérez-Rontomé C, Milla R, Albuixech J, Maestro M (2009) Summer-drought constrains the phenology and growth of two co-existing Mediterranean oaks with contrasting leaf habit: implications for their persistence and reproduction. Trees 23:787–799
Mooney HA, Dunn EL (1970) Convergent evolution of Mediterranean-climate evergreen sclerophyll shrubs. Evolution 24:292–303
Mooney HA (1982) Habitat, plant form, and plant water relations in Mediterranean-climate regions. Ecol Mediterr 8:287–296
Morales F, Abadía A, Abadía J Montserrat G, Gil-Pelegrín E (2002) Trichomes and photosynthetic pigment composition changes: responses of *Quercus ilex* subsp. *ballota* (Desf.) Samp. and *Quercus coccifera* L. to Mediterranean stress conditions. Trees 16:504–510
Morecroft MD, Roberts JM (1999) Photosynthesis and stomatal conductance of mature canopy Oak (*Quercus robur*) and Sycamore *(Acer pseudoplatanus)* trees throughout the growing season. Funct Ecol 13:332–342
Moreno G, Gallardo JF, Vicente MA (2011) How mediterranean deciduous trees cope with long summer drought? The case of *Quercus pyrenaica* forests in western Spain. In: Bredemeier M, Cohen S, Godbold DL, Lode E, Pichler V, Schleppi P (eds) Forest management and the water cycle. An ecosystem-based approach. Ecological Studies 212, Springer, Dordrecht, pp 181–207
Nardini A, Lo Gullo MA, Tracanelli S (1996) Water relations of six sclerophylls growing near Trieste (Northeastern Italy): has sclerophylly a univocal functional significance? Giorn Bot Ital 130:811–828
Nardini A, Lo Gullo MA, Salleo S (1999) Competitive strategies for water availability in two Mediterranean *Quercus* species. Plant, Cell Environ 22:109–116
Nardini A, Pedà G, La Rocca N (2012a) Trade-offs between leaf hydraulic capacity and drought vulnerability: morpho-anatomical bases, carbon costs and ecological consequences. New Phytol 196:788–798
Nardini A, Pedá G, Salleo S (2012b) Alternative methods for scaling leaf hydraulic conductance offer new insights into the structure–function relationships of sun and shade leaves. Funct Plant Biol 39:394–401
Nardini A, Lo Gullo MA, Truifilo P, Salleo S (2014) The challenge of the Mediterranean climate to plant hydraulics: responses and adaptations. Environ Exp Bot 103:68–79
Negi SS, Naithani HB (1995) Oaks of India, Nepal and Bhutan. International Book Distributors, Dehra Dun, India
Nixon KC (2002) The oak (*Quercus*) biodiversity of California and adjacent regions. USDA Forest Service Gen Tech Rep. PSW-GTR-184
Nicotra AB, Cosgrove MJ, Cowling A, Schlichting CD, Jones CS (2008) Leaf shape linked to photosynthetic rates and temperature optima in South African *Pelargonium* species. Oecologia 154:625–635
Niinemets U (2001) Global-scale climatic controls of leaf dry mass per area, density, and thickness in trees and shrubs. Ecology 82:453–469
Niinemets Ü, Cescatti A, Rodeghiero M, Tosens T (2005) Leaf internal conductance limits photosynthesis more strongly in older leaves of Mediterranean evergreen broad-leaved species. Plant, Cell Environ 28:1552–1556
Niinemets Ü, Keenan T (2014) Photosynthetic responses to stress in Mediterranean evergreens: mechanisms and models. Environ Exp Bot 103:24–41
Niinemets Ü (2015) Is there a species spectrum within the world-wide leaf economics spectrum? Major variations in leaf functional traits in the Mediterranean sclerophyll *Quercus ilex*. New Phytol 205:79–96
Niinemets Ü (2016) Does the touch of cold make evergreen leaves tougher? Tree Physiol 36:267–272
Niklas KJ (1999) A mechanical perspective on foliage leaf form and function. New Phytol 143:19–31
Nobel (1991) Physicochemical and environmental plant physiology. Academic Press, San Diego

Noce S, Collalti A, Valentini R, Santini M (2016) Hot spot maps of forest presence in the Mediterranean Basin. iForest 9:766
Oertli JJ (1986) The effect of cell size on cell collapse under negative turgor pressure. J Plant Physiol 124:365–370
Oertli JJ, Lips SH, Agami M (1990) The strength of sclerophyllous cells to resist collapse due to negative turgor pressure. Acta Oecol 11:281–289
Ogaya R, Peñuelas J (2003) Phenological patterns of *Quercus ilex, Phillyrea latifolia*, and *Arbutus unedo* growing under a field experimental drought. Écoscience 11:263–270
Ogaya R, Peñuelas J (2007) Leaf mass per area ratio in *Quercus ilex* leaves under a wide range of climatic conditions. The importance of low temperatures. Acta Oecol 31:168–173
Oliveira G, Peñuelas J (2002) Comparative protective strategies of *Cistus albidus* and *Quercus ilex* facing photoinhibitory winter conditions. Environ Exp Bot 47:281–289
Onoda Y, Westoby M, Adler PB, Choong AMF, Clissold FJ, Cornelissen JHC, Díaz S, Dominy NJ, Elgart A, Enrico L, Fine PVA, Howard JJ, Jalili A, Kitajima K, Kurokawa H, McArthur C, Lucas PW, Markesteijn L, Perez- Harguindeguy N, Poorter L, Richards L, Santiago LS, Sosinski EE, van Bael SA, Warton DI, Wright IJ, Wright SJ, Yamashita N (2011) Global patterns of leaf mechanical properties. Ecol Lett 14:301–312
Paddock WA III, Davis SD, Pratt RB, Jacobsen AL, Tobin MF, López-Portillo J, Ewers FW (2013) Factors determining mortality of adult chaparral shrubs in an extreme drought year in California. Aliso 31:49–57
Panahi P, Jamzad Z, Pourmajidian MR, Fallah A, Pourhashemi M (2012a) Foliar epidermis morphology in *Quercus* (subgenus *Quercus*, section *Quercus*) in Iran. Acta Bot Croat 71:95–113
Panahi P, Jamzad Z, Pourmajidian MR, Fallah A, Pourhashemi M, Sohrabi H (2012b) Taxonomic revision of the *Quercus brantii* complex (*Fagaceae*) in Iran with emphasis on leaf and pollen micromorphology. Acta Bot Hung 54(3–4):355–375
Pasho E, Camarero JJ, de Luis M, Vicente-Serrano SM (2011) Impacts of drought at different time scales on forest growth across a wide climatic gradient in north-eastern Spain. Agr Forest Meteorol 151:1800–1811
Peel MC, Finlayson BL, McMahon TA (2007) Updated world map of the Köppen-Geiger climate classification. Hydrol Earth Syst Sci Discuss 4:439–473
Peeters PJ, Sanson G, Read J (2007) Leaf biomechanical properties and the densities of herbivorous insect guilds. Funct Ecol 21:246–255
Peguero-Pina JJ, Sancho-Knapik D, Cochard H, Barredo G, Villarroya D, Gil-Pelegrín E (2011) Hydraulic traits are associated with the distribution range of two closely related Mediterranean firs, *Abies alba* M. and *Abies pinsapo* Boiss. Tree Physiol 31:1067–1075
Peguero-Pina JJ, Sancho-Knapik D, Barrón E, Camarero JJ, Vilagrosa A, Gil-Pelegrín E (2014) Morphological and physiological divergences within *Quercus ilex* support the existence of different ecotypes depending on climatic dryness. Ann Bot 114:301–313
Peguero-Pina JJ, Sancho-Knapik D, Martín P, Saz MA, Gea-Izquierdo G, Cañellas I, Gil-Pelegrín E (2015) Evidence of vulnerability segmentation in a deciduous Mediterranean oak (*Quercus subpyrenaica* E. H. del Villar). Trees 29:1917–1927
Peguero-Pina JJ, Sisó S, Sancho-Knapik D, Díaz-Espejo Flexas J, Galmés J, Gil-Pelegrín E (2016a) Leaf morphological and physiological adaptations of a deciduous oak (*Quercus faginea* Lam.) to the Mediterranean climate: a comparison with a closely related temperate species (*Quercus robur* L.). Tree Physiol 36:287–299
Peguero-Pina JJ, Sisó S, Fernández-Marín B, Flexas J, Galmés J, García-Plazaola JI, Niinemets Ü, Sancho-Knapik D, Gil-Pelegrín E (2016b) Leaf functional plasticity decreases the water consumption without further consequences for carbon uptake in *Quercus coccifera* L. under Mediterranean conditions. Tree Physiol 36:356–367
Peguero-Pina JJ, Sisó S, Flexas J, Galmés J, García-Nogales A, Niinemets Ü, Sancho-Knapik D, Saz MA, Gil-Pelegrín E (2017a) Cell-level anatomical characteristics explain high mesophyll conductance and photosynthetic capacity in sclerophyllous Mediterranean oaks. New Phytol 214:585–596

Peguero-Pina JJ, Sancho-Knapik D, Gil-Pelegrín E (2017b) Ancient cell structural traits and photosynthesis in today's Environment. J Exp Bot 68:1389–1392

Peñuelas J, Filella I, Llusià J, Siscart D, Piñol J (1998) Comparative field study of spring and summer leaf gas exchange and photobiology of the mediterranean trees *Quercus ilex* and *Phillyrea latifolia*. J Exp Bot 49:229–238

Pérez-Estrada LB, Cano-Santana Z, Oyama K (2000) Variation in leaf trichomes of *Wigandia urens*: environmental factors and physiological consequences. Tree Physiol 20:629–632

Pesoli P, Gratani L, Larcher W (2003) Responses of *Quercus ilex* from different provenances to experimentally imposed water stress. Biol Plantarum 46:577–581

Pinto CA, David JS, Cochard H, Caldeira MC, Henriques MO, Quilhó T, Paço TA, Pereira JS, David TS (2012) Drought-induced embolism in current-year shoots of two Mediterranean evergreen oaks. Forest Ecol Manag 285:1–10

Pratt RB, Jacobsen AL, Golgotiu KA, Sperry JS, Ewers FW Davis SD (2007) Life history type and water stress tolerance in nine California chaparral species (*Rhamnaceae*). Ecol Monogr 77:239–253

Poudyal K, Jha PK, Zobel DB, Thapa CB (2004) Patterns of leaf conductance and water potentila of five Himalayan tree species. Tree Physiol 24:689–699

Poyatos R, Llorens P, Piñol J, Rubio C (2008) Response of Scots pine (*Pinus sylvestris* L.) and pubescent oak (*Quercus pubescens*Willd.) to soil and atmospheric water deficits under Mediterranean mountain climate. Ann For Sci 65:306

Pozo-Vázquez D, Esteban-Parra MJ, Rodrigo FS, Castro-Diez Y (2001) A study of NAO variability and its possible non-linear influences on European Surface temperature. Clim Dyn 17:701–715

Radoglou K (1996) Environmental control of CO_2 assimilation rates and stomatal conductance in five oak species growing under field conditions in Greece. Ann Sci Forest 53:269–278

Read J, Sanson GD (2003) Characterising sclerophylly: the mechanical properties of a diverse range of leaf types. New Phytol 160:81–99

Read J, Sanson GD, Lamont BB (2005) Leaf mechanical properties in sclerophyll woodland and shrubland on contrasting soils. Plant Soil 276:95–113

Read J, Sanson G, De Garine-Wichatitsky M, Tanguy J (2006) Sclerophylly in two contrasting tropical environments: low nutrients versus low rainfall. Am J Bot 93:1601–1614

Read J, Sanson GD, Caldwell E, Clissold F, Chatain A, Peeters P, Lamont BB, De Garine-Wichatitsky M, Jaffré T, Stuart Kerr S (2009) Correlations between leaf toughness and phenolics among species in contrasting environments of Australia and New Caledonia. Ann Bot 103:757–767

Read J, Sanson G, Trautmann MF (2016) Leaf traits in Chilean matorral: Sclerophylly within, among, and beyond matorral, and its environmental determinants. Ecol Evol 6:1430–1446

Reale O, Dirmeyer P (2000) Modeling the effects of vegetation on Mediterranean climate during the Roman classical period—Part I: climate history and model sensitivity. Global Planet 25:163–184

Reille M, Pons A (1992) The ecological significance of sclerophyllous oak forests in the western part of the Mediterranean Basin: a note on pollen analytical data. Vegetatio 99–100:13–17

Retallack GJ (2004) Late Miocene climate and life on land in Oregon within a context of Neogene global change. Palaeogeogr Palaeocl 214:97–123

Ribeiro SP, Basset Y (2007) Gall-forming and free-feeding herbivory along vertical gradients in a lowland tropical rainforest: the importance of leaf sclerophylly. Ecography 30:663–672

Riera-Mora S, Esteban-Amat A (1994) Vegetation history and human activity during the last 6000 years on the central Catalan coast (north-eastern Iberian Peninsula). Veg His Archaebot 3:7–23

Rico M, Gallego HA, Moreno G, Santa Regina I (1996) Stomatal response of *Quercus pyrenaica* Willd to environmental factors in two sites differing in their annual rainfall (Sierra de Gata, Spain). Ann Sci For 53:221–234

Riederer M, Schreiber L (2001) Protecting against water loss: analysis of the barrier properties of plant cuticles. J Exp Bot 52:2023–2032

Ripley BS, Pammenter NW, Smith VR (1999) Function of leaf hairs revisited: the hair layer on leaves of *Arctotheca populifolia* reduces photoinhibition, but leads to higher leaf temperatures caused by lower transpiration rates. J Plant Physiol 155:78–85

Rivas-Martínez S, Rivas-Sáenz S, Penas-Merino A (2011) Worldwide bioclimatic classification system. Global geobotany 1:1–634

Roth-Nebelsick A, Hassiotou F, Veneklaas EJ (2009) Stomatal crypts have small effects on transpiration: a numerical model analysis. Plant Physiol 151:2018–2027

Roth-Nebelsick A, Fernández V, Peguero-Pina JJ, Sancho-Knapik D, Gil-Pelegrín E (2013) Stomatal encryption by epicuticular waxes as a plastic trait modifying gas exchange in a Mediterranean evergreen species (*Quercus coccifera* L.). Plant, Cell Environ 36:579–589

Rubio de Casas R, Vargas P, Pérez-Corona E, Cano E, Manrique E, García-Verdugo C, Balaguer L (2009) Variation in sclerophylly among Iberian populations of *Quercus coccifera* L. is associated with genetic differentiation across contrasting environments. Plant Biology 11:464–472

Rundel PW, Arroyo MTK, Cowling RM, Keeley JE, Lamont BB, Vargas P (2016) Mediterranean biomes: evolution of their vegetation, floras, and climate. Ann Rev Ecol Evol S 47:383–407

Sack L, Scoffoni C, McKown AD, Frole K, Rawls M, Havran JC, Tran H, Tran T (2012) Developmentally based scaling of leaf venation architecture explains global ecological patterns. Nat Commun 3:837

Salleo S, Gullo Lo (1990) Sclerophylly and plant water relations in three Mediterranean *Quercus* species. Ann Bot 65:259–270

Salleo S, Nardini A, Lo Gullo MA (1997) Is sclerophylly of Mediterranean evergreens an adaptation to drought? New Phytol 135:603–612

Salleo S, Nardini A (2000) Sclerophylly: evolutionary advantage or mere epiphenomenon? Plant Biosyst 134:247–259

Salleo S, Pitt F, Nardini A, Hamzé M, Jomaa I (2002) Differential drought resistance of two Mediterranean oaks growing in the Bekaa Valley (Lebanon). Plant Biosyst 136:91–99

Sánchez de Dios R, Benito-Garzón M, Sainz-Ollero H (2009) Present and future extension of the Iberian submediterranean territories as determined from the distribution of marcescent oaks. Plant Ecol 204:189–205

Sancho-Knapik D, Alvarez-Arenas TG, Peguero-Pina JJ, Fernández V, Gil-Pelegrín E (2011) Relationship between ultrasonic properties and structural changes in the mesophyll during leaf dehydration. J Exp Bot 62:3637–3645

Sardans J, Rodá F, Peñuelas J (2004) Phosphorus limitation and competitive capacities of *Pinus halepensis* and *Quercus ilex* subsp. *rotundifolia* on different soils. Plant Ecol 174:305–317

Sardans J, Peñuelas J (2013) Plant-soil interactions in Mediterranean forest and shrublands: impacts of climatic change. Plant Soil 365:1–33

Savé R, Castell C, Terradas J (1999) Gas exchange and water relations. In: Roda F, Retana J, Gracia CA, Bellot J (eds) Ecology of Mediterranean Evergreen Oak forests. Ecological studies 137. Springer, Berlin, pp 135–146

Savé R, Biel C, de Herralde F (2000) Leaf pubescence, water relations and chlorophyll fluorescence in two subspecies of *Lotus creticus* L. Biol Plant 43:239–244

Scarascia-Mugnozza G, Oswald H, Piussi P, Radoglou K (2000) Forests of the Mediterranean region: gaps in knowledge and research needs. Forest Ecol Manag 132:97–109

Scareli-Santos C, Sánchez-Mondragón ML, González-Rodríguez A, Oyama K (2013) Foliar micromorphology of Mexican oaks (*Quercus*: Fagaceae). Acta Bot Mex 104:31–52

Schimper AFW (1903) Plant-geography on a physiological basis. Clarendon Press, Oxford

Schreuder MDJ, Brewer CA, Heine C (2001) Modelled influences of non-exchanging trichomes on leaf boundary layers and gas exchange. J Theor Biol 210:23–32

Scoffoni C, Rawls M, McKown A, Cochard H, Sack L (2011) Decline of leaf hydraulic conductance with dehydration: relationship to leaf size and venation architecture. Plant Physiol 156:832–843

Seddon G (1974) Xerophytes, xeromorphs and sclerophylls: the history of some concepts in ecology. Biol J Linn Soc 6:65–87

Serrano L, Peñuelas J, Ogaya R, Savé R (2005) Tissue-water relations of two co-occurring evergreen Mediterranean species in response to seasonal and experimental drought conditions. J Plant Res 118:263–269

Serrano-Muela MP, Nadal-Romero E, Lana-Renault N, González-Hidalgo JC, López-Moreno JI, Beguería S, Sanjuan Y, García-Ruiz JM (2013) An exceptional rainfall event in the central western Pyrenees: spatial patterns in discharge and impact. Land Degrad Dev 26(3):249–262

Shmida A (1981) Mediterranean vegetation in California and Israel: similarities and differences. Israel J Bot 30:105–123

Shmida A, Whittaker RH (1984) Convergence and non-convergence of Mediterranean type communities in the old and the new world. In: Margaris NS, Arianoustou-Farragitaki M, Oechel WC (eds) Being alive on land. Dr W. Junk, The Hague, pp 5–11

Siam AMJ, Radoglou KM, Noitsakis B, Smiris P (2009) Differences in ecophysiological responses to summer drought between seedlings of three deciduous oak species. Forest Ecol Manag 258:35–42

Singh SP, Zobel DB, Garkoti SC, Tewari A, Negi CMS (2006) Patterns in water relations of central Himalayan trees. Trop Ecol 47:159–182

Sisó S, Camarero JJ, Gil-Pelegrín E (2001) Relationship between hydraulic resistance and leaf morphology in broadleaf *Quercus* species: a new interpretation of leaf lobation. Trees 15:341–345

Sperry JS, Tyree MT (1988) Mechanism of water-stress induced embolism. Plant Physiol 88:581–587

Sperry JS, Christman MA, Torres-Ruiz JM, Taneda H, Smith DD (2012) Vulnerability curves by centrifugation: is there an open vessel artefact, and are 'r' shaped curves necessarily invalid? Plant, Cell Environ 35:601–610

Sobrado MA, Medina E (1980) General Morphology, anatomical structure, and nutrient content of sclerophyllous leaves of the 'Bana' vegetation of amazonas. Oecologia 45:341–345

Suc JP (1984) Origin and evolution of the Mediterranean vegetation and climate in Europe. Nature 307:429–432

Sultan SE, Bazzaz FA (1993) Phenotypic plasticity in *Polygonum persicaria*. II. Norms of reaction to soil-moisture and the maintenance of genetic diversity. Evolution 47:1032–1049

Thompson JD (2005) Plant Evolution in the Mediterranean. Oxford University Press, Oxford

Tobin MF, Pratt RB, Jacobsen AL, De Guzman ME (2013) Xylem vulnerability to cavitation can be accurately characterised in species with long vessels using a centrifuge method. Plant Biol 15:496–504

Tognetti R, Longobucco A, Raschi A (1998) Vulnerability of xylem to embolism in relation to plant hydraulic resistance in *Quercus pubescens* and *Quercus ilex* co-occurring in a Mediterranean coppice stand in central Italy. New Phytol 139:347–448

Tognetti R, Longobucco A, Raschi A (1999) Seasonal embolism and xylem vulnerability in deciduous and evergreen Mediterranean trees influenced by proximity to a carbon dioxide spring. Tree Physiol 19:271–277

Tognetti R, Cherubini P, Marchi S, Raschi A (2007) Leaf traits and tree rings suggest different water-use and carbon assimilation strategies by two co-occurring *Quercus* species in a Mediterranean mixed-forest stand in Tuscany, Italy. Tree Physiol 27:1741–1751

Traiser C, Klotz S, Uhl D, Mosbrugger V (2005) Environmental signals from leaves—a physiognomic analysis of European vegetation. New Phytol 166:465–484

Trigo RM, García-Herrera R, Díaz J, Trigo IF, Valente A (2005) How exceptional was the early August 2003 heatwave in France. Geophys Res Lett 32:L10701

Trouet V, Taylor AH, Carleton AM et al (2009) Interannual variations in fire weather, fire extent, and synoptic-scale circulation patterns in northern California and Oregon. Theor Appl Climatol 95:349

Turner IM (1994) Sclerophylly: primarily protective? Funct Ecol 8:669–675

Tyree MT, Cochard H (1996) Summer and winter embolism in oak. Impact on water relations. Ann Sci Forest 53:173–180

Urbieta IR, Zavala MA, Marañón T (2008) Human and non-human determinants of forest composition in Southern Spain: evidence of shifts towards cork oak dominance as a result of management over the ast century. J Biogeogr 35:1688–1700

Urli M, Porté AJ, Cochard H, Guengant Y, Burlett R, Delzon S (2013) Xylem embolism threshold for catastrophic hydraulic failure in angiosperm trees. Tree Physiol 33:672–683

Valero-Galván J, González-Fernández R, Navarro-Cerrillo R, Gil-Pelegrín E, Jorrín-Novo JV (2013) Physiological and proteomic analyses of drought stress response in holm oak rovenances. J Proteome Res 12:5110–5123

Valero-Galván J, Valledor L, Navarro-Cerrillo R, Gil-Pelegrín E, Jorrín-Novo JV (2011) Studies of variability in Holm oak (*Quercus ilex* subsp. *ballota* [Desf.] Samp.) through acorn protein profile analysis. J Proteomics 74:1244–1255

Valiente-Banuet A, Rumebe AV, Verdú M, Callaway RM (2006) Modern quaternary plant lineages promote diversity through facilitation of ancient tertiary lineages. PNAS 103:16812–16817

Vankat JL (1982) A gradient perspective on the vegetation of Sequoia National Park, California. Madroño 29:200–214

Vaz M, Pereira JS, Gazarini LC, David TS, David JS, Rodrigues A, Maroco J, Chaves MM (2010) Drought-induced photosynthetic inhibition and autumn recovery in two Mediterranean oak species (*Quercus ilex* and *Quercus suber*). Tree Physiol 30:946–956

Vaz M, Cochard H, Gazarini L, Graça J, Chaves MM, Pereira JS (2012) Cork oak (*Quercus suber* L.) seedlings acclimate to elevated CO_2: photosynthesis, growth, wood anatomy and hydraulic conductivity. Trees 26:1145–1157

Venturas MD, Rodriguez-Zaccaro FD, Percolla MI, Crous CJ, Jacobsen AL, Pratt RB (2016a) Single vessel air injection estimates of xylem resistance to cavitation are affected by vessel network characteristics and sample length. Tree Physiol 36:1247–1259

Venturas MD, MacKinnon ED, Dario HL, Jacobsen AL, Pratt RB, Davis SD (2016b) Chaparral shrub hydraulic traits, size, and life history types relate to species mortality during California's historic drought of 2014. PLoS ONE 11(7):e0159145

Verdú M, Dávila P, García-Fayos P, Flores-Hernández N, Valiente-Banuet A (2003) 'Convergent' traits of Mediterranean woody plants belong to pre-Mediterranean lineages. Biol J Linn Soc 78:415–427

Verdú M, Pausas JG, Segarra-Moragues JG, Ojeda F (2007) Burning phylogenies: fire, molecular evolutionary rates, and diversification. Evolution 61:2195–2204

Verheye W, de la Rosa D (2005) Mediterranean soils. In: Land use and land cover from encyclopedia of life support systems (EOLSS), Developed under the Auspices of the UNESCO, Eolss Publishers, Oxford, UK

Vertovec M, Sakçali S, Ozturk M, Salleo S, Giacomich P, Feoli E, Nardini A (2001) Diagnosing plant water status as a tool for quantifying water stress, on a regional basis in Mediterranean drylands. Ann For Sci 58:113–125

Vilagrosa A, Cortina J, Gil-Pelegrín E, Bellot J (2003a) Suitability of drought-preconditioning techniques in Mediterranean climate. Restor Ecol 11:208–216

Vilagrosa A, Bellot J, Vallejo VR, Gil-Pelegrín E (2003b) Cavitation, stomatal conductance, and leaf dieback in seedlings of two co-occurring Mediterranean shrubs during an intense drought. J Exp Bot 54:2015–2024

Vilagrosa A, Morales F, Abadía A, Bellot J, Cochard H, Gil-Pelegrin E (2010) Are symplast tolerance to intense drought conditions and xylem vulnerability to cavitation coordinated? An integrated analysis of photosynthetic, hydraulic and leaf level processes in two Mediterranean drought-resistant species. Environ Exp Bot 69:233–242

Vilagrosa A, Chirino E, Peguero-Pina JJ, Barigah TS, Cochard H, Gil-Pelegrín E (2012) Xylem cavitation and embolism in plants living in water-limited ecosystems. In: Aroca R (ed) Plant responses to drought stress. Springer, Berlin, pp 63–109

Villar R, Merino J (2001) Comparison of leaf construction costs in woody species with differing leaf life-spans in contrasting ecosystems. New Phytol 151:213–226

Villar-Salvador P, Planelles R, Oliet J, Peñuelas-Rubira JL, Jacobs DF, González M (2004) Drought tolerance and transplanting performance of holm oak (*Quercus ilex*) seedlings after drought hardening in the nursery. Tree Physiol 4:1147–1155

Vogel S (2009) Leaves in the lowest and highest winds: temperature, force and shape. New Phytol 183:13–26

Wagner GJ, Wang E, Shepherd RW (2004) New approaches for studying and exploiting an old protuberance, the plant trichome. Ann Bot 93:3–11

Wallace JM, Gutzler DS (1981) Teleconnections in the geopotential height field during the Northern Hemisphere winter. Mon Weather Rev 109:784–812

Wallen CC (1970) Climates of Central and Southern Europe. World survey of climatology Volume 6. Elsevier Scientific Publishing Co, Amsterdam

Walter H (1985) Vegetation of the Earth and ecological systems of the geo-biosphere, 3rd edn. Springer, Berlin, p 318

Westbrook JW, Kitajima K, Burleigh JG, Kress WJ, Erickson DL, Wright SJ (2011) What makes a leaf tough? Patterns of correlated evolution between leaf toughness traits and demographic rates among 197 shade-tolerant woody species in a neotropical forest. Am Nat 177:800–811

Withington JM, Reich PB, Oleksyn j, Eissenstat DM (2006) Comparisons of structure and life Span in roots and leaves among temperate trees. Ecol Monogr 76:381–397

Witkowski ETF, Lamont BB (1991) Leaf specific mass confounds leaf density and thickness. Oecologia 88:486–493

Wright W, Vincent JFV (1996) Herbivory and the mechanics of fracture in plants. Biol Rev Camb Philos Soc 71:401–413

Xu L, Baldocchi DD (2003) Seasonal trends in photosynthetic parameters and stomatal conductance of blue oak (*Quercus douglasii*) under prolonged summer drought and high temperature. Tree Physiol 23:865–877

Yates MJ, Verboom GA, Rebelo AG, Cramer MD (2010) Ecophysiological significance of leaf size variation in *Proteaceae* from the Cape Floristic Region. Funct Ecol 24:485–492

Zhang SB, Zhou ZK, Hu H, Xu K, Yan N, Li SY (2005) Photosynthetic performances of *Quercus pannosa* vary with altitude in the Hengduan Mountains, Southwest China. Forest Ecol Manag 212:291–30

Zhang YJ, Rockwell FE, Graham AC, Alexander T, Holbrook NM (2016) Reversible leaf xylem collapse: a potential "circuit breaker" against cavitation. Plant Physiol 172:2261–2274

Zhu F, Yoh M, Gilliam FS, Lu X, Mo J (2013) Nutrient limitation in three lowland tropical forests in southern China receiving high nitrogen deposition: insights from fine root responses to nutrient additions. PLoS ONE 8(12):e82661

Chapter 6
Coexistence of Deciduous and Evergreen Oak Species in Mediterranean Environments: Costs Associated with the Leaf and Root Traits of Both Habits

Alfonso Escudero, Sonia Mediavilla, Manuel Olmo, Rafael Villar and José Merino

Abstract The geographic distribution of deciduous versus evergreen woody species has been intensively investigated, but the ecological significance of both leaf habits is still far from being fully understood. The purpose of this chapter is to review the factors that are related with the carbon gain of deciduous and evergreen oak species under Mediterranean environmental conditions. We will focus on the morphological, anatomical and chemical adaptations of evergreens necessary to guarantee leaf survival during the unfavorable part of the year. We will review the information available about the construction and maintenance costs associated with the leaf traits of deciduous and evergreen oak species. Moreover, we will compare these traits with those of non-Mediterranean oaks and species belonging to other families. One central leaf trait is the leaf mass per area (LMA), which depends on the leaf anatomy and chemical composition. Differences in LMA are related to photosynthesis and the costs of construction and maintenance. We will assess the differences in these traits between deciduous and evergreen oaks, the aim being to understand the coexistence of both leaf habits in certain environments.

A. Escudero · S. Mediavilla
Departamento de Ecología, Facultad de Biología,
Universidad de Salamanca, Salamanca, Spain

M. Olmo · R. Villar
Área de Ecología, Facultad de Ciencias, Universidad de Córdoba,
14071 Córdoba, Spain

J. Merino (✉)
Departamento Sistemas Físicos, Químicos y Naturales,
Universidad de Pablo Olavide, Sevilla, Spain
e-mail: jamerort@upo.es

© Springer International Publishing AG 2017
E. Gil-Pelegrín et al. (eds.), *Oaks Physiological Ecology. Exploring the Functional Diversity of Genus* Quercus L., Tree Physiology 7,
https://doi.org/10.1007/978-3-319-69099-5_6

6.1 Introduction

The differentiation between the deciduous and evergreen leaf habits in woody species has been a focus of research for many investigators. In particular, the geographic distribution of communities dominated by deciduous or evergreen species suggests that the two habits must have different requirements. However, the patterns of geographic distribution of both habits are complex with apparently inconsistent changes with latitude, which in some cases may seem paradoxical (Givnish 2002). In addition, although there are evident geographic patterns in the dominance of the different leaf habits, spatial and temporal variation in resource supply within a stand are often sufficiently intense to permit the success of different strategies (Reich 2014). Accordingly, deciduous and evergreen species are frequently present in the same sites (Van Ommen Kloeke et al. 2012), which implies that they should be able to resist the competition of species having the other leaf habit.

Different models have been developed to explain the geographic distribution of deciduous and evergreen species. In general, deciduous species are favored wherever the seasonal difference in the net rate of whole plant return from leaves adapted to the favorable season versus the unfavorable season is large (Reich et al. 1992; Givnish 2002). However, in strongly seasonal climates, short favorable seasons and long unfavorable ones should lead to dominance by evergreens with prolonged leaf life span and sclerophyllous structure, because under these circumstances it is difficult for a leaf to pay back its construction costs by photosynthetic gain during a single season since the favorable period is too short (Reich et al. 1992; Kikuzawa 1995). Models also suggest that as the maximum photosynthetic rate increases, leaf longevities should become shorter (Kikuzawa 1995), because high initial C assimilation rates allow for shorter payback times. Accordingly, evergreen communities should dominate regions that are always relatively unfavorable due to low fertility and/or water availability (Reich et al. 1992; Givnish 2002). Finally, the optimal leaf longevity should also decrease as the rate of decrease in photosynthesis with leaf aging becomes higher; whereas when construction costs increase, leaf longevities should increase (Kikuzawa 1995).

Given the above predictions, Mediterranean environments have been considered traditionally as a typical habitat for evergreen sclerophyllous species. Mediterranean climates are characterized by rainfall that occurs mainly in winter and drought during summer, which result in negative correlations of rainfall with temperature and light availability. This peculiarity of Mediterranean climates should lead to relatively low seasonality in the net potential return from photosynthesis (Givnish 2002), because photosynthesis would be limited by low temperatures during cold seasons and by low water availability during warm periods. Under these circumstances, evergreens should be favored. In addition, stomatal limitations in response to low soil water availability might lead to reduced maximum photosynthetic rates. High light intensities during the growing season and low vegetation densities (Flexas et al. 2014) should permit deep penetration of light into the canopy, leading to a slow deterioration of the light environment of the older leaves as new, younger leaf layers are added to the canopy (Ackerly and Bazzaz 1995; Reich et al. 2009). This should also favor the evergreen habit because older leaves may still attain relatively high photosynthetic rates, in comparison with the new leaves. Finally, sclerophylly is a frequent trait in Mediterranean woody species, apparently because it affords protection against drought stress during the Mediterranean summer. Sclerophyllous leaves have a higher leaf mass per area (LMA) than those of malacophylls (Turner 1994) and this should imply higher construction costs, at least when expressed per unit leaf area (Villar and Merino 2001).

However, Mediterranean regions are extremely variable both spatially and temporally, exhibiting different degrees of canopy cover, and a wide range of soil moisture availability and temperature over the course of a year (Baldocchi et al. 2010). In addition, contrary to common assumptions, Mediterranean woody species present photosynthetic capacities similar to those of other biomes (Flexas et al. 2014). Accordingly, the predominance of conditions favorable to sclerophyllous evergreens is not as general as previously supposed, and this probably explains the variety of leaf habits present in the Mediterranean environments. In particular, in both Europe and North America, Mediterranean environments present a rich array of oaks (*Quercus* spp.) with different leaf habits and leaf life spans (Baldocchi et al. 2010), as well as varying photosynthetic capacities.

In our study, we have compiled data from the literature on 115 woody species from different taxonomic groups and geographic origins, to analyze the possible differences between *Quercus* species belonging to Mediterranean and non-Mediterranean environments and between deciduous and evergreen leaf habits. We collected data on leaf traits, leaf chemistry, phenology, and maximum gas-exchange rates together with data on fine roots from different sources (unpublished data and published papers). References for data used in these surveys are listed in the Appendix.

Our main objective was to understand the differences in leaf and root traits between deciduous and evergreen oak species in Mediterranean regions, which may explain the coexistence of both leaf habits. Moreover, we compared these traits as well as leaf phenological patterns with those of oaks belonging to non-Mediterranean areas and with those of species belonging to other families.

One fundamental leaf trait is the LMA, which depends on the leaf anatomy and chemical composition (Fig. 6.1). Differences in LMA are associated with photosynthesis and the costs of construction and maintenance. We will assess the differences in these traits between deciduous and evergreen oaks, with the aim of understanding the coexistence of both leaf habits in certain environments.

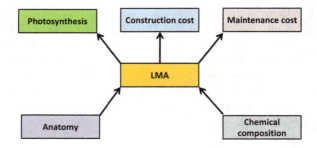

Fig. 6.1 Conceptual diagram of the relationships between the leaf traits that will be discussed in the chapter

6.2 Comparison of Leaf Phenology Among Different Leaf Habits

Despite the well-known existence of a double limitation in Mediterranean climates, arising from low winter temperatures and summer drought (Mitrakos 1980), most analyses of the ecophysiology of woody species in Mediterranean environments have focused on the limitations to plant performance derived from drought and heat stress during summer. Much less attention has been dedicated to the limitations to photosynthesis owing to cold stress during the winter and to the adaptations needed for leaf survival during winter in evergreen Mediterranean species. Only recently, have several papers (Ogaya and Peñuelas 2007; González-Zurdo et al. 2016a) described intense changes in the leaf traits of evergreen species as a response to winter temperature gradients, and this suggests that future research should focus more on the effects of low temperature on leaf traits (Niinemets 2016). Furthermore, among the different leaf habits present in Mediterranean environments, most attention has been devoted to evergreens, despite the fact that in the Mediterranean environments of the Northern Hemisphere numerous deciduous oak species occupy ample extensions, both in Europe and in North America (Table 6.1).

In fact, many of the Mediterranean deciduous species occupy habitats with severe drought during summer. The widespread abundance of winter-deciduous species in warm Mediterranean climates has frequently attracted the attention of researchers, both in the Mediterranean regions of North America: *"Surprisingly, winter-deciduous species are often dominant where summer drought is especially severe"* (Blumler 2015), and in the Mediterranean Basin: *"It could be expected that most of the Mediterranean tree species will be evergreen, but they consist of only about 50% of the tree species in Israel"* (Ne'eman and Goubitz 2000). Given that many oak species exhibit great plasticity in their leaf phenological patterns (Ne'eman 1993; García Nogales et al. 2016), if the selective pressure in favor of the evergreen habit under warm Mediterranean conditions were so intense as generally supposed, the deciduous habit should have been rapidly replaced.

In addition, despite the severe summer drought stress typical of most Mediterranean regions, many deciduous tree species in this area exhibit patterns of leaf phenology similar to those of other deciduous species from temperate climates

Table 6.1 List of Mediterranean oak species from Europe and North America

Evergreen		Deciduous	
North America	Europe	North America	Europe
Q. ajoensis	Q. alnifolia	Q. douglasii	Q. afares
Q. agrifolia	Q. calliprinos	Q. kelloggii	Q. canariensis
Q. berberidifolia	Q. coccifera	Q. lobata	Q. cerris
Q. brandegeei	Q. ilex		Q. faginea
Q. cornelius-mulleri	Q. suber		Q. frainetto
Q. cedrosensis			Q. infectoria
Q. chrysolepis			Q. ithaburensis
Q. dumosa			Q. lusitanica
Q. durata			Q. macrolepis
Q. engelmannii			Q. pubescens
Q. garryana			Q. pyrenaica
Q. john-tuckeri			Q. trojana
Q. palmeri			
Q. pacifica			
Q. parvula			
Q. sadleriana			
Q. turbinella			
Q. tomentella			
Q. wislizeni			

with less severe drought stress and lower winter temperatures. The dates of leaf unfolding and leaf fall of deciduous Mediterranean oaks do not differ excessively from those of their counterparts from non-Mediterranean temperate regions (Table 6.2). In fact, other genera of deciduous trees typical of Mediterranean climates also maintain phenological patterns similar to those of winter-deciduous species from colder climates (Table 6.2). If we assume that the times of leaf emergence and abscission in deciduous plants are selected to maximize the expected whole-plant return per unit leaf mass per season (Givnish 2002), the similitudes between Mediterranean and non-Mediterranean deciduous species in their spring and autumn phenology suggest that in both areas spring and summer conditions for photosynthesis should also be relatively similar.

Actually, leaf unfolding tends to be slightly earlier in Mediterranean climates in comparison with oak species from colder climates, probably as a response to the higher temperatures in spring (Table 6.2). However, oak species often experience embolisms if there are frosts after leaf flush (Tyree and Cochard 1996). As a consequence, species typical of colder sites under Mediterranean conditions in the Iberian Peninsula, such as Q. pyrenaica, tend to have late leaf-out times, very similar to those of Central European species. The mean dates of leaf senescence indicate that Mediterranean oaks are able to keep their leaf biomass during the dry season, whereas other Mediterranean winter-deciduous plants, like many members

Table 6.2 Approximate dates (DOY) of leaf unfolding and leaf fall in deciduous species

Species	Leaf unfolding	Leaf fall	Canopy duration
Mediterranean oak species			
Quercus canariensis	90	365	275
Quercus douglasii	78	320	242
Quercus faginea	94	326	232
Quercus ithaburensis	82	341	260
Quercus pyrenaica	128	329	201
Mean	94	336	242
Non-Mediterranean oak species			
Quercus petraea	120	310	190
Quercus robur	116	304	188
Quercus rubra	129	307	178
Mean	122	307	185
Shrub Mediterranean species			
Pyrus bourgaeana	71	210	139
Crataegus monogyna	87	263	176
Sambucus nigra	78	324	247
Mean	79	266	187
Other Mediterranean trees			
Fraxinus angustifolia	108	329	221
Celtis australis	90	313	223
Pistacia terebinthus	134	312	178
Acer monspessulanum	89	297	208
Mean	105	313	208

Canopy duration in days

of the *Rosaceae* family, tend to lose part of their leaf area during late summer in response to the severe drought stress typical of this part of the year. This results in mean dates of leaf abscission that are earlier than those of deciduous Mediterranean oaks (Table 6.2). This anticipation of leaf loss is compensated by earlier dates of leaf unfolding, owing to the high resistance to late frosts of these species (Duhme and Hinckley 1992). The behavior of the deciduous oak species implies that they are able to obtain a positive carbon (C) balance also during the most stressing part of the drought season. In fact, eddy-flux based measurements in Mediterranean ecosystems dominated by deciduous oaks reveal that significant C assimilation rates are maintained throughout the dry season (Kuglitsch et al. 2008; Ma et al. 2011). Probably, this summer activity is due to the capacity of Mediterranean trees to produce deep roots that obtain water from deep soil layers (Canadell et al. 1996). Even on rocky soils, roots of oak trees have been observed penetrating through fissures and cracks (David et al. 2004). In several species the mechanism of "hydraulic lift" has been demonstrated to release water from deep soil layers into upper soil layers during the night (Prieto et al. 2012). Given that most annual precipitation in Mediterranean environments is concentrated in winter when the low potential evapotranspiration allows a surplus of water to infiltrate into deep soil layers, the mechanism of hydraulic lift may significantly increase the amount of water transpired during summer (Canadell et al. 1996; Kurz-Besson et al. 2006).

In addition, the photosynthetic apparatus of many Mediterranean species seems to be resistant to high summer temperatures and water stress (Daas et al. 2008). A pattern common to different plant species is that metabolic impairment does not limit photosynthesis until the water stress is severe (i.e. maximum daily stomatal conductance below 0.10–0.15 mol H_2O m^{-2} s^{-1}) (Flexas et al. 2004). In addition, although periods with very low maximum daily stomatal conductances are usual under dry Mediterranean conditions, the recovery of photosynthesis after the stress period is usually rapid (Galmés et al. 2007; Bongers et al. 2017).

One evident advantage of the evergreen habit under Mediterranean conditions is that the possession of a year-round green canopy allows significant levels of C assimilation to be attained during winter, when the deciduous species are inactive (Hollinger 1992; Givnish 2002; Van Ommen Kloeke et al. 2012). However, although some instantaneous measurements suggest that the rates of C assimilation may be relatively high during winter months (e.g. Asensio et al. 2007; Bongers et al. 2017), when the amounts of C assimilated are calculated for more extended periods of time, the winter contribution to the annual total is usually modest. For example, even in warm low-altitude sites of the Mediterranean Basin, the contribution of the November–April period amounted to no more than 20% of the total annual gross primary production achieved by an evergreen canopy (calculated from data reported in Allard et al. 2008; Garbulsky et al. 2008; Kuglitsch et al. 2008). Although air temperatures during the Mediterranean winter at low-altitude sites are not especially low, the short photoperiod may be a strong limiting factor at this time of the year (Givnish 2002). Obviously, at colder sites, the winter contribution to total productivity should be even lower.

Accordingly, the extended period for photosynthesis implies only a modest advantage in terms of total annual production. Given that keeping a green leaf canopy during winter has associated maintenance costs and costs of leaf adaptations to freezing (Van Ommen Kloeke et al. 2012), the final balance of the evergreen habit strictly depends on the exact quantification of the costs associated with it.

6.3 The Leaf Economics Spectrum in Deciduous and Evergreen Oaks

The coexistence of deciduous and evergreen species means that both leaf habits must perform similarly under the same environmental conditions. The competition between the two leaf habits can be suitably analyzed in the context of the "leaf economics spectrum" (Wright et al. 2004). In most cases, the evergreenness is achieved by a longer leaf duration that allows the overlapping of several leaf cohorts in the crown. Leaf life span is a pivotal trait in the carbon-fixation 'strategy' of a species (Wright et al. 2002). Species with a long leaf life span usually exhibit high leaf mass per area (LMA), low nutrient concentrations per unit leaf mass, and slow gas-exchange rates (Reich et al. 1999; Poorter et al. 2009). These traits form a spectrum of "leaf economics", which has been analyzed by many researchers (Wright et al. 2004; Wright and Sutton-Grier 2012; Edwards et al. 2014; Reich 2014; Díaz et al. 2016). We can expect that evergreen and deciduous Mediterranean tree species should maintain trends along the leaf economics spectrum similar to those described by the above-mentioned authors. To analyze the position of Mediterranean tree species along the spectrum, we have compiled data available in the literature.

One important limitation for interspecific comparisons that involve differences in leaf habit is that much of the information available in the literature has been obtained with small plants growing in controlled environments (Bassow and Bazzaz 1998; Valladares et al. 2004). This may constitute a problem if we try to extrapolate the results obtained to mature specimens, because many leaf traits tend to change along ontogeny. If the ontogenetic trends were of similar magnitude for different plant functional types, the interspecific differences observed at the seedling stage could be extrapolated to the mature stage. However, the ontogenetic changes tend to be stronger for species with longer leaf life span at maturity (Mediavilla et al. 2014), which means that predictions of plant performances, based on data obtained from studies limited to a part of the life cycle, should be made with caution (Cornelissen et al. 2003). For this reason, we have calculated the species means separately for those studies based on seedling traits. By comparing the trait values obtained for seedlings and mature specimens within each species, we can estimate the ontogenetic change observed for the different leaf traits, for each species, as 100 × (trait value in mature trees − trait value in seedlings)/(trait value in seedlings) (Table 6.3).

As we can see, most leaf traits, as well as the maximum gas-exchange rates, changed along ontogeny. Mature trees produced leaves with greater thickness and LMA. Despite the interspecific trend of high-LMA leaves to have lower nitrogen (N) contents per unit mass (N_{mass}) (Reich et al. 1992; Wright et al. 2004), mature trees tended to have higher N_{mass} values; accordingly, the contents of N per unit area were approximately 50% greater in mature trees compared to their counterparts from the seedling stage. The concentrations of structural carbohydrates also tended to increase along ontogeny, but the lignin concentrations maintained the opposite trend. The larger amounts of N per unit area in mature trees resulted in much higher photosynthetic rates per unit leaf area (A_{area}). However, in the case of evergreens, the effect on A_{area} of the increase in N_{area} with the age of trees was almost compensated by a pronounced decrease in maximum stomatal conductance. Given the different ontogenetic trends, the comparison of the two leaf habits differs depending on the stage addressed. Frequently, comparisons made at the seedling stage provide scarce differences in gas-exchange rates between deciduous and evergreen species (Acherar et al. 1991; Acherar and Rambal 1992; Lo Gullo et al. 2003). By contrast, at the adult stage the differences tend to be much larger (Tretiach 1993; Mediavilla

Table 6.3 Ontogenetic change in several leaf traits for deciduous and evergreen oaks

	LMA	Thickness	Leaf density	N_{mass}	N_{area}	Hemicellulose	Cellulose	Lignin	A_{area}	A_{mass}	g_s
Deciduous	18.7	39.6	8.6	15.0	52.4	19.2	11.2	−15.0	114.8	43.6	2.0
Evergreen	43.8	42.7	16.5	3.5	52.8	75.4	18.0	−10.0	17.3	−32.5	−27.1

LMA Leaf mass per area; N_{mass} N concentration per unit leaf mass; N_{area} N content per unit leaf area; A_{area} Maximum photosynthesis per unit leaf area; A_{mass} Maximum photosynthesis per unit leaf mass; g_s Maximum stomatal conductance. The intensity of the ontogenetic change was evaluated as: 100 × (trait value in mature trees − trait value in seedlings)/(trait value in seedlings) and Escudero 2003; Flexas et al. 2014). For this reason, the results provided in Figs. 6.2, 6.3, and 6.4 refer exclusively to mature specimens.

As observed by many authors (Reich et al. 1992; Wright et al. 2004), the maximum photosynthetic rates of each species were negatively related to leaf life span in the whole set of species. This negative relationship was especially tight when CO_2 assimilation rates were expressed per unit leaf mass and less clear, although still significant, when expressed per unit leaf area (Fig. 6.2a, b).

The negative relationship between leaf life span and instantaneous assimilation rates per unit mass (A_{mass}) followed a power function (linear when both variables were log-transformed). Given this relationship between the two variables, several authors have postulated its slope as a measure of the effects of leaf duration on the CO_2 assimilated throughout the life of the leaf. A slope equal to -1 in the log–linear relationship between leaf longevity and A_{mass} would mean that the product of instantaneous assimilation rate × leaf life span, which may constitute a crude estimation of total assimilation at the end of the life of the leaf, is independent of the leaf life span (Westoby et al. 2000). A slope greater than -1 (less negative) would imply that cumulative C assimilation increases with leaf life span (Reich et al. 1992;

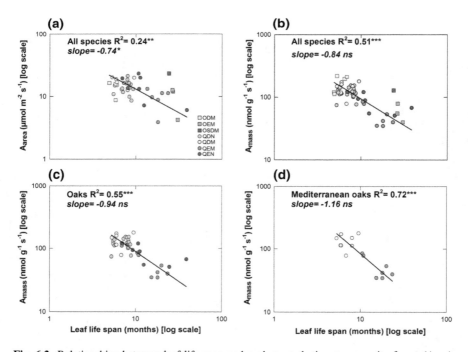

Fig. 6.2 Relationships between leaf life span and **a** photosynthetic rate per unit of area (A_{area}); **b** photosynthetic rate per unit of mass (A_{mass}) for all species; **c** all oaks and **d** Mediterranean oaks. The slopes of the regressions were calculated as Standardized Major Axis Regressions using SMATR (Warton et al. 2006). *Ns* Non significant; **$P < 0.01$; ***$P < 0.001$. The significance of the slope's difference from -1 is also shown

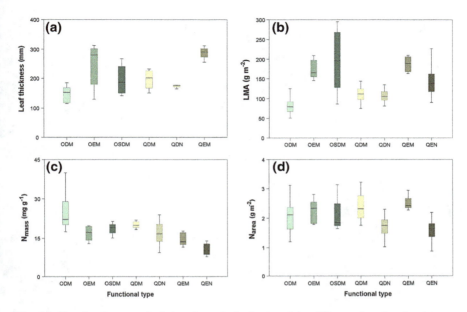

Fig. 6.3 Boxplot diagrams depicting the main leaf traits of the different plant functional types. The box in each box plot shows the median and the lower and upper quartile, and the whiskers show the range of variation. Identification of functional types: *ODM* Mediterranean deciduous non-oak spp; *OEM* Mediterranean evergreen non-oak spp; *OSDM* Mediterranean semideciduous non-oak spp; *QDM* Mediterranean deciduous oak spp; *QDN* Non-Mediterranean deciduous oak spp; *QEM* Mediterranean evergreen oak spp; *QEN* Non-Mediterranean evergreen oak spp

Westoby et al. 2000). The slope calculated for the whole set of species used in this review (Fig. 6.2b) was not significantly different from −1 for the mass-based relationships. Similar relationships between leaf life span and A_{mass} were found when the range of species used was limited to oak species, although the slope of the relationship decreased slightly (Fig. 6.2c), but it was not significantly different from 1. Similarly, the log-log plot of A_{mass} and leaf life span calculated only for Mediterranean oaks provided a slope very close to −1 (Fig. 6.2d), which means that the lengthening leaf duration exactly compensates the concomitant decrease in instantaneous assimilation capacity. These results differ from those obtained by other authors, who found that the photosynthetic rate changed at roughly the −2/3 power of leaf longevity (Reich et al. 1992; Givnish 2002), implying that leaf lifetime C assimilation would increase with leaf life span. Given that the present analysis includes only broadleaf species with a relatively short range of leaf life span, our results suggest that, when we compare species within a limited range of leaf duration, CO_2 assimilation throughout the leaf life is independent from leaf duration, which could explain the coexistence of deciduous and evergreen species in many environments. Instantaneous CO_2 assimilation tends to decrease with leaf age. The inclusion of this effect should contribute to reducing the slope, because the instantaneous assimilation rate averaged for the different leaf cohorts present in the

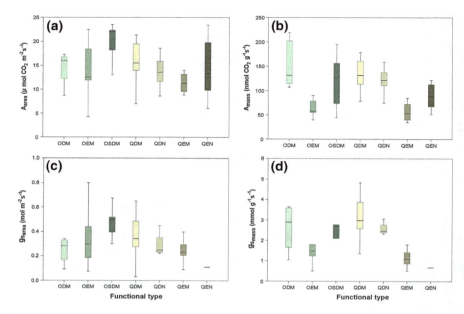

Fig. 6.4 Boxplot diagrams depicting the gas exchange rates of the different plant functional types. The box in each box plot shows the median and the lower and upper quartile, and the whiskers show the range of variation. g_s, stomatal conductance. Identification of functional types as in Fig. 6.3

crown should be lower for species with long leaf life span (Mediavilla and Escudero 2003). However, again the limited range of leaf life spans in our set of species means that even in the evergreen species with the longest leaf life span the proportion of old leaves in the crown is relatively low.

The similar capability for leaf C fixation of Mediterranean species differing in leaf life span (Fig. 6.2) is also apparent at the ecosystem level. Eddy-flux based measurements in Mediterranean ecosystems provide very similar values of gross primary production for oak species such as *Q. douglasii* (Ma et al. 2007), *Q. suber* (Pereira et al. 2007), and *Q. cerris* and *Q. ilex* (Maselli et al. 2006), thus revealing that different leaf habits may perform similarly under Mediterranean conditions. These results are also in line with those found in the dehesas (savanna-like ecosystems) of the Iberian Peninsula, where acorn production has great economic importance. In these systems, evergreen species usually have greater acorn production than deciduous species (Martín Vicente et al. 1998), which gives the impression that the evergreen leaf habit is advantageous under Mediterranean conditions. However, total production (fruits plus litter) per species cover (a surrogate of the species net primary production) is similar for species differing in leaf life span (Martín Vicente et al. 1998 and unpublished data), which could help to explain the coexistence of evergreen and deciduous habits in the same habitats.

The negative relationship between leaf duration and photosynthetic capacity may be interpreted as being the consequence of a trade-off between leaf traits that confer persistence and those that maximize instantaneous productivity (Reich et al. 1997; Warren and Adams 2000; Takashima et al. 2004). Similarly, a strong negative effect of a high LMA on photosynthetic rates per unit leaf mass has been observed in many studies (Reich et al. 1997, 1999). The leaf N concentration is one of the main leaf traits that determine photosynthetic capacity. A high concentration of structural components contributes to increasing leaf resistance and leaf duration (Villar et al. 2006). However, high concentrations of structural components per unit leaf mass contribute to the dilution of N, and this should reduce photosynthetic rates and N use efficiency (Vitousek et al. 1990; Lloyd et al. 1992). Thus, high photosynthetic rates per unit leaf mass basis are necessarily associated with short leaf life span because a high assimilation rate requires high N concentrations and consequently low LMA and thus low concentrations of structural components. All these traits contribute to increasing the vulnerability to herbivory and physical hazards of leaves with a short leaf life span (Wright and Cannon 2001; Shipley et al. 2006; Poorter et al. 2009). Accordingly, several traits of evergreen leaves may help to explain the negative effects of a long leaf life span on instantaneous assimilation rates.

6.4 Leaf Traits Differ Between Deciduous and Evergreen Oaks

We have summarized the main morphological leaf traits typical of different plant functional types in Mediterranean and non-Mediterranean environments (Fig. 6.3). For some traits, we did not find data for several of the species included. Accordingly, the average values for the different plant functional types reported in Fig. 6.3 were obtained with different sets of species. This explains the inconsistencies between the values obtained for some traits. For example, in some cases the N_{area} values are inconsistent with the mean values obtained for N_{mass} and LMA, as $N_{area} = N_{mass} \times$ LMA. Despite this, we have preferred to include the data available for each trait.

The LMA and leaf thickness tended to be greater for evergreen and semideciduous species. The means of the different functional types were significantly different for LMA ($P < 0.0001$) and marginally different ($P = 0.07$) for leaf thickness, according to one-way analysis of variance. However, non-Mediterranean evergreen oak species tended to exhibit lower LMA values in comparison with their counterparts from Mediterranean environments, although we must acknowledge that most of the data on non-Mediterranean evergreen oaks were obtained from a single study on *Quercus* species typical of Florida (Cavender-Bares et al. 2004). By contrast, there were no clear differences in LMA between typically evergreen Mediterranean species and semi-deciduous shrubs (mostly, different species of the genus *Cistus*). Most of the species of this genus are classified as semi-deciduous because they tend to lose a significant part of their leaf biomass during drought periods (Harley et al. 1987; Villar and Merino 2001). A greater LMA has been interpreted as a trait aimed at guaranteeing leaf survival, acting as protection against different environmental factors such as drought or attack by herbivores (Turner 1994; Niinemets 2001). However, according to the results of the present analysis, an elevated LMA in semi-deciduous species does not seem to be an efficient mechanism to guarantee leaf survival during periods of drought stress. In a recent review, Bartlett et al. (2012) conclude that there is no evidence of a mechanistic linkage of LMA with drought tolerance and that sclerophylly is not necessarily a mechanism of drought adaptation. However, Hallik et al. (2009) found a positive correlation between LMA and drought tolerance (according to the habitat of the different species). Also, de la Riva et al. (2016) found that species with high LMA were associated with habitats with low water availability, but the trait most closely associated with water availability was leaf density, which was higher in drier habitats. By contrast, the resistance to leaf drought damage and premature loss during summer was not related to LMA in the study of Günthardt-Goerg et al. (2013), since *Quercus pubescens*, a typical deciduous Mediterranean tree with comparatively high LMA, had more drought injury and leaf mass shedding than the non-Mediterranean deciduous *Q. petraea* under the experimental conditions in the study of these authors.

On the other hand, the tendency of LMA to increase as a response to increased winter cold in evergreen species has been repeatedly observed (Niinemets 2016). Although, usually, Mediterranean environments are assumed to have moderate cold stress, the responses of LMA to winter cold have also been reported for several species in these climates (Ogaya and Peñuelas 2007; González-Zurdo et al. 2016a).

The means of the different functional types were also significantly different for leaf N ($P < 0.0001$), both per unit mass (N_{mass}) and per unit area (N_{area}), with evergreen species showing lower N_{mass}, probably because of dilution of the N content in a larger mass per unit area (Fig. 6.3). Mediterranean oaks tended to maintain greater N concentrations in comparison with oak species with the same leaf habit from non-Mediterranean climates. This difference may reflect the tendency of deciduous species in regions with a shorter favorable period for photosynthesis (i.e. Mediterranean regions) to have higher leaf N concentrations than deciduous species in more favorable environments (Kikuzawa et al. 2013). In turn, this trait of deciduous Mediterranean species constitutes indirect evidence that, despite the similarity in phenological patterns, drought stress contributes to shortening the favorable period for photosynthesis in Mediterranean environments. The differences were especially marked for N_{area}, because of the combination in Mediterranean oaks of elevated N concentrations and high LMA.

The evergreens maintained lower maximum gas-exchange rates (A) than the deciduous and semi-deciduous species (Fig. 6.4). The differences were especially pronounced when gas-exchange rates were expressed per unit leaf mass ($P < 0.0001$). By contrast, the differences in A per unit area between functional types were not statistically significant ($P = 0.446$), similar to the results reported by

Hallik et al. (2009). The Mediterranean deciduous oaks tended to exhibit higher photosynthetic capacity than non-Mediterranean oak species of the same habit, which may constitute compensation for the limitation of total C assimilation derived from the drought stress during part of the Mediterranean summer (Hallik et al. 2009; Baldocchi et al. 2010). Other non-oak deciduous Mediterranean species achieved maximum gas-exchange rates similar to those of the deciduous oaks (Fig. 6.4). The semi-deciduous species, which experience some losses of leaf area during the most-stressed part of the season (Gulías et al. 2009), tended to exhibit maximum photosynthetic rates that were even higher than those of the deciduous oaks, at least when expressed per unit leaf area (Fig. 6.4). The patterns of stomatal conductance (g_s) are similar to those of A_{area}, g_s being higher in deciduous oaks than in evergreen oaks (Fig. 6.4). In fact, the differences in A_{area} are well explained by differences in g_s (Quero et al. 2006).

6.5 Causes of Differences in LMA

Despite its ecological relevance, LMA is still somewhat of a biological black box. We need to understand which traits cause the variation in LMA, and what the functional consequences are in terms of leaf physiology and longevity. We therefore aimed to break down LMA into its underlying anatomical components, following the approach shown in Fig. 6.5. At the whole leaf level, LMA is factorized into two components: (1) leaf thickness and (2) leaf density (LD), the amount of dry mass per unit leaf volume. The LMA is equal to the product of leaf thickness (LT) and leaf density (LD) (Witkowski and Lamont 1991). As we have described before, within the genus *Quercus*, evergreen leaves have a higher LMA than deciduous ones (Fig. 6.3; Table 6.4), as has been found in many studies (Castro-Diez et al. 1997; Villar and Merino 2001). We found that evergreen leaves have significantly higher LT and LD than deciduous leaves (Table 6.4), which accounts for the higher LMA of evergreens. Similar results have been found in other studies (Castro-Diez et al. 1997; Villar and Merino 2001; Prior et al. 2003; Wright et al. 2005; Mediavilla et al. 2008). In fact, LMA was positively correlated with both LT and LD (Fig. 6.6). The relationships between LMA and both LT and LD reveal that for evergreen oaks the variation in LMA was better explained by variation in LD, but for deciduous oaks the variation in LMA was mainly due to variation in LT (Fig. 6.6).

To determine the anatomical causes of the differences in LT and LD between the leaves of deciduous and evergreen oaks (Fig. 6.5) we compiled data of the anatomical structure of different species of *Quercus* (21 deciduous and 14 evergreen), considering the main tissues: epidermis (upper and lower), mesophyll (palisade and spongy), vascular tissue and sclerenchyma, and air spaces (Fig. 6.7).

Our aim was to know which tissues are related to changes in LT and LD, in order to explain the causes of differences in LMA (Castro-Diez et al. 1997; Villar et al. 2013). In the case of LT, its increase is mainly due to an increase in the mesophyll thickness (Fig. 6.8a). Also, the LT increase is related to an increase in all

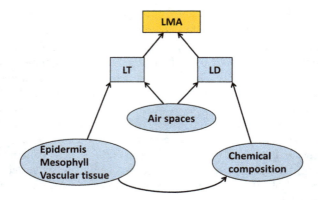

Fig. 6.5 Conceptual diagram of the anatomical characteristics that can affect LT (leaf thickness) and LD (leaf density), which may affect the variation in leaf mass per area (LMA). Vascular tissue includes also the sclerenchymatous tissue

Table 6.4 Mean values ± standard deviation of different leaf traits related to the structure and anatomy in evergreen and deciduous oaks

	Evergreen			Deciduous			Signif.
LMA (g m^{-2})	**147.6**	±	**51.8**	72.8	±	28.5	***
Leaf thickness (μm)	**235.8**	±	**66.1**	143.3	±	39.1	***
LD (g mL^{-1})	**0.6**	±	**0.2**	0.5	±	0.1	***
Epidermis (μm)	29.9	±	10.7	28.2	±	8.0	–
Mesophyll (μm)	169.7	±	61.0	113.3	±	32.7	***
Vasc + Scl. tissue (μm)	**36.1**	±	**29.0**	14.4	±	10.2	a
Air-spaces (μm)	27.0	±	10.0	23.5	±	7.9	–

The statistical significance is shown in bold (a $0.1 > P > 0.05$, ***$P < 0.001$)

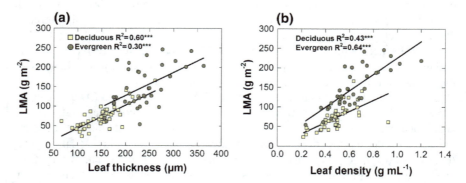

Fig. 6.6 Relationships of leaf mass per area (LMA) with **a** leaf thickness (LT) and **b** leaf density (LD) for Quercus species differing in leaf habit (evergreen and deciduous). The R^2 and the significance of the regression are shown (***$P < 0.001$)

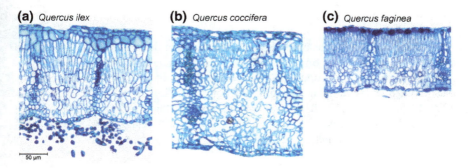

Fig. 6.7 Representative cross sections of the leaves of several species **a** the evergreen *Quercus ilex* ssp. ilex, **b** the evergreen *Quercus coccifera* and **c** the deciduous *Quercus faginea*. The bars represent 50 µm. Photographs from de la Riva et al. (2016)

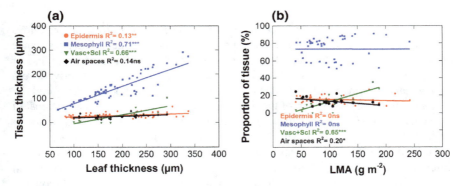

Fig. 6.8 Relationships between **a** leaf thickness and the thickness of the different tissues and **b** leaf mass per area (LMA) and the proportion of the different tissues for oak species. The R^2 and the significance of the regression are shown (*$P > 0.05$; **$P < 0.01$; ***$P < 0.001$)

tissues, except the air spaces -which remain constant throughout the range of LT. These trends were also found in other studies (Castro-Diez et al. 1997; de la Riva et al. 2016).

In the case of LD, it is more appropriate to relate it to the proportion of the tissues than to the absolute values of the thickness of the tissues (Villar et al. 2013). The density of the leaf was related positively to the proportion of vascular and sclerenchyma tissue and negatively to the proportion of air spaces ($P < 0.05$), similar to Villar et al. (2013). This could be due to the specific density of the air spaces (zero) and to the high density of vascular tissue (1.40 g cm^{-3}; indirect estimates of Poorter et al. 2009). Therefore, an increase in the proportion of air spaces will decrease the leaf density, the opposite being true in relation to the vascular and sclerenchymatic tissue.

The effect of leaf anatomy on variation in LMA is due to the joint effect of LT and LD. As LMA increases, the proportion of vascular and sclerenchyma tissue increases and the proportion of air spaces decreases (Fig. 6.8b)—which could be due to the effects of LD as it was explained before.

An increase in vascular and sclerenchyma tissue improves the leaf support and structure, and may confer on the leaf a better structural defense (Poorter et al. 2009). A decrease in the proportion of air spaces can result in lower conductance of the mesophyll and therefore a decrease in photosynthetic activity (Niinemets et al. 2009; Flexas et al. 2012). Thus, these results confirm that variation in LMA can be explained by differences in leaf tissues and that these differences have consequences for the structure and functioning of the leaf.

6.6 Differences in Construction and Maintenance Costs

6.6.1 Construction Costs

The active organs (leaves and roots) of evergreen species differ from the deciduous ones in the putative structural and chemical characteristics which confer resistance to physical stress (i.e. to loose of shape as a result of either summer water loss or low winter temperatures) or biological stress (enzymatic endowments) derived from their greater exposure to herbivores and pathogens. These supplementary endowments required for the growth of the evergreen organs, and the high specific cost of the synthesis of some of the fractions involved (lignin, wax, proteins; Penning de Vries et al. 1974), would inevitably result in high organ construction costs.

However, in the case of leaves, the pattern of the cost of construction of the different species and leaf habits is quite homogeneous, with mean values around 1.50 g glucose g^{-1} (Fig. 6.9b). Possible explanations for this are the following. First, the differences between deciduous and evergreen leaves in the concentration of structural components are not very great (Mediavilla et al. 2008). This suggests that the high toughness of evergreen leaves (around two times that of deciduous ones) is the result not just of their chemical composition (i.e. lignin or other structural fractions) but also of the spatial distribution of cell wall constituents (Gallardo and Merino 1993; Lucas et al. 2000) and their greater thickness (Fig. 6.3; Table 6.4). Second, the concentrations of the different chemical fractions in the organ are not independent but are correlated in relation to their physiological and/or structural roles. Thus, there are positive correlations between expensive and inexpensive fractions (such as phenols and ash) and negative correlations between either inexpensive fractions (such as ash and cellulose) or expensive ones (such as waxes and proteins); all together, these tend to homogenize the construction costs of the different leaf types, keeping them close to the mean values (Chapin 1989; Poorter and Jong 1999; Villar and Merino 2001; Martinez et al. 2002a).

Besides, there exists a negative correlation between the concentration of structural lipids, such as cutin or wax (the most expensive fractions to synthesize) and the concentration of cellulose (the most inexpensive fraction to synthesize), which

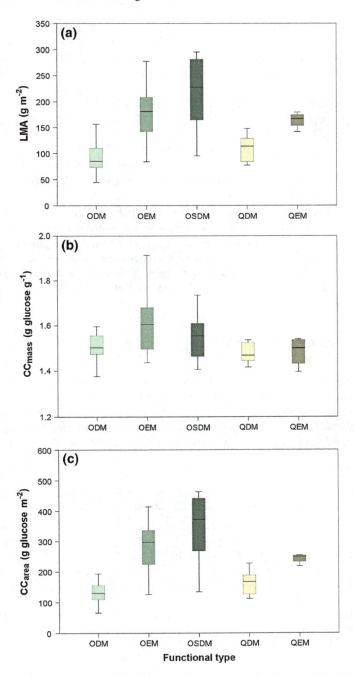

Fig. 6.9 Differences in **a** LMA, **b** leaf construction cost per unit of mass (CC_{mass}) and **c** leaf construction cost per unit of area (CC_{area}) for *Quercus* species or other species differing in leaf habit (deciduous, evergreens and semideciduous). Identification of functional types as in Fig. 6.3

contributes to the existence of differences in the construction cost, in leaves (Gallardo and Merino 1993; Villar and Merino 2001) or roots (Martínez et al. 2002a). A rather high lipid concentration explains the slightly higher construction cost of semideciduous leaves (1.55 g glucose g^{-1}) (Fig. 6.9b) (Villar and Merino 2001).

In accordance with all the above, the differences in construction costs on an area basis are just a reflection of the differences in LMA (Fig. 6.9a) displayed by the leaf types, since $CC_{area} = CC_{mass} \times LMA$. Thus, because of their higher LMA, the leaves of evergreen oaks have higher construction costs than those of deciduous oaks (242 vs. 162 g glucose m^{-2}; Fig. 6.9c); the species other than oaks follow the same pattern, while leaves of semideciduous species show the highest CC_{area} (344 g glucose m^{-2}), mainly because they have the highest LMA.

Besides, fine roots of evergreen oaks, with both a higher wax concentration (about two-times higher) and a lower cellulose concentration than those of deciduous oaks (Table 6.5), show the highest CC_{mass} (1.78 vs. 1.57 g glucose g^{-1}). The fine roots of semideciduous species, with a wax concentration similar to that of the fine roots of evergreen ones but with a higher cellulose concentration exhibit an intermediate value (1.69 g glucose g^{-1}) (Martínez et al. 2002a).

6.6.2 Maintenance Costs

The energy expenses associated with organ maintenance are mainly related to the importance of the enzymatic endowments involved in processes such as gradient maintenance, transport, replacement and reparation of endangered cellular structures, and defense against free radicals, herbivory and pathogens. This explains the significant relationships between N concentration (a surrogate of the concentration

Table 6.5 Comparison of the mean chemical composition (mg g^{-1} dry mass) and construction cost (g glucose g^{-1}) of fine roots, grouped by life form and leaf habit

Life forms	Leaf habit	Pro	Lip	Phe	Cel	Lig	Wax	TNC	Ash	Construction cost
Shrubs	Semi-deciduous	53	27	45	508[a]	129	139	35	62	1.69
Trees	Evergreen	46	23	87	361[b]	167	196[aa]	23	97	1.78[a]
	Deciduous	69	29	80	458[a]	145	86[bb]	30	104	1.57[b]

Data from Martínez et al. (2002a)
Pro Proteins; *Lip* Lipids; *Phe* Phenols; *Cel* Cellulose; *Lig* Lignin; *Wax* Wax; *TNC* Total non-structural carbohydrates
Single letters: $P < 0.05$; double letters: $P < 0.01$

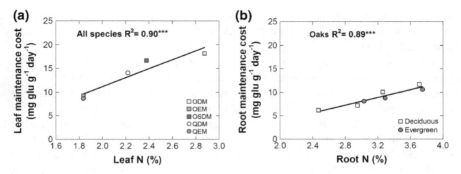

Fig. 6.10 Relationships between maintenance cost (mg glucose g^{-1} day^{-1}) and nitrogen concentration (%) in leaves (**a**) (Villar and Merino, unpublished) and fine roots (**b**) (Martínez et al. 2002b). (*$P > 0.05$; **$P < 0.01$; ***$P < 0.001$)

of enzymatic endowments) and maintenance respiration in leaves and roots (Fig. 6.10).

There was no difference in the fine root maintenance cost between evergreen and deciduous oak species (Table 6.6) (Martínez et al. 2002b). By contrast, the maintenance cost of deciduous oak leaves was almost two-times higher than that of evergreen ones (14.0 vs. 8.7 mg glucose g^{-1} day^{-1}; Fig. 6.11a), probably because of their higher enzymatic endowments (higher N concentration, Fig. 6.10a), with the rest of the species types following the same pattern. However, when expressed on an area basis, the higher LMA of the evergreen leaves compensates for the differences in maintenance cost on a mass basis; all this results in a similar maintenance cost (around 1 g glucose m^{-2} day^{-1}; Fig. 6.11b) for the two leaf habits. The leaves of semideciduous species, with a CM$_{mass}$ similar to that of deciduous species and an LMA similar to that of evergreen ones, show the highest maintenance cost (1.79 g glucose m^{-2} day^{-1}). This agrees with the ecology of this species type, which is characterized by habitats that are water stressed in summer and by very shallow root systems (Martínez et al. 1998), together resulting in severe water stress

Table 6.6 Maintenance cost (mg glucose g^{-1} dry mass d^{-1}) for roots of evergreen and deciduous species

Species	Leaf habit	Maintenance cost (mg glu g^{-1} d^{-1})
Q. pyrenaica	Deciduous	11.7
Q. canariensis	Deciduous	7.2
Q. faginea	Deciduous	10.1
Q. suber	Evergreen	10.6
Q. ilex spp. *ballota*	Evergreen	8.8
Q. coccifera	Evergreen	8.1
Q. fruticosa	Deciduous	6.2

From Martínez et al. (2002b)

during the summer months (Merino et al. 1976; and unpublished) and probably quite high defense endowments; all this explains why they have the highest maintenance cost.

It is important to point out that the maintenance cost values considered in the present discussion were gathered from individuals growing in laboratory conditions, free of temperature, water, or nutrient stress. However, the maintenance respiration of *Quercus* species increases in response to environmental stress (Laureano et al. 2016), as a consequence of greater resource allocation to defensive endowments (i.e. antioxidant enzymatic systems) (Tausz et al. 2007) and homeostatic control (i.e. higher alternative oxidase activity) (Ribas-Carbo et al. 2005).

The respiratory alternative oxidase (AOX) plays a relevant role in defense against reactive oxygen species (Millenaar and Lambers 2003; Challabathula et al. 2016) generated in regular physiological processes or induced by extreme temperatures, high light, and water and nutrient limitations (see, for example, Watanabe et al. 2016). However, its low efficiency results in supplementary energy expenses and, thus, in higher rates of maintenance respiration (maintenance respiration increases as AOX activity does; Florez-Sarasa et al. 2007). In the case of the fine roots of oaks, AOX activity (the fraction of electrons flowing through the AOX pathway in stress-free conditions) is significantly higher in deciduous species than in evergreen species (32% vs. 21% of the total root respiration; Table 6.7) (Martínez et al. 2003), This implies an extra cost for the roots of these species, which is already included in their maintenance respiration (Table 6.6). Besides, the maximum potential value of AOX activity in roots is similar in both leaf habits (around 50% of total respiration, Table 6.7); but, since evergreen species have active organs throughout periods of the year characterized by high degrees of stress, AOX activities close the potential maximum—associated with low winter temperatures or summer water and nutrient limitations—should be expected. A rough calculation considering the low efficiency of AOX, its maximum potential activity, and the root maintenance values of evergreen roots (Fig. 6.10b) shows a 5% maintenance respiration increase; a percentage that, if extended to leaves and

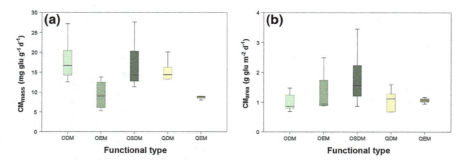

Fig. 6.11 Leaf maintenance cost **a** per unit of mass (CM_{mass}) and **b** per unit of area (CM_{area}). Identification of functional types as in Fig. 6.3

Table 6.7 Fraction (as a percentage of total respiration) of electrons flowing through the cytochrome oxidase (COX) and alternative oxidase (AOX) pathways in stress-free conditions and potential maximum electron flow through the AOX pathway (AOX Potential Activity), in roots of evergreen and deciduous *Quercus* species

Respiratory components			
Species	AOX potential activity (%)	COX activity (%)	AOX activity (%)
Deciduous	48.9	68.5 ± 1.8**	32.0 ± 2.2***
Evergreen	47.4	79.3 ± 0.8**	20.7 ± 0.8**

**$P < 0.01$, ***$P < 0.001$ (ANOVA)

computed along the unfavorable seasons, would increase significantly the maintenance energy expenses of evergreen species.

In a different line, it is important to point out that light inhibits dark respiration in photosynthetic tissues (Kok effect) (Kok 1948), since the utilization of ATP and NADPH generated directly from photosynthesis decreases the energy requirements of catabolic origin, and thus the respiration rates. In Mediterranean deciduous and evergreen leaves, inhibition of respiration by light can reach 100 and 75%, respectively, even at low light intensities (around 400 µmol m^{-2} s^{-1}) (Villar and Merino 1995). This is as expected, since the higher photosynthetic rates of deciduous species (Fig. 6.4b) would result in greater availability of ATP and NADPH and, consequently, in lower demands for respiratory energy. A rough calculation considering these percentages and the leaf maintenance cost values in Fig. 6.11 shows a 12% decrease in maintenance respiration of fully developed deciduous leaves and a decrease of between 9% and (if considering the light extinction in the canopy) 6% in the case of evergreen ones; this indicates higher energy expenses for evergreen leaf maintenance.

In summary, the organs (leaves and roots) of evergreen oaks (and of other evergreen species) have higher construction costs than those of deciduous ones and maintenance costs that are similar; although the latter could be increased for evergreen species, if considering the AOX activity during the unfavorable seasons of the year. Evergreens also display a lower degree of inhibition of respiration by light (lower energy savings). All these factors indicate that the evergreen organs are more expensive in terms of energy requirements.

6.6.3 Payback Time

To have a better idea of the benefit of one leaf strategy over the opposite, we have made simple calculations about the carbon balance for the leaf over its leaf-life span. For that, we have calculated the quotient between the cost of construction (CC) and the net benefit (calculated as the difference between the rate of photosynthesis and respiration). This ratio can give us an idea of the potential of amortization of the investment in the leaves or what is called pay-back time

(Williams et al. 1989; Eamus and Prichard 1998). The construction cost has been considered in a simple way as LMA × 0.5 (an average concentration of C in the leaf of 50%). We recognize that these calculations are very simple, since the rates of photosynthesis and respiration can vary throughout the life of the leaf, as well as with the environmental conditions. However, it can give us a broad idea of the potential amortization of the investment in the leaves.

We found that the leaves with greater longevity have a greater payback time (Fig. 6.12); that is to say, more time is necessary to amortize the cost of construction of these leaves. This relationship is mainly due to the fact that the leaves with the longest life-span have a high cost of construction ($R^2 = 0.46$, $P < 0.001$, Fig. 6.12b), as hypothesized by Kikuzawa (1995). However, a shorter payback time was not related to a higher carbon gain, contrary to the hypothesis of Kikuzawa

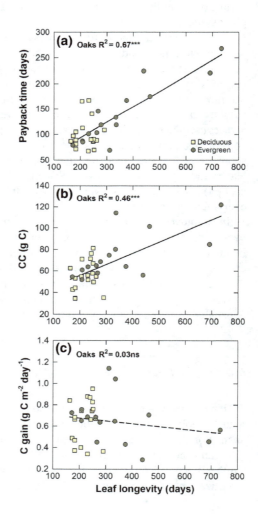

Fig. 6.12 Relationships between **a** leaf life span and payback time, **b** construction costs and **c** estimated C gain per unit leaf area

(1995). This can be explained by the fact that there is not a big change in the photosynthetic rate (A_{area}) between deciduous and evergreen oaks (Fig. 6.4); the change is related mostly to the LMA, which determines a greater change in the construction cost (Fig. 6.9c) for evergreens. However, this higher LMA for evergreens will be an important advantage to resist environmental hazards (e.g. herbivory, pathogens, drought, and cold) (Poorter et al. 2009; González-Zurdo et al. 2016a, b).

6.7 Conclusions

Mediterranean environments present a variety of leaf forms that coexist under the special conditions typical of this climate. The most striking difference among the woody species inhabiting these areas is the dichotomy represented by the deciduous and evergreen habits. As seen by other authors, this dichotomy affects a large number of leaf traits of the different habits, which exhibit pronounced differences in leaf morphology, anatomy, and physiology. Although the information available is still fragmentary, a clear picture emerges that the large differences between leaf habits may finally lead to similar C fixation capabilities for the species that coexist in Mediterranean environments. More research is needed to understand the responses of the different foliar strategies to temporal and geographic changes in these environmental conditions if we want to make predictions about the responses of Mediterranean plant communities to eventual climatic changes in the future.

Appendix: References to the Data Used in the Analysis

Abril M, Hanano R (1998) Ecophysiological responses of three evergreen woody Mediterranean species to water stress. Acta Oecologica 19:377–387
Acherar M, Rambal S (1992) Comparative water relations of four Mediterranean oak species. Vegetatio 99–100:177–184
Acherar M, Rambal S, Lepart J (1991) Évolution du potentiel hydrique foliaire et de la conductance stomatique de quatre chênes méditerranéens lors d'une période de dessèchement. Ann des Sci For 48:561–573
Ackerly DD, Knight CA, Weiss SB, Barton K, Starmer KP (2002) Leaf size, specific leaf area and microhabitat distribution of chaparral woody plants: contrasting patterns in species level and community level analyses. Oecologia 130:449–457
Aranda I, Pardo F, Gil L, Pardos JA (2004) Anatomical basis of the change in leaf mass per area and nitrogen investment with relative irradiance within the canopy of eight temperate tree species. Acta Oecologica 25:187–195

Arend M, Brem A, Kuster TM, Günthardt-Goerg MS (2013) Seasonal photosynthetic responses of European oaks to drought and elevated daytime temperature. Plant Biol 15:169–176

Asensio D, Peñuelas J, Ogaya R, Llusià J (2007) Seasonal soil and leaf CO_2 exchange rates in a Mediterranean holm oak forest and their responses to drought conditions. Atmos Environ 41:2447–2455

Aussenac G, Ducrey M (1977) Etude bioclimatique d'une futaie feuillue (*Fagus sylvatica* L. et *Quercus sessiliflora* Salisb.) de l'Est de la France. I. Analyse des profils microclimatiques et des caractéristiques anatomiques et morphologiques de l'appareil foliaire. Ann des Sci For 34:265–284

Baldocchi DD, Xu L (2007) What limits evaporation from Mediterranean oak woodlands—the supply of moisture in the soil, physiological control by plants or the demand by the atmosphere? Adv Water Resour 30:2113–2122

Baldocchi DD, Ma S, Rambal S, Misson L, Ourcival JM, Limousin JM, Pereira J, Papale D (2010) On the differential advantages of evergreenness and deciduousness in mediterranean oak woodlands: A flux perspective. Ecol Appl 20:1583–1597

Baquedano FJ, Castillo FJ (2007) Drought tolerance in the Mediterranean species *Quercus coccifera, Quercus ilex, Pinus halepensis,* and *Juniperus phoenicea.* Photosynthetica 45:229–238

Bassow S, Bazzaz FA (1998) How environmental conditions affect canopy leaf-level photosynthesis in four deciduous tree species. Ecology 79:2660–2675

Both H, Brüggemann W (2009) Photosynthesis studies on European evergreen and deciduous oaks grown under Central European climate conditions. I: a case study of leaf development and seasonal variation of photosynthetic capacity in *Quercus robur* (L.), *Q. ilex* (L.) and their semideciduous hybrid, *Q.* x *turneri* (Willd.). Trees 23:1081–1090

Bréda N, Cochard H, Dreyer E, Granier A (1993) Water transfer in a mature oak stand (*Quercus petraea*): seasonal evolution and effects of a severe drought. Can J For Res 23:1136–1143

Bussotti F (2008) Functional leaf traits, plant communities and acclimation processes in relation to oxidative stress in trees: A critical overview. Glob Chang Biol 14:2727–2739

Bussotti F, Bettini D, Grossoni P, Mansuino S, Nibbi R, Soda C, Tani C (2002) Structural and functional traits of *Quercus ilex* in response to water availability. Environ Exp Bot 47:11–23

Calatayud V, Cervero J, Calvo E, García-Breijo F-J, Reig Armiñana J, Sanz M (2011) Responses of evergreen and deciduous *Quercus* species to enhanced ozone levels. Environ Pollut 159:55–63

Callaway RM, Nadkarni NM (1991) Seasonal patterns of nutrient deposition in a *Quercus douglasii* woodland in central California. Plant Soil 137:209–222

Callaway RM, Nadkarni NM, Mahall BE (1991) Facilitation and interference of *Quercus douglasii* on understory productivity in Central California. Ecology 72:1484–1499

Castell C, Terradas J, Tenhunen JD (1994) Water relations, gas exchange and growth of dominant and suppressed shoots of *Arbutus unedo* L. and *Quercus ilex* L. Oecologia 98:201–211

Castro-Díez P, Montserrat-Martí G (1998) Phenological pattern of fifteen Mediterranean phanaerophytes from *Quercus ilex* communities of NE-Spain. Plant Ecol 139:103–112

Castro-Díez P, Villar-Salvador P, Pérez-Rontomé C, Maestro-Martínez M, Montserrat-Martí G (1997) Leaf morphology and leaf chemical composition in three *Quercus* (*Fagaceae*) species along a rainfall gradient in NE Spain. Trees - Struct Funct 11:127–134

Castro-Diez P, Puyravaud JP, Cornelissen JHC (2000) Leaf structure and anatomy as related to leaf mass per area variation in seedlings of a wide range of woody plant species and types. Oecologia 124:476–486

Cavender-Bares J, Kitajima K, Bazzaz FA (2004) Multiple trait associations in relation to habitat differentiation among 17 floridian oak species. Ecol Monogr 74:635–662

Cavender-Bares J, Cortes P, Rambal S, Joffre R, Miles B, Rocheteau A (2005) Summer and winter sensitivity of leaves and xylem to minimum freezing temperatures: A comparison of co-occurring Mediterranean oaks that differ in leaf lifespan. New Phytol 168:597–612

Chaves MM, Pereira JS, Maroco J, Rodrigues ML, Ricardo CPP, Osório ML, Carvalho I, Faria T, Pinheiro C (2002) How plants cope with water stress in the field. Photosynthesis and growth. Ann Bot 89:907–916

Chiatante D, Tognetti R, Scippa GS, Congiu T, Baesso B, Terzaghi M, Montagnoli A (2015) Interspecific variation in functional traits of oak seedlings (*Quercus ilex*, *Quercus trojana*, *Quercus virgiliana*) grown under artificial drought and fire conditions. J Plant Res 128:595–611

Cochard H, Lemoine D, Dreyer E (1999) The effect of acclimation to sunlight on the xylem vulnerability to embolism in *Fagus sylvatica* L. Plant, Cell Environ 22:101–108

Cody ML, Prigge B (2003) Spatial and temporal variations in the timing of leaf replacement in a *Quercus cornelius-mulleri* population. J Veg Sci 14:781–798

Corcuera L, Morales F, Abadía A, Gil-Pelegrín E (2005) Seasonal changes in photosynthesis and photoprotection in a *Quercus ilex* subsp. *ballota* woodland located in its upper altitudinal extreme in the Iberian Peninsula. Tree Physiol 25:599–608

Cornelissen JHC, Werger MJA, Castro-Díez P, Van Rheenen JWA, Rowland AP (1997) Foliar nutrients in relation to growth, allocation and leaf traits in seedlings of a wide range of woody plant species and types. Oecologia 111:460–469

Crescente M, Gratani L, Larcher W (2002) Shoot growth efficiency and production of *Quercus ilex* L. in different climates. Flora 197:2–9

Damesin C, Rambal S (1995) Field study of leaf photosynthetic performance by a Mediterranean deciduous oak tree (*Quercus pubescens*) during a severe summer drought. New Phytol 131:159–167

Damesin C, Rambal S, Joffre R (1998) Co-occurrence of trees with different leaf habit: A functional approach on Mediterranean oaks. Acta Oecologica 19:195–204

David TS, Ferreira MI, Cohen S, Pereira JS, David JS (2004) Constraints on transpiration from an evergreen oak tree in southern Portugal. Agric For Meteorol 122:193–205

De La Riva EG, Olmo M, Poorter H, Ubera JL, Villar R (2016) Leaf mass per area (LMA) and its relationship with leaf structure and anatomy in 34 mediterranean woody species along a water availability gradient. PLoS One 11:1–18. doi:10.1371/journal.pone.0148788

Ducousso A, Guyon JP, Kremer A (1996) Latitudinal and altitudinal variation of bud burst in western populations of sessile oak (*Quercus petraea* (Matt) Liebl). Ann des Sci For 53:775–782

Ehleringer JR, Phillips SL (1996) Ecophysiological factors contributing to the distribution of several *Quercus* species in the intermountain west. Ann des Sci For 53:291–302

Epron D, Dreyer (1990) Stomatal and non stomatal limitation of photosynthesis by leaf water deficits in three oak species: a comparison of gas exchange and chlorophyll a fluorescence data. Ann des Sci For 47:435–450

Epron D, Dreyer E (1993) Long-term effects of drought on photosynthesis of adult oak trees [*Quercus petraea* (Matt.) Liebl. and *Quercus robur* L.] in a natural stand. New Phytol 125:381–389

Epron D, Dreyer E, Aussenac G (1993) A comparison of photosynthetic responses to water stress in seedlings from 3 oak species: *Quercus petraea* (Matt) Liebl, *Q. rubra* L and *Q. cerris* L. Ann des Sci For 50:S48–S60

Escudero A, Mediavilla S (2003) Decline in photosynthetic nitrogen use efficiency with leaf age and nitrogen resorption as determinants of leaf life span. J Ecol 91:880–889

Estiarte M, De Castro M, Espelta JM (2007) Effects of resource availability on condensed tannins and nitrogen in two *Quercus* species differing in leaf life span. Ann For Sci 64:439–445

Faria T, GarciaPlazaola JI, Abadia A, Cerasoli S, Pereira JS, Chaves MM (1996) Diurnal changes in photoprotective mechanisms in leaves of cork oak (*Quercus suber*) during summer. Tree Physiol 16:115–123

Faria T, Silverio D, Breia E, Cabral R, Abadia A, Abadia J, Pereira JS, Chaves MM (1998) Differences in the response of carbon assimilation to summer stress (water deficits, high light and temperature) in four Mediterranean tree species. Physiol Plant 102:419–428

Filho JT, Damesin C, Rambal S, Joffre R (1998) Retrieving leaf conductances from sap flows in a mixed Mediterranean woodland: a scaling exercise. Ann des Sci For 55:173–190

Flexas J, Gulías J, Jonasson S, Medrano H, Mus M (2001) Seasonal patterns and control of gas exchange in local populations of the Mediterranean evergreen shrub *Pistacia lentiscus* L. Acta Oecologica 22:33–43

Flexas J, Bota J, Cifre J, Escalona JM, Galmés J, Gulías J, Lefi EK, Martínez-Cañellas SF, Moreno MT, Ribas-Carbó M, Riera D, Sampol B,

Medrano H (2004) Understanding down-regulation of photosynthesis under water stress: Future prospects and searching for physiological tools for irrigation management. Ann Appl Biol 144:273–283

Flexas J, Diaz-Espejo A, Gago J, Gallé A, Galmés J, Gulías J, Medrano H (2014) Photosynthetic limitations in Mediterranean plants: A review. Environ Exp Bot 103:12–23

Ford-Peterson L (2009) Specific leaf area: a key trait to understanding forest structure. University of California, Santa Cruz

Fotelli MN, Radoglou KM, Constantinidou HI (2000) Water stress responses of seedlings of four Mediterranean oak species. Tree Physiol 20:1065–1075

Fusaro L, Mereu S, Brunetti C, Di Ferdinando M, Ferrini F, Manes F, Salvatori E, Marzuoli R, Gerosa G, Tattini M (2014) Photosynthetic performance and biochemical adjustments in two co-occurring Mediterranean evergreens, *Quercus ilex* and *Arbutus unedo*, differing in salt-exclusion ability. Funct Plant Biol 41:391–400

Gallagher DW (2014) Photosynthetic thermal tolerance and recovery to short duration temperature stress in desert and montane plants: a comparative study. California Polytechnic State University

Galmés J, Medrano H, Flexas J (2007) Photosynthetic limitations in response to water stress and recovery in Mediterranean plants with different growth forms. New Phytol 175:81–93

García-Plazaola JI, Faria T, Abadia J, Abadia A, Chaves MM, Pereira JS (1997) Seasonal changes in xanthophyll composition and photosynthesis of cork oak (*Quercus suber* L.) leaves under mediterranean climate. J Exp Bot 48:1667–1674

González-González BD, García-González I, Vázquez-Ruiz RA (2013) Comparative cambial dynamics and phenology of *Quercus robur* L. and *Q. pyrenaica* Willd. in an Atlantic forest of the northwestern Iberian Peninsula. Trees Struct Funct 27:1571–1585

Gordo O, Sanz JJ (2010) Impact of climate change on plant phenology in Mediterranean ecosystems. Glob Chang Biol 16:1082–1106

Goulden ML (1996) Carbon assimilation and water-use efficiency by neighboring Mediterranean-climate oaks that differ in water access. Tree Physiol 16:417–424

Goulden ML, Field CB (1994) Three methods for monitoring the gas exchange of individual tree canopies: ventilated-chamber, sap-flow and Penman-Monteith measurements on evergreen oaks. Funct Ecol 8:125–135

Gratani L, Bombelli A (2001) Differences in leaf traits among Mediterranean broad-leaved evergreen shrubs. Ann Bot Fenn 38:15–24

Gratani L, Foti I (1998) Estimating forest structure and shade tolerance of the species in a mixed deciduous broad-leaved forest in Abruzzo, Italy. Ann Bot Fenn 35:75–83

Gratani L, Varone L (2004) Adaptive photosynthetic strategies of the Mediterranean maquis species according to their origin. Photosynthetica 42:551–558

Gratani L, Pesoli P, Crescente MF (1998) Relationship between photosynthetic activity and chlorophyll content in an isolated *Quercus ilex* L. tree during the year. Photosynthetica 35:445–451

Grulke NE, Dobrowolski W, Mingus P, Fenn ME (2005) California black oak response to nitrogen amendment at a high O_3, nitrogen-saturated site. Environ Pollut 137:536–545

Grünzweig JM, Carmel Y, Riov J, Sever N, McCreary DD, Flather CH (2008) Growth, resource storage, and adaptation to drought in California and eastern Mediterranean oak seedlings. Can J For Res 38:331–342

Gulías J, Flexas J, Abadía A, Medrano H (2002) Photosynthetic responses to water deficit in six Mediterranean sclerophyll species: possible factors explaining the declining distribution of *Rhamnus ludovici-salvatoris*, an endemic Balearic species. Tree Physiol 22:687–697

Gulías J, Cifre J, Jonasson S, Medrano H, Flexas J (2009) Seasonal and inter-annual variations of gas exchange in thirteen woody species along a climatic gradient in the Mediterranean island of Mallorca. Flora 204:169–181

Hallik L, Niinemets Ü, Wright IJ (2009) Are species shade and drought tolerance reflected in leaf-level structural and functional differentiation in Northern Hemisphere temperate woody flora? New Phytol 184:257–274

Handley T, Grulke NE (2008) Interactive effects of O3 exposure on California black oak (*Quercus kelloggii* Newb.) seedlings with and without N amendment. Environ Pollut 156:53–60

Harley PC, Tenhunen JD, Beyschlag W, Lange OL (1987) Seasonal changes in net photosynthesis rates and photosynthetic capacity in leaves of *Cistus salvifolius*, a European mediterranean semi-deciduous shrub. Oecologia 74:380–388

Harley P, Deem G, Flint S, Caldwell M (1996) Effects of growth under elevated UV-B on photosynthesis and isoprene emission in *Quercus gambelii* and *Mucuna pruriens*. Global Change Biol 2:149–154

Hollinger DY (1983) Photosynthesis and water relations of the mistletoe, *Phoradendron villosum*, and its host, the California valley oak, *Quercus lobata*. Oecologia 60:396–400

Hollinger DY (1986) Herbivory and the cycling of nitrogen and phosphorus in isolated California oak trees. Oecologia 70:291–297

Hollinger DY (1992) Leaf and simulated whole-canopy photosynthesis in two co-occurring tree species. Ecology 73:1–14

Hunter JC (1997) Correspondence of environmental tolerances with leaf and branch attributes for six co-occurring species of broadleaf evergreen trees in northern California. Trees - Struct Funct 11:169–175

Infante JM, Damesin C, Rambal S, Fernández-Alés R (1999) Modelling leaf gas exchange in holm-oak trees in southern Spain. Agric For Meteorol 95:203–223

Infante JM, Domingo F, Fernández Alés R, Joffre R, Rambal S (2003) *Quercus ilex* transpiration as affected by a prolonged drought period. Biol Plant 46:49–55

Jacobsen AL, Pratt RB, Davis SD, Ewers FW (2008) Comparative community physiology: Nonconvergence in water relations among three semi-arid shrub communities. New Phytol 180:100–113

Johnson DM, Woodruff DR, McCulloh KA, Meinzer FC (2009) Leaf hydraulic conductance, measured in situ, declines and recovers daily: Leaf hydraulics, water potential and stomatal conductance in four temperate and three tropical tree species. Tree Physiol 29:879–887

Jonasson S, Medrano H, Flexas J (1997) Variation in leaf longevity of *Pistacia lentiscus* and its relationship to sex and drought stress infered from leaf delta13 C. Funct Ecol 11:282–289

Juárez-López FJ, Escudero A, Mediavilla S (2008) Ontogenetic changes in stomatal and biochemical limitations to photosynthesis of two co-occurring Mediterranean oaks differing in leaf life span. Tree Physiol 28:367–374

Kaplan D, Gutman M (1999) Phenology of *Quercus ithaburensis* with emphasis on the effect of fire. For Ecol Manage 115:61–70

Karam F, Doulis A, Ozturk M, Dogan Y, Sakcali S (2011) Eco-physiological behaviour of two woody oak species to combat desertification in the east Mediterranean-A case study from Lebanon. Procedia Soc Behav Sci 19:787–796

Karavin N (2013) Effects of leaf and plant age on specific leaf area in deciduous tree species *Quercus cerris* L. var. *cerris*. Bangladesh JBot 42:301–306

Keenan TF, García R, Friend AD, Zaehle S, Gracia C, Sabate S (2009) Improved understanding of drought controls on seasonal variation in Mediterranean forest canopy CO_2 and water fluxes through combined in situ measurements and ecosystem modelling. Biogeosciences 6:1423–1444

Keenan T, Sabate S, Gracia C (2010) The importance of mesophyll conductance in regulating forest ecosystem productivity during drought periods. Glob Chang Biol 16:1019–1034

Kikuzawa K, Lechowicz MJ (2006) Toward synthesis of relationships among leaf longevity, instantaneous photosynthetic rate, lifetime leaf carbon gain, and the gross primary production of forests. Am Nat 168:373–383

Klein T, Shpringer I, Fikler B, Elbaz G, Cohen S, Yakir D (2013) Relationships between stomatal regulation, water-use, and water-use efficiency of two coexisting key Mediterranean tree species. For Ecol Manage 302:34–42

Knops JMH, Koenig WD (1997) Site fertility and leaf nutrients of sympatric evergreen and deciduous species of *Quercus* in central coastal California. Plant Ecol 130:121–131

Kyparissis A, Grammatikopoulos G, Manetas Y (1997) Leaf demography and photosynthesis as affected by the environment in the drought semi-deciduous Mediterranean shrub *Phlomis fruticosa* L. Acta Oecologica 18:543–555

Lechowicz MJ (1984) Why do temperate deciduous trees leaf out at different times? Adaptation and ecology of forest communities. Am Nat 124:821–842

Liakoura V, Manetas Y, Karabourniotis G (2001) Seasonal fluctuations in the concentration of UV-absorbing compounds in the leaves of some Mediterranean plants under field conditions. Physiol Plant 111:491–500

Limousin JM, Misson L, Lavoir AV, Martin NK, Rambal S (2010) Do photo-synthetic limitations of evergreen *Quercus ilex* leaves change with long-term increased drought severity? Plant, Cell Environ 33:863–87

Llusia J, Peñuelas J (2000) Seasonal patterns of terpene content and emission from seven Mediterranean woody species in field conditions. Am J Bot 87:133–140

Llusia J, Roahtyn S, Yakir D, Rotenberg E, Seco R, Guenther A, Peñuelas J (2016) Photosynthesis, stomatal conductance and terpene emission response to water availability in dry and mesic Mediterranean forests. Trees Struct Funct 30:749–759

Lo Gullo MA, Salleo S, Rosso R, Trifilò P (2003) Drought resistance of 2-year-old saplings of Mediterranean forest trees in the field: Relations between water relations, hydraulics and productivity. Plant Soil 250:259–272

Ma S, Baldocchi DD, Xu L, Hehn T (2007) Inter-annual variability in carbon dioxide exchange of an oak/grass savanna and open grassland in California. Agric For Meteorol 147:157–171

Ma S, Baldocchi DD, Mambelli S, Dawson TE (2011) Are temporal variations of leaf traits responsible for seasonal and inter-annual variability in ecosystem CO_2 exchange? Funct Ecol 25:258–270

Mahall BE, Tyler CM, Cole ES, Mata C (2009) A comparative study of oak (*Quercus, Fagaceae*) seedling physiology during summer drought in southern California. Am J Bot 96:751–761

Manes F, Astorino G, Vitale M, Loreto F (1997a) Morpho-functional characteristics of *Quercus ilex* L. leaves of different age and their ecophysiological behaviour during different seasons. Plant Biosyst 131:149–158

Manes F, Seufert G, Vitale M (1997b) Ecophysiological studies of Mediterranean plant species at the Castelporziano estate. Atmos Environ 31:51–60

Manes F, Vitale M, Donato E, Giannini M, Puppi G (2006) Different ability of three Mediterranean oak species to tolerate progressive water stress. Photosynthetica 44:387–393

Martínez-Vilalta J, Prat E, Oliveras I, Piñol J (2002) Xylem hydraulic properties of roots and stems of nine Mediterranean woody species. Oecologia 133:19–29

Matzner SL, Rice KJ, Richards JH (2001) Intra-specific variation in xylem cavitation in interior live oak (*Quercus wislizenii* A. DC.). J Exp Bot 52:783–789

Mauffette Y, Oechel WC (1989) Seasonal variation in leaf chemistry of the coase live oak *Quercus agrifolia* and implications for the California oak moth *Phryganidia californica*. Oecologia 79:439–445

Mediavilla S, Escudero A (2003a) Stomatal responses to drought at a Mediterranean site : a comparative study of co-occurring woody species differing in leaf longevity. Tree Physiol 23:987–996

Mediavilla S, Escudero A (2003b) Relative growth rate of leaf biomass and leaf nitrogen content in several mediterranean woody species. Plant Ecol 168:321–332

Mediavilla S, Escudero A (2003c) Photosynthetic capacity, integrated over the lifetime of a leaf, is predicted to be independent of leaf longevity in some tree species. New Phytol 159:203–211

Mediavilla S, Escudero A (2004) Stomatal responses to drought of mature trees and seedlings of two co-occurring Mediterranean oaks. For Ecol Manage 187:281–294

Mediavilla S, Escudero A (2009) Ontogenetic changes in leaf phenology of two co-occurring Mediterranean oaks differing in leaf life span. Ecol Res 24:1083–1090

Mediavilla S, Garcia-Ciudad A, Garcia-Criado B, Escudero A (2008) Testing the correlations between leaf life span and leaf structural reinforcement in 13 species of European Mediterranean woody plants. Funct Ecol 22:787–793

Mediavilla S, González-Zurdo P, García-Ciudad A, Escudero A (2011) Morphological and chemical leaf composition of Mediterranean evergreen tree species according to leaf age. Trees - Struct Funct 25:669–677

Mediavilla S, Herranz M, González-Zurdo P, Escudero A (2014) Ontogenetic transition in leaf traits: A new cost associated with the increase in leaf longevity. J Plant Ecol 7:567–575

Mediavilla S, González-Zurdo P, Babiano J, Escudero A (2016) Responses of photosynthetic parameters to differences in winter temperatures throughout a temperature gradient in two evergreen tree species. Eur J For Res 135:1–13

Medlyn B, Badeck F-W, De Pury D, Barton C, Broadmeadow M, Ceulemans R, De Angelis P, Forstreuter M, Jach M, Kellomäki S, Laitat E, Marek M, Philippot S, Rey A, Strassemeyer J, Laitinen K, Liozon R, Portier B, Robertntz P, Wang K, Jarvis P (1999) Effects of elevated [CO_2] on photosynthesis in European forest species: a meta-analysis of model parameters. Plant Cell Environ 22:1475–1495

Mereu S, Salvatori E, Fusaro L, Gerosa G, Muys B, Manes F (2009) A whole plant approach to evaluate the water use of mediterranean maquis species in a coastal dune ecosystem. Biogeosciences Discuss 6:2599–2610

Mészáros I, Veres S, Kanalas P, Oláh V, Szollosi E, Sárvári E, Lévai L, Lakatos G (2007) Leaf growth and photosynthetic performance of two co-existing oak species in contrasting growing seasons. Acta Silv Lign Hung 3:7–20

Milla R, Reich PB (2011) Multi-trait interactions, not phylogeny, fine-tune leaf size reduction with increasing altitude. Ann Bot 107:455–465

Montserrat-Martí G, Camarero JJ, Palacio S, Pérez-Rontomé C, Milla R, Albuixech J, Maestro M (2009) Summer-drought constrains the phenology and growth of two coexisting Mediterranean oaks with contrasting leaf habit: Implications for their persistence and reproduction. Trees - Struct Funct 23:787–799

Morecroft MD, Stokes VJ, Morison JIL (2003) Seasonal changes in the photosynthetic capacity of canopy oak (*Quercus robur*) leaves: The impact of slow development on annual carbon uptake. Int J Biometeorol 47:221–226

Morin X, Roy J, Sonie L, Chuine I (2010) Changes in leaf phenology of three European oak species in response to experimental climate change. New Phytol 186:900–910

Mouillot F, Rambal S, Joffre R (2002) Simulating climate change impacts on fire frequency and vegetation dynamics in a Mediterranean-type ecosystem. Glob Chang Biol 8:423–437

Munné-Bosch S, Jubany-Marí T, Alegre L (2003) Enhanced photo- and antioxidative protection, and hydrogen peroxide accumulation in drought-stressed *Cistus clusii* and *Cistus albidus* plants. Tree Physiol 23:1–12

Nardini A, Lo Gullo MA, Salleo S (1999) Competitive strategies for water availability in two Mediterranean *Quercus* species. Plant, Cell Environ 22:109–116

Nardini A, Salleo S, Lo Gullo M, Pitt F (2000) Different responses to drought and freeze stress of *Quercus ilex* L. growing along a latitudinal gradient. Plant Ecol 148:139–147

Nardini A, Pedà G, La Rocca N (2012) Trade-offs between leaf hydraulic capacity and drought vulnerability: Morpho-anatomical bases, carbon costs and ecological consequences. New Phytol 196:788–798

Navas ML, Ducout B, Roumet C, Richarte J, Garnier J, Garnier E (2003) Leaf life span, dynamics and construction cost of species from Mediterranean old-fields differing in successional status. New Phytol 159:213–228

Ne'eman G (1993) Variation in leaf phenology and habit in *Quercus ithaburensis*, a Mediterranean deciduous tree. J Ecol 81:627–634

Niinemets Ü, Valladares F, Ceulemans R (2003) Leaf-level phenotypic variability and plasticity of invasive *Rhododendron ponticum* and non-invasive *Ilex aquifolium* co-occurring at two contrasting European sites. Plant, Cell Environ 26:941–956

Niinemets Ü, Díaz-Espejo A, Flexas J, Galmés J, Warren CR (2009) Role of mesophyll diffusion conductance in constraining potential photosynthetic productivity in the field. J Exp Bot 60:2249–2270

Nikolopoulos D, Liakopoulos G, Drossopoulos I, Karabourniotis G (2002) The relationship between anatomy and photosynthetic performance of heterobaric leaves. Plant Physiol 129:235–243

Nitta I, Ohsawa M (1997) Leaf dynamics and shoot phenology of eleven warm-temperate evergreen broad-leaved trees near their northern limit in central Japan. Plant Ecol 130:71–88

Nunn AJ, Wieser G, Reiter IM, Häberle K-H, Grote R, Havranek WM, Matyssek R (2006) Testing the unifying theory of ozone sensitivity with mature trees of Fagus sylvatica and *Picea abies*. Tree Physiol 26:1391–1403

Oechel WC (1982) Carbon balance studies in chaparral shrubs: implications for biomass production. Proc Symp Dyn Manag Mediterr Ecosyst General Te:158–166

Ogaya R, Peñuelas J (2003) Comparative field study of *Quercus ilex* and *Phillyrea latifolia*: Photosynthetic response to experimental drought conditions. Environ Exp Bot 50:137–148

Ogaya R, Peñuelas J (2006) Contrasting foliar responses to drought in *Quercus ilex* and *Phillyrea latifolia*. Biol Plant 50:373–382

Oliveira G, Correia OA, Martins-Louçao MA, Catarino FM (1992) Water relations of cork-oak (*Quercus suber* L.) under natural conditions. Vegetatio 99–100:199–208

Osuna JL, Baldocchi DD, Kobayashi H, Dawson TE (2015) Seasonal trends in photosynthesis and electron transport during the Mediterranean summer drought in leaves of deciduous oaks. Tree Physiol 35:485–500

Otieno DO, Schmidt MWT, Kurz-Besson C, Lobo Do Vale R, Pereira JS, Tenhunen JD (2007) Regulation of transpirational water loss in *Quercus suber* trees in a Mediterranean-type ecosystem. Tree Physiol 27:1179–1187

Parker VT (1984) Correlation of physiological divergence with reproductive mode in chaparral shrubs. Madrono 31:231–242

Peguero-Pina JJ, Sancho-Knapik D, Morales F, Flexas J, Gil-Pelegrín E (2009) Differential photosynthetic performance and photoprotection mechanisms of three

Mediterranean evergreen oaks under severe drought stress. Funct Plant Biol 36:453–462

Peñuelas J, Filella I, Llusia J, Siscart D, Pinol J (1998) Comparative field study of spring and summer leaf gas exchange and photobiology of the mediterranean trees *Quercus ilex* and *Phillyrea latifolia*. J Exp Bot 49:229–238

Peñuelas J, Filella I, Comas P (2002) Changed plant and animal life cycles from 1952-2000 in the Mediterranean region. Glob Chang Biol 8:531–544

Pérez Latorre AV, Cabezudo B (2006) Phenomorphology and eco-morphological characters of *Rhododendron lauroid* forests in the Western Mediterranean (Iberian Peninsula, Spain). Plant Ecol 187:227–247

Phillips N, Bond BJ, McDowell NG, Ryan MG, Schauer A (2003) Leaf area compounds height-related hydraulic costs of water transport in Oregon White Oak trees. Funct Ecol 17:832–840

Pivovaroff AL, Sack L, Santiago LS (2014) Coordination of stem and leaf hydraulic conductance in southern California shrubs: A test of the hydraulic segmentation hypothesis. New Phytol 203:842–850

Pivovaroff AL, Pasquini SC, De Guzman ME, Alstad KP, Stemke JS, Santiago LS (2016) Multiple strategies for drought survival among woody plant species. Funct Ecol 30:517–526

Quero JL, Villar R, Marañón T, Zamora R (2006) Interactions of drought and shade effects on seedlings of four *Quercus* species: physiological and structural leaf responses. New Phytol 170:819–834

Quero JL, Villar R, Marañón T, Zamora R, Vega D, Sack L (2008) Relating leaf photosynthetic rate to whole-plant growth: drought and shade effects on seedlings of four *Quercus* species. Funct Plant Biol 35:725–737

Quero JL, Sterck FJ, Martínez-Vilalta J, Villar R (2011) Water-use strategies of six co-existing Mediterranean woody species during a summer drought. Oecologia 166:45–57

Radoglou KM (1996) Environmental control of CO_2 assimilation rates and stomatal conductance in five oak species growing under field conditions in Greece. Ann des Sci For 53:269–278

Ramírez-Valiente JA, Valladares F, Sánchez-Gómez D, Delgado A, Aranda I (2014) Population variation and natural selection on leaf traits in cork oak throughout its distribution range. Acta Oecologica 58:49–56

Reich PB, Walters MB, Ellsworth DS, Vose JM, Volin JC, Gresham C, Bowman WD (1998) Relationships of leaf dark respiration to leaf nitrogen, specific leaf area and leaf life span: a test across biomes and functional groups. Oecologia 114:471–482

Rhizopoulou S, Mitrakos K (1990) Water relations of evergreen sclerophylls. I. Seasonal changes in the water relations of eleven species from the same environment. Ann Bot 65:171–178

Ripullone F, Rivelli AR, Baraldi R, Guarini R, Guerrieri R, Magnani F, Peñuelas J, Raddi S, Borghetti M (2011) Effectiveness of the photochemical reflectance index to track photosynthetic activity over a range of forest tree species and plant water statuses. Funct Plant Biol 38:177–186

Rodríguez-Calcerrada J, Pardos JA, Gil L, Aranda I (2007) Acclimation to light in seedlings of *Quercus petraea* (Mattuschka) Liebl. and *Quercus pyrenaica* Willd. planted along a forest-edge gradient. Trees Struct Funct 21:45–54

Rodríguez-Calcerrada J, Pardos JA, Gil L, Reich PB, Aranda I (2008) Light response in seedlings of a temperate (*Quercus petraea*) and a sub-Mediterranean species (*Quercus pyrenaica*): Contrasting ecological strategies as potential keys to regeneration performance in mixed marginal populations. Plant Ecol 195:273–285

Rundel PW (1986) Origins and Adaptations of California Hardwoods. In: Multiple-Use Management of California's Hardwood Resources. pp 11–17

Sala A, Tenhunen JD (1994) Site-specific water relations and stomatal responses of *Quercus ilex* L. in a Mediterranean watershed. Tree Physiol 14:601–617

Sala A, Tenhunen JD (1996) Simulations of canopy net photosynthesis and transpiration in *Quercus ilex* L under the influence of seasonal drought. Agric For Meteorol 78:203–222

Serrano L, Peñuelas J, Ogaya R, Savé R (2005) Tissue-water relations of two co-occurring evergreen Mediterranean species in response to seasonal and experimental drought conditions. J Plant Res 118:263–269

Siam AMJ, Radoglou KM, Noitsakis B, Smiris P (2008) Physiological and growth responses of three Mediterranean oak species to different water availability regimes. J Arid Environ 72:583–592

Siam AMJ, Radoglou KM, Noitsakis B, Smiris P (2009) Differences in ecophysiological responses to summer drought between seedlings of three deciduous oak species. For Ecol Manage 258:35–42

Simonin KA, Limm EB, Dawson TE (2012) Hydraulic conductance of leaves correlates with leaf lifespan: Implications for lifetime carbon gain. New Phytol 193:939–947

Stpaul NKM, Limousin JM, Rodríguez-Calcerrada J, Ruffault J, Rambal S, Letts MG, Misson L (2012) Photosynthetic sensitivity to drought varies among populations of *Quercus ilex* along a rainfall gradient. Funct Plant Biol 39:25–37

Stylinski CD, Oechel WC, Gamon JA, Tissue DT, Miglietta F, Raschi A (2000) Effects of lifelong [CO_2] enrichment on carboxylation and light utilization of *Quercus pubescens* Willd. examined with gas exchange, biochemistry and optical techniques. Plant, Cell Environ 23:1353–1362

Stylinski CD, Gamon JA, Oechel WC (2002) Seasonal patterns of reflectance indices, carotenoid pigments and photosynthesis of evergreen chaparral species. Oecologia 131:366–374

Tenhunen JD, Lange OL, Harley PC, Beyschlag W, Meyer A (1985) Limitations due to water stress on leaf net photosynthesis of *Quercus coccifera* in the Portuguese evergreen scrub. Oecologia 67:23–30

Tenhunen JD, Serra AS, Harley PC, Dougherty RL, Reynolds JF (1990) Factors influencing carbon fixation and water use by mediterranean sclerophyll shrubs during summer drought. Oecologia 82:381–393

Tognetti R, Longobucco A, Miglietta F, Raschi A (1998a) Transpiration and stomatal behaviour of *Quercus ilex* plants during the summer in a Mediterranean carbon dioxide spring. Plant, Cell Environ 21:613–622

Tognetti R, Longobucco S, Raschi A (1998b) Vulnerability of xylem to embolism in relation to plant hydraulic resistance in *Quercus pubescens* and *Quercus ilex* co-occurring in a Mediterranean coppice stand in central Italy. New Phytol 139:437–447

Tognetti R, Cherubini P, Marchi S, Raschi A (2007) Leaf traits and tree rings suggest different water-use and carbon assimilation strategies by two co-occurring *Quercus* species in a Mediterranean mixed-forest stand in Tuscany, Italy. Tree Physiol 27:1741–1751

Tretiach M (1993) Photosynthesis and transpiration of evergreen Mediterranean and deciduous trees in an ecotone during a growing season. Acta Oecologica 14:341–360

Valentini R, Epron D, Deangelis P, Matteucci G, Dreyer E (1995) In-situ estimation of net CO_2 assimilation, photosynthetic electron flow and photorespiration in turkey oak (*Q. cerris* L) leaves—diurnal cycles under different levels of water-supply. Plant Cell Environ 18:631–640

Valladares F, Sánchez-Gómez D (2006) Ecophysiological traits associated with drought in Mediterranean tree seedlings: Individual responses versus interspecific trends in eleven species. Plant Biol 8:688–697

Vaz M, Pereira JS, Gazarini LC, David TS, David JS, Rodrigues A, Maroco J, Chaves MM (2010) Drought-induced photosynthetic inhibition and autumn recovery in two Mediterranean oak species (*Quercus ilex* and *Quercus suber*). Tree Physiol 30:946–956

Vilagrosa A, Bellot J, Vallejo VR, Gil-Pelegrín E (2003) Cavitation, stomatal conductance, and leaf dieback in seedlings of two co-occurring Mediterranean shrubs during an intense drought. J Exp Bot 54:2015–2024

Villar R, Merino J (2001) Comparison of leaf construction costs in woody species with differing leaf life-spans in contrasting ecosystems. New Phytol 151:213–226

Villar R, Robleto JR, De Jong Y, Poorter H (2006) Differences in construction costs and chemical composition between deciduous and evergreen woody species are small as compared to differences among families. Plant, Cell Environ 29:1629–1643

Villar R, Ruiz-Robleto J, Ubera JL, Poorter H (2013) Exploring variation in leaf mass per area (LMA) from leaf to cell: An anatomical analysis of 26 woody species. Am J Bot 100:1969–1980

Vitasse Y, Delzon S, Dufrene E, Pontailler JY, Louvet JM, Kremer A, Michalet R (2009a) Leaf phenology sensitivity to temperature in European trees: Do within-species populations exhibit similar responses? Agric For Meteorol 149:735–744

Vitasse Y, Porté AJ, Kremer A, Michalet R, Delzon S (2009b) Responses of canopy duration to temperature changes in four temperate tree species: Relative contributions of spring and autumn leaf phenology. Oecologia 161:187–198

von Caemmerer S, Evans JR (2015) Temperature responses of mesophyll conductance differ greatly between species. Plant, Cell Environ 38:629–637

Warren CR, Dreyer E (2006) Temperature response of photosynthesis and internal conductance to CO_2: Results from two independent approaches. J Exp Bot 57:3057–3067

Welter S, Bracho-Nuñez A, Mir C, Zimmer I, Kesselmeier J, Lumaret R, Schnitzler JP, Staudt M (2012) The diversification of terpene emissions in Mediterranean oaks: Lessons from a study of *Quercus suber*, *Quercus canariensis* and its hybrid *Quercus afares*. Tree Physiol 32:1082–1091

Williams DG, Ehleringer JR (2000) Carbon isotope discrimination and water relations of oak hybrid populations in southwestern Utah. West North Am Nat 60:121–129

Withington JM, Reich PB, Oleksyn J, Eissenstat DM (2006) Comparisons of structure and life span in roots and leaves among temerate trees. Ecol Monogr 76:381–397

Wolkerstorfer SV, Wonisch A, Stankova T, Tsvetkova N, Tausz M (2011) Seasonal variations of gas exchange, photosynthetic pigments, and antioxidants in Turkey oak (*Quercus cerris* L.) and Hungarian oak (*Quercus frainetto* Ten.) of different age. Trees Struct Funct 25:1043–1052

Xu L, Baldocchi DD (2003) Seasonal trends in photosynthetic parameters and stomatal conductance of blue oak (*Quercus douglasii*) under prolonged summer drought and high temperature. Tree Physiol 23:865–77

References

Acherar M, Rambal S (1992) Comparative water relations of four Mediterranean oak species. Vegetatio 99–100:177–184

Acherar M, Rambal S, Lepart J (1991) Évolution du potentiel hydrique foliaire et de la conductance stomatique de quatre chênes méditerranéens lors d'une période de dessèchement. Ann des Sci For 48:561–573

Ackerly DD, Bazzaz IFA (1995) Leaf dynamics, self-shading and carbon gain in seedlings of a tropical pioneer tree. Oecologia 101:289–298

Allard V, Ourcival J, Rambal S, Joffre R, Rocheteau A (2008) Seasonal and annual variation of carbon exchange in an evergreen Mediterranean forest in southern France. Global Change Biol 14:714–725

Asensio D, Peñuelas J, Ogaya R, Llusià J (2007) Seasonal soil and leaf CO_2 exchange rates in a Mediterranean holm oak forest and their responses to drought conditions. Atmos Environ 41:2447–2455

Baldocchi DD, Ma S, Rambal S, Misson L, Ourcival JM, Limousin JM, Pereira J, Papale D (2010) On the differential advantages of evergreenness and deciduousness in mediterranean oak woodlands: a flux perspective. Ecol Appl 20:1583–1597

Bartlett MK, Scoffoni C, Sack L (2012) The determinants of leaf turgor loss point and prediction of drought tolerance of species and biomes: a global meta-analysis. Ecol Lett 15:393–405

Bassow S, Bazzaz FA (1998) How environmental conditions affect canopy leaf-level photosynthesis in four deciduous tree species. Ecology 79:2660–2675

Blumler MA (2015) Deciduous woodlands in Mediterranean California. In: Box EO, Fujiwara K (eds) Warm-temperate deciduous forests around the Northern Hemisphere. Springer, Switzerland, pp 257–266

Bongers FJ, Olmo M, Lopez-Iglesias B, Anten N, Villar R (2017) Drought responses, phenotypic plasticity and survival of Mediterranean species in two different microclimatic sites. Plant Biol 19:386–395

Canadell J, Jackson R, Ehleringer J, Mooney HA, Sala OE, Schulze E-D (1996) Maximum rooting depth of vegetation types at the global scale. Oecologia 108:583–595

Castro-Díez P, Villar-Salvador P, Perez-Rontome P, Maestro-Martinez M, Montserrat-Marti G (1997) Leaf morphology and leaf chemical composition in three *Quercus* (*Fagaceae*) species along a rainfall gradient in NE Spain. Trees 11:127–134

Cavender-Bares J, Kitajima K, Bazzaz FA (2004) Multiple trait associations in relation to habitat differentiation among 17 floridian oak species. Ecol Monogr 74:635–662

Challabathula D, Vishwakarma A, Raghavendra AS, Padmasree K (2016) Alternative oxidase pathway optimizes photosynthesis during osmotic and temperature stress by regulating cellular ROS, malate valve and antioxidative systems. Front Plant Sci 7: 68. Published online 2016 Feb 9. doi:10.3389/fpls.2016.00068

Chapin FS III (1989) The cost of tundra plant structures: evaluation of concepts and currencies. Amer Nat 133:1–19

Cornelissen J, Cerabolini B, Castro-Díez P, Villar-Salvador P, Montserrat-Martí G, Puyravaud JP, Maestro M, Werger MJA, Aerts R (2003) Functional traits of woody plants: correspondence of species rankings between field adults and laboratory-grown seedlings? J Veg Sci 14:311–322

Daas C, Montpied P, Hanchi B, Dreyer E (2008) Responses of photosynthesis to high temperatures in oak saplings assessed by chlorophyll-a fluorescence: inter-specific diversity and temperature-induced plasticity. Ann For Sci 65:305

David TS, Ferreira MI, Cohen S, Pereira JS, David JS (2004) Constraints on transpiration from an evergreen oak tree in southern Portugal. Agric For Meteorol 122:193–205

de la Riva EG, Olmo M, Poorter H, Ubera JL, Villar R (2016) Leaf Mass per Area (LMA) and its relationship with leaf structure and anatomy in 34 Mediterranean woody species along a water availability gradient. PLoS ONE. doi:10.1371/journal.pone.0148788

Díaz S, Kattge J, Cornelissen JHC, Wright IJ, Lavorel S, Dray S, Reu B, Kleyer M, Wirth C, Colin Prentice I, Garnier E, Bönisch G, Westoby M, Poorter H, Reich PB, Moles AT, Dickie J, Gillison AN, Zanne AE, Chave J, Joseph Wright S, Sheremet'ev SN, Jactel H, Baraloto C, Cerabolini B, Pierce S, Shipley B, Kirkup D, Casanoves F, Joswig JS, Günther A, Falczuk V, Rüger N, Mahecha MD, Gorné LD (2016) The global spectrum of plant form and function. Nature 529:167–171

Duhme F, Hinckley TM (1992) Daily and seasonal variation in water relations of macchia shrubs and trees in France (Montpellier) and Turkey (Antalya). Vegetatio 99–100:185–198

Eamus D, Prichard H (1998) A cost-benefit analysis of leaves of four Australian savanna species. Tree Physiol 18:537–545

Edwards EJ, Chatelet DS, Sack L, Donoghue MJ (2014) Leaf life span and the leaf economic spectrum in the context of whole plant architecture. J Ecol 102:328–336

Flexas J, Bota J, Cifre J, Escalona JM, Galmés J, Gulías J, Lefi EK, Martínez-Cañellas SF, Moreno MT, Ribas-Carbó M, Riera D, Sampol B, Medrano H (2004) Understanding down-regulation of photosynthesis under water stress: future prospects and searching for physiological tools for irrigation management. Ann Appl Biol 144:273–283

Flexas J, Barbour MM, Brendel O, Cabrera HM, Carriquí M, Díaz-Espejo A, Douthe C, Dreyer E, Ferrio JP, Gago J, Gallé A, Galmés J, Kodama N, Medrano H, Niinemets Ü, Peguero-Pina JJ, Pou A, Ribas-Carbó M, Tomás M, Tosens T, Warren CR (2012) Mesophyll diffusion conductance to CO_2: an unappreciated central player in photosynthesis. Plant Sci 193–194: 70–84

Flexas J, Diaz-Espejo A, Gago J, Gallé A, Galmés J, Gulías J, Medrano H (2014) Photosynthetic limitations in Mediterranean plants: a review. Environ Exp Bot 103:12–23

Florez-Sarasa ID, Bouma TJ, Medrano H et al (2007) Contribution of the cytochrome and alternative pathways to growth respiration and maintenance respiration in Arabidopsis thaliana. Physiol Plant 129:143–151

Gallardo A, Merino J (1993) Leaf decomposition in two Mediterranean ecosystems of southwest Spain: influence of substrate quality. Ecology 74:152–161

Galmés J, Medrano H, Flexas J (2007) Photosynthetic limitations in response to water stress and recovery in Mediterranean plants with different growth forms. New Phytol 175:81–93

Garbulsky MF, Peñuelas J, Papale D, Filella I (2008) Remote estimation of carbon dioxide uptake by a Mediterranean forest. Global Change Biol 14:2860–2867

García Nogales A, Linares JC, Laureano RG, Seco JI, Merino J (2016) Range-wide variation in life-history phenotypes: spatio temporal plasticity across the latitudinal gradient of the evergreen oak *Quercus ilex*. J Biogeogr 43:2366–2379

Givnish TJ (2002) Adaptive significance of evergreen vs. deciduous leaves: solving the triple paradox. Silva Fenn 36:703–743

González-Zurdo P, Escudero A, Babiano J, García-Ciudad A, Mediavilla S (2016a) Costs of leaf reinforcement in response to winter cold in evergreen species. Tree Physiol 36:273–286

González-Zurdo P, Escudero A, Nuñez R, Mediavilla S (2016b) Losses of leaf area owing to herbivory and early senescence in three tree species along a winter temperature gradient. Int J Biometeorol 60:1661–1674

Gulías J, Cifre J, Jonasson S, Medrano H, Flexas J (2009) Seasonal and inter-annual variations of gas exchange in thirteen woody species along a climatic gradient in the Mediterranean island of Mallorca. Flora 204:169–181

Günthardt-Goerg MS, Kuster TM, Arend M, Vollenweider P (2013) Foliage response of young central European oaks to air warming, drought and soil type. Plant Biol 15:185–197

Hallik L, Niinemets Ü, Wright IJ (2009) Are species shade and drought tolerance reflected in leaf-level structural and functional differentiation in Northern Hemisphere temperate woody flora? New Phytol 184:257–274

Harley PC, Tenhunen JD, Beyschlag W, Lange OL (1987) Seasonal changes in net photosynthesis rates and photosynthetic capacity in leaves of Cistus salvifolius, a European mediterranean semi-deciduous shrub. Oecologia 74:380–388

Hollinger DY (1992) Leaf and simulated whole-canopy photosynthesis in two co-occurring tree species. Ecology 73:1–14

Kikuzawa K (1995) Leaf phenology as an optimal strategy for carbon gain in plants. Can J Bot 73:158–163

Kikuzawa K, Onoda Y, Wright IJ, Reich PB (2013) Mechanisms underlying global temperature-related patterns in leaf longevity. Glob Ecol Biogeogr 22:982–993

Kok B (1948) A critical consideration of the quantum yield of Chlorella-photosynthesis. Enzimologia 13:1–56

Kuglitsch FG, Reichstein M, Beer C, Carrara A, Ceulemans R, Granier A, Janssens IA, Koestner B, Lindroth A, Loustau D, Matteucci G, Montagnani L, Moors EJ, Papale D, Pilegaard K, Rambal S, Rebmann C, Schulze ED, Seufert G, Verbeeck H, Vesala T, Aubinet M, Bernhofer C, Foken T, Grünwald T, Heinesch B, Kutsch W, Laurila T, Longdoz B, Miglietta F, Sanz MJ, Valentini R (2008) Characterisation of ecosystem water-use efficiency of european forests from eddy covariance measurements. Biogeosciences Discuss 5:4481–4519

Kurz-Besson C, Otieno D, Lobo Do Vale R, Siegwolf R, Schmidt M, Herd A, Nogueira C, David TS, David JS, Tenhunen J, Pereira JS, Chaves M (2006) Hydraulic lift in cork oak trees in a savannah-type Mediterranean ecosystem and its contribution to the local water balance. Plant Soil 282:361–378

Laureano RG, García-Nogales A, Seco JI, Linares JC, Martínez F, Merino J (2016) Plant maintenance and environmental stress. Summarising the effects of contrasting elevation, soil, and latitude on *Quercus ilex* respiration rates. Plant Soil 409:389–403

Lloyd J, Syvertsen JP, Kriedemann PE, Farquhar GD (1992) Low conductances for CO_2 diffusion from stomata to the sites of carboxylation in leaves of woody species. Plant Cell Environ 15:873–899

Lo Gullo MA, Salleo S, Rosso R, Trifilò P (2003) Drought resistance of 2-year-old saplings of Mediterranean forest trees in the field: relations between water relations, hydraulics and productivity. Plant Soil 250:259–272

Lucas PW, Turner IM, Dominy TN, Yamasitha N (2000) Mechanical defences to herbivory. Ann Bot 86:913–920

Ma S, Baldocchi DD, Xu L, Hehn T (2007) Inter-annual variability in carbon dioxide exchange of an oak/grass savanna and open grassland in California. Agric For Meteorol 147:157–171

Ma S, Baldocchi DD, Mambelli S, Dawson TE (2011) Are temporal variations of leaf traits responsible for seasonal and inter-annual variability in ecosystem CO_2 exchange? Funct Ecol 25:258–270

Martín Vicente A, Infante JM, García Gordo J, Merino J, Fernández Alés R (1998) Producción de bellotas en montes y dehesas del suroeste español. Pastos 28:237–248

Martínez F, Merino O, Martín A, García Martín D, Merino J (1998) Belowground structure and production in a Mediterranean scrub community. Plant Soil 201:209–216

Martinez F, Lazo YO, Fernández-Galiano RM, Merino J (2002a) Chemical composition and construction cost for roots of Mediterranean trees, shrub species and grasslands communities. Plant Cell Environ 25:601–608

Martinez F, Lazo YO, Fernández-Galiano JM, Merino J (2002b) Root respiration and associated costs in deciduous and evergreen species of Quercus. Plant Cell Environ 25:1271–1278

Martínez F, Laureano RG, Merino J (2003) Alternative respiration in seven Quercus spp of SW Spain. J Mediterr Ecol 4:9–14

Maselli F, Barbati A, Chiesi M, Chirici G, Corona P (2006) Use of remotely sensed and ancillary data for estimating forest gross primary productivity in Italy. Remote Sens Environ 100:563–575

Mediavilla S, Escudero A (2003) Photosynthetic capacity, integrated over the lifetime of a leaf, is predicted to be independent of leaf longevity in some tree species. New Phytol 159:203–211

Mediavilla S, Garcia-Ciudad A, Garcia-Criado B, Escudero A (2008) Testing the correlations between leaf life span and leaf structural reinforcement in 13 species of European Mediterranean woody plants. Funct Ecol 22:787–793

Mediavilla S, Herranz M, González-Zurdo P, Escudero A (2014) Ontogenetic transition in leaf traits: a new cost associated with the increase in leaf longevity. J Plant Ecol 7:567–575

Merino J, García Novo F, Sánchez Díaz M (1976) Annual fluctuation of water potential in the xerophytic shrub of the Doñana biological reserve. Oecologia Plant 11:1–11

Millenaar FF, Lambers H (2003) The alternative oxidase: in vivo regulation and fuctions. Plant Biol 5:2–15

Mitrakos K (1980) A theory for Mediterranean plant life. Acta Oecologica 1:245–252

Ne'eman G (1993) Variation in leaf phenology and habit in *Quercus ithaburensis*, a Mediterranean deciduous tree. J Ecol 81:627–634

Ne'eman G, Goubitz S (2000) Phenology of east-Mediterranean vegetation. In: Trabaud L (ed) Life and environment in the Mediterranean. WIT Press, UK, pp 155–202

Niinemets Ü (2001) Global-scale climatic controls of leaf dry mass per area, density, and thickness in trees and shrubs. Ecology 82:453–469

Niinemets Ü (2016) Does the touch of cold make evergreen leaves tougher? Tree Physiol 36:267–272

Niinemets Ü, Díaz-Espejo A, Flexas J, Galmés J, Warren CR (2009) Role of mesophyll diffusion conductance in constraining potential photosynthetic productivity in the field. J Exp Bot 60:2249–2270

Ogaya R, Peñuelas J (2007) Leaf mass per area ratio in Quercus ilex leaves under a wide range of climatic conditions. The importance of low temperatures. Acta Oecologica 31:168–173

Penning de Vries FWT, Brusting AHM, Van Laar HH (1974) Products, requirements and efficiency of biosynthesis: a quantitative approach. J Theor Biol 45:339–377

Pereira JS, Mateus JA, Aires LM, Pita G, Pio C, David JS, Andrade V, Banza J, David TS, Paço TA, Rodrigues A (2007) Net ecosystem carbon exchange in three contrasting Mediterranean ecosystems—the effect of drought. Biogeosciences 4:791–802

Poorter HA, Jong R (1999) A comparison of specific leaf area, chemical composition and leaf construction cost of field plants from 15 habitats differing in productivity. New Phytol 143:163–176

Poorter H, Niinemets Ü, Poorter L, Wright IJ, Villar R (2009) Causes and consequences of variation in leaf mass per area (LMA): a meta-analysis. New Phytol 82:565–588

Prieto I, Armas C, Pugnaire FI (2012) Water release through plant roots: new insights into its consequences at the plant and ecosystem level. New Phytol 193:830–841

Prior LD, Eamus D, Bowman DMJS (2003) Leaf attributes in the seasonally dry tropics: a comparison of four habitats in northern Australia. Funct Ecol 17:504–515

Quero JL, Villar R, Marañon T, Zamora R (2006) Interactions of drought and shade effects on seedlings of four *Quercus* species: physiological and structural leaf responses. New Phytol 170:819–834

Reich PB (2014) The world-wide "fast-slow" plant economics spectrum: a traits manifesto. J Ecol 102:275–301

Reich PB, Walters MB, Ellsworth DS (1992) Leaf life-span in relation to leaf, plant, and stand characteristics among diverse ecosystems. Ecol Monogr 62:365–392

Reich PB, Walters MB, Ellsworth DS (1997) From tropics to tundra: global convergence in plant functioning. Proc Natl Acad Sci U S A 94:13730–13734

Reich PB, Ellsworth DS, Walters MB, Vose JM, Gresham C, Volin JC, Bowman WD (1999) Generality of leaf trait relationships: a test across six biomes. Ecology 80:1955–1969

Reich PB, Falster DS, Ellsworth DS, Wright IJ, Westoby M, Oleksyn J, Lee TD (2009) Controls on declining carbon balance with leaf age among 10 woody species in Australian woodland: do leaves have zero daily net carbon balances when they die? New Phytol 183:153–166

Ribas-Carbo M, Taylor NL, Giles L, Busquets S, Finnegan PM, Day DA, Lambers H, Medrano H, Berry JA, Flexas J (2005) Effects of water stress on respiration in soybean leaves 1. Plant Physiol 139:466–473

Shipley B, Lechowicz MJ, Wright I, Reich PB (2006) Fundamental tradeoffs generating the worldwide leaf economics spectrum. Ecology 87:535–541

Takashima T, Hikosaka K, Hirose T (2004) Photosynthesis or persistence: nitrogen allocation in leaves of evergreen and deciduous *Quercus* species. Plant Cell Environ 27:1047–1054

Tausz M, Grulke NE, Wieser G (2007) Defense and avoidance of ozone under global change. Environ Pollut 147:525–531

Tretiach M (1993) Photosynthesis and transpiration of evergreen Mediterranean and deciduous trees in an ecotone during a growing season. Acta Oecologica 14:341–360

Turner IM (1994) Sclerophylly: primarily protective? Funct Ecol 8:669–675

Tyree MT, Cochard H (1996) Summer and winter embolism in oak: impact on water relations. Ann des Sci For 53:173–180

Valladares F, Vilagrosa A, Peñuelas J, Ogaya R, Julio J, Corcuera L, Sisó S (2004) Estrés hídrico: ecofisiología y escalas de la sequía. In: Valladares F (ed) Ecología del bosque mediterráneo en un mundo cambiante, Ministerio de Medio Ambiente, EGRAF, S. A., Madrid, pp 163–190

Van Ommen Kloeke AEE, Douma JC, Ordoñez JC, Reich PB, Van Bodegom PM (2012) Global quantification of contrasting leaf life span strategies for deciduous and evergreen species in response to environmental conditions. Global Ecol Biogeogr 21:224–235

Villar R, Merino J (1995) Dark leaf respiration in light and darkness of an evergreen and a deciduous plant species. Plant Physiol 107:421–427

Villar R, Merino J (2001) Comparison of leaf construction costs in woody species with differing leaf life-spans in contrasting ecosystems. New Phytol 151:213–226

Villar R, Robleto JR, De Jong Y, Poorter H (2006) Differences in construction costs and chemical composition between deciduous and evergreen woody species are small as compared to differences among families. Plant Cell Environ 29:1629–1643

Villar R, Ruíz-Robleto J, Ubera JL, Poorter H (2013) Exploring variation in leaf mass per area (LMA) from leaf to cell: an anatomical analysis of 26 woody species. Am J Bot 100:1969–1980

Vitousek PM, Field CB, Matson PA (1990) Variation in foliar $\delta 13C$ in Hawaiian *Metrosideros polymorpha*: a case of internal resistance? Oecologia 84:362–370

Warren CR, Adams MA (2000) Trade-offs between the persistence of foliage and productivity in two Pinus species. Oecologia 124:487–494

Warton D, Wright I, Falster D, Westoby M (2006) Bivariate line-fitting methods for allometry. Biol Rev 81:259–291

Watanabe CKA, Yamori W, Takahashi S, Terashima I, Noguchi KXLM (2016) Mitochondrial alternative pathway-associated photoprotection of Photosystem II is related to the photorespiratory pathway. Plant Cell Physiol 57:1426–1431

Westoby M, Warton D, Reich PB (2000) The time value of leaf area. Am Nat 155:649–656

Williams K, Field CB, Mooney HA (1989) Relationships among leaf construction cost, leaf longevity, and light environment in rain-forest plants of the genus *Piper*. Am Nat 133:198–211

Witkowski ETF, Lamont BB (1991) Leaf specific mass confounds leaf density and thickness. Oecologia 88:486–493

Wright IJ, Cannon K (2001) Relationships between leaf lifespan and structural defences in a low-nutrient, sclerophyll flora. Funct Ecol 15:351–359

Wright JP, Sutton-Grier A (2012) Does the leaf economic spectrum hold within local species pools across varying environmental conditions? Funct Ecol 26:1390–1398

Wright IJ, Westoby M, Reich PB (2002) Convergence towards higher leaf mass per area in dry and nutrient-poor habitats has different consequences for leaf life span. J Ecol 90:534–543

Wright IJ, Westoby M, Reich PB, Oleksyn J, Ackerly DD, Baruch Z, Bongers F, Cavender-Bares J, Chapin T, Cornellissen JHC, Diemer M, Flexas J, Gulias J, Garnier E, Navas ML, Roumet C, Groom PK, Lamont BB, Hikosaka K, Lee T, Lee W, Lusk C, Midgley JJ, Niinemets Ü, Osada H, Poorter H, Pool P, Veneklaas EJ, Prior L, Pyankov VI, Thomas SC, Tjoelker MG, Villar R (2004) The worldwide leaf economics spectrum. Nature 428:821–827

Wright IJ, Reich PB, Cornelissen JHC, Falster DS, Groom PK, Hikosaka K, Lee W, Lusk CH, Niinemets Ü, Oleksyn J, Osada N, Poorter H, Warton DI, Westoby M (2005) Modulation of leaf economic traits and trait relationships by climate. Glob Ecol Biogeogr 14:411–421

Chapter 7
The Role of Hybridization on the Adaptive Potential of Mediterranean Sclerophyllous Oaks: The Case of the *Quercus ilex* x *Q. suber* Complex

Unai López de Heredia, Francisco María Vázquez and Álvaro Soto

Abstract Gene flow among closely related species is a not so unusual event, especially in plants. Hybridization and introgression have probably played a relevant role in the evolutionary history of the genus *Quercus*, for instance in the post-glacial northwards migration of European white oaks. In the same way, hybridization between the Mediterranean sclerophyllous oaks *Q. ilex* and *Q. suber* could have been determinant for the survival of the latter species during glaciations. In this chapter, evidences of the ancient introgression between these two species are revised, as well as estimations of current hybridization rates, which are very likely underrated. Pre-zygotic and post-zygotic limitations to introgression between *Q. suber* and *Q. ilex* are described. Finally, the effects of hybridization and introgression on cork quality, and the suitability of *Q. ilex*—*Q. suber* as a model system for the study of introgression and the maintenance/restoration of species boundaries within the genus are briefly discussed.

7.1 Hybridization and Introgression in Oaks

Hybridization, i.e. the combination between different species by means of sexual reproduction, is a common process in nature that can play a key role in the evolution of plants, where is more spread than in animals: 25% of plant taxa versus 10% of animal taxa are thought to be subjected to hybridization processes

---

U. López de Heredia (✉) · Á. Soto
GI Genética y Fisiología Forestal. Dpto. Sistemas y Recursos Naturales. ETSI Montes, Forestal y Del Medio Natural, Universidad Politécnica de Madrid.
Ciudad Universitaria s/n, 28040 Madrid, Spain
e-mail: unai.lopezdeheredia@upm.es

F. M. Vázquez
Grupo de Investigación HABITAT, Instituto de Investigaciones
Agrarias "Finca La Orden-Valdesequera", CICYTEX, Junta de Extremadura,
Autovía a-5 Km 372, 06187, Guadajira Badajoz, Spain

© Springer International Publishing AG 2017
E. Gil-Pelegrín et al. (eds.), *Oaks Physiological Ecology. Exploring the Functional Diversity of Genus* Quercus L., Tree Physiology 7,
https://doi.org/10.1007/978-3-319-69099-5_7

(Mallet 2005). In many plant systems, hybridization has even led to the apparition of new species, via polyploidy (Stebbins 1974), or, in other cases, due to the inability of hybrids to backcross with the parent species because unequal pairing of the chromosomes (Arnold 1997), has caused reproductive isolation of hybrids.

In other species, however, interspecific barriers are more diffuse, and backcrossing of hybrids is possible. Introgression is the transfer of genes from one species to another, as a consequence of interspecific hybridization followed by backcrossing with one of the parental species. Gene transfer has relevant implications in the evolution, because the diversity produced by introgression far exceeds the one produced by mutations (Anderson 1949). The effects of hybridization and introgression may be relevant for low-size populations or for species at risk of extinction, or may increase the ability to colonize new habitats (Potts and Reid 1988; Petit et al. 2004). Moreover, hybridization and introgression are evolutionary forces that have facilitated the stabilization or emergence of species, and, occasionally, have produced populations with more competitive genomes and with a greater ecological range (Sexton et al. 2009), thus providing adaptive solutions to environmental changes (Rieseberg et al. 2003).

For hybridization to occur in nature, it must exist reproductive compatibility between the pairs of species that produce hybrids, which is only possible if species are phylogenetically close. In addition, hybridization normally occurs in areas where species with compatible genomes coexist, and environmental and phenological factors enable effective gene flow (Sork and Smouse 2006).

Quercus L. is a relevant genus in the debate about the extent of hybridization and introgression on the evolution of plants (Anderson 1949; Whittemore and Schaal 1991; Rieseberg and Wendell 1993), on the assessment of the ecological factors limiting hybridization in nature (Stebbins 1950; Muller 1952; Barton and Hewitt 1985; Rushton 1993), and served as a model to re-define the species concept based upon ecological criteria (Burger 1975; Van Valen 1976). From an adaptive point of view, introgression in oaks could have played an important role in northwards migration of European white oak species from southern refugia after the last Würmian glacial period (Petit et al. 1997, 2004). In the same way, in adverse environmental conditions, hybrids of some Mediterranean sclerophyllous oaks could have shown better fitness that at least one of the parental species, while subsequent backcrossing with this species would have led to the recovery of the "pure" character of the species (López de Heredia et al. 2007a). This process, also known as adaptive introgression, consists in allele transfer between different species through subsequent backcrosses; therefore, one species acquires specific traits from another, enabling adaptation to adverse environmental conditions and increasing the ecological specific range of the species (Arnold and Martin 2006).

Hybridization does not always generate obvious hybrids, and complicates the traditional taxonomy treatment of genus *Quercus*. In many sympatric areas where several oak species coexist, the concept of multispecies has been adopted (Dodd and Rafii 2004; Peñaloza-Ramírez et al. 2010). Hence, some authors considered a hybrid origin for some species such as *Q. subpyrenaica* (*Q. faginea* × *Q. pubescens*)

(Himrane et al. 2004), *Q. afares* (*Q. suber* × *Q. canariensis*) (Mir et al. 2006), or *Q. marianica* C.Vicioso (*Q. canariensis* × *Q. faginea*) (Vila-Viçosa et al. 2014).

The presence of fertile oak hybrids resulting from interspecific crossing between the more than 500 species recognized in the genus is well-known, and has been extensively documented (Govaerts and Frodin 1998). The inter-sectional hybrids in *Quercus* L., are frequently sterile, such as *Q. robur* (Sect. *Robur*) cross to *Q. suber* (Sect. *Cerris*) or *Q. faginea* (Sect. *Galliferae*) cross to *Q. suber* (Vázquez et al. 2015). Among the more than 300 oak hybrids (nothotaxa) that have been acknowledged, 70% of them are fertile, capable to generate viable offspring, and showing ability to backcross. Therefore, new generations of fertile individuals with viable offspring (F2, F3…) are produced in mixed populations, to the extent of not being able to distinguish introgressed from pure progenies, depending on the degree of kinship (Staudt et al. 2004).

Hybridization in oaks is facilitated or supported by various environmental and functional factors, but, at the same time, other factors may limit its effects on the populations, and the integrity of the species is largely maintained. In the one hand, oaks are anemophilous and allogamous, and frequently show phenological overlap between species, facilitating inter-species fertilization. On the other hand, there are also barriers in oaks that prevent hybridization, such as physical limitations, climatic and environmental conditions during the breeding process (Cecich and Haenche 1995; Sever et al. 2012), or gametic incompatibility between individuals of the same or of other species (Yacine and Bouras 1997).

7.2 Hybridization and Introgression in the *Q. ilex* x *suber* Complex

Cork oak (*Q. suber*) and holm oak (*Q. ilex*) are two ecologically, economically, environmentally and socially relevant species that are key elements in open oak woodlands from the western Mediterranean basin ("dehesas" in Spain; "montados" in Portugal). Here, cork oak is exploited to extract cork, while the sweet acorns from holm oak constitute an essential portion of the pig livestock feed. Both species are included into distinct infrageneric groups (*Ilex* and *Cerris*), and can be easily distinguished according to morphological traits of bark, cupules of the acorns, or leaves (Amaral Franco 1990). Cork oak is named after its very thick, corky bark (see below, Sect. 7.3), and it presents also medium to large leaves of up to 8.5 cm, with annual and biennial fruit production. The high diversity of phenotypic characters of holm oak in the Mediterranean has led some authors to identify two different species/subspecies. Continental holm oaks with frequently spinescent short leaves, and sweet acorns from Western Mediterranean, have been integrated into *Quercus rotundifolia* Lam. (= *Quercus ilex* L., subsp. *ballota* (Desf.) Samp.) (Amaral Franco 1990; Vázquez et al. 1993), while *Quercus ilex* L. (= *Quercus ilex* subsp. *Ilex* L)., represents the holm oak forests of temperate climates with oceanic influence from the

Mediterranean Basin, and shows longer leaves (more than seven secondary pairs of nerves), and bitter acorns (Rafii et al. 1991; Peguero-Pina et al. 2014).

In addition to phenotypic differences, both species have contrasting ecological requirements. Cork oak is more hygrophilous and thermophilous than holm oak (at least than *Q. rotundifolia*), with the former occurring in areas with 400 mm of annual precipitation, and mean temperatures above 0 °C, and the latter inhabiting areas with hardly 350 mm of annual precipitation, and marked temperature oscillations. Probably the most determinant ecological factor to explain the current distribution of both species is the acidophilus character of cork oak, which hampers the occurrence of the species on calcareous soils. On the contrary, holm oak is tolerant to many growing conditions. Thus, cork and holm oak ecological requirements overlap to a great extent, and both species form mixed forests in the siliceous or decarbonated areas from the Western Mediterranean. The presence of individuals with intermediate morphological characteristics in these regions has long been known (Colmeiro and Boutelou 1854; Laguna 1881; Borzi 1881; Natividade 1936; Camus 1936–1954). These hybrids, that received the names of *Quercus* × *morisii* Borzí, or *Quercus* × *avellaniformis* Colmeiro & E. Boutelou, are described as having strongly cracked bark, glabrescent leaves of light green tone, fruits with conical cupules with free bracts, and viable flavour acorns. More detailed micro-morphological and anatomical characters are related to the presence/absence and distribution of foliar trichomes (Vázquez 2013), the characteristics of floral organs, with hairs typical of cork oaks in the anthers' teaks. However, the only way to confirm the hybrid character of the species is by analyzing the genetic composition of the candidate hybrids with molecular methodologies.

7.3 Hybridization at Regional and Local Scales. Molecular Approaches

Detection, quantification and evaluation of hybrids in the *Q. ilex* × *suber* system using molecular methodologies was approached in the first decade of the twentieth century by several independent research groups. Phylogeographic patterns based upon chloroplast DNA (cpDNA), which is maternally inherited in oaks, and has a low mutation rate, suggested the existence of past hybridization events. These early studies showed higher cpDNA diversity in *Q. ilex* than in *Q. suber*. According to these studies, *Q. ilex* had three cpDNA lineages shared with *Q. coccifera* (López de Heredia et al. 2007b) that have been extended to five in a recent study that sampled the easternmost population of the species, along with other oak members of the Group Ilex (*Q. coccifera* (Mediterranean Basin), *Q. alnifolia* (Mediterranean Basin), *Q. aucherii* (Near East), *Q. baloot* (Asia), and *Q. floribunda* (Asia)) more intensively (Vitelli et al. 2017). For *Q. suber*, however, a single cpDNA lineage was found, showing little variation between eastern and western populations (Jiménez et al. 2004; Lumaret et al. 2005; López de Heredia et al. 2007b; Magri et al. 2007; Pulido Neves da Costa 2011). The consistence of cpDNA lineages for both species is shown when analyzing the correspondence of some of the haplotypes found by López de Heredia et al. (2007b) with mitochondrial DNA (mtDNA), which is also maternally inherited, and has even lower mutation rate than cpDNA (Fig. 7.1). The four cpDNA haplotypes found for *suber* lineage showed a single mitotype, while the more diverse *i-c I* lineage (lineage Euro-Med V from Vitelli et al. 2017), which is shared by *Q. ilex* and *Q. coccifera* of the western range of the species, presented four mitotypes. The remaining *ilex* cpDNA lineages, also shared with *Q. coccifera*, corresponded with a single mitotype each.

While many sympatric populations from the western (i.e. Portugal, western Spain, Morocco, southwestern France) and eastern distribution ranges of *Q. suber* (i.e. southeastern France, the Italian peninsula, Sardinia, Corsica, Sicily, Algeria, Tunisia) showed specific plastidial DNA lineages consistent with the taxonomic identification, the aforementioned studies detected extensive cpDNA lineage sharing in sympatric populations from eastern Iberian Peninsula (Jiménez et al. 2004), and the Balearic Islands (López de Heredia et al. 2005) (Fig. 7.2). Specifically, these *Q. suber* populations, which are mainly located in sub-optimal sites for this acidophilus species, showed only *ilex* cpDNA lineages. An ancient hybridization of *Q. suber* with *Q. ilex* followed by backcrossing and introgression,

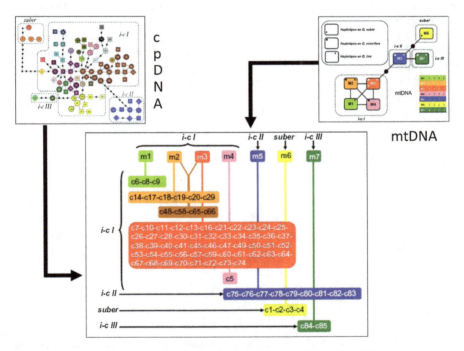

Fig. 7.1 Correspondence of cpDNA and mtDNA haplotypes. Chloroplast DNA haplotypes are depicted in López de Heredia et al. (2007b). *Suber* lineage cpDNA haplotypes correspond with a single mitotype (M6), while *Ilex* cpDNA lineages, also present in *Q. coccifera*, are more diverse, and show correspondence with six mitotypes (one for each of the eastern lineages *i-c II* and *III*, and four for the western lineage *i-c I*)

may explain the complete displacement of the *suber* lineage exactly in the central range of *Q. suber* distribution. As we have said before, *Q. ilex* is a much more tolerant species than *Q. suber* in terms of cold, drought resistance and soil composition; therefore, extensive hybridization in an adverse environmental context, for instance in the glacial pulses of the Holocene, may have had an adaptive role for *Q. suber*. Hybridization is more likely to occur with *Q. suber* acting as pollen donor (see Sect. 7.1), so that hybrids would carry *ilex* chloroplast. Successive pollinations of these presumably better adapted hybrids by *Q. suber* would have yielded introgressed cork oaks with *ilex* chloroplast. However, simulation studies suggested that the complete disappearance of *suber* haplotypes is possible even in lack of selection due to stochastic demographic processes when there is a lower relative effective population size of *Q. suber* versus *Q. ilex*, and/or strong asymmetries to gene flow (Soto et al. 2008). In the rest of the distribution range of *Q. suber*, evidences of ancient introgression with *Q. ilex* are not so clear as in the case of the populations from eastern Spain. However, together with *Q. suber* individuals bearing a *suber* cpDNA lineage, *ilex* cpDNA lineages were also found in some *Q. suber* populations from western Iberia, but in a significantly lower proportion. Among other possible causes, the fact that lineage *suber* has not been completely displaced by lineage *ilex* in these western populations could be due to a more recent hybridization.

But, have the traces of hybridization found for cpDNA a parallel in nuclear DNA? The answer to this question is not trivial. The analysis of nuclear ribosomal ITS sequences and dominant markers such as AFLPs (López de Heredia et al. 2007b; Coelho et al. 2006) showed that both species were clustered in robust

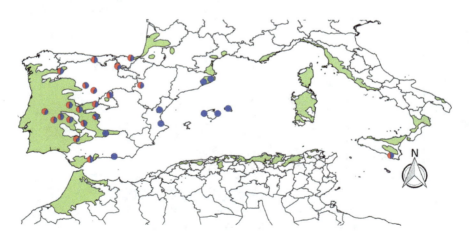

Fig. 7.2 Chloroplast DNA lineage proportion in 23 sympatric populations of *Q. suber* and *Q. ilex* (modified from López de Heredia et al. (2007b)). Notice that some populations in eastern Spain and the Balearic Islands lack lineage *suber* (in red), while other populations show a variable presence of lineage *ilex* (in blue). In light green, the distribution range of *Q. suber*. These results are consistent with those obtained for nuclear EST2T13, and TrnS/PsbC cpDNA haplotypes from Pulido Neves da Costa (2011)

separate groups, even those samples that showed evidence of ancient hybridization and introgression at the cpDNA level. However, if contemporary hybridization is going on, as suggested by hybrids detected in the field and in those mixed populations where *suber* and *ilex* lineages co-exist, a more detailed analysis of nuclear hybridization at a local scale using co-dominant molecular markers would provide new insights in the extent of introgression.

Early studies using allozymes on hybrids (Elena-Rosselló et al. 1992; Oliveira et al. 2003) had found very limited levels of introgression. The use of higher resolution markers, such as nuclear microsatellites, i.e. simple repeated motifs consisting of 1–6 base pairs that can be found in both coding and non-coding regions, were used together with Bayesian clustering methodologies to quantify the extent of hybridization in mixed stands of *Q. suber* and *Q. ilex*. For instance, Soto et al. (2003) used a set of six nuclear microsatellites (nSSR) developed in other oak species for the identification of F1 hybrids obtained in controlled crosses. Later on, the same authors added new nSSRs to the set of markers used for population genetics and gene flow analysis in holm and cork oak (Soto et al. 2007). Unfortunately, those molecular markers revealed a high proportion of shared alleles between both species, and a much lower level of diversity in cork oak than in holm oak, yielding just a moderate exclusion power. Nevertheless, these markers, combined with Bayesian approaches, have been applied for the identification of hybrid and introgressed individuals in the field (Burgarella et al. 2009). The analysis of seven mixed and five cork oak pure populations, using as reference the allele frequencies for the whole distribution range of cork oak, provided an estimation of a very low introgression rate, below 2%. This result was consistent with the results obtained with allozymes (Elena-Rosselló et al. 1992).

More interestingly, Burgarella et al. (2009) pointed out to the possibility of bidirectional introgression, identifying presumably introgressed individuals also among the trees phenotypically classified as holm oaks. These results suggest that phenological, and even pre- and post-zygotic barriers (see below), do not avoid completely backcrosses and introgression towards *Q. ilex*.

Simulation analysis using the allele frequencies for this set of markers, and further application of Bayesian approaches for the detection of introgression in virtual individuals with known pedigree, have revealed that this set of markers is insufficient for a reliable estimation of advanced introgression (individuals resulting from one or several backcrosses with the parental species) (Pérez Rodríguez 2008; López de Heredia and Soto 2017), and that current introgression very likely could have been underestimated in previous studies.

A microsatellite analysis of the progeny from four hybrid adult trees identified in a mixed stand in Central Spain revealed that three of them were preferentially pollinated by cork oak, while for the remaining hybrid the probable pollen donors were holm oaks (López de Heredia et al. 2017a). This result is consistent with the species that occurs more frequently in the vicinity of the hybrid trees. Thus, backcrossing direction could be driven by pollen availability and abundance, although backcrossing "preferences" due to genomic and epigenetic incompatibilities cannot be discarded.

7.4 Reproductive Success and Hybrid Phenotypes

In the previous section, we have seen how hybridization and introgression in mixed forests of cork and holm oak could have been underestimated up to date. We will focus now on the impact of hybridization on the resulting phenotypes of first generation hybrids and backcrosses, at individual levels, and on the integrity of the species. Hybrid phenotypes have been traditionally identified by taxonomists as the phenotypes of individuals that show intermediate characters, based upon the premise of the polygenic control of morphological traits with simple additive effects (Rieseberg et al. 2007). However, establishing hybrid ancestry may not be straightforward considering only intermediate characters by several reasons (López-Camaal and Tovar-Sánchez 2014): (1) the correlation of morphological traits that reduces the number of characters to identify different species; (2) the mode of inheritance and segregation of genes associated to morphological traits; (3) the effect of the environment on morphological expression, i.e. plasticity and epigenetics; and (4) the retention of plesiomorphic character states of ancestral populations in closely related species, that may originate intermediate phenotypes as well (Rieseberg 1995; Judd et al. 2002; Arnold and Martin 2006).

The limitations derived from the correlation of morphological traits may be overcome by selecting the best characters discriminating between species using, for instance, multivariate analysis on a sufficient number of individuals. However, a proper analysis of genetic and epigenetic determinants of morphological expression can only be accomplished in controlled conditions, by analyzing full- or half-sib progenies. However, it is not easy to obtain such progenies, since the path from flowering of the parents to the adult reproductive stage of *Q. ilex x suber* hybrids and backcrosses presents pre-zygotic and post-zygotic barriers that will compromise the success of the hybridization event.

7.4.1 Pre-zygotic Barriers. Pollination and Fertilization in Hybrids

Several endogenous and exogenous factors may prevent zygote formation, and therefore compromise subsequent acorn production. Among these factors, the most evident ones occur to prevent pollination, and are those related with the phenological overlap of flower production between cork and holm oaks, being both of them monoecious protandrous species. Environmental conditions are fundamental in the dispersion, viability and germination of the pollen grains. The most effective conditions for the viability and germination of pollen grains are 80% of relative humidity, and temperatures of 25–29 °C for holm and cork oak, respectively (Vázquez et al. 1996).

Flowering phenology of the species may vary between sites and years, and consequently, the overlapping of male and female flowers of both species may also vary. *Quercus ilex* flowers from mid-February to late-March in southwestern Iberian Peninsula (Boavida et al. 2001; Gómez-Casero et al. 2007; Varela et al. 2008), extending this period until the end of May in some areas from north-eastern Spain (Castro-Díez and Montserrat-Martí 1998; Ogaya and Peñuelas 2004; Montserrat-Martí et al. 2009), and including the possibility of a second flowering period in autumn (Vázquez et al. 1992; Ducousso et al. 1993). *Quercus suber* flowers from early-April to mid-May in Portugal (Boavida et al. 2001; Varela et al. 2008), and may flower earlier in central Spain (Díaz-Fernández et al. 2004; Gómez-Casero et al. 2007). In some areas, such as northwestern Iberia, cork oak's leaf burst and flowering may be delayed until June (Jato et al. 2002). Occasionally, autumn flushes may produce a secondary flowering event in cork oak, but there is no evidence that this leads to acorn production (Eriksson et al. 2017). To complicate the picture, biennial acorn maturation is also possible in cork oak, even in the same tree that produces annual maturation acorns (Díaz-Fernández et al. 2004).

It must be stressed out that the flowering behavior of the two species in mixed populations is usually marked by great differences in time. Therefore, and for western Iberia, early cork oak male flowers can pollinate late female holm oak flowers, but the probability of the opposite process is reduced (Varela and Valdiviesso 1996; Elena-Rosselló and de la Cruz 1998). However, comparative studies for flower phenology on mixed stands in Portugal have shown that the opportunity for cross-pollination could have occurred in both directions (Varela et al. 2008). In the case of mixed populations in north-eastern Spain, flowering seems to overlap for longer periods, increasing the chances for hybridization to occur. To our knowledge, the only study that monitored the flowering phenology of a hybrid individual compared to cork and holm oaks from the same population was performed in central Spain in spring 2006 (Perea García-Calvo 2006). The results of this study showed that male catkin production took place before in holm oak (before May 15th) than in cork oak and the hybrids (between May 15th and June 1st), pointing to a more similar phenological behavior of the hybrid and cork oaks. However, although these results support the preferential hybridization of cork oak acting as pollen donor and suggest the possibility of backcrossing to produce advanced generation introgressed individuals, they should be interpreted with caution because they were obtained from only one hybrid individual in a single year.

However, the discordance between pollen release and stigma receptivity periods is not the only pre-zygotic barrier to hybridization, and relevant post-pollination barriers to interspecific crosses have been suggested. To complete the fertilization process, the pollen tube progresses through the style after the pollen grain arrives the stigma of the male flower. Boavida et al. (1999), while studying the progamic phase of cork oak, reported a sequential arrest of pollen tubes along the style, thus providing preliminary evidence for a pollen tube competition mechanism. For *Q. ilex*-*Q. suber* intraspecific crosses, although some prezygotic interaction may occur at the style, the most important interaction takes place at the ovary, where additional phenomena related to the inability of pollen tubes to penetrate the transmitting tissue after microspore germination were shown by Boavida et al. (2001). These authors reported significantly higher success rate in the interspecific crosses for *Q. suber* acting as pollen donor rather than as female parent due to differential growth of the pollen tubes of both species. Notwithstanding, as we have seen before, there is gaining evidence that suggests that the success of bi-directional hybridization could be more spread than previously thought.

7.4.2 Post-zygotic Barriers: Acorn Germination and Seedling Development

When fertilized female flowers do not abort, acorns are formed and mature in the tree the very same autumn. First generation hybrid acorns are constituted normally and are able to germinate; however, adult hybrids have been identified at extremely low frequency in natural populations (Oliveira et al. 2003; Burgarella et al. 2009; López de Heredia et al. 2017a). Analysis of the progeny of hybrid individuals localized in natural populations has shown that viable acorns are produced by hybrids. However, germination percentages of hybrid progenies are significantly lower (<60%), and show larger variation between families for hybrid progenies than for pure species, where germination rate is in general >80% (Fig. 7.3).

The processes underlying lack of germination are difficult to decipher. The main abiotic cause of acorn mortality is in situ desiccation in winter, as suggested for *Q. ilex* (Joët et al. 2016). For cork oak, it was shown that genes related to response to drought and water transport were mostly represented during early and last stages of acorn development, when tolerance to water desiccation is possibly critical for acorn viability (Miguel et al. 2015). Although differential behavior of *Q. ilex* and *Q. suber* has been regarded in terms of germination and epicotyl emergence rates at diverse target moisture content of the embryos (Salomón et al. 2012), there is no reason to assert that the recalcitrant behavior of acorns could be more or less pronounced in hybrid backcrosses than in pure progenies. According to our experimental observations, some acorns that germinate and produce taproots are

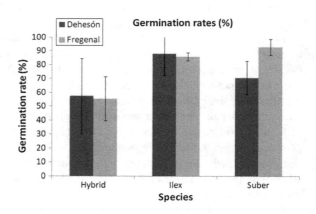

Fig. 7.3 Germination rates of half-sib progenies of hybrids, *Q. ilex* and *Q. suber* from four populations in Spain

however unable to initiate stem differentiation, and die few months after germination. A significant effect of stressful environmental conditions that result in fluctuating asymmetry of acorns was argued to explain the strong developmental selection that resulted in fertilization failure and abortion in holm oak (Díaz et al. 2003). In our controlled experiments with hybrid progenies, however, the stress is likely produced by interspecific genetic incompatibilities that probably lead to the modification of the epigenetic systems controlling germination and initial developmental stages, as has been suggested for hybrids of *Arabidopsis* (Zhu et al. 2017).

At initial stages of seedling development, a relationship between acorn size and seedling growth and survival has been reported for both cork (Ramírez-Valiente et al. 2009) and holm oak (Bonito et al. 2011). For hybrid progenies, however, low correlation was found between acorn size (estimated as the volume of an ellipsoid) and seedling dimensions (height and slenderness ratio) one and two years after sowing (Fig. 7.4).

Hybrid seedlings present very diverse phenotypes. In the aerial part, a percentage of individuals are constituted normally from the shoot apical meristems and the leaf primordia. Seedlings may resemble the phenotypes of pure species, or present transgressive traits possibly due to epistasis, heterosis, and/or developmental instability. Seedlings of abnormal phenotypes in hybrid progenies of *Q. ilex x suber* are frequent. Among these, we can cite seedlings with multiple stems or strong ramification patterns, seedlings that lack leaves or present small micro-leaves, dwarf seedlings without apparent apical growth, and seedlings showing extremely asymmetric leaves (Fig. 7.5a–f). In other cases, heterosis is manifested in leaf size

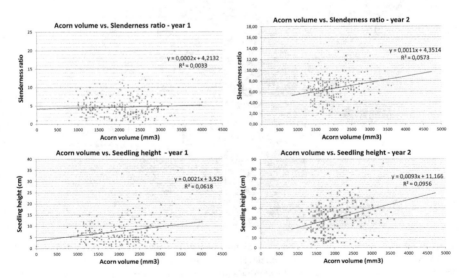

Fig. 7.4 Low correlation between acorn volume and slenderness ratio and height of one and two years-old backcrossed seedlings, respectively

or growth, with hybrids showing superior performance than pure species (Fig. 7.5g). These results are consistent with findings for other species complexes. The phenotypic analysis of F1 hybrids in 46 plant taxa showed 44.6% of intermediate characters, while 45.2% of the characters in hybrids were similar to either parental species, and 10.2% presented transgressive characters (Rieseberg and Ellstrand 1993).

The mechanisms underlying abnormal phenotype formation in hybrids are related to genetic incompatibilities that modify the epigenetic system that controls the differentiation of the apical shoot meristem and the leaf primordia (Lodha et al. 2013; Iwasaki et al. 2013). Axis development and polarity specification of flat bifacial leaves require a network of genes for transcription factor-like proteins and small RNAs. For instance, the development of asymmetric leaves in *Arabidopsis* is produced by repression of key genes in the regulation of the proximal–distal leaf length (class 1 KNOX homeobox genes -BP, KNAT2-), and in the establishment of adaxial–abaxial polarity (ETT/ARF3 and ARF4) (Machida et al. 2015; Matsumura et al. 2016). Possibly, similar disruptions of such regulatory networks due to genomic incompatibilities are behind the production of abnormal phenotypes in the *Q. ilex* x *suber* complex.

The number of seedlings showing malformations varies with ontogenic changes. Although is not a generalized process, seedlings lacking leaves during the first year of life are able to produce leaves at more advanced life-stages, while extremely

Fig. 7.5 Illustrative examples of abnormal phenotypes in half-sib progenies of *Q. ilex* x *suber*. **a** two-guides seedling with abnormal stem development from the shoot apical meristem. **b** seedling lacking leave development from leaf primordia. **c, d, e** seedling with several degrees of abnormal leaf development. **f** seedling showing strong ramification patterns, and small asymmetric leaves. **g** seedling outperforming for growth and leaf size, possibly due to heterosis

Fig. 7.6 Number of normal and malformed individuals in half-sib progenies of four hybrids (IS, LG1, LG2, ZLR), four holm oaks (I1, I2, I3, I4), and four cork oaks (S1, S2, S3, S4), in four- (**b**) and two-years-old (**a**) seedlings. Lower proportions of individuals showing abnormal phenotypes are scored for older seedlings

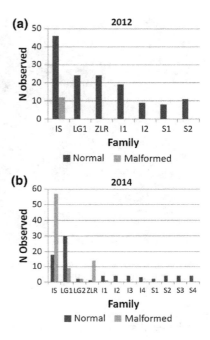

asymmetric leaves seem to have shorter life-spans than normal leaves, thus falling from the seedlings earlier (López de Heredia et al. 2017b) (Fig. 7.6). The effect on the hybrid population is that malformed phenotypes are filtered from the population. Although not verified in field experiments, early expressed abnormality probably results in lower fitness of the hybrid seedlings, and in higher mortality rates than pure species.

7.4.3 Effects of Hybridization on Cork

Bark is probably the most discriminant phenotypic character between cork and holm oaks in the adult stage. The most common model of bark development in *Quercus* species is based upon the rhytidome concept. A rhytidome is formed when successive phellogens differentiate to produce intricate suberized cell layers (phellem) that enclose heterogeneous cortical tissues (parenchyma, fibers, etc.) and collapsed phloem cells, thus producing irregular outer bark (Fig. 7.7). This type of outer bark has been described for American oak species (Howard 1977), *Q. robur* (Trockenbrodt 1991), *Q. petraea* (Gricar et al. 2015), *Q. faginea* (Quilhó et al. 2013), and *Q. cerris* (Sen et al. 2011), and is also present in *Q. ilex*. For *Q. suber*, however, this bark model is not valid, and the outer bark is due to the differentiation of a single phellogen that rather than producing a rhytidome, generates new layers of suberized cells each year (Graça and Pereira 2004). This bark differentiation

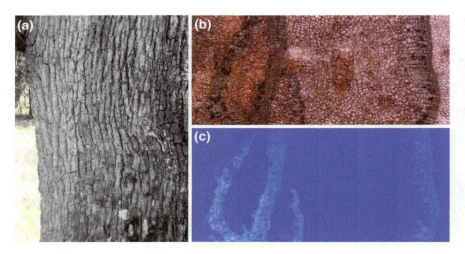

Fig. 7.7 *Quercus ilex* L. bark. **a** Detail of the trunk of a holm oak. **b** and **c**. Microphotographs of *Q. ilex* rhytidome showing the successive, thin periderms under visible (**b**) and UV light (**c**), where autofluorescence of suberin is detected

model results in a thick regular suberized material with extraordinary protective and insulation properties that may contain also lignified cells with thick walls, called phelloids (Fig. 7.8).

Cork is a unique feature of *Q. suber*, and is one of the most important products exploited in the Mediterranean countries due to its high economic and social value in rural areas. Cork is a hydrophobic renewable raw material that protects the tree against water loss, pathogen attacks or fire, and presents good technological properties with industrial application (Pausas et al. 2009). It is constituted mainly by dead, empty cells with their cell walls covered with suberin, a substance partially similar to lignin and cutin. Although suberin composition has been profusely studied (Franke and Schreiber 2007; Schreiber 2010; Serra et al. 2014), and several genes involved in cell wall suberization have been described (Barberon et al. 2016; Soler et al. 2007; Pollard et al. 2008; Serra et al. 2010; Soler et al. 2011), the developmental factors leading to the formation of a long-living phellogen and a thick, continuous phellem remain elusive.

Traditionally, it has been considered that bark of F1 hybrids is not suitable for cork extraction, and that bad quality cork comes from individuals with a certain degree of introgression (Natividade 1936; Llensa de Gelcén 1943; Varela et al. 2008). Early studies of the bark anatomy of *Q. ilex x suber* hybrids revealed segregation of bark characters, with some hybrids showing barks resembling those of *Q. ilex*, others with corky barks, and some others with intermediate characteristics (Natividade 1936). The same study depicted the phellem of hybrids as forming a rhytidome, like in *Q. ilex*, but with considerable thickness of the successive phellem layers, like in *Q. suber*. In addition, the author reported significant intra-individual variability in bark anatomy of hybrids. Differences in bark are

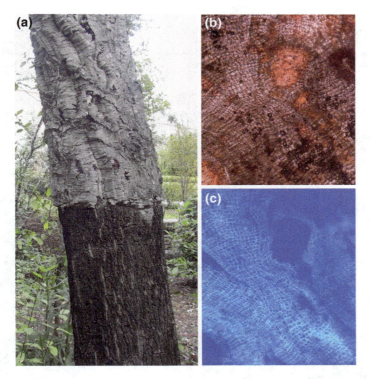

Fig. 7.8 *Quercus suber* L. bark. **a** Detail of the trunk of a cork oak, with the lower part debarked few years ago. **b** and **c** Microphotographs of *Q. suber* thick periderm, with inclusions of lignified phelloids, under visible (**b**) and UV light (**c**)

hardly noticeable during the first years of life of seedlings, when a single phellogen is differentiated. In three years-old seedlings of hybrids, no significant differences in phelloderm development, nor in the phelloderm/phellem thickness ratio were found, while only slight differences in phellem thickness were scored. In older individuals, however, the results by Natividade (1936), have been recently confirmed by microscope bark analysis in branches of different age of hybrids and pure species (Díez Morales 2016) (Table 7.1).

However, a preliminary analysis of a higher number of adult hybrid individuals shows noticeable differences in their barks. While most individuals show a rhytidome with successive, thin phellems, more similar to holm oaks, other individuals present barks mainly formed by a thick phellem, similar to that of cork oak (Soto et al. unpublished). Interestingly, a high variability in bark structure has been observed even within the same individual. In individuals with rhytidome we can find, within the same section of the bark, areas with very thin phellems and other areas with thicker phellems, usually near the lignified radii (Fig. 7.9). It is difficult to determine if these thicker phellems are derived from more active and/or from longer-living phellogens. Anyway, differences in the development of phellems are

Table 7.1 Characteristics of the outer bark in *Q. ilex*, *Q. suber*, and four hybrid individuals. Modified from Díez Morales (2016)

	Q. ilex	Q. suber	Ilicioid hybrid	Suberoid hybrid	Ilicioid hybrids	Suberoid hybrid
Successive separate phelogenes	y	n	y	n	y	n
Phellem thickness	Thin	Massive	Thin	Thick	Thin	Thick
Pheloids	Isolated	Grouped	Isolated	Grouped	Isolated	Grouped
Phellem annual rings	n	y	n	y	n	y
Lignified phellem rings	n	y	n	y	n	y
Other structures	Independent periderms	n	Llenticels, independent periderms	n	Independent periderms	n

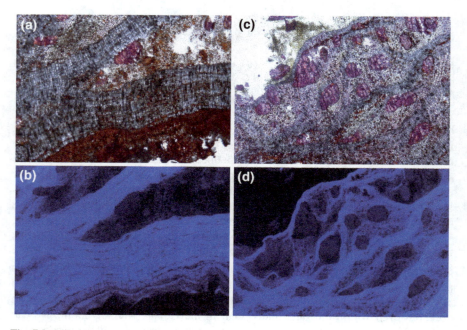

Fig. 7.9 Microphotographs of the bark of a hybrid individual under visible (**a, c**) and UV (**b, d**) light. **a** and **b** show an area with thick periderms. **c** and **d** show an area with successive, thin periderms. The non-lignified cells comprised between successive periderms also present suberin

very likely due to epigenetic modifications. It is also noteworthy the partial suberization of the cells comprised between successive phellems in these individuals. This fact has not been reported before, and does not appear in cork or in holm oaks.

7.5 Conclusions

In conclusion, hybridization and introgression between *Q. ilex* and *Q. suber* are natural processes that, although not being the rule, have probably been underestimated, and which could have played a key role in the evolution of these species. *Quercus ilex* and *Q. suber* conform a very good system to study hybridization in genus *Quercus*, because, even if they may share the same habitat, they are sufficiently divergent in morphological, phenological, and phylogenetic terms to provide better estimates of the effect of hybridization in the phenotypes than other more phylogenetically inter-related oak systems, such as the European white oaks (Kremer 2002). The application of next generation sequencing (NGS) technologies, and more powerful bioinformatic analysis pipelines to explore the genomes of the F1 individuals and their progenies will shed light on the way of inheritance of genomic blocks, the species limits, the genomic incompatibilities and the reversion to parental phenotypes. In the same way, these individuals constitute an excellent material for the analysis of the genomic and epigenetic basis of cork formation, and could provide a good starting point for breeding programs focused on cork quality and adaptability of the species.

References

Amaral Franco J (1990) *Quercus* L. In: Castroviejo S et al. (eds) Flora Ibérica 2:15–36
Anderson E (1949) Hybridization of the habitat. Evol 2:1–9
Arnold ML (1997) Natural hybridization and evolution. Oxford University Press, New York, Oxford Series
Arnold ML, Martin NH (2006) Adaptation by introgression. J Biol 8:82
Barberon M, Vermeer JEM, De Bellis D, Wang P, Naseer S, Andersen TG, Humbel BM, Nawrath C, Takano J, Salt DE, Geldner N (2016) Adaptation of root function by nutrient-induced plasticity of endodermal differentiation. Cell 164:1–13
Barton NH, Hewitt GM (1985) Analysis of hybrid zones. Ann Rev Ecol Syst 16:113–148
Boavida LC, Varela MC, Feijó J (1999) Sexual reproduction in the cork oak (*Quercus suber* L.). I. The progamic phase. Sex Plant Reprod 11(6):347–353
Boavida LC, Silva JP, Feijó JA (2001) Sexual reproduction in the cork oak (*Quercus suber* L). II. Crossing intra- and interspecific barriers. Sex Plant Reprod 14(3):143–152
Bonito A, Varone L, Gratani L (2011) Relationship between acorn size and seedling morphological and physiological traits of *Quercus ilex* L. from different climates. Photosynthetica 49:75–86
Borzi A (1881) L'Ilixi-Suergiu (*Quercus morisii*-Borzi), nuova Querce della Sardegna. Nuov Gior Bot Ital 13(1):3–10

Burgarella C, Lorenzo Z, Jabbour-Zahab R, Lumaret R, Guichoux E, Petit RJ, Soto A, Gil L (2009) Detection of hybrids in nature: application to oaks (*Quercus suber* and *Q. ilex*). Heredity 102:442–452

Burger W (1975) The species-concept in *Quercus*. Taxon 24:45–50

Camus A (1936–1954) Les Chênes: monographie du genre *Quercus* [et Lithocarpus]. Editions Paul Lechevalier, Paris

Castro-Díez P, Montserrat-Martí G (1998) Phenological pattern of fifteen Mediterranean phanerophytes from *Quercus ilex* communities of NE-Spain. Plant Ecol 139(1):103–112

Cecich RA, Haenche WW (1995) Pollination biology of northern red and black oak. 10th Central Hardwood Forest Conference:238–246

Coelho AC, Lima MB, Neves D, Cravador A (2006) Genetic Diversity of Two Evergreen Oaks [*Quercus suber* (L.) and *Quercus ilex* subsp. *rotundifolia* (Lam.)] in Portugal using AFLP Markers. Silvae Genet 55(3):105–118

Colmeiro M, Boutelou E (1854) Examen de las Encinas y demás arboles de la Peninsula Iberica que producen bellotas, con la designación de los que se llaman mestos. Imprenta de D. Jose M, Geofrin

Díaz M, Møller AP, Pulido FJ (2003) Fruit abortion, developmental selection and developmental stability in *Quercus ilex*. Oecologia 135:378–385

Díaz-Fernández PM, Climent J, Gil L (2004) Biennial acorn maturation and its relationship with flowering phenology in Iberian populations of *Quercus suber*. Trees-Struct Funct 18(6):615–621

Díez Morales E (2016) Estudio anatómico de la corteza de los híbridos *Quercus ilex* x *Q. suber*. Dissertation, Universidad Politécnica de Madrid

Dodd RS, Rafii ZA (2004) Selection and dispersal in a multispecies oak hybrid zone. Evolution 58:261–269

Ducousso A, Michaud H, Lumaret R (1993) Reproduction and gene flow in the genus *Quercus*. Ann Sci For 50(S1):91s–106s

Elena-Rosselló JA, de la Cruz PJ (1998) Levels of genetic diversity in natural, mixed populations of oak species. In: Steiner KC (ed.) Diversity and Adaptation in Oak Species. Proceedings of the 2nd meeting of Working Party 2.08.05, Genetics of *Quercus*, of the IUFRO. University Park, USA

Elena-Rossello JA, Lumaret R, Cabrera E, Michaud H (1992) Evidence for hybridization between sympatric holm-oak and cork oak in Spain based on diagnostic enzyme markers. Vegetatio 99–100:115–118

Eriksson G (2017) *Quercus suber*—recent genetic research. European Forest Genetic Resources Programme (EUFORGEN), Bioversity International, Rome, Italy

Franke R, Schreiber L (2007) Suberin-a biopolyester forming apoplastic plant interfaces. Curr Opin Plant Biol 10:1–8

Gómez-Casero MT, Galán C, Domínguez-Vilches E (2007) Flowering phenology of Mediterranean *Quercus* species in different locations (Córdoba, SW Iberian Peninsula). Acta Bot Mal 32:127–146

Govaerts R, Frodin DG (1998) World checklist and bibliography of Fagales (Betulaceae, Corylaceae, Fagaceae and Ticodendraceae). Royal Bot Gardens, London

Graca J, Pereira H (2004) The periderm development in *Quercus suber*. IAWA J 25:325–335

Gricar J, Jagodic S, Prislan P (2015) Structure and seasonal changes in the bark of sessile oak (*Quercus petraea*). Trees 29:747–757

Himrane H, Camarero JJ, Gil-Pelegrín E (2004) Morphological and ecophy-siological variation of the hybrid oak *Quercus subpyrenaica* (*Q. faginea* x *Q. pubescens*). Trees 18:566–575

Howard ET (1977) Bark structure of the Southern Upland Oaks. Wood Fiber Sci 9:172–183

Iwasaki M, Takahashi H, Iwakawa H, Nakagawa A, Ishikawa T, Tanaka H, Matsumura Y, Pekker I, Eshed Y, Vial-Pradel S Ito T, Watanabe Y, Ueno Y, Fukazawa H, Kojima S, Machida Y, Machida C (2013) Dual regulation of *ETTIN* (*ARF3*) gene expression by AS1-AS2, which maintains the DNA methylation level, is involved in stabilization of leaf adaxial-abaxial partitioning in *Arabidopsis*. Development 140:1958–1969

Jato V, Rodríguez-Rajo FJ, Méndez J, Aira MJ (2002) Phenological behaviour of Quercus in Ourense (NW Spain) and its relationship with the atmospheric pollen season. Int J Biometeorol 46:176–184

Jiménez MP, López de Heredia U, Collada C, Lorenzo Z, Gil L (2004) High variability of chloroplast DNA in three Mediterranean evergreen oaks indicates complex evolutionary history. Heredity 93:510–515

Joët T, Ourcival JM, Capelli M, Dussert S, Morin X (2016) Explanatory ecological factors for the persistence of desiccation-sensitive seeds in transient soil seed banks: *Quercus ilex* as a case study. Ann Bot-London 117(1):165–176

Judd WS, Campbell CS, Kellogg EA, Stevens PF, Donoghue MJ (2002) Plant systematics, a phylogenetic approach. Sinauer Associates, Massachusetts

Kremer A (ed) (2002) Range wide distribution of chloroplast DNA diversity and pollen deposits in European white oaks: inferences about colonisation routes and management of oak genetic resources. For Ecol Manage 156:1–223

Laguna M (1881) Un mesto italiano y varios mestos espanoles. Rev Montes 114:477–486

Llensa de Gelcen S (1943) Sistemática, fitogeografía y utilidad forestal del hibrido x *Quercus Morisii* Borzi. Anales de la Escuela de peritos agrícolas y superior de agricultura y de los servicios técnicos de agricultura, vol 3. ESAB, Barcelona, pp 315–328

Lodha M, Marco CF, Timmermans MCP (2013) The ASYMMETRIC LEAVES complex maintains repression of *KNOX* homeobox genes via direct recruitment of Polycomb-repressive complex2. Genes Dev 27:596–601

López de Heredia U, Soto A (2017) Simulation analysis suggest *Quercus suber* x *Quercus ilex* hybridization could be underestimated. International Congress on cork oak trees and woodlands—conservation, management, products and challenges for the future, May 25–26, Sassari, Italy

López de Heredia U, Jiménez P, Díaz-Fernández P, Gil L (2005) The Balearic Islands: a reservoir of cpDNA genetic variation for evergreen oaks. J Biogeogr 32:939–949

López de Heredia U, Carrion JS, Jimenez P, Collada C, Gil L (2007a) Molecular and palaeoecological evidence for multiple glacial refugia for evergreen oaks on the Iberian Peninsula. J Biogeog 34(9):1505–1517

López de Heredia U, Jiménez P, Collada C, Simeone MC, Bellarosa R, Schirone B, Cervera MT, Gil L (2007b) Multi-Marker Phylogeny of Three Evergreen Oaks Reveals Vicariant Patterns in the Western Mediterranean. Taxon 56(4):1209–1220

López de Heredia U, Sánchez H, Soto A (2017a) Molecular evidence of bidirectional introgression between *Quercus suber* and *Quercus ilex*. International Congress on cork oak trees and woodlands—conservation, management, products and challenges for the future, May 25–26, Sassari, Italy

López de Heredia U, Duro MJ, Soto A (2017b) Leaf morphology of progenies of *Q. suber*, *Q. ilex*, and their hybrids using multivariate and geometric morphometric analysis. International Congress on cork oak trees and woodlands—conservation, management, products and challenges for the future, May 25–26, Sassari, Italy

López-Caamal A, Tovar-Sánchez E (2014) Genetic, morphological, and chemical patterns of plant hybridization. Rev Chil Hist Nat 87:1–14

Lumaret R, Tryphon-Dionnet M, Michaud H, Sanuy A, Ipotesi E, Born C, Mir C (2005) Phylogeographical Variation of Chloroplast DNA in Cork Oak (*Quercus suber*). Ann Bot-London 96(5):853–861

Machida C, Nakagawa A, Kojima S, Takahashi H, Machida Y (2015) The complex of ASYMMETRIC LEAVES (AS) proteins plays a central role in antagonistic interactions of genes for leaf polarity specification in Arabidopsis *WIREs*. Dev Biol. 4:655–671

Magri D, Fineschi S, Bellarosa R, Buonamici A, Sebastiani F, Schirone B, Simeone MC, Vendramin GG (2007) The distribution of Quercus suber chloroplast haplotypes matches the palaeogeographical history of the western Mediterranean. Mol Ecol 16(24):5259–5266

Mallet J (2005) Hybridization as an invasion of the genome. Trends Ecol Evol 20(5):229–237

Matsumura Y, Ohbayashi I, Takahashi H, Kojima S, Ishibashi N, Keta S, Nakagawa A, Hayashi R, Saéz-Vázquez J, Echeverria M, Sugiyama M, Nakamura K, Machida C, Machida Y (2016) A genetic link between epigenetic repressor AS1-AS2 and a putative small subunit processome in leaf polarity establishment of Arabidopsis. Biol Open 5(7):942–954

Miguel A, de Vega-Bartol J, Marum L, Chaves I, Santo T, Leitão J, Varela MC, Miguel CM (2015) Characterization of the cork oak transcriptome dynamics during acorn development. BMC Plant Biol 15:158

Mir C, Toumi L, Jarne P, Sarda V, Di Giusto F, Lumaret R (2006) Endemic North African *Quercus afares* Pomel originates from hybridisation between two genetically very distant oak species (*Q. suber* L. and *Q. canariensis* Willd.): evidence from nuclear and cytoplasmic markers. Heredity 96:175–184

Montserrat-Martí G, Camarero JJ, Palacio S, Pérez-Rontomé C, Milla R, Albuixech J, Maestro M (2009) Summer-drought constrains the phenology and growth of two coexisting Mediterranean oaks with contrasting leaf habit: implications for their persistence and reproduction. Trees 23(4):787–799

Muller C (1952) Ecological control of hybridization in *Quercus*: A factor in the mechanism of evolution. Evolution 6:147–161

Natividade JV (1936) Estudo histologico das peridermes do hibrido *Quercus ilex* X *suber*, P Cout Publ Dir G Serv Flor III(1)

Ogaya R, Peñuelas J (2004) Phenological patterns of *Quercus ilex*, *Phillyrea latifolia*, and *Arbutus unedo* growing under a field experimental drought. Ecoscience 11(3):263–270

Oliveira P, Custódio AC, Branco C, Reforço I, Rodrigues F, Varela MC, Meierrose C (2003) Hybrids between cork oak and holm oak: isoenzyme analysis. For Genet 10(4):283–297

Pausas JG, Pereira JS, Aronson J (2009) The Tree. In: Aronson J, Pereira JS, Pausas JG (eds) Cork Oak Woodlands on the Edge. Ecology, Adaptive Management and Restoration. Island Press, Washington, USA

Peguero-Pina JJ, Sancho-Knapik D, Barron E, Camarero JJ, Vilagrosa A, Gil-Pelegrín E (2014) Morphological and physiological divergences within *Quercus ilex* support the existence of different ecotypes depending on climatic dryness. Ann Bot-London 114:301–313

Peñaloza-Ramírez JM, González-Rodríguez A, Mendoza-Cuenca L, Caron H, Kremer A, Oyama K (2010) Interspecific gene flow in a multispecies oak hybrid zone in the Sierra Tarahumara of Mexico. Ann Bot-London 105:389–399

Perea García-Calvo R (2006) Estudio de la estructura de masa de una dehesa de encina con alcornoque en "El Deheson del Encinar" (Toledo). Master Thesis, Universidad Politécnica de Madrid

Perez Rodriguez A (2008) Modelizacion e introgresion en encina y alcornoque: Efecto de los parametros demograficos y adaptativos. Master Thesis, Universidad Politecnica de Madrid

Petit RJ, Pineau E, Demesure B, Bacilieri R, Ducousso A, Kremer A (1997) Chloroplast DNA footprints of postglacial recolonization by oaks. Proc Natl Acad Sci USA 94:9996–10001

Petit RJ, Bodénès C, Ducousso A, Roussel G, Kremer A (2004) Hybridization as a mechanism of invasion in oaks. New Phytol 161:151–164

Pollard M, Beisson F, Li Y, Ohlrogge JB (2008) Building lipid barriers: biosynthesis of cutin and suberin. Trends Plant Sci 13:236–246

Potts B, Reid JB (1988) Hybridization as a dispersal mechanism. Evolution 42:1245–1255

Pulido Neves da Costa JS (2011) Differentiation and genetic variability in cork oak populations (*Quercus suber* L.). Dissertation, Universidade de Lisboa

Quilho T, Sousa V, Tavares F, Pereira H (2013) Bark anatomy and cell size variation in *Quercus faginea*. Turk J Bot 37:561–570

Rafii ZA, Zavarin E, Pelleau Y (1991) Chemosystematic differentation of *Quercus ilex* and *Q. rotundifolia* based on Acorn Fatty Acids. Biochem System Ecol 19(2):163–166

Ramírez-Valiente JA, Valladares F, Gil L, Aranda I (2009) Population differences in juvenile survival under increasing drought are mediated by seed size in cork oak (*Quercus suber* L.). For Ecol Manage 257:1676–1683

Rieseberg LH (1995) The role of hybridization in evolution: old wine in new skins. Am J Bot 82(7):944–953
Rieseberg LH, Ellstrand NC (1993) What can molecular and morphological markers tell us about plant hybridization? Crit Rev Plant Sci 12:213–241
Rieseberg L, Wendel J (1993) Introgression and its consequences in plants. In: Harrison R (ed) Hybrid zones and the evolutionary process. Oxford University Press, New York, pp 70–109
Rieseberg LH, Raymond O, Rosenthal DM et al (2003) Major ecological transitions in wild sunflowers facilitated by hybridization. Science 301:1211–1216
Rieseberg LH, Kim S, Randell RA, Whitney KD, Gross BL, Lexer C, Clay K (2007) Hybridization and the colonization of novel habitats by annual sunflowers. Genetica 129:149–165
Rushton BS (1993) Natural hybridization within the genus *Quercus* L. Ann Sci For 50(suppl 1):73s–90s
Salomón R, Lorenzo Z, Valbuena-Carabaña M, Nicolás JL, Gil L (2012) Seed recalcitrant behavior of Iberian *Quercus*: a multispecies comparison. Aust J For Sci 129(3–4):182–201
Schreiber L (2010) Transport barriers made of cutin, suberin and associated waxes. Trends Plant Sci 15:546–553
Sen A, Quilho T, Pereira H (2011) Bark anatomy of *Quercus cerris* L. var. *cerris* from Turkey. Turk J Bot 35:45–55
Serra O, Hohn C, Franke R, Prat S, Molinas M, Figueras M (2010) A feruloyl transferase involved in the biosynthesis of suberin and suberin-associated wax is required for maturation and sealing properties of potato periderm. Plant J 62:277–290
Serra O, Chatterjee S, Figueras M, Molinas M, Stark RE (2014) Deconstructing a plant macromolecular assembly: Chemical architecture, molecular flexibility, and mechanical performance of natural and engineered potato suberins. Biomacromolecules 15:799–811
Sever K, Škvorc Z, Bogdan S, Franjić J, Krstonošić D, Alešković I, Kereša S, Fruk G, Jemrić T (2012) In vitro pollen germination and pollen tube growth differences among *Quercus robur* L. clones in response to meteorological conditions. Grana 51(1):25–34
Sexton JP, McIntyre PJ, Angert AL, Rice KJ (2009) Evolution and ecology of species range limits. Annu Rev Ecol Evol Syst 40:415–436
Soler M, Serra O, Molinas M, Huguet G, Fluch S, Figueras M (2007) A genomic approach to suberin biosynthesis and cork differentiation. Plant Physiol 144:419–431
Soler M, Serra O, Fluch S, Molinas M, Figueras M (2011) Potato skin SSH library yields new candidate genes for suberin biosynthesis and periderm formation. Planta 233:933–945
Sork VL, Smouse PE (2006) Genetic analysis of landscape connectivity in tree populations. Landscape Ecol 21(6):821–836
Soto A, Lorenzo Z, Gil L (2003) Nuclear microsatellite markers for the identification of *Quercus ilex* L. and *Q. suber* L. hybrids. Silvae Genet 52(2):63–66
Soto A, Lorenzo Z, Gil L (2007) Differences in fine-scale genetic structure and dispersal in *Quercus ilex* L. and *Q. suber* L.: consequences for regeneration of Mediterranean open woods. Heredity 99(6):601–607
Soto A, López de Heredia U, Robledo-Arnuncio JJ, Gil L (2008) Simulation analysis of *Quercus ilex-Quercus suber* chloroplast introgression. In: Vázquez-Piqué J, Pereira H, González-Pérez A (eds) Suberwood—New challenges for the integration of cork oak forests and products. Universidad de Huelva Publicaciones, Huelva, pp 91–98
Staudt M, Mir C, Joffre R, Rambal S, Bonin A, Landais D, Lumaret R (2004) Isoprenoid emissions of *Quercus* spp. (*Q. suber* and *Q. ilex*) in mixed stands contrasting in interspecific genetic introgression. New Phytol 163:573–584
Stebbins GL (1950) Variation and evolution in plants. Columbia University Press, New York
Stebbins GL (1974) Flowering plants: evolution above the species level. Belknap Press of Harvard University Press
Trockenbrodt M (1991) Qualitative structural changes during bark development in *Quercus robur, Ulmus glabra, Populus tremula* and *Betula pendula*. IAWA Bull 11:141–166
Van Valen L (1976) Ecological species, multispecies, and oaks. Taxon 25:233–239
Varela MC, Valdiviesso T (1996) Phenological phases of *Quercus suber* L. flowering. For Genet 3:93–102

Varela MC, Bras R, Barros IR, Oliveira P, Meierrose C (2008) Opportunity for hybridization between two oak species in mixed stands as monitored by the timing and intensity of pollen production. Forest Ecol Manag 256:1546–1551

Vázquez FM (2013) Micromorphological and Anatomical Characters Used to Differentiate Mediterranean Oaks. J Int Oak Soc 24:122–129

Vazquez FM, Perez MC, Esparrago F, Burzaco A (1993) Hibridos del género *Quercus* en Extremadura. I Congreso Forestal Español 1:459–465

Vázquez FM, Esparrago F, López-Marquez JA, Jaraquemada F (1992) Flowering of *Quercus rotundifolia* Lam. In: International workshop quercus ilex l. ecosystems: fuction, dynamics and management. Montpellier-Barcelona, CEPE/CNRS:81

Vázquez FM, Suárez MA, Baselga MP (1996) Nota sobre el efecto de la temperatura y la humedad en la germinación "in vitro" del grano de polen de *Quercus rotundifolia* Lam., y *Q. suber* L. *Inv Agr Sist* Rec. Forest 5(2):351–359

Vázquez FM, Sánchez-Gullón E, Pinto-Gomes C, Pineda MA, García D, Márquez F, Guerra MJ, Blanco J, Vilaviçosa C (2015) Three New Oak Hybrids from Southwest Iberia (Spain and Portugal). J Int Oak Soc 26:43–55

Vila-Viçosa CM, Vázquez FM, Meireles C, Pinto-Gomes C (2014) Taxonomic peculiarities of marcescent oaks (*Quercus*) in southern Portugal. Lazaroa 35:139–153

Vitelli M, Vessella F, Cardoni S, Pollegioni P, Denk T, Grimm GW, Simeone MC (2017) Phylogeographic structuring of plastome diversity in Mediterranean oaks (*Quercus* Group Ilex, Fagaceae). Tree Genet Genomes 13(1):1

Whittemore AT, Schaal BA (1991) Interspecific gene flow in sympatric oaks. Proc Natl Acad Sci USA 88:2540–2544

Yacine A, Bouras F (1997) Self- and cross-pollination effects on pollen tube growth and seed set in Holm oak *Quercus ilex* L (Fagaceae). Ann Sci For 54(5):447–462

Zhu A, Greaves IK, Dennis ES, Peacock WJ (2017) Genome-wide analyses of four major histone modifications in *Arabidopsis* hybrids at the germinating seed stage. BMC Genomics 18:137

Chapter 8
The Anatomy and Functioning of the Xylem in Oaks

Elisabeth M. R. Robert, Maurizio Mencuccini and Jordi Martínez-Vilalta

Abstract Because of its economic and ecological importance, the genus *Quercus* has been relatively intensively studied for its anatomical and hydraulic characteristics, having often been testing ground for development of methods and hypotheses related to tree structure and function. However, despite long-withstanding interest, we are still far from having obtained a clear understanding of the hydraulic functioning of the species within this genus, the occurrence of trade-offs among various xylem properties and the prevalence of syndromes of characters under different environmental conditions. We conducted a review of the xylem anatomical literature of the genus *Quercus*, an undertaking that does not appear to have been carried out before. We also updated existing quantitative databases of vessel diameter and density, volumetric fractions of parenchyma, wood density and xylem hydraulic properties, to synthesise the main patterns of variation in the hydraulic architecture and functioning of the genus. We found that ring-porous (deciduous) species have lower wood density, higher hydraulic conductivity, xylem that is more vulnerable to embolism and lower Huber values compared to diffuse-porous (evergreen) species. We also report systematic differences among taxonomic groups, with species of sections *Quercus* and *Lobatae* having smaller but more numerous vessels, lower wood density, more vulnerable xylem, higher conductivity and lower Huber values as opposed to species of section *Cerris*. Many of these trends appeared to map onto environmental differences across the three main biomes where *Quercus* species are found, i.e. the temperate, the Mediterranean/semi-arid and the tropical biomes. Although limited by the coverage of the empirical data, our compilation contributes to characterise the hydraulic architecture and functioning of the genus as a function of taxonomic grouping, biome, ring-porosity and leaf phenology. Future investi-

E. M. R. Robert (✉) · M. Mencuccini · J. Martínez-Vilalta
CREAF, 08193 Cerdanyola del Vallès, Barcelona, Spain
e-mail: e.robert@creaf.uab.cat

M. Mencuccini
ICREA, Pg. Lluís Companys 23, 08010 Barcelona, Spain

J. Martínez-Vilalta
Universitat Autònoma de Barcelona, 08193 Cerdanyola del Vallès, Barcelona, Spain

gations can benefit by the identification of the main factors responsible for these patterns and their likely ecological significance.

8.1 Introduction

The xylem is the internal water transport system of plants that links water-absorbing roots to the pores inside the leaves where evaporation takes place. The discipline studying the biophysical and physiological processes controlling how water moves inside the xylem is named hydraulics. Despite having a primary focus on water movement in the xylem, hydraulics also often examines jointly aspects related to the functioning of other hydraulic systems in the stem or in other organs, i.e. root water uptake, radial water transport across the stele, radial water transfer in leaves outside the last xylem conduits up to the sites of evaporation and transport of carbohydrates in the phloem (Tyree and Ewers 1991). Because of its fundamental focus on structural features, hydraulics shares a lot of ground with studies of anatomy and it is often at the interface between these two disciplines that scientific progress has been made.

Studies of the wood anatomy and the hydraulics of the genus *Quercus* have been instrumental in advancing our understanding of the physiological ecology of plant water transport, vulnerability to embolism and plant hydraulic architecture across the plant kingdom. Some of the very early experimental measurements of the physiological bases of plant hydraulic performance were carried out on species of the genus *Quercus*, thereby allowing inferences to be made regarding the realized ecological niches of various species in the field. The first vulnerability curves to drought stress published for a species of the genus *Quercus* was for current-year twigs and petioles of *Quercus rubra* (Cochard and Tyree 1990). The authors employed both hydraulic and acoustic techniques and did not find systematic differences in vulnerability between the two studied organs. Older stems were not examined because "*stems had to be cut longer then the longest vessel, and in older stems this meant that we would have to dehydrate stems several meters in length*", an issue we are still grappling with today (see Sect. 8.3.1). More vulnerability curves in response to drought stress followed suit for European oaks (*Quercus petraea, Quercus robur* and *Quercus pubescens*—Cochard et al. 1992).

In another relevant early paper, Sperry and Sullivan (1992) reported curves of vulnerability to embolism after freeze-thaw cycles in *Quercus gambelii*. In it, the authors adopted a comparative approach (contrasting diffuse-porous, ring-porous and coniferous species) and highlighted how small tensions above −0.2 MPa were sufficient to embolize 90% of the xylem in the ring-porous *Q. gambelii* following a single freeze-thaw cycle, whereas embolism in diffuse-porous *Betula* and *Populus* species was much lower and almost non-existent in several conifers under the same conditions. Lo Gullo and Salleo (1993) were the first researchers to employ acoustic emissions in Mediterranean *Quercus ilex* to detect loss of hydraulic conductance, documenting subsequent recovery following an irrigation event overnight. They also investigated the relative sensitivities of this species to freezing and summer drought stresses using hydraulic techniques. Lo Gullo et al. (1995) combined hydraulic techniques with anatomical and staining approaches to determine the relative sensitivities of conduit size in twigs of *Quercus cerris* (larger conduits tended to be more vulnerable), an issue examined also earlier on by Cochard and Tyree (1990) in *Q. rubra*. The concepts of 'hydraulic constriction zones' and 'hydraulic segmentation' proposed by Zimmermann only a few years before (1983) were also tested experimentally in the early' 90s. Already in 1996, issues related to potential methodological artefacts began to be discussed. In a comparative early analysis of the hydraulics of the genus, Tyree and Cochard (1996) reported that they could not replicate the vulnerability curve for *Q. ilex* published by Lo Gullo and Salleo (1993). Their (Tyree and Cochard 1996) figure 1 gives Ψ_{50} values (the water potential at which 50% of the maximum hydraulic conductivity is lost) of −2.9 and −5.7 MPa showing a difference between the two studies of almost 3 MPa. Similar differences have been reported later on by other authors (see Sect. 8.3.1).

It is now generally accepted that studies of xylem anatomy and of hydraulic architecture are integral components of the characterization of a species' ecological niche. A comparative review of the hydraulics of the genus *Quercus* has not been attempted since the early work already cited by Tyree and Cochard (1996), which was limited to six species, while xylem anatomy of oaks has not been reviewed recently. Regarding hydraulics, Tyree and Cochard (1996) concluded their review by stressing the existence of a correlation across species between vulnerability to drought-induced embolism and other indices of drought stress tolerance, while at the same time stating that significant levels of summer embolism are probably avoided in most situations, thanks to an efficient coordination between hydraulic transport system and stomatal control of water loss in leaves. It is therefore obvious already from this initial review (Tyree and Cochard 1996) that a full interpretation of the significance of anatomical and hydraulic relationships in trees needs to incorporate an understanding also of leaf physiology, particularly photosynthesis and stomatal conductance. We therefore recommend reading this chapter in conjunction with the relevant other chapters in this book.

Our objectives for this review were four-fold. Firstly, we reviewed the main elements of vessel anatomy that impinged on xylem function and that allowed a comparative analysis of the genus *Quercus* relative to other angiosperms. We focussed on vessel diameter and its relationship with vessel density, the length of oak water-transporting vessels, the degree of vessel isolation in the wooden matrix (or its reverse, i.e. vessel grouping) and the structure and micro-anatomy of pits (because of their central role in controlling the spread of air emboli) after which we briefly discussed the occurrence of tyloses and their structural and physiological significance. Secondly, we examined tissue-level properties outside of vessel anatomy, i.e. the occurrence and significance of radial and axial parenchyma, and of tracheids and fibres. We ended this section with an overview of the main patterns of variation across the genus in wood density, a central variable that often relates to other physiological and ecological properties. Thirdly, because of the occurrence of very long vessels in oaks, we dedicated a whole section to a critical examination of the major methodological issues related to the measurement of hydraulic efficiency and hydraulic safety in the genus *Quercus*. Here, we touched on various potential artefacts occurring during hydraulic measurements and we discussed the evidence that these artefacts may have occurred in *Quercus* studies. Fourthly and lastly, we examined the major patterns that have been found across the genus in hydraulic safety (as quantified by xylem Ψ_{50}), hydraulic efficiency (as quantified by the sapwood-specific hydraulic conductivity, $K_{S,max}$), minimum values of leaf water potentials encountered in the field (Ψ_{min}, which partly is an indicator of the maximum levels of soil drought stress encountered at the peak of the seasonal droughts and partly depends on rooting strategies and stomatal behaviour) and the Huber value (H_V, the ratio between cross-sectional sapwood and leaf area distal to the section, a measure of relative allocation between xylem and leaves).

Because of heterogeneity in data availability, we adopted an opportunistic approach to the content and structuring of the chapter sections. In some cases, our observations primarily have a qualitative nature, attempting to summarise and interpret the variability encountered in the primary literature and place it in context of the species' broader characteristics. In some instances, however, collation of data from the primary literature has already been carried out and global databases are available. In these more fortunate cases (e.g. vessel diameter and density, parenchyma content, wood density, xylem Ψ_{50}, $K_{S,max}$, Ψ_{min} and H_V), we carried out a quantitative analysis across the whole genus, focusing on the existence of broad inter-specific patterns. In all these cases, we recognise that our conclusions have a preliminary nature and are primarily limited by the quality and quantity of the available data in the literature.

We also examined the hypothesis that traits are coordinated with one another, thereby supporting the idea that hydraulic strategies represented by syndromes of coordinated traits can be identified within the genus. We examined trends in these various traits in relation to other biological properties of the species, i.e. taxonomic grouping (sub-genera *Cyclobalanopsis* and *Quercus*, and within this last sub-genus, sections *Quercus*, *Lobatae*, *Cerris*, *Mesobalanus* and *Protobalanus*), ring-porosity (classed into ring-porous and diffuse-porous) and leaf phenology (evergreen versus deciduous leaf habit). Finally, we classified all the species we examined into three broad biome classes (tropical, temperate and Mediterranean/semi-arid), following an early biome classification employed for hydraulic traits (Choat et al. 2012). Classification into taxonomic groups and leaf phenology largely follows the Wikipedia list of *Quercus* species (Wikipedia 2017). We recognise that all these classifications are problematic. For ring-porosity, we followed the literature and for dubious cases (semi-ring-porousness), we attempted a classification based on the prevailing opinions from the literature. Where appropriate, we highlighted potential problems associated with the classification we adopted. The following online databases were consulted to resolve borderline cases: The Plant List (2013), Encyclopedia of Life, Wikipedia, Oaks of the World (Hélardot 1987 onwards), eFloras (2008), Tropicos and The Wood Database. Species that The Plant List (2013) classified as belonging to the genera *Cyclobalanus*, *Cyclobalanopsis* or *Lithocarpus* were excluded from the analyses. In Table 8.1, we list the taxonomic section, biome, ring-porosity type, leaf phenology type and mean values of the selected traits of all the oak species for which we show data in this chapter.

8.2 Functional Wood Anatomy

For this section, we explored the variability of the major functional wood traits within the genus *Quercus* and compared structural trait characteristics with other angiosperm genera. Vessel anatomy (ring-porosity, vessel dimensions, vessel packing, network connectivity and tyloses—Sect. 8.2.1) and tissue properties (parenchyma, tracheids and fibres and wood density—Sect. 8.2.2) are discussed.

8.2.1 Vessel Anatomy

8.2.1.1 Ring-Porosity

Whether deciduous or evergreen, *Quercus* wood often shows distinct growth rings allowing for dendrochronological analyses. The longest continuous tree-ring chronology in the world is an oak chronology from Southern Germany dating back till 8480 BC (Friedrich et al. 2004; Haneca et al. 2009; Wilson 2010). Although the clearness of the ring boundaries is generally very high for deciduous oak trees from temperate regions due to very pronounced ring-porousness (i.e. large vessels at the beginning of the growing season in contrast to small vessels at the end of it—Wheeler et al. 1989), it might be far less evident how to correctly distinguish annual ring borders for evergreen oaks from Mediterranean or subtropical climates (see some examples in Fig. 8.1). However, several successful attempts have been made (e.g. Cherubini et al. 2003; Gea-Izquierdo et al. 2009) and it has been proven by Campelo et al. (2010) that time series of vessel lumen size in *Q. ilex* trees from Catalonia (Spain) bear climatic signals that can be used for dendrochronology and climate reconstruction in combination with the more classical tree-ring width data, creating opportunities for oak species falling within the gradient from semi-ring-porousness (i.e. intermediate condition between ring-porousness and diffuse-porousness—Wheeler et al. 1989) to diffuse-porousness (i.e. homogenous vessel size distribution over the growth rings—Wheeler et al. 1989). The relationships deciduousness/ring-porousness and evergreenness/diffuse-porousness are often but not always true within the *Quercus* genus (see some examples in Table 8.1). In the quantitative analyses conducted for this chapter, the relation of ring-porosity to other hydraulic traits is examined since it is a crucial aspect of oak xylem structure and thus hydraulics.

Table 8.1 List of all the *Quercus* species employed in the quantitative analyses presented in this chapter

Species	Section	Biome	Porosity	Phenology	Ψ_{50}	$K_{S,max}$	H_V	WD	Ψ_{min}	D_{mean}	N_{cond}	P_{rad}	P_{ax}	P_{tot}
Quercus acuta	Cyclobalanopsis	TMS	Diffuse-porous	E								23.30		
Quercus acutissima	Cerris	TRS	Ring-porous	D	−3.39	2.42		0.74		72.48	7.43	21.90		
Quercus agrifolia	Lobatae	WDS	Diffuse-porous	E	−1.89	4.05		0.62	−3.12	50.88			12.48	35.32
Quercus alba	Quercus	TMS	Ring-porous	D	−1.37	1.37	1.45E-04	0.62	−1.14	52.06	43.40	22.83		
Quercus aliena	Quercus	TMS	Ring-porous	D	−2.09	6.30		0.69						
Quercus aquifolioides	Quercus	TRS		E				0.79						
Quercus argentata	Cyclobalanopsis	TRS		E				0.74						
Quercus arizonica	Quercus	WDS	Ring-porous	E				0.59						
Quercus asymetrica	Cyclobalanopsis	TRS		E				0.88						
Quercus austrina	Quercus	TMS	Ring-porous	D		1.90	1.83E-04	0.75	−2.99					
Quercus berberidifolia	Quercus	WDS		E	−2.16	1.83	8.04E-04	0.70	−4.35	44.56	36.65			
Quercus bicolor	Quercus	TMS	Ring-porous	D				0.65				29.72		
Quercus blakei	Cyclobalanopsis	TRS		E				0.78						
Quercus calliprinos	Cerris	WDS	Diffuse-porous	E						78.00	15.00			
Quercus canariensis	Quercus	WDS	Ring-porous	D						43.63	49.85			
Quercus candicans	Lobatae	TRS	Ring-porous	D								21.90		
Quercus castaneifolia	Cerris	TMS	Ring-porous	D						121.50	7.14			
Quercus cerris	Cerris	WDS	Ring-porous	D				0.70	−2.10	122.45	7.19			
Quercus chapmanii	Quercus	TMS		D		2.50	1.35E-04	0.78						
Quercus chevalieri	Cyclobalanopsis	TRS		E				0.69						
Quercus chrysolepis	Protobalanus	WDS	Diffuse-porous	E				0.70						
Quercus chungii	Cyclobalanopsis	TRS		E				0.78						
Quercus coccifera	Cerris	WDS	Diffuse-porous	E	−6.96	0.29	7.59E-04		−10.30	48.70	56.60			

(continued)

Table 8.1 (continued)

Species	Section	Biome	Porosity	Phenology	Ψ_{50}	$K_{S,max}$	H_V	WD	Ψ_{min}	D_{mean}	N_{cond}	P_{rad}	P_{ax}	P_{tot}
Quercus coccinea	Lobatae	TMS	Diffuse-porous	E				0.61				18.15	23.30	41.45
Quercus cornelius-mulleri	Quercus	WDS		E	−0.70	2.03	5.06E-04	0.68	−4.50					
Quercus costaricensis	Lobatae	TRS		E				0.61						
Quercus crenata	Cerris	WDS		E								9.05		
Quercus delavayi	Cyclobalanopsis	TRS		E				0.78						
Quercus douglasii	Quercus	WDS	Diffuse-porous	D				0.59		65.41	11.41			
Quercus ellipsoidalis	Lobatae	TMS		D				0.59						
Quercus emoryi	Lobatae	WDS	Ring-porous	D	−1.40	1.61	4.97E-05	0.59	−2.36					
Quercus engelmannii	Quercus	WDS		E				0.59						
Quercus engleriana	Quercus	WDS		E				0.72		38.03				
Quercus fabrei	Quercus	TRS	Ring-porous	D	−5.14	3.15		0.67		61.04	94.07			
Quercus faginea	Quercus	WDS	Ring-porous	D			2.16E-04			45.82		18.75	27.60	46.30
Quercus falcata	Lobatae	TMS	Ring-porous	D	−0.92	2.75	1.87E-04	0.59	−2.46					
Quercus fleuryi	Cyclobalanopsis	TRS		E				0.81						
Quercus frainetto	Mesobalanus	WDS	Ring-porous	D	−4.56				−3.13	31.86	107.00			
Quercus fusiformis	Quercus	WDS	Diffuse-porous	E	−0.50	1.70				45.40				
Quercus gambelii	Quercus	WDS	Ring-porous	D	−0.58	2.00	3.90E-05	0.62	−2.16	30.93		13.50	11.00	24.50
Quercus garryana	Quercus	TMS	Diffuse-porous	D	−3.61			0.64		48.47	36.90			
Quercus gemelliflora	Cyclobalanopsis	TRS		E				0.70						
Quercus geminata	Quercus	TMS	Ring-porous	E		1.70	4.95E-04	0.87	−3.23					
Quercus gilva	Cyclobalanopsis	TRS		E				0.80						
Quercus glauca	Cyclobalanopsis	TRS	Diffuse-porous	E				0.73		128.00	4.00			

(continued)

Table 8.1 (continued)

Species	Section	Biome	Porosity	Phenology	Ψ_{50}	$K_{S,max}$	H_V	WD	Ψ_{min}	D_{mean}	N_{cond}	P_{rad}	P_{ax}	P_{tot}
Quercus glaucoides	Quercus	WDS	Diffuse-porous	E				0.80						
Quercus graciliformis	Lobatae	WDS		D				0.59						
Quercus griffithii	Quercus	TRS		D				0.68						
Quercus grisea	Quercus	WDS	Ring-porous	D	-1.69	1.73	4.98E-05	0.59	-3.41					
Quercus hemisphaerica	Lobatae	TMS	Ring-porous	E		2.80	2.25E-04	0.69	-2.13					
Quercus hypoleucoides	Lobatae	WDS	Ring-porous	E	-0.93	2.16	3.83E-05	0.59	-2.39					
Quercus ilex	Quercus	WDS	Diffuse-porous	E	-4.66	0.89	3.64E-04	0.75	-3.75	39.47	72.83	34.10		
Quercus ilicifolia	Lobatae	TMS		D				0.59						
Quercus imbricaria	Lobatae	TMS		D				0.59						
Quercus incana	Lobatae	TMS	Ring-porous	D		3.20	2.04E-04	0.67	-2.86					
Quercus infectoria	Cerris	WDS	Diffuse-porous	E						115.00	40.00			
Quercus ithaburensis	Cerris	WDS	Diffuse-porous	D						125.00	8.00			
Quercus kelloggii	Lobatae	TMS	Ring-porous	D				0.51						
Quercus laevis	Lobatae	TMS	Ring-porous	D	-1.89	1.60	2.09E-04	0.64	-2.72	51.34	4.00			
Quercus lamellosa	Cyclobalanopsis	TRS	Ring-porous	E						138.00	4.00			
Quercus lanata	Quercus	TRS		E						113.00	7.00			
Quercus langbianensis	Cyclobalanopsis	TRS		E				0.81						
Quercus laurifolia	Lobatae	TMS	Ring-porous	E		5.20	1.38E-04	0.58	-2.70			18.85	26.35	45.20
Quercus laurina	Lobatae	WDS	Ring-porous	E				0.70						
Quercus leucotrichophora	Quercus	TRS		E						128.00	7.00			
Quercus lobata	Quercus	WDS	Ring-porous	D				0.55		70.23	31.97			

(continued)

8 The Anatomy and Functioning of the Xylem in Oaks

Table 8.1 (continued)

Species	Section	Biome	Porosity	Phenology	Ψ_{50}	$K_{S,max}$	H_V	WD	Ψ_{min}	D_{mean}	N_{cond}	P_{rad}	P_{ax}	P_{tot}
Quercus lusitanica	Quercus	WDS	Ring-porous	D				0.70						
Quercus lyrata	Quercus	TMS	Ring-porous	D				0.57						
Quercus macrocarpa	Quercus	TMS	Ring-porous	D				0.58		39.54	37.28	20.60	12.00	32.60
Quercus margarettae	Quercus	TMS		D		1.70	2.40E-04	0.70	−3.34					
Quercus marilandica	Lobatae	TMS	Ring-porous	D				0.69				21.55	23.75	45.30
Quercus michauxii	Quercus	TMS	Ring-porous	D	−1.70	4.40	1.11E-04	0.64	−2.91	41.83		25.30		
Quercus minima	Quercus	TMS		E		4.20	3.09E-04	0.79						
Quercus mongolica	Quercus	TMS	Ring-porous	D		0.50	2.28E-04	0.60	−0.96	129.84	18.24	12.80		
Quercus montana	Quercus	TMS	Ring-porous	D				0.57						
Quercus morii	Cyclobalanopsis	TMS	Diffuse-porous	E				0.80						
Quercus muehlenbergii	Quercus	TMS		D				0.59						
Quercus multinervis	Cyclobalanopsis	TMS		E				0.76						
Quercus myrsinifolia	Cyclobalanopsis	TRS		E				0.77						
Quercus myrtifolia	Lobatae	TMS		E	−0.75	1.70	2.64E-04	0.81	−2.46			21.75		
Quercus nigra	Lobatae	TMS	Ring-porous	E	−1.31	2.70	1.82E-04	0.63	−1.95	43.91		23.45	25.95	44.55
Quercus oblongata	Cyclobalanopsis	TRS	Diffuse-porous	E				0.68						
Quercus oblongifolia	Quercus	TMS		E				0.59						
Quercus oglethorpensis	Quercus	TMS	Ring-porous	D				0.59						
Quercus oleoides	Quercus	TRS	Ring-porous	E	−3.03	2.10	1.18E-04	0.86	−3.47					

(continued)

Table 8.1 (continued)

Species	Section	Biome	Porosity	Phenology	Ψ_{50}	$K_{S,max}$	H_V	WD	Ψ_{min}	D_{mean}	N_{cond}	P_{rad}	P_{ax}	P_{tot}
Quercus oxyodon	Cyclobalanopsis	TRS		E						118.00	7.00			
Quercus pachyloma	Cyclobalanopsis	TMS		E				0.73				17.80	24.05	41.85
Quercus pagoda	Lobatae	TMS	Ring-porous	D				0.61						
Quercus palustris	Lobatae	TMS	Ring-porous	D				0.58						
Quercus petraea	Quercus	TMS	Ring-porous	D	−3.06			0.57	−3.17					
Quercus phellos	Lobatae	TMS	ring-porous	D	−1.42	1.35	1.81E-04	0.60	−1.97	49.17				
Quercus phillyraeoides	Cerris	TRS	Diffuse-porous	E				0.69		41.48	14.65			
Quercus prinoides	Quercus	TMS		D				0.59						
Quercus pubescens	Quercus	WDS	Ring-porous	D	−2.88	1.58		0.71	−4.20					
Quercus pumila	Lobatae	TMS		E		3.00	1.68E-04	0.84						
Quercus pyrenaica	Mesobalanus	TMS	Ring-porous	D										
Quercus robur	Quercus	TMS	Ring-porous	D	−1.90	1.30		0.57	−2.94	43.12	39.37	17.54	17.20	33.95
Quercus rubra	Lobatae	TMS	Ring-porous	D	−2.23	1.33	1.75E-04	0.60	−1.30	45.11	11.62	16.88	14.05	31.01
Quercus rugosa	Quercus	TRS		E				0.60						
Quercus salicifolia	Lobatae	TRS		E				0.67						
Quercus sebifera	Quercus	WDS		E			2.45E-04	0.71	−5.58					
Quercus semecarpifolia	Cerris	TRS		E	−5.50	0.71				173.00	5.00			
Quercus semiserrata	Cyclobalanopsis	TRS	Diffuse-porous	E				0.71						
Quercus serrata	Quercus	TRS	Ring-porous	D	−2.70	2.87		0.68		27.34				
Quercus sessilifolia	Cyclobalanopsis	TMS		E				0.68						
Quercus shumardii	Lobatae	TMS	Ring-porous	D		3.70	1.32E-04	0.63	−2.40			17.90	23.60	41.50
Quercus sideroxyla	Quercus	TRS						0.63						
Quercus similis	Quercus	TMS		D				0.59						
Quercus sinuata	Quercus	WDS	Ring-porous	D				0.59		37.00				

(continued)

Table 8.1 (continued)

Species	Section	Biome	Porosity	Phenology	Ψ_{50}	$K_{S,max}$	H_V	WD	Ψ_{min}	D_{mean}	N_{cond}	P_{rad}	P_{ax}	P_{tot}
Quercus stellata	Quercus	TMS	Ring-porous	D	−0.82	1.86	1.77E-04	0.68		47.24		19.15	29.75	48.90
Quercus suber	Cerris	WDS	Diffuse-porous	E	−3.68	1.03		0.77	−2.97	54.52	99.27			
Quercus subsericea	Cyclobalanopsis	TRS		E				0.77						
Quercus texana	Lobatae	TMS		D				0.59						
Quercus turbinella	Quercus	TMS	Ring-porous	D	−0.50									
Quercus variabilis	Cerris	TMS	Ring-porous	D				0.76		110.58	5.36	25.15	21.80	46.95
Quercus velutina	Lobatae	TMS	Ring-porous	D				0.56				32.20		
Quercus virginiana	Quercus	TMS	Diffuse-porous	E		3.00	3.63E-04	0.86	−2.17					
Quercus wislizeni	Lobatae	WDS	Diffuse-porous	E	−1.69	1.56		0.66	−3.57	39.61	37.30			

Selected anatomical, tissue-level and hydraulic traits were summarised from available databases enlarged with additional more recent literature. See text for further details

Notes Nomenclature follows The Plant List (2013). Codes for biomes: *TMS* temperate; *TRS* tropical; *WDS* Mediterranean/semi-arid. For leaf phenology: *D* deciduous, *E* evergreen. Ψ_{50}, water potential at 50% loss of hydraulic conductivity (MPa); $K_{S,max}$, maximum sapwood-specific hydraulic conductivity (kg m^{-1} MPa^{-1} s^{-1}); H_V, Huber value; *WD*, wood density (g cm^{-3}); Ψ_{min}, minimum water potential experienced in the field (MPa); D_{mean}, mean vessel diameter (μm); N_{cond}, mean vessel density (mm^{-2}); P_{rad}, % radial parenchyma; P_{ax}, % axial parenchyma; P_{tot}, % axial plus radial parenchyma

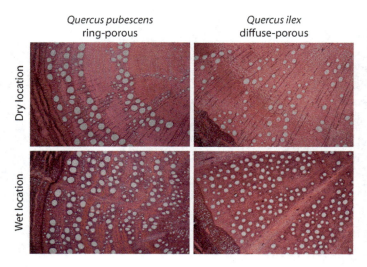

Fig. 8.1 Exemplary transverse sections of ring-porous *Quercus pubescens* (left) and diffuse-porous *Quercus ilex* (right) branch wood taken from trees growing at a dry (top) and a wet (bottom) location in Catalonia, Spain. Classification of the study sites into dry or wet was based on species-specific terciles in the range of average annual precipitation over potential evapotranspiration ratios. Sections are about 20 micrometres thick and were stained with a safranine–astra blue mixture. Images are 1.1 by 0.7 mm (top, left) and 2.2 by 1.4 mm (top, right and bottom)

8.2.1.2 Vessel Size and Density

Conduit size and density are perhaps the two most widely measured anatomical characteristics of trees. These two traits have a direct impact on xylem hydraulic efficiency, particularly through the fourth-power relationship between lumen hydraulic conductivity and conduit diameter (Tyree and Zimmermann 2002). Vessel size and density define the lumen fraction of a wood section (via their multiplication) and the variation in vessel composition within the potential transport space (their ratio). A global dataset complied by Zanne et al. (2010) revealed that hydraulic conductivity is more sensitive to changes in the total lumen fraction than to changes in the size to number ratio, all other factors being equal. However, the vessel composition ratio was the driving factor for potential conductivity differences across species.

To assess vessel characteristics of oaks in comparison with other angiosperm species, we extended the dataset of Zanne et al. (2010) with data from Jacobs (2013) and with anatomy data present in an updated version of the global xylem traits database (Choat et al. 2012) (Table 8.1). Values were averaged at the species level prior to analysis. Only measurements conducted on stems were considered, as measurements on other organs were extremely rare. It should be noted, however, that the few studies measuring vessel dimensions in stems and roots of the same oak individuals have reported much wider conduits in the latter organ, particularly in deep roots, consistent with results for other plant groups (Martínez-Vilalta et al. 2002; McElrone et al. 2004). The mean vessel diameter in deep (7–20 m) roots of *Quercus fusiformis* and *Quercus sinuata* was around 100 µm, with maximum values over 200 µm that allowed seeing the conduits with the naked eye (McElrone et al. 2004).

According to our dataset, average vessel diameter (D_{mean}) in oak stems is 62 ± 5 µm ($N = 42$ species) (mean ± SE; calculated in all cases after back-transformation from log-transformed data) and the average vessel density (N_{cond}) is 19 ± 4 conduits·mm^{-2} ($N = 30$ species). The former value is close to the average for non-oak angiosperm species (68 ± 1 µm, $N = 2204$ non-oak species in Zanne et al. 2010), whereas vessel density is clearly lower for oaks relative to other angiosperms (38 ± 1 conduits·mm^{-2}, $N = 2204$ non-oak species). However, these comparisons must be interpreted with care, particularly for vessel diameters, as there are substantial methodological differences in how this variable can be measured, which were not accounted for in our compilation from disparate literature sources. If only the data from the Zanne et al. (2010) database are considered, including 13 oak species, the low average vessel density for oaks was supported (19 ± 4 conduits·mm^{-2} for oaks), whereas vessel diameter appeared to be larger for oaks (99 ± 18 µm) than for other angiosperms.

We assessed the effect of taxonomic grouping, species biome, ring-porosity and leaf phenology on $\log(D_{mean})$ and $\log(N_{cond})$ using linear models with one explanatory variable at a time (as the strong association among them prevented including them in a single, multiple regression model). For vessel diameter, only taxonomic section had a significant effect ($P < 0.001$), with species from the *Cyclobalanopsis* ($N = 3$) and *Cerris* ($N = 11$) groups having larger vessels than those in sections *Quercus* ($N = 20$) and *Lobatae* ($N = 7$). The only species in section *Mesobalanus* (*Quercus frainetto*) had the smallest vessels. The effect of taxonomic section was also significant for vessel density ($P = 0.015$), with sections *Mesobalanus* ($N = 1$) and *Quercus* ($N = 13$) having higher values than *Cyclobalanopsis* ($N = 3$), whereas *Cerris* ($N = 11$) and *Lobatae* ($N = 2$) had intermediate values. The only other variable with a significant effect on oak vessel density was biome ($P < 0.001$). Temperate ($N = 8$) and Mediterranean/Semi-arid species ($N = 14$) had higher conduit density than tropical species ($N = 8$). Larger vessels in the *Cyclobalanopsis* and *Cerris* groups are probably associated with higher vulnerability to freezing-induced embolism (Davis et al. 1999), although these differences are not reflected in clear bioclimatic segregation among phylogenetic sections (Cavender-Bares et al. 2004).

We assessed the relationship between vessel size and density after converting vessel diameter into area, for consistency with previous work and because the product of vessel area and density is a direct measure of the lumen fraction (Zanne et al. 2010). Oak species occupy a substantial range of the angiosperm space in the relationship between vessel area and density, although they tend to cluster towards the lower vessel density end (Fig. 8.2). Interestingly, oaks tend to have small vessels for a given vessel density, relative to other angiosperms, implying a lower lumen fraction. The fact that oak species tend to be located far from the packing

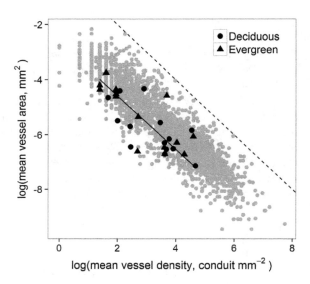

Fig. 8.2 Relationship between mean vessel area and mean vessel density for oak species represented on top of the scatterplot for all angiosperms (the latter obtained from Zanne et al. 2010). Both variables are (natural) log-transformed. Symbol shapes distinguish between deciduous and evergreen oak species. The solid line indicates the standardized major axis regression between the two variables (slope = −0.97). The dashed line indicates the packing limit, corresponding to a lumen fraction of 1

limit indicates the importance of the ground tissue, whether genetically defined or as an adaptation for survival (Jupa et al. 2016), and the tendency of the genus towards xylem structures that provide high mechanical support. The relationship between vessel area and density for oaks was not affected by taxonomic grouping, species biome, ring-porosity or leaf phenology.

8.2.1.3 Vessel Length

Vessel length is a key trait in woody plant hydraulics due to its defining role in affecting both hydraulic safety and efficiency (Lens et al. 2011; Jacobsen et al. 2012). Jacobsen et al. (2012) found average vessel length in the Fagaceae to be smaller than 0.2 m, as was the case for most of the 31 studied species. However, oaks are often mentioned as being long-vesselled (e.g. *Q. gambelii* and *Quercus prinus*—Hacke et al. 2006; *Q. robur*—Cochard et al. 2010; *Q. ilex*—Martínez-Vilalta et al. 2002 and Martin-StPaul et al. 2014; *Quercus variabilis*—Pan et al. 2015). Comparative studies show indeed that *Q. gambelii* has much longer vessels than *Betula occidentalis* and *Populus tremuloides* (Sperry and Sullivan 1992) and that *Q. robur* had the longest vessels in a group of ten woody angiosperms (Cochard et al. 2010). Sperry and Sullivan (1992) show a length distribution of *Q. gambelii* vessels in which almost 40% falls in the 5–10 cm length class and 5% in the 65–70 cm maximum class. But *Quercus* vessels are not always the longest. The maximum vessel length of *Q. ilex* (0.45 ± 0.01 m) is only second in a list of nine studied species, after *Eucaluptus camaldulensis* (0.52 ± 0.06 m) (Trifilò et al. 2015). *Q. rober* had similar maximum vessel length compared to *Vitis vinifera* and *Populus trichocarpa* in a study where its mean vessel length was also the same as for *V. vinifera* but six times larger than in *P. trichocarpa* (Venturas et al. 2016). In Table 8.2, we extended the mean and maximum vessel length data coverage provided in Appendix S1 of Jacobsen et al. (2012). It must be noticed that different measurement methods can generate different outcomes even in the same individuals (Pan et al. 2015; Oberle et al. 2016) and that within-tree and within-species variability might be larger than considered so far (Wang et al. 2014; Zhang and Holbrook 2014; Pan et al. 2015).

8.2.1.4 Vessel Grouping

Oak wood is known for having a high degree of solitary vessels, arranged in a diagonal/oblique (intermediate between radially and tangentially) or a dendritic pattern (Wheeler et al. 1989; Johnson et al. 2014; InsideWood 2004 onwards; Ellmore et al. 2006; Sano et al. 2008; Mencuccini et al. 2010; Kim and Daniel 2016a; Venturas et al. 2016). Mencuccini et al. (2010) classified *Q. petraea* and *Q. robur* as species with an overall random vessel distribution (as opposed to aggregated or uniform) with the note that there is a difference between the early-wood (random) and the late-wood (clustered) vessel distribution in these ring-porous species. Three-dimensional analysis of roots of *Q. fusiformis* showed that only 1 (shallow roots) to 6 (deep roots) % of the vessels were connected to other vessels, a finding that is likely related to the high resistance to embolism of this species' roots relative to other species with more vessel connections (Johnson et al. 2014). In addition, dye experiments in saplings of *Q. rubra* showed high sectoriality and low potential radial flow, attributed to the low degree in radial and tangential vessel interconnections (Ellmore et al. 2006). Lateral connections between vessels and their surrounding non-vascular cells such as ray parenchyma and vasicentric tracheids might thus ensure network connectivity, as has been observed for *Quercus suber* (Sousa et al. 2009) and other Mediterranean oak species (*Q. ilex*, *Quercus pyrenaica* and *Quercus faginea*—cf. Sousa et al. 2009).

8.2.1.5 Pits

The structural variability of intervessel pits—the tiny cavities in the secondary cell walls that allow water and air passage between adjacent vessels (Sperry and Tyree 1988; Choat et al. 2008; Jansen et al. 2009; Schenk et al. 2015)—has not been

Table 8.2 Mean and maximum vessel length data of *Quercus* species as found in literature

Species	Sample information	Mean vessel length (m)	Maximum vessel length (m)	References
Quercus agrifolia	Stems from adult shrubs		1.96	Jacobsen et al. (2007)
Quercus alba	Branches from adult trees	0.1153		Cochard and Tyree (1990)
	Model estimate	0.025		Oberle et al. (2016)
Quercus berberidifolia	Stems from adult Shrubs	0.1779	1.60	Jacobsen et al. (2007), Hacke et al. (2009)
Quercus gambelii	Branches from adult trees	0.1656	0.65 – 0.70	Sperry and Sullivan (1992)
	Stems from adult trees			Hacke et al. (2006)
Quercus ilex	Stems from adult trees		0.96 ± 0.07	Martínez-Vilalta et al. (2002)
	Current year resprouts	0.86	0.75 – 1.00	Martin-StPaul et al. (2014)
	Stems of young trees (>15y)		0.45 ± 0.01	Trifilò et al. (2015)
Quercus prinus	Stems from adult trees	0.1571		Hacke et al. (2006)
Quercus robur	Terminal shoots		1.34 ± 0.38	Cochard et al. (2010)
	Branches from young trees	0.109 ± 0.019	0.830 ± 0.186	Venturas et al. (2016)
Quercus rubra	Branches from adult trees	0.1159		Cochard and Tyree (1990)
	Model estimate	0.075		Oberle et al. (2016)
Quercus variabilis	Stem from mature trees (method 1a)	0.1059 ± 0.0077		Pan et al. (2015)
	Stem from mature trees (method 1b)	0.2106 ± 0.0075		Pan et al. (2015)
	Stem from mature trees (method 2)	0.4549 ± 0.0065		Pan et al. (2015)
Quercus velutina	Model estimate	0.06		Oberle et al. (2016)
Quercus wislizenii	Stems of adult shrubs		1.45	Jacobsen et al. (2007)

The table is an extended version from Jacobsen et al. (2012) (Appendix S1)

extensively studied in general, with oaks not being an exception. This is mainly due to the minuscule nature of pits and the difficulty to access them (mostly by electron microscopy). Table 8.3 gives an overview of the published data on oak pit features (intervessel and other pit types) that play a role in determining how easily liquid or air can pass from one cell to another within the transport network (Lens et al. 2011).

Table 8.3 Overview of *Quercus* pit structural characteristics

Species	Sample information	Pit diameter—horizontal (µm)	Pit diameter—longitudinal (µm)	Pit aperture area (µm^2)	Pit membrane thickness (nm)	Total inter-vessel cell wall thickness (nm)	References
Vessel connections							
Quercus fusiformis	Shallow roots			3.59 ± 0.29	239 ± 30		Johnson et al. (2014)
	Deep roots			18.85 ± 2.71	410 ± 40		
Quercus ilex		5.73		2.01	689		Choat et al. (2012) (updated)
Quercus robur	Last growth ring of young branches	5.886 ± 0.75			278 ± 87	2503 ± 562	Jansen et al. (2009)
Quercus rubra	Braches of mature trees	6–10					Ellmore et al. (2006)
Quercus serrata	Stems of mature trees—early wood			5.67 ± 2.16	163.79		Ahmed et al. (2011)
	Stems of mature trees—late wood			1.95 ± 0.88			
Tracheid connections							
Quercus crispula	Outer sapwood	6.0 ± 0.47	5.9 ± 0.57				Sano and Jansen (2006)
Quercus hypoleucoides		11.3 by 26.5					Adaskaveg et al. (1995)

(continued)

Table 8.3 (continued)

Species	Sample information	Pit diameter—horizontal (μm)	Pit diameter—longitudinal (μm)	Pit aperture area (μm²)	Pit membrane thickness (nm)	Total inter-vessel cell wall thickness (nm)	References
Vessel-tracheid connections							
Quercus crispula	Stem of mature trees—early wood	5.7 ± 0.61	5.2 ± 0.80				Sano et al. (2008)
	Stem of mature trees—late wood	5.8 ± 0.77	5.7 ± 0.79				
Fibre connections							
Quercus crispula	Outer sapwood	2.2 ± 0.40	2.5 ± 0.36				Sano and Jansen (2006)
Quercus serrata	Stems of mature trees—early wood			1.35 ± 0.57			Ahmed et al. (2011)
	Stems of mature trees—late wood			1.08 ± 0.51			
Ray connections							
Quercus serrata	Stems of mature trees—early wood			0.90 ± 0.72	312.48		Ahmed et al. (2011)
	Stems of mature trees—late wood			0.68 ± 1.53			

Intervessel Pits—Arrangement and Size

According to the available information, intervessel pits in oaks mostly seem to be alternately organised, i.e. in diagonal rows (Wheeler et al. 1989; InsideWood 2004 onwards; Jansen et al. 2009). However, Ellmore et al. (2006) reported the oval and widely scattered pits of *Q. rubra* to only be vaguely alternate and intervessel pits in the roots of *Q. fusiformis* were observed to be mostly alternate but occasionally also scalariform or gash-like (Johnson et al. 2014). In a comparative study on 26 hardwood species, average pit diameter of *Q. robur* fell within the top half of all observed values (5.9 μm, species range: 2.1–7.6 μm) (Jansen et al. 2009) and in comparison to poplar (*Populus tomentiglandulosa*), *Quercus serrata* had wider pit apertures (Ahmed et al. 2011). Vessel walls of *Q. rubra* branches were for 12% occupied by intervessel pits (Ellmore et al. 2006) and roots of *Q. fusiformis* for 6–7% (Johnson et al. 2014). However, Christman et al. (2012) found only 7% of the intervessel pits in *Q. gambelii* to be air-seeding. It has been suggested by these authors that ring-porous species such as *Q. gambelii* compensate their large leakiness probability at the pit level that comes with the highly efficient, large vessels in their early-wood, with a low degree of vessel connectivity, thicker pit membranes or with refilling (Christman et al. 2012).

Intervessel Pits—Pit Membranes

Pore sizes in intervessel pit membranes have a large effect on both the hydraulic conductivity of the xylem and its vulnerability to embolism (Wheeler et al. 2005). Dye experiments in *Q. rubra* confirmed that pitting alone cannot explain intervascular liquid transfer pointing to the role of pit membrane porosity (Ellmore et al. 2006). Jansen et al. (2009) found pit membranes of *Q. robur* to be without visible pores on Scanning Electron Microscopy (SEM) images, in contrast to most other studied hardwood species, and observed a species characteristic pit membrane thickness (Table 8.3) that fell within the average thickness of 100–300 nm. In the roots of *Quercus fusilis*, pit membranes did not show any visible pores either (Johnson et al. 2014). Theoretically estimated pit membrane pore diameters were 0.134 ± 0.02 μm for shallow and 0.167 ± 0.02 μm for deep roots (Johnson et al. 2014). Pit membranes were thicker in deep than in shallow roots, accounting for a larger portion of the overall xylem hydraulic resistance (Johnson et al. 2014).

Other Pit Types

Sano et al. (2008) studied other pit types and found that pit pairs between vessels and fibres and between vessels and vasicentric tracheids were frequently present in *Quercus crispula* (vessel-tracheid connections: 88.2 pairs per 10,000 μm^2 on the radial wall for early-wood and 49.0 for late-wood), with the pit membranes observed as sheet-like, homogeneous and without visible pores. Bordered pits were observed between vessels and tracheids in *Q. robur* by Kim and Daniel (2016a). The high density of vessel-fibre/tracheid connections in *Q. crispula*, in contrast to other studied species, was attributed to the species' high degree of solitary vessels, with tracheary elements thus supposed to contribute to the transpiration flow (Sano et al. 2008).

There is a consistent structural difference in pits between fibres and pits between tracheids that can be attributed to their functional roles in support and conduction, respectively (Sano and Jansen 2006). Interfibre pits were found to be smaller than 4 μm and very often showed simple or multiple perforations (studies on *Q. crispula* and *Q. robur*, Sano and Jansen 2006; Kim and Daniel 2016a). Intertracheid pits in *Q. crispula* on the other hand were larger than 4 μm and their membranes were densely and evenly packed with microfibrils, only for 10% showing sparsely packed microfibril regions at the membrane periphery (Sano and Jansen 2006). Thickened walls have also been observed in bordered tracheid pits of *Quercus hypoleucoides* (Adaskaveg et al. 1995).

Pits in *Q. robur* were found to be half-bordered between tracheary elements and parenchyma cells and simple between parenchyma cells (Kim and Daniel 2016a). Roots of *Q. fusiformis* showed scalariform vessel-parenchyma pitting, interspersing and surrounding inter-vessel and vessel-tracheid alternate pitting (Johnson et al. 2014). Heavily encrusted parenchyma pit membranes were observed in *Q. serrata* (Ahmed et al. 2011). Early- and late-wood of *Q. robur* were contrasting in the abundance of pits associated with axial and radial parenchyma, with the former being more abundant in the earlywood and the latter being more abundant in the late-wood (Kim and Daniel 2016a).

Pit membrane chemistry showed large variability, both between pit types and between early- and late-wood in *Q. robur* (Kim and Daniel 2016a, b). However, the functional significance of this variability is yet to be elucidated. It was suggested that the presence of hemicelluloses in the inter-tracheid and tracheid-vessel pits could play a similar role as pectin, i.e. mediate pit membrane porosity through changes in ion concentrations (Kim and Daniel 2016a, b).

8.2.1.6 Tyloses

Tyloses are oily secondary metabolites that contain phenolic substances and terpenoids (Kuroda 2001). In *Quercus macrocarpa* tyloses, a lignin content of *ca.* 28% has been measured (Obst et al. 1988). Tyloses are produced by the ray parenchyma and oxidize and polymerize in the vessel lumen in absence of water, thus in vessels that already are air-filled (Kuroda 2001). Cochard and Tyree (1990) studied tylose formation in ring-porous oaks (*Quercus alba* and *Q. rubra*) and discovered that early-wood vessels start getting filled with tyloses in the first winter after their formation, i.e. at the end of the growing season, to be fully blocked by tyloses by the next summer. Late-wood vessels on the other hand remained free of complete blockage for several years (Cochard and Tyree 1990). In *Q. robur* a wide protective layer, thought to play a role in tyloses formation, was only observed inside axial and ray parenchyma of early- but not of late-wood (Kim and Daniel 2016b). Tylose-filled early-wood vessels after the growing season have been reported for several ring-porous species (Kitin and Funada 2016) and for the following oaks: *Quercus castaneaefolia* (Safdari et al. 2008) and *Q. variabilis* (Kim and Hanna 2006; Pan et al. 2015). Tyloses were moreover commonly present in the wood of *Q. suber* (Sousa et al. 2009) and in the heartwood of *Q. robur* (Fromm et al. 2001; Sorz and Hietz 2006), being used as a potential criterion to define sapwood depth (Fromm et al, 2001; Sohara et al. 2012). Babos (1993) found tyloses in early-wood vessels to be increasing from stump (16 cm: 5.02%) upwards (12 m: 12.58%). Tylose formation can also be linked to wounding and pathogen infestation. In *Q. petraea* more sealed early-wood vessels were found in diseased tree trunks as compared to healthy ones (Babos 1993). Tyloses formation has also been observed in *Q. crispula* and *Q. serrata* trees after infestation by the fungus *Raffaelea* sp. (Kuroda 2001). Although supposed to be protective, the tyloses could not prevent the expansion of the fungus in the study of Kuroda (2001).

8.2.2 Tissue Properties

8.2.2.1 Parenchyma

Aggregate rays, i.e. large units of clustered ray parenchyma, are not unique to oaks but are uncommon in trees and they typically occur in Fagaceae (Carlquist 2001). This typical wood structure element can often be spotted with the naked eye in oak-made furniture or building parts. Together with the small rays and the axial parenchyma tissue, the large aggregated rays form the long-living (2–200 years) (Spicer and Holbrook 2007), elastic fraction of oak wood volumes, likely providing oxygen to the cambial zone and the bark (Spicer and Holbrook 2005) and defining the start of heartwood formation upon death (Spicer and Holbrook 2007). Internal water storage and water release has been linked to these elastic storage compartments, with the thickness of parenchyma cell walls playing an important role in the storage capacity (Jupa et al. 2016). However, in juvenile xylem from excised plant parts of five temperate tree species among which *Q. robur*, the importance of capillary compartments in water storage and release was much larger than that of parenchyma cells (Jupa et al. 2016). It is thus far unclear if this also stands for adult trees in natural conditions (Jupa et al. 2016). Jupa et al. (2016) compared the ring-porous *Q. robur* to diffuse-porous non-*Quercus* species and concluded that, contrary to the expectations, ring-porous wood does not seem to stand out in water storage and release capacity.

We employed the global dataset assembled by Morris et al. (2015) to examine existing data for radial, axial and radial plus axial parenchyma volumetric fractions for the genus *Quercus*. We found 24, 14 and 14 species of the genus *Quercus* for which data were available for the volumetric fractions of radial parenchyma, axial parenchyma and radial plus axial parenchyma, respectively. Given the small sample size, it is impossible to draw any definitive conclusions about large-scale patterns. Overall, fractions of radial parenchyma in oaks varied between a minimum of 9% and a maximum of 34% (median of 21%), axial parenchyma varied between 11 and 26% (median of 21%) and total radial plus axial parenchyma varied between 24 and 46% (median of 42%). This compares with equivalent figures for the rest of the angiosperms of 2 and 68% (median of 17%), 0 and 74% (median of 11%) and 6 and 99% (median of 35%), respectively. Hence, the little available data do not show any unusual patterns for oaks relative to other angiosperms. The only discernible pattern in the dataset specifically for *Quercus* was that section *Lobatae* had significantly greater fractions of axial and radial plus axial parenchyma (but not radial parenchyma on its own) compared to *Quercus*, but given the limited sample size, caution needs to be exercised regarding the significance of this conclusion.

Few studies compare oak parenchyma content between organs or between trees growing under different environmental conditions, and if within-genus structural differences are observed, experimental proof for their possible functional significance is generally lacking. In saplings of *Q. robur*, *Q. petraea* and *Q. pubescens*, the density of non-ray parenchyma cells was slightly increased in drought-exposed individuals (Fonti et al. 2013). The authors suggest a role in embolism repair, more needed in dry conditions. Stokke and Manwiller (1994) found higher ray parenchyma content in roots, followed by branches and stems in *Quercus velutina*, proposing it to be related to the carbohydrate storage function of roots.

Besides in day-to-day tree physiology, xylem parenchyma also plays an important role in wood decay (e.g. Deflorio et al. 2009), wound reactions (Schmitt and Liese 1993, 1995) and pathogen infestations (Morris et al. 2016). The parenchyma tissue is often the place where trees are attacked by pathogens (or parasites) while at the same time parenchyma tissue is responsible for tyloses formation in reaction to infestation. Some example regarding oak species can be found in Kuroda (2001, *Q. serrata* and *Q. crispula*—ambrosia beetle *Platypus quercivorous*), Brummer et al. (2002, *Q. robur*—*Phytophthora quercina*), Miric and Popovic (2006, *Q. robur* and *Q. petraea*—fungus *Chondrostereum purpureum*), Medeira et al. (2012, *Q. suber*—*Phytophthora cinnamomi*), Ebadzad et al. (2015, *Q. ilex* and *Q. suber*—*P. cinnamomi*) and Cocoletzi et al. (2016, *Quercus germana*—mistletoe *Psittacanthus schiedeanus*).

8.2.2.2 Tracheids and Fibres

Wood fibres are known to provide mechanical support and are supposed to offer protection against vessel collapse (Metcalfe and Chalk 1983; Hacke et al. 2001; Jacobsen et al. 2005; Jupa et al. 2016). They also play a role in the trees' water household (Jupa et al. 2016). Fibres as well as fibre-tracheids and tracheids can store water and thus contribute to the sapwood water storage and release capacity as has been proven for *Q. robur* by Jupa et al. (2016). This makes fibre and tracheid lumen size and pit characteristics important structural traits in tree hydraulics. Together with the vessel and parenchyma fraction, the fibre fraction defines the density of a wood volume (Preston et al. 2006), mainly through the ratio of wall thickness to lumen size (Martínez-Cabrera et al. 2009; Zieminska et al. 2013). Fibre dimensions in oaks are found to vary with tree age, location within the tree and growth conditions (e.g. Lei et al. 1996; Leal et al. 2006; Yilmaz et al. 2008).

8.2.2.3 Wood Density

Wood density (WD, dry mass per fresh volume, g cm^{-3}) is the basic measurement of the content of dry biomass within the green (or fresh) volume of a tree. It provides important ecological and physiological insights for foresters, ecologists and physiologists. Wood density varies between 0 and an upper limit of about 1.5 g cm^{-3}, which is the density of the wooden matrix alone (Whitehead and Jarvis 1981; Siau 1984). Different tree species can allocate different amount of carbon to produce their xylem structure, and an obvious trade-off emerges between the allocation to 'cheap' wood with lower construction costs to produce greater volumes of wood versus the allocation of 'expensive' wood with higher costs which might result in lower volumes being produced. Physiologically, across-species differences in wood density have been linked to differences in mechanical properties of wood (Young's modulus, stiffness, resistance to splitting, etc.), in hydraulic properties of the xylem (vulnerability to embolism, conductive efficiency, hydraulic capacitance), in defence against attacks by pathogens, in canopy architecture and in the ratio of leaf area to stem cross-sectional area (all of which are comprehensively reviewed by Chave et al. 2009). Ecologically, studies have often found negative relationships between wood density and either growth rates and/or likelihood of mortality in the field (e.g. Martínez-Vilalta et al. 2010). Therefore, it is important to examine general trends in wood density across the genus *Quercus* as a whole to determine whether broad patterns can be identified.

To do so, we employed existing compilations (Zanne et al. 2009; Chave et al. 2009) and enlarged that dataset by examining recent papers that collated wood density values for species not represented there (i.e. Cavender-Bares et al. 2004; Aiba and Nakashikuza 2009; Miles and Smith 2009; Návar 2009). Multiple entries for each species were averaged and all subsequent analyses were carried out using only species means. We recognise that significant limitations are present when wood density values are pooled across from a heterogeneous literature. Nonetheless, we agree with earlier authors (Chave et al. 2009; Zanne et al. 2010; and discussions therein) that these consolidation exercises have value in themselves and can provide novel insights.

Overall, we found wood density values for 99 oak species, with a relatively even spread across the three biomes (26 tropical, 50 temperate and 23 Mediterranean/semi-arid). The spread across sections of the genus was less even (19 in *Cyclobalanopsis*, 43 in *Quercus*, 29 in *Lobatae*, 5 in *Cerris*, 1 in both *Mesobalanus* and *Protobalanus*; we could not find any reference allowing classing *Quercus sideroxyla* within the genus). For the genus as a whole, wood density varied between 0.51 and 0.88 g cm^{-3}, with a median value of 0.67 g cm^{-3}.

The distribution of wood density values for the rest of the angiosperms has a minimum of 0.08, a maximum of 1.39 and a median of 0.60 g cm^{-3}, therefore suggesting that the genus *Quercus* occupies a fairly central position within the overall distribution of angiosperm wood density values, perhaps shifted towards values slightly above the mean. We log-transformed wood density values for all subsequent analyses to achieve normality of distribution. A very highly significant difference in log(*WD*) was found according to phenology ($P = 0.1.3$ e^{-8}, $N = 98$; Fig. 8.3), with evergreen species ($N = 49$) having on average higher wood density values than deciduous ones ($N = 49$ as well; back-transformed means of 0.72 versus 0.62 g cm^{-3}, respectively). A highly significant difference was also found with regard to wood porosity. Ring-porous species ($N = 43$) had a significantly lower wood density compared to diffuse-porous (0.64 vs. 0.71 g cm^{-3} respectively, $P = 0.009$, $N = 15$, Fig. 8.3). Compared to section *Cerris* (*WD* = 0.73), section *Quercus* and, especially *Lobatae* had significantly lower wood density (0.66 and 0.62 g cm^{-3}, $P = 0.037$ and $P = 0.002$, respectively). Finally, a significant difference was also found among biomes ($P = 0.0002$, $N=99$). *Quercus* species from the temperate biome (*WD* = 0.65) had significantly lower wood density compared to

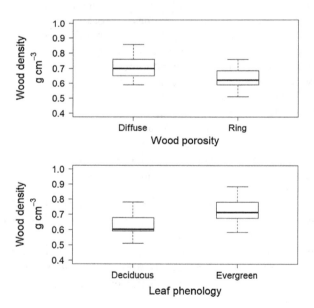

Fig. 8.3 Wood density box-and-whisker plots for two main grouping criteria, i.e. ring-porosity (diffuse- versus ring-porous) and leaf phenology (deciduous vs. evergreen). The dataset consisted of 98 *Quercus* species. Data were obtained from the primary literature and from existing compilations (see text for further details). In both cases, the differences between the two groups were highly significant ($P = 0.009$ and $P = 0.1.3$e^{-8}, respectively). The thick line gives the median, the edges of the box are the lower and upper quartile, while the two whiskers extend up to 1.5 times the interquartile range from the top/bottom of the box to the furthest datum within that distance

those from the tropical ($WD = 0.73$), but not the Mediterranean/semi-arid biomes ($WD = 0.65$).

Overall, the data support the idea that evergreen sub-tropical species with diffuse-porous wood, especially of section *Cerris*, have higher wood density values than deciduous ring-porous oaks of the temperate and Mediterranean biomes, especially those of the *Lobatae* section. It is likely that these differences in wood density also map onto additional hydraulic features (see Sect. 8.3), such that variability in wood density across oaks species better equip those with higher values for life under drier conditions.

8.3 Xylem Hydraulics

The xylem provides a low resistance pathway for water movement from the roots to the evaporation sites in leaves (Tyree and Zimmermann 2002). By supplying water to the leaves, it is a key element controlling plant water relations and stomatal responses, determining plant transpiration and assimilation rates (Sperry et al. 2017) and, ultimately, plant productivity and resistance to major stress factors such as freezing and droughts (Tyree and Sperry 1989; Brodribb 2009). The hydraulic properties of the xylem are usually summarized using two parameters, characterizing hydraulic efficiency and safety. Hydraulic efficiency corresponds to maximum transport capacity, and it is usually expressed as the sapwood-specific hydraulic conductivity under fully hydrated conditions ($K_{S,max}$). $K_{S,max}$ is quantified as the water flow through a wood segment per unit pressure gradient driving the flow, divided by the cross-sectional area of wood (Melcher et al. 2012). Hydraulic safety refers to the susceptibility to (drought-induced) embolism, as measured from vulnerability curves expressing how K_S declines as the water potential becomes more negative, indicating more stressful conditions. Several methods have been used to establish vulnerability curves, the most widely used being bench dehydration, air injection and centrifugation (Cochard et al. 2013). Vulnerability curves are frequently summarized using one single parameter: the xylem water potential inducing a 50% loss of hydraulic conductivity (Ψ_{50}).

Here, we assembled a database of oak xylem hydraulic measurements starting with all oak entries in the Choat et al. (2012) functional xylem traits database, completed with Google Scholar searches of papers published after the compilation of the original dataset. Our analyses focus on $K_{S,max}$ and Ψ_{50} values measured in stems, as data from other organs (roots and leaves) are scarce. In addition, we only considered direct hydraulic measures of $K_{S,max}$ and Ψ_{50}; that is, indirect measures based on conduit sizes (for $K_{S,max}$) or acoustic emissions (for Ψ_{50}) were discarded. Overall, we compiled data on stem $K_{S,max}$ and P_{50} for 41 and 32 oak species, respectively (Table 8.1), corresponding to a total of 89 individual vulnerability curves.

8.3.1 Measuring Xylem Hydraulic Properties in Oaks

The measurement of xylem hydraulic properties in oaks is problematic for several reasons. Firstly, as already mentioned in the introduction, long vessels result in a substantial proportion of open vessels in measured wood segments. This leads to overestimations of both $K_{S,max}$ (by not accounting for the resistance of pit membranes in open vessels; Melcher et al. 2012) and vulnerability to embolism (by exaggerating the percent loss of conductivity at a given pressure). The latter effect is not only caused by the overestimation of $K_{S,max}$ but also because open, water-filled vessels are likely to be emptied immediately when subjected to any positive pressure, thus underestimating individual K_S measurements. Although this potential issue was recognized more than 20 years ago (Cochard and Tyree 1990; Sperry and Saliendra 1994; Martínez-Vilalta et al. 2002), it has received renewed attention in recent years (e.g. Cochard et al. 2010, 2013; Sperry et al. 2012; Hacke et al. 2015). The resulting artefact in vulnerability curves is supposed to affect mostly the air-injection and centrifugation techniques, and can be minimized by measuring $K_{S,max}$ once open vessels have been emptied (through slight pressurization or gentle spinning; Sperry and Saliendra 1994; Alder et al. 1997). However, a limitation of this approach is that wider and longer vessels tend to be more vulnerable to embolism (Hargrave et al. 1994; Cai and Tyree 2010) and thus the population of vessels remaining once open conduits have been emptied may not be representative of the whole xylem. The bench dehydration technique is much less affected by these effects if sampled branches are longer than the longest vessels. However, even this technique, frequently considered the 'gold standard' to establish vulnerability curves, may be prone to related artefacts when measurement segments are cut under tension (Wheeler et al. 2013; Torres-Ruiz et al. 2015; but see Trifilò et al. 2014; Venturas et al. 2015).

A second issue, particularly relevant for ring-porous species and, therefore, many oaks, concerns the reference starting point for vulnerability curves. Initial $K_{S,max}$ estimation is frequently done after flushing to remove any previously embolized vessels. However, flushing may refill conduits that were not functional in vivo, particularly in ring-porous species where only the most recent growth ring is functional. It has been recommended to work only with current-year shoots or exclude older growth rings (e.g. by gluing) before measurements are made in ring-porous species (Cochard et al. 2013). Another option is to sample in the wet season, before substantial embolism has developed, and consider the native K_S obtained under this conditions as a reasonable estimate of $K_{S,max}$.

The combination of the methodological issues described in the previous paragraphs is likely to have resulted in a huge variability in Ψ_{50} estimates for oaks, even within species. A large portion of this variability is associated to the shape of vulnerability curves: 'sigmoidal' curves generally having lower P_{50} (more resistant xylem) than 'exponential' curves (Cochard et al. 2013). Our data for oaks mirror the results obtained by Cochard et al. (2013) in that the vast majority of vulnerability curves obtained with the dehydration technique are reasonably sigmoidal in shape

(36 out of 39), whereas this is not the case for other methods (3 out of 35 for the static centrifuge, 1 out of 4 for the cavitron, 7 out of 11 for air-injection). Our database includes eight oak species (ring-porous and diffuse-porous) for which both shapes of curves have been measured. There is a striking difference in the estimated Ψ_{50} values between curve types for these species, with Ψ_{50} values being always more negative for sigmoidal curves, by as much as 1.8 MPa on average ($P < 0.001$; Fig. 8.4). Differences between curve shapes correspond to differences among methods, with vulnerability curves obtained by dehydration generally showing more negative Ψ_{50} values (Fig. 8.4). Although $K_{S,max}$ values were only reported for both curve types in four of the previous species, there was no consistent difference between curve types for this variable, suggesting that the differences in Ψ_{50} were not necessarily associated with biased $K_{S,max}$ estimates.

While we share the view that exponential vulnerability curves are not necessarily wrong (Sperry et al. 2012), it seems hard to believe that the huge variability observed within species in the shape of vulnerability curves and associated Ψ_{50} values is real. In addition, some of the extreme Ψ_{50} values obtained from exponential vulnerability curves (11 curves with $P_{50} > -0.5$ MPa, many for *Q. gambelii*) seem difficult to reconcile with the ecophysiology of the species, for which midday leaf water potentials are likely to be substantially lower than this value even under well-watered conditions (e.g. minimum water potential of -2.2 MPa reported

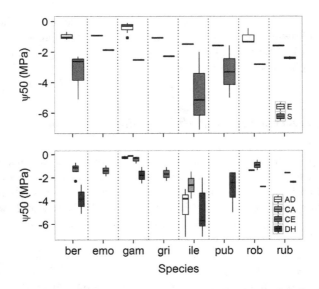

Fig. 8.4 Effect of vulnerability curve shape (upper panel: E, exponential; S, sigmoidal) and vulnerability curve method (lower panel: AD, air injection; CA, cavitron; CE, static centrifuge; DH, dehydration) on estimated vulnerability to embolism (Ψ_{50}) for eight species in which both types of vulnerability curve shapes have been obtained. Species: ber, *Quercus berberidifolia*; emo, *Quercus emoryi*; gam, *Quercus gambelii*; gri, *Quercus grisea*; ile, *Quercus ilex*; pub, *Quercus pubescens*; rob, *Quercus robur*; rub, *Quercus rubra*

in Schwilk et al. 2016; but see Taneda and Sperry 2008). Direct imaging techniques such as X-ray computed microtomography (microCT) are promising to solve this vexing issue (Cochard et al. 2015), and have recently been applied to measure the vulnerability to embolism of *Q. robur*, a ring-porous oak (Choat et al. 2016). The Ψ_{50} obtained for living, intact *Q. robur* plants using microCT was −4.2 MPa, much lower than the Ψ_{50} values obtained in the same study using the cavitron or static centrifuge techniques (−1.4 and −0.5 MPa, respectively), and also much lower than values reported in other studies using the centrifuge (−1.4 MPa, Cochard et al. 1992) and even the bench dehydration method (−2.8 MPa, Venturas et al. 2016). However, even microCT seems to give highly variable vulnerability curves within species (Nardini et al. 2017) and concerns have been raised about the correct interpretation of microCT results (Hacke et al. 2015). In addition, we note that as promising as the microCT technique is by being able to measure and visualize embolized conduits non-invasively, it does not directly measure the hydraulic impact of embolism. Studies modelling water transport in the xylem network show that conductivity is lost faster than the number of vessels (or the estimated theoretical conductivity in cross-sections) as water potential declines (Loepfe et al. 2007; Martínez-Vilalta et al. 2012), which could result in overestimating the resistance to embolism when using microCT (relative to purely hydraulic methods).

Before all these methodological issues are completely settled and standardized protocols are developed (Jansen et al. 2015), it can only be recommended that extreme care is taken at applying methodological protocols to measure hydraulic properties of oaks, and that different methodologies are compared whenever possible. In the following sections, we have taken an agnostic approach, using all data included in our database but always separating the Ψ_{50} values obtained from sigmoidal and exponential vulnerability curves.

8.3.2 An Overview of Oak Vulnerability to Embolism

The average stem vulnerability to embolism (Ψ_{50}) for the 32 oak species included in our dataset is −2.5 ± 0.3 MPa (mean ± SE), ranging from −0.5 MPa for *Q. fusiformis* (McElrone et al. 2004) and *Quercus turbinella* (Hacke et al. 2006) to −7.0 MPa for *Quercus coccifera* (Vilagrosa et al. 2003). This overall mean is close to the general average for angiosperms as reported in Choat et al. (2012) (−2.9 ± 0.1 MPa, N = 361 species excluding oaks). However, if only sigmoidal vulnerability curves are included Ψ_{50} declines to −3.6 ± 0.3 MPa (N = 18 species), which suggests that the oak genus may be a relatively resistant one within angiosperms. Root vulnerability to embolism has been only measured in seven oak species, with an average Ψ_{50} of −1.1 ± 0.2 MPa. Note, however, that the shape of all reported root vulnerability curves is exponential, except for two curves measured on *Q. ilex* by Limousin et al. (2010). When root and stem P_{50} were compared pairing the values by species and considering only studies in which both organs had been measured (mixed-effects model), the difference between organs was only marginally significant ($P = 0.05$), with root Ψ_{50} being more vulnerable by 0.4 MPa on average.

We studied how stem Ψ_{50} depends on taxonomic grouping, species biome, ring-porosity and leaf phenology. Due to limitations in sample size and close correspondence between some of these variables (e.g. most ring-porous oaks are deciduous) we assessed the effect of each variable separately, and thus our analyses are only meant to uncover broad associations. In all analyses the shape of the vulnerability curve (sigmoidal vs. exponential) was considered as a co-variate. Our results show that section *Cerris* has significantly lower (more resistant) Ψ_{50} than the *Quercus* or *Lobatae* sections ($P = 0.04$ in both cases), whereas the only species in the section *Mesobalanus* (*Q. frainetto*) showed intermediate values. This is consistent with two of the three species in section *Cerris* being Mediterranean evergreen oaks adapted to dry habitats (*Q. coccifera* and *Q. suber*). No significant difference was observed among biomes, although Mediterranean/semi-arid species tended to have lower (more negative) Ψ_{50}. Ring-porous oaks were more vulnerable to embolism (modelled $\Psi_{50} = -2.1 \pm 0.2$ MPa, $N = 22$; compared to -3.2 ± 0.4 MPa for diffuse-porous species, $N = 6$). Similar differences were observed between deciduous (-2.1 ± 0.2 MPa, $N = 19$) and evergreen species (-2.8 ± 0.3 MPa, $N = 13$), although in this case the effect was only marginally significant ($P = 0.06$). Overall, the data are consistent with evergreen, diffuse-porous oaks occupying drier habitats being able to withstand more negative water potentials (being more drought tolerant) than species from wetter environments.

The results we showed are in line with earlier studies showing an association between oak species traits and their distribution along gradients of water availability (Monk et al. 1989; Villar-Salvador et al. 1997; Corcuera et al. 2002; Cavender-Bares et al. 2004), albeit some studies report contrasting associations for sympatric species (e.g. Knops and Koenig (1994) found that deciduous species were more drought tolerant).

Freezing-induced xylem embolism is an additional element that needs to be considered when relating xylem traits with oak distribution. The relatively wide xylem vessels of oaks, together with the well-known relationship between conduit size and vulnerability to freezing-induced embolism has been linked with the absence of oaks from cold climates prone to late frosts (Tyree and Cochard 1996). Within oaks, several studies have shown that deciduous species, which tend to have wider xylem vessels, are more vulnerable to freezing-induced embolism than evergreen species (Cavender-Bares and Holbrook 2001; Cavender-Bares et al. 2005), mimicking the patterns obtained for drought-induced embolism. Differences in freezing-induced embolism are also influenced by evolutionary lineage, with white oaks (section *Quercus*) being more vulnerable to freezing than live oaks (all evergreen) than red oaks (section *Lobatae*, including both evergreen and deciduous species) (Cavender-Bares and Holbrook 2001).

8.3.3 Hydraulic Conductivity and the Hydraulic Safety–Efficiency Trade-Off

The average stem $K_{S,max}$ for the species included in our dataset is 2.0 ± 0.3 kg m^{-1} MPa^{-1} s^{-1} ($N = 41$ species; values calculated after back-transformation from log-transformed data). The individual species values range from 0.3 kg m^{-1} MPa^{-1} s^{-1} (*Q. coccifera*; Vilagrosa et al. 2003) to 6.3 kg m^{-1} MPa^{-1} s^{-1} (*Quercus aliena*; Fan et al. 2011), and have a markedly right-skewed distribution. Consistent with the relatively large vessels of oaks (see Sect. 8.2.1), their average $K_{S,max}$ is larger than the overall mean for angiosperms (1.5 ± 0.1 kg m^{-1} MPa^{-1} s^{-1}, $N = 573$ species excluding oaks; data from Gleason et al. 2016). Root $K_{S,max}$ is only available for seven species, with an overall average of 7.1 ± 0.9 kg m^{-1} MPa^{-1} s^{-1}, significantly higher than the value measured on stems for these same species ($P = 0.004$; mixed-effects model of log($K_{S,max}$)).

As before, we also modelled log($K_{S,max}$) as a function of taxonomic grouping, species biome, ring-porosity and leaf phenology. The results reveal significantly higher stem $K_{S,max}$ for species in section *Lobatae* ($N = 15$) than in section *Cerris* ($N = 3$) ($P = 0.01$), with *Quercus* species ($N = 23$) showing intermediate values (no data available for section *Mesobalanus*). However, these results should be taken with care because only three representatives of the section *Cerris* were included (*Quercus acutissima, Q. coccifera* and *Q. suber*). Temperate species ($N = 23$) have higher $K_{S,max}$ than Mediterranean/semi-arid species ($N = 14$) ($P = 0.02$), with the four measured tropical species showing intermediate values. We also observed marginally higher $K_{S,max}$ in ring-porous ($N = 26$) relative to diffuse-porous species ($N = 7$) ($P = 0.08$). There was no difference in $K_{S,max}$ between deciduous and evergreen species ($N = 22$ and 19 species, respectively).

Data for oaks show clear indications of a hydraulic safety–efficiency trade-off, with a positive exponential relationship between Ψ_{50} and $K_{S,max}$ (Fig. 8.5). This effect is highly significant ($P = 0.001$, overall $R^2 = 0.42$ when the effect of vulnerability curve type is accounted for in the model, $N = 29$ species) and particularly clear for species with sigmoidal vulnerability curves, whereas no relationship between Ψ_{50} and $K_{S,max}$ was detected for species with exponential vulnerability curves (Fig. 8.5). This clear trade-off (at least when considering sigmoidal curves) contrasts with the poor relationship observed when all angiosperms are considered ($R^2 = 0.05$ according to Gleason et al. 2016). We did not find evidence for the clustering of oak species in the low efficiency–low safety corner, contrary to what has been found in previous studies covering a wider range of species (Gleason et al. 2016).

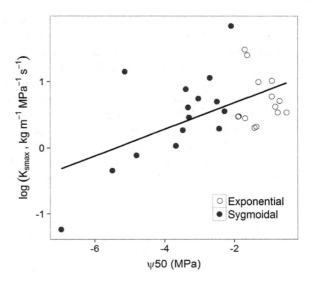

Fig. 8.5 Relationship between hydraulic safety (Ψ_{50}) and efficiency (natural log of $K_{S,max}$) for 29 oak species. Symbol colour indicates differences between vulnerability curve shapes (exponential vs. sigmoidal). The positive relationship between the two variables is highly significant ($P = 0.001$, overall $R^2 = 0.42$ when the effect of vulnerability curve type is accounted for in the model)

8.3.3.1 Huber Value and Relationships Between Allocation and Hydraulic Conductivity

The balance between water supply by the xylem and water demand by the canopy is mediated primarily by the relative allocation to these two tissues, a quantity variably named leaf-sapwood area ratio (leaf area divided by sapwood area) or Huber value (H_V, sapwood area divided by leaf area, Tyree and Ewers 1991). This allocation ratio is central to the interpretation of all measurements of plant hydraulics and water relations, because any estimate of the capacity of the xylem to efficiently conduct water per unit of sapwood area (i.e. $K_{S,max}$) depends on the total cross-sectional area through which water flows and the area of the leaves distal to this section that transpires this water. In addition to the balance between water supply and demand, the Huber value affects also indirectly the balance between vulnerability to embolism Ψ_{50} and the minimum water potential Ψ_{min} that plants are willing to sustain, because at constant values of $K_{S,max}$ and Ψ_{50}, greater allocation to leaf area relative to cross-sectional sapwood area will increase the demand on the xylem and the tension that the conductive system is expected to work under (e.g. Gleason et al. 2016). Several compensating mechanisms may allow plants to avoid or moderate the emerging trade-off between hydraulic safety and efficiency that was discussed in the previous section and which appears to be present also in *Quercus*. One important mechanism of compensation is the one involving the Huber value.

We considered therefore the relationships between the hydraulic traits mentioned above and Huber value. Consistent with a global relationship found elsewhere for angiosperms (Rosas et al. *unpubl.*), a negative relationship was also found here between Huber value and maximum specific conductivity $K_{S,max}$, although the relationship for the genus *Quercus* was only marginally significant ($P < 0.001$ for the global relationship and $P = 0.10$ for the 31 oak species documented here, Fig. 8.6). Although only marginally significant, the negative relationship in oaks suggests the occurrence of compensation mechanisms between high allocation to supporting sapwood area and formation of a highly conductive xylem. Neither of the two slopes were significantly different from $b = -1.00$ (i.e. 95% CI = −0.94 to −1.13 for all angiosperms and −0.83 to −1.69 for the oaks). Given that the leaf-specific conductivity K_L equals the product between $K_{S,max}$ and H_V, a slope of −1.0 is expected if all oak species had on average a constant K_L resulting in a perfect compensation. Obviously, more data across a larger sample of *Quercus* species are required to determine the degree of robustness of the relationship given here and therefore whether the genus *Quercus* also show compensation between water transport efficiency per unit of cross-sectional area and relative allocation between leaf area and sapwood area.

We then modelled H_V as a function of various grouping factors, as done in other sections above. Logarithmic transformation into $\log(H_V)$ was required to achieve normality. A significantly lower stem H_V was found for species in section *Lobatae* ($N = 13$) and marginally *Quercus* ($N = 18$) than in section *Cerris* ($P = 0.03$ and

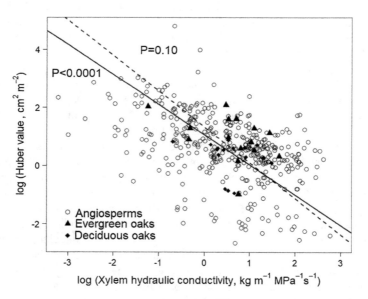

Fig. 8.6 Relationship between the (natural) logarithm of Huber value and the (natural) logarithm of the maximum specific hydraulic conductivity for 375 non-oak angiosperms (Rosas et al. *unpubl.*) and for 31 *Quercus* species. Oak species are separated according to their phenology (evergreen vs. deciduous). The *P*-values refer to the significance of the negative relationships shown for the two datasets. Lines were fitted by standardised major axis regression

$P = 0.08$, respectively), but only one representative of the section *Cerris* was present (*Q. coccifera*), so this relationship is very weak despite its level of significance. No difference was found between temperate species ($N = 21$), tropical ($N = 1$) and Mediterranean/semi-arid species ($N = 10$). A significantly lower H_V ($P = 0.006$) was found in ring-porous ($N = 21$) than in diffuse-porous species ($N = 3$), consistent with the (marginally) higher $K_{S,max}$ found in ring-porous species (Sect. 8.3.3). Deciduous species ($N = 17$) had significantly lower H_V ($P = 0.013$) than evergreens ($N = 15$).

8.3.3.2 Hydraulic Safety Margins and Drought Responses

Our data for oaks shows a strong coordination between the minimum leaf water potential measured for a given species (Ψ_{min}) and its vulnerability to embolism (P_{50}) ($P < 0.001$, overall $R^2 = 0.67$ when the effect of vulnerability curve type is accounted for in the model, $N = 25$ species) (Fig. 8.7), consistent with previous reports at community (e.g. Pockman and Sperry 2000; Martínez-Vilalta et al. 2002; Jacobsen et al. 2007) and global levels (Choat et al. 2012). Interestingly, the slope of this relationship for oaks appears to be similar to the global slope for angiosperms, whereas the intercept is smaller (Fig. 8.7), implying narrower hydraulic

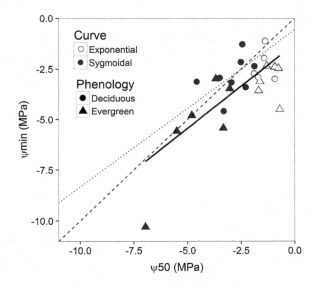

Fig. 8.7 Relationship between hydraulic safety (Ψ_{50}) and minimum leaf water potential (Ψ_{min}) for 25 oak species. Symbol colour indicates differences between vulnerability curve shapes (exponential vs. sigmoidal) and symbol shape distinguishes between leaf phenology (deciduous vs. evergreen). The overall linear fit to the data is shown as a continuous black line. The dotted line indicates the global fit for angiosperms from Choat et al. (2012), and the dashed line is the 1:1 relationship

safety margins (*HSM*) for oaks. The relatively narrow hydraulic safety margins for oaks are robust to differences in the shape of vulnerability curves. The slope of the relationship, however, must be interpreted with caution, as our data shows no significant relationship between Ψ_{min} and Ψ_{50} for exponential curves, whereas the slope of the relationship for sigmoidal curves is highly significant and steeper (1.3, 95% confidence intervals: 0.8–1.8) than the overall slope for angiosperms (0.8; Choat et al. 2012). A slope >1 implies that the hydraulic safety margin is narrower for species having lower Ψ_{50} or experiencing more negative Ψ_{min}.

The relationship between Ψ_{min} and Ψ_{50} was similar for temperate and Mediterranean/semi-arid oak species, even though the latter species tended to have lower Ψ_{min} and Ψ_{50} and, thus, be located towards the bottom-left corner of the graph (not shown). On the other hand, the compiled oak data show a marginal effect of leaf phenology on the relationship between Ψ_{min} and Ψ_{50} ($P = 0.071$), with both species types having similar slopes but evergreen species having narrower hydraulic safety margins (0.9 MPa on average) (Fig. 8.7). This result and particularly the fact that hydraulic safety margins are nearly always negative for evergreens (regardless of the shape of the vulnerability curve) is rather counter-intuitive. Note, however, that these narrower hydraulic safety margins for evergreens do not arise from greater vulnerability to embolism (rather the opposite, as we have seen before) but from lower minimum water potentials at a given xylem vulnerability. This pattern suggests either a more risk-averse strategy for deciduous species or adaptations in other parts of the soil-plant-atmosphere-continuum allowing evergreen species to sustain proportionally greater xylem stress. These adaptations may include a higher Huber value and, thus, higher leaf-specific hydraulic conductivity K_L in evergreens (e.g. Cavender-Bares and Holbrook 2001) or systematic differences in stomatal or rooting behaviour. As shown in the previous section, across 32 species a trend of increased Huber value in evergreen species was indeed found, supporting the explanation above.

8.4 Concluding Remarks

Oak species are very diverse in their ecology and life history, and this diversity is also reflected in their xylem anatomy and function. In this chapter, we have reviewed the main aspects of the xylem structural properties at vessel and tissue level, as well as the hydraulics of oaks. We have done so by combining qualitative reviews from the primary literature when data availability is relatively low or disperse, with quantitative syntheses for the variables for which consolidated databases already existed, some of which have been expanded substantially. We admit that by working with heterogeneous data sources generally covering a small percentage of overall oak diversity, as well as by ignoring intra-specific variability (our analyses are focused at the species level), we are making important assumptions that may limit our capacity to draw general conclusions. Despite these caveats, some general patterns emerged that are summarized in the following paragraphs.

Some of these patterns appeared to be specific for oaks and, in our opinion, merit further investigation.

From a wood anatomical perspective, oaks are characterized by relatively low vessel density and medium to high vessel sizes, which result in low lumen fractions and, everything else being equal, high mechanical strength. Although hydraulic conductivity is highly sensitive to the lumen fraction (Zanne et al. 2010), oaks do not seem to pay an important price in terms of transport efficiency, as they are characterized by relatively high sapwood-specific hydraulic conductivity ($K_{S,max}$) relative to other angiosperms. This is likely related to the presence of long vessels in many oak species and it might be related to relatively high permeability of their inter-vessel pit connections. Long vessels contribute towards increasing the connectivity of the xylem network (sensu Loepfe et al. 2007). However, the high percentage of solitary vessels, characteristic of oaks, has the opposite influence. The net effect of these two patterns on the connectivity of the xylem network remains to be elucidated as is the role of tracheids in conduction. High $K_{S,max}$ in oaks tends to be associated with high investment in leaf area per unit of cross-sectional sapwood area (low Huber value), similar to the pattern observed when all angiosperms are considered.

Oak vulnerability to embolism (Ψ_{50}) appears to be extremely variable, and its measurement is plagued with methodological issues that limit our capacity to draw general conclusions. Many published values appear hard to recognize with the known ecology of the species, but even when all data were pooled together (while accounting for methodological effects in models) some clear patterns emerged. Firstly, a strong safety–efficiency trade-off between $K_{S,max}$ and Ψ_{50} was observed for oaks, much clearer than the pattern reported when all angiosperm species were considered. Secondly, oaks followed a similar relationship between Ψ_{50} and the leaf minimum water potential experienced in the field (Ψ_{min}) to the one reported for angiosperms (Choat et al. 2012) but with even narrower hydraulic safety margins. Importantly, this pattern held even when only sigmoidal vulnerability curves were considered. This result has important implications in the context of increased frequency and intensity of extreme drought events under ongoing climate change, which has already affected several oak species worldwide (see Chapter on drought-induced oak decline in this book). At the same time, however, many oak species show high resprouting capacity, frequently resulting in fast recovery after canopy dieback (e.g. Lloret et al. 2004).

With regards to patterns of variation within the *Quercus* genus, we report consistent differences among taxonomic groups, particularly between sections *Quercus* and *Lobatae*, on the one hand, and *Cerris*, on the other. The former sections had smaller but more numerous vessels, lower wood density, more vulnerable xylem and generally higher conductivity ($K_{S,max}$). These differences do not necessarily agree with expected relationships between wood anatomy and function, as high densities of smaller vessels are normally interpreted as an indicator of high resistance to embolism (e.g. Pfautsch et al. 2016), illustrating the complexities of making this type of generalizations even when comparing phylogenetically close species within the same genus. These consistent differences among taxonomic groups were partially reflected in broad scale biogeographic patterns, with temperate oak species having higher conduit density and lower wood density than tropical species. Temperate oak species had also higher conductivity ($K_{S,max}$) than Mediterranean/semi-arid species, and tended to be more vulnerable to embolism, consistent with the expectations based on differential drought exposure in these different biomes.

Ring-porosity and leaf phenology are highly associated in oaks, with most evergreen species being diffuse-porous and most deciduous species being ring-porous. In agreement with this, we frequently observed consistent patterns when assessing the effects of these two variables. Although we did not observe clear patterns in vessel anatomy between these groups, probably reflecting limited sample size and methodological heterogeneity, we found clear differences in wood density and hydraulic traits. Ring-porous (deciduous) species had lower wood density, higher hydraulic conductivity ($K_{S,max}$) and were more vulnerable to xylem embolism, consistent with expectations based on the fact that these species do not generally occupy dry environments. How these patterns relate with other stress factors (e.g. low temperatures) and with overall plant resource-use strategies is an important question that merits further investigation.

Acknowledgements We thank Kristoffel Jacobs for his help with data preparation. M.M. and J.M.V. are supported by the Spanish Government (grant CGL2013-46808-R), E.M.R.R. by the EU (Marie Skłodowska-Curie Fellowship—No 659191) and by the Research Foundation—Flanders (FWO, Belgium). J.M.V. benefits from an ICREA Academia award.

References

Adaskaveg JE, Gilbertson RL, Dunlap MR (1995) Effects of incubation time and temperature on in vitro selective delignification of silver leaf oak by *Ganoderma colossum*. App Env Microbiol 61:138–144

Ahmed SA, Chun SK, Miller RB, Chong SH, Kim AJ (2011) Liquid penetration in different cells of two hardwood species. J Wood Sci 57:179–188

Aiba M, Nakashikuza T (2009) Architectural differences associated with adult stature and wood density in 30 temperate tree species. Funct Ecol 23:265–273

Alder NN, Pockman WT, Sperry JS, Nuismer S (1997) Use of centrifugal force in the study of xylem cavitation. J Exp Bot 48:665–674

Babos K (1993) Tyloses formation and the state of health of *Quercus petraea* trees in Hungary. IAWA J 14:239–243

Brodribb TJ (2009) Xylem hydraulic physiology: the functional backbone of terrestrial plant productivity. Plant Sci 177:245–251

Brummer M, Arend M, Fromm J, Schlenzig A, Oßwald WF (2002) Ultrastructural changes and immunocytochemical localization of the elicitin quercinin in *Quercus robur* L. roots infected with *Phytophthora quercina*. Phys Mol Plant Path 61:109–120

Cai J, Tyree MT (2010) The impact of vessel size on vulnerability curves: data and models for within-species variability in saplings of aspen, Populus tremuloides Michx. Plant Cell Environ 33:1059–1069

Campelo F, Nabais C, Gutiérrez E, Freitas H, García-González I (2010) Vessel features of *Quercus ilex* L. growing under Mediterranean climate have a better climatic signal than tree-ring width. Trees 24:463–470
Carlquist S (2001) Comparative wood anatomy. Systematic, ecological, and evolutionary aspects of dicotyledon wood. Springer-Verlag, Berlin Heidelberg, p 448
Cavender-Bares J, Holbrook NM (2001) Hydraulic properties and freezing-induced cavitation in sympatric evergreen and deciduous oaks with contrasting habitats. Plant Cell Environ 24:1243–1256
Cavender-Bares J, Kitajima K, Bazzaz FA (2004) Multiple trait association in relation to habitat differentiation among 17 Floridian oak species. Ecol. Monog. 74:635–662
Cavender-Bares J, Cortes P, Rambal S, Joffre R, Miles B, Rocheteau A (2005) Summer and winter sensitivity of leaves and xylem to minimum freezing temperatures: a comparison of co-occurring Mediterranean oaks that differ in leaf lifespan. New Phytol 168:597–612
Chave J, Coomes D, Jansen S, Lewis SL, Swenson NG, Zanne AE (2009) Towards a worldwide wood economics spectrum. Ecol Lett 12:351–366
Cherubini P, Gartner BL, Tognetti R, Braker OU, Schoch W, Innes JL (2003) Identification, measurement and interpretation of tree rings in woody species from mediterranean climates. Biol Rev 78:119–148
Choat B, Cobb AR, Jansen S (2008) Structure and function of bordered pits: new discoveries and impacts on whole-plant hydraulic function. New Phytol 177:608–626
Choat B, Jansen S, Brodribb TJ, Cochard H, Delzon S, Bhaskar R, Bucci SJ, Feild TS, Gleason SM, Hacke UG et al (2012) Global convergence in the vulnerability of forests to drought. Nature 491:752–755
Choat B, Badel E, Burlett R, Delzon S, Cochard H, Jansen S (2016) Noninvasive measurement of vulnerability to drought-induced embolism by X-Ray microtomography. Plant Physiol 170:273–282
Christman MA, Sperry JS, Smith DD (2012) Rare pits, large vessels and extreme vulnerability to cavitation in a ring-porous tree species. New Phytol 193:713–720
Cochard H, Tyree MT (1990) Xylem dysfunction in *Quercus*: vessel sizes, tyloses, cavitation and seasonal changes in embolism. Tree Physiol 6:393–407
Cochard H, Bréda N, Granier A, Aussenac G (1992) Vulnerability to air embolism of three European oak species (*Quercus petraea* (Matt) Liebl, *Q. pubescens* Willd, *Q. robur* L). Ann For Sci 49:225–233
Cochard H, Herbette S, Barigah T, Badel E, Ennajeh M, Vilagrosa A (2010) Does sample length influence the shape of xylem embolism vulnerability curves? A test with the Cavitron spinning technique. Plant Cell Environ 33:1543–1552
Cochard H, Badel E, Herbette S, Delzon S, Choat B, Jansen S (2013) Methods for measuring plant vulnerability to cavitation: a critical review. J Exp Bot 64:4779–4791
Cochard H, Delzon S, Badel E (2015) X-ray microtomography (micro-CT): a reference technology for high-resolution quantification of xylem embolism in trees. Plant Cell Environ 38:201–206
Cocoletzi E, Angeles G, Ceccantini G, Patrón A, Ornelas JF (2016) Bidirectional anatomical effects in a mistletoe–host relationship: *Psittacanthus schiedeanus* mistletoe and its hosts *Liquidambar styraciflua* and *Quercus germana*. Am J Bot 103:986–997
Corcuera L, Camarero JJ, Gil-Pelegrín E (2002) Functional groups in *Quercus* species derived from the analysis of pressure–volume curves. Trees 16:465–472
Davis SD, Sperry JS, Hacke UG (1999) The relationship between xylem conduit diameter and cavitation caused by freeze-thaw events. Am J Bot 86:1367–1372
Deflorio G, Franz E, Fink S, Schwarze FWMR (2009) Host responses in the xylem of trees after inoculation with six wood-decay fungi differing in invasiveness. Botany 87:26–35
Ebadzad G, Medeira C, Maia I, Martins J, Cravador A (2015) Induction of defence responses by cinnamomins against *Phytophthora cinnamomi* in *Quercus suber* and *Quercus ilex* subs. *rotundifolia*. Eur J Plant Pathol 143:705–723

eFloras (2008). Missouri Botanical Garden, St. Louis, MO, USA and Harvard University Herbaria, Cambridge, MA, USA. Published on the Internet. Accessed in August 2017 at http://www.efloras.org

Ellmore GS, Zanne AE, Orians CM (2006) Comparative sectoriality in temperate hardwoods: hydraulics and xylem anatomy. Bot J Linn Soc 150:61–71

Encyclopedia of Life. Published on the Internet. Accessed in August 2017 at http://www.eol.org

Fan D-Y, Jie S-L, Liu C-C, Zhang X-Y, Xu X-W, Zhang S-R, Xie Z-Q (2011) The trade-off between safety and efficiency in hydraulic architecture in 31 woody species in a karst area. Tree Physiol 31:865–877

Fonti P, Heller O, Cherubini P, Rigling A, Arend M (2013) Wood anatomical responses of oak saplings exposed to air warming and soil drought. Plant Biol 15:210–219

Friedrich M, Remmele S, Kromer B, Hofmann J, Spurk M, Kaiser KF, Orcel C, Küppers M (2004) The 12,460-year Hohenheim oak and pine tree-ring chronology from central Europe—a unique annual record for radiocarbon calibration and paleoenvironment reconstructions. Radiocarbon 46:1111–1122

Fromm JH, Sautter I, Matthies D, Kremer J, Schumacher P, Ganter C (2001) Xylem water content and wood density in spruce and oak trees detected by high-resolution Computed Tomography. Plant Physiol 127:415–426

Gea-Izquierdo G, Martín-Benito D, Cherubini P, Cañellas I (2009) Climate-growth variability in *Quercus ilex* L. west Iberian open woodlands of different stand density. Ann For Sci 66:802–814

Gleason SM, Westoby M, Jansen S, Choat B, Hacke UG, Pratt RB, Bhaskar R, Brodribb TJ, Bucci SJ, Cao K-F et al (2016) Weak tradeoff between xylem safety and xylem-specific hydraulic efficiency across the world's woody plant species. New Phytol 209:123–136

Google Scholar. Published on the Internet. Accessed in August 2017 at https://scholar.google.com

Hacke UG, Sperry JS, Pockman WT, Davis SD, McCulloh KA (2001) Trends in wood density and structure are linked to prevention of xylem implosion by negative pressure. Oecol 126:457–461

Hacke UG, Sperry JS, Wheeler JK, Castro L (2006) Scaling of angiosperm xylem structure with safety and efficiency. Tree Physiol 26:689–701

Hacke UG, Jacobsen AL, Pratt RB (2009) Xylem function of arid-land shrubs from California, USA: an ecological and evolutionary analysis. Plant Cell and Environ 32:1324–1333

Hacke UG, Venturas MD, MacKinnon ED, Jacobsen AL, Sperry JS, Pratt RB (2015) The standard centrifuge method accurately measures vulnerability curves of long-vesselled olive stems. New Phytol 205:116–127

Haneca K, Cufar K, Beeckman H (2009) Oaks, tree-rings and wooden cultural heritage: a review of the main characteristics and applications of oak dendrochronology in Europe. J Arch Sc 36:1–11

Hélardot (1987 onwards) Published on the Internet. Accessed in August 2017 at http://oaks.of.the.world.free.fr/liste.htm

Hargrave KR, Kolb KJ, Ewers FW, Davis SD (1994) Conduit diameter and drought induced embolism in *Salvia mellifera* Greene (Labiatae). New Phytol 126:695–705

InsideWood. 2004-onwards. Published on the Internet. Accessed in August 2017 at http://insidewood.lib.ncsu.edu/search

Jacobs K (2013) The hydraulic system of *Quercus* species in relation to water availability. MSc thesis (promoter: Koedam N, co-promoters: Robert EMR, Beeckman H), Vrije Universiteit Brussel, Brussels, Belgium, 113 pp

Jacobsen AL, Ewers FW, Pratt RB, Paddock WA, Davis SD (2005) Do xylem fibers affect vessel cavitation resistance? Plant Physiol 139:546–556

Jacobsen AL, Pratt RB, Ewers FW, Davis SD (2007) Cavitation resistance among 26 chaparral species of southern California. Ecol Monogr 77:99–115

Jacobsen AL, Brandon Pratt R, Tobin MF, Hacke UG, Ewers FW (2012) A global analysis of xylem vessel length in woody plants. Am J of Bot 99:1583–1591

Jansen S, Choat B, Pletsers A (2009) Morphological variation of intervessel pit membranes and implications to xylem function in angiosperms. Am J Bot 96:409–419

Jansen S, Schuldt B, Choat B (2015) Current controversies and challenges in applying plant hydraulic techniques. New Phytol 205:961–964

Johnson DM, Brodersen CR, Reed M, Domec JC, Jackson RB (2014) Contrasting hydraulic architecture and function in deep and shallow roots of tree species from a semi-arid habitat. Ann Bot 13:617–627

Jupa R, Plavcová L, Gloser V, Jansen S (2016) Linking xylem water storage with anatomical parameters in five temperate tree species. Tree Phys 36:756–769

Kim JS, Daniel G (2016a) Distribution of phenolic compounds, pectins and hemicelluloses in mature pit membranes and its variation between pit types in English oak xylem (*Quercus robur*). IAWA J 37:402–419

Kim JS, Daniel G (2016b) Variations in cell wall ultrastructure and chemistry in cell types of earlywood and latewood in English oak (*Quercus robur*). IAWA J 37:383–401

Kim NH, Hanna RB (2006) Morphological characteristics of *Quercus variabilis* charcoal prepared at different temperatures. Wood Sci Technol 40:392–401

Kitin P, Funada R (2016) Earlywood vessels in ring-porous trees become functional for water transport after bud burst and before the maturation of the current-year leaves. IAWA J 37:315–331

Knops JMH, Koenig WD (1994) Water use strategies of five sympatric species of *Quercus* in central coastal California. Madroño 41:290–301

Kuroda K (2001) Responses of *Quercus* sapwood to infection with the pathogenic fungus new wilt disease vectored by the ambrosia beetle *Platypus quercivorus*. J Wood Sci 47:425–429

Leal S, Sousa VB, Vicelina B, Pereira H (2006) Within and between-tree variation in the biometry of wood rays and fibres in cork oak (*Quercus suber* L.). Wood Sci Technol 40:585–597

Lei H, Milota MR, Gartner BL (1996) Between- and within-tree variation in the anatomy and specific gravity of wood in Oregon white oak (*Quercus garryana* Dougl). IAWA J 17:445–461

Lens F, Sperry JS, Christman MA, Choat B, Rabaey D, Jansen S (2011) Testing hypotheses that link wood anatomy to cavitation resistance and hydraulic conductivity in the genus *Acer*. New Phytol 190:709–723

Limousin J-M, Longepierre D, Huc R, Rambal S (2010) Change in hydraulic traits of Mediterranean *Quercus ilex* subjected to long-term throughfall exclusion. Tree Physiol 30:1026–1036

Lloret F, Siscart D, Dalmases C (2004) Canopy recovery after drought dieback in holm-oak Mediterranean forests of Catalonia (NE Spain). Global Change Biol 10:2092–2099

Loepfe L, Martinez-Vilalta J, Piñol J, Mencuccini M (2007) The relevance of xylem network structure for plant hydraulic efficiency and safety. J Theoret Biol 247:788–803

Lo Gullo MA, Salleo S (1993) Different vulnerabilities of *Quercus ilex* L. to freeze- and summer drought-induced xylem embolism: an ecological interpretation. Plant Cell Environ 16:511–519

Lo Gullo MA, Salleo S, Piaceri EC, Rosso R (1995) Relations between vulnerability to xylem embolism and xylem conduit dimensions in young trees of *Quercus cerris*. Plant Cell Environ 18:661–669

Martínez-Cabrera HI, Jones CS, Espino S, Schenk HJ (2009) Wood anatomy and wood density in shrubs: responses to varying aridity along transcontinental transects. Am J Bot 96:1388–1398

Martínez-Vilalta J, Prat E, Oliveras I, Piñol J (2002) Xylem hydraulic properties of roots and stems of nine Mediterranean woody species. Oecologia 133:19–29

Martínez-Vilalta J, Mencuccini M, Vayreda J, Retana J (2010) Interspecific variation in functional traits, not climatic differences among species ranges, determines demographic rates across 44 temperate and Mediterranean tree species. J Ecol 98:1462–1475

Martínez-Vilalta J, Mencuccini M, Álvarez X, Camacho J, Loepfe L, Piñol J (2012) Spatial distribution and packing of xylem conduits. Am J Bot 99:1189–1196

Martin-StPaul NK, Longepierre D, Huc R, Delzon S, Burlett R, Joffre R, Rambal S, Cochard H (2014) How reliable are methods to assess xylem vulnerability to cavitation? The issue of 'open vessel' artifact in oaks. Tree Phys 34:894–905

McElrone AJ, Pockman WT, Martínez-Vilalta J, Jackson RB (2004) Variation in xylem structure and function in stems and roots of trees to 20 m depth. New Phytol 163:507–517

Medeira C, Quartin V, Maia I, Diniz I, Matos MC, Semedo JN, Scotti-Campos P, Ramalho JC, Pais IP, Ramos P, Melo E (2012) Cryptogein and capsicein promote defence responses in *Quercus suber* against *Phytophthora cinnamomi* infection. Eur J Plant Pathol 134:145–159

Melcher PJ, Michele Holbrook N, Burns MJ, Zwieniecki MA, Cobb AR, Brodribb TJ, Choat B, Sack L (2012) Measurements of stem xylem hydraulic conductivity in the laboratory and field. Methods Ecol Evol 3:685–694

Mencuccini M, Martínez-Vilalta J, Piñol J, Loepfe L, Burnat M, Alvarez X, Camacho J, Gil D (2010) A quantitative and statistically robust method for the determination of xylem conduit spatial distribution. Am J Bot 97:1247–1259

Metcalfe CR, Chalk L (1983) Anatomy of the dicotyledons, vol II. Clarendon Press, Oxford, UK, Wood structure and conclusions of the general introduction, p 1500

Miles PD, Smith WB (2009) Specific gravity and other properties of wood and bark for 156 tree species found in North America. USDA For Serv, Northern Exp. St., Research Note NRS-38, p 35

Mirić M, Popović Z (2006) Structural damages of oak-wood provoked by some stereales–basidiomycetes decaying fungi. Wood Structure and Properties '06. Arbora Publishers, Zvolen, Slovakia, pp 111–115

Monk CD, Imm DW, Potter RL, Parker GG (1989) A classification of the deciduous forest of eastern North America. Vegetatio 80:167–181

Morris H, Plavcová L, Cvecko P, Fichtler E, Gillingham MAF, Martínez-Cabrera HJ, McGlinn DJ, Wheeler E, Zheng J, Ziemińska K, Jansen S (2015) A global analysis of parenchyma tissue fractions in secondary xylem of seed plants. New Phytol 209:1553–1565

Morris H, Brodersen C, Francis WMR, Steven Jansen S (2016) The Parenchyma of secondary Xylem and its critical role in tree defense against fungal decay in relation to the CODIT model. Frontiers in Plant Sci 7

Nardini A, Savi T, Losso A, Petit G, Pacilè S, Tromba G, Mayr S, Trifilò P, Lo Gullo MA, Salleo S (2017) X-ray microtomography observations of xylem embolism in stems of Laurus nobilis are consistent with hydraulic measurements of percentage loss of conductance. New Phyt 213:1068–1075

Návar J (2009) Allometric equations for tree species and carbon stocks for forests of northwestern Mexico. For. Ecol. Manag. 257:427–434

Oberle B, Ogle K, Penagos Zuluaga JC, Sweeney J, Zanne AE (2016) A Bayesian model for xylem vessel length accommodates subsampling and reveals skewed distributions in species that dominate seasonal habitats. J Plant Hydr 3:e-003

Obst JR, Sachs IB, Kuster TA (1988) The quantity and type of lignin in tyloses of bur oak (*Quercus macrocarpa*). Holzforschung 42:229–231

Pan R, Geng J, Jing Cai J, Tyree MT (2015) A comparison of two methods for measuring vessel-length in woody plants. Plant Cell Environ 38:2519–2526

Pfautsch S, Harbusch M, Wesolowski A, Smith R, Macfarlane C, Tjoelker MG, Reich PB, Adams MA (2016) Climate determines vascular traits in the ecologically diverse genus *Eucalyptus*. Ecol Lett 19:240–248

Pockman WT, Sperry JS (2000) Vulnerability to xylem cavitation and the distribution of Sonoran Desert vegetation. Am J Bot 87:1287–1299

Preston KA, Cornwell WK, DeNoyer JL (2006) Wood density and vessel traits as distinct correlates of ecological strategy in 51 California coast range angiosperms. New Phyt 170:807–818

Safdari V, Ahmed M, Palmer J, Baig MB (2008) Identification of Iranian commercial wood with hand lens. Pak J Bot 40:1851–1864

Sano Y, Jansen S (2006) Perforated pit membranes in imperforate tracheary elements of some angiosperms. Ann Bot 97:1045–1053

Sano Y, Ohta T, Jansen S (2008) The distribution and structure of pits between vessels and imperforate tracheary elements in angiosperm woods. IAWA J 29:1–15

Schenk HJ, Steppe K, Jansen S (2015) Nanobubbles: a new paradigm for air-seeding in xylem. Trends Plant Sci 20:199–205

Schmitt U, Liese W (1993) Response of xylem parenchyma by suberization in some hardwoods after mechanical injury. Trees 8:23–30

Schmitt U, Liese W (1995) Wound reactions in the xylem of some hardwoods. Drevarsky Vyskum 40:1–10

Schwilk DW, Brown TE, Lackey R, Willms J (2016) Post-fire resprouting oaks (genus: *Quercus*) exhibit plasticity in xylem vulnerability to drought. Plant Ecol 217:697–710

Siau JF (1984) Transport processes in wood. Springer Verlag, Berlin-Heidelberg-New York, p 245

Sohara K, Vitasb A, Läänelaid A (2012) Sapwood estimates of pedunculate oak (*Quercus robur* L.) in eastern Baltic. Dendrochronologia 30:49–56

Sorz J, Hietz P (2006) Gas diffusion through wood: implications for oxygen supply. Trees 20:34–41

Sousa VB, Leal S, Quilhó T, Pereira H (2009) Characterization of cork oak (*Quercus suber*) wood anatomy. IAWA J 30:149–161

Sperry J, Saliendra NZ (1994) Intra- and inter-plant variation in xylem cavitation in Betula occidentalis. Plant Cell Environ:1233–1241

Sperry JS, Sullivan JEM (1992) Xylem embolism in response to freeze-thaw cycles and water stress in ring-porous, diffuse-porous, and conifer species. Plant Physiol 100:605–613

Sperry JS, Tyree MT (1988) Mechanism of water stress-induced xylem embolism. Plant Physiol 88:581–587

Sperry JS, Christman MA, Torres-Ruiz JM, Taneda H, Smith DD (2012) Vulnerability curves by centrifugation: is there an open vessel artefact, and are 'r' shaped curves necessarily invalid?: Vulnerability curves by centrifugation. Plant Cell Environ 35:601–610

Sperry JS, Venturas MD, Anderegg WRL, Mencuccini M, Mackay DS, Wang Y, Love DM (2017) Predicting stomatal responses to the environment from the optimization of photosynthetic gain and hydraulic cost. Plant Cell Environ 40:816–830

Spicer R, Holbrook NM (2005) Within-stem oxygen concentration and sap flow in four temperate tree species: does long-lived xylem parenchyma experience hypoxia? Plant Cell Environ 28:192–201

Spicer R, Holbrook NM (2007) Parenchyma cell respiration and survival in secondary xylem: does metabolic activity decline with cell age? Plant Cell Environ 30:934–943

Stokke DD, Manwiller FG (1994) Proportions of wood elements in stem, branch, and root wood of black oak (*Quercus velutina*). IAWA J 15:301–310

Taneda H, Sperry JS (2008) A case-study of water transport in co-occurring ring- versus diffuse-porous trees: contrasts in water-status, conducting capacity, cavitation and vessel refilling. Tree Physiol 28:1641–1651

The Plant List. Version 1.1 (2013). Published on the Internet. Accessed in August 2017 at http://www.theplantlist.org

The Wood Database. Published on the Internet. Accessed in August 2017 at http://www.wood-database.com

Torres-Ruiz JM, Jansen S, Choat B, McElrone A, Cochard H, Brodribb TJ, Badel E, Burlett R, Bouche PS, Brodersen C et al (2015) Direct micro-CT observation confirms the induction of embolism upon xylem cutting under tension. Plant Physiol 167:40–43

Trifilò P, Raimondo F, Lo Gullo MA, Barbera PM, Salleo S, Nardini A (2014) Relax and refill: xylem rehydration prior to hydraulic measurements favours embolism repair in stems and generates artificially low PLC values. Plant Cell Environ 37:2491–2499

Trifilò P, Nardini A, Gullo MA, Barbera PM, Savi T, Raimondo F (2015) Diurnal changes in embolism rate in nine dry forest trees: relationships with species-specific xylem vulnerability, hydraulic strategy and wood traits. Tree Phys 35:694–705

Tropicos. Published on the Internet. Missouri Botanical Garden. Wikimedia Foundation. Accessed in August 2017 at http://www.tropicos.org

Tyree MH, Cochard H (1996) Summer and winter embolism in oaks: impact on water relations. Ann Sc For 53:173–180

Tyree MH, Ewers FW (1991) The hydraulic architecture of trees and other woody plants. New Phytol 119:345–360

Tyree MT, Sperry JS (1989) Vulnerability of xylem to cavitation and embolism. Annu Rev Plant Biol 40:19–36
Tyree MT, Zimmermann MH (2002) Xylem structure and the ascent of sap. Springer, New York
Venturas MD, Mackinnon ED, Jacobsen AL, Pratt RB (2015) Excising stem samples underwater at native tension does not induce xylem cavitation. Plant Cell Environ 38:1060–1068
Venturas MD, Rodriguez-Zaccaro FD, Percolla MI, Crous CJ, Jacobsen AL, Pratt RB (2016) Single vessel air injection estimates of xylem resistance to cavitation are affected by vessel network characteristics and sample length (F Meinzer, Ed.). Tree Physiol 36:1247–1259
Vilagrosa A, Bellot J, Vallejo VR, Gil-Pelegrin E (2003) Cavitation, stomatal conductance, and leaf dieback in seedlings of two co-occurring Mediterranean shrubs during an intense drought. J Exp Bot 54:2015–2024
Villar-Salvador P, Castro-Díez P, Pérez-Rontomé C, Montserrat-Martí G (1997) Stem xylem features in three *Quercus* (Fagaceae) species along a climatic gradient in NE Spain. Trees 12:90–96
Wang R, Zhang L, Zhang S, Cai J, Tyree MT (2014) Water relations of *Robinia pseudoacacia* L.: do vessels cavitate and refill diurnally or are R-shaped curves invalid in *Robinia*? Plant Cell Environ 37:2667–2678
Wikipedia: The free encyclopedia. Published on the Internet. Wikimedia Foundation, Inc. Accessed in August 2017 at https://www.wikipedia.org
Wikipedia—List of *Quercus* species. Published on the Internet. Accessed in August 2017 at https://en.wikipedia.org/wiki/List_of_Quercus_species
Wheeler EA, Baas P, Gasson PE (1989) IAWA list of microscopic features for hardwood identification. IAWA Bulletin n.s. 10:219–332
Wheeler JK, Sperry JS, Hacke UG, Hoang N (2005) Inter-vessel pitting and cavitation in woody Rosaceae and other vesselled plants: a basis for a safety versus efficiency trade-off in xylem transport. Plant Cell Environ 28:800–812
Wheeler JK, Huggett BA, Tofte AN, Rockwell FE, Holbrook NM (2013) Cutting xylem under tension or supersaturated with gas can generate PLC and the appearance of rapid recovery from embolism. Plant Cell Environ 36:1938–1949
Whitehead D, Jarvis PG (1981) Coniferous forests and plantations. In: Kozlowski TT (ed) Water deficits and plant growth, Vol. VI, Academic Press, New York, pp. 49–152
Wilson R (2010) The longest tree-ring chronologies in the world, summarised by Rob Wilson. Published on the Internet. Accessed in August 2017 at http://lustiag.pp.fi/data/Advance/LongChronologies.pdf
Yilmaz M, Serdar B, Lokman A, Usta A (2008) Relationships between environmental variables and wood anatomy of *Quercus pontica* C. Koch (Fagaceae). Fresenius Environ Bull 17:902–910
Zanne AE, Lopez-Gonzalez G, Coomes DA, Ilic J, Jansen S, Lewis SL, Miller RB, Swenson NG, Wiemann MC, Chave J (2009) Global wood density database. Dryad. Identifier: http://hdl.handle.net/10255/dryad.235
Zanne AE, Westoby M, Falster DS, Ackerley DD, Loarie SR, Arnold SEJ, Coomes DA (2010) Angiosperm wood structure: global patterns in vessel anatomy and their relation to wood density and potential conductivity. Am J Bot 97:207–215
Zhang YJ, Holbrook NM (2014) The stability of xylem water under tension: a long, slow spin proves illuminating. Plant, Cell Environ 37:2652–2653
Zieminska K, Butler DW, Gleason SM, Wright IJ, Westoby M (2013) Fibre wall and lumen fractions drive wood density variation across 24 Australian angiosperms. AoB Plants 5: plt046
Zimmermann MH (1983) Xylem structure and the ascent of sap. Springer-Verlag, Berlin, Heidelberg, New York, Tokio, p 143

Chapter 9
The Role of Mesophyll Conductance in Oak Photosynthesis: Among- and Within-Species Variability

José Javier Peguero-Pina, Ismael Aranda, Francisco Javier Cano,
Jeroni Galmés, Eustaquio Gil-Pelegrín, Ülo Niinemets,
Domingo Sancho-Knapik and Jaume Flexas

Abstract Oak species show a wide range of variation in key foliage traits determining the leaf economics spectrum, including the leaf dry mass per unit area (LMA) and photosynthetic capacity. Though it is well known that stomatal conductance plays a major role in determining maximum rates of carbon assimilation, other factors such as mesophyll conductance to CO_2 (g_m) can constrain the rate of photosynthesis and, under certain conditions, be the most significant photosynthetic limitation. First, this chapter addresses the differences in the photosynthetic limitations imposed by g_m between deciduous and evergreen oak species, covering the role of variations in several leaf anatomical traits determining the variability in g_m and photosynthetic capacity. This analysis emphasizes that cell-wall thickness of mesophyll cells and the chloroplast surface facing intercellular air spaces are the two main anatomical traits contributing to changes in g_m, and as consequence have a high relevance in the carbon fixing capacity of leaves within the genus *Quercus*. The second part of the chapter analyses the within-species variation of g_m and

---

J. J. Peguero-Pina (✉) · E. Gil-Pelegrín · D. Sancho-Knapik
Unidad de Recursos Forestales, Centro de Investigación y Tecnología
Agroalimentaria de Aragón, Gobierno de Aragón, Avda. Montañana 930,
50059 Zaragoza, Spain
e-mail: jjpeguero@aragon.es

I. Aranda
Department of Forest Ecology and Genetics, Forest Research Centre, INIA,
Avda. A Coruña km 7.5, 28040 Madrid, Spain

I. Aranda
Instituto de Investigaciones Agroambientales y de Economía del Agua (INAGEA),
Palma de Mallorca, Islas Baleares, Spain

F. J. Cano
ARC Centre of Excellence for Translational Photosynthesis, Canberra, Australia

F. J. Cano
Hawkesbury Institute for the Environment, Western Sydney University, Science Road,
Richmond, NSW 2753, Australia

© Springer International Publishing AG 2017
E. Gil-Pelegrín et al. (eds.), *Oaks Physiological Ecology. Exploring the Functional Diversity of Genus* Quercus L., Tree Physiology 7,
https://doi.org/10.1007/978-3-319-69099-5_9

photosynthesis rate in oaks as related to long-term variations in site climate (genetic and plastic variability) and to shorter-term variation in environmental drivers (e.g. drought stress and light availability) and during leaf ontogeny. The results of this analysis demonstrate a very high variability within-species across species range and in response to shorter-term environmental drivers, ultimately underlying the success of several *Quercus* species in many ecosystems worldwide.

9.1 Introduction

9.1.1 A Synopsis About the Photosynthesis in C_3 Plants

The vast majority of plants and, ultimately, the ecosystem primary productivity on Earth depend on the fixation of atmospheric CO_2 from the atmosphere by green tissues, in the process termed photosynthesis. Photosynthesis needs the energy from the sun to reduce the molecule of CO_2 into phosphorylated sugars that may be exported out of the leaf. Significantly, the study of photosynthesis is often divided into the processes that are directly regulated by the light absorption (light reactions), which involve the production of ATP and NADPH in the thylakoid membrane, and those in the chloroplastic stroma catalysed by enzymes of the Calvin-Benson cycle that use the products of the former for the fixation of CO_2 (dark reactions) (Emerson and Arnold 1932). Photosynthesis relies on sunlight, but it can also be considered as a diffusion process, because CO_2 from the atmosphere has to diffuse into the leaf as the CO_2 concentration in the air spaces of the leaf mesophyll and further in the chloroplast stroma where photosynthesis takes place is smaller than that in the external air. The enzyme ribulose 1,5-bisphosphate (RuBP) carboxylase/oxygenase (Rubisco) that is responsible for CO_2 fixation into the first organic carbon intermediate is the driving force depleting chloroplastic CO_2 concentration, and thus, generating this gradient. Other photosynthetic organisms with active CO_2 concentrating mechanisms, e.g. C4 grasses, are able to increase the CO_2 concentration

---

J. Galmés · J. Flexas
Research Group on Plant Biology under Mediterranean conditions—Instituto de Investigaciones Agroambientales y de Economía del Agua (INAGEA), Palma de Mallorca, Spain

J. Galmés · J. Flexas
Departament de Biologia, Universitat de les Illes Balears,
Carretera de Valldemossa km 7.5, 07122 Palma de Mallorca, Spain

Ü. Niinemets
Institute of Agricultural and Environmental Sciences,
Estonian University of Life Sciences, Kreutzwaldi 1, Tartu 51014, Estonia

Ü. Niinemets
Estonian Academy of Sciences, Kohtu 6, Tallinn 10130, Estonia

inside the chloroplasts above the ambient (Raven and Beardall 2016), but oaks belong to the older group of plants performing C3 photosynthesis where as in any enzymatic process the rate of carbon fixation depends on the availability of substrate. Furthermore, O_2 competes for the active site of Rubisco with CO_2 and for any O_2 fixation 0.5 CO_2 is emitted as consequence of photorespiration, fostering the decrease of net carbon fixation at low concentration of CO_2 inside the chloroplasts (Laing et al. 1974).

According to the biochemical model of leaf photosynthesis (Farquhar et al. 1980), the net CO_2 assimilation rate (A_N) is hyperbolically dependent on the availability of CO_2 at the chloroplast stroma (C_c). In plants with the C_3 photosynthetic mechanism, including all *Quercus* species, two different parts are described along the A_N–C_c response curve: at low C_c, A_N is limited by Rubisco activity and C_c, while at high CO_2 where the A_N–C_c response starts to saturate, A_N is limited by the regeneration of the RuBP, to which CO_2 is bound. RuBP regeneration, in turn, is limited by the rate of photosynthetic electron transport that increases hyperbolically with increasing light intensity. Under the current CO_2 atmospheric concentration (C_a) and saturating light, A_N is limited by the activity of Rubisco and is mathematically expressed as:

$$A_N = \frac{B(C_c - 0.5O/S_{C/O})k_{cat}^c}{C_c + K_c^{air}} - R_d \qquad (9.1)$$

where B is the concentration of active Rubisco sites, O is the concentration of oxygen in the chloroplast stroma (assumed to be in equilibrium with that in the atmosphere), $S_{C/O}$ is the Rubisco specificity factor for CO_2/O_2, k_{cat}^c is the carboxylase maximum turnover rate of Rubisco, K_c^{air} is the Rubisco Michaelis-Menten affinity constant for CO_2 under atmospheric conditions, and R_d is the rate of mitochondrial respiration. Apart from R_d, Eq. 9.1 demonstrates that the photosynthetic CO_2 assimilation rate in C_3 plants depends on two main components: (i) the diffusive component: C_c, which is determined by the efficiency of transfer of CO_2 from the atmosphere to the sites of carboxylation and the leaf capacity for CO_2 fixation, and (ii) the biochemical component (B, $S_{C/O}$, k_{cat}^c and K_c^{air}), related to the abundance, activation state and catalytic traits of Rubisco. The present chapter deals with the role of mesophyll conductance (g_m) governing the CO_2 transfer from the leaf sub-stomatal cavities to the chloroplasts and, thus, strongly controlling photosynthesis in oaks.

9.1.2 The Diffusion Factor: Leaf Mesophyll Conductance to CO_2

Two resistances dominate the pathway from the atmosphere to the sites of carboxylation. The first is the stomata, which regulates the diffusion of CO_2 from the atmosphere to the sub-stomatal cavities and the diffusion of H_2O in the opposite direction. The following equation, derived from Fick's first law of diffusion, relates the stomatal conductance (g_s) and the photosynthetic CO_2 assimilation rate:

$$A_N = (C_a - C_i)g_s \qquad (9.2)$$

where C_a and C_i are the atmospheric and sub-stomatal CO_2 concentrations, respectively.

The second part of the CO_2 pathway within the leaves is the diffusion of CO_2 from the sub-stomatal cavities to the sites of carboxylation, and consists of resistances in both the gaseous and liquid phases through intercellular air spaces, cell walls, plasmalemma and chloroplast envelope, and cytosol and chloroplast stroma. All these resistances are collectively referred as the mesophyll resistance and its inverse, the mesophyll conductance (g_m). Analogously to Eq. 9.2, the relationship between g_m and A_N is expressed as:

$$A_N = (C_i - C_c)g_m \qquad (9.3)$$

Evidence has accumulated that g_m is finite and variable, and that spans widely among plant functional groups with contrasting leaf structural properties (Flexas et al. 2008, 2014). For instance, g_m is typically higher in annual plants, intermediate in deciduous plants and lower in evergreens (Flexas et al. 2014). Actually, g_m correlates with leaf structural properties such as the leaf dry mass per unit area (LMA) (Flexas et al. 2008; Niinemets et al. 2009a, b), but also, at a lower structural scale, with the mesophyll and/or chloroplast surface area directly exposed to intercellular air spaces (Tosens et al. 2012a; Tomás et al. 2013; Peguero-Pina et al. 2017a), and due to the structural effects, g_m can also change with chloroplast movements that promotes higher exposed chloroplast surface (Tholen et al. 2008).

Nevertheless, g_m also shows a plastic behavior independently of structural controls, both in the long (days, weeks) and short (minutes, hours) term in response to many environmental variables, including light, temperature, water and CO_2 availability in a similar manner than g_s does (Flexas et al. 2008). Consequently, while the structure sets the physical limit for maximum values of g_m, rapid adjustments of g_m can be attributable to rapid biochemical processes. The most likely candidates for the most dynamic g_m changes are carbonic anhydrase and aquaporins, which facilitate the dissolution and diffusion of CO_2 in the mesophyll compartments (Gillon and Yakir 2000; Heckwolf et al. 2011; Pérez-Martín et al. 2014; Sade et al. 2014; Niinemets et al. 2017).

A finite and variable g_m has important implications for plant ecology and resource use efficiency. From a global perspective, the leaf economics spectrum (LES) describes the coordinated variations in leaf structural, chemical and photosynthetic characteristics (Wright et al. 2004). LES runs from the low return end, characterized by low nitrogen content per dry mass (N_m), low photosynthetic capacity per leaf dry mass (A_{mass}), and high LMA and high leaf longevity, to the high return end characterized by opposite variation in these key leaf traits (Fig. 9.1; see also Wright et al. 2004). Moreover, the LES underpinned by leaf anatomical changes is associated with trade-offs in leaf functional performance (reviewed recently in Onoda et al. 2017). Among these trade-offs are the structural controls on g_m, in particular, leaf thickness and density, i.e. the components of LMA (Niinemets 1999), which set a limit to the maximum g_m in agreement with LES (Onoda et al. 2017). Therefore, interspecific differences in g_m are aligned with the spectrum of ecological strategies arising from LES: fast-growing, high photosynthetic capacity and high g_m species from resource-rich environments versus slow-growing, low photosynthetic capacity and low g_m species with a high persistence under environmental stress and low resource availabilities.

As CO_2 diffusion into the leaf is inevitably bound to the loss of water through the stomata, use of water is intimately related to CO_2 fixation. From Eqs. 9.2 and 9.3 it follows that A_N can be increased by rising g_s, g_m or both. However, increasing g_s would decrease the ratio A_N/g_s, and therefore the intrinsic water use efficiency, and thus, be the strategy for the species in the fast end of the economics spectrum. In contrast, increasing g_m/g_s would favor A_N/g_s and improve the performance in drought-prone environments (Flexas et al. 2013, 2016).

9.2 Oak Photosynthesis in the Context of the Worldwide Leaf Economics Spectrum

Apart from global variation in leaf traits across all species discussed above, there are important variations in foliage characteristics among species within given genus driven by evolutionary adjustments to environment. Furthermore, there are major trait variations within species driven by both plastic and ecotypic variations to site climate (Vasseur et al. 2012; Violle et al. 2012; Niinemets 2015, 2016a). Due to

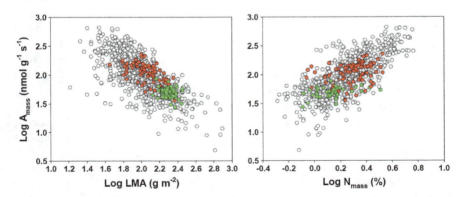

Fig. 9.1 The log-scale relationships between photosynthetic capacity per leaf dry mass (A_{mass}) and leaf dry mass per unit area (LMA) (left panel) and nitrogen content per dry mass (N_m) (right panel). White dots represent the original dataset in Wright et al. (2004), with more than 1000 species from all over the world. Red dots are data for different *Quercus* species, and green dots data for *Quercus ilex* only

differences in selection pressures and environmental heterogeneity, trait variation in within-species economic spectra might somewhat differ from the global trends in trait variation in the worldwide LES (Niinemets 2015; Harayama et al. 2016).

A meta-analysis of leaf trait variation among 39 *Quercus* species (altogether 203 observations obtained from the dataset of Maire et al. 2015) demonstrated a major variation in all key leaf traits with LMA varying 6.3-fold, N_m by 4.2-fold and A_{mass} by 8.2-fold (Fig. 9.1). In comparison with the global LES, different *Quercus* data primarily positioned to the intermediate to lower return end of the spectrum (Fig. 9.1), covering still a high span within the full LES, but consistent with the low growth rate and conservative resource use of *Quercus* species (Reich 2014). This result provides a nice demonstration of how evolutionary modifications within a single genus have resulted in coordinated variation in leaf traits, suggesting that LES is a result of convergent evolution of leaf traits.

Apart from convergence, species ecological strategies, e.g. drought and shade tolerance, can significantly affect LES patterns (Hallik et al. 2009). In fact, analysis of leaf trait data for the Mediterranean evergreen *Q. ilex*—the species with the best data coverage within the *Quercus* genus—indicates that there is still room for differentiation within the general global trends of LES. Even within this single species, LMA varied 2.1-fold, N_m 4-fold and A_{mass} 3.1-fold, and the trait variations were apparently part of the global LES (Fig. 9.1), i.e. similar variations as those described by Niinemets (2015). Yet, the position of leaf traits in bivariate relationships indicates that *Q. ilex* clearly operates at the lower return end of *Quercus* genus. Furthermore, A_{mass} versus N_m response in *Q. ilex* was shallower than that for the global response and for all *Quercus* considered together (Fig. 9.1), suggesting that increases in nutrient availability only moderately enhance photosynthesis for this lower return end species. This evidence collectively indicates that LES is a valuable concept summarizing the overall variation in leaf traits within *Quercus* genus, but also that more species-level analyses are needed to gain insight into the effects of ecological strategies on species deviations from genus-level and global LES.

The recent study by Onoda et al. (2017) has highlighted some mechanistic traits underlying the observed LES, including nitrogen allocation to Rubisco versus cell walls, anatomical traits and mesophyll conductance to CO_2 (g_m). These authors concluded that, globally, gas diffusion limitations associated with low g_m and thick cell walls in species with large LMA explained low photosynthesis in such species better than concomitantly lower N allocation to Rubisco. The fact that the A_{mass} versus N_m response was shallower in *Q. ilex* than the A_{mass} versus LMA response also suggests that, at least for this species, g_m is more limiting for photosynthesis than enzymatic processes associated with carboxylation. Indeed, Peguero-Pina et al. (2017b) have confirmed that g_m is by far the most limiting factor for photosynthesis in a collection of *Q. ilex* provenances, and Peguero-Pina et al. (2017a) have shown that this is a general pattern across *Quercus* species with different leaf life span. The next sections of this chapter are devoted to explain how mesophyll conductance restricts photosynthesis in *Quercus* species, and how g_m in *Quercus* species is regulated at the leaf anatomical level.

9.3 Mesophyll Conductance Constraining Photosynthetic Activity: Evergreen Versus Deciduous Oak Species

The increasing interest of plant physiologists on the role of mesophyll conductance to CO_2 (g_m) in the photosynthetic process is reflected in the number of studies published in recent years addressing the ecophysiological significance of g_m and its regulatory mechanisms (see Flexas et al. 2012 and references therein). Up to now, g_m has been estimated for more than 100 species from all major plant groups, mainly for Spermatophytes (angiosperms and gymnosperms), with only very few data for ferns, liverworts and hornworts (Flexas et al. 2012; Carriquí et al. 2015; Tosens et al. 2016).

The number of studies concerning g_m in *Quercus* is also relatively limited, in spite of the great ecological importance of this genus in the Northern Hemisphere (Breckle 2002). Specifically, g_m has been estimated in ca. 20 evergreen and deciduous oak species, from different sections, climates and geographic locations (see Table 9.1 and references therein). Regarding the deciduous species, several authors have studied g_m variations in some of the most representative oaks widely distributed under temperate-nemoral climate both in Europe (*Q. robur* and *Q. petraea*) and North America (*Q. rubra*). Moreover, the role of g_m in the photosynthetic activity has been also analyzed in deciduous oaks occurring under Mediterranean-type climates, such as *Q. faginea*, *Q. canariensis*, *Q. pyrenaica*, *Q. pubescens* and *Q. garryana*. On the other hand, regarding evergreen oaks, a large number of studies have looked at g_m in *Q. ilex* (see Table 9.1 and references therein), a keystone Circum-Mediterranean tree species which displays a huge morphological and functional variability (Peguero-Pina et al. 2014, 2017b; Niinemets 2015). Recently, Peguero-Pina et al. (2017a) have compared the photosynthetic traits and the different photosynthetic limitations in seven representative Mediterranean oaks from Europe/North Africa and North America. Besides Mediterranean oaks, g_m has been also estimated in other evergreen oak species from warm temperate or tropical mountain climates from East Asia (Table 9.1). Overall, these studies collectively indicate that g_m plays a key role in the photosynthetic process of *Quercus* species, being in many cases the most limiting factor for carbon assimilation.

LMA is one of the most important functional traits that clearly separates evergreen and deciduous oaks (Abrams 1990; Corcuera et al. 2002). There is a general consensus that an increased LMA would negatively affect the mass-based photosynthetic performance of the plant (see Sect. 9.1.2 in this chapter) (Wright et al. 2004, 2005), which has been associated to a higher investment in non-photosynthetic tissues and/or a lower efficiency of the photosynthetically active mesophyll (Niinemets et al. 2009a, b; Hassiotou et al. 2010). However, when considered globally, the relationship between LMA and area-based net CO_2 assimilation (A_N) is less clear and could be modulated by the influence of both diffusive and biochemical factors (e.g. Peguero-Pina et al. 2017a, b).

Table 9.1 List of *Quercus* species with published data for mesophyll conductance to CO_2 (g_m)

Leaf phenology	Species	Section	Origin	Climate	References
Deciduous	*Q. robur*	Quercus	Europe	Temperate	Roupsard et al. (1996), Zhou et al. (2014), Peguero-Pina et al. (2016a)
	Q. petraea	Quercus	Europe	Temperate	Roupsard et al. (1996), Cano et al. (2011, 2013)
	Q. rubra	Lobatae	N America	Temperate	Loreto et al. (1992), Manter and Kerrigan (2004)
	Q. garryana	Quercus	N America	Mediterranean	Manter and Kerrigan (2004)
	Q. faginea	Quercus	Europe	Mediterranean	Peguero-Pina et al. (2016a)
	Q. canariensis	Quercus	Europe/N Africa	Mediterranean	Warren and Dreyer (2006)
	Q. pyrenaica	Quercus	Europe	Mediterranean	Cano et al. (2011)
	Q. pubescens	Quercus	Europe	Mediterranean	Zhou et al. (2014), Sperlich et al. (2015)
Evergreen	*Q. ilex*	Ilex	Europe/N Africa	Mediterranean	Loreto et al. (1992), Roupsard et al. (1996), Peña-Rojas et al. (2005), Niinemets et al. (2005, 2006), Limousin et al. (2010), Fleck et al. (2010), Gallé et al. (2011), Zhou et al. (2014), Sperlich et al. (2015), Varone et al. (2016), Peguero-Pina et al. (2017a, 2017b)
	Q. coccifera	Ilex	Europe/N Africa	Mediterranean	Peguero-Pina et al. (2016b, 2017a)
	Q. suber	Cerris	Europe/N Africa	Mediterranean	Peguero-Pina et al. (2017a)
	Q. chrysolepis	Protobalanus	N America	Mediterranean	Peguero-Pina et al. (2017a)
	Q. agrifolia	Lobatae	N America	Mediterranean	Peguero-Pina et al. (2017a)
	Q. wislizeni	Lobatae	N America	Mediterranean	Peguero-Pina et al. (2017a)
	Q. phillyraeoides	Ilex	E Asia	Warm temperate	Hanba et al. (1999)
	Q. aquifolioides	Ilex	E Asia	Tropical mountain	Feng et al. (2013)
	Q. spinosa	Ilex	E Asia	Tropical mountain	Shi et al. (2015)
	Q. guyavifolia	Ilex	E Asia	Tropical mountain	Huang et al. (2016)
	Q. glauca	Cyclobalanopsis	E Asia	Warm temperate	Hanba et al. (1999)

Leaf phenological type (winter-deciduous, evergreen), infrageneric classification (according to Denk and Grimm 2010), geographic origin and climate are also included. Bibliography has been incorporated in the references list at the end of the chapter

In oaks, a meta-analysis of photosynthetic traits indicated that there were no significant differences in A_N between evergreen and deciduous species (altogether 19 *Quercus* species), despite higher LMA in the evergreens (Fig. 9.2, $P < 0.01$). Stomatal conductance (g_s) did not differ between evergreen and deciduous oak species (Fig. 9.2), but g_m was lower for evergreen oak species (Fig. 9.2, $P < 0.1$). On the other hand, the maximum velocity of carboxylation ($V_{c,max}$) and maximum capacity for photosynthetic electron transport (J_{max}) were slightly higher for evergreens, although the differences were not statistically significant (Fig. 9.2). Flexas et al. (2014) found that Mediterranean evergreen plants compensate their lower g_m with a larger $V_{c,max}$ to reach similar A_N values to other plant groups.

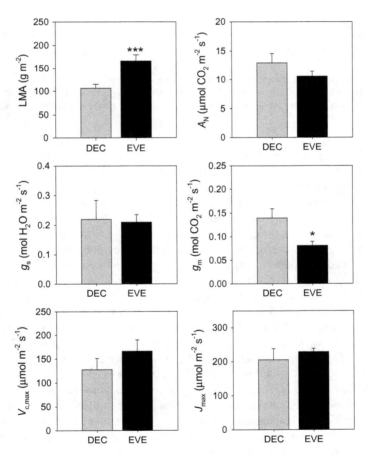

Fig. 9.2 Leaf dry mass per unit area (LMA), net CO_2 assimilation rate (A_N), stomatal conductance for H_2O (g_s), mesophyll conductance (g_m), maximum velocity of carboxylation ($V_{c,max}$) and maximum capacity for photosynthetic electron transport (J_{max}) for evergreen (EVE) and deciduous (DEC) *Quercus* species (see Table 9.1 for data sources). Data are mean ± SE. Asterisk and triple asterisk indicate significant differences between groups at $P < 0.1$ and $P < 0.01$, respectively

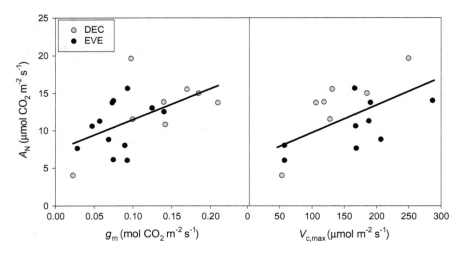

Fig. 9.3 Relationships between mesophyll conductance (g_m) and net CO_2 assimilation rate (A_N) (left panel) and between maximum velocity of carboxylation ($V_{c,max}$) and net CO_2 assimilation rate (A_N) (right panel) for deciduous (DEC) and evergreen (EVE) *Quercus* species (data sources in Table 9.1)

Moreover, these authors also stated that both lower g_m and higher $V_{c,max}$ were associated with larger LMA in Mediterranean evergreen plants. The predominant role of both diffusive and biochemical factors in oak photosynthesis was also evidenced by the existence of statistically significant relationships between A_N and g_m ($R^2 = 0.29$, $P < 0.05$) and between A_N and $V_{c,max}$ ($R^2 = 0.36$, $P < 0.05$) across the compiled set of data (Fig. 9.3).

Besides these general trends, the striking absence of differences in A_N between evergreen and deciduous oaks despite the great differences in LMA can be explained by considering the contrasting relationships found between LMA and photosynthetic traits for each group (Fig. 9.4). In evergreen oaks, we found that A_N positively scaled with LMA ($R^2 = 0.44$, $P < 0.05$), which could be related to the existence of a synergistic effect caused by (i) the absence of influence of LMA on g_m and (ii) the strong positive influence of LMA on $V_{c,max}$ ($R^2 = 0.79$, $P < 0.05$) (Fig. 9.4). Therefore, although g_m is the most limiting factor for A_N in many evergreen species (Roupsard et al. 1996; Piel 2002; Flexas et al. 2014; Galmés et al. 2014; Niinemets and Keenan 2014), $V_{c,max}$ would allow evergreen oaks reaching similar values of net assimilation rate than those in deciduous species. The explanation for the compensating greater $V_{c,max}$ with increasing foliage robustness in evergreen oaks could be a higher content of nitrogen per area, probably as a result of overall greater number of cell layers and greater leaf volume in thicker leaves (Niinemets 1999, 2015; Flexas et al. 2014; Peguero-Pina et al. 2017b), or a higher proportional investment of nitrogen in photosynthetic enzymes within leaves than in other evergreen woody species (Harayama et al. 2016). However, given that the fractional investment of N in Rubisco is typically less in leaves with greater

LMA, the latter possibility seems less likely (Niinemets 1999; Onoda et al. 2017). On the other hand, in deciduous oaks, a negative relationship was found between LMA and g_m ($R^2 = 0.40$, $P < 0.1$; Fig. 9.4) By contrast, no significant relationships were found between LMA and $V_{c,max}$ and between LMA and A_N (Fig. 9.4). In other words, the possible "benefits" of increased LMA observed in evergreen oak species were apparently absent in deciduous ones, further underscoring the contrasting trait correlation networks.

In summary, deciduous oaks showed a negative scaling between g_m and LMA, which could be explained by (i) an increase in the resistance of the CO_2 gas-phase conductance associated with greater leaf thickness and density as LMA increases (Niinemets et al. 2009a, b; Hassiotou et al. 2010; Tosens et al. 2012a; Tomás et al. 2013) and (ii) an increase in the resistance of the CO_2 liquid-phase conductance due to the thicker cell walls observed in species with high leaf density (Peguero-Pina et al. 2012; Tosens et al. 2012a; Tomás et al. 2013). However, although these generic limitations to g_m would be also present in evergreen oaks as LMA increases, we did not find a relationship between LMA and g_m for this group of species. This fact seems to indicate the existence of other anatomical compensatory factors at cell-level that can counteract the negative influence of LMA in g_m in evergreen oaks as discussed in the next section.

9.4 The Role of Leaf Anatomy in the Variability of g_m in *Quercus*

As stated above, g_m is often the most significant photosynthetic limitation, and its variability is strongly driven by different leaf structural traits, such as LMA (Flexas et al. 2012; Galmés et al. 2014; Niinemets and Keenan 2014). Besides LMA, there are other leaf anatomical traits that quantitatively determine the variability in g_m and photosynthetic capacity among species and that can modulate the relationships of g_m with LMA, leaf thickness, leaf density and cell-wall thickness (see Fig. 9.5). Among these traits, different studies have demonstrated major roles of the packing of mesophyll cells relative to the distance and position of stomata, the mesophyll and chloroplast surface area exposed to intercellular air space per unit leaf area (S_m/S and S_c/S, respectively), and the chloroplast size (Terashima et al. 2011; Tosens et al. 2012b; Tomás et al. 2013).

Regarding oaks, a few studies have analysed how leaf anatomy and ultrastructure affect interspecific variability in g_m and photosynthetic capacity. Hanba et al. (1999) analyzed the influence of mesophyll thickness on g_m for several evergreen tree species in Japanese warm-temperate forests, including *Q. glauca* and *Q. phillyraeoides*. These authors found that leaves with thicker mesophyll (e.g. *Q. phillyraeoides*) tended to have larger S_m/S values and smaller volume ratio of intercellular air spaces (i.e. lower

Fig. 9.4 Relationships between leaf dry mass per unit area (LMA) and (i) net CO_2 assimilation rate (A_N) (upper panel), (ii) mesophyll conductance (g_m) (medium panel) and (iii) maximum velocity of carboxylation ($V_{c,max}$) (lower panel) for deciduous (DEC) and evergreen (EVE) *Quercus* species (Table 9.1 for data sources). Grey and black lines indicate significant relationships at $P < 0.05$ for DEC and EVE, respectively

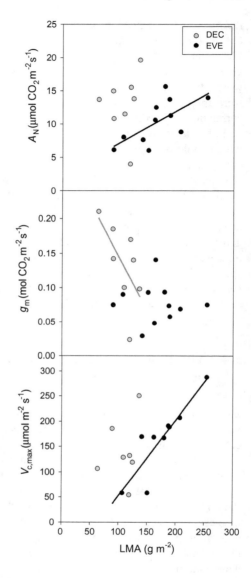

mesophyll porosity), which implied a greater g_m in thicker leaves with higher LMA values. In this context, Peguero-Pina et al. (2017a) compared the morphological, anatomical and photosynthetic traits and the share of different photosynthetic limitations in seven Mediterranean oaks from Europe/North Africa (*Q. coccifera*, *Q. ilex* subsp. *rotundifolia*, *Q. ilex* subsp. *ilex* and *Q. suber*) and North America (*Q. agrifolia*, *Q. chrysolepis* and *Q. wislizeni*) with contrasting LMA values. They observed that g_m was the most limiting factor for carbon assimilation in these species, accordingly, the variation in A_N across species mainly resulted from the interspecific differences in g_m.

Moreover, they also found a good correlation between measured g_m from combined gas exchange and chlorophyll fluorescence measurements (according to Harley et al. 1992) and simulated g_m estimated from anatomical measurements (Niinemets and Reichstein 2003; Tosens et al. 2012b). This confirms the main role of structural traits in determining photosynthetic differences among these *Quercus* species. The detailed analysis of the contributions of ultracellular and cellular components indicated that even the most limiting component for CO_2 diffusion was the cell wall thickness in all species, but differences in g_m and A_N could be primarily attributed to differences in S_c/S (see Fig. 6 in Peguero-Pina et al. 2017a). An increase in S_c/S was achieved through higher values of S_c/S_m and/or S_m/S, and the increase of the latter was due to an increase in mesophyll thickness and LMA (Peguero-Pina et al. 2017a) similarly to the study of Hanba et al. (1999).

The leading role of leaf anatomy in determining g_m and A_N has also been reported for deciduous *Quercus* species (Peguero-Pina et al. 2016a). Comparing the anatomical and photosynthetic traits of two deciduous oaks from western Eurasia (*Q. robur* and *Q. faginea*), they found that the higher photosynthetic capacity of *Q. faginea* was partly explained by variation in several leaf anatomical traits that decreased the resistance to CO_2 diffusion in the liquid phase (higher S_m/S and S_c/S and lower chloroplast thickness) (Peguero-Pina et al. 2016a).

9.5 Within-Species Variation of Mesophyll Conductance and Photosynthesis in *Quercus*

9.5.1 Genetic Factors

Only few studies have looked at the intraspecific variability in g_m among cultivars of several agronomic (Barbour et al. 2010; Galmés et al. 2013; Muir et al. 2014), among mutants or genotypes of plant model species such as *Arabidopsis thaliana* (Easlon et al. 2014), and among different clones or natural genotypes of forest tree species (Soolanayakanahally et al. 2009; Théroux-Rancourt et al. 2015; Milla-Moreno et al. 2016). This intraspecific variability according to different genetic backgrounds within the same species could be relevant as observed for other functional traits such as water use efficiency inferred from carbon isotopic discrimination in response to drought (Roussel et al. 2009; Ramírez-Valiente et al. 2010), respiration of roots and leaves (Boolstad et al. 2003; Laureano et al. 2008), functionality of photochemistry in response to freezing or water stress (Aranda et al. 2005; Cavender-Bares 2007; Koehler et al. 2012; Camarero et al. 2012) or even plasticity in morpho-functional traits to light environment (Balaguer et al. 2001). However, information on g_m variability within *Quercus* is much scarcer, and up to the present, only a few studies have compared variation in g_m across naturally-occurring genotypes in the Mediterranean evergreen species *Q. ilex* and in the temperate deciduous species *Q. robur*.

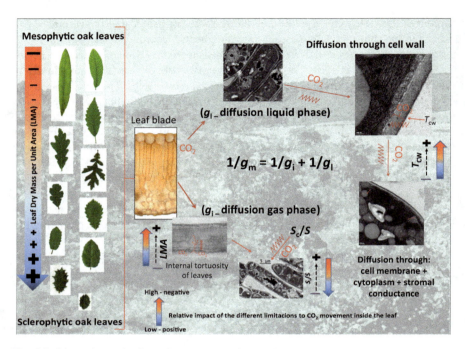

Fig. 9.5 Leaves in species from *Quercus* genus show a high variability in leaf mass per area that increases from typical malacophyllous to sclerophyllous species. This leaf morphological trait impacts in the function of leaves by affecting mesophyll CO_2 resistances in the liquid and gas phases throughout the leaf blade (resistance understood as inverse of the most common term conductance). Increase in LMA has as direct consequence in internal CO_2 movement as result of a higher internal tortuosity of the diffusion pathway. This can be compensated by a higher S_c/S as result of higher tissue density and higher proportion of chloroplasts versus internal surface facing to intercellular spaces. Diffusion in the liquid phase begins by movement of CO_2 across the cell wall matrix where thickness has a fundamental impact in the overall g_m. Cytoplasm and stromal conductance are the two last steps that CO_2 has to overcome before reaching the sites of carbon fixation within chloroplast. Specific aquaporins and carbonic anhydrases have a fundamental role in the kinetic of the CO_2 movement in the liquid phases

Different studies have underscored the great variability exhibited by *Q. ilex* across its circum-Mediterranean distribution, both in terms of morphological and ecophysiological traits such as xylem resistance to drought-induced embolism (see Peguero-Pina et al. 2014 and references therein). In fact, as stated in Sect. 9.2, this species has the best functional data coverage within the *Quercus* genus, showing a major variation in foliage structural, chemical and photosynthetic traits (Niinemets 2015). Thus, Varone et al. (2016) found a range in g_m spanning between 0.09 and 0.06 mol m^{-2} s^{-1} across five ecotypes of *Q. ilex* from Italy grown in the same environmental conditions. In line with this, Peguero-Pina et al. (2017b) observed that the ecotypic variation of photosynthetic characteristics within seven *Q. ilex* provenances from Spain and Italy was driven by changes in ultrastructural/biochemical mesophyll traits. Specifically, their study demonstrated that within-species differences in A_N in *Q. ilex* could be attributed to a synergistic effect between (i) the variation in g_m, by means of changes in several leaf anatomical traits, mainly cell wall thickness, chloroplast thickness and S_c/S, and (ii) the variation in $V_{c,max}$ associated with changes in nitrogen content per area. On the other hand, in contrast to the results found for *Q. ilex*, the genetic control of g_m was not observed by Roussel et al. (2009) in a full-sib family of *Q. robur*.

9.5.2 Environmental Factors: Light Availability and Water Stress

9.5.2.1 Light Availability

The variation in light availability is among the most conspicuous features of plant canopies (Niinemets et al. 2015) and leads to significant acclimatory changes in leaf anatomy that translate to modifications in leaf functional traits such as g_m and A_N (Terashima et al. 2011; Tosens et al. 2012b). Leaves growing in the forest understory usually present lower g_m values than sun leaves (Hanba et al. 2002; Warren et al. 2007; Flexas et al. 2008) and the decreased photosynthetic capacity is considered a common phenomenon in plants growing in the understory (Montpied et al. 2009). Regarding oaks, within-canopy variation in foliage structural and photosynthetic traits in *Q. ilex* has been studied by Niinemets et al. (2006), who found that g_m scaled positively with growth irradiance (Q_{int}) for current-year and 1-year-old leaves of *Q. ilex*. The same results were observed by Cano et al. (2013) when studied the deciduous *Q. petraea* growing in a dense mixed forest with *Fagus sylvatica*. These authors hypothesized that this phenomenon would be explained by a light-dependent increase in S_m/S and S_c/S associated to increases in leaf thickness. Such adaptive modifications in response to light availability were observed by Peguero-Pina et al. (2016c), who compared g_m and anatomical traits of *Abies pinsapo* needles grown in the open field and in the understory, and by Tosens et al. (2012b), who looked at g_m and associated anatomical traits in *Populus tremula* grown under different light availabilities. Similarly, Cano et al. (2011) found that leaves of *Q. petraea* and *Q. pyrenaica* grown under high light showed greater A_N and g_m values than those grown under low light. On the contrary, A_N and g_m in *Q. coccifera* did not exhibit a plastic response to changes in Q_{int}, despite the plasticity showed in several anatomical leaf traits (Peguero-Pina et al. 2016b). This non-plastic response might be explained by a trade-off between (i) an increased conductance of the gas phase (due to the higher fraction of the mesophyll tissue occupied by intercellular air spaces and lower mesophyll thickness) and (ii) a reduced conductance of the liquid phase (due to the lower S_m/S) in *Q. coccifera* leaves grown under low Q_{int} when compared with leaves grown under high Q_{int}.

9.5.2.2 Water Stress

Different studies have suggested that reduced g_m is one of the main factors limiting photosynthetic activity under water stress conditions (Flexas et al. 2004, 2012; Cano et al. 2014; Niinemets and Keenan 2014), including different *Quercus* species. Regarding deciduous oaks, this topic has been studied in *Q. robur* and *Q. petraea*, two of the most representative oaks widely distributed in Europe under temperate-nemoral climate (www.euforgen.org). Grassi and Magnani (2005) observed that g_m declined in response to drought in *Q. robur*, pointing out the importance of this photosynthetic component in the response of photosynthesis to water stress in oaks. In line with this, Cano et al. (2013) showed that g_m decreased due to summer drought in *Q. petraea* and hypothesized that this fact could be related to changes in aquaporin conductance and/or carbonic anhydrase expression instead of anatomical changes. It should be noted that, to the best of our knowledge, this is the only study addressing the impact of water stress on g_m according to leaf position with the canopy. These authors concluded that the effect of water stress on A_N were more important in shaded leaves than in sun-exposed leaves, and although the intensity of water stress was moderate a clear trend revealed that stomatal limitation was more important in leaves from the top canopy, but mesophyll and biochemical limitations dominated the decrease of A_N in the shaded canopy. Regarding evergreen oaks, some studies have dealt with the response of g_m to drought in *Q. ilex*. For instance, Limousin et al. (2010) and Gallé et al. (2011) found that g_m and A_N decreased markedly with soil water stress during summer, with a high degree of co-regulation between both traits in this species. More recently, Zhou et al. (2014) compared the response of photosynthetic traits of *Q. ilex* to water stress with other Mediterranean (*Q. pubescens*) or temperate-nemoral (*Q. robur*) deciduous oaks. This study found that the decrease of g_m associated with water stress was influenced by the different climate of origin, as species from more mesic habitats (*Q. robur*) experienced larger rates of decline of g_m than those from more xeric habitats (*Q. ilex* and *Q. pubescens*). In any case, although drought-induced g_m decrease is a common response in *Quercus* species, more research is needed to elucidate the underlying anatomical and/or biochemical factors that would explain the drought-dependent reduction in g_m.

9.5.3 Leaf Ageing

Major changes in leaf structure and physiology occur during leaf growth and senescence (Niinemets et al. 2012). In particular, in young developing leaves, g_m is limited by low S_c/S_m and tight packing of leaf gas phase, and g_m further increases with leaf maturation (Tosens et al. 2012b). With leaf senescence, leaf photosynthetic capacities declines due to coordinated dismantling of the components of leaf photosynthetic machinery and this is associated with decreased g_m (Evans and Vellen 1996), but changes in g_m during leaf growth and senescence have not yet been studied in oaks. On the other hand, in evergreen species, there are important modifications in foliage functional traits through leaf life span (Niinemets et al. 2012; Niinemets 2016b; Kuusk et al. 2017). To the best of our knowledge, only Niinemets et al. (2005, 2006) have considered the age-dependent decline in g_m in older leaves of evergreen oak species. They demonstrated that g_m is much lower in older (0.015–0.04 mol CO_2 m^{-2} s^{-1}) than in younger leaves (0.05–0.1 mol CO_2 m^{-2} s^{-1}) of *Q. ilex*, and concluded that net CO_2 assimilation is more constrained by g_m in older leaves (2–5-year old) compared with current-year and one-year-old leaves. As discussed by Niinemets et al. (2005), the mechanistic explanation for this decrease in g_m and photosynthetic potential with leaf age may be related to (i) a decrease in S_c/S and (ii) an increase in cell wall lignification and total amount of cell wall. Because older leaves comprise often more than 50% of canopy leaves in evergreen species, such age-dependent changes in the functional leaf activity in evergreens are important to consider in predicting the whole canopy photosynthesis.

9.6 Concluding Remarks

This chapter demonstrates large inter- and intraspecific variation in foliage photosynthesis potential and g_m in *Quercus* species. The variation in leaf functional traits within *Quercus* species is broadly consistent with the worldwide leaf economics spectrum (Wright et al. 2004), but *Quercus* species typically allocate to the lower return end of the leaf economic spectrum. This is consistent with the evidence that most *Quercus* species are early- to mid-successional and moderately to very drought tolerant (Niinemets and Valladares 2006). Nevertheless, consistent with this broad ecological niche range, there is evidence of functional differentiation of *Quercus* species within the worldwide leaf economics spectrum.

The genetic (ecotypic) variation among genotypes from different parts of species range appeared to be a relatively moderate determinant of g_m in *Quercus*, but data are very limited, and clearly, more common garden studies on variation of foliage photosynthetic traits among different genotypes across species range is needed to gain an insight into the extent of genetic variation in g_m and associated structural and physiological traits in oaks. Data coverage of plastic changes in g_m and associated traits in response to differences in light availability, to water stress and in leaves of different age is also limited, although a major variation, several-fold for g_m, has been demonstrated for several oak species. Yet, the sources of this plastic variation remain larger unknown and we conclude that more information is needed about how the plastic modifications in g_m to changes in environmental drivers and leaf age in oak species are driven by changes in leaf anatomy and ultrastructure. Moreover, how far the plasticity of different anatomical factors responsible for plastic modifications in g_m is species-dependent is a high-priority topic for further research.

Acknowledgements The work of Domingo Sancho-Knapik is supported by a DOC INIA contract co-funded by INIA and the European Social Fund (ESF). This research was supported by the grants RTA2015-00054-C02-01 of the INIA (Spain) and SEDIFOR-AGL2014-57762-R of the MINECO (Spain).

References

Abrams MD (1990) Adaptations and responses to drought in *Quercus* species of North America. Tree Physiol 7:227–238

Aranda I, Castro L, Alía R, Pardos JA, Gil L (2005) Low temperature during winter elicits differential responses among populations of the Mediterranean evergreen cork oak (*Quercus suber*). Tree Physiol 25:1085–1090

Balaguer L, Martínez-Ferri E, Valladares F, Pérez-Corona ME, Baquedano FJ, Castillo FJ, Manrique E (2001) Population divergence in the plasticity of the response of *Quercus coccifera* to the light environment. Funct Ecol 15:124–135

Barbour MM, Warren CR, Farquhar GD, Forrester G, Brown H (2010) Variability in mesophyll conductance between barley genotypes, and effects on transpiration efficiency and carbon isotope discrimination. Plant Cell Environ 33:1176–1185

Boolstad PV, Reich P, Lee T (2003) Rapid temperature acclimation of leaf respiration rates in *Quercus alba* and *Quercus rubra*. Tree Physiol 23:969–976

Breckle SW (2002) Walter's vegetation of the earth. Springer, Berlin

Camarero JJ, Olano JM, ArroyoAlfaro SJ, Fernández-Marín B, Becerril JM, García-Plazaola JI (2012) Photoprotection mechanisms in *Quercus ilex* under contrasting climatic conditions. Flora 207:557–564

Cano FJ, Sánchez-Gómez D, Gascó A, Rodríguez-Calcerrada J, Gil L, Warren CR, Aranda I (2011) Light acclimation at the end of the growing season in two broadleaved oak species. Photosynthetica 49:581–592

Cano FJ, Sánchez-Gómez D, Rodríguez-Calcerrada J, Warren CR, Gil L, Aranda I (2013) Effects of drought on mesophyll conductance and photosynthetic limitations at different tree canopy layers. Plant Cell Environ 36:1961–1980

Cano FJ, López R, Warren CR (2014) Implications of the mesophyll conductance to CO_2 for photosynthesis and water-use efficiency during long-term water stress and recovery in two contrasting *Eucalyptus* species. Plant Cell Environ 37:2470–2490

Carriquí M, Cabrera HM, Conesa MÀ, Coopman RE, Douthe C, Gago J, Gallé A, Galmés J, Ribas-Carbó M, Tomás M, Flexas J (2015) Diffusional limitations explain the lower photosynthetic capacity of ferns as compared with angiosperms in a common garden study. Plant Cell Environ 38:448–460

Cavender-Bares J (2007) Chilling and freezing stress in live oaks (*Quercus* section Virentes): intra- and inter-specific variation in PSII sensitivity corresponds to latitude of origin. Photosynth Res 94:437–453

Corcuera L, Camarero JJ, Gil-Pelegrín E (2002) Functional groups in *Quercus* species derived from the analysis of pressure-volume curves. Trees 16:465–472

Denk T, Grimm GW (2010) The oaks of western Eurasia: traditional classifications and evidence from two nuclear markers. Taxon 59:351–366

Easlon HM, Nemali KS, Richards JH, Hanson DT, Juenger TE, McKay JK (2014) The physiological basis for genetic variation in water use efficiency and carbon isotope composition in *Arabidopsis thaliana*. Photosynth Res 119:119–129

Emerson R, Arnold W (1932) The photochemical reaction in photosynthesis. J Gen Physiol 16:191–205

Evans JR, Vellen L (1996) Wheat cultivars differ in transpiration efficiency and CO_2 diffusion inside their leaves. In: Ishii R, Horie T (eds) Crop research in Asia: achievements and perspectives. Proceedings of the 2nd Asian crop science conference "toward the improvement of food production under steep population increase and global environment change", 21–23 Aug 1995, the Fukui Prefecture University, Fukui, Japan. Crop Science Society of Japan, Asian Crop Science Association (ACSA), Fukui, pp 326–329

Farquhar GD, von Caemmerer S, Berry JA (1980) A biochemical model of photosynthetic CO_2 assimilation in leaves of C_3 species. Planta 149:78–90

Feng Q, Centritto M, Cheng R, Liu S, Shi Z (2013) Leaf functional trait responses of *Quercus aquifolioides* to high elevations. Int J Agric Biol 15:69–75

Fleck I, Peña-Rojas K, Aranda X (2010) Mesophyll conductance to CO_2 and leaf morphological characteristics under drought stress during *Quercus ilex* L. resprouting. Ann For Sci 67:308

Flexas J, Bota J, Loreto F, Cornic G, Sharkey TD (2004) Diffusive and metabolic limitations to photosynthesis under drought and salinity in C_3 plants. Plant Biol 6:269–279

Flexas J, Ribas-Carbó M, Díaz-Espejo A, Galmés J, Medrano H (2008) Mesophyll conductance to CO_2: current knowledge and future prospects. Plant Cell Environ 31:602–621

Flexas J, Barbour MM, Brendel O, Cabrera HM, Carriquí M, Díaz-Espejo A, Douthe C, Dreyer E, Ferrio JP, Gago J, Gallé A, Galmés J, Kodama N, Medrano H, Niinemets Ü, Peguero-Pina JJ, Pou A, Ribas-Carbó M, Tomás M, Tosens T, Warren CR (2012) Mesophyll conductance to CO_2: an unappreciated central player in photosynthesis. Plant Sci 193–194:70–84

Flexas J, Niinemets Ü, Gallé A, Barbour MM, Díaz-Espejo A, Galmés J, Ribas-Carbó M, Rodríguez PL, Rosselló F, Soolanayakanahally R, Tomàs M, Wright IJ, Farquhar GD, Medrano H (2013) Diffusional conductances to CO_2 as a target for increasing photosynthesis and photosynthetic water-use efficiency. Photosynth Res 117:45–59

Flexas J, Díaz-Espejo A, Gago J, Gallé A, Galmés J, Gulías J, Medrano H (2014) Photosynthetic limitations in Mediterranean plants: a review. Environ Exp Bot 103:12–23

Flexas J, Díaz-Espejo A, Conesa MA, Coopman RE, Douthe C, Gago J, Gallé A, Galmés J, Medrano H, Ribas-Carbo M, Tomàs M, Niinemets Ü (2016) Mesophyll conductance to CO_2 and Rubisco as targets for improving intrinsic water use efficiency in C_3 plants. Plant Cell Environ 39:965–982

Gallé A, Flórez-Sarasa I, El Aououad H, Flexas J (2011) The Mediterranean evergreen *Quercus ilex* and the semi-deciduous *Cistus albidus* differ in their leaf gas exchange regulation and acclimation to repeated drought and re-watering cycles. J Exp Bot 14:5207–5216

Galmés J, Ochogavía JM, Gago J, Roldán EJ, Cifre J, Conesa MÀ (2013) Leaf responses to drought stress in Mediterranean accessions of *Solanum lycopersicum*: anatomical adaptations in relation to gas exchange parameters. Plant Cell Environ 36:920–935

Galmés J, Andralojc PJ, Kapralov MV, Flexas J, Keys J, Molins A, Parry MAJ, Conesa MÀ (2014) Environmentally driven evolution of Rubisco and improved photosynthesis and growth within the C_3 genus *Limonium* (Plumbaginaceae). New Phytol 203:989–999

Gillon JS, Yakir D (2000) Internal conductance to CO_2 diffusion and $C^{18}OO$ discrimination in C_3 leaves. Plant Physiol 123:201–213

Grassi G, Magnani F (2005) Stomatal, mesophyll conductance and biochemical limitations to photosynthesis as affected by drought and leaf ontogeny in ash and oak trees. Plant Cell Environ 28:834–849

Hallik L, Niinemets Ü, Wright IJ (2009) Are species shade and drought tolerance reflected in leaf-level structural and functional differentiation in Northern Hemisphere temperate woody flora? New Phytol 184:257–274

Hanba YT, Miyazawa SI, Terashima I (1999) The influence of leaf thickness on the CO_2 transfer conductance and leaf stable carbon isotope ratio for some evergreen tree species in Japanese warm temperate forests. Funct Ecol 13:632–639

Hanba YT, Kogami H, Terashima I (2002) The effect of growth irradiance on leaf anatomy and photosynthesis in *Acer* species differing in light demand. Plant Cell Environ 25:1021–1030

Harayama H, Ishida A, Yoshimura J (2016) Overwintering evergreen oaks reverse typical relationships between leaf traits in a species spectrum. R Soc Open Sci 3:160276

Harley PC, Loreto F, Di Marco G, Sharkey TD (1992) Theoretical considerations when estimating the mesophyll conductance to CO_2 flux by the analysis of the response of photosynthesis to CO_2. Plant Physiol 98:1429–1436

Hassiotou F, Renton M, Ludwig M, Evans JR, Veneklaas EJ (2010) Photosynthesis at an extreme end of the leaf trait spectrum: how does it relate to high leaf dry mass per area and associated structural parameters? J Exp Bot 61:3015–3028

Heckwolf M, Pater D, Hanson DT, Kaldenhoff R (2011) The *Arabidopsis thaliana* aquaporin AtPIP1;2 is a physiologically relevant CO_2 transport facilitator. Plant J 67:795–804

Huang W, Hu H, Zhang S-B (2016) Photosynthesis and photosynthetic electron flow in the alpine evergreen species *Quercus guyavifolia* in winter. Front Plant Sci. doi:10.3389/fpls.2016.01511

Koehler K, Center A, Cavender-Bares J (2012) Evidence for a freezing tolerance-growth rate trade-off in the live oaks (*Quercus* series *Virentes*) across the tropical-temperate divide. New Phytol 193:730–744

Kuusk V, Niinemets Ü, Valladares F (2017) A major trade-off between structural and photosynthetic investments operative across plant and needle ages in three Mediterranean pines. Tree Physiol (in press)

Laing WA, Ogren WL, Hageman RH (1974) Regulation of soybean net photosynthetic CO_2 fixation by the interaction of CO_2, O_2, and ribulose 1,5-diphosphate carboxylase. Plant Physiol 54:678–685

Laureano RG, Lazo YO, Linares JC, Luque A, Martínez F, Seco JI, Merino J (2008) The cost of stress resistance: construction and maintenance costs of leaves and roots in two populations of *Quercus ilex*. Tree Physiol 28:1721–1728

Limousin JM, Misson L, Lavoir AV, Martin NK, Rambal S (2010) Do photosynthetic limitation of evergreen *Quercus ilex* leaves change with long-term increased drought severity? Plant Cell Environ 33:863–875

Loreto F, Harley PC, Di Marco G, Sharkey TD (1992) Estimation of mesophyll conductance to CO_2 flux by three different methods. Plant Physiol 98:1437–1443

Maire V, Wright IJ, Prentice IC, Batjes NH, Bhaskar R, van Bodegom PM, Cornwell WK, Ellsworth D, Niinemets Ü, Ordoñez A, Reich PB, Santiago LS (2015) Global effects of soil and climate on leaf photosynthetic traits and rates. Global Ecol Biogeogr 24:706–717

Manter DK, Kerrigan J (2004) A/C_i curve analysis across a range of woody plant species: influence of regression analysis parameters and mesophyll conductance. J Exp Bot 55:2581–2588

Milla-Moreno EA, McKown AD, Guy RD, Soolanayakanahally RY (2016) Leaf mass per area predicts palisade structural properties linked to mesophyll conductance in balsam poplar (*Populus balsamifera* L.). Botany 94:225–239

Montpied P, Granier A, Dreyer E (2009) Seasonal time-course of gradients of photosynthetic capacity and mesophyll conductance to CO_2 across a beech (*Fagus sylvatica* L.) canopy. J Exp Bot 60:2407–2418

Muir CD, Hangarter RP, Moyle LC, Davis PA (2014) Morphological and anatomical determinants of mesophyll conductance in wild relatives of tomato (*Solanum* sect. *Lycopersicon*, sect. *Lycopersicoides*; Solanaceae). Plant Cell Environ 37:1415–1426

Niinemets Ü (1999) Research review. Components of leaf dry mass per area—thickness and density—alter leaf photosynthetic capacity in reverse directions in woody plants. New Phytol 144:35–47

Niinemets Ü (2015) Is there a species spectrum within the world-wide leaf economics spectrum? Major variations in leaf functional traits in the Mediterranean sclerophyll *Quercus ilex*. New Phytol 205:79–96

Niinemets Ü (2016a) Does the touch of cold make evergreen leaves tougher? Tree Physiol 36:267–272

Niinemets Ü (2016b) Leaf age dependent changes in within-canopy variation in leaf functional traits: a meta-analysis. J Plant Res 129:313–318

Niinemets Ü, Keenan TF (2014) Photosynthetic responses to stress in Mediterranean evergreens: mechanisms and models. Environ Exp Bot 103:24–41

Niinemets Ü, Reichstein M (2003) Controls on the emission of plant volatiles through stomata: a sensitivity analysis. J Geophys Res 108:4211

Niinemets Ü, Valladares F (2006) Tolerance to shade, drought and waterlogging in the temperate dendroflora of the Northern hemisphere: tradeoffs, phylogenetic signal and implications for niche differentiation. Ecol Monogr 76:521–547

Niinemets Ü, Cescatti A, Rodeghiero M, Tosens T (2005) Leaf internal conductance limits photosynthesis more strongly in older leaves of Mediterranean evergreen broad-leaved species. Plant Cell Environ 28:1552–1556

Niinemets Ü, Cescatti A, Rodeghiero M, Tosens T (2006) Complex adjustments of photosynthetic potentials and internal diffusion conductance to current and previous light availabilities and leaf age in Mediterranean evergreen species *Quercus ilex*. Plant Cell Environ 29:1159–1178

Niinemets Ü, Díaz-Espejo A, Flexas J, Galmés J, Warren CR (2009a) Role of mesophyll diffusion conductance in constraining potential photosynthetic productivity in the field. J Exp Bot 60:2249–2270

Niinemets Ü, Wright IJ, Evans JR (2009b) Leaf mesophyll diffusion conductance in 35 Australian sclerophylls covering a broad range of foliage structural and physiological variation. J Exp Bot 60:2433–2449

Niinemets Ü, García-Plazaola JI, Tosens T (2012) Photosynthesis during leaf development and ageing. In: Flexas J, Loreto F, Medrano H (eds) Terrestrial photosynthesis in a changing environment. A molecular, physiological and ecological approach. Cambridge University Press, Cambridge, pp 353–372

Niinemets Ü, Keenan TF, Hallik L (2015) A worldwide analysis of within-canopy variations in leaf structural, chemical and physiological traits across plant functional types. New Phytol 205:973–993

Niinemets Ü, Berry JA, von Caemmerer S, Ort DR, Parry MAJ, Poorter H (2017) Photosynthesis: ancient, essential, complex, diverse ... and in need of improvement in a changing world. New Phytol 213:43–47

Onoda Y, Wright IJ, Evans JR, Hikosaka K, Kitajima K, Niinemets Ü, Poorter H, Tosens T, Westoby M (2017) Physiological and structural tradeoffs underlying the leaf economics spectrum. New Phytol 214:1447–1463

Peguero-Pina JJ, Flexas J, Galmés J, Niinemets Ü, Sancho-Knapik D, Barredo G, Villarroya D, Gil-Pelegrín E (2012) Leaf anatomical properties in relation to differences in mesophyll conductance to CO_2 and photosynthesis in two related Mediterranean *Abies* species. Plant Cell Environ 35:2121–2129

Peguero-Pina JJ, Sancho-Knapik D, Barrón E, Camarero JJ, Vilagrosa A, Gil-Pelegrín E (2014) Morphological and physiological divergences within *Quercus ilex* support the existence of different ecotypes depending on climatic dryness. Ann Bot 114:301–313

Peguero-Pina JJ, Sisó S, Sancho-Knapik D, Díaz-Espejo A, Flexas J, Galmés J, Gil-Pelegrín E (2016a) Leaf morphological and physiological adaptations of a deciduous oak (*Quercus faginea* Lam.) to the Mediterranean climate: a comparison with a closely related temperate species (*Quercus robur* L.). Tree Physiol 36:287–299

Peguero-Pina JJ, Sisó S, Fernández-Marín B, Flexas J, Galmés J, García-Plazaola JI, Niinemets Ü, Sancho-Knapik D, Gil-Pelegrín E (2016b) Leaf functional plasticity decreases the water consumption without further consequences for carbon uptake in *Quercus coccifera* L. under Mediterranean conditions. Tree Physiol 36:356–367

Peguero-Pina JJ, Sancho-Knapik D, Flexas J, Galmés J, Niinemets Ü, Gil-Pelegrín E (2016c) Light acclimation of photosynthesis in two closely related firs (*Abies pinsapo* Boiss. and *Abies alba* Mill.): the role of leaf anatomy and mesophyll conductance to CO_2. Tree Physiol 36:300–310

Peguero-Pina JJ, Sisó S, Flexas J, Galmés J, García-Nogales A, Niinemets Ü, Sancho-Knapik D, Saz MÁ, Gil-Pelegrín E (2017a) Cell-level anatomical characteristics explain high mesophyll conductance and photosynthetic capacity in sclerophyllous Mediterranean oaks. New Phytol 214:585–596

Peguero-Pina JJ, Sisó S, Flexas J, Galmés J, Niinemets Ü, Sancho-Knapik D, Gil-Pelegrín E (2017b) Coordinated modifications in mesophyll conductance, photosynthetic potentials and leaf nitrogen contribute to explain the large variation in foliage net assimilation rates across *Quercus ilex* provenances. Tree Physiol. doi:10.1093/treephys/tpx057

Peña-Rojas K, Aranda X, Joffre R, Fleck I (2005) Leaf morphology, photochemistry and water status changes in resprouting *Quercus ilex* during drought. Funct Plant Biol 32:117–130

Pérez-Martín A, Michelazzo C, Torres-Ruiz JM, Flexas J, Fernández JE, Sebastiani L, Díaz-Espejo A (2014) Regulation of photosynthesis and stomatal and mesophyll conductance under water stress and recovery in olive trees: correlation with gene expression of carbonic anhydrase and aquaporins. J Exp Bot 65:3143–3156

Piel C (2002) Diffusion du CO_2 dans le mésophylle des plantes à métabolisme C_3. Ph.D. thesis, Université Paris XI Orsay, Paris, France

Ramírez-Valiente JA, Valladares F, Gil L, Aranda I (2010) Phenotypic plasticity and local adaptation in leaf ecophysiological traits of 13 contrasting cork oak populations under different water availabilities. Tree Physiol 31:1–10

Raven JA, Beardall J (2016) The ins and outs of CO_2. J Exp Bot 67:1–13

Reich PB (2014) The world-wide 'fast-slow' plant economics spectrum: a traits manifesto. J Ecol 102:275–301

Roupsard O, Gross P, Dreyer E (1996) Limitation of photosynthetic activity by CO_2 availability in the chloroplasts of oak leaves from different species and during drought. Ann Sci For 53:243–254

Roussel M, Dreyer E, Montpied P, Le-Provost G, Guehl JM, Brendel O (2009) The diversity of ^{13}C isotope discrimination in a *Quercus robur* full-sib family is associated with differences in intrinsic water use efficiency, transpiration efficiency, and stomatal conductance. J Exp Bot 60:2419–2431

Sade N, Shatil-Cohen A, Ziv AZ, Christophe MC, Boursiac Y, Kelly G, Granot D, Yaaran A, Lerner S, Moshelion M (2014) The role of plasma membrane aquaporins in regulating the bundle sheath mesophyll continuum and leaf hydraulics. Plant Physiol 166:1609–1620

Shi Z, Haworth M, Feng Q, Cheng R, Centritto M (2015) Growth habit and leaf economics determine gas exchange responses to high elevation in an evergreen tree, a deciduous shrub and a herbaceous annual. AoB Plants 7:plv115

Soolanayakanahally RY, Guy RD, Silim SN, Drewes EC, Schroeder WR (2009) Enhanced assimilation rate and water use efficiency with latitude through increased photosynthetic capacity and internal conductance in balsam poplar (*Populus balsamifera* L.). Plant Cell Environ 32:1821–1832

Sperlich D, Chang CT, Peñuelas J, Gracia C, Sabaté S (2015) Seasonal variability of foliar photosynthetic and morphological traits and drought impacts in a Mediterranean mixed forest. Tree Physiol 35:501–520

Terashima I, Hanba YT, Tholen D, Niinemets Ü (2011) Leaf functional anatomy in relation to photosynthesis. Plant Physiol 155:108–116

Théroux-Rancourt G, Ethier G, Pepin S (2015) Greater efficiency of water use in poplar clones having a delayed response of mesophyll conductance to drought. Tree Physiol 35:172–184

Tholen D, Boom C, Noguchi K, Ueda S, Katase T, Terashima I (2008) The chloroplast avoidance response decreases internal conductance to CO_2 diffusion in *Arabidopsis thaliana* leaves. Plant Cell Environ 31:1688–1700

Tomás M, Flexas J, Copolovici L, Galmés J, Hallik L, Medrano H, Ribas-Carbó M, Tosens T, Vislap V, Niinemets Ü (2013) Importance of leaf anatomy in determining mesophyll diffusion conductance to CO_2 across species: quantitative limitations and scaling up by models. J Exp Bot 64:2269–2281

Tosens T, Niinemets Ü, Vislap V, Eichelmann H, Castro-Díez P (2012a) Developmental changes in mesophyll diffusion conductance and photosynthetic capacity under different light and water availabilities in *Populus tremula*: how structure constrains function. Plant Cell Environ 35:839–856

Tosens T, Niinemets Ü, Westoby M, Wright IJ (2012b) Anatomical basis of variation in mesophyll resistance in eastern Australian sclerophylls: news of a long and winding path. J Exp Bot 63:5105–5119

Tosens T, Nishida K, Gago J, Coopman RE, Cabrera HM, Carriquí M, Laanisto L, Morales L, Nadal M, Rojas R, Talts E, Tomás M, Hanba Y, Niinemets Ü, Flexas J (2016) The photosynthetic capacity in 35 ferns and fern allies: mesophyll CO_2 diffusion as a key trait. New Phytol 209:1576–1590

Varone L, Vitale M, Catoni R, Gratani L (2016) Physiological differences of five Holm oak (*Quercus ilex* L.) ecotypes growing under common growth conditions were related to native local climate. Plant Species Biol 31:196–210

Vasseur F, Violle C, Enquist BJ, Granier C, Vile D (2012) A common genetic basis to the origin of the leaf economics spectrum and metabolic scaling allometry. Ecol Lett 15:1149–1157

Violle C, Enquist BJ, McGill BJ, Jiang L, Albert CH, Hulshof C, Jung V, Messier J (2012) The return of the variance: intraspecific variability in community ecology. Trends Ecol Evol 27:244–252

Warren CR, Dreyer E (2006) Temperature responses of photosynthesis and internal conductance to CO_2: results from two independent approaches. J Exp Bot 57:3057–3067

Warren CR, Löw M, Matyssek R, Tausz M (2007) Internal conductance to CO_2 transfer of adult *Fagus sylvatica*: variation between sun and shade leaves and due to free-air ozone fumigation. Environ Exp Bot 59:130–138

Wright IJ, Reich PB, Westoby M, Ackerly DD, Baruch Z, Bongers F, Cavender-Bares J, Chapin T, Cornelissen JHC, Diemer M, Flexas J, Garnier E, Groom PK, Gulias J, Hikosaka K, Lamont BB, Lee T, Lee W, Lusk C, Midgley JJ, Navas ML, Niinemets Ü, Oleksyn J, Osada N, Poorter H, Poot P, Prior L, Pyankov VI, Roumet C, Thomas SC, Tjoelker MG, Veneklaas E, Villar R (2004) The world-wide leaf economics spectrum. Nature 428:821–827

Wright IJ, Reich PB, Cornelissen JHC, Falster DS, Groom PK, Hikosaka K, Lee W, Lusk CH, Niinemets Ü, Oleksyn J, Osada N, Poorter H, Warton DI, Westoby M (2005) Modulation of leaf economic traits and trait relationships by climate. Global Ecol Biogeogr 14:411–421

Zhou S, Medlyn B, Sabaté S, Sperlich D, Prentice IC (2014) Short-term water stress impacts on stomatal, mesophyll and biochemical limitations to photosynthesis differ consistently among tree species from contrasting climates. Tree Physiol 34:1035–1046

Chapter 10
Carbon Losses from Respiration and Emission of Volatile Organic Compounds—The Overlooked Side of Tree Carbon Budgets

Roberto L. Salomón, Jesús Rodríguez-Calcerrada and Michael Staudt

Abstract The balance between photosynthetic carbon (C) assimilation and C loss via respiration (R), emission of volatile organic compounds (VOCs), and rhizodeposition determines plant net primary production and controls to a large extent ecosystem C budgets. Compared to photosynthesis, the physiology, environmental control and ecological importance of processes involving C release from trees have been less studied; it is the purpose of this review to address these questions in oak trees with special focus on R and VOC emissions. Mass-based leaf dark R scales positively with specific leaf area, nitrogen content and photosynthetic capacity, and it is normally greater in deciduous species than evergreen sclerophyllous ones. Leaf dark R increases with temperature, and is constrained by water shortages; however, the magnitude of these responses may vary at different temporal scales. Similarly, R in woody tissues increases with temperature, although in a hysteretic manner during a diel period. On a seasonal basis, besides temperature, water availability becomes the main abiotic driver of woody tissue R as drought stress down-regulates maintenance and growth metabolic processes in stems and roots. Respiration in foliar and woody tissues is expected to account for about half of photosynthesis; nevertheless, R can largely fluctuate with ontogenetic, biotic and abiotic factors independently of C uptake. Volatile organic compounds have multiple roles in plant-environment interactions and plant-plant signalling. Oak genus is one of the

---

R. L. Salomón (✉)
Laboratory of Plant Ecology, Department of Applied Ecology
and Environmental Biology, Faculty of Bioscience Engineering,
Ghent University, Coupure links 653, 9000 Ghent, Belgium
e-mail: RobertoLuis.SalomonMoreno@UGent.be

R. L. Salomón · J. Rodríguez-Calcerrada
Forest Genetics and Ecophysiology Research Group,
E.T.S. Forestry Engineering, Technical University of Madrid,
Ciudad Universitaria s/n, 28040 Madrid, Spain

M. Staudt
Centre d'Ecologie Fonctionnelle et Evolutive CEFE, CNRS,
UMR 5175, 1919 route de Mende, 34293 Montpellier Cedex 5, France

© Springer International Publishing AG 2017
E. Gil-Pelegrín et al. (eds.), *Oaks Physiological Ecology. Exploring the Functional Diversity of Genus* Quercus *L.*, Tree Physiology 7,
https://doi.org/10.1007/978-3-319-69099-5_10

strongest emitter of isoprenoids, which are the most important VOCs released from plants. Most oak species release isoprene constitutively; however, several oak species distributed around the Mediterranean (mostly evergreen) do not produce isoprene, but alternatively emit monoterpenes or lack constitutive emissions of VOCs. The rate of emission of VOCs from leaves increases with leaf temperature and irradiance, being the derived C loss relative to photosynthesis about 1%, except during heat waves when this percentage may increase up to 5%. Emission of VOCs is constrained by drought-stress to a lesser extent than leaf photosynthesis, thus the relative C loss through VOCs also increases with drought severity. Overall, the hypothesis of homeostatic ratios between plant C gain and C loss, an artefact of our better understanding of photosynthesis in comparison to all these processes that encompass tree C loss, should be revisited to better understand C cycling in oaks and to better predict oak physiological performance under climate change scenarios.

10.1 Introduction

The advance in the understanding of our environment is largely driven by the development of suitable technologies to quantify and explain the subject of study. Research in plant carbon (C) cycling clearly illustrates how methodological feasibility has driven knowledge in a particular direction (Körner 2015): As leaves can be easily enclosed in sealed chambers and gas exchange measured with an array of sophisticated systems (Hunt 2003), research on leaf C assimilation has traditionally held a predominant role in studying tree C budgets to the detriment of C efflux from other tree organs. Atmospheric CO_2 is assimilated by plants through photosynthesis (P) and part of it is released back to the atmosphere and soil through respiration (R), emission of volatile organic compounds (VOCs) and rhizodeposition. The difference between C assimilation and C loss is known as net primary production (NPP) and represents the net gain of C to be invested in plant growth, maintenance, defence, reproduction or storage. Net primary production is a key output of dynamic global vegetation models to predict C exchange between terrestrial ecosystems and the atmosphere. Net primary production is commonly calculated as the difference between P and R (e.g. Waring et al. 1998; Luyssaert et al. 2007; Piao et al. 2010; Rambal et al. 2014):

$$NPP = P - R \qquad (10.1)$$

In this equation, P is a well-known process mechanistically described more than 35 years ago (Farquhar et al. 1980), whereas R is a comparatively understudied C flux, despite its predominant role in ecosystem C balance (Valentini et al. 2000; Amthor 2000). Moreover, VOCs and rhizodeposits are commonly ignored in this conceptual framework despite evidence of a non-negligible contribution to tree C loss (Jones et al. 2004; Bracho-Nunez et al. 2013; Sindelarova et al. 2014). Equation (10.1) should be therefore further developed to integrate all C fluxes between the plant and its environment:

$$NPP = P - R - VOCs - rhizodeposits \qquad (10.2)$$

The biological significance of processes involving C release for survival and regeneration is unquestionable. Briefly, the mitochondrial oxidation of C substrates in all plant living cells produces reducing power [e.g. NAD(P)H from NAD(P)$^+$], C skeleton intermediates, and usable energy (ATP) from ADP and inorganic phosphate to fulfil metabolic requirements. During the numerous reactions of mitochondrial respiration CO_2 is formed, a fraction of which is released to the atmosphere, and another recycled within chloroplasts. VOCs are organic chemicals emitted from all plant organs that play multiple roles in plant reproduction, plant protection, and plant-plant signalling; whereas rhizodeposits consist of a wide range of compounds involved in plant nutrition, plant defence, and signalling between plant roots and surrounding organisms of the rhizosphere.

The genus *Quercus* comprises more than 600 species including deciduous and evergreen trees and shrubs adapted to a broad range of environmental conditions, from semi-arid Mediterranean evergreen woodlands to sub-boreal, temperate and subtropical deciduous forests (Mabberley 2008). Due to its wide distribution in the Northern hemisphere, *Quercus* spp. are extensively surveyed in physiological research and constitute an excellent taxonomic group to study the variability in R and the emission of VOCs among different plant functional types and across environmental conditions. In this chapter we want to draw the attention to an important side of plant C budget that has largely been overlooked in oak species: the release of C from the plant to the environment. We will focus on R and VOCs emission due to the scarce literature on oak rhizodeposition. We aim at (i) explaining the variability in R and VOCs emission among different oak species, along gradients of environmental conditions, and at different temporal scales; (ii) highlighting the relative contribution of R and VOCs emission to oak C budgets; (iii) and summarizing information on the chemical typology, mechanisms of synthesis and release, and ecophysiological significance of VOCs among oak species.

10.2 Plant Respiration

Plant R consists on the mitochondrial oxidation of C substrates to produce usable energy, reducing power, and C skeleton intermediates with the consequent release of CO_2 as a reaction product. Plant R is commonly simplified by a single equation developed in the early 1970s, in which R is partitioned into growth and maintenance processes (McCree 1970; Thornley 1970):

$$R = R_G + R_M = g_R G + m_R W \tag{10.3}$$

where R is respiration rate (mol CO_2 s^{-1}), R_G and R_M are growth and maintenance respiration rates (mol CO_2 s^{-1}), G is growth rate (g new biomass s^{-1}), W is living biomass (g dry mass), g_R is growth respiration coefficient (mol CO_2 (g new biomass)$^{-1}$) and m_R is maintenance respiration coefficient (mol CO_2 (g dry biomass)$^{-1}$ s^{-1}). Despite the magnitude of R, ca. 35 to 80% of P (Amthor 2000), substantial improvements in the mechanistic understanding of respiration are still lacking since the early 1970s (Cannell and Thornley 2000; Amthor 2000; Thornley 2011). Respiratory processes continue to be simplified by the growth-and-maintenance-respiration paradigm, and we are far from understanding respiration at the same detail as we do for photosynthesis (Farquhar et al. 1980). Gifford (2003) stated that *"plant respiratory regulation is too complex for a mechanistic representation in current terrestrial productivity models for carbon accounting and global change research"*. This idea seems to be tacitly accepted and given that the rates of enzymatic reactions involved in R are temperature-dependent, plant R is commonly estimated from a single equation derived from Arrhenius kinetics (Davidson et al. 2006):

$$R = R_b \times Q_{10}(T - T_b)/10 \tag{10.4}$$

where R is the respiration rate at temperature T, R_b is the respiration rate at a basal temperature T_b, and Q_{10} is the relative increase in respiration rate corresponding to a 10 °C temperature rise. Hence dynamic global vegetation models often estimate plant R from temperature data—neglecting other biotic and abiotic regulators of plant R such as water availability, C and nutrient supply, and energy demand—to quantify C fluxes and pools at the global scale (Smith and Dukes 2013; Fatichi et al. 2014). Alternatively, R is sometimes assumed to be a constant fraction of P at the tree scale (Waring et al. 1998; Van Oijen et al. 2010). Nevertheless, there is growing evidence that C cycling is not uniquely driven by C assimilation ("source-driven"), but also by abiotic constraints to plant growth and cell maintenance processes ("sink-driven") (see Körner 2015 for a review), as recently observed in *Q. ilex* (Lempereur et al. 2015), which challenges the assumption of homeostatic R:P ratios. Consequently, we are unable to accurately estimate plant R at different temporal and spatial scales, and to comprehensively understand the regulation of the plant respiratory physiology despite its primary role in tree C cycling, productivity and survival (Atkin and Macherel 2009).

In this section, we distinguish between leaf and woody tissues to better summarize current knowledge on oak R. The physiological functioning of an organ is determined by its particular anatomy and structure, and thus leaves and woody tissues have different energy requirements and C-related expenditures derived from respiratory processes. Moreover, dissimilar methodological approaches to quantify R in different organs have contributed to further distance R research between leaves and woody tissues.

10.2.1 Leaf Respiration

10.2.1.1 Physiology and Variability Among Species

Leaf R is mostly dependent on the availability of ADP, C substrates—lipids, amino acids and mostly carbohydrates—and the amount, position and protein content of mitochondria within leaf cells, particularly mesophyll cells, whose contribution to total leaf R in some species is >90% (Long et al. 2015). Data compilations from plant species across the globe evidence that mass-based leaf dark respiration (R_d) rates—more easily measured and typically higher than leaf R rates in light conditions (Zaragoza-Castells et al. 2007)—are positively related with specific leaf area (SLA), leaf nitrogen (N) concentration and photosynthetic capacity (P_{max}) (Reich et al. 1998; Wright et al. 2006; Reich et al. 2008; Atkin et al. 2015). In the first case, the relationship is due to the lower proportion of structural components in high-SLA leaves; in the second, to the higher amount of respiratory enzymes present in leaves with high N concentrations. Finally, the correlation of R_d with P_{max} can reflect the role of P_{max} in providing the mitochondria with respiratory C substrates, or the importance of N for synthesizing both respiratory and photosynthetic enzymes.

Because oak species encompass a wide gradient of leaf functional characteristics, primarily abridged in the separation of broad-leaf evergreen and deciduous oaks, variability in leaf R_d rates among oak species is also large and partly related with climate conditions. Taking as example the 13 oak species included in the global database of respiration GlobResp (Kattge et al. 2011; Atkin et al. 2015), both area- and mass-based leaf R_d varied by approximately sevenfold. Global patterns across broad climatic gradients evidence that plant species from cold sites exhibit higher rates of leaf R_d than those from warmer sites at a comparable measurement temperature, a pattern that holds for oak species (Kattge et al. 2011; Atkin et al. 2015). Further, plant species from dry sites tend to have higher leaf R_d rates than those from mesic sites (Atkin et al. 2015), with this difference being only partly explained by differences in SLA or P_{max} (Wright et al. 2006). Some adaptations to stress, such as high rates of metabolite turnover can result in high respiratory costs in processes of cell maintenance and repair in cold, high-light or arid environments (Wright et al. 2006; Atkin et al. 2015). Similarly, high N investment in repair compounds could also explain why individuals from colder and drier populations of *Q. ilex* exhibited higher leaf maintenance respiratory costs than those from warmer and wetter populations when grown under the same conditions (Laureano et al. 2008). However, different evolutionary selection pressures, including abiotic and biotic stress factors and competition for resources have shaped the respiratory metabolism in different ways, which precludes a straightforward relationship of leaf R_d with stress resistance. As such, some leaf structural and chemical adaptations to stress, such as low N concentrations and low biomass allocation to metabolic components, can result in low leaf R_d rates in long-lived, sclerophyllous leaves of some oaks growing in nutrient-poor or drought-prone sites. At a comparable temperature of 25 °C, mass-based leaf R_d was lowest in

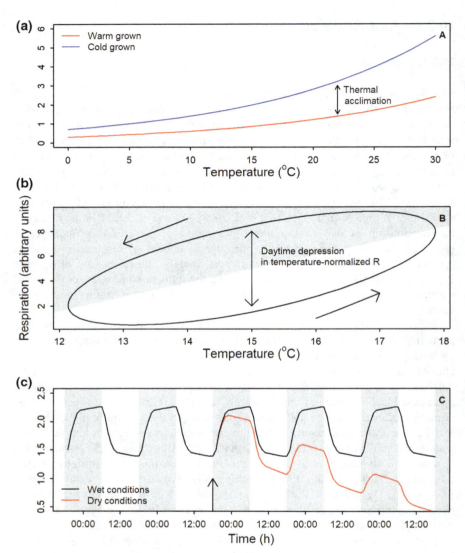

Fig. 10.1 Theoretical representation of the plant respiration (R) response to shifts in temperature (**a**, **b**) and water availability (**c**) at different temporal scales. Respiration exponentially increases with temperature at any temporal scale; however, thermal acclimation (commonly observed in leaves) leads to lower respiratory capacity of tissues grown under warm conditions (**a**, adapted from Atkin and Tjoelker 2003). Hysteresis between R and temperature over diel cycles has been observed for stem R, but not for leaf dark R (even if leaf R at a given temperature is usually lower in light than dark conditions) (**b**, adapted from Salomón et al. 2016b). Improved water status at night-time increases stem R under constant temperature on a diel basis, whereas stem R progressively decreases when water availability becomes limiting on a seasonal basis, as theoretically illustrated during the drought event (the beginning of the drought is represented by the vertical arrow) (**c**, adapted from Saveyn et al. 2007b). Shaded areas indicate night-time

the Mediterranean evergreen *Q. ilex* (4.0 nmol CO_2 g^{-1} s^{-1}) and highest in the deciduous *Q. alba* and *Q. rubra* (25–27 nmol CO_2 g^{-1} s^{-1}). This is consistent with low P_{max} of sclerophyllous leaves, and the positive relationship between P_{max} and leaf R_d (Wright et al. 2006).

10.2.1.2 Response to Environmental Changes at Different Time Scales

Within species, R_d varies in relation to ontogeny and environmental changes. As oak leaves progressively stop growing, both in lamina area and thickness, R_d decreases rapidly and reaches a phase in which fluctuations are not due to ontogenetic changes but mostly to climatic shifts; eventually, when leaves start to senesce, R_d starts to decline again markedly due to remobilization of leaf N and degradation of the respiratory machinery (Collier and Thibodeau 1995; Miyazawa 1998; Xu and Baldocchi 2003; Rodríguez-Calcerrada et al. 2012). Ontogenetic variations over the leaf life span make that, for comparative purposes, measurements of R_d are typically made in non-senescent mature leaves that have fully expanded. However, the responses of leaf physiology and R_d to climatic shifts can vary as affected by ontogeny and age, something that clearly merits more research to improve C balance models (Niinemets 2014).

The plasticity of R_d to air temperature, irradiance or water and nutrient availability is considerable in oak species. Multiple changes occur in the respiratory metabolism in response to the need of the tree to adjust the production of respiratory products to shifting demands imposed by environmental changes. This plasticity in the respiratory metabolism—reflected in varying rates of R_d—allows trees to orchestrate whole-plant plasticity and overcome periods of sub-optimal growing conditions. Two of the most important drivers of leaf physiology over the leaf life span are temperature and water availability.

Changes in temperature elapsed over hours, days, months or years affect the rates of R_d. However, the magnitude of the change in R_d rates depends on the time scale of temperature changes. Short-term raises in temperature provoke an exponential increase in leaf R_d that is typically higher than that occurred over longer term warming periods due to the thermal acclimation of the respiratory metabolism (Reich et al. 2016). This process of acclimation involves a reversible decline in the activity of respiratory enzymes. Frequently, the respiratory capacity (i.e. intercepts of respiratory temperature response curves) differs across temporal scales, probably due to shifts in the amount of mitochondria or mitochondrial enzymes. The consequence of this is that R_d measured at prevailing ambient temperature barely changes over broad, long-term changes in temperature (see Fig. 10.1a for an illustration of this phenomenon, and Slot and Kitajima 2015 for a recent review of the process across biomes and experimental conditions). The thermal acclimation of leaf R (more easily and frequently examined in dark than light conditions, i.e. R_d) is a general response of healthy, non-growing oak leaves (e.g. Bolstad et al. 2003; Lee et al. 2005; Zaragoza-Castells et al. 2008; Rodríguez-Calcerrada et al. 2011) that may accompany different thermal photosynthetic adjustments to balance leaf net C gain with C needs in the new environment

(Way and Yamori 2014; Slot and Kitajima 2015). The extent of the thermal acclimation of R_d varies among species and with leaf developmental status, magnitude of temperature change (Slot and Kitajima 2015) and interaction with other abiotic factors such as irradiance (Bolstad et al. 1999) or soil water availability (Turnbull et al. 2001; Rodríguez-Calcerrada et al. 2011).

Despite the importance of C losses for leaf and plant C balance, very few studies have examined the impacts of temporal soil water fluctuations on R_d in oak species. Most of these studies have been conducted on the drought-tolerant, widespread Mediterranean oak *Q. ilex* (e.g. Rodríguez-Calcerrada et al. 2011; Varone and Gratani 2015). The results of these studies are consistent with the drought-induced decline in leaf R_d and increase in leaf R_d/P ratio that is generally reported for other plant species (see review of Atkin and Macherel 2009). The main reasons behind this short-term decline in leaf R_d are: (i) a reduction in the amount of mitochondrial protein and (ii) a reduction in enzymatic activity due to limited turnover of ATP to ADP (associated to down-regulation of energy consumption processes) or limited flow of triose phosphate from chloroplasts into mitochondria (associated to impaired P). The complex regulation of leaf R makes that, as it happens in response to temperature, leaf R_d does not necessarily exhibit the same response to drought over short- and long-term time scales. However, few studies have examined how long-term decreases in soil water availability affect leaf R and C balance in oak trees (Turnbull et al. 2001; Rodríguez-Calcerrada et al. 2011; Sperlich et al. 2016). In southwestern Europe, two parallel throughfall-reduction experiments have been set up in two *Q. ilex* forest stands to study the long-term effects of increased drought on foliar respiratory rates. Rodríguez-Calcerrada et al. (2011) observed that leaf R_d decreased in response to seasonal decline of leaf water potential similarly in trees subjected to normal and 7-year reduced throughfall, so that leaf R_d was lower in the trees that had experienced a reduction in throughfall and a greater decline of leaf water potential during the dry season, but did not differ between treatments at optimal soil water conditions. These results and those of Limousin et al. (2010) on the nature of photosynthetic limitations in the same species and experimental site suggested that 7 years of increased drought had not modified the physiology of leaf mesophyll cells. In another *Q. ilex* stand subjected to a longer period of rain reduction (14 years) of similar intensity, Sperlich et al. (2016) found the same lack of treatment effect on leaf R_d, but a significant increase in leaf R during daytime in trees receiving less rain, suggesting that the reorganization of the respiratory metabolism depends on the duration of increased drought. Collectively, the results suggest that drought-induced declines in leaf R over short time periods seemingly change to drought-induced increases in leaf R as trees adapt to increased drought (Turnbull et al. 2001; Rodríguez-Calcerrada et al. 2011; Atkin et al. 2015; Sperlich et al. 2016), however, more studies are needed to understand long-term responses of leaf R to drought.

10.2.2 Stem and Root Respiration

10.2.2.1 How to Estimate It? Methodological Constrains to Measure Woody Respiration

Although stems and roots constitute the largest fraction of biomass in woody species, especially in large trees (Poorter et al. 2012), our knowledge of woody R is by far less advanced compared with that of leaf R. The main obstacle to understand R in woody tissues remains in the difficulty to accurately measure it. Radial CO_2 efflux to the atmosphere (E_A) from stems and roots, which can be measured with cuvettes surrounding the monitored organ, is commonly assumed to equal the rate of R of these organs. Nevertheless, locally respired CO_2 in roots and stems can either diffuse to the soil or the atmosphere, respectively, or alternatively accumulate within woody tissues due to substantial barriers to radial gas diffusion offered by outer tissues. Accordingly, concentrations of internal CO_2 in xylem (xylem [CO_2]) range from <1 to 26%, values one to two orders of magnitude higher than atmospheric [CO_2]. As xylem [CO_2] builds up inside the tree, it dissolves in the sap solution until equilibrium between gaseous and liquid phases is reached according to Henry's law; respired CO_2 moves upward in the transpiration stream, and eventually diffuses to the soil or the atmosphere elsewhere (see Teskey et al. 2008; Rodríguez-Calcerrada et al. 2015b for reviews). Internal transport of respired CO_2 has therefore resulted in significant misestimation of woody R from measurements of E_A, as consistently observed in several oak species (McGuire and Teskey 2002; Teskey and McGuire 2002; Bloemen et al. 2014; Salomón et al. 2016b).

An additional constraint that hinders direct measurements of woody R is the re-assimilation of internal CO_2 by chloroplast-containing woody tissues (see Ávila et al. 2014 for a review). For instance, recycling of respired CO_2 transported through the xylem offset 19 and 70% of C respiratory losses in branches of *Q. alba* and stems of *Q. robur*, respectively (Coe and McLaughlin 1980; Berveiller et al. 2007). To solve this issue, woody P is commonly disabled by using opaque cuvettes; nevertheless, woody P above and below the cuvette might induce axial diffusion of internal CO_2 that would decrease E_A within the monitored segment, as observed in *Q. robur* stems during the dormant season (Saveyn et al. 2008). Additional difficulties arise when measuring root R due to inaccessibility of root systems and the unclear discrimination between autotrophic and heterotrophic respiration from measurements of soil CO_2 efflux (Hanson et al. 2000). All these limitations hinder the establishment of a widely accepted methodological approach to systematically measure woody R.

10.2.2.2 Response to Environmental Changes at Different Time Scales

Due to the few studies on woody R in oaks, the influence of abiotic drivers (mainly temperature and water availability) on woody R is summarized independently of any intrageneric classification. In a study with seven oak species grown under uniform conditions, Martinez et al. (2002) did not find any intrinsic difference in root R attributable to the evergreen or deciduous character of the species. Likewise, potential differences in woody R ascribed to the ring-porous or diffuse-porous wood anatomy of oaks remain untested, despite their differential wood phenology and growth (Pérez-de-Lis et al. 2016).

The temperature sensitivity of woody R, expressed as the change in R that occurs over 10 °C (Q_{10}) ranges from 1.4 to 3.1 in oak species, with mean values close to two (i.e., R rates double for an increase in temperature of 10 °C, Table 10.1), as similarly observed for a variety of species in leaves and roots (see Atkin and Tjoelker 2003 and references therein). On a diel basis, woody R increases along the day with increasing temperatures and decreases at night-time exhibiting a characteristic hysteresis (Table 10.1; Fig. 10.1b). Several factors have been suggested to cause the day-time depression in temperature-normalized stem R observed in oaks: (i) internal transport of respired CO_2 with the transpiration stream at day-time (Negisi 1982; McGuire and Teskey 2002; Teskey et al. 2008), (ii) enhanced metabolic activity of woody tissues owing to improved water status at night-time (Negisi 1982; Saveyn et al. 2007a; Salomón et al. 2016b), (iii) lagged temperature transmission and/or delayed radial CO_2 diffusion due to physical barriers presented by peripheral tissues (Rodríguez-Calcerrada et al. 2014), and (iv) refixation of respired CO_2 nearby the darkened monitored stem segment at day-time (Saveyn et al. 2008). At a seasonal scale, the down-regulation of temperature-normalized R with increasing temperatures across the year (Atkin and Tjoelker 2003) has the potential to reduce C loss through woody R. The thermal acclimation of leaf R (Fig. 10.1a) is a well-documented phenomenon in oaks (see previous sub-section) that has been less studied in woody tissues. A meta-analysis across 44 forested ecosystems, six of them dominated by oak species, supports the hypothesis of thermal acclimation of R in roots (Burton et al. 2008): An attenuated rate of temperature-driven increase in root R across ecosystems (Q_{10} = 1.6) was observed in comparison to short-term fluctuations within individual stands (Q_{10} = 2–3). Likewise, lower rates of temperature-normalized root R were registered in experimentally heated plots in mixed hardwood forests co-dominated by *Q. velutina*, although the concomitant effect of soil drying along with soil heating could not be discarded as a driver of R reductions (Burton et al. 2008). Thermal acclimation of R rates in stems may also occur in oak species such as *Q. ilex* (Rodríguez-Calcerrada et al. 2014), but literature is much scarcer in this regard. Again, the concomitant increase in temperature with summer drought in some ecosystems hinders to unequivocally ascribe seasonal reductions in stem R to thermal acclimation, given that water shortages constrain stem growth and associated respiratory costs in this widespread Mediterranean oak (Lempereur et al. 2015). The different temperature sensitivity of metabolic processes involved in maintenance and growth R (Amthor 2000) complicates the study of acclimation of maintenance R to abiotic stress in sites where secondary growth varies amply throughout the year. Accordingly, higher Q_{10} values observed in stems of *Q. accutisima* during the colder non-growing season relative to the warmer growing

Table 10.1 Temperature sensitivity of stem and root respiration (Q_{10}) in *Quercus* species

Organ	Species	Q_{10}	Hysteresis[a]	References
Stem	*Q. accutisima*	2.2		Yang et al. (2012b)
	Q. alba	1.5–2.4	✓	Edwards and Hanson (1996), Li et al. (2012)
	Q. ilex	1.5–2.5	✓	Rodríguez-Calcerrada et al. (2014)
	Q. mongolica	2.1–2.4		Wang et al. (2010), Yang et al. (2012a)
	Q. petraea	1.6–2.1		Rodríguez-Calcerrada et al. (2015a)
	Q. prinus	2.4	✓	Edwards and Hanson (1996)
	Q. pyrenaica	1.4–2	✓	Rodríguez-Calcerrada et al. (2015a), Salomón et al. (2016b)
	Q. robur	1.9–2.8	✓	Saveyn et al. (2007a, b, 2008)
	Q. serrata	2.1	✓	Miyama et al. (2006)
	Q. velutina	1.6		Li et al. (2012)
Roots[b]	*Q. accutisima*	2.8		Luan et al. (2011)
	Q. serrata mixed stand	2.4		Dannoura et al. (2006)
	Q. cerris	2.2		Rey et al. (2002)
	Quercus-Carya stand	3.1		Burton et al. (2002)
	Mixed *Quercus* stand	2.4		Burton et al. (2002)

[a]Studies in which a hysteretic relationship between temperature and respiration has been reported (✓). Empty spaces denote studies in which this phenomenon was not evaluated. [b]Root respiration integrates CO_2 originated from fine and coarse roots as well as root-associated microorganisms present in the rhizosphere

season were attributed to the greater temperature sensitivity of maintenance processes (Yang et al. 2012b) rather than to the potential effect of thermal acclimation of R.

Oak species subjected to drought stress commonly show reduced R as a consequence of constrained growth and metabolic activity. This effect has been observed at the ecosystem (Reichstein et al. 2002; Unger et al. 2009; Rambal et al. 2014), organ (Saveyn et al. 2007b; Rodríguez-Calcerrada et al. 2014), and cellular level (Saveyn et al. 2007a). During a diel cycle, night-time reduction in the vapour pressure deficit and transpiration lead to replenishment of water reservoirs within woody tissues (Steppe et al. 2006), as observed in *Q. ilex* (Salomón et al. 2017). Increase in cell turgor facilitates cell expansion and growth (Lockhart 1965), which in turn may lead to enhanced rates of overall R. This hypothesis of growth respiration mainly confined to night-time hours is supported by substantial night-time increases in both E_A and xylem [CO_2] observed in *Q. robur* stems under relatively constant temperature across 24 h (Fig. 10.1c; Saveyn et al. 2007a, b). During the course of a year, progressive soil drying in summer was found to constrain fine root turnover (López et al. 2001) and stem growth (Lempereur et al. 2015) in *Q. ilex*. Impeded growth and down-regulation of maintenance processes likely explains typical reductions in woody R when water becomes limiting (Fig. 10.1c), as observed in roots of *Q. cerris* (Rey et al. 2002), *Q. robur* (Molchanov 2009), and mixed oak stands (Burton et al. 2002), as well as in stems of *Q. ilex* (Rodríguez-Calcerrada et al. 2014) and *Q. robur* (Saveyn et al. 2007b). Drought-induced reductions in R suggest a threshold in soil water content below which woody R becomes largely driven by water availability and independent of temperature (e.g. Reichstein et al. 2002; Rey et al. 2002). Likewise, sharp increases in xylem [CO_2] observed in *Q. robur* and *Q. pyrenaica* stems after rain events following dry periods (Saveyn et al. 2007b; Salomón et al. 2016a) provide further evidence of drought-driven constraints to woody R. On the other hand, reduced resistance to radial CO_2 diffusion due to reduced water content of peripheral woody tissues as the soil dries out (Teskey et al. 2008; Salomón et al. 2016b) may partially explain increases in stem and root CO_2 efflux during mild drought in oak trees (e.g. Edwards and Hanson 1996; Dannoura et al. 2006; Molchanov 2009). Such conflicting results evidence our deficient understanding of drought effects on woody R. At an inter-annual timescale, acclimation of stem R to long-term increased drought was not observed in *Q. ilex* after eight years of experimental throughfall-reduction (Rodríguez-Calcerrada et al. 2014). Further research in this line would be necessary to test potential down-regulation of woody R to prolonged drought in order to better predict C cycling at the whole-tree level under changing climate regimes.

10.2.3 Relative Importance of R for Tree Carbon Budgets

Ecosystem R (R_{ECO}) determines ecosystem C balance in a wide range of environmental conditions (Valentini et al. 2000). Ecosystem R can be biometrically partitioned into leaf, stem and soil R by measuring samples of each component in chambers and upscaling the measurements to the stand level. The broad range of variation in the contribution of each component to R_{ECO} is partly due to stand structure, composition and age, but also to uncertainties in calculations. First, estimations of annual leaf respiratory C losses at the stand scale range from 3 to 37% of R_{ECO} in oak forests (Table 10.2). At an intra-annual scale, the largest contribution of leaf R to R_{ECO} generally occurs when new leaves expand; the lowest contribution can occur at peak summer drought in some evergreen Mediterranean forests (Rodríguez-Calcerrada et al. 2012). Second, the high proportion of parenchyma in the wood of oak species can make C losses from stems potentially important in oak-dominated ecosystems. The contribution of stem R to R_{ECO} in pure and mixed oak stands ranges between 5 and 38% on an annual basis, with mean values of ca. 15% (Table 10.2). The highest contribution of stem R to

R_{ECO} occurs in spring coincident with high stem growth rates in temperate and continental climates (Curtis et al. 2005; Miyama et al. 2006), and before water becomes limiting in the case of drought-prone Mediterranean regions (Rodríguez-Calcerrada et al. 2014). Third, soil R is the largest respiratory C flux to the atmosphere (Valentini et al. 2000) and accounts for 48–85% of R_{ECO} in oak forests (Table 10.2). Ambiguous discrimination between heterotrophic (microbes and soil fauna) and autotrophic sources of soil CO_2 efflux hinders accurate estimations of root R. Assuming an average contribution of soil R to R_{ECO} of 67% (Table 10.2), and a mean contribution of root R to soil R of 50% (Hanson et al. 2000; Burton et al. 2008), root R would account for 33% of R_{ECO}. More conservative contributions of root R to soil R observed in oak stands—ranging from 15 to 40% (Reichstein et al. 2002; Rey et al. 2002; Unger et al. 2009; Luan et al. 2011)—would reduce the contribution of root R to R_{ECO} to 10–27%, respectively. Furthermore, it is worth noting that neglecting internal fluxes of root respired CO_2 through xylem results in substantial underestimation of root R rates when these are estimated via soil CO_2 efflux measurements; underestimation ranges from 2 to 18% in *Quercus* species (Bloemen et al. 2014; Salomón et al. 2015) and reach up to 50% in other taxa (Aubrey and Teskey 2009). Overall, these estimates evidence the important role of autotrophic R in plant and ecosystem C budgets and further highlight the need of more experimental research on plant R to improve the accuracy of dynamic global vegetation models.

10.3 Volatile Organic Compounds (VOCs)

Plants produce a large array of metabolites whose vapor pressures are high enough (approx. ≥ 0.01 kPa) to become volatilized under ambient temperature conditions. All plant organs, namely flowers and fruits, foliage, stem and roots can release VOCs. Flower and leaf emissions are by far the best investigated ones. However, in the last decade increasing research has been afforded to root emissions, whose ecological roles in soil biotic interactions are only in the beginning to be appreciated (see e.g. Weissteiner et al. 2012; Delory et al. 2016). VOCs are emitted from plant organs either constitutively or temporarily following induction by stress factors. This classification is however not straightforward, because the emissions of constitutive VOCs are also up and down-regulated by environmental factors including stress events (Peñuelas and Staudt 2010).

Phytogenic VOCs are mainly composed of C and hydrogen plus occasionally other elements such as oxygen, nitrogen and sulphur, or more rarely halogens. Once emitted the C skeleton of VOCs reacts gradually with oxidants in the atmosphere to form ultimately CO_2, thus closing the carbon cycle. However, a substantial portion of intermediate products may be removed from the atmosphere via dry and wet deposition. Products from VOC oxidation can condense with each other and other atmospheric constituents and contribute to the formation of secondary organic

Table 10.2 Ecosystem respiration (R_{ECO}) in pure and mixed oak stands, and its partitioning (%) into leaf (R_L), stem (R_S[a]) and soil (R_{SOIL}) respiration

Stand	Site and climate	Stand characteristics	R_{ECO} (g C m^{-2} year^{-1})	R_L	R_S	R_{SOIL}	Reference
Q. ilex coppice	Puéchabon (France) Mediterranean	5100 stems ha^{-1} 27.4 m^2 ha^{-1} 70 years-old	977	37	13	>50	Four studies[b]
Mixed oak[c]	Ozark (MO, USA) Continental	997 stems ha^{-1} >70 years-old	1684	10	15	>71	Li et al. (2012)
Mixed broadleaved[d]	Yamashiro (Japan) Temperate	5953 stems ha^{-1} 19.6 m^2 ha^{-1}	1080	21	16	63	Miyama et al. (2006)
Mixed harwood[e]	Michigan (USA) Continental	2214 stems ha^{-1} 29.7 m^2 ha^{-1}	1425	18	11	71	Curtis et al. (2005)
Temperate deciduous[e]	Massachusetts (USA)	20–24 m height 50–70 years-old	2460	28	5	68	Goulden et al. (1996)
Temperate mixed[f]	Changbai (China) Temperate	1556 stems ha^{-1} 43.2 m^2 ha^{-1} Uneven-aged	1490	31	21	48	Wang et al. (2010)
Mixed plantation[g]	Akou (Japan) Temperate	11,150 stems ha^{-1} 34 m^2 ha^{-1} 9 years-old	1144	36	Neglected	64	Kosugi et al. (2005)
Evergreen oak savanna[h]	Mitra (Portugal) Mediterranean	7.5 m height 21% tree canopy cover	568–1325	19–40	Neglected	60–81	Unger et al. (2009)
Deciduous-broadleaved[i]	Takayama (Japan) Cool-temperate	1379 stems ha^{-1} 50 years-old 20 m height	1397	24	10	67	Ito et al. (2007)
Q. robur forest steppe	Voronezh (Russia) Continental	Not reported	1205–1478	3–5	12–38	60–85	Molchanov (2009)

Data is presented on a yearly basis per unit of soil area. [a]R_S estimated from CO_2 efflux measurements; R_S commonly integrates branch respiration. [b]Data gathered from four studies in the same experimental site: R_{ECO} from Rambal et al. (2002); R_L from Rodríguez-Calcerrada et al. (2012); R_S from Rodríguez-Calcerrada et al. (2014); R_{SOIL} from Reichstein et al. (2002). [c]*Q. alba*, *Q. velutina* and *Q. cocinea* are dominant species; uniquely data from the unmanaged control plot is shown here. [d]*Q. serrata* is one of the dominant species. [e]*Q. rubra* is one of the dominant species. [f]*Q. mongolica* is one of the dominant species. [g]*Q. serrata*, *Q. glauca* and *Q. phillyraeoides* comprise ¼ of total basal area. [h]*Q. ilex* and *Q. suber* are dominant species. [i]*Q. crispula* is one of the dominant species aerosols, which have important impacts on climate forcing and human health (Hallquist et al. 2009).

10.3.1 Metabolic Origins and Ecological Importance

Of the thousands existing plant VOCs, the majority belongs to three biosynthetic classes, fatty acid derived volatiles, volatile aromatic compounds and volatile terpenoids (isoprenoids). The most prominent fatty acid-derived VOCs are Green Leaf Volatiles (GLVs) that are formed from the breakdown of free fatty acids by lipoxygenase and hydroperoxide-lyase enzymes (Matsui 2006). GLVs comprise mainly mono-unsaturated C6 alcohols, aldehydes and acetate esters that are potentially emitted from all plant species. GLV are emitted in high amounts only after exposure to stresses such as wounding, herbivore attack, extreme temperatures or acute ozone exposure. Upon stress GLVs are formed and emitted within seconds to minutes and rapidly disappear when the stress ceases (e.g. Staudt et al. 2010). Volatile aromatic compounds derive from the shikimate pathway (benzenoids and phenylpropanoids). They are most common in flower scents (Schiestl 2010). However, recent studies made at plant and canopy levels emphasize that considerable amounts of benzenoids are also emitted from vegetative tissues in particular under stress conditions (Misztal et al. 2015). Volatile isoprenoids are classified according to the number of C5 units they have: C5 hemiterpenes, C10 monoterpenoids and C15 sesquiterpenoids. *In planta*, isoprenoids are synthesized within two distinct pathways, the 2-methyl-erythritol-4-phosphate pathway operating in plastids and the mevalonate pathway operating in the cytosol, endoplasmic reticulum and peroxisomes (Lu et al. 2016 and references therein). Volatile isoprenoids can also be secondarily formed from the breakdown of higher isoprenoids such as homoterpenes (Tholl et al. 2011) and apocarotenoids (Walter et al. 2010). The quantity and quality of emitted isoprenoids and aromatic compounds inherently differ among plant taxa. The hemiterpene isoprene is the most important VOC released from terrestrial vegetation. Globally ca. 600 Tg (10^{12} g) of isoprene are annually emitted, which is about 2/3 of the total biogenic VOC release (Sindelarova et al. 2014). However, only about 30% of vascular plants emit isoprene from their foliage (Fineschi et al. 2013). Oaks, poplars and willows figure among the strongest isoprene emitters (Kesselmeier and Staudt 1999).

In addition to the volatiles of these three major classes, a number of short-chain oxygenated volatiles (collectively called OVOCs) are frequently observed in plant emissions such as methanol, formaldehyde, formic acid, ethanol, acetaldehyde, acetic acid, methyl acetate, acetone, ethylene and methylglyoxal. These have different metabolic origins and occur rather universally in plants. For instance methanol is formed during cell wall formation and is therefore particularly released during the growing period (Bracho-Nunez et al. 2011; Brilli et al. 2016). With an estimated annual emission of ca. 130 Tg, methanol is the second important phytogenic VOC worldwide (Sindelarova et al. 2014). Emissions of ethanol and the equivalent C2 aldehyde and acid are mostly associated with fermentation during hypoxia (Kreuzwieser and Rennenberg 2013). Many of these OVOCs are also secondarily formed during the oxidation of primary emitted higher VOCs such as terpenes (e.g. Gaona-Colmán et al. 2016).

Diverse ecological functions have been attributed to VOC production in plants. The attraction of pollinators and seed dispersers by flower and fruit scents is an essential driver of sexual reproduction and evolution in many plant species (Schiestl 2010). Further, numerous studies have shown that plant volatiles induced by herbivore or pathogen attacks have toxic, repellent or aposematic effects, or attract predators and parasitoids of the attackers and thereby contribute to limit damages to plants (see e.g. Van Loon et al. 2000). So far, such functions are largely unknown for oaks. Oaks are wind pollinated and a potential role of volatiles in sustaining acorn dispersal has rarely been documented (e.g. Verdú et al. 2007). Only few studies have reported that herbivory or pathogen attack affect quantitatively and/or qualitatively the volatile production from oaks (Staudt and Lhoutellier 2007; Paris et al. 2010; Copolovici et al. 2014). A study with *Q. robur* suggests potential beneficial effects of herbivory-induced VOCs against the European oak leaf roller (Ghirardo et al. 2012). On the other hand, Vuts et al. (2016) demonstrated that volatiles of *Q. robur* can attract the two-spotted oak buprestid, a bark beetle that causes severe damages in European oak forests.

The possible function of the constitutive isoprene production in chloroplasts (or analog monoterpenes) is still a matter of debate. With regard to biotic stress, a study using transgenic isoprene emitting *Arabidopsis* plants has shown that isoprene can disturb the attraction of a parasitic wasp to volatiles from herbivore-infested plants. Thus, by troubling VOC-mediated trophic interactions of neighbouring species, isoprene emitters could promote their interspecific competitiveness (Loivamäki et al. 2008). However, since isoprene is one of the most common background-VOC in the atmosphere, insects may rapidly adapt to avoid such interferences in host recognition (Müller et al. 2015). The major function of isoprene is thought to reinforce the resistance to abiotic stresses, in particular to oxidative and high temperature stress (see e.g. Loreto and Schnitzler 2010). The most pertinent results supporting these hypotheses were achieved with transgenic plants, either isoprene synthase over-expressing mutants of the non-isoprene emitting species *Arabidopsis* (Sasaki et al. 2007; Velikova et al. 2012) and tobacco (Vickers et al. 2009b; Tattini et al. 2014) or isoprene synthase silenced mutants of the isoprene emitting species poplar (Behnke et al. 2007, 2010). However, several studies using the same or other transgenic mutants reported conflicting results with respect to these hypotheses (Behnke et al. 2009, 2012; Palmer-Young et al. 2015). Further, it is uncertain that the improved resistance to abiotic stress in isoprene emitters is, as initially hypothesized, due to direct effect of isoprene by scavenging oxidants or stabilizing membrane structure. In fact, isoprene dissolves too little in bio-membranes to efficiently change membrane conformation (Harvey et al. 2015) and is only fairly suitable to scavenge oxidants due to its moderate reactivity (Atkinson and Arey 2003; Palmer-Young et al. 2015) and the high toxicity of reaction products (Cappellin et al. 2017; Matsui 2016 and references therein). More likely, the physiological effects of isoprene are indirect by taking part in the plant's stress signalling network (Vickers et al. 2009a; Vanzo et al. 2016). In any case, the genetically engineered absence or presence of isoprene synthase in mutants causes transcriptomic, proteomic and metabolic changes in various metabolic pathways even under non-stress conditions, including the biosynthesis of phenyl-propanoids and lipids that affects the composition and ultrastructure of chloroplast membranes (Velikova et al. 2015; Harvey and Sharkey 2016). Numerous other VOCs than isoprene have shown to be involved in within-plant and between-plant stress signalling, among which diverse GLVs, aromatic compounds and higher isoprenoids (Havaux 2014; Delory et al. 2016; Matsui 2016).

10.3.2 Diversity of VOCs in Oak Species

The great majority of oak species hitherto screened for VOC emission has been found to release isoprene constitutively at high rates from its foliage (up to several tenths of nmol $m^{-2} s^{-1}$). However, several oak species with distribution around the Mediterranean do not produce isoprene, but produce either monoterpenes in high amounts or shown no constitutive VOC emissions. These exceptions are mostly but not exclusively evergreen oaks and belong all to the two very closely groups *Cerris* and *Ilex* (Welter et al. 2012; Monson et al. 2013). Diversification of isoprene emission in oaks has been mainly observed at species and higher taxon levels and more rarely at population level. So far, inherent intra and/or inter population variability in the quantity or quality of constitutively produced VOC has not been reported within isoprene emitting oak species (Welter et al. 2012; Steinbrecher et al. 2013). By contrast in monoterpene emitting species, compositional diversification (i.e. chemotypes) has been frequently observed within and among populations (Staudt et al. 2001b, 2004; Loreto et al. 2009; Welter et al. 2012). In addition, a few low or non-emitting individuals were detected in some monoterpene emitting populations. Conversely, dual isoprene/monoterpene emitting oak individuals seem to be extremely rare in natural oak populations (Staudt et al. 2004). This is somewhat surprising, because many oaks can hybridize with each other resulting in widespread genetic introgression or rise of new species (e.g. Valbuena-Carabaña et al. 2007). For example, the endemic oak *Q. afares* (Algerian oak) is considered to be a stabilized hybrid between *Q. suber* and *Q. canariensis*. Yet, *Q. afares* produces exclusively monoterpenes (Welter et al. 2012) as one of its parent species (*Q. suber*), but seems to have completely lost or suppressed the VOC production capacity of its second parent *Q. canariensis*, which produces only isoprene. These findings indicate that qualitative diversification in the monoterpene production capacity occurs frequently and has no or only minor consequences for the competitiveness of the trees. By contrast, the loss or gain of isoprene synthesis happens more rarely during species evolution, possibly because it requires co-evolving compensatory mutations to overcome failures in metabolic homeostasis.

10.3.3 Response to Environmental Changes at Different Time Scales

Unlike for CO_2 and water vapor, the foliar exchange of VOCs is less constrained by stomatal conductance. This is due to their generally low gas phase concentrations inside the leaves, which allow changes in the diffusive resistance by stomata movements to be compensated by concomitant changes in the concentration gradients between the substomatal cavity and the outer atmosphere. Nevertheless, many OVOCs are at least partly under stomatal control either because these VOCs are transported with the transpiration stream and/or because these VOCs dissolve efficiently in liquid phase, which in turn delays the re-equilibrium of the gaseous concentration in the apoplast in response to stomata movements (Niinemets et al. 2002). In any case, all VOC emissions are strongly modulated by external factors, above all temperature, which governs the VOC's vapor pressures and diffusion velocities, in addition to metabolic processes involved in the VOC release (Staudt et al. 2017b). In oaks, the major bulk of the emitted VOCs (constitutive isoprene and monoterpenes) is directly regulated by their synthesis rate inside the chloroplasts, which in turn depends on the availability of primary substrates coming from light-dependent photosynthetic processes. Recent studies on poplar suggest that most of the short-term variation of isoprene emission is due to limitations in the availability of reduction power from photosynthetic electron transport (Rasulov et al. 2016 and references therein). Thus, constitutive isoprene and monoterpene emissions from oaks, and perhaps also stress-induced de novo synthesized emissions of sesquiterpenes, are strongly and almost instantaneously modulated by both temperature and light (e.g. Staudt and Lhoutellier 2011). The shape of the emission response to light resembles that of net photosynthesis; i.e. a rectangular hyperbola approaching an asymptote at high light values (Staudt and Bertin 1998). By contrast, the temperature response exhibits a shape typical for the temperature dependence of enzyme-catalyzed reactions (Fig. 10.2). In addition to these fast responses, light and temperature modulates the oak's overall emission capacity over the seasons with changing weather conditions (e.g. Pier and McDuffie 1997; Staudt et al. 2002, 2003, 2017a) mostly via the expression of genes of rate-limiting enzymes (Schnitzler et al. 1997; Fischbach et al. 2002; Lavoir et al. 2009). Constitutive isoprenoid emissions from oaks are also influenced by other environmental factors such as ozone (Velikova et al. 2005), atmospheric CO_2 concentration (Loreto and Sharkey 1990; Staudt et al. 2001a) or shortage and excess of soil water (Bertin and Staudt 1996; Staudt et al. 2002, 2008; Rodríguez-Calcerrada et al. 2013; Bourtsoukidis et al. 2014; Saunier et al. 2017).

10.3.4 Relative Importance for Tree Carbon Budgets

Given that in oaks almost all primary C substrate used in the biosynthesis of volatile isoprenoids comes from ongoing photosynthesis, the C loss associated with emission can be expressed as the ratio of mol emitted C to mol assimilated C. As shown in Fig. 10.2, the instantaneous C loss varies much with changing temperature but little with light. Under most conditions the instantaneous C loss of

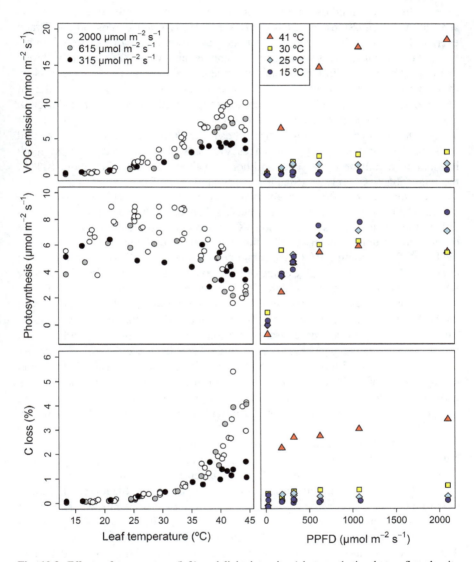

Fig. 10.2 Effects of temperature (left) and light intensity (photosynthetic photon flux density (PPFD), right) on monoterpene emission (upper graphs), photosynthesis (middle graphs) and the resulting relative carbon loss from monoterpene emission (C loss, lower graphs) in *Q. ilex* leaves. Temperature effect was measured at a PPFD of about 315, 615 and 2000 µmol m^{-2} s^{-1}. Light effect was measured at a temperature of about 15, 25, 30 and 41 °C. Data were compiled from Staudt and Bertin (1998)

assimilated C by constitutive isoprenoid emissions will be less than 1%. However, during heat spells, the C loss can substantially increase, especially when combined with drought events. In fact, isoprenoid emissions from oaks are less sensitive to water shortage than photosynthesis. The emissions decrease only when drought is severe while under mild drought they remain stable or even increase (Fig. 10.3). As a result, VOC emissions from oaks may significantly drain photosynthetic C and energy during conditions in which other sinks associated with growth are inhibited. In addition to constitutively produced isoprenoids, new volatiles induced by abiotic or biotic stresses may further exhaust the tree's C resources. The exact C loss by these emissions is difficult to assess since stress usually induces a wealth of high-molecular, very reactive trace compounds (many of which remain undetected) that are emitted sporadically. Generally, the emission rates reported for stress-induced VOCs are lower than those for strong constitutive isoprenoid emissions. For instance, in *Q. ilex* leaves VOC emissions induced by gypsy moth infestation (mainly sesquiterpenes and homoterpenes) accounted for about 10% of the total foliar VOC release (Staudt and Lhoutellier 2007).

The emissions of OVOCs from foliage have often been neglected when estimating C losses by VOC emission, because their accurate measurement requires different techniques to that of common VOCs. However, as mentioned above, many of them are ubiquitous plant volatiles that can be emitted at quite high rates, such as methanol. By combining PTR-MS technique with classical GC-MS, Bracho-Nunez et al. (2013) determined a large range of VOC exchange in 28 plant species, among which three oak species. At standard light and temperature conditions, mean C loss of net-photosynthesis by VOC emission ranged between 1.4 and 3.7%, with methanol and acetone contributing between 5 and 66% to the total VOC release. The quantities of VOCs released from roots and stem tissues are only poorly known and to our knowledge have never been reported for oaks. Nevertheless, Weissteiner et al. (2012) identified more than 60 VOCs in the headspace of washed healthy and damaged roots of young *Q. petraea* × *Q. robur* trees, of which 13 compounds were consistently released. Asensio et al. (2007) investigated the VOC exchange of soil in a *Q. ilex* forest and concluded that it represented 0.003% of the total C emitted by soil as CO_2.

On the other hand, there is increasing awareness that VOC exchanges can be bidirectional at least for some OVOCs and hence compensation points in VOC concentration exist, above which VOCs are taken up by plants (Niinemets et al. 2014; Matsui 2016 and references therein). For example, Staudt et al. (2000) observed bidirectional exchanges of acetic acid from diverse plant species, with net emissions dominating during day-time and net depositions dominating during night-time. For *Q. ilex*, the average deposition rate observed in darkness was more than half of the average emission rate observed under illumination (0.41 and 0.72 ng C m^{-2} s^{-1}, respectively). Furthermore, secondary VOCs produced during atmospheric oxidation can be taken up by the vegetation and metabolized, thus possibly recovering parts of the C lost by emission (Karl et al. 2010; Park et al. 2013). Bidirectional above-canopy VOC fluxes have been recently measured by Schallhart et al. (2016) in a mixed oak forest during the early summer season. These

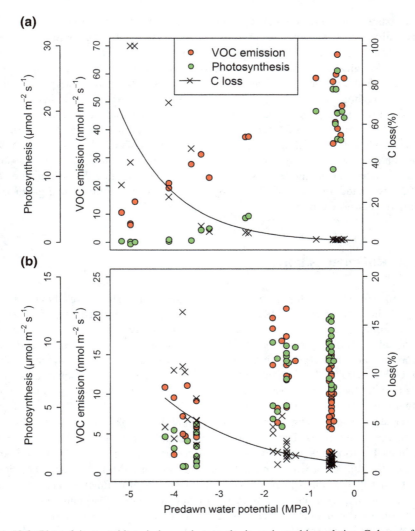

Fig. 10.3 Plot of isoprenoid emissions, photosynthesis and resulting relative C losses from isoprenoid emissions against predawn water potential during two drought experiments on *Quercus pubescens* (**a**, isoprene emitter) and *Quercus suber* (**b**, monoterpene emitter). Oak saplings were exposed to one or two consecutive drying cycles for *Q. pubescens* and *Q. suber*, respectively. Gas exchange measurements were made under the same standard light and temperature conditions (1000 μmol m^{-2} s^{-1} PPFD; 30 °C). Lines are best fits on C-loss data assuming an exponential relationship for *Q. pubescens* (R^2 = 0.92; Rodríguez-Calcerrada et al. 2013) and *Q. suber* (R^2 = 0.67; Staudt et al. 2008)

authors observed an average net VOC efflux to the atmosphere of 41.8 nmol C m^{-2} s^{-1} which accounted for a bit less than 2% of the net uptake of CO_2. This number likely represented the upper limit of VOC-related C loss in that study site, because the measurements were conducted during the season when the emission capacity for constitutive isoprenoids reaches usually its maximum. Continuous online year-long measurements of total VOC exchanges at plant or canopy levels are still sparse due to methodological constraints. As an exception, Brilli et al. (2016) have monitored by eddy covariance both VOC and CO_2 exchanges over a temperate poplar plantation throughout a whole growing season. Although poplar is, similar to most temperate oak species, a strong isoprene emitter, they observed a relatively small net VOC flux of ca. 1 g m^{-2} per growing season accounting for about 0.8% of the net ecosystem CO_2 exchange. Earlier studies extrapolated discontinuous VOC emission/flux measurements by means of generic emission models to assess their weight relative to annual ecosystem C budget. For example Kesselmeier et al. (2002) estimated the annual C loss by VOC emission from an evergreen oak Mediterranean forest being 0.45% of their annual GPP.

10.4 Rhizodeposition

The rhizosphere is a highly populated environment. There are thousands of non-volatile organic compounds released by roots that mainly consist of carbohydrates, amino acids, vitamins, lipids, and a wide variety of secondary metabolites and proteins. Rhizodeposits can alter the physico-chemical soil properties and play important roles in the interactions of the plant with microbes or competing plant species (Bais et al. 2004; Bashir et al. 2016). As example, phenolic compounds help roots to deter the attack of pathogens (Lanoue et al. 2010); flavonoids facilitate the mutualistic symbiosis with mycorrhizal fungi (Nagahashi et al. 2010); the synthesis and release of some enzymes increase the availability of phosphorus forms that are absorbable by the roots (Dakora and Phillips 2002); while the exudation of phospholipids by root tips can reduce the surface tension of the soil solution and enhance the uptake of water and nutrients (Read et al. 2003). Similar interactions might exist in the rhizosphere of oak trees. However, the function and chemical profiling of rhizodeposits have been rarely studied within this genus. One of the scarce studies documents the effect of herbivory on C rhizodeposition in 2-year-old *Q. rubra* seedlings (Frost and Hunter 2008): It was observed that foliar herbivory reduced C allocation to fine roots whilst root exudation was actively regulated to maintain constant rates of C rhizodeposition, likely to sustain nutrient supply to microbes.

Rhizodeposition is affected by edaphic and environmental factors (reviewed by Nguyen 2009). The abundance of soil microorganisms substantially enhances the allocation of C assimilates to the rhizosphere. Besides, soil texture affects microbial activity, nutrient cycling, and soil physical properties, so that rhizodeposition increases in loam and clay soils due to their higher fertility and the smaller size of soil pores that facilitate the flow of organic compounds. Regarding climatic conditions, rhizodeposition is expected to increase during drought stress (Henry et al.

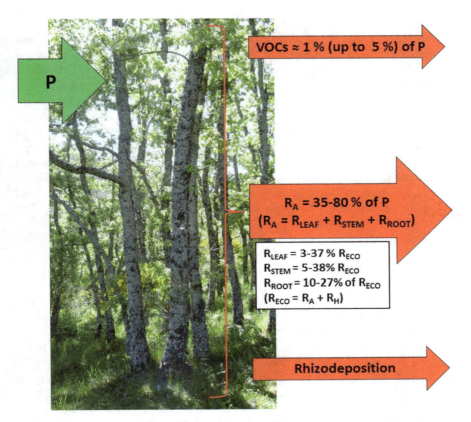

Fig. 10.4 Simplified schematic of tree carbon (C) budget in an oak forest. Net primary production can be estimated as the difference between photosynthesis (P) and overall C loss. Tree C loss occurs via emission of volatile organic compounds (VOCs), autotrophic respiration (R_A), and rhizodeposition of organic compounds. Autotrophic respiration is partitioned into leaf, stem and root respiration (R_{LEAF}, R_{STEM} and R_{ROOT}, respectively) and is expressed as a fraction of ecosystem respiration (R_{ECO}), which additionally integrates heterotrophic respiration (R_H) of living organisms in the soil

2007), and seems to be unaffected by changes in atmospheric [CO_2] (Nguyen 2009).

For experimental simplicity, research on rhizodeposition has focussed on herbaceous plants and young tree seedlings (<2 months old; reviewed by Jones et al. 2009), so that any extrapolation to large trees would be biased by the potential effect of plant age on C allocation patterns (Nguyen 2009). Assessments on the contribution of C rhizodeposits to plant C budgets is an experimentally elusive task, mainly because of the technical limitation that the soil imposes for the quantification of C flow through the rhizosphere, and the natural abundance of soil microorganisms that promptly assimilate rhizodeposits. We are aware of only one study in which the contribution of C rhizodeposits to tree C budgets has been surveyed in oak species. In mature trees of *Q. serrata* and *Q. glauca*, C loss via root exudation was proportional to that of root R (10%) on a root-weight basis, and accounted for 3% of NPP (Sun et al. 2017). Similarly, studies on annual herbs and tree seedlings using isotopic tracers estimate that the portion of C assimilated through photosynthesis and lost via rhizodeposition ranges between 2 and 11% (Jones et al. 2009; Preece and Peñuelas 2016). However, these rough estimates should be taken with caution due to the uncertain origin and fate of C within the rhizosphere (Jones et al. 2009). Finally, there is increasing evidence that roots of autotrophic plants can take up amino-acids hence assimilating organic C and nutrients (reviewed by Schmidt et al. 2013). This mixotrophic behaviour has been observed in *Q. petraea* roots during spring, when the strong C demand for growth before budburst cannot completely rely in new assimilates (Bréda et al. 2013), thus adding further complexity to the estimation of net C loss belowground.

10.5 Conclusions

Carbon assimilation traditionally occupies a predominant role in the study of tree C cycling, whereas processes involved in the C release from the plant to the atmosphere are comparatively understudied. Respiration, emission of VOCs and rhizodeposition constitute therefore the overlooked side of tree C budgets, included those of oak trees. The ecological importance, physiology and environmental control of R and VOC emissions in oak species have been reviewed; a simplified schematic of the tree C budget in oak—dominated stands is presented in Fig. 10.4. Autotrophic R is expected to consume half of the assimilated C, with respiration—to—photosynthesis ratios ranging between 35 and 80%. The relative contribution of leaves, stems and roots to the overall C respiratory expenditure largely fluctuates according to stand structure, composition and age (Table 10.2). The C loss associated with VOC emissions accounts for about 1% of gross P. This percentage may increase up to 5% during heat waves and under drought stress (Figs. 10.2 and 10.3), and even more if VOCs emitted from flowers, fruits, stems and roots are taken into account. Rhizodeposits represent an additional and non-negligible source of tree C loss. However, scarce literature on oak rhizodeposition discourages attempts to provide a rough quantitative estimation. Furthermore, environmental—induced fluctuations in oak R and VOC emissions are not necessarily proportional to fluctuations in C assimilation. Thus, the assumption of homeostatic ratios between C loss and C gain should be revisited as it might lead to erroneous predictions on the strength of oak stands as C sinks in a climate changing world, conclusion that could be extrapolated to other tree taxa. A comprehensive understanding of oak C loss comparable to that of photosynthesis would be therefore necessary to accurately assess oak C cycling in scenarios of climate change.

Acknowledgements This book chapter has received funding from the FWO and the European Union's Horizon 2020 research and innovation programme under the Marie Skłodowska-Curie grant agreement No 665501, and from the "Legado de González-Esparcia" granted to RLS. J R-C acknowledges the support of the Spanish Ministry of Economy and Competitiveness via the "Ramón y Cajal" programme.

References

Amthor J (2000) The McCree–de wit-penning de Vries-Thornley respiration paradigms: 30 years later. Ann Bot 86:1–20
Asensio D, Penuelas J, Ogaya R, Llusià J (2007) Seasonal soil VOC exchange rates in a Mediterranean holm oak forest and their responses to drought conditions. Atmos Environ 41:2456–2466
Atkin OK, Macherel D (2009) The crucial role of plant mitochondria in orchestrating drought tolerance. Ann Bot 103:581–597
Atkin OK, Tjoelker MG (2003) Thermal acclimation and the dynamic response of plant respiration to temperature. Trends Plant Sci 8:343–351
Atkin OK, Bloomfield KJ, Reich PB et al (2015) Global variability in leaf respiration in relation to climate, plant functional types and leaf traits. New Phytol 206:614–636
Atkinson R, Arey J (2003) Gas-phase tropospheric chemistry of biogenic volatile organic compounds: a review. Atmos Environ 37:197–219
Aubrey DP, Teskey RO (2009) Root-derived CO_2 efflux via xylem stream rivals soil CO_2 efflux. New Phytol 184:35–40
Ávila E, Herrera A, Tezara W (2014) Contribution of stem CO_2 fixation to whole-plant carbon balance in nonsucculent species. Photosynthetica 52:3–15
Bais HP, Park SW, Weir TL et al (2004) How plants communicate using the underground information superhighway. Trends Plant Sci 9:26–32
Bashir O, Khan K, Hakeem K et al (2016) Soil microbe diversity and root exudates as important aspects of rhizosphere ecosystem. In: Hakeem K, Akhtar M (eds) Plant, soil and microbes. Springer International Publishing, Switzerland, pp 337–357
Behnke K, Ehlting B, Teuber M et al (2007) Transgenic, non-isoprene emitting poplars don't like it hot. Plant J 51:485–499
Behnke K, Kleist E, Uerlings R et al (2009) RNAi-mediated suppression of isoprene biosynthesis in hybrid poplar impacts ozone tolerance. Tree Physiol 29:725–736
Behnke K, Loivamäki M, Zimmer I et al (2010) Isoprene emission protects photosynthesis in sunfleck exposed Grey poplar. Photosynth Res 104:5–17
Behnke K, Grote R, Brüggemann N et al (2012) Isoprene emission-free poplars—a chance to reduce the impact from poplar plantations on the atmosphere. New Phytol 194:70–82
Bertin N, Staudt M (1996) Effect of water stress on monoterpene emissions from young potted holm oak (*Quercus ilex* L.) trees. Oecologia 107:456–462
Berveiller D, Kierzkowski D, Damesin C (2007) Interspecific variability of stem photosynthesis among tree species. Tree Physiol 27:53–61
Bloemen J, Agneessens L, Van Meulebroek L et al (2014) Stem girdling affects the quantity of CO_2 transported in xylem as well as CO_2 efflux from soil. New Phytol 201:897–907
Bolstad P, Mitchell K, Vose J (1999) Foliar temperature-respiration response functions for broad-leaved tree species in the southern Appalachians. Tree Physiol 19:871–878
Bolstad PV, Reich P, Lee T (2003) Rapid temperature acclimation of leaf respiration rates in *Quercus alba* and *Quercus rubra*. Tree Physiol 23:969–976
Bourtsoukidis E, Kawaletz H, Radacki D et al (2014) Impact of flooding and drought conditions on the emission of volatile organic compounds of *Quercus robur* and *Prunus serotina*. Trees Struct Funct 28:193–204

Bracho-Nunez A, Welter S, Staudt M, Kesselmeier J (2011) Plant-specific volatile organic compound emission rates from young and mature leaves of Mediterranean vegetation. J Geophys Res 116:D16304

Bracho-Nunez A, Knothe NM, Welter S et al (2013) Leaf level emissions of volatile organic compounds (VOC) from some Amazonian and Mediterranean plants. Biogeosciences 10:5855–5873

Bréda N, Maillard P, Montpied P et al (2013) Isotopic evidence in adult oak trees of a mixotrophic lifestyle during spring reactivation. Soil Biol Biochem 58:136–139

Brilli F, Gioli B, Fares S et al (2016) Rapid leaf development drives the seasonal pattern of volatile organic compound (VOC) fluxes in a "coppiced" bioenergy poplar plantation. Plant Cell Environ 39:539–555

Burton A, Pregitzer KS, Ruess R et al (2002) Root respiration in North American forests: effects of nitrogen concentration and temperature across biomes. Oecologia 131:559–568

Burton AJ, Melillo JM, Frey SD (2008) Adjustment of forest ecosystem root respiration as temperature warms. J Integr Plant Biol 50:1467–1483

Cannell MGR, Thornley JHM (2000) Modelling the components of plant respiration: some guiding principles. Ann Bot 85:45–54

Cappellin L, Algarra Alarcon A, Herdlinger-Blatt I et al (2017) Field observations of volatile organic compound (VOC) exchange in red oaks. Atmos Chem Phys 17: 4218-4207

Coe JM, McLaughlin SB (1980) Winter season corticular photosynthesis in *Cornus florida, Acer rubrum, Quercus alba*, and *Liriodendron tulipifera*. For Sci 26:561–566

Collier DE, Thibodeau BA (1995) Changes in respiration and chemical content during autumnal senescence of *Populus tremuloides* and *Quercus rubra* leaves. Tree Physiol 15:759–764

Copolovici L, Vaartnou F, Estrada MP, Niinemets U (2014) Oak powdery mildew (*Erysiphe alphitoides*)-induced volatile emissions scale with the degree of infection in *Quercus robur*. Tree Physiol 34:1399–1410

Curtis PS, Vogel CS, Gough CM et al (2005) Respiratory carbon losses and the carbon-use efficiency of a northern hardwood forest, 1999–2003. New Phytol 167:437–456

Dakora FD, Phillips DA (2002) Root exudates as mediators of mineral acquisition in low-nutrient environments. Plant Soil 245:35–47

Dannoura M, Kominami Y, Tamai K et al (2006) Development of an automatic chamber system for long-term measurements of CO_2 flux from roots. Tellus 58:502–512

Davidson EA, Janssens IA, Luo Y (2006) On the variability of respiration in terrestrial ecosystems: moving beyond Q_{10}. Glob Change Biol 12:154–164

Delory BM, Delaplace P, Fauconnier ML, du Jardin P (2016) Root-emitted volatile organic compounds: can they mediate belowground plant-plant interactions? Plant Soil 402:1–26

Edwards NT, Hanson PJ (1996) Stem respiration in a closed-canopy upland oak forest. Tree Physiol 16:433–439

Farquhar GD, Von Caemmerer S, Berry JA (1980) A biochemical model of photosynthestic CO_2 assimilation in leaves of C3 species. Planta 90:78–90

Fatichi S, Leuzinger S, Korner C (2014) Moving beyond photosynthesis: from carbon source to sink-driven vegetation modeling. New Phytol 201:1086–1095

Fineschi S, Loreto F, Staudt M, Peñuelas J (2013) Diversification of volatile isoprenoid emissions from trees: evolutionary and ecological perspectives. In: Niinemets Ü, Monson RK (eds) Biology, controls and models of tree volatile organic compound emissions. Tree Physiol vol 5. Springer, Dordrecht, pp 1–20

Fischbach RJ, Staudt M, Zimmer I et al (2002) Seasonal pattern of monoterpene synthase activities in leaves of the evergreen tree *Quercus ilex*. Physiol Plant 114:354–360

Frost CJ, Hunter MD (2008) Herbivore-induced shifts in carbon and nitrogen allocation in red oak seedlings. New Phytol 178:835–845

Gaona-Colmán E, Blanco MB, Barnes I, Teruel MA (2016) Gas-phase ozonolysis of β-ocimene: temperature dependent rate coefficients and product distribution. Atmos Environ 147:46–54

Ghirardo A, Heller W, Fladung M et al (2012) Function of defensive volatiles in pedunculate oak (*Quercus robur*) is tricked by the moth *Tortrix viridana*. Plant, Cell Environ 35:2192–2207

Gifford RM (2003) Plant respiration in productivity models: conceptualisation, representation and issues for global terrestrial carbon-cycle research. Funct Plant Biol 30:171

Goulden ML, Munger W, Fan S et al (1996) Measurements of carbon sequestration by long-term eddy covariance: methods and a critical evaluation of accuracy. Glob Change Biol 2:169–182

Hallquist M, Wenger JC, Baltensperger U et al (2009) The formation, properties and impact of secondary organic aerosol: current and emerging issues. Atmos Chem Phys 9:5155–5236

Hanson PJ, Edwards NT, Garten CT, Andrews JA (2000) Separating root and soil microbial contributions to soil respiration: a review of methods and observations. Biogeochemistry 48:115–146

Harvey CM, Sharkey TD (2016) Exogenous isoprene modulates gene expression in unstressed *Arabidopsis thaliana* plants. Plant Cell Environ 39:1251–1263

Harvey CM, Li Z, Tjellström H et al (2015) Concentration of isoprene in artificial and thylakoid membranes. J Bioenerg Biomembr 47:419–429

Havaux M (2014) Carotenoid oxidation products as stress signals in plants. Plant J 79:597–606

Henry A, Doucette W, Norton J, Bugbee B (2007) Changes in crested wheatgrass root exudation caused by flood, drought, and nutrient stress. J Environ Qual 36:904–912

Hunt S (2003) Measurements of photosynthesis and respiration in plants. Physiol Plant 117: 314–325

Ito A, Inatomi M, Mo W et al (2007) Examination of model-estimated ecosystem respiration using flux measurements from a cool-temperate deciduous broad-leaved forest in central Japan. Tellus 59:616–624

Jones DL, Hodge A, Kuzyakov Y (2004) Plant and mycorrhizal regulation of rhizodeposition. New Phytol 163:459–480

Jones DL, Nguyen C, Finlay RD (2009) Carbon flow in the rhizosphere: carbon trading at the soil-root interface. Plant Soil 321:5–33

Karl T, Harley P, Emmons L et al (2010) Efficient atmospheric cleansing of oxidized organic trace gases by vegetation. Science 330:816–819

Kattge J, Díaz S, Lavorel S et al (2011) TRY—a global database of plant traits. Glob Change Biol 17:2905–2935

Kesselmeier J, Staudt M (1999) Biogenic volatile organic compounds (VOC): an overview on emission, physiology and ecology. J Atmos Chem 33:23–88

Kesselmeier J, Ciccioli P, Kuhn U et al (2002) Volatile organic compound emissions in relation to plant carbon fixation and the terrestrial carbon budget. Global Biogeochem Cycles 16:1–9

Körner C (2015) Paradigm shift in plant growth control. Curr Opin Plant Biol 25:107–114

Kosugi Y, Tanaka H, Takanashi S et al (2005) Three years of carbon and energy fluxes from Japanese evergreen broad-leaved forest. Agric For Meteorol 132:329–343

Kreuzwieser J, Rennenberg H (2013) Flooding-driven emissions from trees. In: Niinemets Ü, Monson RK (eds) Biology, controls and models of tree volatile organic compound emissions. Springer, Dordrecht, pp 237–252

Lanoue A, Burlat V, Henkes GJ et al (2010) De novo biosynthesis of defense root exudates in response to *Fusarium* attack in barley. New Phytol 185:577–588

Laureano RG, Lazo YO, Linares JC et al (2008) The cost of stress resistance: construction and maintenance costs of leaves and roots in two populations of *Quercus ilex*. Tree Physiol 28:1721–1728

Lavoir AV, Staudt M, Schnitzler JP et al (2009) Drought reduced monoterpene emissions from the evergreen Mediterranean oak *Quercus ilex*: results from a throughfall displacement experiment. Biogeosciences 6:1167–1180

Lee TD, Reich PB, Bolstad PV (2005) Acclimation of leaf respiration to temperature is rapid and related to specific leaf area, soluble sugars and leaf nitrogen across three temperate deciduous tree species. Funct Ecol 19:640–647

Lempereur M, Martin-StPaul NK, Damesin C et al (2015) Growth duration is a better predictor of stem increment than carbon supply in a Mediterranean oak forest: implications for assessing forest productivity under climate change. New Phytol 207:579–590

Li Q, Chen J, Moorhead DL (2012) Respiratory carbon losses in a managed oak forest ecosystem. For Ecol Manage 279:1–10

Limousin JM, Misson L, Lavoir AV et al (2010) Do photosynthetic limitations of evergreen *Quercus ilex* leaves change with long-term increased drought severity? Plant Cell Environ 33:863–875

Lockhart JA (1965) An analysis of irreversible plant cell elongation. J Theor Biol 8:264–275

Loivamäki M, Mumm R, Dicke M, Schnitzler J-P (2008) Isoprene interferes with the attraction of bodyguards by herbaceous plants. Proc Natl Acad Sci 105:17430–17435

Long BM, Bahar NHA, Atkin OK (2015) Contributions of photosynthetic and non-photosynthetic cell types to leaf respiration in *Vicia faba* L. and their responses to growth temperature. Plant Cell Environ 38:2263–2276

López B, Sabate S, Gracia CA (2001) Annual and seasonal changes in fine root biomass of a *Quercus ilex* L. forest. Plant Soil 230:125–134

Loreto F, Schnitzler J-P (2010) Abiotic stresses and induced BVOCs. Trends Plant Sci 15: 154–166

Loreto F, Sharkey TD (1990) A gas-exchange study of photosynthesis and isoprene emission in *Quercus rubra* L. Planta 182:523–531

Loreto F, Bagnoli F, Fineschi S (2009) One species, many terpenes: matching chemical and biological diversity. Trends Plant Sci 14:416–420

Lu X, Tang K, Li P (2016) Plant metabolic engineering strategies for the production of pharmaceutical terpenoids. Front Plant Sci. https://doi.org/10.3389/fpls.2016.0164

Luan J, Liu S, Wang J et al (2011) Rhizospheric and heterotrophic respiration of a warm-temperate oak chronosequence in China. Soil Biol Biochem 43:503–512

Luyssaert S, Inglima I, Jung M et al (2007) CO_2 balance of boreal, temperate, and tropical forests derived from a global database. Glob Change Biol 13:2509–2537

Mabberley DJ (2008) Mabberley's plant book: a portable dictionary of plants, their classifications and uses. Cambridge University Press, UK

Martinez F, Lazo YO, Fernandez-Galiano RM, Merino JA (2002) Chemical composition and construction cost for roots of Mediterranean trees, shrub species and grassland communities. Plant Cell Environ 25:601–608

Matsui K (2006) Green leaf volatiles: hydroperoxide lyase pathway of oxylipin metabolism. Curr Opin Plant Biol 9:274–280

Matsui K (2016) A portion of plant airborne communication is endorsed by uptake and metabolism of volatile organic compounds. Curr Opin Plant Biol 32:24–30

McCree KJ (1970) An equation for the rate of respiration of white clover plants grown under controlled conditions. In: Prediction and measurement of photosynthetic productivity. Proceedings of the IBP/PP technical meeting, Trebon, 14–21 Sept 1969, pp 221–229

McGuire MA, Teskey RO (2002) Microelectrode technique for in situ measurement of carbon dioxide concentrations in xylem sap of trees. Tree Physiol 22:807–811

Misztal PK, Hewitt CN, Wildt J et al (2015) Atmospheric benzenoid emissions from plants rival those from fossil fuels. Sci Rep 5:12064

Miyama T, Kominami Y, Tamai K et al (2006) Components and seasonal variation of night-time total ecosystem respiration in a Japanese broad-leaved secondary forest. Tellus 58:550–559

Miyazawa S (1998) Slow leaf development of evergreen broad-leaved tree species in Japanese warm temperate forests. Ann Bot 82:859–869

Molchanov AG (2009) Effect of moisture availability on photosynthetic productivity and autotrophic respiration of an oak stand. Russ J Plant Physiol 56:769–779

Monson RK, Jones RT, Rosenstiel TN, Schnitzler JP (2013) Why only some plants emit isoprene. Plant Cell Environ 36:503–516

Müller A, Kaling M, Faubert P et al (2015) Isoprene emission by poplar is not important for the feeding behaviour of poplar leaf beetles. BMC Plant Biol 15:1

Nagahashi G, Douds D, Ferhatoglu Y (2010) Functional categories of root exudate compounds and their relevance to AM fungal growth. In: Koltai H, Kapulnik Y (eds) Arbuscular mycorrhizas: physiology and function. Springer, Berlin, pp 33–56

Negisi K (1982) Diurnal fluctuations of the stem bark respiration in relationship to the wood temperature in standing young *Pinus densiflora*, *Chamaecyparis obtusa* and *Quercus myrsinaefolia* trees. J Japanese For Soc 64:315–319

Nguyen C (2009) Rhizodeposition of organic C by plant: mechanisms and controls. In: Lichtfouse E, Navarrete M, Debaeke P et al (eds) Sustainable agriculture. Springer, Netherlands, pp 97–123

Niinemets U (2014) Cohort-specific tuning of foliage physiology to interacting stresses in evergreens. Tree Physiol 34:1301–1304

Niinemets U, Reichstein M, Staudt M et al (2002) Stomatal constraints may affect emission of oxygenated monoterpenoids from the foliage of *Pinus pinea*. Plant Physiol 130:1371–1385

Niinemets Ü, Fares S, Harley P, Jardine KJ (2014) Bidirectional exchange of biogenic volatiles with vegetation: emission sources, reactions, breakdown and deposition. Plant Cell Environ 37:1790–1809

Palmer-Young EC, Veit D, Gershenzon J, Schuman MC (2015) The Sesquiterpenes(E)-ß-Farnesene and (E)-α-Bergamotene quench ozone but fail to protect the wild tobacco *Nicotiana attenuata* from Ozone, UVB, and drought stresses. PLoS ONE 10:e0127296

Paris CI, Llusia J, Peñuelas J (2010) Changes in monoterpene emission rates of *Quercus ilex* infested by aphids tended by native or invasive *Lasius* ant species. J Chem Ecol 36:689–698

Park J-H, Goldstein AH, Timkovsky J et al (2013) Active atmosphere-ecosystem exchange of the vast majority of detected volatile organic compounds. Science 341:643–647

Peñuelas J, Staudt M (2010) BVOCs and global change. Trends Plant Sci 15:133–144

Pérez-de-Lis G, García-González I, Rozas V, Olano JM (2016) Feedbacks between earlywood anatomy and non-structural carbohydrates affect spring phenology and wood production in ring-porous oaks. Biogeosci Discuss 35:1–19

Piao S, Luyssaert S, Ciais P et al (2010) Forest annual carbon cost: a global-scale analysis of autotrophic respiration. Ecology 91:652–661

Pier PA, McDuffie C (1997) Seasonal isoprene emission rates and model comparisons using whole-tree emissions from white oak. J Geophys Res 102:23963–23971

Poorter H, Niklas KJ, Reich PB et al (2012) Biomass allocation to leaves, stems and roots: meta-analysis of interspecific variation and environmental control. New Phytol 193:30–50

Preece C, Peñuelas J (2016) Rhizodeposition under drought and consequences for soil communities and ecosystem resilience. Plant Soil 409:1–17

Rambal S, Lempereur M, Limousin JM et al (2014) How drought severity constrains gross primary production (GPP) and its partitioning among carbon pools in a *Quercus ilex* coppice? Biogeosciences 11:6855–6869

Rasulov B, Talts E, Niinemets Ü (2016) Spectacular oscillations in plant isoprene emission under transient conditions explain the enigmatic CO_2 response. Plant Physiol 172:2275–2285

Read DB, Bengough AG, Gregory PJ et al (2003) Plant roots release phospholipid surfactants that modify the physical and chemical properties of soil. New Phytol 157:315–326

Reich PB, Walters MB, Ellsworth DS et al (1998) Relationships of leaf dark respiration to leaf nitrogen, specific leaf area and leaf life-span: a test across biomes and functional groups. Oecologia 114:471–482

Reich PB, Tjoelker MG, Pregitzer KS et al (2008) Scaling of respiration to nitrogen in leaves, stems and roots of higher land plants. Ecol Lett 11:793–801

Reichstein M, Tenhunen J, Roupsard O et al (2002) Ecosystem respiration in two Mediterranean evergreen Holm Oak forests: drought effects and decomposition dynamics. Funct Ecol 16:27–39

Reich et al (2016) Boreal and temperate trees show strong acclimation of respiration to warming. Nature531(7596): 633–636

Rey A, Pegoraro E, Tedeschi V et al (2002) Annual variation in soil respiration and its components in a coppice oak forest in Central Italy. Glob Change Biol 8:851–866

Rodríguez-Calcerrada J, Jaeger C, Limousin JM et al (2011) Leaf CO_2 efflux is attenuated by acclimation of respiration to heat and drought in a Mediterranean tree. Funct Ecol 25:983–995

Rodríguez-Calcerrada J, Limousin JM, Martin-Stpaul NK et al (2012) Gas exchange and leaf aging in an evergreen oak: causes and consequences for leaf carbon balance and canopy respiration. Tree Physiol 32:464–477

Rodríguez-Calcerrada J, Buatois B, Chiche E et al (2013) Leaf isoprene emission declines in *Quercus pubescens* seedlings experiencing drought—any implication of soluble sugars and mitochondrial respiration? Environ Exp Bot 85:36–42

Rodríguez-Calcerrada J, Martin-StPaul NK, Lempereur M et al (2014) Stem CO_2 efflux and its contribution to ecosystem CO_2 efflux decrease with drought in a Mediterranean forest stand. Agric For Meteorol 195–196:61–72

Rodríguez-Calcerrada J, López R, Salomón R et al (2015a) Stem CO_2 efflux in six co-occurring tree species: underlying factors and ecological implications. Plant Cell Environ 38:1104–1115

Rodríguez-Calcerrada J, Salomón R, Gil L (2015b) Transporte y reciclaje de CO_2 en el interior del árbol: factores que complican la estimación de la respiración leñosa a través de la emisión radial de CO_2. Bosque 36:5–14

Salomón R, Valbuena-Carabaña M, Rodríguez-Calcerrada J et al (2015) Xylem and soil CO_2 fluxes in a *Quercus pyrenaica* Willd. coppice: root respiration increases with clonal size. Ann For Sci 72:1065–1078

Salomón R, Valbuena-Carabaña M, Teskey R et al (2016a) Seasonal and diel variation in xylem CO_2 concentration and sap pH in sub-Mediterranean oak stems. J Exp Bot 67:2817–2827

Salomón RL, Valbuena-Carabaña M, Gil L et al (2016b) Temporal and spatial patterns of internal and external stem CO_2 fluxes in a sub-Mediterranean oak. Tree Physiol 36:1409–1421

Salomón RL, Limousin JM, Ourcival JM et al (2017) Stem hydraulic capacitance decreases with drought stress: implications for modelling tree hydraulics in the Mediterranean oak *Quercus ilex*. Plant Cell Environ 40:1379–1391

Sasaki K, Saito T, Lämsä M et al (2007) Plants utilize isoprene emission as a thermotolerance mechanism. Plant Cell Physiol 48:1254–1262

Saunier A, Ormeño E, Wortham H et al (2017) Chronic drought decreases anabolic and catabolic BVOC emissions of *Quercus pubescens* in a Mediterranean Forest. Front Plant Sci 8:71

Saveyn A, Steppe K, Lemeur R (2007a) Daytime depression in tree stem CO_2 efflux rates: is it caused by low stem turgor pressure? Ann Bot 99:477–485

Saveyn A, Steppe K, Lemeur R (2007b) Drought and the diurnal patterns of stem CO_2 efflux and xylem CO_2 concentration in young oak (*Quercus robur*). Tree Physiol 27:365–374

Saveyn A, Steppe K, Lemeur R (2008) Report on non-temperature related variations in CO_2 efflux rates from young tree stems in the dormant season. Trees 22:165–174

Schallhart S, Rantala P, Nemitz E et al (2016) Characterization of total ecosystem-scale biogenic VOC exchange at a Mediterranean oak-hornbeam forest. Atmos Chem Phys 16:7171–7194

Schiestl FP (2010) The evolution of floral scent and insect chemical communication. Ecol Lett 13:643–656

Schmidt S, Raven JA, Paungfoo-Lonhienne C (2013) The mixotrophic nature of photosynthetic plants. Funct Plant Biol 40:425–438

Schnitzler J, Lehning A, Steinbrecher R (1997) Seasonal pattern of isoprene synthase activity in *Quercus robur* leaves and its significance for modeling isoprene emission rates. Bot Acta 110:240–243

Sindelarova K, Granier C, Bouarar I et al (2014) Global data set of biogenic VOC emissions calculated by the MEGAN model over the last 30 years. Atmos Chem Phys 14:9317–9341

Slot M, Kitajima K (2015) General patterns of acclimation of leaf respiration to elevated temperatures across biomes and plant types. Oecologia 177:885–900

Smith NG, Dukes JS (2013) Plant respiration and photosynthesis in global-scale models: incorporating acclimation to temperature and CO_2. Glob Change Biol 19:45–63

Sperlich D, Barbeta A, Ogaya R et al (2016) Balance between carbon gain and loss under long-term drought: impacts on foliar respiration and photosynthesis in *Quercus ilex* L. J Exp Bot 67:821–833

Staudt M, Bertin N (1998) Light and temperature dependence of the emission of cyclic and acyclic monoterpenes from holm oak *Quercus ilex* L. leaves. Plant Cell Environ 21:385–395

Staudt M, Lhoutellier L (2007) Volatile organic compound emission from holm oak infested by gypsy moth larvae: evidence for distinct responses in damaged and undamaged leaves. Tree Physiol 27:1433–1440

Staudt M, Lhoutellier L (2011) Monoterpene and sesquiterpene emissions from *Quercus coccifera* exhibit interacting responses to light and temperature. Biogeosciences 8:2757–2771

Staudt M, Wolf A, Kesselmeier J (2000) Influence of environmental factors on the emissions of gaseous formic and acetic acids from orange (*Citrus sinensis* L.) foliage. Biogeochemistry 48:199–216

Staudt M, Joffre R, Rambal S, Kesselmeier J (2001a) The effect of elevated CO_2 on terpene emission, leaf structure and related physiological parameters of young *Quercus ilex*. Tree Physiol 21:437–445

Staudt M, Mandl N, Joffre R, Rambal S (2001b) Intraspecific variability of monoterpene composition emitted by *Quercus ilex* leaves. Can J For Res 31:174–180

Staudt M, Rambal S, Joffre R (2002) Impact of drought on seasonal monoterpene emissions from *Quercus ilex* in southern France. J Geophys Res 107:4602

Staudt M, Joffre R, Rambal S (2003) How growth conditions affect the capacity of *Quercus ilex* leaves to emit monoterpenes. New Phytol 158:61–73

Staudt M, Mir C, Joffre R et al (2004) Isoprenoid emissions of *Quercus* spp. (*Q. suber* and *Q. ilex*) in mixed stands contrasting in interspecific genetic introgression. New Phytol 163:573–584

Staudt M, Ennajah A, Mouillot F, Joffre R (2008) Do volatile organic compound emissions of Tunisian cork oak populations originating from contrasting climatic conditions differ in their responses to summer drought? Can J For Res 38:2965–2975

Staudt M, Jackson B, El-aouni H et al (2010) Volatile organic compound emissions induced by the aphid *Myzus persicae* differ among resistant and susceptible peach cultivars and a wild relative. Tree Physiol 30:1320–1334

Staudt M, Morin X, Chuine I (2017a) Contrasting direct and indirect effects of warming and drought on isoprenoid emissions from Mediterranean oaks. Reg Environ Change 17:2121–2133

Staudt M, Bourgeois I, Al Halabi R et al (2017b) New insights into the parametrization of temperature and light responses of mono-and sesquiterpene emissions from Aleppo pine and rosemary. Atmos Environ 152:212–221

Steinbrecher R, Contran N, Gugerli F et al (2013) Inter- and intra-specific variability in isoprene production and photosynthesis of Central European oak species. Plant Biol 15:148–156

Steppe K, De Pauw DJW, Lemeur R, Vanrolleghem A (2006) A mathematical model linking tree sap flow dynamics to daily stem diameter fluctuations and radial stem growth. Tree Physiol 26:257–273

Sun et al (2017) Relationship between fine-root exudation and respiration of two Quercus species in a Japanese temperate forest. Tree Physiol 37(8): 1011–1020

Tattini M, Velikova V, Vickers C et al (2014) Isoprene production in transgenic tobacco alters isoprenoid, non-structural carbohydrate and phenylpropanoid metabolism, and protects photosynthesis from drought stress. Plant Cell Environ 37:1950–1964

Teskey RO, McGuire MA (2002) Carbon dioxide transport in xylem causes errors in estimation of rates of respiration in stems and branches of trees. Plant Cell Environ 25:1571–1577

Teskey RO, Saveyn A, Steppe K, McGuire MA (2008) Origin, fate and significance of CO_2 in tree stems. New Phytol 177:17–32

Tholl D, Sohrabi R, Huh J-H, Lee S (2011) The biochemistry of homoterpenes—common constituents of floral and herbivore-induced plant volatile bouquets. Phytochemistry 72:1635–1646

Thornley JHM (1970) Respiration, growth and maintenance in plants. Nature 227:304–305

Thornley JHM (2011) Plant growth and respiration re-visited: maintenance respiration defined—it is an emergent property of, not a separate process within, the system—and why the respiration: photosynthesis ratio is conservative. Ann Bot 108:1365–1380

Turnbull MH, Whitehead D, Tissue DT et al (2001) Responses of leaf respiration to temperature and leaf characteristics in three deciduous tree species vary with site water availability. Tree Physiol 21:571–578

Unger S, Máguas C, Pereira JS et al (2009) Partitioning carbon fluxes in a Mediterranean oak forest to disentangle changes in ecosystem sink strength during drought. Agric For Meteorol 149:949–961

Valbuena-Carabaña M, González-Martínez SC, Hardy OJ, Gil L (2007) Fine-scale spatial genetic structure in mixed oak stands with different levels of hybridization. Mol Ecol 16:1207–1219

Valentini R, Matteucci G, Dolman AJ et al (2000) Respiration as the main determinant of carbon balance in European forests. Nature 404:861–865

Van Loon JJA, De Boer JG, Dicke M (2000) Parasitoid-plant mutualism: parasitoid attack of herbivore increases plant reproduction. Entomol Exp Appl 97:219–227

Van Oijen M, Schapendonk A, Hoglind M (2010) On the relative magnitudes of photosynthesis, respiration, growth and carbon storage in vegetation. Ann Bot 105:793–797

Vanzo E, Merl-Pham J, Velikova V et al (2016) Modulation of protein S-nitrosylation by isoprene emission in poplar. Plant Physiol 170:1945–1961

Varone L, Gratani L (2015) Leaf respiration responsiveness to induced water stress in Mediterranean species. Environ Exp Bot 109:141–150

Velikova V, Tsonev T, Pinelli P et al (2005) Localized ozone fumigation system for studying ozone effects on photosynthesis, respiration, electron transport rate and isoprene emission in field-grown Mediterranean oak species. Tree Physiol 25:1523–1532

Velikova V, Sharkey TD, Loreto F (2012) Stabilization of thylakoid membranes in isoprene-emitting plants reduces formation of reactive oxygen species. Plant Signal Behav 7:139–141

Velikova V, Müller C, Ghirardo A et al (2015) Knocking down of isoprene emission modifies the lipid matrix of thylakoid membranes and influences the chloroplast ultrastructure in poplar. Plant Physiol 168:859–870

Verdú JR, Lobo JM, Numa C et al (2007) Acorn preference by the dung beetle, *Thorectes lusitanicus*, under laboratory and field conditions. Anim Behav 74:1697–1704

Vickers CE, Gershenzon J, Lerdau MT, Loreto F (2009a) A unified mechanism of action for volatile isoprenoids in plant abiotic stress. Nat Chem Biol 5:283–291

Vickers CE, Possell M, Cojocariu CI et al (2009b) Isoprene synthesis protects transgenic tobacco plants from oxidative stress. Plant Cell Environ 32:520–531

Vuts J, Woodcock CM, Sumner ME et al (2016) Responses of the two-spotted oak buprestid, *Agrilus biguttatus* (Coleoptera: Buprestidae), to host tree volatiles. Pest Manag Sci 72:845–851

Walter MH, Floss DS, Strack D (2010) Apocarotenoids: hormones, mycorrhizal metabolites and aroma volatiles. Planta 232:1–17

Wang M, Guan D-X, Han S-J, Wu J-L (2010) Comparison of eddy covariance and chamber-based methods for measuring CO_2 flux in a temperate mixed forest. Tree Physiol 30:149–163

Waring R, Landsberg J, Williams M (1998) Net primary production of forests: a constant fraction of gross primary production? Tree Physiol 18:129–134

Way DA, Yamori W (2014) Thermal acclimation of photosynthesis: on the importance of adjusting our definitions and accounting for thermal acclimation of respiration. Photosynth Res 119:89–100

Weissteiner S, Huetteroth W, Kollmann M et al (2012) cockchafer larvae smell host root scents in soil. PLoS One. doi:10.1371/journal.pone.0045827

Welter S, Bracho-Nuñez A, Mir C et al (2012) The diversification of terpene emissions in Mediterranean oaks: lessons from a study of *Quercus suber*, *Quercus canariensis* and its hybrid *Quercus afares*. Tree Physiol 32:1082–1091

Wright IJ, Reich PB, Atkin OK et al (2006) Irradiance, temperature and rainfall influence leaf dark respiration in woody plants: evidence from comparisons across 20 sites. New Phytol 169:309–319

Xu L, Baldocchi DD (2003) Seasonal trends in photosynthetic parameters and stomatal conductance of blue oak (*Quercus douglasii*) under prolonged summer drought and high temperature. Tree Physiol 23:865–877

Yang JY, Teskey RO, Wang CK (2012a) Stem CO_2 efflux of ten species in temperate forests in Northeastern China. Trees 26:1225–1235

Yang Q, Xu M, Chi Y et al (2012b) Temporal and spatial variations of stem CO_2 efflux of three species in subtropical China. J Plant Ecol 5:229–237

Zaragoza-Castells J, Sánchez-Gómez D, Valladares F et al (2007) Does growth irradiance affect temperature dependence and thermal acclimation of leaf respiration? Insights from a Mediterranean tree with long-lived leaves. Plant Cell Environ 30:820–833

Zaragoza-Castells J, Sánchez-Gómez D, Hartley IP et al (2008) Climate-dependent variations in leaf respiration in a dry-land, low productivity Mediterranean forest: the importance of acclimation in both high-light and shaded habitats. Funct Ecol 22:172–184

Chapter 11
Photoprotective Mechanisms in the Genus *Quercus* in Response to Winter Cold and Summer Drought

José Ignacio García-Plazaola, Antonio Hernández,
Beatriz Fernández-Marín, Raquel Esteban, José Javier
Peguero-Pina, Amy Verhoeven and Jeannine Cavender-Bares

Abstract The photosynthetic apparatus must cope with the excess energy when light exceeds what plant can use. Under these conditions, plants, including oaks, can activate an array of "photoprotection mechanisms", which are crucial to understand the relationships between plants and their environment. First, this chapter gives a general description of the different photoprotection mechanisms that operate at several levels: (i) the reduction of light collection by chlorophylls, (ii) the enhancement of the metabolic use of light energy absorbed, (iii) the enhancement of the dissipation of the absorbed energy as heat, and (iv) the mechanisms for preventing and repairing oxidative damage (Sect. 11.1). These photoprotection mechanisms are subsequently analyzed in detail for evergreen oaks exposed to winter stress (Sect. 11.2) and for both deciduous and evergreen oaks under drought-stress conditions (Sect. 11.3), with particular emphasis on the role of free and enzymatic antioxidants, xanthophyll cycles and sustained engagement of dissipation. Afterwards, the chapter addresses with the need of photoprotection in deciduous oaks during autumn senescence associated to the risks of chlorophyll degradation and reactive oxygen species (ROS) generation (Sect. 11.4).

J. I. García-Plazaola (✉) · A. Hernández · B. Fernández-Marín · R. Esteban
Department of Plant Biology and Ecology, University of the Basque Country (UPV/EHU),
Apdo 644, Bilbao 48080, Spain
e-mail: joseignacio.garcia@ehu.es

J. J. Peguero-Pina
Unidad de Recursos Forestales, Centro de Investigación y
Tecnología Agroalimentaria de Aragón, Gobierno de Aragón,
Avda. Montañana 930, 50059 Saragossa, Spain

A. Verhoeven
Biology Department (OWS352), University of St Thomas,
2115 Summit Ave, St. Paul, MN, USA

J. Cavender-Bares
Department of Ecology, Evolution & Behavior, University of Minnesota,
1479 Gortner Ave, St. Paul, MN, USA

© Springer International Publishing AG 2017
E. Gil-Pelegrín et al. (eds.), *Oaks Physiological Ecology. Exploring the Functional Diversity of Genus Quercus L.*, Tree Physiology 7,
https://doi.org/10.1007/978-3-319-69099-5_11

11.1 Photoprotection in Leaves, the Basics

Whenever light exceeds what plants can use, the photosynthetic apparatus has the unavoidable need to get rid of the excess energy by activating an array of "photoprotection mechanisms". Photoprotection is foundational to understand the relationships between plants and their environment and it has been recently reviewed by several authors (e.g. Goss and Lepetit 2015; Ruban 2016; Wobbe et al. 2015). Basically, photoprotection mechanisms operate at several hierarchic levels that first, reduce light collection by chlorophylls (Sect. 11.1.2); second, enhance the metabolic use of light energy absorbed (Sect. 11.1.3); third, enhance the dissipation of the absorbed energy as heat (Sect. 11.1.4); and fourth, prevent and repair oxidative damage (Sect. 11.1.5).

11.1.1 Dealing with Light Excess in Quercus Canopies

Applying a shade tolerance index that ranges from 0-untolerant to 5-maximum tolerance, Niinemets and Valladares (2006) classified 57 oak species as intermediate shade-tolerant or shade-intolerant species (index 1–3.5), reaching scores lower than those of the traditionally considered shade-tolerant genus, such as *Abies*, *Picea*, *Acer* or *Fagus* (index 4–5) (Fig. 11.1). As a consequence of this limited shade-tolerance, *Quercus* seedlings usually regenerate in the understory of relatively open canopies, but cannot compete under deep shade with seedlings of more shade-tolerant species, such as beech (Hansen et al. 2002). To survive in such open understories, oak seedlings have to be able to dynamically adjust photosynthetic activity and photoprotection mechanisms to changes in light environment, such as those generated by canopy opening, while maintaining a high photosynthetic gain (Hansen et al. 2002; Naidu and De Lucia 1997). However, even among *Quercus* species, the degree of such plasticity differs, with the more shade-tolerant being less plastic and more susceptible to high light (Rodríguez-Calcerrada et al. 2007).

11.1.2 Decreasing Light Absorption: Morphological and Biochemical Adjustments

When light interception by photosynthetic cells is in excess of what plants can use, it can be reduced efficiently through the development of plastic morphological modifications. Most of them are slowly reversible in the short-term and they can be grouped in two main categories: (1) changes in leaf anatomy and morphology (leaf and petiole angle, lamina size, leaf rolling, chloroplast movements) and (2) changes in leaf reflectance (pubescence, wax deposition, accumulation of red pigments) and/or regulation of antenna size. A comparative examination of the leaves developed in

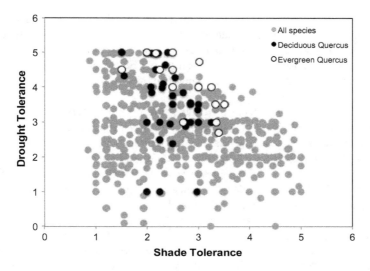

Fig. 11.1 Relationship between shade tolerance and drought tolerance indexes in *Quercus* species in the context of northern hemisphere trees and shrubs. Database published by Niinemets and Valladares (2006), (see the original publication for details on index calculation). Light grey symbols show data from all species included in the database, black symbols correspond to deciduous *Quercus* species, and white symbols to evergreen species

the outer (sun) and inner (shade) crown of a well-developed *Quercus* tree immediately reveals most of these characters, particularly in the more plastic species such as *Q. velutina* or *Q. coccifera*. Thus, when developed in a sunny position, leaves are thicker, with more layers of palisade parenchyma and higher epidermal cell thickness (Ashton and Berlyn 1994). Leaf angle also responds to canopy position and, for example in *Q. coccifera*, there is considerable variation in this character, with steeper leaves in the upper canopy and more horizontal shade leaves (Rubio de Casas et al. 2007). Leaf rolling is also common among Mediterranean *Quercus* species and it increases with irradiance within the canopy, reducing the effective light-interception area (Niinemets 2007). At the biochemical level, irradiance acclimation within *Quercus* canopies is reflected by a shift to higher values of chlorophyll (Chl) *a/b* and Carotenoid/Chl ratios in more exposed positions (Hansen et al. 2002). As Chl *b* is only present in light harvesting complexes (LHCs) and carotenoids are mostly involved in photoprotection, these trends are indicative of a change from efficient light harvesting under low irradiance to high capacity for energy dissipation in sunny positions.

Other foliar traits that greatly affect leaf interaction with the environment are pubescence (presence of hairs) and glaucescence (presence of waxes). The presence of a hairy epidermis increases both the thickness of the boundary layer (Bacelar et al. 2004) and the leaf hydrophobicity (Brewer et al. 1991) and reduces the susceptibility to herbivory (Karioti et al. 2011). In *Q. ilex*, adaxial trichomes can also contribute to water absorption thanks to their high hydrophilicity (Fernández et al. 2014). However, the exact selective pressure that makes a plant species pubescent is difficult to demonstrate. Independently of what is the main function of trichomes, pubescence always generates a change in the spectral properties of leaves (Holmes and Keiller 2002) increasing light reflectance at all wavelengths of the spectrum and consequently, reducing leaf absorptance up to 50% in some species (Ehleringer 1982).

In the case of *Quercus* species, trichomes are mainly present in the abaxial leaf surface, but some species (i.e.: *Q. ilex*, *Q. pyrenaica*, *Q. pubescens*, *Q. macranthera*) have trichomes in the adaxial side of their leaves, at least during the early stages of their development (Bussotti and Grossoni 1997; Hardin 1979). An unavoidable consequence of the presence of leaf trichomes is that the uniform photon scattering at all wavelengths caused by the hairy surface reduces light reaching the mesophyll (Karabourniotis and Bornman 1999). For example, when trichomes were mechanically removed from *Q. ilex* leaves, reflectance decreased by 5% (Morales et al. 2002). Furthermore, this treatment also enhanced susceptibility to photoinhibition, demonstrating the photoprotective role of leaf pubescence for this species (Morales et al. 2002). Similarly, reflectance from hairless species such as *Q. coccifera* is on average 5% lower than in *Q. ilex* (Morales et al. 2002). In agreement with a protective role, it has been shown that in *Q. ilex* pubescence is higher in xeric ecotypes compared to mesic ones (Camarero et al. 2012). Apart from their role in protection against excessive radiation, trichomes also contribute in *Q. ilex* to attenuate UVA and UVB protection thanks to the presence of flavonoids (Karabourniotis and Bornman 1999). However, this is probably not the main factor determining pubescence, as shown by the decrease in leaf hair density with increasing elevation (and consequently UV exposure) in *Q. ilex* (Filella and Peñuelas 1999).

In some *Quercus* species, leaf reddening is conspicuous at certain developmental stages. The question of the functional role of leaf reddening (caused by the accumulation of anthocyanins, carotenoids or betacyanins) has been tackled by a surprisingly high number of studies (reviewed in Chalker-Scott 1999; Hughes 2011; Steyn et al. 2002). Two main hypotheses have been tested in those studies: red pigments act as light filters reducing the irradiance reaching the photosynthetic cell layers, or alternatively, red molecules act as antioxidants. An alternative hypothesis, not related to photoprotection is that red colouration in leaves represents a type of communication signal between plants and insects (Archetti et al. 2008). This hypothesis has been experimentally tested in *Q. coccifera* by Karageorgou and Manetas (2006). These authors showed that green leaves were more damaged by herbivore insects with little differences on photoinhibition between both phenotypes. Despite the role in bio-communication, the photoprotective hypothesis is also supported in *Quercus* species by the fact that anthocyanins appear during critical phenological periods such as leaf expansion in *Q. coccifera* (Manetas et al. 2003) or *Q. ilex* (Brossa et al. 2009) and autumn senescence in *Q. rubra* (Lee et al. 2003) or *Q. palustris* (Boyer et al. 1988) (see also Sect. 11.4.2); but also in response to low temperature stress in evergreen *Quercus* species (Ramírez-Valiente et al. 2015) (see Sect. 11.2.1). Such anthocyanin accumulation is triggered by light, and consequently is not observed when senescing leaves are artificially shaded (Lee et al. 2003). Furthermore, when comparing young leaves from red and green phenotypes of *Q. coccifera*, the higher photochemical efficiency of first, together with their shade-like phenotype (Manetas et al. 2003), support that shading generated by the accumulation of such pigments enhances photoprotection during leaf expansion.

11.1.3 Metabolic Dissipation

Another mechanism that can reduce over-excitation of the photosynthetic machinery is to enhance the metabolic use of the light energy absorbed by processes as photorespiration, the cyclic electron transport, the water-water cycle or/and the chlororespiration. All these processes act as alternative electron sinks, thus preventing damage to the photosynthetic apparatus (Ort and Baker 2002).

Photorespiration results in a light-driven loss of CO_2 from cells that are simultaneously fixing CO_2 by the Calvin cycle. Photorespiratory reactions can dissipate the energy excess either directly (by ATP, NAD(P)H and reduced ferredoxin) or indirectly (*e.g.*, via alternative oxidase and providing an internal CO_2 pool) (Voss et al. 2013). This process has been described also for species of the genus *Quercus* in response to different stress factors (see Sect. 11.3.1) as: (i) water stress in *Q. ilex* (Tsonev et al. 2014), (ii) drought stress in *Q. suber*, *Q. ilex* and *Q. coccifera* (Peguero-Pina et al. 2009), (iii) low temperature in *Q. guyavifolia* (Huang et al. 2016a), (iv) high temperature in *Q. ilex* (Peñuelas and Llusia 2002) and (v) high temperature combined with high CO_2 in resprouts from *Q. ilex* (Pintó-Marijuan et al. 2013). In the latter, photorespiration was showed to be dependent on CO_2 concentration (Pintó-Marijuan et al. 2013). These data suggest that photorespiration provides a "safety-valve" for excess energy to avoid photochemical damage when CO_2 assimilation is inhibited (Peñuelas and Llusia 2002), thus preventing accumulation of reactive oxygen species (ROS) (Voss et al. 2013). The leaf internal CO_2 pool provided by photorespiration supports the Calvin cycle and isoprenoid biosynthesis (Peñuelas and Llusia 2002).

Chlororespiration also might play a role in the regulation of photosynthesis by modulating the activity of cyclic electron flow around photosystem (PS) I (Peltier and Cournac 2002). Besides, it has been described as a protective mechanism based on the induction of chloroplast NAD(P)H dehydrogenase by stressful treatments (e.g. Öquist and Huner 2003; Streb et al. 2005). This protective role was studied in winter-acclimated *Q. ilex* and *Q. suber*, whose leaves produced zeaxanthin (Z) in darkness (possibly by creation of a chlororespiratory pH-gradient), contributing to winter hardening. (Brüggemann et al. 2009). Cyclic electron flow around PSI and water-water cycle have been found to be involved in winter photoprotection of evergreen Asian oaks *Q. douglasii* and *Q. guyavifolia* (Huang et al. 2016b).

The Mehler reaction, that reduces oxygen to superoxide anion (O_2^-), and subsequently to water by the ascorbate-glutathione pathway (described in Sect. 11.1.5.) is other major sink for the electrons from the photosynthetic electron transport chain.

11.1.4 Enhancing Dissipation

Once light has been absorbed, the first site of photoprotection is within the LHCs themselves. If light is excessive and excited Chl is unable to drive photochemistry, then the lifetime of the singlet state is extended, resulting in a higher yield of triplet state formation. This is undesirable because energy transfer from triplet Chl to oxygen generates singlet oxygen ($^1O_2$), a highly reactive type of ROS. However, in addition to driving photochemistry, excited Chl can return to the ground state by the emission of light (chlorophyll fluorescence) or by the harmless emission of heat (safe thermal dissipation of excess absorbed light energy), a process that is modulated by the xanthophyll cycles. The latter route is a major component of photoprotection, also termed non-photochemical quenching (NPQ) as it results in an easily measurable quenching of chlorophyll fluorescence (Müller et al. 2001; Murchie and Niyogi 2011). Two xanthophyll cycles associated to the thermal energy dissipation have been described in higher plants so far: the ubiquitous violaxanthin-zeaxanthin cycle (VAZ-cycle) and the taxonomically restricted lutein epoxide-lutein cycle (LxL-cycle) (Esteban and García-Plazaola 2014).

The regulation of energy dissipation through NPQ associated to the VAZ-cycle is one of the main photoprotective mechanisms described in higher plants (Niyogi 1999). The cycle involves inter-conversions between three carotenoids in the thylakoid membrane: violaxanthin (V), antheraxanthin (A) and zeaxanthin (Z). Under excess light, the efficient PSII light-harvesting antenna is switched into a photoprotected state in which the potentially harmful excess of absorbed energy is thermally dissipated. Changes occur rapidly and reversibly, enhanced by the de-epoxidation of V to Z via A. This mechanism has been described in many *Quercus* species in response to different stress factors (see Sects. 11.2.2 and 11.3.2 for a more detailed description of the regulation of NPQ by these stresses): (i) the excess of light (e.g. *Q. suber*, García-Plazaola et al. 1997; *Q. ilex*, Corcuera et al. 2005a; *Q. alba*, Wang and Bauerle 2006; *Q. petraea*, Rodríguez-Calcerrada et al. 2007; *Q. coccifera*, Peguero-Pina et al. 2013); (ii) summer drought (e.g. *Q. coccifera*, Peguero-Pina et al. 2008; *Q. ilex* and *Q. suber*, Peguero-Pina et al. 2009), and (iii) winter stress due to low temperatures (e.g. *Q. ilex*, Corcuera et al. 2005b; Camarero et al. 2012; *Q. myrsinaefolia*, Yamazaki et al. 2011).

The light-driven inter-conversions between lutein epoxide (Lx) and lutein (L) in the thylakoid membrane, constitutes the LxL-cycle (Fig. 11.2). This cycle operates concurrently with the VAZ-cycle (Esteban and García-Plazaola 2014). Lutein epoxide de-epoxidation enhances the already large pool of L in leaves, giving rise to newly formed L molecules (ΔL) (Nichol et al. 2012). In darkness, the epoxidation back from ΔL to Lx operates at two modes: one "completed", in which the initial Lx pool is restored in minutes or hours, and one "truncated", in which ΔL remains for a longer period (days or weeks). The latter, has been widely described in leaves of the genus *Quercus*: *Q. ilex*, (Llorens et al. 2002); *Q. ilex*, *Q. coccifera*, *Q. robur*, *Q. faginea* and *Q. suber*, (García-Plazaola et al. 2002); *Q. rubra* (García-Plazaola et al. 2003a). In addition, the "truncated" LxL-cycle has been also described in the enclosed buds of some woody plants, as *Q. robur*, when bud-burst takes place (García-Plazaola et al. 2004). The regulation of energy dissipation through NPQ is, as in the case for the VAZ-cycle, associated to the operation of LxL-cycle (Esteban and García-Plazaola 2014). Indeed, faster engagement of NPQ in leaves with ΔL not containing A + Z prior to light exposure was described in *Q. rubra* under light stress (García-Plazaola et al. 2003a). The ecophysiological significance of this "truncated" LxL-cycle in species of the genus *Quercus* relies on the fact that Lx de-epoxidation may represent an emergency mechanism of special relevance for long-term downregulation of photosynthetic efficiency, supplementing retention of Z + A and their sustained engagement in energy dissipation in response to prolonged environmental stress, in cases of winter or summer acclimation (Llorens et al. 2002; García-Plazaola et al. 2002) or sudden exposure to high light, as the generation of forest gaps or budbreak (Esteban and García-Plazaola 2014). Thus, the combination of both cycles increases the plasticity of photoprotective responses in *Quercus* species.

11.1.5 Antioxidant and Repair Mechanisms (Free and Enzymatic Antioxidants)

Environmental stress conditions that result in restricted CO_2 fixation rates can induce an imbalance between the generation and utilization of photosynthetic electrons. Thus, whenever excess excitation energy is not safely removed, photo-oxidative damage can occur due to an enhanced formation of ROS in the chloroplasts (Hernández et al. 2012). Besides the oxidation of different molecules (i.e. lipids, proteins, sugars, nucleic acids), ROS may also impair the PSII repair process through the inhibition of the de novo synthesis of PSII proteins (primarily the D1 protein) (Takahashi and Badger 2011). To cope with this situation, chloroplasts possess non-enzymatic and enzymatic antioxidant defense mechanisms. These mechanisms detoxify ROS and maintain an adequate cellular redox balance (Mittler 2002), alleviating the inhibition of PSII repair (Takahashi and Badger 2011).

The general operating mechanism of non-enzymatic, low molecular weight, antioxidants consists in the donation of electrons or hydrogen atoms to the oxidizing agent. For example, carotenoids as xanthophylls and carotenes can act as direct scavengers of ROS and stabilize the light-harvesting complexes within the thylakoid membranes (Bassi and Caffarri 2000). Tocopherols (Toc) are also

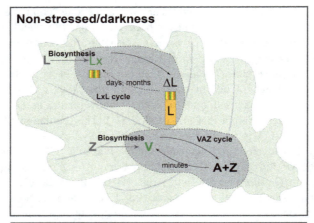
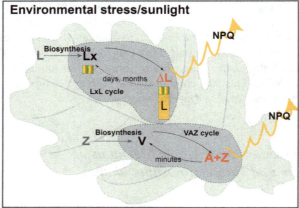
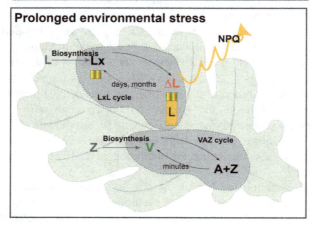

◀**Fig. 11.2** Concurrent operation of LxL- and VAZ-cycles in *Quercus*. In non-stressed leaves, the LxL- and VAZ-cycles remained in the epoxidated forms, Lx and V shown (shown in green). These carotenoids can be synthesized by *novo* from L and Z respectively (shown in grey). Under environmental stress or sunlight, the light harvesting antenna is switched into a photoprotected state, in which Lx and V are de-epoxidated to the de-epoxidated forms (L and A + Z respectively; shown in orange). In the case of LxL-cycle, Lx enhances the already large pool of L in leaves, giving rise to newly formed L molecules (ΔL). The potentially harmful excess of absorbed energy can be then dissipated thermally (NPQ). When the leaf is again without stress or in darkness, A + Z is epoxidated-back to V, restoring the initial V pool (minutes or hours). However, in the LxL-cycle, ΔL remains for a longer period (days, months), giving rise to the "truncated" LxL-cycle (dashed arrow). If the stress is sustained for a long time, a faster engagement of NPQ occurs then in leaves, activated by the truncated LxL-cycle (ΔL) and without any dissipation through A + Z. The combination of both cycles, with different mode and kinetics of operation increases the plasticity of NPQ in *Quercus* species

lipophilic antioxidants, and are able to donate single electrons to lipid peroxyl radicals, preventing the propagation of lipid peroxidation chains in thylakoids (Munné-Bosch 2005). Other plant molecules such as lipoic acid, anthocyanins, (poly)phenols, melatonine and tocotrienols have also been described to have antioxidant activity, but their relevance in vivo has not been fully determined. Ascorbate is the most abundant antioxidant in *Quercus* leaves (García-Plazaola et al. 1999a, b, 2000) and a substrate for ascorbate peroxidase (APX) that detoxifies H_2O_2. It is also involved in the regeneration of (oxidized) tocopheryl radical to Toc, and is a co-substrate for V de-epoxidase in the VAZ-cycle.

Enzymatic antioxidants are proteins that use electron donors, mainly antioxidants and NAD(P)H, either to eliminate ROS or to regenerate "burned" (oxidized) antioxidants (Hernández et al. 2012). Among others, antioxidative enzymes include glutathione reductase (GR), monodehydroascorbate reductase (MDAR), catalase (CAT), APX and other peroxidases. The interaction with non-enzymatic antioxidants such as ascorbate (Asc) and glutathione (GSH) plays a central role in the chemical and metabolic destruction of ROS. Ascorbate is considered as the key compound of O_2^- and H_2O_2 removal in the chloroplast, and the ascorbate-glutathione cycle constitutes a powerful pathway to maintain Asc in its reduced form by using GSH as an electron donor (Noctor and Foyer 1998). Another large group of non-enzymatic antioxidants are the superoxide dismutases (SODs), which constitute the first line of defense against ROS, catalyzing the decomposition of O_2^- in chloroplasts and other organelles (Alscher et al. 2002).

Several studies have dealt with the role of non-enzymatic and enzymatic antioxidants in different *Quercus* species. In particular, the changes in antioxidant concentration under several stress factors have been analyzed, such as drought in deciduous (e.g. *Q. robur*, Schwanz and Polle 2001; *Q. pubescens*, Gallé et al. 2007) and evergreen species (e.g. *Q. coccifera* and *Q. ilex*, Baquedano and Castillo 2006) (see also Sect. 11.3.3), winter stress in evergreens (e.g. *Q. ilex*, Corcuera et al. 2005a; *Cyclobalanopsis helferiana*, Zhu et al. 2009) (see also Sect. 11.2.3), and other factors such as pollutants (e.g. *Q. ilex*, Munné-Bosch et al. 2004). In general positive responses of antioxidant pools to those stress factors confirm the important role of this strategy in the battery of photoprotective defense mechanisms in *Quercus*.

11.2 Photoprotection in Evergreen Oaks During Winter

11.2.1 When and Where Evergreen Oaks Are Exposed to Winter Stress?

Many oak species are evergreen, sub-evergreen or brevideciduous, meaning that they maintain leaf function for most or all of the year. Evergreen species occur in all of the major lineages and across all the continents of their distribution in the Northern Hemisphere, including in Asia, Europe, North Africa and North and Meso America. In the Americas, in the southeastern United States, California and in northern, high elevation regions of Mexico, in some areas of the Mediterranean basin, and in high mountains in Asia, numerous oaks are exposed to chilling and freezing temperatures. While these different regions experience contrasting climatic regimes, they are similar in having thermal seasonality marked by warm summers and relatively cold winters. Plants with evergreen leaves, which are not programmed to senesce and abscise in response to cold temperatures, require mechanisms to protect the photosynthetic apparatus during freezing. This allows them to benefit in terms of carbon gain by maintaining function under mild freezing stress. Mediterranean oaks have been the subject of considerable study in the Mediterranean basin and California; these regions experience cold wet winters and hot dry summers, in contrast to the southeastern US and northeastern Mexico, where summers are hot and wet and winters are cold and dry. Nights in the Mediterranean region of southern Europe and North Africa frequently reach freezing temperatures during winter months (December, January and February) but only rarely extend below −10 °C. Leaf photochemistry is known to be impaired by night-time freezing temperatures as a result of impairment of enzymatic processes involved in photosynthesis. Acclimation to cold temperatures in overwintering evergreen oaks species has been linked to increases in antioxidants and xanthophyll pigments (García-Plazaola et al. 1997, 1999a, b; Brüggemann et al. 2009) (see Sect. 11.2.2), as well as changes in the composition of PS II antenna and increases in cyclic electron transport that allow increased quenching of absorbed light (Öquist and Huner 2003). In a common garden in the Mediterranean region of France, where two evergreen (*Q. ilex* and *Q. suber*) and two deciduous (*Q. afares* and *Q. faginea*) were grown for almost two decades, evergreen species showed important differences in cold acclimation ability and protection of the photosynthetic apparatus (Cavender-Bares et al. 2005). In the two evergreen species, photosynthetic function in leaves showed a strong acclimation response during winter, which protected the leaves even at −15 °C. This did not occur in the deciduous species.

Starch and lipid content increased in the evergreen species, and sugar content increased in *Q. ilex*, the most freezing tolerant species, consistent with changes associated with cold acclimation to stabilize membrane structure and function. These changes appeared to be coordinated with hydraulic function, such that species with long-lived leaves had greater protection of both the photosynthetic apparatus and xylem transport.

In general, species responses to chilling and freezing were predicted by their climates of origin. Thus for example, in another common garden experiments with the American live oaks, which maintain green leaves throughout the year, divergences in response to cold were tested for by examining PS II photosynthetic yield ($\Delta F/F_M'$) and NPQ under chilling or warm growing conditions after short-term exposure to three temperatures (6, 15 and 30 °C) and under moderate light (400 µmol m^{-2} s^{-1}). Without cold acclimation (tropical treatment), the most northern population of the species occurring in areas with cold winters, *Q. virginiana*, showed the highest photosynthetic yield in response to chilling temperatures (6 °C). With cold acclimation, *Q. virginiana* populations showed greater NPQ under chilling temperatures than the tropical *Q. oleoides* populations, suggesting enhanced mechanisms of photoprotective energy dissipation in the species adapted to cold winters. In a subsequent experiment that included more species from this lineage, species from climates with cold winters again showed greater leaf-level freezing tolerance than the tropical species, *Q. oleoides*, as indicated by changes in maximal photochemical efficiency of PSII (F_V/F_M) under continuous dark environments after freezing at −10 °C (Koehler et al. 2012; Cavender-Bares et al. 2011). At the population level, the degree of their loss of photosynthetic function depended on the mean minimum temperature of their climate of origin. Interestingly, seedlings originating from warmer climates had higher anthocyanin concentration in leaves when grown under cold winter conditions but did not exhibit a higher de-epoxidation state (Ramírez-Valiente et al. 2015).

Photoprotection mechanisms against chilling in Asian oaks have only been studied in a couple of species native to savanna-valleys and high elevations (above 3000 m.a.s.l.) of Southwest China: *C. helferiana* and *Q. guyavifolia* (Zhu et al. 2009; Huang et al. 2016a, b). These species are exposed to below zero temperatures during winter (mostly at night). When freezing night temperatures reduce or inhibit photosynthesis, energy dissipation, photorespiration and alternative electron flow acquire a key role in the photoprotection of photosynthetic apparatus (Huang et al. 2016a). All together, these mechanisms allow an efficient performance and protection of PSII of Asian oaks during winter.

In summary, the presence of a cold season in their habitat of origin together with a proper period of acclimation represent the two main factors explaining when and where *Quercus* evergreen species are able to successfully deal with winter stress.

11.2.2 Sustained Energy Dissipation Under Winter Stress

Evergreen oaks growing in areas where temperatures drop below freezing have been shown to employ sustained thermal energy dissipation during the winter months (García-Plazaola et al. 1999b, 2003b; Martínez-Ferri et al. 2004). Sustained thermal energy dissipation is characterized by reductions in maximal photochemical efficiency that correlates with overnight retention of the de-epoxidized forms of the VAZ and LxL-cycles: Z, A and ΔL (García-Plazaola et al. 2002). It is thought that this mechanism represents a sustained form (or forms) of thermal energy dissipation that protects the photosynthetic apparatus from excess excitation pressure during conditions of high light and low temperatures that occur during winter (Adams et al. 2004; Verhoeven 2014). This type of sustained dissipation has been widely observed in other evergreens acclimated to winter conditions. The mechanism(s) of sustained thermal dissipation are not fully understood, however dark retention of A and Z, as well as reorganization of photosynthetic proteins including increases in early light induced proteins (ELIP) are likely involved (Verhoeven 2014).

Studies on Mediterranean evergreen oak species have reported winter values of F_V/F_M ranging from 0.4 to 0.7, correlating with retention of A + Z, such that values for AZ/VAZ range from 0.3 to 0.7 (García-Plazaola et al. 1999a, 2003b; Martínez-Ferri et al. 2004), while in alpine Asian oaks winter values of F_V/F_M can be even lower (around 0.1) (Zhu et al. 2009; Huang et al. 2016b). This pattern is consistent with winter induced sustained thermal dissipation occurring in these species. In fact, dramatic increases in sustained dissipation were demonstrated to occur upon exposure to sudden drops in temperature (García-Plazaola et al. 2003b). However, observations that during winter, F_V/F_M values continued to decrease while AZ/VAZ remained the same, suggest that processes other than sustained thermal dissipation are likely also causing winter declines in F_V/F_M, possibly including some sustained photo-damage (Martínez-Ferri et al. 2004). Additionally, pool sizes of xanthophyll cycle pigments as well as L and β-carotene (β-Car) have also been shown to increase during winter in Mediterranean oak species, suggesting an increased capacity for photoprotection (García-Plazaola et al. 1999a, 2003b; Martínez-Ferri et al. 2004).

In a study comparing co-occurring deciduous and evergreen oaks in northern Florida, a region with cold but mild winters, short-term chilling stress (without prior cold acclimation) resulted in greater than 50% reduction in maximum photosynthesis, 60–70% reduction in electron transport rate and irreversible quenching of PSII fluorescence in both species (Cavender-Bares et al. 1999). However, the kinetics of recovery after combined high light exposure and chilling showed that the evergreen species exhibited greater photoprotective quenching (qE) and less irreversible quenching (qI) than the deciduous species. Higher photoprotective capacity may be inherent in evergreen oaks compared to deciduous oaks even without cold acclimation.

11.2.3 Antioxidants: Responses of Free and Enzymatic Antioxidants to Winter Stress

Studies examining the antioxidant responses of cold tolerant *Quercus* species to low temperatures have demonstrated that responses vary considerably depending upon both species and the particular environmental conditions encountered (García-Plazaola et al. 1999a, 2000, 2003b; Corcuera et al. 2005a). In a study comparing winter to spring antioxidant content in *Q. ilex*, a high synthesis of Asc (without significant effects on α-Toc and GSH) occurred in winter (2–3 fold over spring on a leaf area basis, García-Plazaola et al. 1999a). However, in another study comparing a cold with a mild winter, the most highly induced antioxidant in the coldest winter was α-Toc, being 400 and 60% higher (on an area basis) than the mild winter in *Q. coccifera* and *Q. ilex* respectively (García-Plazaola et al. 2003b). In this study, a decrease in Asc content was observed in both species in the coldest winter relative to the mild one.

A study examining enzymatic antioxidants in *Q. ilex* dealing with winter stress reported an induction in GR and MDAR activity in winter, but a weak response of other antioxidant enzymes (CAT, SOD, APX, García-Plazaola et al. 1999a). Interestingly, both GR and MDAR use NADPH as electron donor, which is a sink for photosynthetic electrons, minimizing overexcited photosynthetic electron chain and ROS production. The authors concluded that MDAR activity plays a central role in the Asc regeneration. A weak response of APX, GR, SOD and guaiacol peroxidase has been also reported for *Q. ilex* in its upper altitudinal extreme in the Iberian Peninsula (Corcuera et al. 2005a), which could mean that the constitutive activity of these enzymes is enough to cope with oxidative stress at low temperatures. Similarly, enhanced activity of GR, SOD and particularly of MDAR have also been described in Asian oak *C. helferiana* during chilling period (Zhu et al. 2009).

Antioxidant contents in response to winter stress can change, not only depending on chilling severity, but also on internal (ecotype, leaf ontogeny) and external factors (time of the day, irradiance). In this sense, García-Plazaola et al. (1999b) reported that the antioxidant content in *Q. ilex* (Asc, GSH, α-Toc) was constitutively higher in sun than in shade leaves during winter. Besides, it decreased sensitively along a sunny day, showing antioxidant content is dynamic and it changes depending on consumption and regeneration. Furthermore, an age-dependent tocopherol accumulation has been observed in woody plants including oaks (Hansen et al. 2002), but its physiological meaning is unknown.

Additionally, provenances of ecotypes also condition antioxidant content, and commonly, ecotypes adapted to colder climates evolve higher antioxidant content. In a common garden study with *Q. ilex* from three contrasting Mediterranean climatic provenances (semiarid, cold and oceanic), Camarero et al. (2012) observed that the highest content of α-Toc occurred in *Q. ilex* seedlings from the coldest provenance, which supports the role of this antioxidant in cold adaptation. García-Plazaola et al. (2000) also reported an increase in VAZ and β-Car (in a Chl ratio) in Mediterranean *Q. ilex* growing in an altitudinal gradient in winter, but the response in other antioxidants was not so evident.

Overall, studies suggest that climate has been a key factor in shaping species and population differences in winter stress antioxidant response in *Quercus*. Despite the remarkable diversity in the antioxidant strategies followed by different species, generally, enhanced hydrophilic antioxidants (particularly MDAR) together with increased Z (and α-Toc in some cases) could summarize oak antioxidant response to winter stress.

11.3 Photoprotection in Drought-Stressed *Quercus*

Drought leads to water deficit in the leaf tissue, which affects many physiological processes such as photosynthesis. Stomatal closure is a common response to drought stress in *Quercus* species (e.g. Mediavilla and Escudero 2003; Peguero-Pina et al. 2009), as a way of minimizing water loss at the expense of reducing net CO_2 assimilation. Under this situation, when light incident on the leaf surface exceeds largely the amount that can be used for photosynthesis, different mechanisms allow the protection of the photosynthetic apparatus both dissipating excess of light as heat (Demmig-Adams and Adams 2006) or decreasing ROS formation. Photoprotective mechanisms have been described for both evergreen and deciduous *Quercus* species in response to moderate or severe drought stress conditions, although with some differences among them.

11.3.1 Functional Differences Between Deciduous and Evergreen Species Under Drought

The photoprotective mechanisms of deciduous *Quercus* species under drought stress have been analyzed since the 1990s. Thus, several studies have dealt with this topic in *Q. robur* and *Q. petraea*, two of the most representative oaks widely distributed in Europe under temperate-nemoral climate (www.euforgen.org). Both *Q. robur* and *Q. petraea* have developed effective photoprotective mechanisms to withstand mild water deficit (Epron et al. 1992; Schwanz et al. 1996), although both species seem unable to effectively cope with severe drought stress in terms of photoprotection (Epron and Dreyer 1992; Schwanz and Polle 2001). Nevertheless, deciduous oak species are not exclusive of the humid climates, but they are also present in more xeric habitats with summer drought, i.e. the so-called "nemoro-Mediterranean oaks" (Corcuera et al. 2002). Among them,

Q. pubescens, which has a wide distribution range including most of central and southern Europe (www.euforgen.org), is able to withstand and survive extreme summer droughts (Damesin and Rambal 1995). The tolerance of this species to water stress has been related to the existence of efficient photoprotective mechanisms, as showed by Marabottini et al. (2001), Gallé et al. (2007), Contran et al. (2013) and Hu et al. (2013). Other "nemoro-Mediterranean oaks" such as *Q. cerris* and *Q. frainetto* also exhibited efficient photoprotective mechanisms in response to drought stress (ca. -3 MPa, Wolkerstorfer et al. 2011).

The photoprotection mechanisms have also been studied in drought-stressed evergreen oaks, mainly in those species occurring under Mediterranean-type climates (i.e. the so-called "Mediterranean oaks", Corcuera et al. 2002). Although photoprotective mechanisms are strong enough in these species, the performance of their photosynthetic machinery in response to an intense summer stress period varied markedly among them, as showed by Peguero-Pina et al. (2009) for *Q. coccifera*, *Q. ilex* and *Q. suber*. Apart from the "Mediterranean oaks", few studies have dealt with photoprotective mechanisms in other evergreen oaks. Zhu et al. (2009) reported that *Cyclobalanopsis helferiana*—a resilient species that can survive in the savannas in the hot-dry valleys in SW China—was highly tolerant to severe drought stress (ca. -4 MPa at predawn) due to the existence of photoprotective mechanisms that resembled the performance of "Mediterranean oaks" explained below (see Sects. 11.3.2 and 11.3.3 for details). Recently, Ramírez-Valiente et al. (2015) found that four evergreen oaks (*Q. virginiana*, *Q. geminata*, *Q. fusiformis* and *Q. oleoides*) included in a group of species called *Quercus* series *Virentes* from southern USA, Mexico and Central America, living under contrasting climatic conditions, developed photoprotective mechanisms when exposed to drought stress.

In conclusion, both deciduous and evergreen *Quercus* species occurring under contrasting climates implement different photoprotective mechanisms under drought stress. Regardless of this common performance, some evergreen (e.g. *Q. coccifera* and *Q. ilex*) and deciduous (e.g. *Q. pubescens*) oaks living under Mediterranean-type climates seem to be better adapted for withstanding severe drought periods. By contrast, deciduous oaks from temperate-nemoral climates have developed effective photoprotective mechanisms only to withstand mild water deficit, being unable to effectively cope with the severe water scarcity experienced by evergreen species. This differential physiological performance under water stress might play an important role in tree mortality and landscape formation in the context of future climate projections, which suggest that the proportion of land surface under extreme drought can be dramatically increased by the end of the present century (Xu et al. 2013).

11.3.2 Energy Dissipation and Xanthophyll Cycles Under Drought

During a drought episode, the dissipation as heat (thermal dissipation) of part of the absorbed energy acquires a crucial role in the photoprotection of the photosynthetic apparatus preventing the accumulation of ROS. Independent of the leaf strategy, both deciduous and evergreen oaks use thermal dissipation as an important alleviation mechanism under water limitation conditions. The vast majority of works, where thermal energy dissipation was studied in drought-stressed oaks, has been conducted in Europe but a few examples from North America are found in the bibliography. Despite the existence of two xanthophyll cycles (VAZ and LxL-cycles) in *Quercus* species (García-Plazaola et al. 2002), very little knowledge is available regarding the relevance of LxL-cycle and thermal dissipation during drought in oaks (Llorens et al. 2002).

Deciduous oaks normally display lower values of NPQ under drought conditions than evergreen species do. An example is provided by Mahall et al. (2009) when comparing two Mediterranean oak species under field conditions in southern California: seedlings of *Q. agrifolia* (evergreen) showed NPQ of 3.09 at the end of summer 2002, while co-occurring *Q. lobata* (deciduous) seedlings showed values of 1.91. Similar examples can be found in the Mediterranean Basin: i.e. compare *Q. ilex* NPQ values of 3.8 against values of the deciduous *Q. humilis* (NPQ = 3.0) under severe drought (Gulías et al. 2002). Compared to Mediterranean oaks, "nemoro-Mediterranean species" generally show lower NPQ values under moderate water stress. In this regard, *Q. pubescens* increased NPQ only from 0.7 (when well watered) to 1.7 under imposed drought during summer 2004 (Gallé et al. 2007). NPQ values, however, can be enhanced after previous drought events as it is shown in the same work: a drought episode during the next summer (2005) in the same trees induced an increase of NPQ that reached values of up to 5 (Gallé et al. 2007). Also in *Q. petraea*, a reversible decrease of photochemical efficiency (probably related to an enhanced AZ/VAZ) was shown to prevent photoihnnhibitory damage under moderate water stress with a predawn water potential below −2 MPa (Epron et al. 1992).

In evergreen *Quercus* species, sustained de-epoxidation of xanthophylls can be induced during a prolonged drought event: i.e. 3 weeks without irrigation during the summer led to very high AZ/VAZ morning-levels c.a. 0.8 in *Q. coccifera* seedlings (Peguero-Pina et al. 2008). This photoprotective mechanism could be related to a low intra-thylakoid lumenal pH and efficiently prevented photoinhibitory damage. Both dynamic and chronic photoinhibition have been observed during water stress in *Q. coccifera* and *Q. ilex* (Baquedano and Castillo 2006). And although the responsiveness of photoprotective mechanisms will partly depend on the climate of origin, heterogeneity of responses can also be found within evergreen oaks of the same climatic region (i.e. Mediterranean). Thus, *Q. ilex* and *Q. coccifera* tended to sustain a chronic photoinhibition evidenced by a decrease in predawn F_V/F_M values to 0.3–0.4 and an overnight retention of A + Z at water potentials below −6 MPa (Peguero-Pina et al. 2009). This was interpreted by these authors as an additional photoprotective mechanism that preserved the photosynthetic pigment machinery after a long summer stress period. By contrast, F_V/F_M in *Q. suber* remains at high values around 0.7 and most of the midday A + Z were converted into V during the night, irrespective of the degree of water stress (Peguero-Pina et al. 2009). In line with this, García-Plazaola et al. (1997) did not find changes in predawn F_V/F_M and the VAZ pool was maintained in a highly epoxidated state at predawn under drought stress in *Q. suber*. A similar performance was found by Zhu et al. (2009) for the evergreen *C. helferiana* in response to severe drought stress, i.e. a down-regulation of PSII activity characterized by gradually NPQ increases with an overnight retention of A + Z.

In summary, drought generally triggers an increase of NPQ in oaks when leaf predawn water potential falls below −2 to −3 MPa. Commonly, this rise in NPQ is mainly, although not completely, related to the de-epoxidation of xanthophylls: i.e. A + Z and NPQ correlate well in sun leaves (García-Plazaola et al. 1997). The highest NPQ values have been recorded for Mediterranean evergreen oaks: i.e. NPQ of up to 10 was measured in adult *Q. suber* trees subjected to three consecutive years of severe drought (ca. −4 MPa at predawn) (Grant et al. 2010). Also acclimation, due to either provenance (i.e. xericity of the site) or seasonal acclimation (i.e. end of dry season) and hardening after repeated periods of drought, are able to induce a progressive increase in VAZ pool, de-epoxidation state of xanthophylls (i.e. AZ/VAZ values) and NPQ (Grant et al. 2010; Camarero et al. 2012; Ramírez-Valiente et al. 2015). All these mechanisms efficiently reduce the risks of photodamage under drought conditions in oaks.

11.3.3 Role of Free and Enzymatic Antioxidants Against Drought

Non-enzymatic and enzymatic antioxidant defense mechanisms detoxify ROS and maintain an adequate cellular redox balance under drought stress conditions in order to avoid photo-oxidative damage (Hernández et al. 2012). The role of antioxidants in drought-stressed *Quercus* has been studied mainly in conjunction with other photoprotective mechanisms, such as VAZ and LxL-cycle, both in deciduous and evergreen species. Specifically, several studies have dealt with interspecific variations in the concentration and/or activity of different antioxidants and its influence on the differential physiological performance of oak species under water stress.

The first studies about this topic were published during the 1990s, in which *Q. robur* experienced a reduction in SOD and catalase activities when subjected to mild drought stress (ca. −1 MPa) (Schwanz et al. 1996). Furthermore, even under severe drought (ca. −3 MPa at predawn), Schwanz and Polle (2001) did not find a photoprotective response of antioxidants in *Q. robur* because the key enzymes involved in antioxidant protection (i.e. SOD) declined and oxidation of Asc and GSH increased under these conditions. Contrary to these findings, Hu et al. (2013) stated that, under similar conditions, this species enhanced its leaf Asc and thiol levels as the most drought-sensitive species in response to an increase in ROS production when compared with *Q. petraea* and, specially, *Q. pubescens*. The latter species maintained high amounts of antioxidants (mainly Asc and α-Toc), minimizing oxidative stress and irreversible damage in leaves under severe drought conditions (ca. −4 MPa at predawn) (Gallé et al. 2007). Similar results were found by Contran et al. (2013), who stated that *Q. pubescens* reacted to water deficit by increasing antioxidant enzyme activity, avoiding ROS toxic effects. The tolerance of this species to water shortage in terms of foliar antioxidant status has also been evidenced by Marabottini et al. (2001) and Hu et al. (2013). For these reasons, *Q. pubescens* is considered as a drought-tolerant species when compared with *Q. robur* and *Q. petraea*. In this regard, *Q. cerris* and *Q. frainetto*—two deciduous oaks that co-occur with *Q. pubescens*—increased three times the α-Toc content during summer in response to drought stress (ca. −3 MPa, Wolkerstorfer et al. 2011). However, these authors could not explain whether the observed accumulation of α-Toc contributed to the protection against the photo-oxidation.

Evergreen oaks, specially those species occurring under Mediterranean-type climates, show efficient antioxidant defenses against drought stress. Faria et al. (1996) found that SOD and APX activities in *Q. suber* were high enough to cope with the increase in ROS under reversible stressful conditions of midday, providing an additional mechanism for energy dissipation. Baquedano and Castillo (2006) found midday dynamic photoinhibition in water-stressed *Q. coccifera* and *Q. ilex* plants (ca. −1.5 MPa at predawn) coupled with a significantly increase in the total antioxidant activity and in the Asc pool. In an additional study on *Q. ilex*, Nogués et al. (2014) found a sustained increase in non-enzymatic (total Asc and phenolic compounds) and enzymatic antioxidants (APX and GR) in response to drought stress (ca. −2.5 MPa at midday). According to these authors, this may indicate the activation of defense responses for scavenging ROS produced under increasing limitations of primary metabolism, and to ultimately avoid stronger oxidative damage in the photosynthetic apparatus of *Q. ilex*. Besides Mediterranean oaks, Zhu et al. (2009) found that SOD and glutathione peroxidase showed a sustained high activity during the driest period of the year (ca. −4 MPa at predawn) for

C. helferiana, a drought-resistant evergreen oak occurring in SW China. This high activity of antioxidant system could efficiently scavenge ROS and protect the photosynthetic apparatus from oxidative injury.

In conclusion, both free and enzymatic antioxidants play an important role in photoprotection against drought in *Quercus* species. However, although some evergreen species seem to display higher antioxidant protection, more comparative studies are needed to elucidate the role of environmental factors in antioxidant activity in oaks.

11.3.4 Synergic Effect of Drought and Heat Waves

Climatic observation of the last century reveals a trend to a higher frequency and intensity of "extreme climatic events" such as severe drought episodes, intense rainfall events or heat waves (IPCC 2014). The term "extreme climatic event" is debatable and its meaning may depend on the organism, the issue or the scale of analysis (see Smith 2011 for an ecological revision). From a meteorological point of view, a heat wave could be defined as an extraordinary event of abnormally high temperatures above the 90th percentile (for a location and season), persisting for at least 3 days (based on the definition from Pezza et al. 2012). Heat waves can have a high impact in ecosystems, particularly when co-occurring with a period of drought, due to their potential to directly or indirectly trigger irreversible changes in them. One of the most severe heat waves in the last century (see "list of heat waves, 2016" for a complete list of most significant heat waves over the last century) was the summer heat wave of 2003, which affected western and central Europe. This was the hottest episode in the last 180 years, reaching temperatures up to 6 °C above the long-term average and, additionally occurred on a year of considerable drought, with annual precipitation 50% below the average (Luterbarcher et al. 2004; Stott et al. 2004). Besides its extremity in magnitude, the European heat wave of 2003 affected different "Temperate-Nemoral" and Mediterranean ecosystems dominated by *Quercus* species. Thus, many of the works available in the literature correspond to this scenario. Nevertheless, the unpredictability of heat waves, limits the availability of field data to those obtained during the course of studies that were already in course when a heat wave event occurred (referred to as "opportunistic studies" in Smith 2011). Hence, in this section, manipulative experiments dealing with the interaction of heat and drought and their effects on photoprotective responses of *Quercus* species have also been included to build up a complete overview of the synergistic effects of heat waves and drought over *Quercus* photoprotection strategies.

Water availability is the main factor determining the effects of a heat wave over a population of *Quercus*. This statement could be expected from a merely physical point of view since transpirative cooling alleviates heating, preventing leaves to reach lethal temperatures. This premise is reinforced by field works: i.e. even extraordinary hot episodes as the European heat wave of 2003 did not affect the growth of *Q. robur* if water supply from the soil was available (Wilkinson et al. 2012). Similarly, *Q. rubra* seedlings are able to increase their biomass under imposed artificial heat waves of up to +6 °C above control whenever water supply is assured, while low soil moisture content itself induces a decrease of biomass (Ameye et al. 2012; Bauweraerts et al. 2013). In that sense, biogeographical and topological location (type of substrate and its capability for water retention, elevation, orography and its interaction with fog or cloud retention, wind, etc.) as well as species-dependent tolerance to drought, determine the consequences of extremely hot and dry episodes (Bertini et al. 2011; Contran et al. 2013). Thus, higher mortality and reduced growth experienced by *Q. petraea* changed oak forests towards a new stand composition where the more drought-tolerant Turkey oak *Q. cerris* has became dominant after the summer of 2003 in Italy (Bertini et al. 2011).

In addition to a proper water supply, some other factors can buffer or mitigate the effect of episodes of simultaneous heat and drought in oaks. This is the case of canopy buffering-effect over the understory plants and seedlings. As an example, during August 2003, maximum temperatures were on average up to 3 °C cooler under the oak forest canopy of the Southern Swiss Alps than in open areas. When compared to other forest types, only the beech forests produced a greater cooling effect (Renaud and Rebetez 2009). Also elevated CO_2 (i.e. 700 ppm) seems to mitigate the effect of heat waves and drought stress in *Quercus* in a supposed scenario of future atmospheric conditions (Ameye et al. 2012).

In a broad sense, *Quercus* species could be generally considered as tolerant to the combined effect of heat waves and drought (Gallé et al. 2007; García-Plazaola et al. 2008; Haldimann et al. 2008; Ameye et al. 2012; Bauweraerts et al. 2013; Contran et al. 2013). Photoprotective barriers of many *Quercus* species start with structural passive-protection of leaves: i.e. many species are covered by a more or less dense hairy surface which reduces light absorption by photosynthetic cell layers (see Sect. 11.1.2 for details). Leaves of *Q. ilex*, which is typically considered as a plastic species, show higher density of adaxial trichomes in xeric habitats than when growing on continental sites or in mesic sites (Camarero et al. 2012). As a constitutive and less dynamic leaf property, density and structure of foliar hairs could have its main meaning in terms of long-term acclimation and ecological stress memory, more than in the immediate response to an acute and quick event such as a heat wave. Nevertheless, morphological acclimation to a more xeric and/or hot environment may provide advantages against a sudden episode of extreme temperature (such as heat wave).

At the chloroplast level, most species down-regulate the photochemical efficiency to prevent photo-oxidative damage. Thus, $\Delta F/F_M'$ decreased 88% at midday: i.e. $\Delta F/F_M'$ values fell from 0.8 at sunrise to 0.1 at midday during the heat wave 2003 in *Q. pubescens* (Haldimann et al. 2008). The decrease is more acute if trees were previously exposed to a heat wave as it has been shown in controlled heat wave experiments of consecutive +12 °C episodes with *Q. rubra* (Bauweraerts et al. 2014). Also predawn F_V/F_M can experience a progressive decrease when trees are exposed for several days to unusually high temperatures combined with low water availability (Gallé et al. 2007). This down-regulation of photochemistry can be attributed, at least in part, to enhanced de-epoxidation and higher pools of VAZ and LxL-cycles during the stress (García-Plazaola et al. 2008; Haldimann et al. 2008) and it seems to be related to reorganization of thylakoid membranes but not to changes in the amount of LHC II proteins, at least in *Q. pubescens* (Haldimann et al. 2008). Under these conditions, heat dissipation is enhanced and high NPQ kept during stress. Additionally, excitation pressure can decrease by a combined Chl loss and Chl a/b increase (García-Plazaola et al. 2008; Haldimann et al. 2008; Contran et al. 2013), although these effects are species-dependent and related to the severity and extent of the heat wave. The antioxidant system can also enhance, in particular α-Toc, for which a dramatic enhancement has been described in response to extreme heat (García-Plazaola et al. 2008). In some species as is the case of *Q. ilex*, some monoterpenes as α-pinene, seem to be involved in the thermotolerance of the leaves through the antioxidative protection of membranes (Copolovici et al. 2005) while other volatiles, such as isoprene, can confer thermotolerance through mechanisms independent of the antioxidant response (Peñuelas et al. 2005).

In sum, an efficient down regulation of photochemical efficiency together with an up regulation of NPQ, and an enhanced antioxidative response, allow most *Quercus* species to effectively protect their photosynthetic apparatus against the combined effects of heat and drought, although the effect of drought superimposed with a severe heat wave can kill many *Quercus* trees within a population in a short time-lapse of days or weeks (Haldimann et al. 2008; Bertini et al. 2011). The additional die-back of further individuals in the following years due to weakness and predisposition to succumb to biotic or abiotic menace must be also considered, and both immediate and delayed effects of heat waves plus drought are able to strongly affect the composition of the ecosystem (Breda and Badeau 2008; Bertini et al. 2011).

11.4 Photoprotection During Autumn Senescence

11.4.1 *The Risks of Chlorophyll Degradation and Leaf Senescence*

In winter-deciduous oak species, leaf senescence involves a highly coordinated process of remobilization of nutrients that is triggered by the shortening photoperiod and decreasing temperatures characteristic of autumn conditions (Rosenthal and Camm 1996; Hoch et al. 2001; Keskitalo et al. 2005; Fracheboud et al. 2009).

The most visible sign of autumn is the dramatic change in leaf pigment content resulting from chlorophyll degradation, while carotenoids are retained longer, and in some species anthocyanins are produced. Pigment changes are accompanied by decreases in photosynthesis, however, maximal photochemical efficiencies are retained at high levels until very late in autumn when chlorophyll contents are quite low, suggesting that the chlorophyll that is retained during senescence resides in functional photosynthetic centers (Adams et al. 1990; Keskitalo et al. 2005; Moy et al. 2015). In contrast to photosynthesis, respiration is maintained later into autumn in order to provide energy needed for remobilization of leaf nutrients prior to leaf abscission (Adams et al. 1990; Collier and Thibodeau 1995; Hoch et al. 2001; Keskitalo et al. 2005). The process of disassembly and degradation of the photosynthetic apparatus must occur in a manner that prevents the accumulation of damaging reactive oxygen species that might preclude optimal nutrient resorption (Matile et al. 1999; Lee et al. 2003). Therefore, photoprotective strategies are an important component of the process of leaf senescence.

11.4.2 Why Leaves Turn Red?: Anthocyanic and Acyanic Oaks

Much of the beauty of autumn resides in the immense variation in color of leaf foliage among species, with colors varying from reds to yellows. According to Hoch et al. (2001) among oaks there are 9 species that produce high amounts of anthocyanins during autumn (all in North America) and 10 species in which anthocyanin production is nonexistent, low or infrequent (split between North America and Europe). The synthesis of anthocyanins during autumn occurs after chlorophyll degradation has begun and the pigments accumulate in the vacuoles of upper palisade cells (Hoch et al. 2001; Lee et al. 2003). The anthocyanins are hypothesized to serve as a light screen during autumn senescence (see Sect. 11.1.2.), which protects the photosynthetic apparatus during the period of nutrient resorption (Hoch et al. 2001, 2003; Lee and Gould 2002; Lee et al. 2003).

11.4.3 Coordinate Operation of Photoprotection Mechanisms in Senescing Leaves

The mechanisms of photoprotection during autumn senescence have not been particularly well studied. Pigment studies have demonstrated that carotenoids are degraded more slowly than chlorophylls affecting in a photoprotective capacity during the degradation process (Adams et al. 1990; Lee et al. 2003; Keskitalo et al.

2005). Additionally, the accumulation of anthocyanins, discussed above, likely serves a photoprotective role. A study monitoring both carotenoids and anthocyanins found no difference in the rate of degradation of carotenoids in anthocyanic and acyanic species (Lee et al. 2003), suggesting that carotenoid based photoprotection likely functions in all species. Few studies have been conducted using methods that differentiate individual carotenoids. The available data show that in aspen and sugar maple, the VAZ-cycle pigments were retained in higher abundance than other carotenoids, while in an oak species (*Q. bicolor*) the VAZ-cycle pigments were retained at relatively high levels only in early stages of autumn senescence, while L was retained in higher abundance than other carotenoids in late autumn (Keskitalo et al. 2005; Moy et al. 2015). Additionally, the oak species was shown to accumulate the PsbS protein in early autumn, which did not occur in the maple, suggesting a role for increased xanthophyll associated energy dissipation in early autumn in the oak (Moy et al. 2015). These studies suggest that there is variation among species in the strategies used for photoprotection during autumn senescence.

11.5 Concluding Remarks

This chapter has outlined that oaks are remarkably plastic and diverse (both inter- and intra-specifically) in terms of morphological and biochemical photoprotective mechanisms (Fig. 11.3), providing tolerance to winter cold and summer drought. These unfavourable climatic conditions are indeed, the key factor in shaping *Quercus* distribution and stress responses, being particularly Mediterranean evergreen oaks more resilient to them. Future climate change scenarios predict warmer and dryer environmental conditions in most of the distribution range of *Quercus*, detailed physiological studies are, therefore, essential to anticipate to the ecological responses. However, most of the information available nowadays comes from a few Mediterranean species, being holm oak (*Q. ilex*) the one most intensively studied. We conclude then that several knowledge gaps should be filled into get a more complete and global perspective of the group in all its distribution range: (i) photoprotective mechanisms and their ecological significance in Asian oaks are understudied, (ii) the role of L (considering the LxL-cycle) in the development of NPQ has received little attention in species of the genus and (iii) few is still known regarding thylakoid proteins and energy dissipation mechanisms in *Quercus*.

Leaf phenology	Species	Origin Geography	Origin Climate	Morphological photoprotection	COLD Tolerance	COLD Response	DROUGHT Tolerance	DROUGHT Response	HEAT+DROUGHT Tolerance	HEAT+DROUGHT Response
Deciduous	Q. cerris	Eu	Med	Ab hairs			u	↑Toc		
	Q. frainetto	Eu	Med	Ab hairs				↑Toc		
	Q. humilis (sin. Q. pubescens)	Eu	Med	Ab hairs			✓	↑NPQ, Asc, Toc	✓	↑NPQ, VAZ, AZ/VAZ
	Q. pyrenaica	Eu	Med	Ab hairs			u	↑NPQ		
	Q. lobata	N Am	Med	Ab hairs				↑NPQ		
	Q. bicolor	N Am	Temp	Ab hairs						
	Q. macranthera	W Asia	Temp	Ab hairs					▷	
	Q. palustris	N Am	Temp	none			▷▷	↑NPQ, AZ/VAZ		
	Q. velutina	N Am	Temp	Ab hairs				↓Photoresp, SOD, Asc red, GSH red		
	Q. petraea	Eu	Temp	none			u			
	Q. robur	Eu	Temp	none						
	Q. alba	N Am	Temp	none						
	Q. rubra	N Am	Temp	none						
Evergreen	Q. coccifera	Eu	Med	Ab hairs	✓	↑Toc, ↓Asc	✓	↑AZ/VAZ, Asc	✓	↑Toc
	Q. ilex (subsps. rotundifolia & ilex)	Eu	Med	Ab hairs		↑Toc, Asc, GR, MDHAR, Lut, β-Car, VAZ	✓	↑NPQ, AZ/VAZ, Asc, GR, APX	✓	↑VOCs, Toc
	Q. suber	Eu	Med	Ab hairs			✓	↑NPQ, SOD, APX	✓	↑Toc
	Q. agrifolia	N Am	Med				✓	↑NPQ		
	Q. douglasii	N Am	Med							
	Q. fusiformis	N Am	Med		✓	↑Anthocyanins, NPQ, AZ/VAZ	✓	↑NPQ, AZ/VAZ		
	Q. geminata	N Am	Med			↑Anthocyanins, NPQ		↑NPQ, AZ/VAZ		
	Q. virginiana	N Am	Temp	Ab hairs		↑Anthocyanins, NPQ, AZ/VAZ	✓	↑NPQ, AZ/VAZ		
	Q. myrsinaefolia	E Asia	Temp & Trop	none	u					
	Q. oleoides	N & Cent Am	Temp & Trop		u	↑NPQ, AZ/VAZ, MDAR, Photoresp				
	Q. guyavifolia	SE Asia	Trop (alpine)		u	↑NPQ, AZ/VAZ, MDAR, Photoresp				
	C. helferiana	SE Asia	Trop (dry-hot)				✓	↑NPQ, AZ/VAZ, SOD		

Fig. 11.3 Summary of the oak species mentioned in the present chapter, their geographic origin, main functional traits, degree of tolerance to abiotic stressors and main photoprotective mechanisms employed by each species. Abbreviations: *Am* America, *Cent* central, *Eu*: Europe, *Med* Mediterranean, *Temp*: temperate, *Trop* tropical

Acknowledgements R.E. and B.F.M received a "Juan de la Cierva-Incorporación" grant (IJCI-2014-21452 and IJCI-2014-22489, respectively). This research was supported by projects CTM2014-53902-C2-2-P from the Ministry of Education and Science of Spain and the ERDF (FEDER), RTA2015-00054-C02-01 from the Spanish National Institute for Agriculture and Food Research and Technology (INIA) and research project UPV/EHU IT-1018-16.

References

Adams WWIII, Winter K, Schreiber U, Schramel P (1990) Photosynthesis and chlorophyll fluorescence characteristics in relationship to changes in pigment and element composition of leaves of *Platanus occidentalis* L. during autumnal leaf senescence. Plant Physiol 93: 1184–1190

Adams WWIII, Zarter CR, Ebbert V, Demmig-Adams B (2004) Photoprotective strategies of overwintering evergreens. Bioscience 54:41–50

Alscher RG, Erturk N, Heath LS (2002) Role of superoxide dismutases (SODs) in controlling oxidative stress in plants. J Exp Bot 53:1331–1341

Ameye M, Wertin TM, Bauweraerts I, McGuire A, Teskey RO, Steppe K (2012) The effect of induced heat waves on *Pinus taeda* and *Quercus rubra* seedlings in ambient and elevated CO_2 atmospheres. New Phytol 196:448–461

Archetti M, Döring TF, Hagen SB, Hughes NM, Leather SR, Lee DW, Lev-Yadun S, Manetas Y, Ougham HJ, Schaberg PG, Thomas H (2008) Unravelling the evolution of autumn colours: an interdisciplinary approach. Trends Ecol Evol 24:166–173

Ashton PMS, Berlyn GP (1994) A comparison of leaf physiology and anatomy of *Quercus* (Section *Erythrobalanus*-Fagaceae) species in different light environments. Am J Bot 81: 589–597

Bacelar EA, Correia CM, Moutinho-Pereira JM, Gonçalves BC, Lopes JI, Torres-Pereira JMG (2004) Sclerophylly and leaf anatomical traits of five field-grown olive cultivars growing under drought conditions. Tree Physiol 24:233–239

Baquedano FJ, Castillo FJ (2006) Comparative ecophysiological effects of drought on seedlings of the Mediterranean water-saver *Pinus halepensis* and water-spenders *Quercus coccifera* and *Quercus ilex*. Trees 20:689–700

Bassi R, Caffarri S (2000) Lhc proteins and the regulation of photosynthetic light harvesting function by xanthophylls. Photosynth Res 64:243–256

Bauweraerts I, Ameye M, Wertin TM, McGuire MA, Teskey RO, Steppe K (2014) Acclimation effects of heat waves and elevated [CO_2] on gas exchange and chlorophyll fluorescence of northern red oak (*Quercus rubra* L.) seedlings. Plant Ecol 215:733–746

Bauweraerts I, Wertin TM, Ameye M, McGuire MA, Teskey RO, Steppe K (2013) The effect of heat waves, elevated [CO_2] and low soil water availability on northern red oak (*Quercus rubra* L.) seedlings. Global Change Biol 19:517–528

Bertini G, Amoriello T, Fabbio G, Pivosi M (2011) Forest growth and climate change: evidences from the ICP-Forests intensive monitoring in Italy. iForest-Biogeosci Forestry 4:262–267

Boyer M, Miller J, Belanger M, Hare E (1988) Senescence and spectral reflectance in leaves of northern pin oak (*Quercus palustris* Muenchh). Rem Sens Env 25:71–87

Breda N, Badeau V (2008) Forest tree responses to extreme drought and some biotic events: Towards a selection according to hazard tolerance? CR Geosci 340:651–662

Brewer CA, Smith WK, Vogelmann TC (1991) Functional interaction between leaf trichomes, leaf wettability and the optical properties of water droplets. Plant Cell Environ 14:955–962

Brossa R, Casals I, Pinto-Marijuan M, Fleck I (2009) Leaf flavonoid content in *Quercus ilex* L. resprouts and its seasonal variation. Trees 23:401–408

Brüggemann W, Bergmann M, Nierbauer K-U, Plug E, Schmidt C, Weber D (2009) Photosynthesis studies on European evergreen and deciduous oaks grown under Central

European climate conditions: II. Photoinhibitory and light-independent violaxanthin deepoxidation and downregulation of photosystem II in evergreen, winter-acclimated European *Quercus* taxa. Trees 23:1091–1100

Bussotti F, Grossoni P (1997) European and Mediterranean oaks (*Quercus* L.; Fagaceae): SEA4 characterization of the micromorphology of the abaxial leaf surface. Bot J Linn Soc 124:183–199

Camarero JJ, Olano JM, Arroyo Alfaro SJ, Fernández-Marín B, Becerril JM, García-Plazaola JI (2012) Photoprotection mechanisms in *Quercus ilex* under contrasting climatic conditions. Flora 207:557–564

Cavender-Bares J, Apostol S, Moya I, Briantais JM, Bazzaz FA (1999) Chilling-induced photoinhibition in two oak species: are evergreen leaves inherently better protected than deciduous leaves? Photosynthetica 36:587–596

Cavender-Bares J, Cortes P, Rambal S, Joffre R, Miles B, Rocheteau A (2005) Summer and winter sensitivity of leaves and xylem to minimum freezing temperatures: a comparison of co-occurring Mediterranean oaks that differ in leaf lifespan. New Phytol 168:597–612

Cavender-Bares J, González-Rodríguez A, Pahlich A, Koehler K, Deacon N (2011) Phylogeography and climatic niche evolution in live oaks (*Quercus* series *Virentes*) from the tropics to the temperate zone. J Biogeogr 38:962–981

Chalker-Scott L (1999) Environmental significance of anthocyanins in plant stress responses. Photochem Photobiol 70:1–9

Collier DE, Thibodeau BA (1995) Changes in respiration and chemical content during autumnal senescence of *Populus tremuloides* and *Quercus rubra* leaves. Tree Physiol 15:759–764

Contran N, Gunthardt-Goerg MS, Kuster TM, Cerana R, Crosti P, Paoletti E (2013) Physiological and biochemical responses of *Quercus pubescens* to air warming and drought on acidic and calcareous soils. Plant Biol 15:157–168

Copolovici LO, Filella I, Llusia J, Niinemets U, Peñuelas J (2005) The capacity for thermal protection of photosynthetic electron transport varies for different monoterpenes in *Quercus ilex*. Plant Physiol 139:485–496

Corcuera L, Camarero JJ, Gil-Pelegrín E (2002) Functional groups in *Quercus* species derived from the analysis of pressure-volume curves. Trees 16:465–472

Corcuera L, Morales F, Abadía A, Gil-Pelegrín E (2005a) Seasonal changes in photosynthesis and photoprotection in a *Quercus ilex* subsp. *ballota* woodland located in its upper altitudinal extreme in the Iberian Peninsula. Tree Physiol 25:599–608

Corcuera L, Morales F, Abadía A, Gil-Pelegrín E (2005b) The effect of low temperatures on the photosynthetic apparatus of *Quercus ilex* subsp. *ballota* at its lower and upper altitudinal limits in the Iberian peninsula and during a single freezing-thawing cycle. Trees 19:99–108

Damesin C, Rambal S (1995) Field-study of leaf photosynthetic performance by a Mediterranean deciduous oak tree (*Quercus pubescens*) during a severe summer drought. New Phytol 131:159–167

Demmig-Adams B, Adams WW III (2006) Photoprotection in an ecological context: the remarkable complexity of thermal energy dissipation. New Phytol 172:11–21

Ehleringer JR (1982) The influence of water stress and temperature on leaf pubescence development in *Encelia farinosa*. Am J Bot 69:670–675

Epron D, Dreyer E (1992) Effect of a severe dehydration on leaf photosynthesis in *Quercus petraea* (Mtt.) Liebl.: photosystem II and electrolyte leakage. Tree Physiol 10:273–284

Epron D, Dreyer E, Bréda N (1992) Photosynthesis of oak trees [*Quercus petraea* (Matt.) Liebl.] during drought under field conditions: diurnal course of net CO_2 assimilation and photochemical efficiency of photosystem II. Plant Cell Environ 15:809–820

Esteban R, García-Plazaola JI (2014) Involvement of a second xanthophyll cycle in non-photochemical quenching of chlorophyll fluorescence: the lutein epoxide story. In: Demmig-Adams B, Garab G, Adams III W, Govindjee (eds) Non-photochemical quenching and energy dissipation in plants, algae and cyanobacteria, Springer Netherlands, pp 277–295

Faria T, García-Plazaola JI, Abadía A, Cerasoli S, Pereira JS, Chaves MM (1996) Diurnal changes in photoprotective mechanisms in leaves of cork oak (*Quercus suber*) during summer. Tree Physiol 16:115–123

Fernández V, Sancho-Knapik D, Guzmán P, Peguero-Pina JJ, Gil L, Karabourniotis G, Khayet M, Fasseas C, Heredia-Guerrero JA, Heredia A, Gil-Pelegrín E (2014) Wettability, polarity, and water absorption of holm oak leaves: effect of leaf side and age. Plant Physiol 166:168–180

Filella I, Peñuelas J (1999) Altitudinal differences in UV absorbance. UV reflectance and related morphological traits of *Quercus ilex* and *Rhododendron ferrugineum* in the Mediterranean region. Plant Ecol 145:157–165

Fracheboud Y, Luquez V, Bjorken L, Sjodin A, Tuominen H, Jansson S (2009) The control of autumn senescence in european aspen. Plant Physiol 149:1982–1991

Gallé A, Haldimann P, Feller U (2007) Photosynthetic performance and water relations in young pubescent oak (*Quercus pubescens*) trees during drought stress and recovery. New Phytol 174:799–810

García-Plazaola JI, Faria T, Abadía J, Abadía A, Chaves MM, Pereira JS (1997) Seasonal changes in xanthophyll composition and photosynthesis of cork oak (*Quercus suber* L.) leaves under mediterranean climate. J Exp Bot 48:1667–1674

García-Plazaola JI, Artetxe U, Dunabeitia MK, Becerril JM (1999a) Role of photoprotective systems of holm-oak (*Quercus ilex*) in the adaptation to winter conditions. J Plant Phys 155:625–630

García-Plazaola JI, Artexe U, Becerril JM (1999b) Diurnal changes in antioxidant and carotenoid composition in the Mediterranean schlerophyll tree *Quercus ilex* (L) during winter. Plant Sci 143:125–133

García-Plazaola JI, Hernandez A, Becerril JM (2000) Photoprotective responses to winter stress in evergreen Mediterranean ecosystems. Plant Biol 2:530–535

García-Plazaola JI, Errasti E, Hernandez A, Becerril JM (2002) Occurrence and operation of the lutein epoxide cycle in *Quercus* species. Funct Plant Biol 29:1075–1080

García-Plazaola JI, Hernandez A, Olano JM, Becerril JM (2003a) The operation of the lutein epoxide cycle correlates with energy dissipation. Funct Plant Biol 30:319–324

García-Plazaola JI, Olano JM, Hernández A, Becerril JM (2003b) Photoprotection in evergreen Mediterranean plants during sudden periods of intense cold weather. Trees 17:285–291

García-Plazaola JI, Hormaetxe K, Hernandez A, Olano JM, Becerril JM (2004) The lutein epoxide cycle in vegetative buds of woody plants. Funct Plant Biol 31:815–823

García-Plazaola JI, Esteban R, Hormaetxe K, Fernández-Marín B, Becerril JM (2008) Photoprotective responses of Mediterranean and Atlantic trees to the extreme heat-wave of summer 2003 in Southwestern Europe. Trees 22:385–392

Goss R, Lepetit B (2015) Biodiversity of NPQ. J Plant Physiol 172:13–32

Grant OM, Tronina L, Ramalho JC, Besson CK, Lobo-do-Vale R, Pereira JS, Jones HM, Chaves MM (2010) The impact of drought on leaf physiology of *Quercus suber* L. trees: comparison of an extreme drought event with chronic rainfall reduction. J Exp Bot 61:4361–4371

Gulías J, Flexas J, Abadía A, Medrano H (2002) Photosynthetic responses to water deficit in six Mediterranean sclerophyll species: possible factors explaining the declining distribution of *Rhamnus ludovici-salvatoris*, an endemic Balearic species. Tree Physiol 22:687–697

Haldimann P, Galle A, Feller U (2008) Impact of an exceptionally hot dry summer on photosynthetic traits in oak (*Quercus pubescens*) leaves. Tree Physiol 28:785–795

Hansen U, Fiedler B, Rank B (2002) Variation of pigment composition and antioxidative systems along the canopy light gradient in a mixed beech/oak forest: a comparative study on deciduous tree species differing in shade tolerance. Trees 16:354–364

Hardin JW (1979) Patterns of variation in foliar trichomes of eastern North American *Quercus*. Am J Bot 66:576–585

Hernández I, Cela J, Alegre L, Munné-Bosch S (2012) Antioxidants defences against drought stress. In: Aroca R (ed) Plant responses to drought stress. Springer, Berlin, pp 231–258

Hoch WA, Zeldin EL, McCown BH (2001) Physiological significance of anthocyanins during autumn leaf senescence. Tree Physiol 21:1–8

Hoch WA, Singsaas EL, McCown BH (2003) Resorption protection. Anthocyanins facilitate nutrient recovery in autumn by shielding leaves from potentially damaging light levels. Plant Physiol 133:1296–1305

Holmes MG, Keiller DR (2002) Effects of pubescence and waxes on the reflectance of leaves in the ultraviolet and photosynthetic wavebands: a comparison of a range of species. Plant Cell Environ 25:85–93

Hu B, Simon J, Rennenberg H (2013) Drought and air warming affect the species-specific levels of stress-related foliar metabolites of three oak species on acidic and calcareous soil. Tree Physiol 33:489–504

Huang W, Hu H, Zhang SB (2016a) Photosynthesis and photosynthetic electron flow in Alpine evergreen species *Quercus guyavifolia* in winter. Front Plant Sci 7:1511

Huang W, Yang YJ, Hu H, Zhang SB (2016b) Huang W, Yang YJ, Hu H, Zhang SB (2016b) Seasonal variations in photosystem I compared with photosystem II of three alpine evergreen broad-leaf tree species. J Photochem Photobiol B-Biol 165:71–79

Hughes N (2011) Winter leaf reddening in "evergreen" species. New Phytol 190:573–581

Karabourniotis G, Bornman JF (1999) Penetration of UV-A, UV-B and blue light through the leaf trichome layers of two xeromorphic plants, olive and oak, measured by optical fibre microprobes. Physiol Plant 105:655–661

IPCC (2014) Climate change 2014: synthesis report. Contribution of working groups I, II and III to the fifth assessment report of the intergovernmental panel on climate change. In: Core Writing Team, Pachauri RK, Meyer LA (eds). IPCC, Geneva, Switzerland

Karageorgou P, Manetas Y (2006) The importance of being red when young: anthocyanins and the protection of young leaves of *Quercus coccifera* from insect herbivory and excess light. Tree Physiol 26:613–621

Karioti A, Tooulakoc G, Bilia AR, Psaras GK, Karabourniotis G, Skaltsa H (2011) Erinea formation on *Quercus ilex* leaves: anatomical, physiological and chemical responses of leaf trichomes against mite attack. Phytochem 72:230–237

Keskitalo J, Bergquist G, Gardestrom P, Jansson S (2005) A cellular timetable of autumn senescence. Plant Physiol 139:1635–1648

Koehler K, Center A, Cavender-Bares J (2012) Evidence for a freezing tolerance-growth rate trade-off in the live oaks (*Quercus* series *Virentes*) across the tropical-temperate divide. New Phytol 193:730–744

Lee DW, Gould KS (2002) Why Leaves Turn Red. Amer Sci 90:524–531

Lee DW, O'Keefe J, Holbrook NM, Feild TS (2003) Pigment dynamics and autumn leaf senescence in a New England deciduous forest, eastern USA. Ecol Res 18:677–694

List of heat waves (2016) In Wikipedia, The Free Encyclopedia. Retrieved 13:18, 20 Sept 2016, from https://en.wikipedia.org/w/index.php?title=List_of_heat_waves&oldid=740342075

Llorens L, Aranda X, Abadia A, Fleck I (2002) Variations in *Quercus ilex* chloroplast pigment content during summer stress: involvement in photoprotection according to principal component analysis. Funct Plant Biol 29:81–88

Luterbarcher J, Dietrich D, Xoplaki E, Grosjean M, Wanner H (2004) European seasonal and annual temperature variability, trends, and extremes since 1500. Science 303:1499–1503

Mahall BE, Tyler CM, Cole ES, Mata C (2009) A comparative study of oak (*Quercus*, Fagaceae) seedling physiology during summer drought in southern California. Am J Botany 96:751–761

Manetas Y, Petropoulou Y, Psaras GK, Drinia A (2003) Exposed red (anthocyanic) leaves of *Quercus coccifera* display shade characteristics. Funct Plant Biol 30:265–270

Marabottini R, Schraml C, Paolacci AR, Sorgona A, Raschi A, Rennenberg H, Badiani M (2001) Foliar antioxidant status of adult Mediterranean oak species (*Quercus ilex* L. and *Q. pubescens* Willd.) exposed to permanent CO_2-enrichment and to seasonal water stress. Environ Pollut 113:413–423

Martínez-Ferri E, Manrique E Valladares F, Balaguer L (2004) Winter photoinhibition in the field involves different processes in four co-occurring Mediterranean tree species. Tree Physiol 24:981–990

Matile P, Hörnsteiner S, Thomas H (1999) Chlorophyll degradation. Ann Rev Plant Physiol Plant Molec Biol 50:67–95

Mediavilla S, Escudero A (2003) Stomatal responses to drought at a Mediterranean site: a comparative study of co-occurring woody species differing in leaf longevity. Tree Physiol 23:987–996

Mittler R (2002) Oxidative stress, antioxidants and stress tolerance. Trends Plant Sci 7:405–410

Morales F, Abadía A, Abadía J, Montserrat G, Gil-Pelegrín E (2002) Trichomes and photosynthetic pigment composition changes: responses of *Quercus ilex* subsp. *ballota* (Desf.) Samp. and *Quercus coccifera* L. to Mediterranean stress conditions. Trees 16:504–510

Moy A, Le S, Verhoeven AS (2015) Different strategies for photoprotection during autumn senescence in maple and oak. Physiol Plant 155:205–216

Müller P, Li X-P, Niyogi KK (2001) Non-photochemical quenching. A response to excess light energy. Plant Physiol 125:1558–1566

Munné-Bosch S (2005) The role of alpha-tocopherol in plant stress tolerance. J Plant Physiol 162:743–748

Munné-Bosch S, Peñuelas J, Asensio D, Llusià J (2004) Airborne ethylene may alter antioxidant protection and reduce tolerance of holm oak to heat and drought stress. Plant Physiol 136:2937–2947

Murchie EH, Niyogi KK (2011) Manipulation of photoprotection to improve plant photosynthesis. Plant Physiol 155:86–92

Naidu SL, De Lucia EH (1997) Acclimation of shade-developed leaves on saplings exposed to late-season canopy gaps. Tree Physiol 17:367–376

Nichol CJ, Pieruschka R, Takayama K, Forster B, Kolber Z, Rascher U, Grace J, Robinson SA, Pogson B, Osmond CB (2012) Canopy conundrums: building on the Biosphere 2 experience to scale measurements of inner and outer canopy photoprotection from the leaf to the landscape. Funct Plant Biol 39:1–24

Niinemets U (2007) Photosynthesis and resource distribution through plant canopies. Plant Cell Environ 30:1052–1071

Niinemets U, Valladares F (2006) Tolerance to shade, drought, and waterlogging of temperate Northern Hemisphere trees and shrubs. Ecol Monogr 76:521–547

Niyogi KK (1999) Photoprotection revisited: genetic and molecular approaches. Annu Rev Plant Phys 50:333–359

Noctor G, Foyer CH (1998) Ascorbate and glutathione: keeping active oxygen under control. Annu Rev Plant Physiol Plant Mol Biol 49:249–279

Nogués I, Llusià J, Ogaya R, Munné-Bosch S, Sardans J, Peñuelas J, Loreto F (2014) Physiological and antioxidant responses of *Quercus ilex* to drought in two different seasons. Plant Biosyst 148:268–278

Öquist G, Huner NPA (2003) Photosynthesis of overwintering evergreen plants. Annu Rev Plant Biol 54:329–355

Ort DR, Baker NR (2002) A photoprotective role for O_2 as an alternative electron sink in photosynthesis? Curr Opin Plant Biol 5:193–198

Peguero-Pina JJ, Gil-Pelegrín E, Morales F (2013) Three pools of zeaxanthin in *Quercus coccifera* leaves during light transitions with different roles in rapidly reversible photoprotective energy dissipation and photoprotection. J Exp Bot 64:1649–1661

Peguero-Pina JJ, Morales F, Flexas J, Gil-Pelegrín E, Moya I (2008) Photochemistry, remotely sensed physiological reflectance index and de-epoxidation state of the xanthophyll cycle in *Quercus coccifera* under intense drought. Oecologia 156:1–11

Peguero-Pina JJ, Sancho-Knapik D, Morales F, Flexas J, Gil-Pelegrín E (2009) Differential photosynthetic performance and photoprotection mechanisms of three Mediterranean evergreen oaks under severe drought stress. Funct Plant Biol 36:453–462

Peltier G, Cournac L (2002) Chlororespiration. Ann Rev. Plant Biol 53:523–550

Peñuelas J, Llusia J (2002) Linking photorespiration, monoterpenes and thermotolerance in *Quercus*. New Phytol 155:227–237

Peñuelas J, Llusia J, Asensio D, Munne-Bosch S (2005) Linking isoprene with plant thermotolerance, antioxidants and monoterpene emissions. Plant Cell Environ 28:278–286

Pezza AB, Rensch P, Cai WJ (2012) Severe heat wave sin Southern Australia: synoptic climatology and large scale connections. Clim Dynam 38:209–224

Pintó-Marijuan M, Joffre R, Casals I, De Agazio M, Zacchini M, García-Plazaola JI, Esteban R, Aranda X, Guàrdia M, Fleck I (2013) Antioxidant and photoprotective responses to elevated CO_2 and heat stress during holm-oak regeneration by resprouting evaluated by NIRS (Near Infrared Reflectance Spectroscopy). Plant Biol 15:5–17

Ramírez-Valiente JA, Koehler K, Cavender-Bares J (2015) Climatic origins predict variation in photoprotective leaf pigments in response to drought and low temperatures in live oaks (*Quercus* series *Virentes*). Tree Physiol 35:521–534

Renaud V, Rebetez M (2009) Comparison between open-site and below-canopy climatic conditions in Switzerland during the exceptionally hot summer of 2003. Agr Forest Meteorol 149:873–880

Rodríguez-Calcerrada J, Pardos JA, Gil L, Aranda I (2007) Acclimation to light in seedlings of *Quercus petraea* (Mattuschka) Liebl. and *Quercus pyrenaica* Willd. planted along a forest-edge gradient. Trees 21:45–54

Rosenthal SI, Camm EL (1996) Effects of air temperature, photoperiod and leaf age on foliar senescence of western larch (*Larix occidentalis* Nutt.) in environmentally controlled chambers. Plant Cell Environ 19:1057–1065

Ruban AV (2016) Nonphotochemical chlorophyll fluorescence quenching: mechanism and effectiveness in protecting plants from photodamage. Plant Physiol 170:1903–1916

Rubio de Casas R, Vargas P, Pérez-Corona E, Manrique E, Quintana JR, García-Verdugo C, Balaguer L (2007) Field patterns of leaf plasticity in adults of the long-lived evergreen *Quercus coccifera*. Ann Bot 100:325–334

Schwanz P, Picon C, Vivin P, Dreyer E, Guehl JM, Polle A (1996) Responses of antioxidative systems to drought stress in pendunculate oak and maritime pine as modulated by elevated CO_2. Plant Physiol 110:393–402

Schwanz P, Polle A (2001) Differential stress responses of antioxidative systems to drought in pedunculate oak (*Quercus robur*) and maritime pine (*Pinus pinaster*) grown under high CO_2 concentrations. J Exp Bot 52:133–143

Smith MD (2011) An ecological perspective on extreme climatic events: a synthetic definition and framework to guide future research. J Ecol 99:656–663

Stott A, Stone DA, Allen MR (2004) Human contribution to the European heatwave of 2003. Nature 432:610–614

Steyn WJ, Wand SJE, Holcroft DM, Jacobs G (2002) Anthocyanins in vegetative tissues: a proposed unified function in photoprotection. New Phytol 155:349–361

Streb P, Josse EM, Gallouet E, Baptist F, Kuntz M, Cornic G (2005) Evidence for alternative electron sinks in the photosynthetic carbon assimilation ion the high mountain plant species *Ranunculus glacialis*. Plant Cell Environ 28:1123–1135

Takahashi S, Badger MR (2011) Photoprotection in plants: a new light on photosystem II damage. Trends Plant Sci 16:53–60

Tsonev T, Wahbi S, Sun P, Sorrentino G, Centritto M (2014) Gas exchange, water relations and their relationships with photochemical reflectance index in *Quercus ilex* plants during water stress and recovery. Int J Agric Biol 16:335–341

Verhoeven AS (2014) Sustained energy dissipation in winter evergreens. New Phytol 201:57–65

Voss I, Sunil B, Scheibe R, Raghavendra AS (2013) Emerging concept for the role of photorespiration as an important part of abiotic stress response. Plant Biol 15:713–722

Wang GG, Bauerle WL (2006) Effects of light intensity on the growth and energy balance of photosystem II electron transport in *Quercus alba* seedlings. Ann For Sci 63:111–118

Wilkinson M, Eaton EL, Broadmeadow MSJ, Morison JIL (2012) Inter-annual variation of carbon uptake by a plantation oak woodland in south-eastern England. Biogeosci 9:5373–5389

Wobbe L, Bassi R, Kruse O (2015) Multi-Level Light Capture Control in Plants and Green Algae. Trends Plant Sci 21:55–68

Wolkerstorfer SV, Wonisch A, Stankova T, Tsvetkova N, Tausz M (2011) Seasonal variations of gas exchange, photosynthetic pigments, and antioxidants in Turkey oak (*Quercus cerris* L.) and Hungarian oak (*Quercus frainetto* Ten.) of different age. Trees 25:1043–1052

Yamazaki J-Y, Kamata K, Maruta E (2011) Seasonal changes in the excess energy dissipation from Photosystem II antennae in overwintering evergreen broad-leaved trees *Quercus myrsinaefolia* and *Machilus thunbergii*. J Photoch Photobio B 104:348–356

Xu C, McDowell NG, Sanna-Sevanto S, Fisher RA (2013) Our limited ability to predict vegetation dynamics under water stress. New Phytol 200:298–300

Zhu J-J, Zhang J-L, Liu H-C, Cao K-F (2009) Photosynthesis, non-photochemical pathways and activities of antioxidant enzymes in a resilient evergreen oak under different climatic conditions from a valley-savanna in Southwest China. Physiol Plant 135:62–72

Chapter 12
Growth and Growth-Related Traits for a Range of *Quercus* Species Grown as Seedlings Under Controlled Conditions and for Adult Plants from the Field

Rafael Villar, Paloma Ruiz-Benito, Enrique G. de la Riva, Hendrik Poorter, Johannes H. C. Cornelissen and José Luis Quero

Abstract Forests and shrublands occupy a large area in the world (*c.* 31% of the total continental area) and in Spain (*c.* 36% of the area), in which around 30% of forests are formed by *Quercus* species. Therefore, the ecosystem services provided by *Quercus* species are critical to human well-being. Thus, it is essential to understand how *Quercus* species grow and how they will respond to global change. Bringing together data of comparative growth experiments with seedlings, field data and allometric equations developed for adult plants, our main objectives for this chapter are: (1) to quantify the relative growth rates (RGR) and growth components of seedlings of *Quercus* species and compare them to values of woody species belonging to other families; (2) to characterise biomass allocation patterns in leaves, stem and roots and RGR in *Quercus* adults; (3) to understand how temperature, precipitation, tree size and tree density affect the RGR of adult *Quercus* species; and (4) to compare the RGR of seedlings and adults, and identify which functional traits can explain the differences in RGR. Compared to woody species from other families, seedlings of *Quercus* species were characterized by low RGR and specific leaf area (SLA), a high proportion of biomass invested in roots (RMF, root mass

---

R. Villar (✉) · E. G. de la Riva
Área de Ecología, Facultad de Ciencias, Universidad de Córdoba, 14071 Córdoba, Spain
e-mail: rafael.villar@uco.es

P. Ruiz-Benito
Grupo de Ecología y Restauración Forestal, Departamento de Ciencias de la Vida, Universidad de Alcalá, Edificio de Ciencias, Campus Universitario, 28805 Alcalá de Henares, Madrid, Spain

H. Poorter
Plant Sciences (IBG2), Forschungszentrum Jülich GmbH, D-52425 Jülich, Germany

J. H. C. Cornelissen
Systems Ecology, Department of Ecological Science, Vrije Universiteit, De Boelelaan 1085, 1081 HV Amsterdam, The Netherlands

J. L. Quero
Dpto. Ingeniería Forestal, Universidad de Córdoba, 14071 Córdoba, Spain

© Springer International Publishing AG 2017
E. Gil-Pelegrín et al. (eds.), *Oaks Physiological Ecology. Exploring the Functional Diversity of Genus Quercus L.*, Tree Physiology 7,
https://doi.org/10.1007/978-3-319-69099-5_12

fraction) and a large seed mass. One of the most important traits explaining differences in RGR among seedlings of *Quercus* species was the leaf area ratio (LAR, total leaf area per unit of total biomass). In *Quercus* species, the fraction of biomass in leaves (LMF) and roots (RMF) decreased with tree size, while the proportion of biomass in stems (SMF) increased. Thus, for a tree with 20 cm diameter at breast height, the values of LMF were only between 0.01 and 0.05 (i.e. 1–5% of total biomass invested in leaves) and SMF ranged from 0.50 to 0.80. RGR values of adult *Quercus* species were highly variable, due to differences in tree size, stand density and abiotic factors. Tree size and density negatively affected RGR, so bigger trees tend to grow more slowly. However, the variation in RGR explained by temperature and/or precipitation was relatively low (<7% of total variation). We observed a positive relationship between the RGR of seedlings in controlled conditions and those of adults in the field. Furthermore, median RGR values of adult plants for *Quercus* species were positively related to SLA and leaf nitrogen. To sum up, *Quercus* species differ in RGR and key leaf traits from other woody species and the RGR of adult trees depend on tree size, density, temperature and precipitation. Our results suggest that climate change synchronised with density might affect future trends on the growth of *Quercus* species.

Abbreviations

DBH	Diameter at breast height
LAR	Leaf area ratio
LMF	Leaf mass fraction
MAP	Mean annual precipitation
MAT	Mean annual temperature
NAR	Net assimilation rate
RGR	Relative growth rate
RGR_B	Relative growth rate based on biomass
RGR_{DBH}	Relative growth rate based on diameter
RGR_H	Relative growth rate based on height
RMF	Root mass fraction
SLA	Specific leaf area
SMF	Stem mass fraction

12.1 Introduction

Plant growth is a complex process, which depends on the balance between capture and loss of carbon and nutrients (Lambers et al. 1998; Poorter et al. 2013). Growth can be measured in absolute or relative units (Paine et al. 2012): the absolute growth rate is the increase of biomass per time (e.g. g day^{-1}), and the relative growth rate (RGR) is the increase of biomass per unit of biomass and time (e.g. mg g^{-1} day^{-1}). In general, it is accepted that for comparative purposes it is more appropriate to use RGR as this allows to compare different species on a relative scale (Villar et al. 2008; Rees et al. 2010; Ruiz-Benito et al. 2015). In absolute terms, a large plant increases more in biomass than a small plant just because it is larger, although the growth rate per unit biomass (RGR) is probably lower (Poorter and Garnier 1999). Absolute growth values can be particularly interesting to study the role of plants in storage of C and nutrients, for example in the dynamics of the carbon cycle.

As primary producers, plants determine the main energy inputs into an ecosystem, energy on which all other trophic levels depend. Because the light energy fixed by plants is stored in carbon-based molecules, plants play an important role in the C-cycle, acting over time both as a sink and a source (Dixon et al. 1994). Human activity is strongly increasing the concentration of CO_2 in the atmosphere, which has led to fertilisation and episodes with increased growth during 20th century (Nabuurs et al. 2003). However, there is great concern that the positive effects of carbon fertilisation might be offset by the negative effects of climate change (Ciais et al. 2005; IPCC 2014; Ruiz-Benito et al. 2014b; Nabuurs et al. 2013). As forests can be an important sink for C at the intermediate time scale, it is important to understand the role of forests as carbon sources and sinks (Pan et al. 2011).

Forests and shrublands are critical ecosystems, which occupy a large area in the world (31% of the total continental area, FAO 2012) as well as in Spain (36% of the area, Ruiz de la Torre 1990). In Spain, circa 30% of the total basal area of tree trunks is formed by *Quercus* species (Spanish Forest Inventory, SFI). *Quercus* forests in the Iberian peninsula vary from Mediterranean evergreen broadleaved forests dominated by *Quercus ilex* (holm oak) and *Q. suber* (cork oak) to Sub-Mediterranean deciduous broadleaved forests dominated by *Q. faginea* and *Q. pyrenaica* and Atlantic deciduous broad-leaved forests dominated by *Q. robur* and *Q. petraea* (Fig. 12.1) (Gómez-Aparicio et al. 2011). *Quercus* forests are a structural component to Iberian vegetation and if the environmental conditions are favourable they are the successional endpoint of many pine forests with which they coexist, alternate and segregate (see e.g. Costa et al. 1997; Zavala and Zea 2004).

In the last decades, an enormous effort has been made to characterise the structure, density and composition of Spanish forests through the Spanish Forest Inventories (SFI, Villanueva 2004). The SFI is performed on permanent plots every decade, allowing to estimate individual tree growth under field conditions at large spatial scales (see e.g. Gómez-Aparicio et al. 2011). There are few studies that have estimated tree growth rates based on biomass in the field, which is mainly due to the difficulties of estimating above- and belowground biomass which requires destructive methods (Poorter et al. 2012a). Therefore, most studies on forests estimate tree growth by using proxies based on increases in height and/or diameter or stem volume (e.g. Chaturvedi et al. 2011). The extensive nature of the SFI measurements (e.g. 20,080 individuals of *Quercus ilex* have been measured in 6914 different plots in the second census) allow to study the role of underlying abiotic factors (e.g. temperature and precipitation), tree size and tree density in how they affect the growth and mortality of individuals (see e.g. Gómez-Aparicio et al. 2011; Ruiz-Benito et al. 2013).

Fig. 12.1 Examples of oak forests in the Iberian Peninsula: **a** *Quercus suber* (cork oak) forest under use; notice the lack of understory as management strategy to facilitate traditional cork extraction. **b** *Quercus ilex* subsp. *ballota* (holm oak) under a "dehesa" regime (Mediterranean-like savanna), where a low oak density and open spaces without shrubs are managed to promote pasture. **c** *Quercus canariensis* (Algerian oak) in the Natural Park of Los Alcornocales. **d** *Quercus suber* (cork oak) and *Quercus canariensis* (Algerian oak) mixed forest where the later is restricted to the bottom part of the valley due to a lower drought-tolerance. (Credits: all pictures by JL. Quero)

In contrast to the few data of RGR obtained in the field for adult plants, many studies have been done with seedlings grown under controlled conditions with close-to-optimal levels of water and nutrients (Cornelissen et al. 1996; Antúnez et al. 2001; Ruíz-Robleto and Villar 2005; Lopez-Iglesias et al. 2014). The main objective of these studies was to understand the intrinsic causes of variation in RGR between different species. Other studies aimed to understand how the growth of *Quercus* seedlings was affected by different abiotic factors (light, water and nutrients) (Quero et al. 2006; Sánchez-Gómez et al. 2006; Pérez-Ramos et al. 2010; González-Rodríguez et al. 2011). It can be expected that RGR values obtained for seedlings grown under controlled conditions differ from those in adults under field conditions. These growth differences might be due to differences in ontogeny and/or the approach used. For example, it is generally known, that RGR values decrease with plant size (Evans 1972; Hunt 1982; Metcalf et al. 2003; Rees et al. 2010) because plants become increasingly inefficient when they are larger. This may be due to several factors such as shelf-shading, tissue aging and increased allocation to structural components (Rees et al. 2010). Nonetheless, RGR values of adult plants may still be useful for comparative purposes. Field data at large scales as forest inventories provides a high representativeness and, therefore, relevance for existing forests, whereas controlled experiments are generally orthogonal ensuring causality and capture the process of interest (Ratcliffe et al. 2016). However, if the ranking of growth and growth-related traits from comparative studies under controlled conditions is well correlated with those obtained from plants in the field, we could advantageously use the wealth of data from controlled studies for careful and informed extrapolation (Poorter et al. 2015).

In the case that we find differences in RGR between *Quercus* and non-*Quercus* woody species it is of high relevance to understand the intrinsic causes of these differences. RGR has traditionally been decomposed into two underlying components: the leaf area ratio (LAR, the ratio between the total leaf area and total dry biomass) and the net assimilation rate (NAR, biomass gain per unit area and time; Evans 1972; Poorter et al. 2013). Thus, RGR = LAR × NAR. In turn, LAR can be decomposed into the specific leaf area (SLA, leaf area per leaf dry biomass) and leaf mass fraction (LMF), being LAR = SLA × LMF. Many studies have concluded that variation in the morphological factor (LAR) is more important than the physiological factor (NAR) to explain interspecific variation of RGR (Poorter and van der Werf 1998; Galmes et al. 2005). However, some studies have found the variation in NAR to be more important than in LAR (Reich et al. 1998; Poorter and Garnier 1999; Shipley 2002). In addition, many studies with woody species have concluded that a very easy-to-measure variable, such as the specific leaf area (SLA), is a good predictor of maximum RGR (Lambers and Poorter 1992; Cornelissen et al. 1996; Khurana and Singh 2000; Antúnez et al. 2001). However, other studies point out that light conditions can influence which variable is more related to RGR (LAR or NAR). Thus, Veneklaas and Poorter (1998) found that in low light, LAR explained most of the interspecific differences in RGR, whereas at high light, differences in NAR were more important. Gibert et al. (2016) have found that there is a shift in the direction of correlation between growth rates and traits across ontogenetic stages. Thus, for example, specific leaf area was found to be correlated with RGR in seedlings but not in adult plants (Gibert et al. 2016). The variation in RGR between species will also be affected by other traits such the leaf nitrogen concentration (N) (Huante et al. 1995; Cornelissen et al. 1997; Villar et al. 2006, 2008). All these characteristics (morphological, physiological and/or phenological) of the species that directly or indirectly influence growth, reproduction and survival are broadly called functional traits (Violle et al. 2007). So far, we have insufficient insight in what causes the differences in RGR and functional traits among *Quercus* species as compared to other woody species.

Here, our aim is to characterise RGR patterns of seedlings and adults for the most common *Quercus* species in the Iberian Peninsula and further investigate the relationships with functional traits and the underlying drivers. We used individual data of seedlings under controlled conditions and adults from the Spanish National Forest Inventory. Our specific objectives were: (1) to identify the differences in

RGR and functional traits between *Quercus* species and other non-*Quercus* species under controlled conditions, (2) to characterise the biomass allocation patterns (leaf, stem and root) and RGR in *Quercus* species under field conditions; (3) to analyse how abiotic factors (mean annual temperature and/or annual precipitation) and tree size affect the growth of adults of *Quercus* species under field conditions, and (4) to compare the ranking of the RGR of seedlings across a range of *Quercus* species with those of adults grown in the field and to understand which functional traits explain the differences in RGR.

12.2 Differences in RGR and Functional Traits Between *Quercus* and Non-*Quercus* Species Under Controlled Conditions

We used data from two studies in which a range of *Quercus* and woody non-*Quercus* species were grown under controlled conditions, with the aim to compare RGR and several functional traits between these two groups (Cornelissen et al. 1996; Villar et al. 2008). For the comparisons we separately analysed deciduous and evergreens species, as leaf habit has a strong influence on functional traits (Cornelissen et al. 1996; Villar et al. 2006; Poorter et al. 2009). The study of Cornelissen et al. (1996), which we name *study 1* here, comprised a total of 97 woody species, of which 10 belonged to the genus *Quercus*. In this experiment, seedlings were grown for 21 days after the first true leaves had unfolded in a growth chamber at 135 µmol m^{-2} s^{-1} for 14 h per day (daily light integral 6.8 mol m^{-2} day^{-1}). The study of Villar et al. (2008), which we call *study 2*, combined data from two comparable studies (Antúnez et al. 2001; Ruíz-Robleto and Villar 2005), together comprising 24 woody species of which seven are *Quercus* species. In study 2, the seedlings were grown in a greenhouse with a total daily light integral of 31.6 mol m^{-2} day^{-1} during 124 days.

Quercus species differed in a number of functional traits compared to other woody species (Fig. 12.2). *Quercus* species, both evergreen and deciduous, had lower values of RGR and SLA than other evergreen and deciduous woody species (Fig. 12.2a, b, c, d). *Quercus* species were also characterized by a high fraction of biomass present in roots (RMF) and a large seed mass (Fig. 12.2e, f, g, h). Low RGR and SLA are typical of resource-conservative species (Diaz et al. 2004; Wright et al. 2004). The low values of RGR may be due to the low LAR of *Quercus* species (Fig. 12.3a, b), as a result of both low leaf mass fraction (LMF) and low specific leaf area (SLA). SLA and LMF are important factors that contribute to the regulation of plant photosynthesis and growth (Poorter and Remkes 1990; Antúnez et al. 2001). Thus, lower values of these traits imply a low leaf area displayed per unit total plant mass invested and consequently a low photosynthetic capacity per unit total plant mass (Reich et al. 1997). The advantage of plants with an inherently low SLA is that it promotes long-lived foliage, which have longer time to pay-off the construction cost of leaves (Wright et al. 2004; Escudero et al. 2017). This strategy, which promotes the duration and nutrient use efficiency over potential growth rates, is energetically and competitively more efficient in stressed environments (Reich et al. 1997). In addition, the low leaf area/total plant mass ratio (LAR) on a whole-plant basis and net assimilation (NAR) of *Quercus* species can also be explained by the high proportion of biomass present in roots (around 40%) compared to other woody species (around 20%) (Fig. 12.2e, f). The strong investment in roots makes the investment in leaves very low compared to that of non-*Quercus* species (Figs. 12.2e, f and 12.3g, h). A greater proportion of roots may constitute an advantage to resist drought, allowing access to a greater amount of water in the soil, which can confer higher survival rates under adverse conditions (Lloret et al. 1999).

One of the most important factors explaining the differences in RGR among *Quercus* species was the leaf area ratio (LAR) in both studies (Fig. 12.3a, b). Also, for non-*Quercus* species RGR was positively related to LAR. However, the relationship between RGR and the physiological component (NAR) was different for both studies (it was positively related in study 2, but no significant relationship was found in study 1 (Fig. 12.3c, d), neither for *Quercus* nor for non-*Quercus* species. These different results may in part be due to the different growth conditions in the two experiments (for example the high light conditions in the study 2). Interspecific variation in RGR was nearly significantly related to variation in in study 2, but it was not significant for study 1 for species. In the case of the species, the variable that better explained variation in RGR was LAR (Fig. 12.3a, b) and SLA (Fig. 12.3e, f) for both studies.

Another important functional trait in which *Quercus* differed with respect to other species was the seed mass. *Quercus* species have seeds that are four to six times greater than those of non-*Quercus* woody species with which they coexist (Fig. 12.2g, h). The seed mass determines the carbohydrate and nutrient reserves that the seedling has when it germinates. A large seed mass has certain advantages during the early stages of the seedling, since it allows to survive situations with a low carbon gain, such as shade conditions (Quero et al. 2007). A large seed will also imply that the seedlings have larger biomass (Quero et al. 2007; González-Rodríguez et al. 2010; Pérez-Ramos et al. 2010). Large seeds can confer a competitive advantage because it allows access to limiting resources (e.g. water, nutrients, light) and it will promote higher survival (Lloret et al. 1999).

It is important to notice that also within *Quercus* species, there is a high variability in functional traits, mainly because of clear differences between deciduous and evergreen (Fig. 12.2; Antúnez et al. 2001; Wright et al. 2004; Villar et al. 2006). In fact, in *Quercus* species of the Iberian Peninsula there is an economics spectrum of functional traits (de la Riva et al. 2014). On the one hand, there are species with a strategy of conservative-use of resources, which present slow growth but are tolerant to environmental stresses such as drought. This is the case of *Q. ilex*, which has leaves with low SLA and low concentration of chlorophyll and nitrogen, high wood density and roots with low specific root area (de la Riva et al. 2014). At the other extreme, there are species with a strategy of acquisitive use of resources, with greater growth capacity but

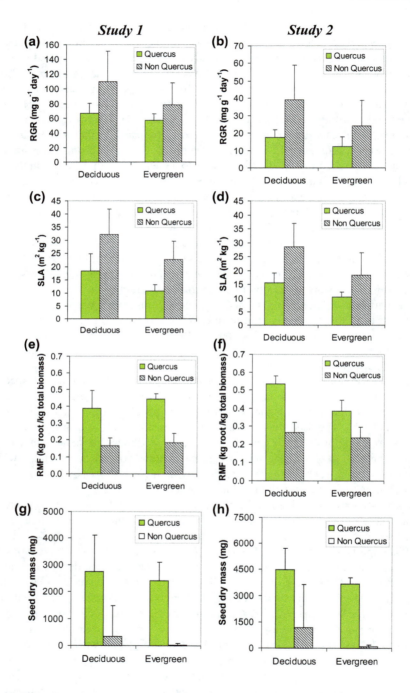

Fig. 12.2 Mean values ± standard deviation for **a, b** relative growth rate (RGR), **c, d** specific leaf area (SLA), **e, f** root mass fraction (RMF), and **g, h** seed dry mass of *Quercus* species and other non-*Quercus* woody species measured in seedlings under controlled conditions. Figures on the left (*study* 1) are from the experiments of Cornelissen et al. (1996) and figures on the right (*study* 2) are from Villar et al. (2008). Species were classified according to their leaf habit (deciduous and evergreen)

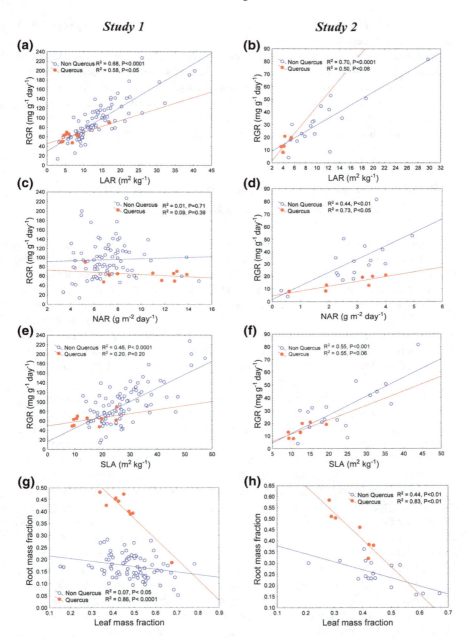

Fig. 12.3 Relationship between relative growth rate (RGR) and **a**, **b** the leaf area ratio (LAR), **c**, **d** the net assimilation rate (NAR), and **e**, **f** the specific leaf area (SLA); and the relationship between the root mass fraction and leaf mass fraction **g**, **h** of *Quercus* species and non-*Quercus* woody species measured in seedlings under controlled conditions. Figures on the left (*study* 1) are from Cornelissen et al. (1996) and figures on the right (*study* 2) are from Villar et al. (2008)

with a lower tolerance to environmental stresses (e.g. drought), as it is the case of *Q. robur* (de la Riva et al. 2014).

In summary, compared to woody species from other families, seedlings of *Quercus* species showed a low RGR and specific leaf area (SLA), a high proportion of biomass invested in roots (RMF, root mass fraction) and a large seed mass. One of the most important traits explaining differences in RGR among seedlings of *Quercus* species was the leaf area ratio (LAR, total leaf area per unit of total biomass), which is the product of SLA and LMF (the leaf proportion).

12.3 Biomass Allocation and Relative Growth Rate of *Quercus* Species Under Field Conditions

Field data of biomass allocation and RGR in adults are very scarce, due to the difficulties to obtain the root biomass. To know how biomass allocation to leaves, stems and roots differed between *Quercus* species under field conditions and how it can be affected by the tree size, we used the allometric equations of Montero et al. (2005). These equations are based on several adults plants of each species (between 23 to 141), in which the aboveground (leaves and stems) and belowground (roots) biomass was measured, along with tree height and stem diameter at breast height. The explained variation of the allometric relationships between total biomass (also for leaf, stems and roots) and tree diameter was very high for all *Quercus* species (R^2 between 0.90 to 0.97, Montero et al. 2005).

We used these allometric equations for trees present in permanent and unmanaged plots from the Spanish Forest Inventory (SFI) belonging to each of the six *Quercus* species to calculate the relative growth rate of adult trees. The SFI does a systematic sampling throughout the entire forest area of Spain approximately every ten years (Villaescusa and Díaz 1998; Villanueva 2004). The plots measured during the second census (SFI_2, in the years 1986–96), were revisited during the third SFI (SFI_3, 1997–2007), with a mean difference between both sampling of 11 years. For each individual, the species is recorded, and measurements of their height and diameter at breast height (DBH) are taken. We aimed to understand how biomass allocation to leaves, stems and roots changes with tree diameter. We found that the leaf and root mass fraction (LMF and RMF, respectively) decreased with tree size (Fig. 12.4a, c), while the proportion of stem (SMF) increased with tree size (Fig. 12.4b; Villar et al. 2014), in accordance with general observations (Veneklaas and Poorter 1998; Poorter et al. 2015). Apart from this overall pattern, we found that the dynamics of the various species were not fully similar (Fig. 12.4). For example, the decrease in LMF with tree size was very pronounced for *Q. faginea*, but less noticeable for *Q. canariensis*. When compared to other *Quercus* species, *Q. canariensis* had opposite slope for SMF (negative) and RMF (positive) with tree size. We currently do not have any explanation for this pattern. If we consider an average tree of 20 cm in diameter, the LMF values are between 0.01 and 0.05 (this implies 1–5% of total biomass invested in leaves), those of SMF between 0.50 and 0.80 (50–80% in stems) and those of RMF between 0.20 and 0.40 (20 and 40% in roots) (Fig. 12.4).

Data of RGR for adult individuals of *Quercus* under field conditions were calculated using the census data of the SFI. In the SFI all the trees are individually identified and remeasured and, therefore, it is possible to calculate a relative growth rate based on the diameter and/or height for each individual. To calculate RGR values, we selected individuals of the six most common *Quercus* species (*Q. canariensis, Q. faginea, Q. ilex, Q. pyrenaica, Q. robur* and *Q. suber*) with the following criteria: (i) trees should be alive in the two SFI censusus; (ii) DBH in SFI_2 was greater than 200 mm, (iii) changes in DBH and height between consecutive inventories should be greater than or equal to 0; and (iv) trees should come from unmanaged plots, with no evidence of cutting or thinning and no evidence of being plantations between the consecutive inventories. Since the plots cover the entire area of Spain, the environmental conditions of the plots are very different in terms of temperature, precipitation and soil type. For example, the mean annual temperature for plots with *Quercus* species varies between 5.7 °C (Logroño, North Spain) and 18 °C (Seville, South Spain) and the annual precipitation ranges from 200 mm (Almería, South Spain) to 2500 mm (Galicia, North of Spain). Therefore, it is expected that these different conditions and resources affect the growth of *Quercus* individuals under field conditions. We found that mean annual precipitation (MAP) showed a negative relationship with mean annual temperature (MAT), but the relationship was weak ($R^2 = 0.06$; $P < 0.001$).

Estimates of relative growth rate in mature trees in field conditions are relatively complicated. To measure growth, the best unit of measure is biomass but, given the difficulties in obtaining biomass data from leaves, stems and especially roots, there are not many observational studies that measure RGR based on biomass. Therefore, non-destructive, more practical measures are used in the field, such as the tree height and/or diameter. These variables have been measured systematically in the Spanish National Forest Inventories and other studies, so that by comparing consecutive inventories we can obtain individual estimates of RGR based on height and diameter. We used data of the second and third inventory (SFI_2 and SFI_3) to calculate the RGR of each individual in three different ways: (i) based on the increase in height (RGR_H, m m^{-1} year^{-1}); (ii) based on the increase in DBH (RGR_{DBH}, cm cm^{-1} year^{-1}), and (iii) based on the estimated whole plant biomass (RGR_B, mg g^{-1} year^{-1}). RGR was calculated using the following formula: $RGR = [Ln\ (X_2) - Ln\ (X_1)]/[time\ 2 - time\ 1]$, with X being the variable in which RGR is estimated (height, DBH or biomass), subscripts 2 and 1 refer to measurements at times 2 and 1 and the difference in *time* (*time* 2 − *time* 1) is the period in years between SFI_2 and SFI_3 (generally 11 years).

To estimate the total biomass (above and belowground) of the individual trees in each plot, we applied the allometric equations of Montero et al. (2005). In spite of the limitations of these equations (e.g. they do not take into account that the biomass allocation to the different organs of the tree may be affected by the

Fig. 12.4 a Leaf mass fraction (LMF), **b** stem mass fraction (SMF) and **c** root mass fraction (RMF) as a function of tree diameter at breast height for six species of *Quercus*. Data from Montero et al. (2005)

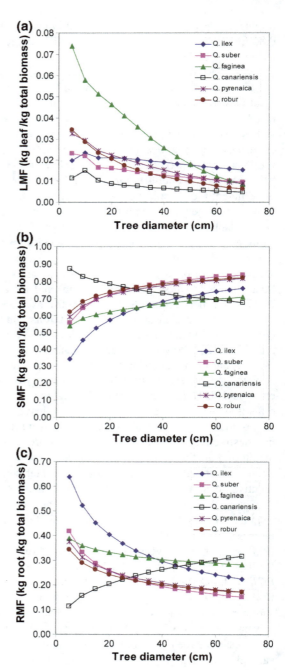

environmental factors, Poorter et al. 2015), this approach allows the comparison between the six *Quercus* species.

We expected that RGR values expressed in different dimensions would be similar. However, the correlation between RGR_H and RGR_{DBH} found for the six *Quercus* species was very low (Fig. 12.5). Although there was a significant relationship ($P < 0.0001$) between RGR_H and RGR_{DBH} for all species except for *Q. canariensis* ($P = 0.08$), the percentage of variation in RGR_H explained by RGR_{DBH} was very low (between 1.2 and 2%, Fig. 12.5). This suggests that there are large differences between growth estimates based on height and diameter, and, therefore,

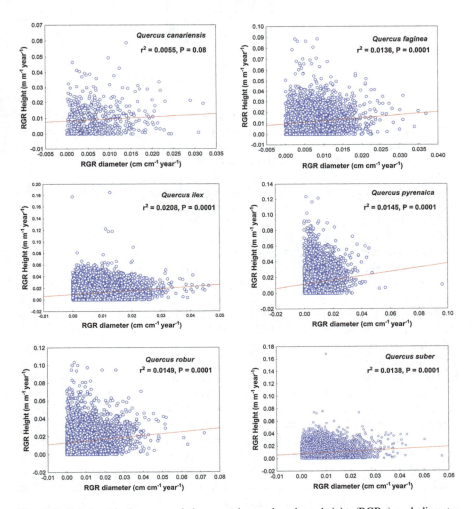

Fig. 12.5 Relationship between relative growth rate based on height (RGR_H) and diameter (RGR_{DBH}) for the six species of *Quercus*

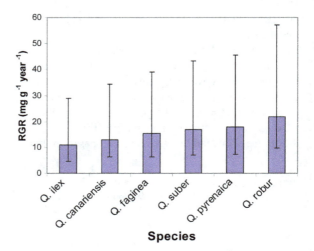

Fig. 12.6 Median values of relative growth rate (RGR) based on estimated biomass increment, in six species of *Quercus*. Data are from the Spanish Forest Inventory and the allometric equations from Montero et al. (2005). Bars indicate the upper and lower quartile

the conclusions based on one or another variable could be rather different. This result could partly be determined by competition, especially for light, and the fact that climatic conditions affect the allometric relationships in trees (e.g. Lines et al. 2012). On the one hand, some individuals could invest more in height and not in diameter, and vice versa depending on the environmental conditions. On the other hand, the precision in the estimation of the height in species with irregular crowns makes the calculation of the RGR_H a parameter with a degree of error greater than the error for RGR_{DBH}. Also, the changes of tree height and tree diameter along tree age could differ. In this sense, Poorter et al. (2012b) found that tree height reached an asymptote with age for all Quercus species. Overall, our results indicate the importance of estimating the tree growth considering both tree height and diameter or stem volume, and that these RGR values could strongly vary depending on the climatic and structural conditions of the plot.

As the RGR values in terms of biomass are calculated on the basis of the diameter using the equations of Montero et al. (2005), the correspondence of the RGR_{DBH} and RGR_B is exactly the same. We found that RGR_B in the field were very variable within species (from 0 to 90 mg g^{-1} year^{-1}), showing a wide overlap across the six species considered (Fig. 12.6). This large variability may be due to the large number of factors that can affect RGR, such abiotic factors (e.g. temperature, precipitation or nutrient availability) or biotic factors (e.g. competition, herbivory, management) (Gómez-Aparicio et al. 2011; Paine et al. 2015).

We conclude that in the six *Quercus* species, the fraction of biomass in leaves (LMF) and roots (RMF) decreased with tree size, while the proportion of biomass in stems (SMF) increased. Thus, for a tree with 20 cm diameter at breast height, the values of LMF were only between 0.01 and 0.05 (i.e. 1–5% of total biomass invested in leaves). The estimates of RGR based on height were very different than those based on biomass. RGR values of adult *Quercus* species were highly variable, due to differences in tree size, stand density and abiotic factors.

12.4 Sources of Variation in RGR Under Field Conditions: The Role of Abiotic Factors, Tree Size and Density

Frequently, water availability is one of the major limiting factors for growth in Mediterranean environments (Galmes et al. 2005), but also variation in temperature could be an important factor (Gómez-Aparicio et al. 2011). To understand the influence of these two climatic factors (precipitation and temperature) in RGR, we performed multiple linear regressions (MLR) per each species using RGR_B as dependent variable and mean annual temperature (MAT, °C) and mean annual precipitation (MAP, mm) as independent variables. In general, we found that both abiotic factors had a significant effect on RGR for most *Quercus* species, but the joined explained variance of both factors on RGR was very small (between 0.1 to 7%, Table 12.1). Paine et al. (2015) also found that relative growth rate based on height increment was weakly related to potential evapotranspiration (which summarizes the joint ecological effects of temperature and precipitation). Apparently, other factors as nutrient availability, soil texture, plot (density) and tree attributes (size) also modulate RGR under field conditions.

We also found different patterns among *Quercus* species in relation to the effect of climatic factors on RGR. Water limitation can influence the RGR in each species differently, which seems to be the result of differences in the variation in NAR, LAR, LMR and SLA between species (Galmes et al. 2005). Hence, different responses and adaptation to environmental constraints might exist. While RGR of *Q. faginea* and *Q. canariensis* show no clear or even a negative relationship with temperature and precipitation (Table 12.1), other deciduous species like *Q. pyrenaica* showed clear associations with climatic factors: RGR was ~ 40 mg g^{-1} year^{-1} in areas with overall temperatures higher than 14 °C and precipitation of 1600 mm, and no growth (~ 0) in areas below 14 °C and 700 mm (Fig. 12.7). This result accords with southern distribution limit of these species (Sierra de Cardeña

Table 12.1 Multiple linear regressions of relative growth rate based on biomass (RGR_B) with mean annual temperature (MAT) and mean annual precipitation (MAP) for the six *Quercus* species growing in field conditions

Species	n	Intercept	b MAT	b MAP	R^2
Quercus canariensis	171	77.036***	−2.274**	−0.021**	0.067 ***
Quercus faginea	1121	18.805***	0.178 ns	−0.0022 ns	0.001 ns
Quercus ilex	6913	14.563***	−0.384***	0.0083***	0.034***
Quercus pyrenaica	1612	−2.149 ns	1.787***	0.0058***	0.061***
Quercus robur	1113	21.407***	1.579***	−0.0077***	0.014***
Quercus suber	2285	13.521***	0.383*	0.0018 ns	0.002 *a*

The values the number of plots (n), the intercept, the slope of MAT (b MAT) and of MAP (b MAP), the R^2 and the significance are shown. ns, non significant, *a* $0.01 > P > 0.05$; *$P < 0.05$; **$P < 0.01$; ***$P < 0.001$

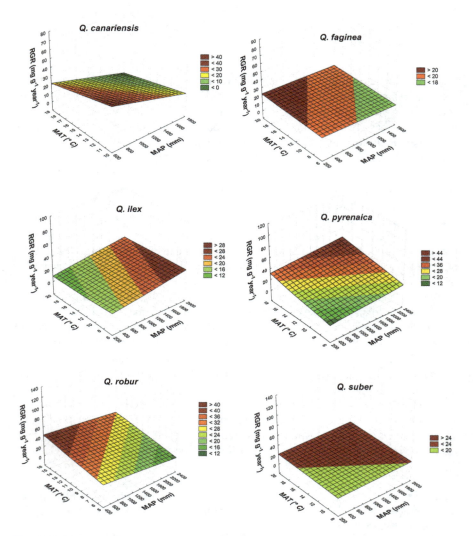

Fig. 12.7 Multiple linear regressions of relative growth rate based on biomass (RGR_B) with mean annual temperature (MAT) and mean annual precipitation (MAP) for the six *Quercus* species growing in field conditions

and Montoro Natural Park and Sierra Nevada National Park), where regeneration can become a bottleneck of forest dynamics (Ninyerola et al. 2009). Indeed, many recent efforts of reforestation of *Q. pyrenaica* seedlings in Sierra de Cardeña had failed probably due to the sever summer drought of the last years (José Manuel Quero, personal communication). For the evergreen species we found different responses to climate. On the one hand, *Q. ilex* had a wide range of temperature and precipitation where it can grow well (Fig. 12.7). In any case, the low RGR of *Q. ilex* even under better conditions of MAT and MAP (just to 20 mg g^{-1} year^{-1}) may be related to its conservative traits (low SLA, high wood density). On the other hand, *Q. suber,* is restricted to acidic soils (Hidalgo et al. 2008), generally distributed in the western part of Iberian Peninsula, where there are higher values of MAP than in the eastern part (Instituto Geográfico Nacional 2017); suggesting that precipitation is not restricting growth of this species (Table 12.1; Fig. 12.7).

In summary, we found significant effects of MAT and MAP on RGR but the explanatory power of these two environmental factors was low (smaller than 7%). Other factors such as nutrient availability, soil texture and plot density and tree attributes (size) can also be related to variation in RGR under field conditions.

In this sense, RGR decreased significantly ($P < 0.001$) with tree size in all species. Larger trees tend to grow more slowly (Fig. 12.8), similar to that found in other studies (Ryan et al. 2004; Coates et al. 2009, Gómez-Aparicio et al. 2011). Noteworthy is the low leaf mass fraction of the trees (for a tree of 20 cm in diameter, LMF values range from 0.01–0.03, thus 1–3%). In addition, the proportion of leaves decreases with tree size for all *Quercus* species (Fig. 12.4). In the review of Poorter et al. (2015) a similar pattern was found for numerous tree species, with a low percentage of leaf and with an almost linear decrease of LMF with plant size. Ryan et al. (2004) found that the decline in *Eucalyptus* forest production with age was due to the combined effect of a decrease in photosynthesis and a greater carbon cost due to increased allocation to roots and leaf respiration. Our results suggest that the decrease in RGR with the increase of tree size could be caused by a decrease in LMF and that, therefore, there is proportionally less leaf

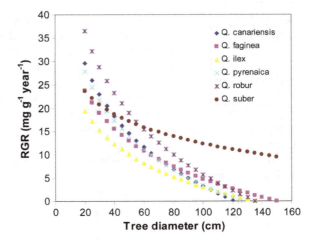

Fig. 12.8 Variation of the relative growth rate based on biomass (RGR$_B$) and stem diameter (DBH, diameter at breast height, cm) for adult trees of six species of *Quercus*, obtained from observational data from the Spanish Forest Inventory and the allometric equations of Montero et al. (2005)

Fig. 12.9 Values of the standardized coefficients (beta values) of the multiple linear regressions per species using RGR_B as dependent variable and mean annual temperature (MAT, °C) and mean annual precipitation (MAP, mm), tree size (DBH, cm) and basal area (m^2 ha^{-1}) of the plot as independent variables. The values of R^2 of the multiple linear regressions are also shown

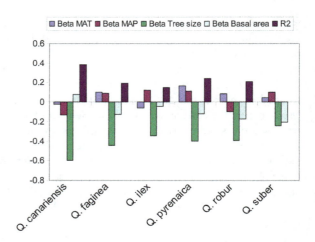

area to perform photosynthesis. An additional important explanatory factor may be that an increasing proportion of the xylem is no longer metabolically active but has a negative effect of RGR by inflating the mass basis on which RGR is expressed. This is especially true for the heartwood, the proportion of which increases with age and size. How much the proportion of heartwood varies and thereby affects RGR among *Quercus* species is an interesting area of future research.

RGR values were also negatively affected by the basal area of the plot (except for *Q. canariensis*), which reflects the negative effect of tree competition. The basal area is the area of a given section of land that is occupied by the cross-section of tree trunks and stems at the breast height, and therefore is a variable related to tree competition. To understand which of the factors exert a higher effect on RGR, we performed multiple linear regressions per species using RGR_B as dependent variable and mean annual temperature (MAT, °C) and mean annual precipitation (MAP, mm), tree size (DBH) and basal area of the plot as independent variables. The comparison of the standardized coefficients (beta) for each factor showed that tree size and basal area of the plot exert a larger and more negative effect on RGR than abiotic factors such as MAT and MAP (Fig. 12.9). We also found that the R^2 of the models was much higher (between 0.08 to 0.38) that those models which only considered abiotic factors (Table 12.1).

In summary, in *Quercus* species growing in the field, tree size and density negatively affected RGR, so bigger trees tend to grow more slowly. However, the variation in RGR explained by temperature and/or precipitation was relatively low (<7% of total variation).

12.5 Comparison of RGR in Adults and Seedlings of *Quercus* Species and Their Relationships with Functional Traits

To test if the values of RGR under controlled conditions are representative of those measured under field conditions, we have compared both measurements. Despite the high variability in RGR under field conditions, we estimated the median values per species. The ranking of RGR for the six species considered was: *Q. ilex* < *Q. canariensis* < *Q. faginea* < *Q. suber* < *Q. pyrenaica* < *Q. robur* (Fig. 12.6).

To compare the RGR under controlled and field conditions, we calculated the maximum RGR under field conditions. To do this, we considered the diametric category of 20–40 cm of DBH, which is the category that shows the maximum RGR values (Fig. 12.8), and calculated the median values for this category. To know if these differences in RGR between species were related to functional traits, we collected data of SLA and leaf N concentration of individuals in the field from Wright et al. (2004) and our own data (de la Riva et al. 2014). We found that RGR in the field was significantly correlated ($P < 0.05$) with SLA and leaf N concentration (Fig. 12.10a, b). The construction of plant tissues of woody Mediterranean species implies a trade-off between mechanical support and storage of water and nutrients (Pratt et al. 2007). Plants allocate nitrogen, according with their requirements, to produce enzymes and pigments that control different plant processes (i.e., photosynthesis, respiration, growth) at a cellular scale (Ghimire et al. 2017). Hence, plants with thinner leaves and higher SLA have high photosynthetic rates per unit mass and high net CO_2 exchange rates, which implies necessarily, from a physiological point of view, a high allocation of leaf N to achieve a high A_{mass} (Poorter 1989; Reich et al. 1997; de la Riva et al. 2016).

When comparing RGRs of seedlings with those of adults (measured in diameter classes of 20–40 cm), a positive trend and nearly significant ($P = 0.11$) was observed for the six species considered (Fig. 12.10c). It is necessary to emphasize the difference of units: in seedlings the RGR range was 10–20 mg g^{-1} day^{-1} and in trees the range was between 16 and 30 mg g^{-1} $year^{-1}$, possibly due to the limitation of resources (water and nutrients) that is common in natural conditions and to the short growth period in field conditions as well as to the larger size of field plants, all of which contribute to a lower RGR (Metcalf et al. 2003; Rees et al. 2010).

In summary, we observed a positive trend between the RGR of seedlings in controlled conditions and those of adults in the field. Median RGR values of adult plants for *Quercus* species were positively related to SLA and leaf nitrogen.

Fig. 12.10 Relationship between the relative growth rate based on biomass (RGR$_B$, mg g^{-1} year^{-1}) of adult individuals measured under field conditions with respect to: **a** specific leaf area (SLA, m^2 kg^{-1}), **b** leaf nitrogen concentration (N field, %), and **c** the relative growth rate measured under controlled conditions and in seedlings (mg g^{-1} day^{-1}). SLA and N data are from individuals growing in the field from Wright et al. (2004) and de la Riva et al. (2014). Note the difference in units of RGR in adults (mg g^{-1} year^{-1}) and seedling (mg g^{-1} day^{-1}). The values of RGR$_B$ were calculated for the trees between 20–40 cm of diameter (DBH) to have the maximum values of RGR$_B$ under field conditions and avoid the effect of tree size on RGR$_B$ of adult individuals measured under field conditions

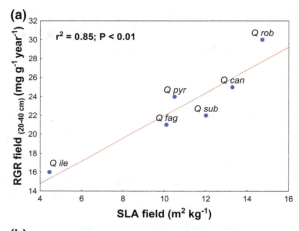

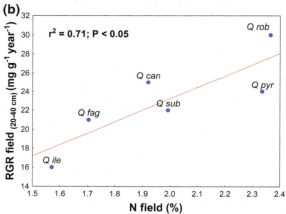

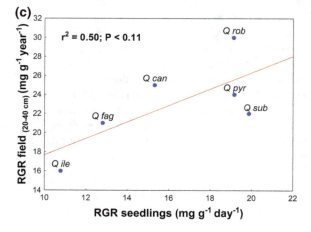

12.6 Conclusions

As compared to other woody species, *Quercus* species are characterized by relatively low relative growth rates, low specific leaf area, high root mass fractions and large seeds. One of the most important factors explaining the differences in relative growth rate between *Quercus* species in both seedlings and adults was the differences in leaf area ratio and the specific leaf area. Data of biomass allocation under field conditions showed that the proportion of leaf and root decreases with tree size, while the proportion of stem increases. Under field conditions, *Quercus* species showed a great variability in the relative growth rates, which is due to differences in tree size and basal area and—to a lesser extent—due to climatic conditions (temperature and precipitation). Larger trees grew more slowly, which could be explained by a decrease in the proportion of leaves with tree size as well as the increasing proportion of non-respiring xylem in the heartwood. In contrast, temperature and precipitation exert a low effect on the relative growth rate. Other environmental factors such as nutrient availability and soil depth can also affect growth under field conditions. Nonetheless, the RGR of seedlings grown under controlled conditions can be partially extrapolated to adult trees under field conditions, which highlights the importance of using both approaches for a better understanding of the causes of differences growth and related traits in *Quercus* species.

Acknowledgements This work was funded by the coordinated project DIVERBOS (CGL2011-30285-C02-02), ECO-MEDIT (CGL2014-53236-R), ANASINQUE (PGC2010-RNM-5782) by Junta de Andalucía, and Life + Biodehesa Project (11/BIO/ES/000726); and FEDER funds. PRB was funded by TALENTO Fellow (Comunidad de Madrid, 2016-T2/AMB-1665).

References

Antúnez I, Retamosa EC, Villar R (2001) Relative growth rate in phylogenetically related deciduous and evergreen woody species. Oecologia 128:172–180
Chaturvedi RK, Raghubanshi AS, Singh JS (2011) Leaf attributes and tree growth in a tropical dry forest. J Veg Sci 22:917–931
Ciais P, Reichstein M, Viovy N, Granier A, Ogee J, Allard V et al (2005) Europe-wide reduction in primary productivity caused by the heat and drought in 2003. Nature 437:529–533
Costa M, Morla C, Sáinz H (1997) Los bosques ibéricos: una interpretación geobotánica. Editorial Planeta, Barcelona.
Coates KD, Canham CD, LePage PT (2009) Above- versus below-ground competitive effects and responses of a guild of temperate tree species. J Ecol 97:118–130
Cornelissen JHC, Castro Díez P, Hunt R (1996) Seedling growth, allocation and leaf attributes in a wide range of woody plant species and types. J Ecol 84:755–765
Cornelissen JHC, Werger MJA, Castro-Díez P, van Rheenen JWA, Rowland AP (1997) Foliar nutrients in relation to growth, allocation and leaf traits in seedlings of a wide range of woody plant species and types. Oecologia 111:460–469
de la Riva EG, Pérez-Ramos IM, Navarro-Fernández CM, Olmo M, Marañón T, Villar R (2014) Estudio de rasgos funcionales en el género *Quercus*: estrategias adquisitivas frente a conservativas en el uso de recursos. Ecosistemas 23(2):82–89

de la Riva EG, Olmo M, Poorter H, Ubera JL, Villar R (2016) Leaf Mass per Area (LMA) and its relationship with leaf structure and anatomy in 34 Mediterranean woody species along a water availability gradient. PLoS ONE. doi:10.1371/journal.pone.0148788

Diaz S, Hodgson JG, Thompson K, Cabido M, Cornelissen JHC et al (2004) The plant traits that drive ecosystems: Evidence from three continents. J Veg Sci 15:295–304

Dixon RK, Solomon AM, Brown S, Houghton RA, Trexier MC, Wisniewski J (1994) Carbon pools and flux of global forest ecosystems. Science 263:185–190

Escudero A, Mediavilla S, Olmo M, Villar R, Merino JA (2017) Coexistence of deciduous and evergreen oak species in Mediterranean environments: costs associated with the leaf traits of both habits. Chapter 6 of this book

Evans GC (1972) The quantitative analysis of plant growth. Blackwell Scientific, Oxford

FAO (2012) State of the world's forests. Food and Agriculture Organization of the United Nations, Rome

Galmes J, Cifre J, Medrano H, Flexas J (2005) Modulation of relative growth rate and its components by water stress in Mediterranean species with different growth forms. Oecologia 145:21

Ghimire B, Riley WJ, Koven CD, Kattge J, Rogers A, Reich PB, Wright IJ (2017) A global trait-based approach to estimate leaf nitrogen functional allocation from observations. Ecol Appl. doi:10.1002/eap.1542

Gibert A, Gray EF, Westoby M, Wright IJ, Falster DS (2016) On the link between functional traits and growth rate: meta-analysis shows effects change with plant size, as predicted. J Ecol 104:1488–1503

Gómez-Aparicio L, García-Valdés R, Ruiz-Benito P, Zavala MA (2011) Disentangling the relative importance of climate, size and competition on tree growth in Iberian forests: implications for management under global change. Glob Change Biol 17:2400–2414

González-Rodríguez V, Navarro Cerrillo R, Villar R (2010) Maternal influences on seed mass effect and initial seedling growth in four *Quercus* species. Acta Oecol 37:1–9

González-Rodríguez V, Villar R, Casado R, Suárez-Bonnet E, Quero JL, Navarro Cerrillo R (2011) Spatio-temporal heterogeneity effects on seedling growth and establishment in four *Quercus* species. Ann For Sci 68:1217–1232

Hidalgo PJ, Marín JM, Quijada J, Moreira JM (2008) A spatial distribution model of cork oak (*Quercus suber*) in southwestern Spain: a suitable tool for reforestation. For Ecol Manag 255 (1):25–34

Huante P, Rincón E, Acosta I (1995) Nutrient availability and growth rate of 34 woody species from a tropical deciduous forest in Mexico. Funct Ecol 9:849–858

Hunt R (1982) Plant growth curves: a functional approach to plant growth analysis. Edward Arnold, London

Instituto Geográfico Nacional (2017) Precipitación media annual. http://www.02.ign.es/espmap/mapas_clima_bach/Mapa_clima_05.htm

IPCC (2014) Climate Change 2014: synthesis report. Contribution of working groups I, II and III to the fifth assessment report of the intergovernmental panel on climate change. In: Core Writing Team, Pachauri RK, Meyer LA (eds). IPCC, Geneva, Switzerland, 151 pp

Khurana E, Singh JS (2000) Influence of seed size on seedling growth of *Albizia procera* under different soil water levels. Ann Bot 86:1185–1192

Lambers H, Poorter H (1992) Inherent variation in growth rate between higher plants: a search for physiological causes and ecological consequences. Adv Ecol Res 23:187–261

Lambers H, Chapin III FS, Pons TL (1998) Plant physiological ecology. Springer

Lines ER, Zavala MA, Purves DW, Coomes DA (2012) Predictable changes in aboveground allometry of trees along gradients of temperature, aridity and competition. Global Ecol Biogeogr 21:1017–1028

Lloret F, Casanovas C, Peñuelas J (1999) Seedling survival of Mediterranean shrubland species in relation to root:shoot ratio, seed size and water and nitrogen use. Funct Ecol 13:210–216

Lopez-Iglesias B, Villar R, Poorter L (2014) Functional traits predict drought performance and distribution of Mediterranean woody species. Acta Oecol 56:10–18

Metcalf JC, Rose KE, Rees M (2003) Evolutionary demography of monocarpic perennials. Trends Ecol Evol 18:471–480.
Montero G, Ruiz-Peinado R, Muñoz M (2005) Producción de biomasa y fijación de CO_2 por los bosques españoles. Instituto Nacional de Investigación y Tecnología Agraria y Alimentaria
Nabuurs GJ, Schelhaas MJ, Mohren GMJ, Field CB (2003) Temporal evolution of the European forest sector carbon sink from 1950 to 1999. Glob Change Biol 9:152–160
Nabuurs GJ, Lindner M, Verkerk PJ, Gunia K, Deda P, Michalak R, Grassi G (2013) First signs of carbon sink saturation in European forest biomass. Nat Clim Change 3:792–796
Ninyerola M, Serra-Díaz JM, Lloret F (2009). Atlas de idoneidad topo-climática de leñosas. http://www.opengis.uab.cat/IdoneitatPI/index.html
Paine CET, Matthews TR, Vogt DR, Purves D, Rees M, Hector A, Turnbull LA (2012) How to fit nonlinear plant growth models and calculate growth rates: an update for ecologists. Methods Ecol Evol 3:245–256
Paine CET, Amissah L, Auge H, Baraloto C, Baruffol M, Bourland N et al (2015) Globally, functional traits are weak predictors of juvenile tree growth, and we do not know why. J Ecol 103:978–989
Pan Y, Birdsey RA, Fang J, Houghton R, Kauppi PE, Kurz WA et al (2011) A large and persistent carbon sink in the world's forests. Science 333:988–993
Pérez-Ramos IM, Gomez-Aparicio L, Villar R, García LV, Marañón T (2010) Seedling growth and morphology of three oak species along field resource gradients and seed mass variation: a seedling age-dependent response. J Veg Sci 21:419–437
Poorter H (1989) Interspecific variation in relative growth rate: on ecological causes and physiological consequences. In: Lambers H (ed) Causes and consequences of variation in growth rate and productivity of higher plants, pp 45–68. SPB Academic Publishing
Poorter H, Garnier E (1999) Ecological significance of inherent variation in relative growth rate and its components. In: Pugnaire F, Valladares F (eds) Handbook of functional plant ecology, vol. 20, p. 81–120
Poorter H, Niinemets U, Poorter L, Wright IJ, Villar R (2009) Causes and consequences of variation in leaf mass per area (LMA): a meta-analysis. New Phytol 182:565–588.
Poorter H, Remkes C (1990) Leaf area ratio and net assimilation rate of 24 wild species differing in relative growth rate. Oecologia 83:553–559
Poorter H, van der Werf A (1998) Is inherent variation in RGR determined by LAR at low irradiance and by NAR at high irradiance? A review of herbaceous species. In: Lambers H, Poorter H, Van Vuuren MI (eds) Inherent Variation in Plant Growth. Backhuys, Leiden, The Netherlands, pp 309–336
Poorter H, Niklas KJ, Reich PB, Oleksyn J, Poot P, Mommer L (2012a) Biomass allocation to leaves, stems and roots: meta-analyses of interspecific variation and environmental control. New Phytol 193:30–50
Poorter L, Lianes E, Moreno-de las Heras M, Zavala MA (2012b) Architecture of Iberian canopy tree species in relation to wood density, shade tolerance and climate. Plant Ecol 213:707–722
Poorter H, Anten NPR, Marcelis LFM (2013) Physiological mechanisms in plant growth models: do we need a supra-cellular systems biology approach? Plant Cell Environ 36:1673–1690
Poorter H, Jagodzinski AM, Ruiz-Peinado R, Kuyah S, Luo Y, Oleksyn J et al (2015) How does biomass distribution change with size and differ among species? An analysis for 1200 plant species from five continents. New Phytol 208:736–749
Pratt RB, Jacobsen AL, Ewers FW, Davis SD (2007) Relationships among xylem transport, biomechanics and storage in stems and roots of nine Rhamnaceae species of the California chaparral. New Phytol 174:787–798.
Quero JL, Villar R, Marañon T, Zamora R (2006) Interactions of drought and shade effects on seedlings of four *Quercus* species: physiological and structural leaf responses. New Phytol 170:819–834
Quero JL, Villar R, Marañon T, Zamora R, Poorter L (2007) Seed mass effects in four mediterranean *Quercus* species (Fagaceae) growing in contrasting light environments. Am J Bot 94:1795–1803

Ratcliffe S, Liebergesell M, Ruiz-Benito P, Madrigal González J, Muñoz Castañeda JM, Kändler G, et al (2016) Modes of functional biodiversity control on tree productivity across the European continent. Glob Ecol Biogeogr 25:251–262.

Rees M, Osborne CP, Woodward FI, Hulme SP, Turnbull LA, Taylor SH (2010) Partitioning the components of relative growth rate: How important is plant size variation? Am Nat 176:E152–E161

Reich PB, Walters MB, Ellsworth DS (1997) From tropics to tundra: global convergence in plant functioning. Proc Natl Acad Sci 94(25):13730–13734

Reich PB, Tjoelker MG, Walters MB, Vanderklein DW, Buschena C (1998) Close association of RGR, leaf and root morphology, seed mass and shade tolerance in seedlings of nine boreal tree species grown in high and low light. Funct Ecol 12:327–338

Ruiz de la Torre J (1990) Distribución y características de las masas forestales Españolas. Ecología, Fuera de Serie 1:11–30.

Ruiz-Benito P, Lines ER, Gómez-Aparicio L, Zavala MA, Coomes DA (2013) Patterns and drivers of tree mortality in Iberian forests: climatic effects are modified by competition. PLoS ONE 8:e56843.

Ruiz-Benito P, Gómez-Aparicio L, Paquette A, Messier C, Kattge J, Zavala MA (2014a) Diversity increases carbon storage and tree productivity in Iberian forests. Global Ecol Biogeogr 23:311–322

Ruiz-Benito P, Madrigal-González J, Ratcliffe S, Coomes DA, Kändler G, Lehtonen A, Wirth C, Zavala MA (2014b) Stand structure and recent climate change constrain stand basal area change in European forests: a comparison across boreal, temperate and Mediterranean biomes. Ecosystems 17:1439–1454

Ruiz-Benito P, Madrigal-González J, Young S, Mercatoris P, Cavin L, Huang T-J et al (2015) Climatic stress during stand development alters the sign and magnitude of age-related growth responses in a subtropical mountain pine. PLoS ONE 10:e0126581

Ruíz-Robleto J, Villar R (2005) Relative growth rate and biomass allocation in ten woody species with different leaf longevity using phyllogenetic independent contrasts (PICs). Plant Biol 7:484–494

Ryan MG, Binkley D, Fownes JH, Giardina CP, Senock RS (2004) An experimental test of the causes of forest growth decline with stand age. Ecol Monogr 74:393–414

Sánchez-Gómez D, Valladares F, Zavala MA (2006) Performance of seedlings of Mediterranean woody species under experimental gradients of irradiance and water availability: trade-offs and evidence for niche differentiation. New Phytol 170:795–806

Shipley B (2002) Trade-offs between net assimilation rate and specific leaf area in determining relative growth rate: relationship with daily irradiance. Funct Ecol 16(5):682–689

Veneklaas EJ, Poorter L (1998) Growth and carbon partitioning of tropical tree seedlings in contrasting light environments. In: Lambers H, Poorter H, VanVuuren M (eds) Inherent variation in plant growth: physiological mechanisms and ecological consequences, Backhuys Publishers, Leiden. ISBN 90-73348-96-X (hardbound), pp 337–361

Villaescusa R, Díaz R (1998) Segundo Inventario Forestal Nacional (1986–1996). España. Ministerio de Medio Ambiente–ICONA. Madrid

Villanueva JA (2004) Tercer Inventario Forestal Nacional (1997–2007). Comunidad de Madrid. Ministerio de Medio Ambiente, Madrid

Villar R, Ruiz-Robleto J, De Jong Y, Poorter H (2006) Differences in construction costs and chemical composition between deciduous and evergreen woody species are small as compared to differences among families. Plant Cell Environ 29:1629–1643

Villar R, Ruiz-Robleto J, Quero JL, Poorter H, Valladares F, Marañón T (2008) Tasas de crecimiento en especies leñosas: aspectos funcionales e implicaciones ecológicas. In: Valladares F (ed) Ecología del bosque mediterráneo en un mundo cambiante (Segunda edición). Ministerio de Medio Ambiente. EGRAF, S. A., Madrid. ISBN: 978-84-8014-738-5. pp 193–230

Villar R, Lopez-Iglesias B, Ruiz-Benito P, Zavala MA, Riva E (2014) Crecimiento de plántulas y árboles de seis especies de *Quercus*. Ecosistemas 23:64–72

Violle C, Navas M-L, Vile D, Kazakou E, Fortunel C, Hummel I, Garnier E (2007) Let the concept of trait be functional! Oikos 116:882–892

Wright IJ, Reich PB, Westoby M, Ackerly DD et al (2004) The worldwide leaf economics spectrum. Nature 428:821–827

Zavala MA, Zea E (2004) Mechanisms maintaining biodiversity in Mediterranean pine-oak forests: insights from a spatial simulation model. Plant Ecol 171:197–207.

Chapter 13
Drought-Induced Oak Decline—Factors Involved, Physiological Dysfunctions, and Potential Attenuation by Forestry Practices

Jesús Rodríguez-Calcerrada, Domingo Sancho-Knapik, Nicolas K. Martin-StPaul, Jean-Marc Limousin, Nathan G. McDowell and Eustaquio Gil-Pelegrín

Abstract The increasing duration and intensity of drought is precipitating widespread episodes of forest decline around the world, possibly in combination with other climatic (e.g. soil water logging) and land-use (e.g. fire and coppicing suppression) changes that have been occurring through the last century. A meta-analysis of physiological data compiled from recent data bases indicates that, comparatively to other species, oaks are mostly tolerant to drought and anisohydric, although also relatively vulnerable to severe water stress. The succession of droughts in unmanaged overcrowded stands is affecting many physiological processes associated with water transport and carbon metabolism and making trees more susceptible to further abiotic and biotic stresses. Although oaks maintain large reserves of nonstructural carbohydrates (NSC) in comparison with other tree species, water restrictions to net carbon gain in declining trees can reduce NSC reserves and favor the sudden or more progressive death of trees. In view of correlative and process-based vegetation model simulations, oaks in the

J. Rodríguez-Calcerrada (✉)
Grupo de Investigación en Genética, Fisiología e Historia Forestal,
Universidad Politécnica de Madrid, Ciudad Universitaria s/n, 28040 Madrid, Spain
e-mail: jesus.rcalcerrada@upm.es

D. Sancho-Knapik · E. Gil-Pelegrín
Unidad de Recursos Forestales, Centro de Investigación y Tecnología
Agroalimentaria de Aragón, Avda. Montañana 930, 50059 Saragossa, Spain

N. K. Martin-StPaul
Ecologie des Forêts Méditerranéennes, URFM, INRA, 84000 Avignon, France

J.-M. Limousin
Centre d'Ecologie Fonctionnelle et Evolutive CEFE, UMR 5175,
CNRS—Université de Montpellier—Université Paul-Valéry Montpellier—EPHE,
1919 route de Mende, 34293 Montpellier, France

N. G. McDowell
Pacific Northwest National Laboratory, Richland, WA 99354, USA

© Springer International Publishing AG 2017
E. Gil-Pelegrín et al. (eds.), *Oaks Physiological Ecology. Exploring the Functional Diversity of Genus* Quercus L., Tree Physiology 7,
https://doi.org/10.1007/978-3-319-69099-5_13

southern-most part of their range can undergo extinctions in the future due to increases in temperature and drought. Hence, managing forests so as to increase acorn production and seedling recruitment is a necessary task to increase adaptability of oak forests to Global Change. Different silvicultural alternatives can improve the physiological status of residual trees and be employed to mitigate the sensitivity of oak forests to drought in the short and long run. However, more studies are needed to understand and mitigate oak decline.

13.1 Oak Decline—Definition and Worldwide Extent

Tree decline refers to a visible lack of "vigor" of mature individuals that, by age, should be exhibiting near optimal performance. The process of decline involves a complex series of phenotypic changes that reflect the response of trees to often multiple interacting abiotic and biotic stress factors. The term refers to a continuum of tree decay, since incipient symptoms characterized by low growth, branch dieback, leaf shedding and reduced foliage area, to more severe conditions of almost complete defoliation, growth stagnation and stem dieback. At the forest scale the term is used to refer to forest stands with a variable proportion of trees showing symptoms of decline or being dead. Climate and land use change, together with an increasing awareness of its consequences are fueling reports of forest decline throughout the world (Allen et al. 2015).

The first records of oak decline that were verified by archival reports started to appear in Central Europe and North eastern United States of America (USA) during the 18th and 19th centuries respectively (Millers et al. 1989; Thomas 2008). Later, from the 1900s to the 1970s, more episodes of oak decline were reported in these and other adjacent areas such as North eastern and western Europe or Central USA. During this time the type of oaks that experienced decline were mostly deciduous oaks, in particular those from groups *Quercus* (white oaks) and *Lobatae* (mainly red, but also black and scarlet oaks) (Millers et al. 1989; Thomas 2008). Since the 1980s, episodes of decline started to be frequently reported over a wide range of forests of the northern hemisphere, including those of deciduous species from group *Cerris* (Halász 2001) and evergreen species such as *Q. ilex* from group *Ilex* (Lloret et al. 2004). Records of oak decline appeared in Southern Europe (Amorini et al. 1996; Desprez-Loustau et al. 2006), North Africa (Brasier 1996), Western USA (Rizzo et al. 2002; Guo et al. 2005; Coleman and Seybold 2008), Mexico (Tainter et al. 2002), Colombia (Ciesla and Donaubauer 1994), China (Gottschalk and Wargo 1997) and Japan (Nakajima and Ishida 2014) (Fig. 13.1). Today, oak decline is a very important issue for society due to the extreme importance of oaks in the culture, history and economy of the affected countries (Gil-Pelegrín et al. 2008; Fei et al. 2011). However, the causes and functional basis of oak decline remain unclear.

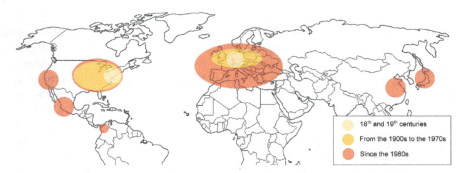

Fig. 13.1 Map showing the areas of oak decline records since the 18th century. References are within the text

Most oak decline episodes have been observed after extreme climatic events (e.g. heat waves or severe droughts), or after consecutive events of winter freezing and waterlogging or drought (Millers et al. 1989; Führer 1998 and references therein; Peñuelas et al. 2001; Lloret et al. 2004). Oak decline has also been associated to different pathogens and site conditions. However, in most cases, none of these factors have acted alone. The simultaneous or subsequent interaction of at least two stress agents, where one of them is often an extreme climatic event, has triggered important outbreaks of decline (Pedersen 1998; Guo et al. 2005; Desprez-Loustau et al. 2006; Thomas 2008; McDowell et al. 2011). Although oaks are displacing less stress-tolerant species in response to climatic changes occurring in some regions (Carnicer et al. 2014; McIntyre et al. 2015), the on-going and projected increase in intensity and frequency of extreme climatic events (Planton et al. 2008) threatens oak performance in many others, especially those where oaks are already in marginal climatic conditions. The problematic sexual regeneration of oak species may further jeopardize adaptability of oak forests to ongoing climatic changes. It is essential that more research is devoted to understand the mechanisms and interactive nature of factors involved in tree decline and regeneration, also in scenarios of elevated air CO_2 concentrations, to project and mitigate future impacts of Global Change on forest health.

13.2 Factors Predisposing to Oak Decline

Different frameworks have been proposed to explain tree death from an ecophysiological perspective (e.g. Selye 1936; Manion 1991; McDowell et al. 2008; Mitchell et al. 2015). All of them argue that growth in sub-optimal conditions predisposes trees to decline and make them more vulnerable to climatic factors such as drought or unusual temperatures that may kill the trees, in association or not with pathogen outbreaks. Edaphoclimatic features (e.g. rainfall, and soil nutrient concentrations, depth or slope), anthropogenic alterations of the habitat (e.g. soil compaction or pollution), and previous/current management of forest stands (e.g. fire suppression or stand overcrowding) all determine tree competition and vigor, and thus tree vulnerability to abiotic and biotic stresses. Trees growing in conditions far from optimal are generally more prone to decline, and consequently high mortality rates have been observed in forests at marginal edaphoclimatic conditions, such as ridges with shallow soils (Thomas et al. 2002; Leuzinger et al. 2005; Fig. 13.2). Meanwhile, the decreasing exploitation of forests for wood products appears to have accelerated tree responses to climate warming and drying in some regions over the 20th century (Doležal et al. 2010). The physiological mechanisms by which both factors—climate and land use change—predispose oaks to decline have some common patterns.

13.2.1 Climate

Repeated and protracted climatic stresses such as drought, waterlogging, freezing, or abnormally elevated temperatures all induce phenotypic changes in trees that may either reduce their subsequent vulnerability to drought, or increase it and

Fig. 13.2 Trees of *Q. pyrenaica* exhibiting reduced growth, little ramification and profuse branch dieback in a steep area with shallow soil in Central Spain

jeopardize tree survival in the long run. The occurrence of extremely cold winter temperatures followed by summer drought stress is a relevant example of how sequential climatic perturbations can trigger oak decline. While large vessels embolize every winter in ring-porous oaks (Zimmermann 1983), extremely cold winters can also cause the embolism of relatively small vessels. When this is followed by a harsh summer drought impeding latewood formation, water transport relies on large vessels of earlywood more prone to embolize, and hydraulic failure can provoke leaf shedding, shoot dieback and eventually death if trees are weakened by other factors (Auclair et al. 1992; Führer 1998; Helama et al. 2016). Another example of how climatic perturbations exacerbate the sensitivity of trees to subsequent stresses is the succession of periods of soil water excess and deficit. The decline of European oak species has been linked to deficient soil aeration caused by waterlogging and flooding; altered soil concentrations of O_2 and CO_2 inhibit root growth (Gaertig et al. 2002), while high soil water content favors root infection by pathogenic fungi (Brasier 1996). These factors weaken the tree and make it more susceptible to subsequent drought than trees thriving in well aerated soils. Moreover, any climatic stress causing permanent photoinhibition of photosynthesis or leaf wilting can demand an increased allocation of resources to repair or replace damaged tissues at the cost of those available for defense (see next section).

13.2.2 Land-Use Changes

The relationship of man with nature has changed dramatically over the 20th century. The emergence of new construction materials and energy sources reduced the consumption of oak products (e.g. charcoal, timber, firewood ...) and modified the exploitation and management of oak forests. Over the past century the frequency and intensity of both harvestings and wildfires have been reduced, with the consequence that the structure and composition of forests have changed (Flatley et al. 2015; Oak et al. 2016). Today, many stands are overcrowded with relatively old oak trees increasingly dominated by more competitive, later-successional species that weaken the vigor of oaks and increase their susceptibility to other stress factors. One example of how land-use changes have favored tree decline is observed in abandoned oak coppices. Because of the typical resprouting ability of oaks, coppicing—the repeated cutting down of stems of trees able to resprout from their roots or stumps—was a widespread practice in oak forests. The abandonment of coppicing has been suggested as a factor causing stem growth stagnation and dieback in these stands. When the stems of coppiced trees of *Q. pyrenaica* exceed a certain age, stems start to display symptoms of decline at anatomical, physiological and morphological levels, e.g., short, little ramified stems with narrow tree rings of absent latewood, higher xylem embolism vulnerability, and higher growth sensitivity to drought than in younger stems (Corcuera et al. 2006; Salomón et al. 2017). Centuries of coppicing have favored the development of multi-centennial, huge, and shallow root systems that result in higher root to shoot ratios in coppiced than non-coppiced trees (Canadell and Rodà 1991; Salomón et al. 2016). The disproportionate root system of coppiced trees entails a high respiratory cost belowground (Salomón et al. 2015) that may reduce carbohydrate availability to growth or fruit production and render trees more susceptible to drought, similarly as to starch-depleted shrubs succumb to drought after resprouting (Pratt et al. 2014). Moreover, the increasing biomass of the root system over decades of coppicing could be related to the observed decline in the amount of taproots and their rapid tapering, which confine the root system to the first 60–100 cm of soil (Zadworny et al. 2014; Salomón et al. 2016). The deep taproots of oaks can facilitate access to deep soil water layers and somewhat attenuate the impact of interannual rainfall fluctuations on tree water status (Rambal 1984; Joslin et al. 2000; Querejeta et al. 2007). Research on the plasticity of root growth and architecture in adult oak trees are scarce (e.g. Canadell and Rodà 1991; Di Iorio et al. 2005; Montagnoli et al. 2012); more research is needed on the way individual roots and entire root systems adjust to land use changes, given the importance of plasticity in vertical rooting patterns in determining oak susceptibility to ongoing changes in rainfall (Joslin et al. 2000).

13.3 Functional Alterations Occurring Through Drought-Induced Oak Decline

Drought is one of the main factors related to oak mortality in recent decades (Allen et al. 2015). Increased warming and drought severity projected for many regions where oaks are dominant could exacerbate ongoing episodes of decline, even in cold and wet regions (Helama et al. 2016). In general, the velocity and magnitude of tree decline depends on the intensity, duration, timing and recurrence of drought periods and on the physiological status of trees. Before dying, some trees exhibit strong phenotypic changes in response to chronic, moderate droughts that progressively weaken the tree, until it finally succumb to drought or any other abiotic or biotic stress (Franklin et al. 1987; Manion 1991). On the other hand, sudden episodes of oak die-off can also occur without trees showing any prior sign of physiological stress. The occurrence of a drought period of extreme intensity and/or duration can exceed the limits of certain physiological attributes key to survival, with possibly less vigorous trees dying from less acute abiotic stresses or pathogen outbreaks (Fig. 13.3). Tree death involves in these cases fewer factors (Mitchell et al. 2015).

The central role that water and carbon play in plant functioning has led to the hypothesis that tree hydraulic and carbon economy are the main factors around which to outline the physiological frameworks of tree mortality. For example, McDowell et al. (2008) proposed a conceptual mechanistic framework in which the importance of hydraulic and carbon related mechanisms in drought-induced tree

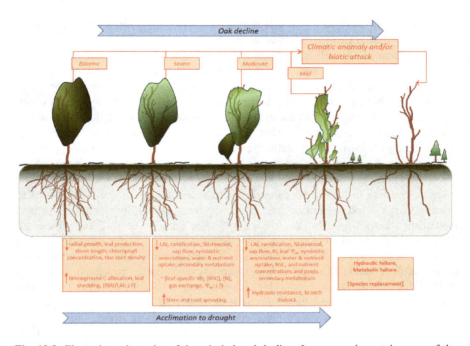

Fig. 13.3 Illustrative schematics of drought-induced decline. *Lower panels* contain some of the phenotypic changes that usually occur in a progressive process of decline. Trees experiencing repeated events of drought start to exhibit symptoms of crown transparency from defoliation and reduced leaf production. This generally limits fine root survival and growth. However, increasing carbon allocation below ground may transiently increase the ratio of fine root area to leaf area ratio (fRAI/LAI). Leaf area index (LAI) can progressively decrease due to leaf shedding and reduced ramification. Carbon limitations to root growth and associated mycorrhiza reduce water and nutrient uptake, which may further reduce carbon available to growth and synthesis of defense compounds. Defoliation and increasing resistance to water uptake reduce canopy transpiration and sap flow. However, this does not necessarily affect leaf physiology or concentration of non-structural carbohydrates (NSC) or nitrogen (N). The decline in leaf area may balance the decline in water transport so that leaf-specific hydraulic conductivity (K_h) and water potential (Ψ_w) may remain relatively unaffected until crown dieback and restrictions to water transport are severe. All along this process of decline, intertwined carbon and water restrictions to plant metabolism may lead to the death of the tree, which can benefit other species able to exploit the residual resources. *Top panels* reflect that sudden oak death can occur at any time along this sequence of decline following climatic and/or biotic perturbations; while relatively vigorous trees will succumb to extreme to severe perturbations, trees weakened by drought can succumb to simultaneous perturbations of just moderate to mild intensity

death depended on the stomatal sensitivity to water stress among species: isohydric species exhibiting rapid stomatal closure in response to water stress would be more susceptible to die from the exhaustion of carbon substrates necessary for respiratory processes, whereas anisohydric species exhibiting a less sensitive stomatal closure would die from hydraulic failure. Further works have provided partial support to this hydraulic framework. In a review of all existing drought manipulation studies, hydraulic failure was a ubiquitous feature, and carbon starvation a frequent feature, of mortality during drought (Adams et al. in revision).

The sequence and relative importance of functional changes leading to tree death is likely to vary among species depending on their tolerance and plasticity to successive or simultaneous stresses. The genus *Quercus* includes hundreds of species with highly variable physiology and ecological requirements. However, there are some features particular to oaks that can shape the mortality spiral in this genus. In general, oaks are tolerant to drought and fire, and poorly tolerant to waterlogging (Niinemets and Valladares 2006; Table 13.1). These ecological features are driven by some typical physiological and anatomical traits, among which the most characteristic are the large tap-root system, root to shoot ratio, NSC reserves and size of the xylem parenchyma. These traits are collectively related to relatively high wood respiration rates and resprouting ability, low stem growth, and infrequent mast crops (Abrams 1990; Dey 1995; Langley et al. 2002; Zhu et al. 2012; Rodríguez-Calcerrada et al. 2015; Salomón et al. 2016; Table 13.1). The need of maintaining sufficient NSC levels belowground to resprout after aerial perturbations is likely to impinge on the responses of oaks to drought. In particular, maintaining a large amount of xylem parenchyma cells alive requires that tissue water stress is minimized under drought or that leaf net carbon gain, phloem transport, and sucrose unload in the roots are all relatively resistant, and/or carbon sink activities and associated respiratory costs relatively sensitive to water stress. Some responses of oaks to drought are consistent with this hypothetical view. For example, the deep root system of oaks allows for the uptake of water from deep wet layers when shallower layers are dry, which helps maintaining leaf water supply, cell turgor and stomatal conductance during drought (Abrams 1990; Epron and Dreyer 1993; Martin-StPaul et al. 2017). This is consistent with the trend of oaks to exhibit higher predawn and midday leaf water potential and higher stem xylem vulnerability to cavitation than other species during drought (Table 13.1). However, a relatively high xylem vulnerability to cavitation (Choat et al. 2012) is at odds with the relatively anisohydric behavior of oaks reported by Klein (2014) and Martinez-Vilalta et al. (2014). Methodological issues when measuring hydraulic conductivity may play a role in this inconsistency (see below).

In any case, the variability in plant functional attributes related with drought tolerance (e.g. turgor loss point and embolism resistance) is rather large across oak species (Corcuera et al. 2002; Bartlett et al. 2012; Martin-StPaul et al. 2017), and there are insufficient observations so as to discriminate how the process of decline in oak differs from that in other botanical groups.

13.3.1 Acclimation to Single Drought Events

Acclimation to drought involves the regulation of numerous metabolic pathways affecting a plethora of processes. One of the very first responses of plants to soil

Table 13.1 Results of t-tests comparing means (with standard deviation) and Mann-Wilcoxon-Whitney tests comparing medians (with interquartile range around the median) of different ecophysiological features between oaks and other species compiled in several global data bases

Variable	N		Mean			Median			Sources
	All	Oak	All	Oak	Diff	All	Oak	Diff	
Tolerance to drought (unitless from 0 to 5)	806	58	2.88 (1.06)	3.76 (1.03)	****	2.88 (1.69)	3.92 (1.61)	****	Niinemets and Valladares (2006)
$\Psi gs50$ (MPa)	69	11	−1.87 (0.76)	−2.87 (0.58)	***	−1.75 (0.88)	−2.90 (0.57)	***	Klein (2014)
P50 (MPa)	480	22	−3.38 (2.50)	−2.34 (1.66)	**	−2.80 (2.93)	−1.93 (1.59)	*	Choat et al. (2012)
Min. Ψpd (MPa)	150	17	−3.52 (2.06)	−2.87 (1.31)	•	−3.01 (3.30)	−2.55 (1.84)	ns	Plaut et al. (in prep.)
Min. Ψmd (MPa)	150	17	−4.35 (1.97)	−3.65 (1.29)	•	4.00 (3.33)	−3.05 (1.44)	ns	Plaut et al. (in prep.)
σ, slope of the midday versus predawn	102	13	0.86 (0.21)	0.89 (0.12)	ns	0.90 (0.20)	0.91 (0.13)	ns	Martinez-Vilalta et al. (2014)
Total NSC in stems (mg/g)	2216	202	74.7 (61.4)	81.0 (31.5)	*	59.5 (67.5)	86.1 (38.6)	****	Martinez-Vilalta et al. (2016)

Variables tolerance to drought [unitless from lowest to maximum value (0–5)]; $\Psi gs50$ midday leaf water potential inducing 50% stomata closure; *P50* xylem water potential inducing a 50% loss of hydraulic conductivity in stems; *Min.* Ψpd minimum predawn leaf water potential ever reported in the literature; *Min.* Ψmd minimum midday leaf water potential ever reported in the literature; σ slope of the midday vs. predawn leaf water potential; and total nonstructural carbohydrates (NSC) in stems. *N* number of taxa, except for Total NSC in stems accounting for the number of observations, with several observations per species
****$p < 0.0001$; ***$p < 0.001$; **$p < 0.01$; *$p < 0.05$; •$p < 0.1$

(and air) drought is the closing of stomata. This reduces water losses, but also the CO_2 concentration inside the leaves and chloroplasts. The primary metabolism is also rapidly altered towards the synthesis of amino acids and soluble sugars, which participate in stress signaling and act as compatible solutes reducing cell osmotic potential. Tissue water relations change through osmotic adjustment, elastic adjustment or water redistribution between symplastic and apoplastic compartments to minimize the loss of turgor and/or volume of living cells (Parker and Pallardy 1988). However, the growth of all plant organs declines with drought. Shoot, leaf and fine root production decreases at different rates due to the different sensitivity of each organ to drought stress and their uncoupled growth patterns (Teskey and Hinckley 1981; Harris et al. 1995; Arend et al. 2011), which can transiently enhance root to shoot or root to leaf ratios under moderate drought. The specific root length and number of tips of fine roots can even increase with increasing seasonal drought due to altered carbon allocation patterns (Coll et al. 2012). However, wood—and particularly latewood in ring porous species—is generally negatively affected by drought (Woodcock 1989; Corcuera et al. 2004, 2006; Doležal et al. 2010); whereas xylem features change towards reducing water transport efficiency (e.g. via smaller vessels—Corcuera et al. 2004; Fonti et al. 2013).

Also the rates of carbon exchange decline with drought due to photosynthetic and respiratory limitations, as well as impaired synthesis of volatile organic compounds (Rodríguez-Calcerrada et al. 2013). The nature of photosynthetic limitations changes from diffusional to biochemical and photochemical as drought stress increases (Limousin et al. 2010; Cano et al. 2013), with permanent photoinhibition of the photosystem II generally occurring only at very low water potentials (Méthy et al. 1996; Rodríguez-Calcerrada et al. 2007), due to the decrease in chlorophyll concentrations and increase in carotenoid concentrations that take place in leaves to balance the energy budget (Haldimann et al. 2008). The nature of respiratory limitations is less clear, with drought-induced decreases in energy demand being a likely factor in ADP limitation of respiration (Rodríguez-Calcerrada et al. 2013). The early reduction in carbohydrate sinks—mainly from growth attenuation—offsets subsequent photosynthetic limitations and helps to maintain NSC reserves nearly unchanged until drought stress is severe (McDowell 2011; Rodríguez-Calcerrada et al. 2017).

Collectively, these changes allow for a relatively fast recovery of water potential and gas exchange after the end of the drought period. However, when the drought period is long and/or acute, and leaf water potential drop to very low values, photoinhibition of the photosystem II and eventually leaf wilting and shedding can ensue, thus delaying or impairing the complete restoration of plant's functionality at the end of the drought period. The magnitude of these short-term responses varies among oak species and populations as a function of their adaptation to drought, with mesic oak species and populations being less tolerant to drought than those inhabiting drier regions (Abrams 1990; Corcuera et al. 2002).

13.3.2 Response to Recurrent Droughts—The Interrelationship of Mechanisms in Tree Decline

As trees experience more frequent and/or more intense periods of drought, molecular and organ-level responses appear to scale up progressively to the whole tree and eventually the stand level. Martin-StPaul et al. (2013), by comparing *Q. ilex* forest stands spanning a regional gradient of aridity with a stand subjected to 8 years of throughfall exclusion, observed increased ratios of sapwood and fine root area relative to leaf area index in the drier stands, which contributed to maintain the minimum water potential within safety margins across all stands; leaf- and stem-level variables such as leaf mass per area or xylem water potential causing

50% loss of hydraulic conductivity were also similar across stands and treatments. More studies are needed to understand long-term responses of oaks to drought. Nonetheless, a reduction in leaf area is a ubiquitous response to drought which results from intertwined carbon and hydraulic limitations (Baldocchi and Xu 2007; Limousin et al. 2009). The leaf area deployed by an oak tree with preformed growth (i.e. having the shoot primordia of the next flush protected in buds) in a given year is in balance with the climate of the previous year via leaf number and with the climate of the current year via leaf area (Alla et al. 2013). When high evaporative demand and transpiration result in depletion of soil water reserves, trees respond to hydraulic limitations to water supply via leaf shedding to avoid catastrophic xylem embolism. In addition, intertwined carbon and hydraulic limitations during bud formation can result in fewer shoots and leaves in response to recurrent droughts (Limousin et al. 2012). Leaf shedding, reduced ramification, and reduced number and area of leaves may contribute to maintain leaf-specific hydraulic conductivity and net photosynthetic CO_2 uptake per unit leaf area rather constant across stands of variable summer drought intensity (Limousin et al. 2009; Martin-StPaul et al. 2013). However, once trees show more obvious symptoms of decline (i.e. severe crown transparency, minimal growth and branch dieback), organ-specific physiology is more likely to be affected. It has been observed that leaf predawn water potential, relative water content, stomatal conductance and net photosynthetic CO_2 uptake are lower in declining than non-declining trees (Thomas and Hartmann 1996; Corcobado et al. 2013). Fine root mortality and increasingly smaller xylem rings prevent the ever-decreasing canopy foliage from receiving enough water to sustain leaf gas exchange. The process of decline does rarely revert; however, foliage recovery has been reported to occur in oaks exhibiting severe crown dieback (Dwyer et al. 2007).

A common response to drought is a reduction in height and diameter growth (although rainfall-exclusion field experiments have evidenced a variable sensitivity of growth; Hanson et al. 2001; Rodríguez-Calcerrada et al. 2011; Rosas et al. 2013). The degree to which plant growth is limited by drought controls the usage of carbon in respiration and the extent of carbon allocated to storage of NSC reserves in stems, roots and lignotubers (Trumbore et al. 2015; Rodríguez-Calcerrada et al. 2017). However, despite metabolic down regulation, the inevitable use of carbohydrates in respiration may eventually result in carbon substrates reaching limiting concentrations for respiration (and life) over long or frequently repeated periods of drought. Concentrations of NSC can remain relatively high in trees with incipient symptoms of decline, and decrease together with NSC pools in severely declining trees (Galiano et al. 2012; Rosas et al. 2013).

On the other hand, restrictions to carbon gain and NSC storage feed forward on limitations to water transport and tolerance to xylem tension by reducing the potential for refilling, the production of latewood, and the thickness of cell walls of vessels relative to their lumen size (Bréda et al. 2006; Eilmann et al. 2009). The reduction (or absence) of latewood reflects the sensitivity of wood production to summer drought stress (Lempereur et al. 2015), and at the same time increases the probability to suffer xylem cavitation, given the positive relationship between vessel size and cavitation risk within species (Zimmerman 1983). Hence, another general acclimation response of oaks to drought is the reduction of vessel size (Eilmann et al. 2009; Fonti et al. 2013). Xylem embolism has a primary role in plant dehydration and is probably one of the latest phenomena occurring in stems before they succumb to death (Rodríguez-Calcerrada et al. 2017; Adams et al. in revision). Embolism resistance is thus a relevant trait to assess the drought resistance of species (Bartlett et al. 2012; Table 13.1). However, recent research has raised doubts regarding embolism resistance measurements in long-vessel species such as oaks (Martin-StPaul et al. 2014). Even if recent methodological progress based on embolism visualization in intact plants now allows monitoring long-vessel species for embolism resistance (Cochard et al. 2015; Venturas et al. 2017), very few data are so far available for oaks (e.g. Bartlett et al. 2012), with a recent study on the topic suggesting that oaks are probably far more embolism resistant than previously expected (Martin-StPaul et al. 2017; Table 13.1). The interrelationship between water- and carbon-related processes is also evidenced by drought-induced carbon restrictions to fine root growth and mycorrhizal associations feeding forward on further limitations to water and nutrient uptake (Jönsson 2006; Coll et al. 2012; Gessler et al. 2016).

Trees can die once drought acclimation limits are exceeded. Intertwined hydraulic and carbon limitations can cause not just the die-back of the stems, but the inability of root living tissues to produce new stems and resprout (Figs. 13.3 and 13.4). Nevertheless, trees rarely die from drought stress alone. The phenotypic adjustments to drought enhance the susceptibility of trees to other biotic (and abiotic) factors that ultimately act as mortality agents (see next section). A trade-off between allocation of carbon (and energy and nutrients) to drought-acclimation *vs* defense can explain the higher susceptibility of drought-acclimated trees to concurrent or successive attacks of pathogens and insects.

13.3.3 The Role of Other Living Organisms in Tree Decline

Pathogens are frequently the proximate cause of massive oak death (see Thomas 2008) and must therefore be taken into consideration when trying to establish an etiology of the process at a global scale.

Fungal species of the genus *Diplodia* have been associated to decline of mediterranean oak species in Europe (Luque and Girbal 1989), North Africa (Linaldeddu et al. 2013) and North America (Lynch et al. 2010). The "Oak Wilt Disease" that has been threatening populations of oaks in many areas of the USA since the early 20th century (Lewis 1978) is essentially considered a vascular wilt produced by *Ceratocystis fagacearum* (Lewis and Oliveira 1979; Wilson 2005; Juzwik et al. 2008 and references therein). This fungus invades the outermost vascular tissues (French and Stienstra 1980), impairing the adequate xylem flow. The pathogeny of this fungus was confirmed since the first studies on this disease

Fig. 13.4 Unidentified dead oak tree in a mixed deciduous oak stand with clear symptoms of decline in central Spain

(Lewis 1978). To explain the wide spread of the problem in a relatively short period of time, Juzwik et al. (2008) consider two main hypotheses: (i) that human-induced promotion of dense stands of highly sensitive red oaks in areas where less sensitive species formerly dominated (McDonald 1995; Wilson 2005) has fueled attacks of *C. fagacearum*, an endemic pathogen in the USA which had minor effects on north American oak forests before, or (ii) that *C. fagacearum* originates from Mexico, Central America, or northern South America where oak species richness is particularly high.

Phytophthora is involved in the decline of a wide range of tree species worldwide, including many *Quercus* spp. *P. ramorum* has affected *Q. agrifolia*, *Q. kelloggii* and *Q. parvula* in California (McPherson et al. 2001; Rizzo et al. 2002); *P. quercina* has affected *Q. robur* and *Q. petraea* in Central Europe (Jung et al. 1999; Jönsson-Belyazio and Rosengren 2006) and *P. cinnamomi*, indigenous to the New Guinea-Celebes region and Southern Africa (Brasier et al. 1993), has been associated with decline of *Q. ilex* and *Q. suber* in southern Europe (Brasier 1996; Gallego et al. 1999; Scanu et al. 2013), of *Q. alba* in USA (Balci et al. 2010; McConnell and Balci 2014; Hubbart et al. 2016) and of *Q. glaucoides*, *Q. peduncularis* and *Q. salicifolia* in Mexico (Tainter et al. 2002). The main damage caused by the different *Phytophthora* species to oak trees is the loss of fine roots and small woody roots (Jung et al. 1999; Jönsson-Belyazio and Rosengren

2006; Balci et al. 2010; Pérez-Sierra et al. 2013), and, therefore, the reduction of the capacity to explore the soil for water and nutrients. This directly implies a weakening of the oak and an increase of the risk of decline (Pérez-Sierra et al. 2013). For example, the inoculation of *Q. ilex* seedlings with *P. cinnamomi* reduced the leaf water content, net CO_2 assimilation and protein abundance (Sghaier-Hammami et al. 2013). Despite the direct involvement of *Phytophthora* spp. in the weakening of oaks through the destruction of the root system, the decline and mortality associated with these pathogens has been described as a complex disease triggered by several interacting environmental factors (Scanu et al. 2013). It is a common observation that this pathogen is more frequently isolated in areas with high soil water content, as this is a critical factor in the infection of the tree root system (Balci et al. 2010). Such evidence has promoted the idea that soil water excess may be as detrimental as severe drought for tree pathogen infection (Hubbart et al. 2016). The most accepted mechanism explaining the incidence of the pathogen in Mediterranean oak stands implies the alternation of periods of high and low soil water content (Brasier et al. 1993; Gallego et al. 1999; Corcobado et al. 2013). The periods of soil wetness would promote the spreading of the infection by favoring the motility of flagellated *Phytophthora* spores, while the infected tree—with a reduced root system—could be less suited for withstanding the subsequent drought (Corcobado et al. 2014a). More recently, a "feed-back loop hypothesis" has been proposed by Corcobado et al. (2015), where the root rot induced by *P. cinnamomi* reduces tree water use, so increasing soil wetness and chances of further fungal infection.

Root rot associated to different fungi of the genus *Armillaria* has been proposed as a possible factor contributing to oak decline (Luisi et al. 1996), particularly in the Missouri Ozark forests (Bruhn et al. 2000; Kelley et al. 2009). High mortality in these forests, particularly of red oaks *Q. coccinea* and *Q. velutina* (Shifley et al. 2006; Kabrick et al. 2007) reflects the high sensitivity of these "red oaks" to *Armillaria* (Bruhn et al. 2000) and their human propagation at the expense of other less sensitive tree species since the early 1900s (Kabrick et al. 2008). Moreover, it is likely that water stress predisposes trees to suffer from *Armillaria* infection because (i) the incidence of oak death in this area is correlated with episodes of drought (Fan et al. 2012), and (ii) *Armillaria* is considered an opportunistic pathogen that invades weakened trees (Marçais and Bréda 2006). Vice versa, the infection by *Armillaria* can feed forward on the susceptibility of trees to further stresses. For example, *Q. robur* trees infected by *Armillaria gallica* were more susceptible to defoliation (Marçais and Bréda 2006).

The fungal species *Biscogniauxia mediterranea* has also been proposed as a factor explaining oak decline in drought-prone areas (Vannini et al. 1996). This endophytic fungus, although considered a facultative saprophyte living on dead tissues (Ju et al. 1998; Anselmi et al. 2004; Vannini et al. 2009), becomes a parasite under some circumstances, especially when the host tree is suffering from water stress (Vannini and Scarascia Mugnozza 1991; Luque et al. 2000; Desprez-Loustau et al. 2006; Capretti and Battisti 2007; Vannini et al. 2009; Linaldeddu et al. 2010). As a mechanistic link between drought and the pathogeny of *B. mediterranea*, Vannini and Valentini (1994) suggested that xylem vessels embolized due to water stress were the path for the spreading of the fungal hyphae.

Plant-eating insects have been directly associated to different processes of massive oak death since the early 20th century in Europe (Gibbs and Greig 1997) and the late 19th century in north-America (Millers et al. 1989). Bark and wood boring insects, mainly beetles belonging to the Cerambycidae, (e.g. *Cerambyx cerdo*), Curculionidae (e.g. *Scolytus intricatus*) and Buprestidae (e.g. *Agrilus biguttatus*), have been recognised as inciting factors in some of the main episodes of oak decline during the last decades (see Sallé et al. 2014 an references therein). The succession of climatic extremes could be responsible for the increasing oak damage by wood boring species of *Agrilus* or *Cerambyx* in Europe and North America that were almost innocuous in the past (see Haavik et al. 2015 for a comprehensive review). These beetles cause a severe reduction in water transport to the crown due to damage and embolization of xylem vessels. Moreover, they often act as vectors of other microorganisms that contribute to oak death. The "Acute Oak Decline", a newly described phenomenon that is threatening different populations of *Q. robur* and *Q. petraea* in the United Kingdom, is explained through the combined action of the buprestid *Agrilus biguttatus* (Brown et al. 2014; Denman et al. 2014) and bacteria (Denman et al. 2012; Brown et al. 2016). The ambrosia beetle *Platypus quercivorus* is considered as the main vector species (Kuroda 2001; Yamasaki et al. 2016) of the fungus *Raffaelea quercivora* (Kubono and Ito 2002). This association has been proposed to explain the high mortality of different *Quercus* species in the so-called "Japanese Oak Wilt" (Nakajima and Ishida 2014). In a similar way, *Platypus cylindrus* has been proposed as a vector of species of the genus *Raffaelea* (Inacio et al. 2008) and of *Biscogniauxia mediterranea* in the Iberian Peninsula (Inacio et al. 2011).

Crown defoliation by insects is also a recurrent causal factor in explaining massive oak decline (Gibbs and Greig 1997; Siwecki and Ufnalski 1998). The gypsy moth *Lymantria dispar* L. has received particular attention due to its impact on large areas of Eurasia (Milanović et al. 2014) and North America (Tobin and Whitmire 2005), where it was introduced in the 19th century (Hunter and Elkinton 2000; Haavik et al. 2015). Thomas et al. (2002) stated that European temperate oak species, such as *Q. robur* or *Q. petraea*, are expected to experience several defoliation episodes through their life cycle. In fact, as is reflected in Gieger and Thomas (2005), oak stands in wide areas of Central Europe can be exposed to full crown defoliations every decade. Isolated events of defoliation may not prevent healthy oaks to produce new leaves from stored reserves. However, consecutive defoliations caused by or in concurrency with climatic perturbations, such as abnormally intense droughts, will drastically weaken the tree and induce massive decline (Thomas et al. 2002).

A positive feed-back is established between tree weakening—by severe defoliations, by climatic stress, or both—and the probability of being colonized and/or damaged by biotic agents (Marçais and Bréda 2006). The nature of this enhanced susceptibility to pathogens of trees weakened by stress has generally an anatomical and chemical origin. Drought-induced reductions in tissue water content and photosynthesis affect the nutritional quality of trees and their attractiveness to fungal pathogens and herbivorous insects depending on their feeding guilds (Huberty and Denno 2004; Rouault et al. 2006). For example, drought-induced shifts in the metabolome of *Q. ilex*, specifically towards osmoprotection and antioxidation via increased leaf concentration of soluble sugars stimulated attraction to herbivores (Rivas-Ubach et al. 2014). Moreover, drought-induced defoliation and increased nitrogen allocation to crown vs. stem can lead to higher soluble nitrogen concentrations in tree leaves (Mattson and Haack 1987; Gessler et al. 2016). As nitrogen is generally limiting for many insects, increased available nitrogen during water stress could result in improved growth of folivorous insects (Jactel et al. 2012 and references within). Besides host attractiveness to insects, water stress can also affect host metabolism and resistance to pests and pathogen damages. Concentrations of secondary metabolites are often higher in foliage of water-stressed trees (Rouault et al. 2006; Sallé et al. 2014), which can affect larval performance of Lepidoptera such as *Operophtera brumata* on *Q. robur* (Buse and Good 1996). In agreement with the 'growth–differentiation balance' hypothesis (Herms and Mattson 1992), without any drought stress, carbohydrates produced by photosynthesis are mainly allocated to growth and development of new foliage, while production of defensive chemicals has a lower priority. Under moderate water stress, the carbohydrates can be redirected to the synthesis of defensive secondary chemicals (such as phenolic and terpenoid compounds), so that trees become more resistant to insect attacks (Herms and Mattson 1992). However, under severe drought, exhaustion of water and carbon reserves can ensue and result in lower constitutive resistance to attacks by bark beetles (Rouault et al. 2006; McDowell et al. 2011; Jactel et al. 2012).

In addition to these biochemical changes, less straightforward changes in response to water stress can occur that increase the probability of an oak to be colonized by a biotic agent. For example, higher leaf temperatures provoked by stomatal closure in response to water stress can increase the rate of insect attack (Mattson and Haack 1987). Finally, it has also been suggested that acoustic signals emitted during the drought-induced cavitation of xylem conduits attract wood boring beetles (Rouault et al. 2006).

Biotic interactions within the context of mortality go beyond the role of insects or pathogens in killing weakened trees. The interaction of bacteria, viruses and endophytic or mycorrhizal fungi with tree physiology is an interesting area of research. These organisms can improve the physiological capacity of plants to cope with other pathogenic microorganisms, herbivores, or even abiotic stresses such as drought (Moricca and Ragazzi 2008; Xu et al. 2008). However, the nature of symbiotic relationships can change from mutualistic to parasitic depending on environmental conditions (Moricca and Ragazzi 2008). The modulation of the abundance and composition of symbiotic microorganisms caused by drought is one unexpected way by which this stress can predispose or lead trees to suffer from decline. Probably in relation to limitations in NSC export and fungal development, the degree of root ectomycorrhizal colonization decreases with decreasing soil water availability (Coll et al. 2012), which may have important consequences for the tree water and nutrient status. For example, Querejeta et al. (2009) observed strong interannual variations in the degree of root colonization by ectomycorrhizal fungi in xeric hill sites of *Q. agrifolia* stands, and discussed that a decrease in rainfall in the future could affect nitrogen uptake by mycorrhizal symbionts of trees in dryland oak woodlands. Although a clear relationship between oak decline and mycorrhizal symbiosis has not always been found (e.g. Lancellotti and Franceschini 2013), some studies have reported higher ectomycorrhizal abundance in the healthy *Q. ilex* trees in Spanish woodlands affected by oak decline (Corcobado et al. 2014b). Corcobado et al. (2014b) considered that a higher crown transparency in declining *Q. ilex* trees should have a direct negative effect on ectomycorrhizal symbiosis by reducing photosynthetic carbon gain and carbon allocation belowground.

Multiple, complex interactions among abiotic and biotic factors modulate the physiology of trees along the spiral of changes that precedes death of declining trees (Manion 1991). In this sense, Corcobado et al. (2014b) suggested that non-mycorrhizal tips might be the "entry points" for *Phytophthora cinnamomi* when invading the root system of *Q. ilex* in Spanish woodlands, establishing in this way a further step in the explanation of such causal complexity.

13.4 Modeling the Future of Oak Forests in a Context of Global Change

The distribution of oak species is expected to change largely and rapidly in the near future due to global change. Correlative and process-based models are the two main types of numerical tools that have been used to assess climate change impacts on forests (Pearson et al. 2004; Morin and Thuiller 2009; Cheaib et al. 2012; Adams et al. 2013). While correlative models simulate the species realized niche, process-based models (PBMs) are built to predict the potential species niche (Morin and Thuiller 2009). Although simulations from correlative models do not always match simulations from PBMs, these agree in the prediction that tree species, including oaks, will migrate toward the north in America and the North-East in Europe to track their niche and escape to warming and increased drought predicted for mid latitudes and the Mediterranean (Czúcz et al. 2011; Cheaib et al. 2012; Hanewinkel et al. 2012).

Correlative models rely on the statistical association between environmental variables (computed from climate and land-use variables) and maps of spatial distributions of species presence; they can be applied to many species to compute their probability of presence in various environmental conditions. Projections performed with correlative models often lead to drastic changes in tree distributions for temperate, sub-boreal and Mediterranean oak species (Morin and Thuiller 2009; Cheaib et al. 2012; Hanewinkel et al. 2012). Results obtained for North American temperate and sub-boreal regions (Morin and Thuiller 2009), different regions of

Europe (Czúcz et al. 2011; Cheaib et al. 2012) and all Europe (Hanewinkel et al. 2012) indicate that oaks will track their current niche by moving toward the north where climate will become warmer, and suffer extinctions in the southern-most part of the range due to increases in temperature and drought. By studying the response to climate change of different European species with correlative models, Hanewinkel et al. (2012) projected an important decline in most temperate tree species in Europe. European beech and pine forests were predicted to decline the most in favor of different oak species, which is line with historic trends reported from forest inventory data (Carnicer et al. 2014). For example, in Europe, *Q. robur* is projected to increase in northern-most latitudes at the cost of more cold-adapted species, while the drought adapted *Q. ilex* is projected to extend over mid-latitudes where water scarcity is expected to increase over the next decades at the cost of more drought-sensitive species.

Alternative to the correlative models, PBMs rely on relatively well-described mechanisms underpinning tree growth, survival, reproduction and migration. PBMs generally predict a lower impact of climate change on oak distribution ranges than correlative models (e.g. Morin and Thuiller 2009; Cheaib et al. 2012). One of the causes of such discrepancy is the effect of raising air CO_2 concentrations, which is not accounted for in correlative models, but that systematically translates into a stimulation of photosynthesis and a decrease in drought stress in PBMs including a photosynthesis algorithm (Davi et al. 2005; Keenan et al. 2011; Cheaib et al. 2012). Another important process driving species niche in PBMs is phenology (Chuine and Beaubien 2001). Oaks are often predicted to leaf out earlier in the future, and thus to have a longer growing season (Vitasse et al. 2011). However these effects may be offset by drought or species chilling requirements, topics which are still poorly understood and accounted for in models (Chuine 2017). Only the most common European and American oak species have been simulated with PBMs, so there is a need to study more species and to improve several physiological processes needed in models. These include hydraulic mechanisms leading to plant dehydration (Sperry and Love 2015), or species chilling requirements (Chuine 2017), migration (Saltré et al. 2015) and genetic adaptation (Oddou-Muratorio and Davi 2014).

13.5 Forestry Practices that Can Mitigate Oak Decline—Functional Bases

Forestry practices are the main tool to alleviate oak decline (Spittlehouse and Stewart 2003; Clatterbuck and Kauffman 2006; Wang et al. 2013; Fig. 13.5). With the aim of improving the vigor of trees and so mitigate the extent or risk of decline, different harvesting alternatives can be applied to oak stands suffering or projected to suffer from decline. In view of the dramatic changes in the climate that are projected for this century, it is important that more research is devoted to understand how different forestry practices affect the vigor of adult trees and enhance sexual regeneration in forest stands of different levels of decline.

13.5.1 Practices Aimed at Improving the Vigor of Adult Trees

(a) Thinning increases irradiance and the availability of soil water and nutrients for residual trees (Aussenac 2000). This increase in resources availability is reflected in a better physiological status of residual trees, which exhibit less negative leaf water potentials during the dry season (Bréda et al. 1995); and higher stem diameter growth (Bréda et al. 1995; Rodríguez-Calcerrada et al. 2011), latewood production (Corcuera et al. 2006), xylem hydraulic conductance (Bréda et al. 1995), transpiration (Bréda et al. 1995; Asbjornsen et al. 2007) and photosynthesis (Moreno and Cubera 2008) than those from unthinned stands. Hence, thinning might be a useful practice to mitigate drought stress, and the impact of oak decline in the future (Johnson et al. 2009). Thinning resulted to be an effective practice in reducing the proportion of sites with moderate to high risk of decline in the Ozark Highlands in central North America at short- to long-term temporal scales, although less effective than other harvesting practices such as clearcutting and group selection cutting causing larger openings in the canopy (Wang et al. 2013). The degree of decline of trees is likely to affect the outcome of the thinning. Because weakened trees are likely to exhibit a less sensitive response to thinning than more vigorous trees, suppressed trees with a high risk of decline should be first in priority of removal (Johnson et al. 2009).

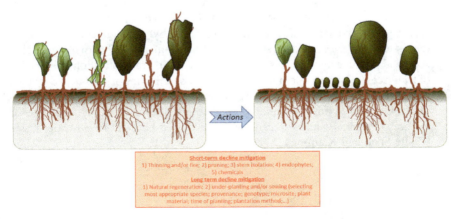

Fig. 13.5 Summary of forestry practices potentially useful to mitigate oak decline

However, the benefits of thinning in preventing or mitigating oak decline at large scales depends in part on its effect on NSC concentrations. The increased photosynthesis rates that follow thinning may be balanced by the increased use of NSC in growth and reproduction. Thus, a decline in NSC levels can ensue and jeopardize the capacity of trees to face perturbations or stress factors for several years after thinning, e.g. via limited NSC available for synthesis of defense compounds (López et al. 2009). Dwyer et al. (2007) did not observe any improvement in decline after 14 years of removing trees with slight to moderate crown dieback; and Johnson et al. (2009) did not observe that thinning reduced the incidence of oak decline in the Missouri Ozark forests. Future research should also consider the clonal relationships among stems in stands of resprouting origin to avoid an unnecessary loss of genetic diversity. Moreover, by removing most stems of a multi-stemmed tree, a root to shoot imbalance could transiently arise and limit (or impair) any physiological improvement in the remaining stems.

(b) Canopy pruning is unlikely to be a suitable practice to slow down oak decline. This practice is commonly applied in some savanna-like woodlands to obtain firewood and enhance acorn production (Alejano et al. 2008). However, pruning does not necessarily result in increased stem growth, and can even reduce growth when applied at severe intensities (Martín et al. 2015). Moreover, cut surfaces opened by pruning demand an extra consumption of carbohydrates for healing that can temporarily enhance susceptibility to other stresses, and are an avenue through which pathogenic microorganisms can easily penetrate into the tree (Dujesiefken et al. 2005; Denman et al. 2010). To the best of our knowledge, pruning has not been extensively tested as a tool to prevent or mitigate oak decline anywhere. Somehow the removal of stems in thinned multi-stemmed trees would be a sort of pruning, with the same benefits and problems associated.

(c) Alternative treatments have to be tested in the management of oak decline. In ancient clonal trees with multiple stems connected by the root system, the isolation of the root system around a single stem could lower its root to shoot imbalance and increase above ground NSC and acorn production (Camarero et al. 2004). The same effect could be achieved by girdling selected stems in multi-stemmed trees; girdled stems would accumulate NSC above the girdled zone and produce more acorns (see Rivas et al. 2007 for application to branches of fruit trees). Furthermore, the inoculation of endophytes or chemical cocktails could be a promising practice to recover the vigor of weakened trees, particularly those of significant cultural or aesthetic value. However, this line of research remains understudied and unadvised on a forest-wide basis (Clatterbuck and Kauffman 2006).

13.5.2 Practices Aimed at Favoring Sexual Regeneration

The profuse asexual regeneration of oak species contrasts with their problematic sexual regeneration (Hodges and Gardiner 1993). Besides measures to prevent or reduce decline of mature trees, measures to promote sexual regeneration are needed to enlarge the genetic pool and adaptability of oak species to ongoing climatic changes (Lefèvre et al. 2014) and so ensure the sustainability of oak forests in the longer run.

Natural regeneration of many oak species has been reported to be problematic: acorn production is subject to intermittent masting events—which might become sparser for oaks as the climate becomes drier (Sork et al. 1993; Dey 1995; Ogaya and Peñuelas 2007; Pérez-Ramos et al. 2015)—and seedling survival is low due to multiple stresses and competition (Pulido et al. 2013; Anninghöffer et al. 2015). Silvicultural treatments such as prescribed burns, pruning and/or thinning favor sexual regeneration by stimulating flower and acorn production, acorn survival, and seedling survival and growth via enhanced photosynthesis (Kruger and Reich 1997a, b; Alejano et al. 2008; Rodríguez-Calcerrada et al. 2011). The elimination of competition from fire-sensitive species by means of prescribed burns could be a feasible option to maintain oak populations in mesic sites otherwise prone to be dominated by fire-sensitive faster growing species (Kruger and Reich 1997a). However, the menace of climate change urges to study the success of classic silvicultural treatments on regeneration under experimentally manipulated conditions of soil water availability, air CO_2 concentration and/or other environmental cues (Leuzinger et al. 2005; Rodríguez-Calcerrada et al. 2011).

Sowing of acorns and planting of young seedlings are viable options to assist sexual regeneration; however, the success of these practices is often low, particularly when failing to exclude mammalian herbivorous, or to account for microhabitat and timing of plantations, or quality of plant material, among other factors. Of particular importance for initial seedling growth and survival is the microsite of sowing/planting. Open microsites generally favor seedling growth, but can negatively affect survival and recruitment. In dry to semi-arid regions, high radiation and belowground competition with herbaceous species exacerbate soil drought stress and may cause massive xylem cavitation, photoinhibition of photosynthesis and metabolic impairment in oak seedlings (Gordon and Rice 2000; Rodríguez-Calcerrada et al. 2007; Pérez-Ramos et al. 2013). In more fertile and humid places, the success of oak regeneration in open sites is jeopardized by competition with fast growing herbaceous species (Harmer and Morgan 2007). Initial survival increases with increasing shading, however, the shade-intolerant nature of most oak species prevents that seedlings establish at low irradiance levels existing below a dense overstory canopy cover. The low morphological (Farque et al. 2001) and physiological (Ponton et al. 2002) plasticity to light availability limits light harvesting, water and nitrogen use efficiency, and carbon reserves available to grow and withstand defoliation and other stresses in dense understories

(Hodges and Gardiner 1993; Kim et al. 1996; Beier et al. 2005). Moreover, in dense forests of sub-humid and dry regions, throughfall interception and soil water uptake by overstory trees can make seedlings to suffer from both shade and drought stress, with multiple field studies reporting lower leaf predawn water potential in uncut than thinned stands (Rodríguez-Calcerrada et al. 2007; Parker and Dey 2008; Prévosto et al. 2011). Hence, in general, sowing of acorns and plantations of seedlings should be conducted in forest understories and gaps where canopy cover is sufficient to moderate radiation and ameliorate soil physico-chemical properties (including root mycorrhizal infection; Dickie et al. 2007), but does not provide excessive shade or competition for belowground resources; this optimal light availability ranges from 20 to 50% of full sunlight among species and populations (Gardiner and Hodges 1998; Gardiner et al. 2001; Kaelke et al. 2001; Rey Benayas et al. 2005). How ongoing climatic changes will alter the role of overstory canopy in facilitating seedling establishment is a question that deserves more research. In a field study in south Europe, throughfall reduction shifted optimal microsites for survival of *Q. ilex* towards denser, more shaded microsites (Pérez-Ramos et al. 2013).

For its importance in plant survival, great attention must be paid to the nutritional status and size of plant material used in reforestation. Multiple works have concluded that nitrogen fertilization of container plants in the nursery increases chances of field survival (e.g. Andivia et al. 2012; Villar-Salvador et al. 2012). In humid regions where interspecific competition for light is an important driver of regeneration, nitrogen fertilization can benefit transplanted plants by increasing their size at the time of transplant, and photosynthesis and growth later on. In seasonally dry regions, water is a limiting resource for plantation success. Some works have suggested that high plant size can be detrimental to transplant success in these regions, so that prior to transplant, it is desirable to reduce watering, air temperatures and/or fertilization to stop growth and avoid an excessive area of transpiring foliage that might affect plant water status after the transplant (Gazal and Kubiske 2004). However, a growing body of literature suggests that fertilized plants having both higher plant size and tissue nutrient concentrations are more likely to overcome the dry season, with fertilization resulting in less negative leaf water potentials during summer drought in some oak species (Villar-Salvador et al. 2012). The higher transplant success of vigorous plants in seasonally dry regions is generally explained by the higher root development and growth capacity compared to smaller, hardened plants (Grossnickle 2005).

Finally, one factor that is being given increasing consideration in reforestation plans is the use of plant material from provenances adapted to near-future projected climatic conditions instead of or mixed with material from local provenances. In relation to warmer and drier conditions projected for some regions, acorns from southern, rear-edge populations might be the most appropriate material to use in reforestations of higher-latitude, more humid populations in the near future (Aitken and Bemmels 2016). The preservation of marginal oak populations could therefore be crucial to species conservation.

Acknowledgements J R-C is thankful to the Spanish Ministry of Economy and Competitiveness for financial support through a "Ramón y Cajal" contract. D S-K is also thankful to the Spanish Ministry of Economy and Competitiveness for financial support through a DOC-INIA contract co-funded by INIA and ESF. NGM was supported by Pacific Northwest National Laboratories LDRD program.

References

Abrams MD (1990) Adaptations and responses to drought in *Quercus* species of North America. Tree Physiol 7:227–238

Adams HD, Williams AP, Xu C, Rauscher SA, Jiang X, McDowell NG (2013) Empirical and process-based approaches to climate-induced forest mortality models. Front Plant Sci 4:438

Adams HD, Zeppel MJB, Anderegg WRL, Hartmann H, Landhäusser SM, Tissue DT, Huxman TE, Hudson PJ, Franz TE, Allen CD, et al (in review) A multi-species synthesis of physiological mechanisms in drought-induced tree mortality. Nat Ecol Evol

Aitken SN, Bemmels JB (2016) Time to get moving: assisted gene flow of forest trees. Evol Appl 9:271–290

Alejano R, Tapias R, Fernández M, Torres E, Alaejos J, Domingo J (2008) Influence of pruning and the climatic conditions on acorn production in holm oak (*Quercus ilex* L.) dehesas in SW Spain. Ann For Sci 65:209

Alla AQ, Camarero JJ, Montserrat-Martí G (2013) Seasonal and inter-annual variability of bud development as related to climate in two coexisting Mediterranean *Quercus* species. Ann Bot 111:261–270

Allen CD, Breshears DD, McDowell NG (2015) On underestimation of global vulnerability to tree mortality and forest die-off from hotter drought in the Anthropocene. Ecosphere 6:129

Amorini E, Baiocca M, Manetti MC, Motta E (1996) A dendroecological study in a declining oak coppice stand. Ann For Sci 53:731–742

Andivia E, Marquez-García B, Vazquez-Piqué CF, Fernández M (2012) Autumn fertilization with nitrogen improves nutritional status, cold hardiness and the oxidative stress response of Holm oak (*Quercus ilex* ssp. *ballota* [Desf.] Samp) nursery seedlings. Trees 26:311–320

Annighöfer P, Beckschäfer P, Vor T, Ammer C (2015) Regeneration patterns of European oak species (*Quercus petraea* (Matt.) Liebl., *Quercus robur* L.) in dependence of environment and neighborhood. PLoS ONE 10(8):e0134935

Anselmi N, Cellerino GP, Franceschini A, Granata G, Luisi N, Marras F, Mazzaglia A, Mutto Accordi S, Ragazzi A (2004) Geographic distribution of fungal endophytes of *Quercus* sp. in Italy. In: Ragazzi A, Moricca S, Dellavalle I (eds) Endophytism in forest trees. Accademia Italiana di Scienze Forestali, Firenze, pp 73–89

Arend M, Kuster T, Günthardt-Goerg MS, Dobbertin M, Abrams M (2011) Provenance-specific growth responses to drought and air warming in three European oak species (*Quercus robur*, *Q. petraea* and *Q. pubescens*). Tree Physiol 31:287–297

Asbjornsen H, Tomer MD, Gomez-Cardenas M, Brudvig LA, Greenan CM, Schilling K (2007) Tree and stand transpiration in a Midwestern bur oak savanna after elm encroachment and restoration thinning. Forest Ecol Manag 247:209–219

Auclair AND, Worrest RC, Lachance D, Martin HC (1992) Climatic perturbation as a general mechanism of forest dieback. In: Manion PD, Lachance D (eds) Forest decline concepts. APS Press, St Paul, pp 38–58

Aussenac G (2000) Interactions between forest stands and microclimate: ecophysiological aspects and consequences for silviculture. Ann For Sci 57:287–301

Balci Y, Long RP, Mansfield M, Balser D, MacDonald WL (2010) Involvement of *Phytophthora* species in white oak (*Quercus alba*) decline in southern Ohio. Forest Pathol 40:430–442

Baldocchi DD, Xu L (2007) What limits evaporation from Mediterranean oak woodlands—The supply of moisture in the soil, physiological control by plants or the demand by the atmosphere? Adv Water Resour 30:2113–2122

Bartlett MK, Scoffoni C, Sack L (2012) The determinants of leaf turgor loss point and prediction of drought tolerance of species and biomes: a global meta-analysis. Ecol Lett 15:393–405

Beier CM, Horton JL, Walker JF, Clinton BD, Nilsen ET (2005) Carbon limitation leads to suppression of first year oak seedling beneath evergreen understory shrubs in Southern Appalachian hardwood forests. Plant Ecol 176:131–142

Brasier CM (1996) *Phytophthora cinnamomi* and oak decline in southern Europe. Environmental constraints including climate change. Ann Sci Forest 53:347–358

Brasier CM, Robredo F, Ferraz JFP (1993) Evidence for *Phytophthora cinnamomi* involvement in Iberian oak decline. Plant Pathol 42:140–145

Bréda N, Granier A, Aussenac G (1995) Effects of thinning on soil and tree water relations, transpiration and growth in an oak forest (*Quercus petraea* (Matt.) Liebl.). Tree Physiol 15:295–306

Bréda N, Huc R, Granier A, Dreyer E (2006) Temperate forest trees and stands under severe drought: a review of ecophysiological responses, adaptation processes and long-term consequences. Ann For Sci 63:625–644

Brown N, Inward DJG, Jeger M, Denman S (2014) A review of *Agrilus biguttatus* in UK forests and its relationship with acute oak decline. Forestry 88:53–63

Brown N, Jeger M, Kirk S, Xu X, Denman S (2016) Spatial and temporal patterns in symptom expression within eight woodlands affected by Acute Oak Decline. Forest Ecol Manag 360:97–109

Bruhn JN, Wetteroff JJ, Mihail JD, Kabrick JM, Pickens JB (2000) Distribution of *Armillaria* species in upland Ozark Mountain forests with respect to site, overstory species composition and oak decline. Eur J Forest Pathol 30:43–60

Buse A, Good J (1996) Synchronization of larval emergence in winter moth (*Operophtera brumata* L.) and budburst in pedunculate oak (*Quercus robur* L.) under simulated climate change. Ecol Entomol 21:335–343

Camarero JJ, Lloret F, Corcuera L, Peñuelas J, Gil-Pelegrín E (2004) Cambio Global y decaimiento del bosque. In: Valladares F (ed) Ecología del bosque Mediterraneo en un mundo cambiante. Ministerio de Medio Ambiente, EGRAF, SA, Madrid, pp 397–423

Canadell J, Rodà F (1991) Root biomass of *Quercus ilex* in a montane Mediterranean forest. Can J For Res 21:1771–1778

Cano FJ, Sánchez-Gómez D, Rodríguez-Calcerrada J, Warren CR, Gil L, Aranda I (2013) Effects of drought on mesophyll conductance and photosynthetic limitations at different tree canopy layers. Plant Cell Environ 36:1961–1980

Capretti P, Battisti A (2007) Water stress and insect defoliation promote the colonization of *Quercus cerris* by the fungus *Biscogniauxia mediterranea*. Forest Pathol 37:129–135

Carnicer J, Coll M, Pons X, Ninyerola M, Vayreda J, Peñuelas J (2014) Large-scale recruitment limitation in Mediterranean pines: The role of *Quercus ilex* and forest successional advance as key regional drivers. Global Ecol Biogeogr 23:371–384

Cheaib A, Badeau V, Boe J, Chuine I, Delire C, Dufrêne E, François C, Gritti ES, Legay M, Pagé C, Thuiller W, Viovy N, Leadley P (2012) Climate change impacts on tree ranges: model intercomparison facilitates understanding and quantification of uncertainty. Ecol Lett 15:533–544

Choat B, Jansen S, Brodribb TJ, Cochard H, Delzon S, Bhaskar R, Bucci SJ, Feild TS, Gleason SM, Hacke UG, Jacobsen AL, Lens F, Maherali H, Martinez-Vilalta J, Mayr S, Mencuccini M, Mitchell PJ, Nardini A, Pittermann J, Pratt RB, Sperry JS, Westoby M, Wright IJ, Zane AE (2012) Global convergence in the vulnerability of forests to drought. Nature 491:752–755

Chuine I (2017) Process-based models of phenology and climate. Annu Rev Ecol Evol Syst 48. doi:10.1146/annurev-ecolsys-110316-022706

Chuine I, Beaubien EG (2001) Phenology is a major determinant of tree species range. Ecol Lett 4:500–510

Ciesla WM, Donaubauer E (1994) Decline and dieback of trees and forests. United Nations Food and Agriculture Organization (FAO) Forestry Paper 120. FAO, Rome, Italy

Clatterbuck WK, Kauffman BW (2006) Managing oak decline. University of Kentucky Cooperative Extension Publication FOR-099; University of Tennessee, Knoxville, TN, USA

Cochard H, Delzon S, Badel E (2015) X-ray microtomography (micro-CT): a reference technology for high-resolution quantification of xylem embolism in trees. Plant Cell Environ 38:201–206

Coleman TW, Seybold J (2008) Previously unrecorded damage to oak, *Quercus* spp., in southern California by the goldspotted oak borer, *Agrilus coxalis* Waterhouse (Coleoptera: Buprestidae). Pan-Pac Entomol 84:288–300

Coll L, Camarero JJ, Martínez de Aragón J (2012) Fine root seasonal dynamics, plasticity, and mycorrhization in 2 coexisting Mediterranean oaks with contrasting aboveground phenology. Ecoscience 19:238–245

Corcobado T, Cubera E, Moreno G, Solla A (2013) *Quercus ilex* forests are influenced by annual variations in water table, soil water deficit and fine root loss caused by *Phytophthora cinnamomi*. Agr Forest Meteorol 169:92–99

Corcobado T, Cubera E, Juárez E, Moreno G, Solla A (2014a) Drought events determine performance of *Quercus ilex* seedlings and increase their susceptibility to *Phytophthora cinnamomi*. Agr Forest Meteorol 192–193:1–8

Corcobado T, Vivas M, Moreno G, Solla A (2014b) Ectomycorrhizal symbiosis in declining and non-declining *Quercus ilex* trees infected with or free of *Phytophthora cinnamomi*. Forest Ecol Manag 324:72–80

Corcobado T, Moreno G, Marisa Azul AM, Solla A (2015) Seasonal variations of ectomycorrhizal communities in declining *Quercus ilex* forests: interactions with topography, tree health status and *Phytophthora cinnamomi* infections. Forestry 88:257–266

Corcuera L, Camarero JJ, Gil-Pelegrín E (2002) Functional groups in *Quercus* species derived from the analysis of pressure-volume curves. Trees 16:465–472

Corcuera L, Camarero JJ, Gil-Pelegrín E (2004) Effects of a severe drought on a *Quercus ilex* radial growth and xylem anatomy. Trees 18:83–92

Corcuera L, Camarero JJ, Sisó S, Gil-Pelegrín E (2006) Radial-growth and wood-anatomical changes in overaged *Quercus pyrenaica* coppice stands: functional responses in a new Mediterranean landscape. Trees 20:91–98

Czúcz B, Gálhidy L, Mátyás C (2011) Present and forecasted xeric climatic limits of beech and sessile oak distribution at low altitudes in Central Europe. Ann For Sci 68:99–108

Davi H, Dufrêne E, Granier A, Le Dantec V, Barbaroux C, François C, Bréda N (2005) Modelling carbon and water cycles in a beech forest: Part II.: validation of the main processes from organ to stand scale. Ecol Model 185:387–405

Denman S, Kirk S, Webber J (2010) Managing acute oak decline. In: Forestry comission practice note, vol 15. Forestry Comission

Denman S, Brady C, Kirk S, Cleenwerck I, Venter S, Coutinho T, De Vos P (2012) *Brenneria goodwinii* sp. nov., associated with acute oak decline in the UK. Int J Syst Evol Microbiol 62:2451–2456

Denman S, Brown N, Kirk S, Jeger M, Webber J (2014) A description of the symptoms of acute oak decline in Britain and a comparative review on causes of similar disorders on oak in Europe. Forestry 87:535–551

Desprez-Loustau ML, Marçais B, Nageleisen LM, Piou D, Vannini A (2006) Interactive effects of drought and pathogens in forest trees. Ann For Sci 63:597–612

Dey DC (1995) Acorn production in red oak. For. Res. Paper No. 127. Ontario Forest Research Institute, Sault Ste. Marie, Canada, p 22

Di Iorio A, Lasserre B, Scippa GS, Chiatante D (2005) Root system architecture of *Quercus pubescens* trees growing on different sloping conditions. Ann Bot 95:351–361

Dickie IA, Schnitzer SA, Reich PB, Hobbie SE (2007) Is oak establishment in old-fields and savanna openings context dependent? J Ecol 95:309–320

Doležal J, Mazůrek P, Klimešoválá J (2010) Oak decline in southern Moravia: the association between climate change and early and late wood formation in oaks. Preslia 82:289–306

Dujesiefken D, Liese W, Shortle W, Minocha R (2005) Response of beech and oaks to wounds made at different times of the year. Eur J Forest Res 124:113–117

Dwyer JP, Kabrick JM, Weteroff J (2007) Do improvement harvests mitigate oak decline in Missouri Ozark forests? North J Appl For 24:123–128

Eilmann B, Zweifel R, Buchmann N, Fonti P, Rigling A (2009) Drought-induced adaptation of the xylem in Scots pine and pubescent oak. Tree Physiol 29:1011–1020

Epron D, Dreyer E (1993) Long-term effects of drought on photosynthesis of adult oak trees [*Quercus petraea* (Matt.) Liebl. and *Quercus robur* L.] in a natural stand. New Phytol 125:381–389

Fan Z, Xiuli Fan X, Crosby MK, Moser WK, He H, Spetich MA, Shifley SR (2012) Spatio-temporal trends of oak decline and mortality under periodic regional drought in the Ozark Highlands of Arkansas and Missouri. Forests 3:614–631

Farque L, Sinoquet H, Colin F (2001) Canopy structure and light interception in *Quercus petraea* seedlings in relation to light regime and plant density. Tree Physiol 21:1257–1267

Fei S, Kong N, Steiner KC, Moser WK, Steiner EB (2011) Change in oak abundance in the eastern United States from 1980 to 2008. Forest Ecol Manag 262:1370–1377

Flatley WT, Lafon CW, Grissino-Amyer HD, LaForest LB (2015) Changing fire regimes and old-growth forest succession along a topographic gradient in the Great Smoky Mountains. Forest Ecol Manag 350:96–106

Fonti P, Heller O, Cherubini P, Rigling A, Arend M (2013) Wood anatomical responses of oak saplings exposed to air warming and soil drought. Plant Biol 15:210–219

Franklin JF, Shugart HH, Harmon ME (1987) Tree death as an ecological process. Bioscience 37:550–556

French DW, Stienstra WC (1980) Oak Wilt. Folder 310. University of Minnesota, Agricultural Extension Service, Minneapolis, MN, 11 p

Führer E (1998) Oak decline in Central Europe: a synopsis of hypotheses. In: McManus ML, Liebhold AM (eds) Proceedings: population dynamics, impacts, and integrated management of forest defoliating insects. USDA Forest Service General Technical Report NE-247, pp 7–24

Gaertig T, Schack-Kirchner H, Hildebrand EE, Wilpert KV (2002) The impact of soil aeration on oak decline in southwestern Germany. Forest Ecol Manag 159:15–25

Galiano L, Martínez-Vilalta J, Sabaté S, Lloret F (2012) Determinants of drought effects on crown condition and their relationship with depletion of carbon reserves in a Mediterranean holm oak forest. Tree Physiol 32:478–489

Gallego FJ, Pérez de Algaba A, Fernández-Escobar R (1999) Etiology of oak decline in Spain. Eur J Forest Pathol 29:17–27

Gardiner ES, Hodges JD (1998) Growth and biomass distribution of cherrybark oak (*Quercus pagoda* Raf.) seedlings as influenced by light availability. Forest Ecol Manag 108:127–134

Gardiner ES, Schweitzer CJ, Stanturf JA (2001) Photosynthesis of Nutall oak (*Quercus nuttalli* Palm.) seedlings interplanted beneath an eastern cottonwood (*Populus deltoides* Bartr. Ex Marsh.) nurse crop. Forest Ecol Manag 149:283–294

Gazal RM, Kubiske ME (2004) Influence of initial root length on physiological responses of cherrybark oak and Shumard oak seedlings to field drought conditions. Forest Ecol Manag 189:295–305

Gessler A, Schaub M, McDowell NG (2016) The role of nutrients in drought-induced tree mortality and recovery. New Phytol. doi:10.1111/nph.14340

Gibbs JN, Greig JW (1997) Biotic and abiotic factors affecting the dying back of pedunculate oak *Quercus robur* L. Forestry 70:399–406

Gieger T, Thomas FM (2005) Differential response of two Central-European oak species to single and combined stress factors. Trees 19:607–618

Gil-Pelegrín E, Peguero-Pina JJ, Camarero JJ, Fernández-Cancio A, Navarro-Cerrillo R (2008) Drought and forest decline in the Iberian Peninsula: a simple explanation for a complex phenomenom? In: Sánchez JM (ed) Droughts: causes, effects and predictions. Nova Science Publishers, New York, pp 27–68

Gordon DR, Rice KJ (2000) Competitive suppression of *Quercus douglasii* (Fagaceae) seedling emergence and growth. Am J Bot 87:986–994

Gottschalk KW, Wargo PM (1997) Oak decline around the world. In: Proceedings USDA interagency gypsy moth research forum 1996

Grossnickle SC (2005) Importance of root growth in overcoming planting stress. New Forest 30:273–294

Guo Q, Kelly M, Graham CH (2005) Support vector machines for predicting distribution of sudden oak death in California. Ecol Modell 182:75–90

Haavik LJ, Billings SA, Guldin JM, Stephen FM (2015) Emergent insects, pathogens and drought shape changing patterns in oak decline in North America and Europe. Forest Ecol Manag 354:190–205

Halász G (2001) Serious damage in turkey oak (*Quercus cerris*) stands in Hungary caused by frost effect (sun scald) and *Armillaria* attack. J For Sci 47:93–96

Haldimann P, Gallé A, Feller U (2008) Impact of an exceptionally hot dry summer on photosynthetic traits in oak (*Quercus pubescens*) leaves. Tree Physiol 28:785–795

Hanewinkel M, Cullmann D, Schelhaas M-J, Nabuurs G-J, Zimmermann NE (2012) Climate change may cause severe loss in the economic value of European forest land. Nat Clim Change 3:203–207

Hanson PJ, Todd DE, Amthor JS (2001) A six-year study of saplings and large-tree growth and mortality responses to natural and induced variability in precipitation and throughfall. Tree Physiol 21:345–358

Harmer R, Morgan G (2007) Development of *Quercus robur* advance regeneration following canopy reduction in an oak woodland. Forestry 80:137–149

Harris JR, Bassuk NL, Zobel RW, Whitlow TH (1995) Root and shoot growth periodicity of green ash, scarlet oak, Turkish hazelnut, and tree lilac. J Am Soc Hortic Sci 120:211–216

Helama S, Sohar K, Läänelaid A, Mäkelä HM, Raisio J (2016) Oak decline as illustrated through plant-climate interactions near the northern edge of species range. Bot Rev 82:1–23

Herms DA, Mattson WJ (1992) The dilemma of plants: to grow or defend. Q Rev Biol 67:283–335

Hodges JD, Gardiner ES (1993) Ecology and physiology of oak regeneration. In: Loftis D, McGee CE (eds) Oak regeneration: serious problems, practical recommendations. USDA Forest Service, Southeastern Forest Experiment Station, General Technical Report SE-84, pp 54–65

Hubbart JA, Guyette R, Muzika RM (2016) More than drought: precipitation variance, excessive wetness, pathogens and the future of the western edge of the eastern deciduous forest. Sci Total Environ 566–567:463–467

Huberty A, Denno R (2004) Plant water stress and its consequences for herbivorous insects: a new synthesis. Ecology 85:1383–1398

Hunter AF, Elkinton JS (2000) Effects of synchrony with host plant on populations of a spring-feeding lepidopteran. Ecology 81:1248–1261

Inacio ML, Henriques J, Lima A, Sousa E (2008) Fungi of *Raffaelea* genus (Ascomycota: Ophiostomatales) associated to *Platypus cylindrus* (Coleoptera: Platypodidae) in Portugal. Rev Cien Agr 31:96–104

Inacio ML, Henriques J, Guerra-Guimaraes L, Azinheira HG, Lima A, Sousa E (2011) *Platypus cylindrus* Fab. (Coleoptera: Platypodidae) transports *Biscogniauxia mediterranea*, agent of cork oak charcoal canker. Bol San Veg Plagas 37:181–186

Jactel H, Petit J, Desprez-Loustau ML, Delzon S, Piou D, Battisti A, Koricheva J (2012) Drought effects on damage by forest insects and pathogens: a meta-analysis. Glob Change Biol 18:267–276

Johnson PS, Shifley SR, Rogers R (2009) The ecology and silviculture of oaks, 2nd ed. CAB International, 580 pp

Jönsson U (2006) A conceptual model for the development of *Phytophthora* disease in *Quercus robur*. New Phytol 171:55–68

Jönsson-Belyazio U, Rosengren U (2006) Can *Phytophthora quercina* have a negative impact on mature pedunculate oaks under field conditions? Ann For Sci 63:661–672

Joslin JD, Wolfe MH, Hanson PJ (2000) Effects of altered water regimes on forest root systems. New Phytol 147:117–129

Ju Y-M, Rogers JD, San Martin F, Granmo A (1998) The genus *Biscogniauxia*. Mycotaxon 66: 1–98

Jung T, Cooke DEL, Blaschke H, Duncan DJ, Osswald W (1999) *Phytophthora quercina* sp. nov., causing root rot of European Oaks. Mycol Res 103:785–798

Juzwik J, Harrington TC, MacDonald WL, Appel DN (2008) The origin of *Ceratocystis fagacearum*, the oak wilt fungus. Annu Rev Phytopathol 46:13–26

Kabrick JM, Fan Z, Shifley SR (2007) Red oak decline and mortality by ecological type in the Missouri Ozarks. e-Gen. Tech. Rep. SRS101. U.S. Department of Agriculture, Forest Service, Southern Research Station, pp 181–186

Kabrick JM, Dey DC, Jensen RG, Wallendorf M (2008) The role of environmental factors in oak decline and mortality in the Ozark Highlands. Forest Ecol Manag 255:1409–1417

Kaelke CM, Kruger EL, Reich PB (2001) Trade-offs in seedling survival, growth, and physiology among hardwood species of contrasting successional status along a light-availability gradient. Can J Forest Res 31:1602–1616

Keenan T, Serra JM, Lloret F, Ninyerola M, Sabate S (2011) Predicting the future of forests in the Mediterranean under climate change, with niche- and process-based models: CO_2 matters! Global Change Biol 17:565–579

Kelley MB, Fierke MK, Stephen FM (2009) Identification and distribution of *Armillaria* species associated with an oak decline event in the Arkansas Ozarks. Forest Pathol 39:397–404

Kim C, Sharik TL, Jurgensen MF, Dickson RE, Buckley DS (1996) Effects of nitrogen availability on northern red oak seedling growth and pine stands in northern Lower Michigan. Can J Forest Res 26:1103–1111

Klein T (2014) The variabilitry of stomatal sensitivity to leaf water potential across tree species indicates a continuum between isohydric and anisohydric species. Funct Ecol 28:1313–1320

Kruger EL, Reich PB (1997a) Responses of hardwood regeneration to fire in mesic forest openings. I. Post-fire community dynamics. Can J Forest Res 27:1822–1831

Kruger EL, Reich PB (1997b) Responses of hardwood regeneration to fire in mesic forest openings. II. Leaf gas exchange, nitrogen concentration, and water status. Can J Forest Res 27:1832–1840

Kubono T, Ito S (2002) *Raffaelea quercivora* sp. nov. associated with mass mortality of Japanese oak, and the ambrosia beetle (*Platypus quercivorus*). Mycoscience 43:255–260

Kuroda K (2001) Responses of *Quercus* sapwood to infection with the pathogenic fungus of a new wilt disease vectored by the ambrosia beetle *Platypus quercivorus*. J Wood Sc 47:425–429

Lancellotti E, Franceschini A (2013) Studies on the ectomycorrhizal community in a declining *Quercus suber* L. stand. Mycorrhiza 23:533–542

Langley JA, Drake BG, Hungate BA (2002) Extensive belowground carbon storage supports roots and mycorrhizae in regenerating scrub oaks. Oecologia 131:542–548

Lefèvre F, Boivin T, Bontemps A, Courbet F, Davi H, Durand-Gillmann M, Fady B, Gauzere J, Gidoin C, Karam M-J, Lalagüe H, Oddou-Muratorio S, Pichot C (2014) Considering evolutionary processes in adaptive forestry. Ann For Sci 71:723–739

Lempereur M, Martin-StPaul NK, Damesin C, Joffre R, Ourcival J-M, Rocheteau A, Rambal S (2015) Growth duration is a better predictor of stem increment than carbon supply in a Mediterranean oak forest: implications for assessing forest productivity under climate change. New Phytol 207:579–590

Leuzinger S, Zotz G, Asshoff R, Körner C (2005) Responses of deciduous forest trees to severe drought in Central Europe. Tree Physiol 25:641–650

Lewis R Jr, Oliveria FL (1979) Live oak decline in Texas. J Arboric 5:241–244

Lewis R Jr (1978) Control of live oak decline in Texas with Lignasan and Arbotech. Proc. of the symposium on systemic chemical treatments in tree culture (9–11 Oct 1978). The Kellogg Center for Continuing Education, Michigan State

Limousin JM, Rambal S, Ourcival JM, Rocheteau A, Joffre R, Rodríguez-Cortina R (2009) Long-term transpiration change with rainfall decline in a Mediterranean *Quercus ilex* forest. Global Change Biol 15:2163–2175

Limousin J-M, Misson L, Lavoir A-V, Martin NK, Rambal S (2010) Do photosynthetic limitations of evergreen *Quercus ilex* leaves change with long-term increased drought severity? Plant Cell Environ 33:863–875

Limousin J-M, Rambal S, Ourcival J-M, Rodríguez-Calcerrada J, Pérez-Ramos IM, Rodríguez-Cortina R, Misson L, Joffre R (2012) Morphological and phenological shoot plasticity in a Mediterranean evergreen oak facing long-term increased drought. Oecologia 169:565–577

Linaldeddu BT, Sirca C, Spano D, Franceschini A (2010) Variation of endophytic cork oak-associated fungal communities in relation to plant health and water stress. Forest Pathol 41:193–201

Linaldeddu BT, Franceschini A, Alves A, Phillips AJL (2013) *Diplodia quercivora* sp. nov.: a new species of Diplodia found on declining *Quercus canariensis* trees in Tunisia. Mycologia 105:1266–1274

Lloret F, Siscart D, Dalmases C (2004) Canopy recovery after drought dieback in holm-oak Mediterranean forests of Catalonia (NE Spain). Global Change Biol 10:2092–2099

López BC, Gracia CA, Sabaté S, Keenan T (2009) Assessing the resilience of Mediterranean holm oaks to disturbances using selective thinning. Acta Oecol 35:849–854

Luisi N, Sicoli G, Lerario P (1996) Observations on *Armillaria* occurrence in declining oak woods of southern Italy. Ann Sci For 53:389–394

Luque J, Girbal J (1989) Dieback of cork oak (*Quercus suber*) in Catalonia (NE Spain) caused by *Botryosphaeria stevensii*. Eur J For Path 19:7–13

Luque J, Parladé J, Pera J (2000) Pathogenicity of fungi isolated from *Quercus suber* in Catalonia (NE Spain). Forest Pathol 30:247–263

Lynch SC, Eskalen A, Zambino P, Scott T (2010) First report of bot banker caused by *Diplodia corticola* on coast live oak (*Quercus agrifolia*) in California. Plant Dis 94:1510

Manion PD (1991) Tree disease concepts, 2nd edn. Prentice-Hall, New Jersey

Marçais B, Bréda N (2006) Role of an opportunistic pathogen in the decline of stressed oak trees. J Ecol 94:1214–1223

Martín D, Vázquez-Piqué J, Alejano R (2015) Effect of pruning and soil treatments on stem growth of holm oak in open woodland forests. Agroforest Syst 89:599–609

Martinez-Vilalta J, Poyatos R, Aguadé D, Retana J, Mencuccini M (2014) A new look at water transport regulation in plants. New Phytol 204:105–115

Martinez-Vilalta J, Sala A, Asensio D, Galiano L, Hoch G, Palacio S, Piper FI, Lloret F (2016) Dynamics of non-structural carbohydrates in terrestrial plants: a global synthesis. Ecol Monogr 86:495–516

Martin-StPaul NK, Limousin JM, Vogt-Schilb H, Rodríguez-Calcerrada J, Rambal S, Longepierre D, Misson L (2013) The temporal response to drought in a Mediterranean evergreen tree: comparing a regional precipitation gradient and a throughfall exclusion experiment. Global Change Biol 19:2413–2426

Martin-StPaul NK, Longepierre D, Huc R, Delzon S, Burlett R, Joffre R, Rambarl S, Cochard H (2014) How reliable are methods to assess xylem vulnerability to cavitation? The issue of 'open vessel' artifact in oaks. Tree Physiol 34:894–905

Martín-StPaul N, Delzon S, Cochard H (2017) Plant resistance to drought relies on early stomata closure. bioRxiv. doi:https://dx.doi.org/10.1101/099531

Mattson W, Haack R (1987) Role of drought in outbreaks of plant-eating insects. Bioscience 37:110–118

McConnell ME, Balci Y (2014) *Phytophthora cinnamomi* as a contributor to white oak decline in mid-Atlantic United States forests. Plant Dis 98:319–327

McDonald WL (1995) Oak wilt: an historical perspective. In: Appel D, Billings R (eds) Oak wilt perspectives: proceedings of national oak wilt symposium. Information Development, Austin, Houston, pp 7–13

McDowell NG (2011) Mechanisms linking drought, hydraulics, carbon metabolism, and vegetation mortality. Plant Physiol 155:1051–1059

McDowell N, Pockman WT, Allen CD, Breshears DD, Cobb N, Kolb T, Plaut J, Sperry J, West A, Williams DG, Yepez EA (2008) Mechanisms of plant survival and mortality during drought: why do some plants survive while others succumb to drought? New Phytol 178:719–739

McDowell NG, Beerling DJ, Breshears DD, Fisher RA, Raffa KF, Stitt M (2011) The interdependence of mechanisms underlying climate-driven vegetation mortality. Trends Ecol Evol 26:523–532

McIntyre PJ, Thorne JH, Dolanc CR, Flint AL, Flint LE, Kelly M, Ackerly DD (2015) Twentieth-century shifts in forest structure in California: Denser forests, smaller trees, and increased dominance of oaks. PNAS 112:1458–1463

McPherson BA, Wood DL, Storer AJ, Kelly NM, Standiford RB (2001) Sudden oak death, a new forest disease in California. Integr Pest Manage Rev 6:243–246

Méthy M, Damesin C, Rambal S (1996) Drought and photosystem II activity in two Mediterranean oaks. Ann For Sci 53:255–262

Milanović S, Milhajlović L, Karadžićzic D, Jankovsky L, Aleksić P, Janković-Tomanić M, Lazarević J (2014) Effects of pedunculate oak tree vitality on gypsy moth preference and performance. Arch Biol Sci Belgrade 66:1659–1672

Millers I, Shriner DS, Rizzo D (1989) History of hardwood decline in the Eastern United States. Gen. Tech. Rep. NE-126. U. S. Department of Agriculture, Forest Service, Northeastern Forest Experiment Station, Broomall, PA, 75 p

Mitchell P, Wardlaw T, Pinkard L (2015) Combined stresses in forests. In: Mahalingam R (ed) Combined stresses in plants. Springer International Publishing, Switzerland

Montagnoli A, Terzaghi M, Di Iorio A, Scippa GS, Chiatante D (2012) Fine-root morphological and growth traits in a Turkey-oak stand in relation to seasonal changes in soil moisture in the Southern Apennines, Italy. Ecol Res 27:1015–1025

Moreno G, Cubera E (2008) Impact of stand density on water status and leaf gas exchange in *Quercus ilex*. Forest Ecol Manag 254:74–84

Moricca S, Ragazzi A (2008) Fungal endophytes in Mediterranean oak forests: a lesson from *Discula quercina*. Phytopathology 98:380–386

Morin X, Thuiller W (2009) Comparing niche- and process-based models to reduce prediction uncertainty in species range shifts under climate change. Ecology 90:1301–1313

Nakajima H, Ishida M (2014) Decline of *Quercus crispula* in abandoned coppice forests caused by secondary succession and Japanese oak wilt disease: stand dynamics over twenty years. Forest Ecol Manag 334:18–27

Niinemets U, Valladares F (2006) Tolerance to shade, drought and waterlogging of temperate northern hemisphere trees and shrubs. Ecol Monogr 76:521–547

Oak SW, Spetich MA, Morin RS (2016) Oak decline in central hardwood forests: frequency, spatial extent, and scale. In: Greenberg H, Collins BS (eds) Natural disturbances and historic range of variation, managing forest ecosystems, vol 32. Springer International Publishing Switzerland, pp 49–71

Oddou-Muratorio S, Davi H (2014) Simulating local adaptation to climate of forest trees with a physio-demo-genetics model. Evol Appl 7:453–467

Ogaya R, Peñuelas J (2007) Species-specific drought effects on flower and fruit production in a Mediterranean holm oak forest. Forestry 80:351–357

Parker WC, Dey DC (2008) Influence of overstory density on ecophysiology of red oak (*Quercus rubra*) and sugar maple (*Acer saccharum*) seedlings in central Ontario shelterwoods. Tree Physiol 28:797–804

Parker WC, Pallardy SG (1988) Leaf and root osmotic adjustment in drought-stressed *Quercus alba*, *Q. macrocarpa*, and *Q. stellata* seedlings. Can J For Res 18:1–5

Pearson R, Dawson T, Liu C (2004) Modelling species distributions in Britain: a hierarchical integration of climate and land-cover data. Ecography 3:285–298

Pedersen BS (1998) The role of stress in the mortality of Midwestern oaks as indicated by growth prior to death. Ecology 79:79–93

Peñuelas J, Lloret F, Montoya R (2001) Severe drought effects on Mediterranean woody flora in Spain. Forest Sci 47:214–218

Pérez-Ramos IM, Rodríguez-Calcerrada J, Ourcival JM, Rambal S (2013) *Quercus ilex* recruitment in a drier world: a multi-stage demographic approach. Perspect Plant Ecol 15:106–117

Pérez-Ramos IM, Padilla-Díaz MC, Koenig WD, Marañón T (2015) Environmental drivers of mast-seeding in Mediterranean oak species: does leaf habit matter? J Ecol 103:691–700

Pérez-Sierra A, López-García C, León M, García-Jiménez J, Abad-Campos P, Jung T (2013) Previously unrecorded low-temperature *Phytophthora* species associated with *Quercus* decline in a Mediterranean forest in eastern Spain. Forest Pathol 43:331–339

Planton S, Déqué M, Chauvin F, Terray L (2008) Expected impacts of climate change on extreme climate events. CR Geosci 340:564–574

Plaut JA, Limousin JM, McDowell NG, Pockman WT (in preparation) The iso-anisohydry continuum and susceptibility of mature woody plants to drought-related mortality

Ponton S, Dupouey J-L, Bréda N, Dreyer E (2002) Comparison of water-use efficiency of seedlings from two sympatric oak species: genotype x environmental interactions. Tree Physiol 22:413–422

Pratt RB, Jacobsen AL, Ramirez AR, Helms AM, Traugh CA, Tobin MF, Heffner MS, Davis SD (2014) Mortality of resprouting chaparral shrubs after a fire and during a record drought: physiological mechanisms and demographic consequences. Global Change Biol 20:893–907

Prévosto B, Monnier Y, Ripert C, Fernández C (2011) Can we use shelterwoods in Mediterranean pine forests to promote oak seedling development? Forest Ecol Manag 262:1426–1433

Pulido F, McCreary D, Cañellas I, McClaran M, Plieninger T (2013) Oak regeneration: ecological dynamics and restoration techniques. In: Campos P et al (eds) Mediterranean oak woodland working landscapes. Landscape series, vol 16. Springer Science+Business Media, Dordrecht

Querejeta JI, Egerton-Warburton LM, Allen MF (2007) Hydraulic lift may buffer rhizosphere hyphae against the negative effects of severe soil drying in a California Oak savanna. Soil Biol Biochem 39:409–417

Querejeta JI, Egerton-Warburton LM, Allen MF (2009) Topographic position modulates the mycorrhizal response of oak trees to interannual rainfall variability. Ecology 90:649–662

Rambal S (1984) Water balance and pattern of root water uptake by a *Quercus coccifera* L. evergreen scrub. Oecologia 62:18–25

Rey Benayas JM, Navarro J, Espigares T, Nicolau JM, Zavala MA (2005) Effects of artificial shading and weed mowing in reforestation of Mediterranean abandoned cropland with contrasting *Quercus* species. Forest Ecol Manag 212:302–314

Rivas F, Gravina A, Agustí M (2007) Girdling effects on fruit set and quantum yield efficiency of PSII in two *Citrus* cultivars. Tree Physiol 27:527–535

Rivas-Ubach A, Gargallo-Garriga A, Sardans J, Oravec M, Mateu-Castell L, Pérez-Trujillo M, Parella T, Ogaya R, Urban O, Peñuelas J (2014) Drought enhances folivory by shifting foliar metabolomes in *Quercus ilex* trees. New Phytol 202:874–885

Rizzo DM, Garbelotto M, Davidson JM, Sloughter GW, Koike ST (2002) *Phytophthora ramorum* as the cause of extensive mortality of *Quercus* spp. and *Lithocarpus densiflorus* in California. Plant Dis 86:5–14

Rodríguez-Calcerrada J, Pardos JA, Gil L, Aranda I (2007) Summer field performance of *Quercus petraea* (Matt.) Liebl and *Quercus pyrenaica* Willd seedlings, planted in three sites with contrasting canopy cover. New Forest 33:67–80

Rodríguez-Calcerrada J, Pérez-Ramos I, Ourcival J-M, Limousin J-M, Joffre R, Rambal S (2011) Is selective thinning an adequate practice for adapting *Quercus ilex* coppices to climate change? Ann For Sci 68:575–585

Rodríguez-Calcerrada J, Buatois B, Chiche E, Shahin O, Staudt M (2013) Leaf isoprene emission declines in *Quercus pubescens* seedlings experiencing drought—any implication of soluble sugars and mitochondrial respiration? Env Exp Bot 85:36–42

Rodríguez-Calcerrada J, López R, Salomón R, Gordaliza GG, Valbuena-Carabaña M, Oleksyn J, Gil L (2015) Stem CO_2 efflux in six co-occurring tree species: underlying factors and ecological implications. Plant Cell Environ 38:1104–1115

Rodríguez-Calcerrada J, Li M, López R, Cano FJ, Oleksyn J, Atkin OK, Pita P, Aranda I, Gil L (2017) Drought-induced shoot dieback starts with massive root xylem embolism and variable depletion of nonstructural carbohydrates in seedlings of two tree species. New Phytol. doi:10.1111/nph.14150

Rosas T, Galiano L, Ogaya R, Peñuelas J, Martínez-Vilalta J (2013) Dynamics of non-structural carbohydrates in three Mediterranean woody species following long-term experimental drought. Front Plant Sci 4:400

Rouault G, Candau J-N, Lieutier F, Nageleisen L-M, Martin J-C, Warzée N (2006) Effects of drought and heat on forest insect populations in relation to the 2003 drought in Western Europe. Ann For Sci 63:613–624

Sallé A, Nageleisen LM, Lieutier F (2014) Bark and wood boring insects involved in oak declines in Europe: current knowledge and future prospects in a context of climate change. Forest Ecol Manag 328:79–93

Salomón R, Valbuena-Carabaña M, Rodríguez-Calcerrada J, Aubrey D, McGuire MA, Teskey R, Gil L, González-Doncel I (2015) Xylem and soil CO_2 fluxes in a *Quercus pyrenaica* Willd. Coppice: root respiration increases with clonal size. Ann For Sci 72:1065–1078

Salomón R, Rodríguez-Calcerrada J, Zafra E, Morales-Molino C, Rodríguez-García A, González-Doncel I, Oleksyn J, Zytkowiak R, López R, Miranda JC, Gil L, Valbuena-Carabaña M (2016) Unearthing the roots of degradation of *Quercus pyrenaica* coppices: a root-to-shoot imbalance caused by historical management? Forest Ecol Manag 363:200–211

Salomón R, Rodríguez-Calcerrada J, González-Doncel I, Gil L, Valbuena-Carabaña M (2017) On the general failure of coppice conversion into high forest in *Quercus pyrenaica* stands: a genetic and physiological approach. Folia Gebot. doi:10.1007/s12224-016-9257-9

Saltré F, Duputié A, Gaucherel C, Chuine I (2015) How climate, migration ability and habitat fragmentation affect the projected future distribution of European beech. Global Change Biol 21:897–910

Scanu B, Linaldeddu BT, Franceschini A, Anselmi N, Vannini A, Vettraino AM (2013) Occurrence of *Phytophthora cinnamomi* in cork oak forests in Italy. Forest Pathol 43:340–343

Selye H (1936) A syndrome produced by diverse nocuous agents. Nature 138:32

Sghaier-Hammami B, Valero-Galvàn J, Romero-Rodríguez MC, Navarro-Cerrillo RM, Abdelly C, Jorrín-Novo J (2013) Physiological and proteomics analyses of Holm oak (*Quercus ilex* subsp. *ballota* [Desf.] Samp.) responses to *Phytophthora cinnamomi*. Plant Physiol Bioch 71:191–202

Shifley SR, Fan Z, Kabrick JM, Jensen RG (2006) Oak mortality risk factors and mortality estimation. Forest Ecol Manag 229:16–26

Siwecki R, Ufnalski K (1998) Review of oak stand decline with special reference to the role of drought in Poland. Eur J Forest Pathol 28:99–112

Sork VL, Bramble J, Sexton O (1993) Ecology of mast-fruiting in three species of north American deciduous oaks. Ecology 74:528–541

Sperry JS, Love DM (2015) Tansley review: what plant hydraulics can tell us about plant responses to climate-change droughts. New Phytol 207:14–17

Spittlehouse DL, Stewart RB (2003) Adaptation to climate change in forest management. BC J Ecosyst Manag 4:1–11

Tainter FH, O'Brien JG, Hernández A, Orozco Fd, Rebolledo O (2002) *Phytophthora cinnamomi* as a cause of oak mortality in the state of Colima, Mexico. Plant Dis 84:394–398

Teskey RO, Hinckley TM (1981) Influence of temperature and water potential on root growth of white oak. Physiol Plantarum 52:363–369

Thomas FM (2008) Recent advances in cause-effect research on oak decline in Europe. CAB Rev Perspect Agric Vet Sci Nutr Nat Resour 3(037):1–12

Thomas FM, Hartmann G (1996) Soil and tree water relations in mature oak stands of northern Germany differing in the degree of decline. Ann For Sci 53:697–720

Thomas FM, Blank R, Hartmann G (2002) Abiotic and biotic factors and their interactions as causes of oak decline in Central Europe. Forest Pathol 32:277–307

Tobin PC, Whitmire SL (2005) Spread of gypsy moth (Lepidoptera: Lymantriidae) and its relationship to defoliation. Environ Entomol 34:1448–1455

Trumbore S, Czimczik CI, Sierra CA, Kuhr J, Xu X (2015) Non-structural carbon dynamics and allocation relate to growth rate and leaf habit in California oaks. Tree Physiol 35:1206–1222

Vannini A, Scarascia Mugnozza G (1991) Water stress: a predisposing factor in the pathogenesis of *Hypoxylon mediterraneum* on *Quercus cerris*. Eur J Forest Pathol 21:193–201

Vannini A, Valentini R (1994) Influence of water relations on *Quercus cerris-Hypoxylon mediterraneum* interaction: a model of drought-induced susceptibility to a weakness parasite. Tree Physiol 14:129–139

Vannini A, Valentini R, Luisi N (1996) Impact of drought and *Hypoxylon mediterraneum* on oak decline in the Mediterranean region. Ann Sci For 53:753–760

Vannini A, Lucero G, Anselmi N, Vettraino AM (2009) Response of endophytic *Biscogniauxia mediterranea* to variation in leaf water potential of *Quercus cerris*. Forest Pathol 39:8–14

Venturas MD, Sperry JS, Hacke UG (2017) Plant xylem hydraulics: what we understand, current research, and future challenges. J Integr Plant Biol. doi:10.1111/jipb.12534

Villar-Salvador P, Puértolas J, Cuesta B, Peñuelas J, Uscola M, Heredia-Guerrero N, Rey Benayas JM (2012) Increase in size and nitrogen concentration enhances seedling survival in Mediterranean plantations. Insights from an ecophysiological conceptual model of plant survival. New Forest 43:755–770

Vitasse Y, François C, Delpierre N, Dufrêne E, Kremer A, Chuine I, Delzon S (2011) Assessing the effects of climate change on the phenology of European temperate trees. Agr Forest Meteorol 151:969–980

Wang WJ, He HS, Spetich MA, Shifley SR, Thompson FR III, Fraser JS (2013) Modeling the effects of harvest alternatives on mitigating oak decline in a central hardwood forest landscape. PLoS ONE 8:e66713

Wilson AD (2005) Recent advances in the control of oak wilt in the United States. Plant Pathol J 4:177–191

Woodcock D (1989) Climate sensitivity of wood-anatomical features in a ring-porous oak (*Quercus macrocarpa*). Can J Forest Res 19:639–644

Xu P, Chen F, Mannas JP, Feldman T, Sumner LW, Roossinck MJ (2008) Virus infection improves drought tolerance. New Phytol 180:911–921

Yamasaki M, Kaneko T, Takayanagi A, Ando M (2016) Analysis of oak tree mortality to predict ambrosia beetle *Platypus quercivorus* movement. Forest Sci 62:377–384

Zadworny M, Jagodziński AM, Łakomy P, Ufnalski K, Oleksyn J (2014) The silent shareholder in deterioration of oak growth: common planting practices affect the long-term response of oaks to periodic drought. Forest Ecol Manag 318:133–141

Zhu W-Z, Xiang J-S, Wang S-G, Li M-H (2012) Resprouting ability and mobile carbohydrate reserves in an oak shrubland decline with increasing elevation on the eastern edge of the Qinghai-Tibet Plateau. Forest Ecol Manag 278:118–126

Zimmermann MH (1983) Xylem structure and the ascent of sap. Springer, Berlin Heidelberg

Chapter 14
Physiological Keys for Natural and Artificial Regeneration of Oaks

Jesús Pemán, Esteban Chirino, Josep María Espelta,
Douglass Frederick Jacobs, Paula Martín-Gómez,
Rafael Navarro-Cerrillo, Juan A. Oliet, Alberto Vilagrosa,
Pedro Villar-Salvador and Eustaquio Gil-Pelegrín

Abstract Oak forests can naturally regenerate from seed or from sprouts. Both strategies result in the establishment of a tree layer, but they involve a crucial difference: i.e. regeneration from seeds affects population genetics while sprouting assures the recovery of biomass after a disturbance but it does not involve sexual reproduction. In addition the two regeneration mechanisms differ in their complexity and are affected by different constraints: i.e. regeneration from seed is a more intricate pathway with several potential bottlenecks (e.g. seed and micro-sites availability, predation, seedling-saplings conflicts) while sprouting is a much more straightforward process benefiting from the presence of an already established root system and more independent from environmental stochasticity. Ultimately, regeneration from seeds or sprouts will result in contrasting forest structures

---

J. Pemán (✉) · P. Martín-Gómez
Department of Crop and Forest Science, University of Lleida-Agrotecnio Center,
Rovira Roure 191, 25198 Lleida, Spain
e-mail: peman@pvcf.udl.cat

E. Chirino
Facultad de Ciencias Agropecuarias, Universidad Laica "Eloy Alfaro" de Manabí,
Ciudadela Universitaria, vía San Mateo s/n., Manta, Manabí, Ecuador

J. M. Espelta
CREAF, 08193 Cerdanyola del Valles, Catalonia, Spain

D. F. Jacobs
Department of Forestry and Natural Resources, Hardwood Tree Improvement and Regeneration Center, Purdue University, West Lafayette, IN 47907, USA

R. Navarro-Cerrillo
Laboratory of Dendrochronology, Selviculture and Climate Change
DendrodatLab-ERSAF, Department of Forestry, School of Agriculture
and Forestry, University of Córdoba, Edif. Leonardo da Vinci,
Campus de Rabanales s/n, 14071 Córdoba, Spain

J. A. Oliet
Departamento de Sistemas y Recursos Naturales, Universidad Politécnica de Madrid,
Ciudad Universitaria s/n, 28040 Madrid, Spain

© Springer International Publishing AG 2017
E. Gil-Pelegrín et al. (eds.), *Oaks Physiological Ecology. Exploring the Functional Diversity of Genus* Quercus L., Tree Physiology 7,
https://doi.org/10.1007/978-3-319-69099-5_14

(respectively, high-forests and coppices) with a different functioning and dynamics and requiring particular forestry practices. When natural regeneration is not possible, oak forest restoration must be done using artificial regeneration by seeding or planting (traditionally, both methods have been recommended), provided that acorn predators are controlled. Although similar results have been obtained with regard to survival, under Mediterranean conditions, shoot growth patterns clearly differ for both methods. Indeed, one-year seedlings often discontinue their shoot elongation shortly after transplanting, especially under drought or competition. At this time, a new taproot and fine lateral roots are formed. This observation suggests that the seeding and planting techniques may bear different consequences with regard to root system development, which may ultimately affect seedling establishment. Survival and growth planted seedlings depends on morphological and physiological attributes (Burdett in Can J For Res 20:415–427, 1990; Villar-Salvador et al. in New For 43:755–770, 2012; Grossnickle in New For 43:711–738, 2012). Cultivation techniques strongly determine the functional attributes of seedlings by manipulating the amount of resources (water, mineral nutrients, light, space) and the conditions (temperature, growing medium pH, photoperiod) for seedling growth. Consequently, how seedlings are cultivated impacts on the performance of forest plantations. Cultivation practices improve the "seedling physiological potential", increasing the chances of survival immediately after field planting. Each of these has an influence and interacts with the others (Ketchum and Rose in Interaction of initial seedling size, fertilization and vegetation control. Redding, CA, pp 63–69, 2000), which should be taken into consideration when evaluating a reforestation proposal; otherwise, artificial forest regeneration often results in unacceptably poor seedling performance. Planting date and site preparation, since they increase water availability, have been shown to be the factors most relevant to the survival of Mediterranean species. However, in less restrictive conditions, the use of less intensive soil preparation, on dates more favorable to the initial growth of the seedlings in the field, might be more efficient. Similarly, the use of tree shelters in oaks plantations is under debate, as its effects are species and environmental dependent. A better understanding of the ecophysiological seedling response under the microenvironment of the tree shelter is needed to improve the

A. Vilagrosa
Mediterranean Center for Environmental Studies (Fundación CEAM),
Joint Research Unit University of Alicante—CEAM, University of Alicante,
PO Box 99, 03080 Alicante, Spain

P. Villar-Salvador
Forest Ecology and Restoration Group, Departamento de Ciencias de la Vida, Apdo. 20, Universidad de Alcalá, 28805 Alcalá de Henares, Madrid, Spain

E. Gil-Pelegrín
Unidad de Recursos Forestales, Centro de Investigación y Tecnología Agroalimentaria de Aragón, Gobierno de Aragón, Avda. Montañana 930, 50059 Zaragoza, Spain

management of this protection tool. On the other side, the effects of cultivation practices can be linked closely to newly established seedlings (the post-planting phenological cycle), and such benefits are ephemeral in nature; thus, the effects of cultivation practices have their greatest importance during the initial growing seasons (1–2 years), diminishing with time.

14.1 Natural Regeneration

14.1.1 Regeneration from Seed

Probably, natural regeneration of oaks (*Quercus* sp.) from seed has received more attention than any other tree species from plant ecologists and forest practitioners. Several studies have highlighted the apparent paradox of a genus dominant in forests of the Northern Hemisphere but with the general perception of some difficulties to regenerate from seeds (see among others: Retana et al. 1999; Marañón et al. 2004; Pérez-Ramos et al. 2012). Factors that constrain the sexual regeneration of trees appear during the sequential interaction among different processes: i.e. seed production, predation, dispersal, and, finally, seedling establishment, and these pathways are particularly intricate in the case of oaks with numerous bottlenecks and ontogenic conflicts (Zavala et al. 2011; Pérez-Ramos et al. 2012).

Concerning seed production, oaks show strong variation in masting behaviour: namely, the synchronous and intermittent production of large and nil acorn crops over wide areas (Espelta et al. 2008; Koenig et al. 2016). Strikingly, although many studies have emphasized that this bizarre reproduction pattern should follow an "economy of scale" principle: i.e. the production of infrequent bumper crop episodes should be more beneficial than moderate and continuous reproductive events (Kelly and Sork 2002), we still lack empirical evidences of its ultimate positive effect on oak recruitment (but see Oddou-Muratorio et al. 2011 for beech). Notwithstanding this, there is consensus on the role of masting to reduce pre-dispersal acorn predation (see for granivorous insects Bonal and Muñoz 2007; Espelta et al. 2008; Xia et al. 2016). Yet losses can still be high (e.g. 60% in a given year per Leiva and Fernández-Alés 2005), and ultimately, contribute to constrain seed availability and seedling recruitment (Espelta et al. 2009a, b; Muñoz et al. 2014). Whatever the evolutionary origin of masting in oaks it has been shown that weather conditions play a major role as proximate cues for synchrony and variability in reproduction (e.g. rainfall in Pérez-Ramos et al. 2010, or even "weather packages" in Fernández-Martínez et al. 2017, see also Koenig et al. 2016). Apart from particular weather cues, oaks require a minimum initial threshold of resources accumulated and an increase in their potential photosynthetic activity prior to the production of a mast event (Sánchez-Humanes et al. 2011; Fernández-Martínez et al. 2015). These bumper crops often result in a decrease in leaf area due to crown self-thinning after the masting episode (Camarero et al. 2010, but see Sánchez-Humanes et al. 2011; Fernández-Martínez et al. 2015).

Once acorns have escaped pre-dispersal predation by insects, even though oaks benefit from seed dispersal mostly by vertebrates, they lack "legitimate" dispersers: i.e. those not destroying part of the seed crop. In fact, both birds (e.g. jays) and scatter-hoarding rodents (e.g. mice) play a dual role as dispersers and consumers when they mobilize acorns (Hollander and Vander Wall 2004). Interestingly, their decision may shift depending on tree location and phenology (Sunyer et al. 2014), seed traits (Muñoz et al. 2012), state (Perea et al. 2012); environmental context such as shrub cover (Perea et al. 2011) and even, seed dispersers' fear of predators (Sunyer et al. 2013). To summarize, it has been shown that likelihood of acorn predation increases around isolated trees, for smaller seeds, in areas with low understory cover and when rodents perceive the risk of predators (Sunyer et al. 2013) or competitors for that food resource (González-Rodríguez and Villar 2012). Ultimately these changes in the behaviour of acorn dispersers towards predation are another source of uncertainty that may also reduce seedling recruitment. In addition, as acorns are highly nutritious they are also the food source for other animals that will further reduce seed availability (e.g. wild boar in Gómez and Hódar 2008; Sunyer et al. 2015).

In the soil, acorns require protection from desiccation and frost damage (Esteso-Martínez and Gil-Pelegrín 2004), and thus germination is enhanced if buried under soil or litter in shady places (e.g. under a nurse shrub), especially in harsh climates (e.g. Mediterranean-type). Thus, for Mediterranean species deciduous, seedlings tend to exhibit higher root length and area, lower leaf area, and higher N content than evergreen conspecifics (Espelta et al. 2005). These differences further enlarge in a gradient of increasing aridity: i.e. evergreen oaks reduce their specific leaf area, while deciduous oaks increase their root-shoot as two different responses to cope with water stress (Espelta et al. 2005, see also Cotillas et al. 2016). Recently established seedlings may benefit from shade, especially in drier sites (e.g. Mediterranean areas in Gómez-Aparicio et al. 2008; Esteso-Martínez et al. 2010), and therefore they survival has been observed to increase when established under the canopy of more pioneer species (e.g. pines in Gómez 2003) during succession. In fact, this evidence has been the basis for the traditional method of planting oaks under the shelter of pine plantations to recover oak forests (Gómez-Aparicio et al. 2009). Once spontaneously established, seedlings can often remain as "stunted seedlings" in a sort of advanced regeneration (see Marañón et al. 2004 for *Q. ilex*). This is due to a seedling-sapling conflict concerning the most suitable conditions for survival and growth, as seedlings require an increase in light intensity (e.g. gap opening) to reach the sapling stage and progress (Zavala et al. 2011).

Probably, the perception that regeneration from acorns is somewhat difficult in oaks is linked to the fact that the reported interactions between dispersers' behaviour and environmental variables for seedling recruitment are highly context-dependent and their output may largely vary in different ecological scenarios (Fig. 14.1). In general, in sparse or fragmented oak forests, with low cover, the role of acorn dispersal towards safe sites becomes more important as it is the presence of shrubs or other trees (e.g. pines) that may facilitate seedling

	Savannah-like landscapes/ oak woodlands		Oak forests/coppices

Importance of different factors constraining natural regeneration from acorns

+	Acorn availability	−
−	Acorn size	+
=	Pre-dispersal predation by insects	=
+	Post-dispersal predation and mobilization by vertebrates	−
+	Presence of nurse shrubs	−
−	Seedling-sapling conflict in the environmental conditions for development	+

Fig. 14.1 Relative importance of factors constraining natural regeneration of oaks from acorns depending on different forest structures and according to the literature. (−) = Low relevance, (+) = High relevance, (=) = Neutral. See the main body text for a detailed explanation of each factor

establishment. Conversely, in mixed or more continuous oak forests, the successful recruitment of seedlings may be more dependent on seed availability, the dynamics of gap formation and the mobilization of acorns to these sites. Clearly, a more comprehensive understanding of the factors that limit regeneration from acorns would benefit from a comparison of regeneration along these environmental and forest structure gradients (see Gómez-Aparicio et al. 2008; Pérez-Ramos et al. 2010, 2012).

14.1.2 Regeneration from Sprouts

Most oaks, as for other species in the Fagaceae family (e.g. *Castanea, Fagus, Corylus*), are able to resprout after an acute stress or physical damage, and for some species, this process is their main regeneration mechanism after disturbances (see for Mediterranean oaks Espelta et al. 2012). Resprouting—i.e. the production of new sprouts from buds protected in the stump, root collar, roots or branches is driven by the combined effect of changes in hormonal levels and environmental conditions (e.g. light, moisture, temperature) after a stress or damage (Verdaguer et al. 2001; Pascual et al. 2002; Espelta et al. 1999). Concerning the role of hormones, resprouting onset has been observed to be related to changes in the levels of growth hormones, in particular auxin [notably indole-3-acetic acid (IAA)] and cytokinins (CTK) that are synthesized primarily in opposite plant compartments: i.e. growing stem tips and root tips, respectively (Cline 1991). As shown by Vogt and Cox (1970) for *Q. alba* and *Q. palustris* the application of IAA in the stump inhibited the development of suppressed buds. Similarly, Zhu et al. (2014) showed for coppiced *Q. aquifolioides* in southwestern China, that CTK concentrations and CTK/IAA ratios in the stump were positively correlated with resprouting ability. Concerning the physical environment, several factors have been observed to constrain the activation of sprouts in oaks (for the negative effects of shading, see Gracia and Retana 2004; Kabeya and Sakai 2005). In addition to a high CTK/IAA ratio, resprouting requires the mobilization of stored reserves to support the initial growth before leaves are fully photosynthetically active. However, there is still some controversy concerning which nutrient or resource is the more limiting for this process. On the one hand, some experiments have indicated that previously accumulated nitrogen is more critical than starch reserves in *Q. ilex* roots to sustain resprouting (El Omari et al. 2003). On the other hand, other studies have shown that resprouting ability is largely determined by the initial carbohydrate pool in roots (e.g. *Q. crispula* in Kabeya and Sakai 2005, *Q. aquifolioides* in Zhu et al. 2012). While onset of resprouting is determined by the levels of growth hormones and the initial mobilization of stored resources, there are a myriad of other factors that will influence the vigour of this process: (i) type, season, intensity and frequency of disturbance, (ii) species-specific and individual characteristics (e.g. age, size), and (iii) local characteristics influencing site quality (climate, soil and topography).

The first factor influencing resprouting vigour is the type of disturbance. In spite of the lack of comparative studies analysing the influence of the type of disturbance on the resprouting process under similar conditions (but see Bonfil et al. 2004), indirect evidence based on the survival and regrowth of oaks suggests that fire is the physical damage that causes the greatest mortality among individuals (Espelta et al. 1999; Bonfil et al. 2004), and the lowest resprouting vigour (i.e. Hmielowski et al. 2014) in comparison to other types, such as clear-cutting or herbivory (Espelta et al. 2003, 2006). The more negative effect of fire has been attributed to the physical destruction it may cause to part of the bud-bank in comparison to other disturbances, rather than to physiological differences. Indeed, the study of the variations in the physiology of resprouts originated after fire and after tree fall in *Q. ilex*, revealed no differences among them in any of the measured variables (Fleck et al. 1996a) and thus, concluded that whatever the cause of topkill, this was not relevant for the physiology of the resprouts (Fleck et al. 1996b). In addition to the type of disturbance, the season when it occurs has often been stressed as a relevant factor that may condition resprouting vigour, with two general conclusions. First, resprouting vigour in oaks is higher after disturbances occurring during the dormant than during the growing-season (Ducrey and Turrel 1992; Hmielowski et al. 2014). Second, for those species inhabiting areas where a harsh period occurs during the growing season—e.g. dry and hot summers in Mediterranean-type climates—resprouting vigour is much lower for "late season disturbances", occurring towards the end of summer (Bonfil et al. 2004). This constrain has been attributed to the lower reserves of resources available to sustain resprout growth (N and P in Peguero and Espelta 2011, see also Kays and Canham 1991; Canadell and López-Soria 1998; Saura-Mas and Lloret 2009; Hmielowski et al. 2014). Therefore, coppicing of oak woodlands has been traditionally conducted during the dormant season (Buckley and Mills 2015). Frequency is the third characteristics of the disturbance regime that affects resprouting, and several studies have emphasized a decrease in resprouting vigour (survival, growth) after repeated disturbances, especially when they occur at a short time interval (see for *Q. suber* in Díaz-Delgado et al. 2003, *Q. ilex* and *Q. cerrioides* in Bonfil et al. 2004). Similar to the effects of seasonality, the negative effects of repeated and frequent disturbances have been attributed to the depletion of the bud bank and the exhaustion of the resource involved in the initial growth of resprouts (Espelta et al. 2012).

In addition to the influence of the characteristics of the disturbance regime (type of disturbance, season and frequency) in resprouting vigour, there have been reported important species-specific differences among oak species and effects of individual characteristics (age, size) in the success of this process. Species-specific differences may appear at very early stages during ontogeny depending on the location of dormant buds. For example, because of the presence of underground dormant buds, *Q. suber* seedlings can resprout even when aerial biomass is removed below the cotyledons, while *Q. ilex* and *Q. humilis* will resprout only from buds located above the cotyledonary node (Verdaguer et al. 2001, see also Pascual et al. 2002). Species-specific differences in the number and size of resprouts are also observed at more mature stages: i.e. evergreen *Q. ilex* trees tend to produce more numerous but shorter resprouts after fire while the co-existing deciduous *Q. cerrioides* produces less but taller resprouts after the same type of disturbance (Espelta et al. 2003). To what extent these particular differences may benefit each species under different disturbance scenarios has not been explored, yet it has been argued that having a larger bud bank (i.e. producing more resprouts) would be more beneficial under more intense and repeated disturbance events while producing less but taller resprouts may help to compete with other tree species during later stages of succession (Bonfil et al. 2004, see also Gracia et al. 2002; Salomón et al. 2013).

Whatever the species-specific means to resprout, the process is also influenced by the size of the individual prior to the disturbance event (Espelta et al. 1999; Cotillas et al. 2009, 2016; Catry et al. 2010; García-Jiménez et al. 2017). In general, resprouting vigour tends to be higher in larger sized individuals, probably because they possess a larger bud-bank and their roots are able to mobilize more resources (Quevedo et al. 2007; Catry et al. 2010; Cotillas et al. 2016) and therefore, larger genets produce resprouts that are thicker and larger (Espelta et al. 2003). Yet the positive effect of size seems to have an upper limit and shifts with ageing toward the production of less resprouts with shorter height and the reduction of genetic diversity in the stands (Ducrey and Boisserie 1992, but see Valbuena-Carabaña and Gil 2017). Stool stagnation and degradation has been attributed to abundant formation of root grafts and the accumulation of a huge amount of living tissues in the root systems of old individuals that are expensive to maintain, consume high amounts of resources, and result in limited growth of the aerial part (Landhäusser and Lieffers 2002; Salomón et al. 2015). Finally, local characteristics shaping site quality (climate, soil and topography) will also influence resprouting ability in oaks. For example, under a Mediterranean-type climate, characterized by severe water shortage, resprouting has been observed to be favoured in north versus south slopes and in deep vs. shallow soils, owing probably to the higher water stress faced in the latter conditions (López-Soria and Castells 1992; Espelta et al. 2003). Similarly, in mountain areas, resprouting ability decreases with increasing altitude owing to the decrease in the root biomass and root non-structural carbohydrates pool size (Zhu et al. 2012). Although several factors can constrain resprouting success, their development is much faster than that of seedlings and this has led to use of coppicing to harvest oak forests. To maintain high growth rates, oak resprouts take advantage of their ability to mobilize resources from roots (N in El Omari et al. 2003), increased water availability (Fleck et al. 1998), enhanced photosynthesis, leaf conductance and Rubisco activity (Fleck et al. 1996a, 2010).

14.2 Artificial Regeneration

Seeding and planting are the artificial regeneration methods used in oak forest restoration programs (Dey et al. 2008; González-Rodríguez et al. 2011b; Oliet et al. 2015). The success of seedling establishment depends on several general factors such as genotype, plant-plant and plant-animal interactions, and site conditions (Grossnickle and Heikurinen 1989; Burdett 1990; Margolis and Brand 1990; Landis et al. 2010) and other factors depending on the regeneration method selected. Seed size and defence against acorn predators are other factors affecting direct seeding success (Quero et al. 2007; Sage et al. 2011; St-denis et al. 2013; Pesendorfer 2014; Castro et al. 2015) while plant attributes, especially root system attributes, nutrition (Davis and Jacobs 2005; Grossnickle 2012) or cold hardiness among other factors (Mckay 1997; Landis et al. 2010) also affect planting success.

When using direct seeding, germination is the key process driving seedling establishment. Mediterranean oaks usually germinate in autumn, developing a strong taproot that normally grows several centimetres in length within a few weeks after germination (Pardos 2000; Johnson et al. 2001; Pemán et al. 2006). Root architecture has been defined as the spatial configuration of the root system and is characterized by its diversity and plasticity (Lynch 1995). Diversity among species and genotypes and plasticity is a consequence of soil environmental factors, physiography or disturbances of shoot architecture (Di Iorio et al. 2005; Siegel-Issem et al. 2005; Tamasi et al. 2005; Chiatante et al. 2006). Root architecture of Holm oak (*Quercus ilex* subsp. *ballota*), after the first growing season under optimal conditions, consists in a dual root system with a strong deep and orthogeotropic taproot, with a length of more than 90 cm (Pemán and Gil-Pelegrín 2008). This main root is branched with an unequal development of lateral roots depending of root depth (Riedacker et al. 1982; Johnson et al. 2001) and has been described as a dual root system. This system allows water uptake from deep soil horizons during drought episodes and from the upper soil horizons in wet seasons, especially when water has not reached deep horizons after the summer drought (Rambal 1984; Canadell and Zedler 1995). The first reference to the dual root system corresponds to Cannon (1914) when he studied three species of oaks in California.

Early development of the taproot in xeric habitats is directly correlated with successful establishment as this ensures water uptake and nutrient supply during the dry periods. Initial daily taproot growth may reach 0.7 cm day^{-1} for *Quercus ilex* (Pemán and Gil-Pelegrín 2008), similar to that described for *Q. robur* and *Q. suber* (Riedacker and Belgrand 1983; Pardos 2000). This high growth rate allows Holm oak taproot to reach 2/3 of the annual length at 60 days after germination (Fig. 14.2). Soil environmental properties control growth and development of root systems, with soil moisture content being one of the most important. Pardos (2000) found, for *Quercus suber*, that daily taproot growth decreased from 0.85 to 0.66 cm day^{-1} according to water availability and Canadell et al. (1999) showed, for *Q. ilex*, the loss of the orthogeotropic taproot pattern in mesic environments.

When planting is used, the key process limiting seedling establishment is to overcome planting stress or transplant shock (Rietveld 1989; Jacobs et al. 2009; Close et al. 2013; Grossnickle and El-Kassaby 2016). One of the main causes of this process is water stress, resulting from: (i) limited exploration capacity of the soil for water uptake by the confinement of the roots to the planting hole, (ii) poor root—soil contact, and/or (iii) low root permeability caused by the reduced number or short length of fine roots (Kozlowski and Davies 1975; Sands 1984; Burdett 1990; Haase and Rose 1993). These deficiencies can be overcome when seedlings

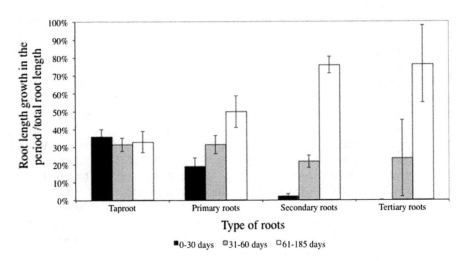

Fig. 14.2 Evolution of holm oak (*Quercus ilex* subsp. *ballota*) root growth from direct seeding in three periods of time after germination (Pemán and Gil-Pelegrín 2008)

initiate new root growth following planting. Therefore, seedling capacity to initiate rapidly new roots will determine seedling survival and growth (Simpson and Ritchie 1997; Grossnickle 2005; Villar-Salvador et al. 2012).

Cultivated seedlings can be classified into bareroot and container stocktypes, which are produced with different cultivation techniques. Bareroot seedlings are cultivated directly in the field and the seedlings must be removed from the soil by cutting the roots at some depth. After seedling lifting, roots are usually free of soil. Container seedlings are grown in pots or cells of variable size, which makes the roots to bind the growing medium into a cohesive structure termed plug (Landis 1995).

Container characteristics affect the architecture of seedling root system. For example, ribs or slits on the cell inner walls prevent root spiralling, and holes in the base facilitate drainage and encourage air pruning of roots (Landis 1990). Continual air pruning limits the main root growth by shortening its length to the depth of the container and preventing the development of replacement taproots. The confining of lateral roots in the container and their downward growth generates an orthogeotropic lateral root system rather than the more usual plagiotropic one that originates after taproot air pruning (Riedacker et al. 1982).

Root development of container seedlings under site conditions during the first growing season after planting keeps the root pattern generated inside the container, developing an orthogeotropic lateral root system and few plagiotropic or sub-horizontal roots at upper of seedling plug. *Quercus ilex* seedlings grown in mini-rhizotrons during the first growing season after planting, showed a variable number of orthogeotropic primary roots, 11–38, reaching a depth between 70 and 80 cm and a horizontal extension between 38 and 45 cm (Fig. 14.3). Therefore, the decision of seeding or planting generates, as first consequence, the development of seedlings with different root architecture as has been described in *Quercus ilex* (Table 14.1). Roots generated by planting are thicker and longer than those from seeding but less efficient for the low values of specific root length (Table 14.1). This parameter is often used either as an overall index of root thickness or as an estimator of the benefit (length) to cost (dry weight) ratio of the root system.

Root architecture is usually linked with water acquisition and nutrient supply (Lynch 1995) and a strong positive correlation between root hydraulic conductance and specific root length is in agreement with the view that hydraulic architecture conforms to the 'energy minimization' principle (West 1999). This strong correlation suggests that root systems characterised by less massive roots per unit length have a higher hydraulic conductance (North and Nobel 1991; Steudle and Meshcheryakov 1996; Rieger and Litvin 1999). According to the composite transport model (Steudle 2000), root water supply to the shoot may change according to the shoot demand owing to an adjustment of root hydraulic conductance. Root hydraulic conductance of *Q. ilex* may vary in response to external (drought or salinity) or internal factors such as nutritional state, water status, water demand and root morphology (Pemán et al. 2006) (Table 14.2).

Seasonal root hydraulic conductance per leaf unit surface area (K_{RL}) changes as reported by Nardini (Nardini et al. 1998b) indicate that *Q. ilex* presents a maximum

Fig. 14.3 Comparasion of the root architecture of *Quercus ilex* subsp. *ballota* developed in mini-rhizotrons, under optimal conditions, after the first growing season, according to the selected artificial regeneration method: seeding (left) and planting (right) (straight line = 10 cm)

Table 14.1 Root morphological parameters of *Quercus ilex* subsp. *ballota* generated by seeding or planting after the first growing season under optimal conditions (mean ± SE)

	Root volume V_R (cm³)	Root diameter D_R (mm)	Total root length L_R (cm)	Dry root weight DRW (g)	Specific root length SRL (m g^{-1})
Seeding	4.3 (0.3)	5.7 (0.3)	617 (59)	5.7 (0.3)	3.7 (0.37)
Planting	9.1 (1.7)	7.3 (0.3)	957.4 (128.47)	10.6 (1.0)	1.2 (0.1)

Pemán and Gil-Pelegrín (2008)

efficiency in water uptake during the spring, when the soil is still wet, and minimum efficiency in November as a consequence of freezing stress. The decrease of these values in summer with respect to spring has been explained by the vulnerability of this species to drought. Changes in root morphology, as a consequence of root length growth restrictions, also implies strong variation in root hydraulic

Table 14.2 *Quercus ilex* root hydraulic conductance per leaf unit surface area (KRL)

Treatment		K_{RL} 10^{-5} kg s^{-1} m^{-2} MPa^{-1}	References
Seasonal changes	May	3.53	Nardini et al. (1998a)
	August	1.9 to 2.3	
	November	0.3	
Freeze stress	Stressed	0.4	Nardini et al. (1998b)
	Non-stressed	3	
Ectomycorrhizas Tuber melanosporum	Inoculate	3.7	Nardini et al. (2000)
	Non-inoculated	2.9	
Root morphology (Root growth restrictions in depth)	No restrictions	3.2	Pemán et al. (2006)
	20 cm	1.2	

conductance, which could be a result of greater suberization of roots limiting their growth up to 20 cm in depth (Pemán et al. 2006). Low permeability of coarser roots, together with an unbalanced root to shoot ratio, is one of the main causes of transplanting stress, which may affect seedling establishment in field conditions (South and Zwolinski 1996). In summary, there are morphological and functional differences in development of seedling root systems as result of establishment techniques (seeding or planting), with root systems developed by seeding being more efficient in water uptake. This result suggests that the modification of root growth patterns brought about by commercial forest containers may influence establishment in the field.

14.2.1 Seed and Plant Quality

14.2.1.1 Seed Quality

Acorns are recalcitrant seeds (Fansworth 2000; Bonner 2008b) and so viability depends on maintaining a high moisture content. Because acorns are large seeds and seedling size is tightly coupled to seed size (Jurado and Westoby 1992), acorns result in large seedlings compared with small-seed species at the same ontogenetic stage. Poor acorn quality reduces the number of emerged seedlings and seedling vigor, which can impair reforestation success if acorns are directly seeded into the field, and reduce the yield of nurseries and the quality of cultivated seedlings. The quality of acorns depends on its moisture content, size and infestation by insects.

Moisture content on a fresh weight basis of ripened acorns ranges between 40 and 55% at maturity depending on the species (Bonner and Vozzo 1987). Desiccation of acorns reduces and delays germination (Connor et al. 1996; Connor and Sowa 2003; Ganatsas and Tsakaldimi 2013) and is an important factor hindering oak recruitment (Joët et al. 2013). Acorn desiccation is directly related to the surface of xylem vascular bundles in the cupule scar and the thickness of the vascularized layer in the pericarp (Xia et al. 2012a). Similarly, oaks whose acorns desiccate rapidly also take up water faster. Neither the presence of cuticular wax nor the thickness of the cuticle or the palisade layer in the pericarp was shown to be related to desiccation of acorns (Xia et al. 2012b). The moisture content at which acorn viability is reduced widely differs among *Quercus* species (Table 14.3). Acorn viability in most oaks is strongly reduced at high moisture (>30%) but some species can tolerate lower moisture levels (20%). In general, acorns are unviable when their moisture content is <15–18% but the oaks of the *Quercus* section lose their viability at higher moisture than the oaks of the *Lobatae* section (red and black oaks) (Bonner 2008b). Consequently, Bonner (2008a) recommends that acorn moisture during storage should be at least 30% for the Section *Lobatae* oaks and 45–50% for the American oaks of the Section *Quercus*.

Table 14.3 Acorn moisture content [(Fresh weight − Dry weight)/(Dry weight) × 100; Bonner 1981] at which maximum germination or viability is reduced to half (M_{50}) and completely lost (M_0) in several *Quercus* species

Species	M_{50} (%)	M_0 (%)	References
Q. alba	30	20–25	Connor and Sowa (2003)
Q. alnus	34	31	Anagiotos et al. (2012)
Q. coccifera	23	16	Ganatsas and Tsakaldimi (2013)
Q. fabric	33	24	Tian and Tang (2010)
Q. humilis	18	13	Ganatsas and Tsakaldimi (2013)
Q. ilex	30–32	18–22	Joët et al. (2013), Villar-Salvador et al. (2013a)
Q. lamellosa	43	26	Xia et al. (2012a)
Q. macrocarpa	38	13–15	Schroeder and Walker (1987)
Q. nigra	15	10–15	Bonner (1996)
	30	22	Connor et al. (1996)
Q. pagoda	13–17	11	Sowa and Connor (2003)
Q. rubra	16–20	11–15	Pritchard (1991)
Q. robur	35 (acorns stored for 100 days) 22 (fresh acorns)	27 15	Finch-Savage et al. (1996)
Q. schottkyana	23	11	Xia et al. (2012a)
Q. vulcanica	19	11	Tilki and Alptekin (2006)
Quercus section oaks (white oaks) *Lobatae* section oaks (black and red oaks)	25–30 15–20		Bonner (2008a)

High moisture content prevents acorns from being desiccated or stored well below freezing temperatures for long term storage. It also causes high acorn respiration even at cool temperature, which prevents acorns from being stored in airtight containers. Although in some studies acorns have been stored for 3–5 years with little vigor loss (Suszka and Tylkowski 1980), acorn storage for more than 2 years under large scale management conditions is impracticable. If acorn moisture is preserved, cold storage can speed seedling emergence and uniformity (Merouani et al. 2001; Bonner 2008a; Doody and O'Reilly 2008). Merouani et al. (2001) reported that cold storage of Q. suber acorns for 6 months also increased seedling tap root length but reduced chlorophyll concentration compared to seedlings cultivated from fresh acorns.

Acorns should be stored at temperatures ranging from −3 to 4 °C. Storing down to −3 °C will not harm most oak species. Q. durandii, Q. nigra and Q. pagoda maintained higher vigor when stored at −2 °C than at 4 °C while the opposite occurred in Q. virginiana (Connor and Sowa 2002). Acorn freezing tolerance importantly differs among species. The deciduous oaks, Q faginea, Q. pyrenaica and Q. petraea had 50% frost injury (LT_{50}) at −5, −6.9 and −8.2 °C, respectively while LT_{50} in the evergreen oaks, Q. ilex and Q. coccifera were −9.2 and −10.6 °C, respectively (Guthke and Spethmann 1993; Esteso-Martínez and Gil-Pelegrín 2004).

Seed lots of around 20 kg can be stored in low density polyethylene bags with a wall thickness of 75–100 μm. This wall thickness allows for relatively low water-vapor permeability but high permeability to gases (Lauridsen et al. 1992). Thinner polyethylene bags should be avoided because they are too weak and permeable to moisture vapor (Bonner 2008a). Acorns should not be stored wet to avoid rotting and special care must be afforded to desiccation during acorn transport and cleaning before storage. Soaking of Q. robur acorns can increase germination especially if acorns suffer slight desiccation after short storage. This increase in germination has been related to water leaching of germination inhibitors such as ABA (Doody and O'Reilly 2008).

Another component of acorn quality is its size. Independently of the nursery cultivation method, increase in acorn size results in larger oak seedlings (Rice et al. 1993; Navarro et al. 2006; Quero et al. 2007; Pesendorfer 2014) and larger acorns tend to have higher germination than small acorns (Pesendorfer 2014). Compared to seedlings emerging from small acorns, seedlings produced from large acorns recover better from herbivory (Bonfil 1998) and usually have higher short-term survival and growth in deep shade (Ke and Werger 1999; Quero et al. 2007), under competition (Rice et al. 1993) and under cold and drought stress conditions (Aizen and Woodcock 1996; Ramírez-Valiente et al. 2009). However, Navarro et al. (2006) did not find any effect of Q. ilex acorn size on field survival in a direct-seeding trial. González-Rodríguez et al. (2011a) concluded that for four Mediterranean oak species the *seedling size effect* rather than the *reserve* and the *metabolic effect* (see Leishman et al. 2000) was the main mechanism explaining the high performance of seedlings grown from large acorns under hazard conditions. A similar result was concluded for *Quercus lobata* (Sage et al. 2011). This means that large oak seedlings are frequently more resistant to stress factors than small seedlings (Ramírez-Valiente et al. 2009; Villar-Salvador et al. 2012). These results may make large acorns more attractive for maximizing forestation success than seeding small acorns. But selecting only large acorns may have long-term drawbacks in forestation as it can narrow genetic diversity of new populations, especially if large acorns are recollected in a few close individuals (Burgarella et al. 2007) or in small forest patches. Burgarella et al. (2007) suggest that the most effective acorn harvest design in large oak stands should include at least 20–30 scattered trees, distributed in a few high-distant groups (separated by hundreds of meters) and the trees within each group separated tens of meters.

Acorns are predated by several insects, most of which develop their larval phase inside the seed (Bonner 2008b). Infested acorns have lower germination and emerged seedlings are smaller than non-infested acorn (Leiva and Fernández-Alés 2005; Xiao et al. 2007; Lombardo and McCarthy 2009). The chance of an infested acorn to germinate and produce a viable seedling increases with acorn size and early germination in the fall (Branco et al. 2002; Xiao et al. 2007). Acorn batches can be floated in water to partially screen out infested and damaged acorns (Bonner and Vozzo 1987). Acorn X-ray imaging can also screen out infested and desiccated low-vigor acorns (Goodman et al. 2005; Lombardo and McCarthy 2009) but has not been used at management scale. Some practitioners immerse the acorns in hot water (48 °C) for 40 min to kill larvae of acorn weevils (Olson 1974).

14.2.1.2 Plant Quality

A seedling is considered of high quality when it has high survival and growth capacity after transplanting in a specific environment. Consequently, plant quality is defined as the capacity of seedlings to outperform in a specific site (Ritchie 1984; Wilson and Jacobs 2006). Performance of outplanted seedlings depends on its carbon, water and nutrient economy, which ultimately depend on the structure and physiological attributes of the seedling (Burdett 1990; Villar-Salvador et al. 2012; Grossnickle 2012). It is not possible to ascribe a universal set of functional attributes with a high-quality plant. Seedling characteristics conferring high performance for one site do not necessarily maximize seedling outplanting performance in another site.

A major aim of forest restoration managers and researchers is to discriminate poor-quality seedling lots and to determine which functional attributes better predict the potential outplanting survival and growth of seedlings. Discrimination of low quality seedling lots is important to attain forest restoration objectives while minimizing post-planting costs. Plant quality assessment cannot predict the actual performances of seedlings because field survival and growth not only depend on plant functional attributes but also on other site-related factors, which can vary unpredictably such as post-planting climatic conditions. Plant quality can be assessed by measuring key functional attributes (material attributes), or by measuring seedling performance under specific environment conditions (performance attributes).

Material attributes

Material attributes encompass both morphological and physiological traits. Morphological attributes are easier to measure than physiological traits and therefore physiological traits are little used in operation to assess the quality or seedling lots. However, morphological attributes have limited predictive capacity of transplanting performance. For instance, morphology will not indicate if tissue mineral nutrient concentration and photosynthesis are suboptimal or if the plant is cold hardened, which are also important outplanting performance drivers. Therefore, physiological quality attributes ideally should complement morphological attributes and frequently are better related to seedling functionality.

Morphological attributes

The morphological quality of plants comprises a set of attributes that measure the structure of the whole plant or any of its parts. Due to their measurement simplicity, root-collar diameter (RCD) and shoot height (from RCD to stem apex) are the most commonly used morphological quantitative attributes for operational quality assessment. For bareroot stock quality assessment, the number of first-order laterals roots and root volume are widely used and provide better predictive capacity than shoot size (Thompson and Schultz 1995; Jacobs et al. 2005; Wilson and Jacobs 2006). From these quantitative morphological traits, several indicators have been proposed to estimate the structural balance of plant organs and their conformation. The most common indexes are the shoot slenderness, which measures the relation between the shoot height and the RCD, and the shoot to root mass ratio, which is a proxy to the transpiration-water uptake balance of the plant (Lloret et al. 1999; Grossnickle 2005).

Plant size has a relatively moderate predictive capacity of out-planting performance whenever plants are not physiologically damaged (Thompson 1985; Navarro et al. 2006; Wilson and Jacobs 2006). Root collar diameter predicts better out-planting performance than shoot height and the relationship tends to be stronger with field growth than with survival (Thompson 1985; Mexal and Landis 1990). The relationships among plant size and survival or growth depends on the interaction of species and climate (Cortina et al. 1997). Among Mediterranean evergreen oaks, outplanting survival and absolute growth frequently increases with shoot or root seedling size at planting for seedlings of the same age (Cortina et al. 1997; Villar-Salvador et al. 2004a, 2013a; Tsakaldimi et al. 2005, 2013; Cuesta et al. 2010a). Very often, this positive effect of plant size on oak survival is apparent under the harshest planting conditions whenever plants are not subjected to very dry conditions at planting (Cuesta et al. 2010a; Villar-Salvador et al. 2013a). However, in semiarid climates with high probability of dry periods without rains or if seedlings are planted in dry soils, the probability of seedling mortality increases with seedling size in oak and in other species (Trubat et al. 2008, 2011; del Campo et al.

2010). Similarly, under similar climatic conditions the relationship between survival and seedling size can differ among oak species. In eastern Spain survival of the tree *Q. ilex* showed clear positive relationship with seedling size while no relation was found for the shrub *Q. coccifera* both in plantations performed in dry and semiarid locations (Cortina et al. 1997). Among deciduous oaks from wet temperate biomes, which usually are cultivated as bareroot stock, both positive (Dey and Parker 1997; Ward et al. 2000) and negative (Hashizume and Han 1993; Thompson and Schultz 1995) relations between field survival and seedling size have been observed.

The number of first order lateral roots (FOLR; structural lateral roots ≥ 1 mm in diameter at junction with the tap root) is also frequently used as an indicator of bareroot seedling quality and performance potential (Wilson and Jacobs 2006). Quantifying FOLR is quick and non-destructive, reflects the structural framework of the root system, and is often positively correlated with survival and growth in the field (Thompson and Schultz 1995; Dey and Parker 1997). However, FOLR has not shown entirely consistent ability to predict field performance in bareroot hardwood seedlings, probably associated with its inability to accurately characterize the root system (Jacobs and Seifert 2004). Other measures such as root volume, lateral root length, or root fibrosity may thus be better correlated with field performance in oaks (Jacobs et al. 2005; Wilson et al. 2007).

The integrity of the plug in containerized plants, which chiefly depends on the growth of fine roots (<1 mm) binding the growing medium is considered an important trait for high quality container stock. Seedlings with loose (poorly consolidated) plugs are prone to fine root breakage during manipulation at planting and this can reduce new root growth capacity after transplanting (Mckay 1997).

Physiological attributes

Many physiological attributes have been proposed for plant quality assessment (Mattsson 1997) but only a limited number of them have been used in large scale operational plant quality assessment. At present, very few regions in the world routinely carry out physiological measurements of stock quality (e.g., mainly limited to USA, UK, Canada, and Sweden per Dunsworth 1997). The most common physiological tests that have been used are the mineral nutrient and non-structural carbohydrate concentration of foliage and fine root electrolyte leakage. However, other physiological measurements such as gas exchange capacity, water use efficiency, photochemical efficiency of the PSII o can be very useful to establish functional limitations or seedling performance (Vilagrosa et al. 2003a; Hernández et al. 2009; Trubat et al. 2011). The main limitation for the operational use of these physiological attributes are the cost of the equipment used in these measurements, the need for highly trained technical staff and the low number of replicates that can be measured in some attributes (Vilagrosa et al. 2005).

The nitrogen (N), phosphorus (P) and potassium (K) status of planted seedlings can affect their outplanting performance (van den Driessche 1982, 1991a). Most published literature has focused on N and for oaks very little information exists for the role of P and K on seedling field performance. A significant part of the new growth in planted seedlings is supported by remobilization of stored N. Around 30–80% of N in new leaves and shoots and 20–60% in new roots in juveniles of broadleaf species comes from mobilization of stored N (Villar-Salvador et al. 2015). As N remobilization is mainly a source-driven process (Millard and Grelet 2010), an increase in N content due to high N fertilization enhances seedling capacity to remobilize N for supporting new growth after transplanting (Grelet et al. 2003; Uscola et al. 2015b). In poor-nutrient soils or high competition environments, remobilized N can have a greater contribution to new organ N than soil N (Salifu et al. 2008; Cuesta et al. 2010a). High outplanting survival and growth in *Quercus* species has been frequently linked to high tissue N concentration and content (Villar-Salvador et al. 2004a; Oliet et al. 2009; Salifu et al. 2009; Andivia et al. 2011). It is possible that the frequent increase in absolute growth after planting with seedling size might be determined by the increase in N content linked to size increase (Cuesta et al. 2010b).

Root growth in *Q. ilex* is especially sensitive to P deficiency (Sardans and Peñuelas 2006), and oak survival and new root growth capacity have been positively related to root P concentration (Villar-Salvador et al. 2004a; Oliet et al. 2009). Although K is involved in drought resistance of plants (van den Driessche 1991b) few studies have shown field performance relationships with seedling K status. However in *Q. ilex* del Campo et al. (2010) showed a positive link between field survival and leaf K concentration relation and concluded that leaf levels lower than 3.4 mg g^{-1} are undesirable for this species.

Non-structural carbohydrates (TNC) comprise starch and a variety of soluble sugars, these latter having a prominent role in cold hardiness and drought tolerance of oaks (Morin et al. 2007; Heredia et al. 2014). TNC support respiration and growth especially when photosynthesis is low and in both deciduous and evergreen oaks early growth in spring depends on TNC remobilization (Vizoso et al. 2008; Uscola et al. 2015b). Poor transplanting performance in cold-stored seedlings has been frequently linked to tissue depletion of TNC during cold storage and low photosynthesis after transplanting in conifers (Puttonen 1986; Grossnickle and South 2014). In *Quercus rubra*, low field survival was reported for saplings with low root TNC induced by repeated defoliation (Canham et al. 1999). Low TNC can jeopardize the capacity of plants to survive under dry conditions.

A healthy and vigorous root system is essential for seedling establishment (Grossnickle 2005) but frost and desiccation prior planting, for instance during seedling storage or transportation can damage fine roots (McKay et al. 1999; Garriou et al. 2000; Radoglou and Raftoyannis 2001). McKay (1992) described a test to assess the vitality of bareroot stock based on fine root electrolyte leakage (REL), which predicts seedling out-planting performance. For deciduous oaks maximum acceptable REL values for bareroot and container stock is 30 and 35%, respectively (Edwards 1998). Species show different sensitivity to fine root desiccation. *Quercus robur* showed lower REL than *Fagus sylvatica* and *Betula pubescens* after fine root desiccation (McKay et al. 1999). In contrast, for a similar reduction in fine root water content, *Q. rubra* and *Q. robur* showed higher REL and lower outplanting performance than *Pinus nigra* (Garriou et al. 2000).

Performance attributes

Performance attributes are assessed by analysing the survival, growth or any other physiological response of seedlings under specific environmental conditions, optimal or not. The most frequently used performance tests are root growth capacity (RGC) and frost tolerance tests.

Root growth capacity test

The RGC, which is also called root growth potential measures the ability of a seedling to produce and elongate new roots under optimal growth conditions within a limited period of time, usually less than two weeks (Ritchie 1985). Seedlings are transplanted into large containers with peat, sand, perlite or vermiculite or in hydroponic cultures. RGC varies seasonally reaching the highest values when seedling have attained highest dormancy and frost tolerance. Accordingly, RGC test have been used to decide the optimum temporal window for cold storage of cultivated seedlings (Ritchie and Dunlap 1980). RGC not only depends on the root physiological status, but also on the global functional characteristics of the seedling. Therefore, the RGC test measures the functional integrity and vigour of seedlings (Ritchie and Dunlap 1980; Simpson and Ritchie 1997). RGC in oak seedlings increases with plant size, nursery fertilization and frost tolerance (Villar-Salvador et al. 2004a, 2005a, 2013a; Mollá et al. 2006; Trubat et al. 2010, 2011; Oliet et al. 2011; Andivia et al. 2014) while water stress conditioning in the nursery (Villar-Salvador et al. 2004b) and desiccation, warming or freezing during storage reduces RGC (Webb and von Althen 1980; Cabral and O'Reilly 2008). RGC has been frequently used to predict seedling outplanting survival and growth and its predictive capacity is highest in harsh sites where field performance is mostly determined by seedling vigour and stress tolerance (Simpson and Ritchie 1997). As for other plant quality tests, the RGC test must be performed as close as possible to the outplanting time to maximize the field performance prediction. Relationships between survival and RGC are often asymptotic (Burdett et al. 1983; Simpson and Vyse 1995), indicating that survival diminishes under a specific RGC threshold because seedlings are damaged or have low vigour. In Mediterranean (Villar-Salvador et al. 2004a, 2013a) and wet temperate deciduous oaks, survival and growth has been positively related to RGC (Garriou et al. 2000; Cabral and O'Reilly 2008). Folk and Grossnickle (1997) showed that RGC tests performed under suboptimal environmental conditions, more similar those that seedlings would encounter after planting seem had higher capacity to predict outplanting performance in conifers than RGC tests performed under optimal growth conditions. However, in the oak shrub *Q. coccifera*, RGC measured outdoors at low air temperature did not discriminate better poor seedling lots than the RGC measured inside a greenhouse under optimal growth conditions (Villar-Salvador et al. 2013b).

14.2.2 Cultivation of Seedlings

At present, container stocktypes are mainly used in regions with harsh climatic conditions, while bareroot stocktypes are used in regions where abiotic environmental conditions after planting are not strongly limiting. Worldwide, both stocktypes are used for planting oaks. Oaks are produced as container stocktypes in the Mediterranean basin (Tsakaldimi et al. 2005; Chirino et al. 2008), while in wet temperate Europe and eastern North America bareroot is the dominant stock-type.

14.2.2.1 The Role of Container for Nursery Culture

The architecture and development of the root system determines plant functioning. Root properties affect the volume and depth of soil explored, the absorption of soil resources and the efficiency in water transport (Landis 1990; Tsakaldimi 2009; Makita et al. 2011). Rooting depth and colonized volume is related to increased survival in drought periods (Domínguez-Lerena 2000; Vilagrosa 2002; Padilla and Pugnaire 2007). As root system development is a species-specific characteristic selecting the right containers to match the morphofunctial characteristics of the species and root system is key for increasing the success of seedling establishment.

Container design determines the morphological and physiological traits of root systems and other seedlings characteristics (Landis 1990; Vilagrosa et al. 1997; Tsakaldimi et al. 2005). Container volume, depth, area of the cell top section, cell spacing, which determines seedling density, and the type of material for manufacturing the containers affect plant quality.

Scientific and technical literature on the effects of container features on seedling morphology and field performance is prolific (Biran and Eliassaf 1980; Landis 1990; Peñuelas and Ocaña 1996; Domínguez-Lerena 1997; South et al. 2005; Dey et al. 2008; Pinto et al. 2011; De Souza et al. 2016; Salto et al. 2016) However, studies on oak species (*Quercus* spp.) are quite scarce compared to other species, and most of the studied oaks are from the Mediterranean basin. Overall, studies of container size show that an increase in container volume increases plant growth and nutrient content (Domínguez-Lerena 2000; Dominguez-Lerena et al. 2006; Mariotti et al. 2015). However, general patterns of the relation between seedling field performance and container characteristics are not straightforward. Domínguez-Lerena (1997) reported that morphological characteristics of *Q. ilex* were little affect by container size whilst the effect was high for several *Pinus* species. Tsakaldimi et al. (2005) reported that *Quercus ilex* and *Q. coccifera* seedlings grown in paper-pot FS 615 (volume: 482 cm^3, depth: 15 cm) showed higher survival than those cultivated in Quick pot T18 (volume: 650 cm^3, depth: 18 cm) and Plantek 35F (volume: 275 cm^3, depth: 13 cm). Other authors pointed out that the using seedlings cultivated in large containers was the main factor determining the success *Q. ilex* plantations. This study found the best results when combining bigger containers with adequate soil preparation (Jelic et al. 2015). In this context, field establishment will depend on the ability of the root system to colonize the soil.

According to Landis (1990), the plants that develop shallow, fibrous root systems, grow better in shorter containers; while the plants with long taproots grow better in longer containers. Most Mediterranean species such as *Rhamnus alaternus, Arbutus unedo, Fraxinus ornus* or *Acer granatense* have dense and shallow root system while *Quercus* species (e.g. *Quercus ilex, Q. faginea, Q. coccifera* or *Q. ithaburensis*) and other species (e.g. *Chamaerops humilis, Retama sphaerocarpa*) develop a deep root system with a strong taproot (Domínguez-Lerena 2000; Chirino et al. 2013; Tsitsoni et al. 2015) (Fig. 14.3).

Use of deep containers may promote the development of deep, strong tap roots in species that produce orthotropic root systems (Fig. 14.4). Cultivating *Quercus* species in deep containers increases the length of the tap root, projected root area and volume, total and fine root length, the specific root length and hydraulic conductance. These plants had higher water potential under drought stress conditions (Pemán et al. 2006; Chirino et al. 2008). Similarly, seedlings of *Q. coccifera, Q. ilex* and *Q. suber* grown in deep containers (depth: 24–30 cm) developed longer taproots than those grown in a paper-pot of standard depth (i.e. 16–18 cm; Chirino et al. 2005). After field transplanting, these seedlings had greater root biomass colonizing the surrounding soil than those cultivated in shorter (standard) containers (Fig. 14.5). Therefore, the use of a large container (volume 400–500 cm^3) or deep container (depth 25–30 cm) (Fig. 14.4) for *Quercus* species produces high quality plants and enhances outplanting performance (Vilagrosa et al. 1997; Domínguez-Lerena 2000; Cortina et al. 2006; Chirino et al. 2008).

14.2.2.2 Fertilization

Fertilization during nursery cultivation affects oak seedling quality and outplanting performance. Most fertilization studies on *Quercus* species have focused either only on nitrogen or on the amount of manufactured fertilizers (i.e. different compositions). Phosphorus (P) nutrition plays an important role in the root growth and function of *Q. ilex* seedlings (Sardans and Peñuelas 2006; Oliet et al. 2011) but very few studies have addressed the role of nutrients other than N on oak seedling quality and outplanting performance (see Sardans and Rodà 2006; Trubat et al. 2010; Sepúlveda et al. 2014 for studies on the specific role of P fertilization on *Quercus* species). Overall, studies on wet temperate and Mediterranean oaks show that moderate to high fertilization in most cases increases field survival and growth

Fig. 14.4 *Quercus suber* seedlings cultivated in a shallow container (depth: 18 cm) and deep container (depth: 30 cm; cylindrical in shape, diameter: 5 cm, depth: 30 cm, volume: 589 cm^3, material: high-density polyethylene, open bottomed, plant density: 318 seedlings m^{-2}) (left). After these experiments, a model patented with the design of a deep container (30 cm depth, 6 cm in diameter and 505 cm^3 of volume) has been registered in Spanish Patent and Trademark Office (OEPM 2014) by the Mediterranean Center for Environmental Studies (Foundation CEAM) and University of Alicante (Spain) (right) for the use in water limited ecosystems (Photos by E. Chirino)

compared to low or unfertilized plants (Villar-Salvador et al. 2004a, 2013b; Salifu et al. 2009; Cuesta et al. 2010b; Oliet et al. 2011). The positive effect of N fertilization on out-planting survival in *Q. ilex* and *Q. coccifera* was more noticeable under harsh planting conditions (Cuesta et al. 2010b; Villar-Salvador et al. 2013a), while the benefit of N fertilization on seedling growth was observed under mild stress conditions. Contrary to the above cited studies, Trubat et al. (2011) reported higher performance of nutrient-deprived *Q. coccifera* seedlings planted under semiarid conditions. To reconcile these conflicting results, Cortina et al. (2013) suggested that the effect of nursery fertilization on seedling out-planting survival may largely depend on the timing and intensity of drought after planting.

Moderate to high N fertilization increases seedling water use efficiency (Guehl et al. 1995; Welander and Ottosson 2000), photosynthesis rate (Hernández et al. 2009), and new root growth (Trubat et al. 2010; Villar-Salvador et al. 2013b), which potentially can increase outplanting performance. N availability may also affect drought and cold stress tolerance but the magnitude and direction of the effect in *Quercus* species shows conflicting results among studies. In *Q. ilex* and *Q. suber*, high N fertilized seedlings that were supplied with nutrients during the growing

Fig. 14.5 Mass of roots protruding out of the root plug one month after field outplanting in seedlings of two Mediterranean *Quercus* species that were cultivated in two types of containers: standard container (CCS-18) and depth container (CCL-30) after. Containers were the same that in Fig. 14.4 (see Chirino et al. 2008 for more details)

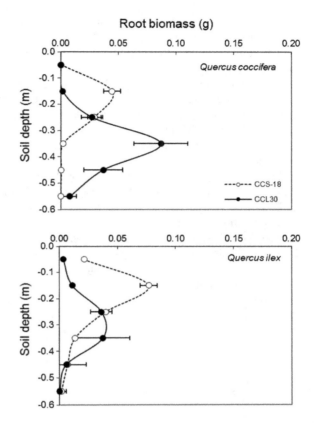

season but not in the fall, had similar osmotic potential to unfertilized plants (Villar-Salvador et al. 2005b). Low or no N fertilization, especially during cold acclimation in the fall hindered frost tolerance in *Q. coccifera* (Villar-Salvador et al. 2013b) and frost and drought tolerance in *Q. ilex* (Andivia et al. 2012a, b, 2014), suggesting that N nutrition during the cold acclimation period is important for hardening in Mediterranean oaks. In contrast, very high N supply rates at the end of the summer reduced the frost and drought tolerance of *Q. ilex* seedlings cultivated in a mild winter sites but not in a cold winter site (Heredia et al. 2014). This suggests that low fall temperature overrides the negative effects of high N fertilization on frost acclimation and drought tolerance, likely explaining the lack of differences among N fertilization treatments observed in some studies. A similar response has been reported for frost tolerance and N fertilization in *Q. petraea* and *Q. robur* (Thomas and Ahlers 1999). All these results indicate that N fertilization timing and fall temperature interact to drive cold and drought stress of oaks.

Nitrogen fertilization increases seedling N content through an increase in seedling mass and tissue N concentration (Salifu and Jacobs 2006; Schmal et al. 2011; Trubat et al. 2011; Villar-Salvador et al. 2013b; Uscola et al. 2015a).

The growth of seedlings after outplanting depends on the remobilization of N and other nutrients. In nutrient poor soils or high competition environments, remobilized N can have a greater contribution to new organ N than soil N (Salifu et al. 2008; Cuesta et al. 2010b). As explained in the plant quality section, an increase in tissue N content involves a greater N remobilization capacity for supporting new growth after transplanting. This explains, in part, the positive effect of N fertilization on new root growth capacity of oak seedlings (Villar-Salvador et al. 2004a; Trubat et al. 2010, 2011; Cuesta et al. 2010a; Oliet et al. 2011), which is crucial for seedling establishment and survival in dry-climate ecosystems (Grossnickle 2005; Padilla and Pugnaire 2007).

Responsiveness of *Quercus* species growth to fertilization is low compared to other species (Valladares et al. 2000; Uscola et al. 2014). Dose-response curves of seedling growth against N availability made from published data shows that the slope of the line in the deficiency phase of the dose-response curves (see Salifu and Timmer 2003) is smaller for *Quercus* species than for other species such as conifers (Fig. 14.6). This indicates that *Quercus* species have low N uptake efficiency at seedling stages and that N must be supplied at high rates to maximize seedling growth and tissue N concentration.

Lower N uptake efficiency in *Quercus* species probably reflects lower N root absorption and post uptake metabolism after uptake but also high dependence on

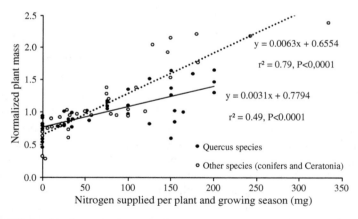

Fig. 14.6 Relationship between the normalized seedling mass and the amount of supplied nitrogen per plant. The data has been separated into *Quercus* species (*Q. ilex, Q. suber, Q. coccifera, Q. faginea, Q.rubra*) and other species (the conifers *Pinus halepensis, P. pinea, P. sylvestris, P. tabulaeformis, Juniperus thurifera*, and the broadleaf *Ceratonia siliqua*). For comparing the data of different studies, the seedling mass was normalized following methodology in Poorter et al. (2010). Studies used for each species are: *Q. ilex* (Ocaña et al. 1997; Villar-Salvador et al. 2004a; Oliet et al. 2009; Uscola et al. 2015a), *Q. coccifera* and *Q. faginea* (Villar-Salvador et al. 2013b), *Q. suber* (Martínez Romero et al. 2001), *Q. rubra* (Salifu and Jacobs 2006), *Pinus halepensis* (Oliet et al. 2004), *P. pinea* and *P. sylvestris* (Ocaña et al. 1997), *P. tabulaeformis* (Wang et al. 2015), *Juniperus thurifera* (Villar-Salvador et al. 2005b) and *Ceratonia siliqua* (Planelles 2004). Each point is a N fertilization treatment

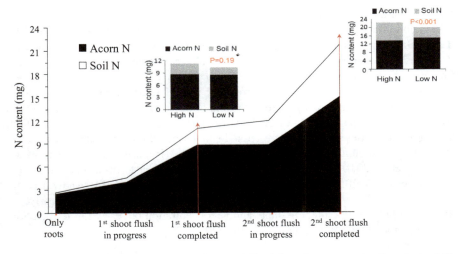

Fig. 14.7 Evolution of the contribution of the N derived from the acorn and soil to the total N content of *Quercus ilex* seedlings through their early ontogeny. Inserted figures compare the total N derived from acorn and soil N in high (10 mM N) and low fertilized (1 mM N) seedlings at the end of the first and second flush of shoot growth. For methodological details see Villar-Salvador et al. (2010). Seedlings completed their first and second shoot flush of growth 53–56 and 83–90 days, respectively, after sowing

acorn N during early ontogeny (Fig. 14.7), which facilitates independence on soil N (García Cebrián et al. 2003; Villar-Salvador et al. 2010; Yi and Wang 2016). *Quercus ilex* seedlings are almost independent of soil N during early development. Acorns made up 62–75% of all seedling N at the end of the first shoot flush and up to this ontogenetic stage, N fertilization did not affect seedling N content. Dependence of seedlings on acorn N progressively decreased during later seedling development and at the end of the second flush of growth differences in N fertilization significantly affected seedling N content (Villar-Salvador et al. 2010). Per these results, an efficient fertilization schedule that maximizes seedling N uptake and minimizes N runoff should delay N supply until the onset of second flush of growth.

The efficiency of fertilization on the cultivation of woody species depends on how nutrients are supplied through time. Contrary to a constant fertilization regime where seedlings are supplied with the same amount of N throughout the cultivation period, the exponential fertilization (Timmer 1997) matches N supply to seedling N demand, which increases exponentially through cultivation period (Birge et al. 2006; Oliet et al. 2013). Compared to constant fertilization, exponential fertilization reduces N run-off in the nursery and in conifer seedlings it results in higher tissue N concentration with few differences in morphology (Salifu and Timmer 2003; Dumroese et al. 2005). Few studies have compared the constant and exponential fertilization regimes on the quality of *Quercus* seedlings. In the evergreen *Q. ilex*

(Oliet et al. 2009; Heredia et al. 2014) and in the deciduous *Q. robur* (Schmal et al. 2011) no differences in growth and seedling N content were observed between both fertilization regimes. However, in the deciduous *Q. rubra* and *Q. alba* exponential fertilization was more effective in promoting nutrient acquisition and storage than the constant fertilization (Birge et al. 2006).

Seedling N loading can also be promoted by fall fertilization. Nitrogen fertilization is usually reduced or ceased at the end of the growing season (Landis 1989). However, an increasing body of evidences points out that moderate N fertilization during the fall after seedlings have ceased shoot elongation significantly promotes N loading with small or no change in morphology (Islam et al. 2009; Zhu et al. 2013; Li et al. 2014). N loading by fall fertilization has also been reported for *Q. ilex* (Oliet et al. 2011; Andivia et al. 2012b; Heredia et al. 2014; Andivia et al. 2014) and for *Q. velutina* (Wang et al. 2016). Timing of fall fertilization (early vs. late in the fall) did not affect tissue N and K concentration but early fall fertilization increased root P concentration (Oliet et al. 2011).

14.2.2.3 Drought and Cold Hardening

Soil water availability and extreme temperatures are major environmental constraints for plant development in many ecosystems (Levitt 1980; Larcher 2003). Low soil water availability and/or high evaporative demand causes plant water stress, which hinders many physiological processes and ultimately plant survival. Cold stress during winter can also damage plants at a structural (whole-plant) (e.g. freezing of xylem tissues) and at a cellular level, which includes alterations in specific metabolites, proteins, and changes in membrane structure (Gusta and Wisniewski 2013). Freezing tolerance is defined as the lowest temperature below the freezing point that a tissue can be exposed to without damage (Grossnickle 2000). Seedling and sapling are especially vulnerable to cold and drought stress. In addition, freezing and drought (i.e. lack of water flow to leaves) can happen together and interactions between both stresses may occur (Pratt et al. 2005; Mayr et al. 2006a, b; Fernández et al. 2007).

Implications for nursery management and restoration practices

Adverse climatic conditions after outplanting are a major limitant for seedling establishment. Much evidence suggests that a key hurdle in plantation success is transplant shock, which is the short-term stress experienced by seedlings when transferred from favourable nursery conditions to the adverse field environment (Burdett 1990). During transplant shock, seedlings may remain stressed until they become acclimated to field conditions. In general, adverse climatic conditions after outplanting are major limiting factors for seedling establishment. Suitable restoration techniques may help the seedlings to establish successfully.

Drought stress is the main cause of seedling mortality in forest plantations during the first year in Mediterranean-climate sites (Vallejo et al. 2006). Drought occurs during the summer in the Mediterranean climate. However, dry conditions during the wet season, when seedlings are planted, are becoming more frequent in the last decades. To increase the chance of seedling establishment and reduce seedling outplanting stress, forest managers have developed nursery techniques to harden (increase the stress tolerance) of the cultivated seedlings. Common drought hardening techniques, also called drought preconditioning, involve irrigation restriction by exposing seedlings to drought cycles, altering and reducing nutrient inputs, and by growing plants outdoors under full sunlight (Landis et al. 1998). The final objective is to promote physiological mechanisms in the seedling that confer stress tolerance to before outplanting.

Cold stress also limits planting performance in natural systems (Mitrakos 1980; Larcher 2003). Cold stress can impose severe restrictions during plant culture in the nursery but also in the field after outplanting. Freezing tolerance is highly variable among species, ranging from species that only tolerate temperatures slightly under 0 °C such as for some mild winter Mediterranean species until to −70 °C or less in boreal and high mountain conifers (Bannister and Neuner 2001; Larcher 2000). In general, forest plantations are established after winter to avoid frosts, but spring frosts are also always a risk. In mild or cool winter regions, planting can be done at the middle or the end of autumn to benefit from autumn and winter rains. However, autumn planted-stock are exposed to some frosts during winter. Therefore, it is important to use frost-tolerant seedlings in plantations made in continental areas or areas with high risk of strong frosts, especially for evergreen woody species (Mollá et al. 2006; Tibbits and Hodge 2003). As cold tolerance has an important genetic basis, using the appropriate provenances is also essential for reducing negative effects of low temperatures on reforestation projects. However, once the right provenance has been selected, nursery treatments for promoting seedling cold tolerance are necessary (Gratani 1995; Gimeno et al. 2009). The general procedure for container seedlings consists of growing seedlings outdoors at the end of the summer and in the fall under a natural temperature and daylight length regime, which will promote cold tolerance (Mollá et al. 2006). This process happens naturally for bareroot seedlings, which are grown under ambient outdoor conditions (Jacobs 2003). Cold hardening strongly depends on nursery location and for *Q. ilex*, hardening was faster and more intense in a cold winter location than in a mild winter location (Mollá et al. 2006). Short-day treatment is another nursery technique to promote hardening that may be applicable to oaks. This involves reducing photoperiods in mid-summer to arrest shoot growth, induce dormancy, and increase cold hardiness (Jacobs et al. 2008). The practice, which is used regularly for production of boreal conifers worldwide, has been shown to also effectively stimulate dormancy and cold tolerance in *Quercus rubra* from temperate North America (Davis 2006) and may have application to hardening of oak species in other regions.

Seedling responses to drought hardening in the nursery

Drought hardening techniques can promote a wide array of plant functional responses at several organization levels (Table 14.4). Species respond to drought

Table 14.4 Comparative morphological and physiological traits affected by drought hardening techniques

Physiological variables	Main effect	Quercus species	Other species[c]	References
Gas exchange, water use efficiency, water flow	Lower residual and cuticular transpiration	Q. ilex[a], Q. coccifera [a], Q. suber[a]	P. halepensis, P. lentiscus, R. officinalis, Pinus pinea. J. oxycedrus[a]	1, 2, 3, 4, 24
	Reduction on stomatal conductance Higher chlorophyll fluorescence	Q. ilex[b], Q. suber, Q. coccifera[a]	P. lentiscus [a]), Rhamnus alaternus, Nerium oleander, P. halepensis[b], P. pinea, J. oxycedrus[a]	1, 2, 3, 4, 5, 6, 7, 8, 24
Cell water relations	Decrease (more negative water potential values) in turgor loss point due to osmotic adjustment and/or changes in cell wall elasticity	Q. ilex[b], Q. coccifera[a]	Olea europaea, P. lentiscus[a], Rh. Alaternus[a], N. oleander, P. halepensis[b], P. pinaster, P. pinea[a], J. oxycedrus[a]	1, 2, 3, 4, 5, 6, 7, 8, 9, 10, 11, 12, 13, 14
Nutrients and other compounds	Accumulation of soluble carbohydrates Increase or no change in mineral nutrient concentration Increase of ABA	Q. ilex, Q. coccifera,	P. halepensis, P. pinea, P. pinaster, R. officinalis, P. lentiscus, Picea mariana, Olea europaea ssp sylvestris, Acacia cyanophylla, Thuja occidentalis,	1, 2, 3, 4, 5, 8, 9, 11, 12, 13, 15, 16, 17, 18, 19
Root functioning and structure	Lower new root growth capacity Higher root hydraulic conductance Higher root-shoot ratio	Q.ilex[b], Q. suber[a, b], Q coccifera,	P. halepensis, P. pinea, P. lentiscus[b], R. officinalis, J. oxycedrus, Lotus creticus	1, 2, 4, 5, 20, 21, 22, 23, 24
Resistance to other stresses	Increased tolerance to freezing and high temperatures	No information	Pinus pinea, Pseudotsuga menziesii,	14, 25

Species and main effects observed
[a]no effect observed
[b]reported variable response, in general opposite response
[c]the majority are co-existing species
References: 1: Villar-Salvador et al. (1999), 2: Villar-Salvador et al. (2000), 3: Vilagrosa et al. (2003a), 4: Rubio et al. (2001), 5: Villar-Salvador et al. (2004b), 6: Bañón et al. (2003), 7: Bañón et al. (2005), 8: Puértolas et al. (2003), 9: Dichio et al. (2003), 10: Larcher et al. (1981), 11: Calamassi et al. (2001), 12: Tognetti et al. (1997), 13: Fernández et al. (1999), 14: Villar-Salvador et al. (2013c), 15: Stewart and Lieffers (1993), 16: Royo et al. (2001), 17: Sánchez-Blanco et al. (2004), 18: Albouchi et al. (1997), 19: Edwards and Dixon (1995), 20: Fonseca (1999), 21: Franco et al. (2002), 22: Chirino et al. (2003), 23: Sánchez-Blanco et al. (2004), 24: Vilagrosa (unpub. data), 25: Blake et al. (1991)

hardening depending on their strategies to cope with drought stress (Valladares et al. 2008 for a review on this topic, Vilagrosa et al. 2003a, 2006; Villar-Salvador et al. 2004a) and their functional characteristics (Vilagrosa et al. 2003b). Consequently, a common response to drought hardening in the nursery cannot be expected for all species. Comparing plant responses among different plant groups, Vilagrosa et al. (2006) observed that boreal and wet temperate woody plants tend to have a more intense acclimation that affects a greater variety of functional traits in response to drought hardening than woody plants from dry or semi-arid ecosystems such as Mediterranean-type ecosystems.

Most studies on nursery drought hardening in *Quercus* species have been applied to oaks of the Mediterranean basin. This is because oaks have been widely used in afforestation programs in southern Europe for the last 20 years. Despite the large amount of studies and different types of applied hardening, Mediterranean oaks have low outplanting performance compared with other planted species (Vilagrosa et al. 1997; Cortina et al. 2004; Villar-Salvador et al. 1999). *Quercus* species tend to have low responsiveness to drought hardening in the nursery (Vilagrosa et al. 2003a), which has been related to low plasticity to variations in other abiotic factors compared to other co-existing species (Vilagrosa et al. 2005).

A common response among oak species to drought hardening is osmotic adjustment (see Abrams 1990; Kleiner et al. 1992; Collet and Ghuel 1997), which allows plants to maintain cell turgor at lower water content values (González et al. 1999). Drought hardening intensity and duration and the period of application seem to determine the effectivity of drought hardening on enhancing drought tolerance in the Mediterranean evergreen oak, *Q. ilex*. Villar-Salvador et al. (2004b) reported that moderate levels of drought stress rather than strong or low water stress were the most effective for enhancing drought tolerance mechanism in *Q. ilex* (Villar-Salvador et al. 2004b). Moreover, drought hardening in the fall had an additive effect on seasonal osmotic adjustment, reducing osmotic potential at full turgor by 50% but did not affect morphological traits. Finally, drought hardening for 3.5 months did not enhance physiological traits related to drought tolerance in *Q. ilex* seedlings compared with hardening for 2.5 months. However, long drought hardening periods (i.e. about six months) especially when applied during the growing season can decrease whole seedling biomass, especially aboveground biomass, but also other functional traits such as increasing the capacity for water transport through xylem (Chirino et al. 2004; Chirino and Vilagrosa 2006).

Moderate drought hardening usually reduces growth of oak seedlings, especially if applied during the growing season, although the effect is species dependent (Sanz-Pérez et al. 2007, 2009). Thus, drought stress had little effect on the growth of seedlings of the oak shrub *Q. coccifera* while it reduced it in *Q. ilex* and *Q. faginea* (Sanz-Pérez et al. 2007). Nutrient and non-structural carbohydrates (NSC) can play an important role in seedling outplanting performance. Drought hardening has varied effects on the accumulation of NSC and mineral nutrients in oak species. Moderate drought had little effect on tissue mineral nutrient concentration in *Q. ilex* and *Q. faginea* while it decreased it in *Q. coccifera* (Villar-Salvador et al. 2004b; Sanz-Pérez et al. 2009). However, strong drought stress reduced soluble sugar concentration and increased starch concentration in *Q. ilex* (Villar-Salvador et al. 2004b). Drought hardening reduces in most cases the new root growth capacity of cultivated seedlings (Fonseca 1999; Trubat 2012), a response that has also been observed among oak species (Villar-Salvador et al. 2004b; Chirino et al. 2009; Trubat 2012). This is an undesired effect of drought hardening because new root growth after planting is an important requisite for seedling establishment (Tinus 1996). Observations of new root growth capacity assessed at the nursery conflict with field observations where the general rule was an increment in root length in hardened seedlings compared to control seedlings (Chirino et al. 2004; Rubio et al. 2001; Franco et al. 2001, 2002).

Seedling responses to cold hardening in the nursery

In cold climate areas cold hardiness is a key factor for survival of planted seedlings. Cold hardening of cultivated seedlings is carried out in the last phases of nursery culture in which seedlings are subjected to progressively low temperatures. In this way, seedlings can develop resistance mechanisms as pointed out in Table 14.4. Cold hardiness can also be stimulated by controlling the photoperiod, irrigation and fertilization, which can induce dormancy and therefore cold hardiness (Villar-Salvador et al. 2004a; Fernández et al. 2007). However, some studies found that N fertilization can decrease cold resistance, thereby avoiding cold hardening or promoting the de-hardening processes in seedlings (Andivia et al. 2011; Heredia et al. 2014).

Pardos et al. (2003) and Mollá et al. (2006) suggesting that temperature was the main factor influencing differences cold resistance. In addition, these authors found a higher resistance to cold conditions in *Q. ilex* than in *Pinus halepensis*. Mollá et al. (2006) compared cold resistance in seedling cultivated in a coastal nursery (warm site) with those cultivated in an inland nursery (cold site). The results pointed out that nursery location did not affect transplanting mortality. However, inland seedlings had greater growth than coastal seedlings when planted in mid-winter planting but not in late fall. This study demonstrated that differences in winter conditions in the nursery have a strong effect on the functional and transplanting performance of *Q. ilex* seedlings.

14.2.3 Outplanting Limitations to the Use of Quercus Species in the Field

Early mortality and low growth rates have been identified as critical constraints to successful plantation establishment in Mediterranean areas (Margolis and Brand 1990; Navarro-Cerrillo et al. 2006b). Often, these seedling losses have been attributed to various environmental or biotic circumstances (e.g. weather conditions, animal damage, disease, insects, etc.) which may act synergistically, influencing plant performance (Burdett 1990; Broncano et al. 1998). However, in many cases, foresters have considered how key cultivation practices influence seedling survival after field planting and how these practices must be managed to make sound forest restoration decisions. The cultivation practices related most to seedling performance are seedling storage and handling, site preparation, planting date and planting practices, microclimatic/microsite conditions, vegetation management and the use of thee shelters (Grossnickle 2000; Landis et al. 2010; Potter 1991). This improved survival has been attributed to the enhancement of drought resistance, due to greater water-uptake ability (through more efficient root systems), seedling nutrition, and growth; these effects increase the speed with which seedlings can overcome competition from other plants (enhanced by vegetation control and microclimatic effects) and become established in the forest restoration site. Without such cultivation practices, seedlings lack the physiological capability to become rapidly established in the face of the stress they suffer after planting in forest-restoration sites. This section has shown that these key cultivation practices can improve the chances of survival and growth of the seedlings after transplanting in such sites.

In Mediterranean areas, the seedling physiological response to transplanting stress is related closely to the plant-root-soil interaction, which limits access to soil water. In these conditions, foresters have used cultivation practices to improve seedling survival and growth (Navarro-Cerrillo et al. 2006a), as well as the ability of seedlings to respond to environmentally stressful conditions that can occur after they have been planted. Adequate cultivation practices enhance the ability of seedlings to survive after being planted in forest restoration sites in Mediterranean areas because they reduce the susceptibility to drought-induced mortality. This positive response is due to changes in the topographic, physical, and microclimatic conditions and the soil water availability at the site, which limit seedling susceptibility to planting stress (i.e. water stress). In Mediterranean areas, the most relevant of these conditions limiting seedling performance are seedling handling, planting date, soil physical and microclimatic conditions, and soil preparation (Navarro-Cerrillo et al. 2006a; Ruthrof et al. 2013) (Table 14.5). Thus, seedling survival in artificial forest restoration is related to how foresters modify the field-site environmental conditions to improve the inherent potential of the seedlings to overcome establishment stress and become coupled into the forest ecosystem (Burdett et al. 1984; Grossnickle 2000; del Campo et al. 2007). This section summarizes our current knowledge about the influence of the main cultivation practices affecting seedling performance after planting (i.e. survival and growth).

Seedling shipping and handling before planting

In the nursery, seedlings are typically grown in optimal conditions (i.e. water availability, plant nutrition, and environmental conditions) that ensure their maximum "seedling functionality" immediately before they are harvested. However, seedlings must be transported before being outplanted; thus, handling and shipping are critical factors in plantation success (Paterson et al. 2001). Inappropriate storage or handling of seedlings, or both, before plantation can decrease their physiological functionality, causing an imbalance in the physiological status of the seedlings that

Table 14.5 Spearman correlation matrix for different cultivation practices and the seedling establishment response (survival and growth) for eight Mediterranean species, based on different reforestation trials in Southern Spain

Variable	Species	Planting date	Soil type	Soil preparation
Survival (n = 56)	0.095	**−0.711****	**0.560****	**0.266***
Height growth (n = 30)	−0.330	−0.085	−0.155	0.202
Diameter growth (n = 27)	−0.290	**−0.397***	0.020	0.329

Species *Ceratonia siliqua, Quercus suber, Olea europaea, Pinus halepensis, P. pinaster, P. pinea, Pistacia lentiscus, Tetraclinis articulata*; Planting date: Early (November–December), Middle (January–February), Late (March–April); Soil type: Agricultural and forest soils; Soil preparation: Plow, manual holing, mechanical holing, subsoiling
*$P < 0.05$, **$P < 0.01$ (Navarro-Cerrillo et al. 2006a)

reduces the survival potential. The most common stresses during this process are extreme temperatures, desiccation, mechanical injuries, and storage molds (Landis et al. 2010) which result in drought and nutritional stress, lower root growth capability, and damage to leaves and lateral roots (Stjernberg 1997; McKay 1997). Therefore, during the transfer from the nursery to the site of transplanting, the quality of the seedlings must be optimized by minimizing both the stress during each stage of the process and the cumulative effect of the various stresses (Landis et al. 2010).

Site preparation treatments

Site preparation (SP) methods are typically oriented to modifying the environment of the site where the seedlings are planted—to improve the microsites for the seedlings by increasing or decreasing the soil water and nutrient contents at the time of planting, as well as by modifying the temperature, soil texture-compaction (e.g. surface layer structure, bulk density, aeration), and soil depth and controlling competing vegetation, among other parameters (South et al. 2001; de Chantal et al. 2004; English et al. 2005). The effects of SP on forest restoration outcomes are well documented and summarized (see Löf et al. 2012), and intensive treatments can improve survival and growth on sites where there is high environmental stress (Querejeta et al. 2001; Palacios et al. 2009).

In general, mechanical soil preparation resulted in higher survival, which agrees with many studies performed under different environmental conditions (Querejeta et al. 2001; Bocio et al. 2004; Barberá et al. 2005; Saquete et al. 2006; Palacios et al. 2009) (Fig. 14.8). One of the major ecological factors acting on afforestation performance under the Mediterranean climate is water availability (Bocio et al. 2004) and mechanical SP effectively increases water penetration into the soil (Querejeta et al. 2001), promoting the development of a deep rooting system which is essential for survival and growth under extreme conditions (Querejeta et al. 2001; Padilla and Pugnaire 2007; Löf et al. 2012). Seedling survival and growth increased in response to more intensive SP on sites with high competition or harsh conditions, although seedlings planted on sites with little environmental stress may show a

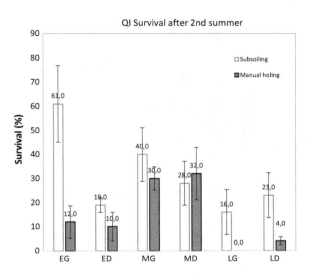

Fig. 14.8 Survival of seedlings of *Quercus ilex* two year after plantations according to planting date (*E* Early *M* Mid-season, *L* Late), site preparation (white = subsoling; grey = holing) and high-fertilized (G) or low-fertilized (D) seedlings (Palacios et al. 2009). Vertical bars (mean ± SE)

positive response to low-intensity SP treatments. More intensive soil preparation permits greater root development and root system fibrosity so that seedlings can explore larger volumes of soil—thus increasing their capability for uptake and transport of water, reducing seedling water stress, and increasing seedling survival (Grossnickle 2005). In particular, in Mediterranean areas, during the summer period, the deep rooting of seedlings becomes increasingly important for their water uptake due to the exhaustion of the topsoil moisture (Palacios et al. 2009).

Planting date and planting practices

An inappropriate planting date has been pointed out as one of the variables with the greatest influence on plant survival. The weather after planting affects soil temperature and moisture—which is mainly determined by the occurrence of rain both before and just after planting—and, consequently, influences seedling establishment (de Chantal et al. 2003). Therefore, an incorrect planting date selection may greatly jeopardize the survival of a plantation. Its effect, combined with SP, has been studied under Mediterranean conditions (Royo et al. 2000; Radoglou et al. 2003; Palacios et al. 2009; Navarro-Cerrillo et al. 2014).

Inappropriate planting dates and planting practices can result in lower survival on sites with limited soil water and greater environmental stress (Palacios et al. 2009; Navarro-Cerrillo et al. 2014) (Fig. 14.8). Under Mediterranean dry conditions, early (November) and middle (January) planting dates, when compared to later (March) dates, result in early shoot growth and increased foliar mass, which can increase the possibility that seedlings overcome the effects of transplanting (Grossnickle 2005). Seedlings planted on the early and mid-season dates had a favourable period of time—5 and 3 months, respectively—with more frequent precipitation for their vegetative growth, mainly for roots, thus reducing water stress and promoting higher rates of photosynthesis (Fig. 14.9). However, these

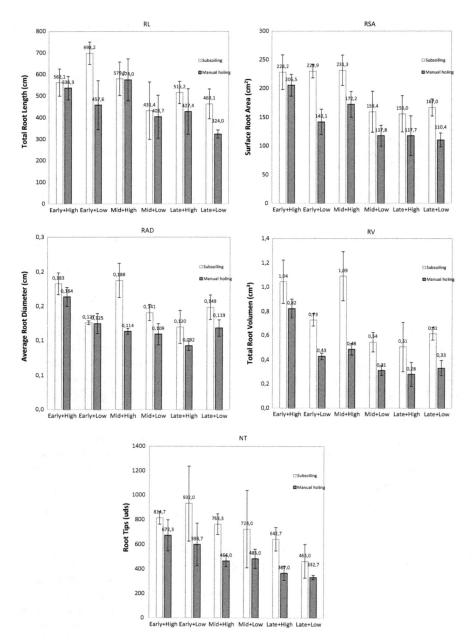

Fig. 14.9 *Quercus ilex* L. root seedlings morphology traits (total root length-RL, cm; surface root area-RSA, cm^2; average root diameter-RAD, cm; total root volume-RV, cm^3, and root tips-NT, uds; ± SE) according to planting date (*E* Early *M* Mid-season, *L* Late), site preparation (white = subsoling; grey = holing) and high-fertilized (G) or low-fertilized (D) seedlings (Palacios-Rodríguez 2015)

results differ from previous studies in Mediterranean environments, which found no statistical differences between planting dates (Royo et al. 2000; Radoglou et al. 2003) or established an optimal an optimal planting period that was too long (November to March) (Jenkinson et al. 1993). Seifert et al. (2006) also reported no significant differences among planting dates for *Q. rubra* and *Q. alba* seedlings across two planting years.

Vegetation management

Numerous reviews have discussed the merits of vegetation control (VC) in relation to seedling performance and thus regeneration success (Nilsson and Allen 2003; Fleming et al. 2006; Navarro-Cerrillo et al. 2005). The reduction of competition is directly related to seedling root growth and functional physiological response and thus provides a better seedling performance potential. Therefore, VC in newly planted sites has long been recognized as important to ensure successful survival and establishment (Navarro-Cerrillo et al. 2005; Jiménez et al. 2007; Ceacero et al. 2012; Pinto et al. 2012). Competition with other vegetation defines the potential drought avoidance of seedlings because seedling water status is directly tied to the vegetation complex. Lack of VC can result in water stress for seedlings, reducing the chance of survival (Jose et al. 2003), while survival increases as VC increases (Navarro-Cerrillo et al. 2005). Ceacero et al. (2012) found that *Quercues ilex* seedlings provided with intensive VC had, on average, a higher level of survival and a better physiological status 21 months after plantation, in particular cultivation + tree shelters (52.5%), herbicide (30%) and herbicide + tree shelters (27.5%). Even though the data showing the importance of VC for seedling survival are compelling, other authors warn of potential problems related to the use of intensive/recurrent VC without also considering other effects on the root system (i.e. lateral roots, root damage, root shape, etc.), which can limit seedling survival in the long-term (Grossnickle 2005). Finally, other reports show that the importance of the "facilitation" effect—with regard to the enhancement of seedling survival and growth, based on total shoot and root weights—may be limited when forecasting survival under harsh field conditions (del Campo et al. 2007).

Outplanting results obtained during practical reforestation work within the forest restoration community have been very relevant to the improvement of reforestation success, through increased seedling survival and growth. The research and practical experience obtained during the past half century has confirmed that the traditional perception of the impact of cultivation procedures on reforestation success was correct. Adequate cultivation practices during the plantation process do not guarantee high survival rates, because foresters work in an "unpredictable" environment, but planting seedlings in combination with desirable cultivation practices increases their chances of survival in dry continental Mediterranean sites.

Tree shelters

Tree shelters have been widely used in forest restoration programs in the world, especially with broadleaved species, since the first studies carried out by Tuley in Great Britain in 1979 (Tuley 1985). The initial objectives of the use of tree shelters were to protect the seedlings from herbivores and to facilitate the application of herbicides for weed control. Oaks have been one of the most used species with tree shelters both in wet temperate and Mediterranean environments (Tuley 1985; Mayhead and Boothman 1997; Welander and Ottosson 1998; Dubois et al. 2000; Bellot et al. 2002; Quilhó et al. 2003; Sharew and Hairston-Strang 2005; Oliet and Jacobs 2007; Dey et al. 2008; Pemán et al. 2010; Puértolas et al. 2010; Ceacero et al. 2014; Vázquez de Castro et al. 2014; Oliet et al. 2015; Mariotti et al. 2015). Along with a physical barrier to predation, effects of tree-shelters on the environmental conditions around the tree were recognized during early studies (Potter 1991). The microclimate inside tree shelters is characterized by a reduction in radiation and wind, an increase in temperature during daytime and marked daily changes in the air humidity, vapour pressure deficit and CO_2 concentration (Dupraz and Bergez 1999; Bergez and Dupraz 2000; Devine and Harrington 2008; Oliet and Jacobs 2007; Vázquez de Castro et al. 2014). The effect of these changes on seedling response is highly dependent upon the interactions between some specific functional traits and environmental conditions of the planting site.

Under mesic conditions in temperate forests, highly assimilating seedlings exhaust air CO_2 concentration within the shelter very early in the morning due to the reduced ventilation rate (Dupraz and Bergez 1999). At midday in summer, high temperatures reduced assimilation to negative values (Mayhead and Jones 1991). Both effects explain why biomass and diameter of sheltered oaks from mesic sites is usually reduced (Devine and Harrington 2008; Mariotti et al. 2015), leading to suggest an increment in ventilation rate via opening holes (Bergez and Dupraz 2009). The opposite effect has been observed. These environmental limitations to growth, however, do not seem to affect survival of sheltered oaks in mesic sites (Dubois et al. 2000; Sharew and Hairston-Strang 2005; Devine and Harrington 2008; Mariotti et al. 2015).

The effect of tree-shelters on survival of planted Mediterranean oaks is complex and deserves a deeper analysis. In a meta-analysis testing different eco-technologies in xeric conditions, Piñeiro et al. (2013) reported an overall positive effect of shelters on survival of different species. These results suggest a protective effect of shelters against abiotic stress. For instance, when planted in a contrasted gradient of drought, *Quercus ilex* survival is enhanced in semiarid areas, while the effect in more humid sites of the Mediterranean is weaker (Fig. 14.10).

These results suggest a moderating effect of shelters on extreme environmental conditions of dry summers. Other experimental plantations of Mediterranean oaks show a consistent improvement of survival (Bellot et al. 2002; Navarro-Cerrillo et al. 2005; Chaar et al. 2008; Padilla et al. 2011; Ceacero et al. 2014) (Fig. 14.11a). High irradiation combined with drought is the main cause of seedlings mortality in Mediterranean oaks (Gómez-Aparicio et al. 2008). Thus, a positive effect against

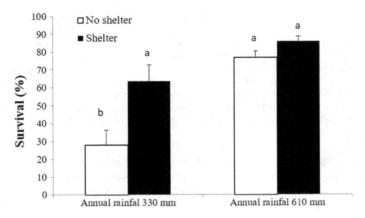

Fig. 14.10 Second year survival (mean ± SE, n = 100) of simultaneously planted holm oak (*Quercus ilex* subsp. *ballota*) in two locations of Andalusia with contrasted rainfall regimes. Within a location, columns with different letters significantly differ (α = 0.05). Modified from Oliet et al. (2003)

excessive light stress of tube shelters due to light reduction is a primary hypothesis (Puértolas et al. 2010; Pemán et al. 2010) to explain improvements of oaks survival, which are late successional-shade tolerant species (Ke and Werger 1999). Protective shade of shelters simultaneously occurs with higher temperatures (up to a 6 °C mean daytime temperature increase in summer in an oak trial with ventilated shelters Vazquez de Castro et al. 2014), which can also be a source of stress (Esteso-Martínez et al. 2006). For instance, higher temperatures combined with intense summer drought explain reductions in assimilation rates (Ceacero et al. 2014) of protected holm oak in summer. Therefore, along with light stress protection, other factors associated to microclimatic modifications of shelters are hypothesised as drivers to survival improvements. For instance, the restricted air movement in the shelters that severely reduced boundary layer conductance (Kjelgren and Rupp 1997) may help to reduce water loss of seedlings and improve hydric status during the dry season (Bergez and Dupraz 1997). This effect can also be overlapped with the reduction in stomatal conductance within tree shelters by light reduction that could also improve water status. The effect of tree shelters on plant physiology of Mediterranean oaks can change along the course of the year. For instance, higher temperatures in spring could trigger vegetative activity and transpiration within shelters, with reductions of water potential. During summer, a positive effect of tree shelters on water status is observed for protected oak seedlings, but only for those grown from seed (Fig. 14.11b, c).

In addition, when irrigating in summer, photochemical efficiency and water status of sheltered Mediterranean holm oaks did not differed from those non-protected (Vazquez de Castro et al. 2014), evidencing the importance of summer drought and its interaction with treeshelter effect in the Mediterranean. In conclusion, we believe that tree shelters exert a complex protective role against the

Fig. 14.11 Effect of the protection method (unprotected, ventilated tree shelter and unventilated ▶ tree shelter) and establishment method (direct acorn seeding vs. one-year-old seedlings) on the survival (**a**) predawn water potential measured at the end of the spring (**b**) and in mid-summer (**c**) and of the first growing season in the field of *Quercus faginea*. Survival was measured three years after planting. Data are means ± one SE. n = 6 and 75 for water potential and mortality data, respectively. The experiment was done in an abandoned wheat cropland in Santorcaz, Madrid, Spain. Both tree shelters were Tubex-Press 0.65 m but the ventilated shelters had two 3 cm in diameter holes cut 14 cm from the lowest end of the shelter. Planted seedlings were cultivated in Forest Pot 300 containers. *Source* Villar-Salvador, unpublished data

multi-stress (light, drought and temperature) effect caused by summer drought in Mediterranean conditions, with overall positive impact on survival.

The effect of shelters on growth of Mediterranean oaks is also influenced by daily or seasonal oscillations of microclimatic conditions. Few studies have provided estimates of gas exchange variables inside tree shelters due to the difficulty of measuring directly inside the shelters (Kjelgren and Rupp 1997; Oliet and Jacobs 2007; Ceacero et al. 2014). Unlike wet temperate oaks, assimilation rates at the plant level are lower, and CO_2 reductions in the shelter are not the main limiting factor (Oliet and Jacobs 2007). Instead, simulations conducted with Y-plant in *Quercus faginea* suggests a positive effect of shelter on assimilation rates in spring but the converse effect in summer, due to a modulating temperature effect within the shelter (Pemán et al. 2010). The way biomass growth of Mediterranean oaks will be finally stimulated or depleted by tree shelters will change with temperature regime and light transmission of the shelter.

The effect of tree-shelters on height growth of oaks from all biomes is mostly positive across published studies, (Kittredge et al. 1992; Burger et al. 1996; Navarro-Cerrillo et al. 2005; Taylor et al. 2006; Chaar et al. 2008; Ceacero et al. 2014; Mariotti et al. 2015), although some of them show null effects (Vazquez de Castro et al. 2014). Light transmission and ratio of red:far-red (r:f-r) of radiation entering the shelter through the wall affect the plant photomorphogenesis: lower r:f-r ratios stimulates length growth of new internodes within the shelter, although this response is associated to shade tolerance of oak species (Sharew and Hairston-Strang 2005; Mariotti et al. 2015). For instance, height growth of holm oak was promoted under shelters with moderate light transmission (55%) but not within highly transmissive shelters (70%, Oliet and Jacobs 2007). On the contrary, too dark tree-shelters (10% light transmission) deplete height growth due to resources limitations (Vázquez de Castro et al. 2014). The use of shelters with an appropriate length that fits herbivore size and light properties that stimulates height growth is considered a good tool to reduce browse damage of young oaks (Gillespie et al. 1996; Taylor et al. 2006; Dey et al. 2008; Chaar et al. 2008).

On the other side, oaks can also be established by direct seeding. Setting acorns within tree shelters below the soil surface has shown good results for both temperate (Löf et al. 2004; Dey et al. 2008; Valkonen 2008) and Mediterranean oaks (Cortina et al. 2009) (Fig. 14.11c). Apart from protecting seeds from predation, improvement of microenvironment at both soil and air levels around sheltered

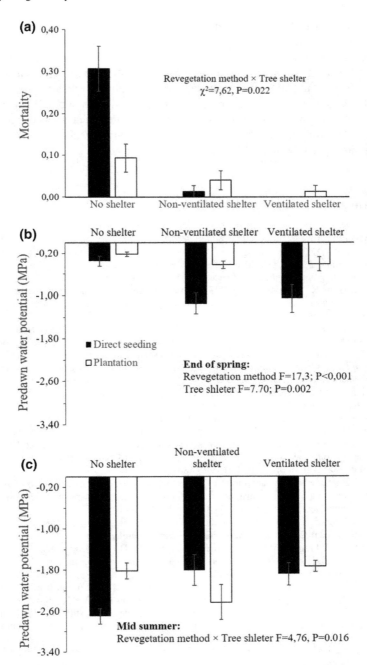

acorns in spring can explain the superior emergence ratios of recalcitrant acorns of oaks (Oliet et al. 2015).

Overall improvements in survival of planted Mediterranean oaks and protection and height growth stimulus of tree shelters of oaks from Mediterranean and wet biomes lead to conclude that tree shelters are good allies in restoration and plantation of oak forests. A better understanding of ecophysiological processes involved during establishment will help to refine shelters features such as wall light transmission to fit requirements of each species and/or environmental conditions of the planting site, specially under Mediterranean conditions.

References

Abrams MD (1990) Adaptations and responses to drought in *Quercus* species of North America. Tree Physiol 7:227–238

Aizen MA, Woodcock H (1996) Effects of acorn size on seedling survival and growth in *Quercus rubra* following simulated spring freeze. Can J Bot 74:308–314. doi:10.1139/b96-037

Albouchi A, Ghrir R, El Aouni MH (1997) Endurcissement à la sécheresse et accumulation de glucides solubles et d'acides aminés libres dans les phyllodes d'*Acacia cyanophylla* Lindl. Ann Sci For 54:155–168

Anagiotos G, Tsakaldimi M, Ganatsas P (2012) Variation in acorn traits among natural populations of *Quercus alnifolia*, an endangered species in Cyprus. Dendrobiology 68:3–10

Andivia E, Fernández M, Vázquez-Piqué J (2011) Autumn fertilization of *Quercus ilex* subsp. *ballota* (Desf.) Samp. nursery seedlings: effects on morpho-physiology and field performance. Ann For Sci 68:543–553. doi:10.1007/s13595-011-0048-4

Andivia E, Fernández M, Vázquez-Piqué J, Alejano R (2012a) Two provenances of *Quercus ilex* subsp. *ballota* (Desf) Samp. nursery seedlings have different response to frost tolerance and autumn fertilization. Eur J For Res 131:1091–1101

Andivia E, Márquez-García B, Vázquez-Piqué J, Córdoba F, Fernández M (2012b) Autumn fertilization with nitrogen improves nutritional status, cold hardiness and the oxidative stress response of holm oak (*Quercus ilex* subsp. *ballota* [Desf.] Samp) nursery seedlings. Trees Struct Funct 26:311–320. doi:10.1007/s00468-011-0593-3

Andivia E, Fernández M, Vázquez-Piqué J (2014) Assessing the effect of late-season fertilization on holm oak plant quality: insights from morpho-nutritional characterizations and water relations parameters. New For 45:149–163. doi:10.1007/s11056-013-9397-1

Bannister P, Neuner G (2001) Frost resistance and the distribution of conifers. In: Bigras FJ, Colombo SJ (eds) Conifer cold hardiness. Springer, Netherlands, pp 3–21

Bañón S, Ochoa J, Franco JA, Sánchez-Blanco MJ, Alarcón JJ (2003) Influence of water deficit and low air humidity in the nursery on survival of *Rhamnus alaternus* seedlings following planting. J Hort Sci Biotech 78:518–522

Bañón S, Ochoa J, Franco JA, Alarcón JJ, Sánchez-Blanco M (2005) Hardening of oleander seedlings by deficit irrigation and low air humidity. Environ Exp Bot 56:36–43

Barberá GG, Martínez-Fernández F, Alvarez-Rogel J, Albaladejo J, Castillo V (2005) Short and intermediate-term effects of site and plant preparation techniques on reforestation of a Mediterranean semiarid ecosystem with Pinus halepensis Mill. New For 29:177–198

Bellot J, Urbina JM, Bonet A, Sánchez JR (2002) The effects of tree shelters on the growth of *Quercus coccifera* L. seedlings in a semiarid environment. Forestry 75:89–106

Bergez J, Dupraz C (1997) Transpiration rate of *Prunus avium* L. seedlings inside an unventilated treeshelter. For Ecol Manage 97:255–264

Bergez J, Dupraz C (2000) Effect of ventilation on growth of *Prunus avium* seedlings grown in tree shelters. Agric For Meteorol 104:199–214

Bergez JE, Dupraz C (2009) Radiation and thermal microclimate in tree shelter. Agric For Meteorol 149:179–186. doi:10.1016/j.agrformet.2008.08.003

Biran I, Eliassaf A (1980) The effect of container shape on the development of roots and canopy of woody plants. Sci Hortic 12(2):183–193

Birge ZKD, Salifu KF, Jacobs DF (2006) Modified exponential nitrogen loading to promote morphological quality and nutrient storage of bareroot-cultured *Quercus rubra* and *Quercus alba* seedlings. Scand J For Res 21:306–316. doi:10.1080/02827580600761611

Blake TJ, Bevilacqua E, Zwiazek JJ (1991) Effects of repeated stress on turgor pressure and cell elasticity changes in black spruce seedlings. Can J For Res 21:1329–1333

Bocio I, Navarro F, Ripoll M, Jiménez M, Simón E (2004) holm oak (*Quercus rotundifolia* Lam.) and Aleppo pine (*Pinus halepensis* Mill.) response to different soil preparation techniques applied to forestation in abandoned farmland. Ann For Sci 61:171–178

Bonal R, Muñoz A (2007) Multi-trophic effects of ungulate intraguild predation on acorn weevil. Oecologia 152(3):533–540

Bonfil C (1998) The effects of seed size, cotyledon reserves, and herbivory on seedling survival and growth in *Quercus rugosa* and *Q. laurina* (Fagaceae). Am J Bot 85:79–87

Bonfil C, Cortés P, Espelta JM, Retana J (2004) The role of disturbance in the co-existence of the evergreen *Quercus ilex* and the deciduous *Quercus cerrioides*. J Veg Sci 15:423–430

Bonner FT (1981) Measurement and management of tree seed moisture. Research paper SO-177. USDA, Forest Service, Southern Forest Experiment Station, New Orleans

Bonner FT (1996) Responses to drying of recalcitrant seeds of *Quercus nigra* L. Ann Bot 78:181–187. doi:10.1006/anbo.1996.0111

Bonner FT (2008a) Storage of seeds. Woody plant seed Man. USDA Forest Service, pp 86–95

Bonner FT (2008b) *Quercus*. In: Bonner FT (ed) Woody plant seed manual, USDA Forest Service, Agriculture handbook 727, pp 928–996

Bonner FT, Vozzo JA (1987) Seed biology and technology of *Quercus*. General Te, New Orleans

Branco M, Branco C, Merouani H, Almeida MH (2002) Germination success, survival and seedling vigour of *Quercus suber* acorns in relation to insect damage. For Ecol Manage 166:159–164

Broncano M, Riba M, Retana J (1998) Seed germination and seedling performance of two Mediterranean tree species, holm oak (*Quercus ilex* L.) and Aleppo pine (*Pinus halepensis* Mill.): a multifactor experimental approach. Plant Ecol 138:17–26

Buckley GP, Mills J (2015) Coppice silviculture: from the Mesolithic to 21st century. In: Kirby KJ, Watkins C (eds) Europe's changing woods and forests: from wildwood to managed landscapes. CABI, Wallingford, pp 77–92

Burdett A (1990) Physiological processes in plantation establishment and the development of specifications for forest planting stock. Can J For Res 20:415–427

Burdett AN, Simpson DG, Thompson CF (1983) Root development and plantation establishment success. Plant Soil 71:103–110

Burdett AN, Herring LJ, Thompson CF (1984) Early growth of planted spruce. Can J For Res 14 (5):644–651. doi:10.1139/x84-116

Burgarella C, Navascues M, Soto A, Lora A, Fici S (2007) Narrow genetic base in forest restoration with holm oak (*Quercus ilex* L.) in Sicily. Ann For Sci 64:757–763

Burger D, Forister GW, Kiehl PA (1996) Height, caliper growth, and biomass response of ten shade tree species to tree shelters. J Arboric 22:161–166

Cabral R, O'Reilly C (2008) Physiological and field growth responses of oak seedlings to warm storage. New For 36:159–170. doi:10.1007/s11056-008-9090-y

Calamassi R, Rocca GD, Falusi M, Paoletti E Strati S (2001) Resistance to water stress in seedlings of eight European provenances of *Pinus halepensis* Mill. Ann For Sci 58: 663–672

Camarero JJ, Albuixech J, López-Lozano R, Casterad MA, Montserrat-Marti G (2010) An increase in canopy cover leads to masting in *Quercus ilex*. Trees 24:909–918

Canadell J, López-Soria L (1998) Lignotuber reserves support regrowth following clipping of two Mediterranean shrubs. Funct Ecol 12:31–38

Canadell J, Zedler PH (1995) Underground structures of woody plants in Mediterranean ecosystems. In: Kalin Arroyo MT, Zedler PH, Fox MD (eds) Ecology and biogeography of Mediterranean ecosystems in Chile, California, and Australia, Springer, New-York, pp 177–210

Canadell J, Djema A, López B, Lloret F, Sabaté S, Siscart D, Gracia CA (1999) Structure and dynamics of the root systems. In: Rodá F, Retana J, Gracia C, Bellot J (eds) Ecology of Mediterranean evergreen oak forests, pp 47–59

Canham CD, Kobe RK, Latty EF, Chazdon RL (1999) Interspecific and intraspecific variation in tree seedling survival effects of allocation to roots versus carbohydrate reserves. Oecologia 121:1–11. doi:10.1007/s004420050900

Cannon W (1914) Specialization in vegetation and in environment in California. Plant World 17:223–237

Castro J, Leverkus AB, Fuster F (2015) A new device to foster oak forest restoration via seed sowing. New For 46:919–929. doi:10.1007/s11056-015-9478-4

Catry FX, Rego F, Moreira F, Fernandes PM, Pausas JG (2010) Post-fire tree mortality in mixed forests of central Portugal. For Ecol Manage 260:1184–1192

Ceacero CJ, Díaz-Hernández JL, del Campo AD, Navarro-Cerrillo RM (2012) Interactions between soil gravel content and neighboring vegetation control management in oak seedling establishment success in Mediterranean environments. For Ecol Manage 271:10–18

Ceacero CJ, Navarro-Cerrillo RM, Díaz-Hernández JL, del Campo AD (2014) Is tree shelter protection an effective complement to weed competition management in improving the morpho-physiological response of holm oak planted seedlings? iForest Biogeosciences For 7:289–299. doi:10.3832/ifor1126-007

Chaar H, Mechergui T, Khouaja A, Abid H (2008) Effects of tree shelters and polyethylene mulch sheets on survival and growth of cork oak (*Quercus suber* L.) seedlings planted in northwestern Tunisia. For Ecol Manage 256:722–731

Chiatante D, Di A, Sciandra S, Scippa GS, Mazzoleni S (2006) Effect of drought and fire on root development in *Quercus pubescens* Willd. and *Fraxinus ornus* L. seedlings. Environ Exp Bot 56:190–197. doi:10.1016/j.envexpbot.2005.01.014

Chirino E, Vilagrosa A (2006) Work package 3, task 3.4. Improvement of nursery techniques. In: conservation and restoration of European cork oak woodlands: a unique ecosystem in the balance (CREOAK; LK5-CT-2002–01594). Final project report UE CREAOK project

Chirino E, Vilagrosa A, Rubio E (2003) Efectos de la reducción del riego y la fertilización en las características morfológicas de *Quercus suber*. Reunión del Grupo de Repoblaciones Forestales de la S.E.C.F., Murcia (Digital format)

Chirino E, Vilagrosa A, Rubio E (2004) Efectos de la reducción del riego y la fertilización en las características morfológicas de *Quercus suber*. Cuad Soc Esp Cien For 17:51–56

Chirino E, Vilagrosa A, Fernández R, Vallejo VR (2005) Uso de contenedor profundo en el cultivo de quercineas. Efectos sobre el crecimiento y distribución de biomasa. In: SECF (eds) Actas del IV Congreso Forestal Nacional, Zaragoza

Chirino E, Vilagrosa A, Hernández E, Vallejo VR (2008) Effects of a deep container on morpho-functional characteristics and root colonization in *Quercus suber* L. seedlings for reforestation in Mediterranean climate. For Ecol Manage 256:779–785. doi:10.1016/j.foreco.2008.05.035

Chirino E, Vilagrosa A, Cortina J, Valdecantos A, Fuentes D, Trubat R, Peñuelas JL (2009) Ecological restoration in degraded drylands: the need to improve the seedling quality and site conditions in the field. For Manage 85–158 (Nova Publisher, New York)

Chirino E, Erades A, Vilagrosa A, Vallejo VR (2013) Dinámica, morfología y topología del sistema radical de seis especies leñosas mediterráneas. In: Martínez-Ruiz C, Lario FJ, Fernández-Santos B (eds) Avances en la restauración de sistemas forestales, Técnicas de implantación, SECF-AEET, pp 177–182

Cline MG (1991) Apical dominance. Bot Rev 57:318–358

Close DC, Beadle CL, Brown PH (2013) The physiological basis of containerised tree seedling "transplant shock": a review. Aust For 68:112–120. doi:10.1080/00049158.2005.10674954

Collet C, Ghuel JM (1997) Osmotic adjustment in sessile oak seedlings in response to drought. Ann Sci For 54:389–394

Connor KF, Sowa S (2002) Recalcitrant behavior of temperate forest tree seeds: storage, biochemistry, and physiology. General technical report SRS 48. U.S. Department of Agriculture, Forest Service, Southern Research Station, Asheville, NC. pp 47–50

Connor KF, Sowa S (2003) Effects of desiccation on the physiology and biochemistry of *Quercus alba* acorns. Tree Physiol 23:1147–1152

Connor KF, Bonner FT, Vozzo JA (1996) Effects of desiccation on temperate recalcitrant seeds: differential scanning calorimetry, gas chromatography, electron microscopy, and moisture studies on *Quercus nigra* and *Quercus alba*. Can J For Res 26:1813–1821

Cortina J, Valdecantos A, Seva JP, Vilagrosa A, Bellot J, Vallejo VR (1997) Relación tamaño-supervivencia en plantones de especies arbustivas y arbóreas mediterráneas producidas en vivero. In: Puertas F (ed) II Congreso Forestal Español, Gobierno de Navarra, Pamplona, pp 22–27

Cortina J, Bellot J, Vilagrosa A, Caturla RN, Maestre FT, Rubio E, Bonet A (2004) Restauración en semiárido. Avances en el estudio de la gestión del monte mediterráneo, Fundación CEAM, Valencia, pp 345–406

Cortina J, Navarro-Cerrillo RM, del Campo AD (2006) Evaluación del éxito de la reintroducción de especies leñosas en ambientes medierráneos. In: Cortina J, Peñuelas JL, Puértolas J, Vilagrosa A, Savé R (eds) Calidad de planta forestal para la restauración en ambientes Mediterráneos. Estado actual de conocimientos, Organismo Autónomo Parques Nacionales, Ministerio de Medio Ambiente, Madrid, pp 11–30

Cortina J, Pérez-Devesa M, Vilagrosa A, Abourouh M, Messaoudene M, Berrahmouni N, Neves Silva L, Almeida MH, Khaldi A (2009) Field techniques to improve cork oak establishment. In: Aronson J, Pereira JS, Pausas JG (eds) Cork oak woodlands on the edge. pp 141–149

Cortina J, Vilagrosa A, Trubat R (2013) The role of nutrients for improving seedling quality in drylands. New For 44(5): 719–732. doi:10.1007/s11056-013-9379-3

Cotillas M, Sabaté S, Gracia C, Espelta JM (2009) Growth response of mixed Mediterranean oak coppices to rainfall reduction: could selective thinning have any influence on it? For Ecol Manage 258:1677–1683

Cotillas M, Espelta JM, Sánchez-Costa E, Sabaté S (2016) Aboveground and belowground biomass allocation patterns in two Mediterranean oaks with contrasting leaf habit: an insight into carbon stock in young oak coppices. Eur J For Res 135:243–252

Cuesta B, Vega J, Villar-Salvador P, Rey-Benayas JM (2010a) Root growth dynamics of Aleppo pine (*Pinus halepensis* Mill.) seedlings in relation to shoot elongation, plant size and tissue nitrogen concentration. Trees 24:899–908. doi:10.1007/s00468-010-0459-0

Cuesta B, Villar-Salvador P, Puértolas J, Jacobs DF, Rey-Benayas JM (2010b) Why do large, nitrogen rich seedlings better resist stressful transplanting conditions? A physiological analysis in two functionally contrasting Mediterranean forest species. For Ecol Manage 260:71–78. doi:10.1016/j.foreco.2010.04.002

Davis AS (2006) Photoperiod manipulation during nursery culture: effects on *Quercus rubra* seedling development and responses to environmental stresses following transplanting. Ph.D. dissertation, Purdue University, 180 p

Davis AS, Jacobs DF (2005) Quantifying root system quality of nursery seedlings and relationship to outplanting performance. New For 30:295–311. doi:10.1007/s11056-005-7480-y

de Chantal M, Leinonen K. Ilvesniemi H, Westman C (2003) Combined effects of site preparation, soil properties, and sowing date on the establishment of *Pinus sylvestris* and *Picea abies* from seeds, Can J For Res 33: 931–945

de Chantal M, Leinonen K. Ilvesniemi H, Westman C (2004) Effects of site preparation on soil properties and on morphology of *Pinus sylvestris* and *Picea abies* seedlings sown at different dates. New For 27(2): 159–173

De Souza H, De Araújo R, De Gois L, Dantas Viana CV, De Almeida M, Jun R (2016) Substrates and containers for the development of *Brassica pekinensis* L. seedlings. Bragantia 75 (3):344–350

del Campo A, Navarro-Cerrillo R, Hermoso J, Ibáñez A (2007) Relationships between site and stock quality in *Pinus halepensis* Mill. reforestation on semiarid landscapes in eastern Spain. Ann For Sci 64:719–731

del Campo AD, Navarro-Cerrillo RM, Ceacero CJ (2010) Seedling quality and field performance of commercial stock lots of containerized holm oak (*Quercus ilex*) in Mediterranean Spain: an approach for establishing a quality standard. New For 39:17–37

Devine W, Harrington CA (2008) Influence of four tree shelter types on microclimate and seedling performance of Oregon white oak and western redcedar. Research paper PNW-RP-576. USDA Forest Service, Pacific North-West Research Station, 54 pp

Dey DC, Parker WC (1997) Morphological indicators of stock quality and field performance of red oak (*Quercus rubra* L.) seedlings underplanted in a central Ontario shelterwood. New For 14:145–156

Dey DC, Jacobs DF, Mcnabb K, Miller G, Baldwin V, Foster G (2008) Artificial regeneration of major oak (*Quercus*) species in the Eastern United States—a review of the literature. For Sci 54 (1):77–106

di Iorio A, Lasserre B, Scippa G, Chiatante D (2005) Root system architecture of *Quercus pubescens* trees growing on different sloping conditions. Ann Bot 95:351–361. doi:10.1093/aob/mci033

Díaz-Delgado R, Lloret F, Pons X (2003) Influence of fire severity on plant regeneration by means of remote sensing imagery. Int J Remote Sens 24:1751–1763

Dichio B, Xiloyannis C, Angelopoulos K, Nuzzo V, Bufo SA, Celano G (2003) Drought-induced variations of water relations parameters in *Olea europaea*. Plant Soil 257:381–389

Domínguez-Lerena S (1997) La importancia del envase en la producción de plantas forestales. Quercus 134:34–37

Domínguez-Lerena S (2000) Influencia de distintos tipos de contenedores en el desarrollo en campo de *Pinus halepensis* y *Quercus ilex*. Reunión de Coordinación I + D, Fundación CEAM

Dominguez-Lerena S, Herrero N, Carrasco I, Ocaña L, Peñuelas JL, Mexal JG (2006) Container characteristics influence *Pinus pinea* seedling development in the nursery and field. For Ecol Manage 221:63–71. doi:10.1016/j.foreco.2005.08.031

Doody CN, O'Reilly C (2008) Drying and soaking pretreatments affect germination in pedunculate oak. Ann For Sci 65:505–509

Dubois MR, Chappelka AH, Robbins E, Somers G, Baker K (2000) Tree shelters and weed control: Effects on protection, survival and growth of cherrybark oak seedlings planted on a cutover site. New For 20:105–118

Ducrey M, Boisserie M (1992) Recrû naturel dans des taillis de chêne vert (*Quercus ilex* L.) à la suite d'explotations partielles. Ann Sci For 49:91–109

Ducrey M, Turrel M (1992) Influence of cutting methods and dates on stump sprouting in holm oak (*Quercus ilex* L.) coppice. Ann Sci For 49:449–464

Dumroese RK, Page-Dumroese DS, Salifu KF, Jacobs DF (2005) Exponential fertilization of *Pinus monticola* seedlings: nutrient uptake efficiency, leaching fractions, and early outplanting performance. Can J For Res 35:2961–2967

Dunsworth GB (1997) Plant quality assessment: an industrial perspective. New For 13:439–448

Dupraz C, Bergez J (1999) Carbon dioxide limitation of the photosynthesis of *Prunus avium* L. seedlings inside an unventilated treeshelter. For Ecol Manage 119:89–97

Edwards C (1998) Testing plant quality. Forestry Commission, Edinburgh

Edwards DR, Dixon MA (1995) Mechanisms of drought response in *Thuja occidentalis* LI Water stress conditioning and osmotic adjustment. Tree Physiol 15:121–127

El Omari B, Aranda X, Verdaguer D, Pascual G, Fleck I (2003) Resource remobilization in *Quercus ilex* L. resprouts. Plant Soil 252:349–357

English NB, Weltzin J, Fravolini A, Thomas L, Williams D (2005) The influence of soil texture and vegetation on soil moisture under rainout shelters in a semi-desert grassland. J Arid Environ 63:324–343
Espelta JM, Sabaté S, Retana J (1999) Resprouting dynamics. In: Rodà F, Retana J, Gracia CA, Bellot J (eds) Ecology of Mediterranean evergreen oak forests. Springer, Berlin, pp 61–73
Espelta JM, Retana J, Habrouk A (2003) Resprouting patterns after fire and response to stool cleaning of two coexisting Mediterranean oaks with contrasting leaf habits on two different sites. For Ecol Manage 179:401–414
Espelta JM, Cortés P, Mangirón M, Retana J (2005) Differences in biomass partitioning, leaf nitrogen content and water use efficiency ($\delta 13C$) result in a similar performance of seedlings of two Mediterranean oaks with contrasting leaf habit. Ecoscience 12:447–454
Espelta JM, Retana J, Habrouk A (2006) Response to natural and simulated browsing of two Mediterranean oaks with contrasting leaf habit after a wildfire. Ann For Sci 63:441–447
Espelta JM, Arnan X, Verkaik I (2008) Evaluación ecológica de diferentes tratamientos silvícolas de mejora de la regeneración natural en zonas afectadas por incendios y sequías extremas. In: Modelos silvícolas en bosques privados Mediterraneos. Colección_Documentos de Trabajo, Serie_Territorio, 5. Diputación de Barcelona, Barcelona, pp 151–179
Espelta JM, Cortés P, Molowny-Horas R, Retana J (2009a) Acorn crop size and pre-dispersal predation determine inter-specific differences in the recruitment of co-occurring oaks. Oecologia 161:559–568
Espelta JM, Bonal R, Sánchez-Humanes B (2009b) Pre-dispersal acorn predation in mixed oak forests: interspecific differences are driven by the interplay among seed phenology, seed size and predator size. J Ecol 97:1416–1423
Espelta JM, Barbati A, Quevedo L, Tárrega R, Navascués P, Bonfil C, Peguero G, Fernández-Martínez M, Rodrigo A (2012) Post-fire management of Mediterranean broadleaved forests. In: Moreira, F, Arianoutsou M, Corona P, de las Heras J (eds) Post-fire management and restoration of Southern European forests. Springer, Netherlands, pp 171–194
Esteso-Martínez J, Gil-Pelegrín E (2004) Frost resistance of seeds in Mediterranean oaks and the role of litter in the thermal protection of acorns. Ann For Sci 61:481–486. doi:10.1051/forest
Esteso-Martínez J, Valladares F, Camarero J, Gil-Pelegrín E (2006) Crown architecture and leaf habit are associated with intrinsically different light-harvesting efficiencies in *Quercus* seedlings from contrasting environments. Ann For Sci 63:511–518
Esteso-Martínez J, Peguero-Pina JJ, Valladares F, Morales F, Gil-Pelegrín E (2010) Self-shading in cork oak seedlings: functional implications in heterogeneous light environments. Acta Oecol 36:423–430
Fansworth E (2000) The ecology and evolution of viviparous and recalcitrant seeds. Annu Rev Ecol Syst 31:107–138. doi:10.1146/annurev.ecolsys.34.011802.132402
Fernández M, Gil L, Pardos JA (1999) Response of *Pinus pinaster* Ait. provenances at early age to water supply. I. Water relation parameters. Ann For Sci 56:179–187
Fernández M, Marcos C, Tapias R, Ruiz F, López G (2007) Nursery fertilisation affects the frost-tolerance and plant quality of *Eucalyptus globulus* Labill. cuttings. Ann For Sci 64:865–873
Fernández-Martínez M, Garbulsky M, Peñuelas J, Peguero G, Espelta JM (2015) Temporal trends in the enhanced vegetation index and spring weather predict seed production in Mediterranean oaks. Plant Ecol 216:1061–1072
Fernández-Martínez M, Vicca S, Janssens IA, Espelta JM, Peñuelas J (2017) The role of nutrients, productivity and climate in determining tree fruit production in European forests. New Phytol 213:669–679
Finch-Savage WE, Blake PS, Clay HA (1996) Desiccation stress in recalcitrant *Quercus robur* L. seeds results in lipid peroxidation and increased synthesis of jasmonates and abscisic acid. J Exp Bot 47:661–667
Fleck I, Grau D, Sanjosé M, Vidal D (1996a) Influence of fire and tree-fell on physiological parameters in *Quercus ilex* resprouts. Ann Sci For 53:337–348

Fleck I, Grau D, Sanjosé M, Vidal D (1996b) Carbon isotope discrimination in *Quercus ilex* resprouts after fire and tree-fell. Oecologia 105:286–292

Fleck I, Hogan KP, Llorens L, Abadia A, Abadia X (1998) Photosynthesis and photoprotection in *Quercus ilex* resprouts after fire. Tree Physiol 18(8–9):607–614

Fleck I, Peñas-Rojas K, Aranda X (2010) Mesophyll conductance to CO_2 and leaf morphological characteristics under drought stress during *Quercus ilex* L. resprouting. Ann For Sci 67(3): 308. doi:10.1051/forest/2009114

Fleming R, Powers R, Foster N, Kranabetter J, Scott D, JrF Ponder, Morris D (2006) Effects of organic matter removal, soil compaction, and vegetation control on 5-year seedling performance: a regional comparison of long-term soil productivity sites. Can J For Res 36(3):529–550

Folk RS, Grossnickle SC (1997) Determining filed performance potential with the use of limiting environmental conditions. New For 13:121–138

Fonseca D (1999) Manipulacion de las características morfo-estructurales de plantones de especies forestales mediterráneas producidas en vivero. Master-Tesis, CIHEAM-IAMZ, Zaragoza (Spain)

Franco JA, Bañón S, Fernández JA, Leskovar DI (2001) Effect of nursery regimes and establishment irrigation on root development of *Lotus creticus* seedlings following transplanting. J Hort Sci Biotech 76:174–179

Franco JA, Cros V, Bañón S, González A, Abrisqueta JM (2002) Effects of nursery irrigation on post-planting root dynamics of *Lotus creticus* in semiarid field conditions. HortScience 37:525–528

Ganatsas P, Tsakaldimi M (2013) A comparative study of desiccation responses of seeds of three drought-resistant Mediterranean oaks. For Ecol Manage 305:189–194. doi:10.1016/j.foreco.2013.05.042

García Cebrián F, Esteso-Martínez J, Gil-Pelegrín E (2003) Influence of cotyledon removal on early seedling growth in *Quercus robur* L. Ann For Sci 60:69–73

García-Jiménez R, Palmero-Iniesta M, Espelta JM (2017) Contrasting effects of fire severity on the regeneration of *Pinus halepensis* Mill. and resprouter species in recently thinned thickets. Forests 8:55. doi:10.3390/f8030055

Garriou D, Girard S, Guehl J, Généré B (2000) Effect of desiccation during cold storage on planting stock quality and field performance in forest species. Ann For Sci 57:101–111

Gillespie AR, Rathfon R, Myers RK (1996) Rehabilitating a young northern red oak planting with tree shelters. NJAF 13(1):24–29

Gimeno TE, Pías B, Lemos-Filhos JP, Valladares F (2009) Plasticity and stress tolerance override local adaptation in the response of Mediterranean holm oak seedlings to drought and cold. Tree Physiol 29:87–98

Gómez JM (2003) Spatial patterns in long-distance dispersal of *Quercus ilex* acorns by jays in a heterogeneous landscape. Ecography 26:573–584

Gómez JM, Hódar JA (2008) Wild boars (*Sus scrofa*) affect the recruitment rate and spatial distribution of holm oak (*Quercus ilex*). For Ecol Manage 256:1384–1389

Gómez-Aparicio L, Pérez-Ramos IM, Mendoza I, Matias L, Quero JL, Castro J, Zamora R, Marañon T (2008) Oak seedling survival and growth along resource gradients in Mediterranean forests: implications for regeneration in current and future environmental scenarios. Oikos 117:1683–1699

Gómez-Aparicio L, Zavala MA, Bonet FJ, Zamora R (2009) Are pine plantations valid tools for restoring Mediterranean forests? An assessment along abiotic and biotic gradients. Ecol Appl 19:2124–2141

González RA, Jiménez MS, Morales D, Aschan G, Losch R (1999) Physiological responses of *Laurus azorica* and *Viburnum rigidum* to drought stress: osmotic adjustment and tissue elasticity. Phyton 39:251–263

González-Rodríguez V, Villar R (2012) Post-dispersal seed removal in four Mediterranean oaks: species and microhabitat selection differ depending on large herbivore activity. Ecol Res 27:587–594

González-Rodríguez V, Villar R, Navarro-Cerrillo RM (2011a) Maternal influences on seed mass effect and initial seedling growth in four *Quercus* species. Acta Oecol 37:1–9. doi:10.1016/j.actao.2010.10.006

González-Rodríguez V, Navarro-Cerrillo RM, Villar R (2011b) Artificial regeneration with *Quercus ilex* L. and *Quercus suber* L. by direct seeding and planting in southern Spain. Ann For Sci 68:637–646. doi:10.1007/s13595-011-0057-3

Goodman RC, Jacobs DF, Karrfalt RP (2005) Evaluating desiccation sensitivity of *Quercus rubra* acorns using X-ray image analysis. Can J For Res 35(12):2823–2831. doi:10.1139/X05-209

Gracia M, Retana J (2004) Effect of site quality and shading on sprouting patterns of holm oak coppices. For Ecol Manage 188:39–49

Gracia M, Retana J, Roig P (2002) Mid-term successional patterns after fire of mixed pine–oak forests in NE Spain. Acta Oecol 23:405–411

Gratani L (1995) Structural and ecophysiological plasticity of some evergreen species of the Mediterranean maquis in response to climate. Photosynthetica 31:335–343

Grelet GA, Alexander IJ, Millard P, Proe MF (2003) Does morphology or the size of the internal nitrogen store determine how *Vaccinium* spp. respond to spring nitrogen supply? Funct Ecol 17:690–699

Grossnickle SC (2000) Ecophysiology of northern spruce species: the performance of planted seedlings. NRC Research Press, Canada, 407 pp

Grossnickle SC (2005) Importance of root growth in overcoming planting stress. New For 30:273–294. doi:10.1007/s11056-004-8303-2

Grossnickle SC (2012) Why seedlings survive: influence of plant attributes. New For 43:711–738. doi:10.1007/s11056-012-9336-6

Grossnickle S, El-Kassaby Y (2016) Bareroot versus container stocktypes: a performance comparasion. New For 47(1):1–51. doi:10.1007/s11056-015-9476-6

Grossnickle SC, Heikurinen J (1989) Site preparation: water relations and growth of newly planted jack pine and white spruce. New For 3:99–123. doi:10.1007/BF00021576

Grossnickle S, South D (2014) Fall acclimation and the lift/store pathway: effect on reforestation. Open For Sci J 7:1–20

Guehl JM, Fort C, Ferhi A (1995) Differential response of leaf conductance, carbon isotope discrimination and water-use efficiency to nitrogen deficiency in maritime pine and pedunculate oak plants. New Phytol 131:149–157. doi:10.1111/j.1469-8137.1995.tb05716.x

Gusta LV, Wisniewski M (2013) Understanding plant cold hardiness: an opinion. Physiol Plant 147:4–14

Guthke J, Spethmann W (1993) Physiological and pathological aspects of long-term storage of acorns. Ann For Sci 50:384–387

Haase DL, Rose R (1993) Soil moisture stress induces transplant shock in stored and seedlings of varying root volumes transplant shock in stored and unstored 2 + 0 Douglas-fir seedlings of varying root volumes. For Sci 39(2):275–294

Hashizume H, Han H (1993) A study on forestation using large-size *Quercus acutissima* seedlings. Hardwood Res 7:1–22

Heredia N, Oliet JA, Villar-Salvador P, Benito LF, Peñuelas JL (2014) Fertilization regime interacts with fall temperature in the nursery to determine the frost and drought tolerance of the Mediterranean oak *Quercus ilex* subsp. *ballota*. For Ecol Manage 331:50–59. doi:10.1016/j.foreco.2014.07.022

Hernández EI, Vilagrosa A, Luis VC, Llorca M, Chirino E, Vallejo VR (2009) Root hydraulic conductance, gas exchange and leaf water potential in seedlings of *Pistacia lentiscus* L. and *Quercus suber* L. grown under different fertilization and light regimes. Environ Exp Bot 67:269–276. doi:10.1016/j.envexpbot.2009.07.004

Hmielowski TL, Robertson KM, Platt WJ (2014) Influence of season and method of topkill on resprouting characteristics and biomass of *Quercus nigra* saplings from a southeastern US pine-grassland ecosystem. Plant Ecol 215:1221–1231

Hollander JL, Vander Wall SB (2004) Effectiveness of six species of rodents as dispersers of singleleaf pinon pine (*Pinus monophylla*). Oecologia 138:57–65

Islam MA, Apostol KG, Jacobs DF, Dumroese RK (2009) Fall fertilization of *Pinus resinosa* seedlings: nutrient uptake, cold hardiness, and morphological development. Ann For Sci 66:704–712

Jacobs DF (2003) Nursery production of hardwood seedlings. Purdue University Cooperative Extension Service, FNR-212. 8 p

Jacobs DF and Seifert JR (2004) Re-evaluating the significance of the first-order lateral root grading criterion for hardwood seedlings. In: Proceedings of the fourteenth central hardwood forest conference, USDA Forest Service North Central Experiment Station, General technical report NE-316, pp 382–388

Jacobs DF, Salifu KF, Seifert JR (2005) Relative contribution of initial root and shoot morphology in predicting field performance of hardwood seedlings. New For 30:235–251

Jacobs DF, Davis AS, Wilson BC, Dumroese RK, Goodman RC, Salifu KF (2008) Short-day treatment alters Douglas-fir seedling dehardening and transplant root proliferation at varying rhizosphere temperatures. Can J For Res 38:1526–1535

Jacobs DF, Salifu KF, Davis AS (2009) Drought susceptibility and recovery of transplanted *Quercus rubra* seedlings in relation to root system morphology. Ann For Sci 66(5):1–12

Jelic G, Topic V, Butorac L, Jazbec A, Orsanic M, Durdevic Z (2015) The impact of the container size and soil preparation on afforestation success of one year old holm oak (*Quercus ilex*) seedlings in Croatian Mediterranean area. Period Biol 117(4):493–503. doi:10.18054/pb.2015.117.4.3532

Jenkinson J, Nelson J, Huddleston M (1993) Improving planting stock quality. The Humbolt experience, USDA. Forest service, General technical report

Jiménez M, Fernández-Ondoño E, Ripoll M, Navarro F, Gallego E, de Simón E, Lallena A (2007) Influence of different post-planting treatments on the development in holm oak afforestation. Trees Struct Funct 21:443–455

Joët T, Ourcival JM, Dussert S (2013) Ecological significance of seed desiccation sensitivity in *Quercus ilex*. Ann Bot 111:693–701. doi:10.1093/aob/mct025

Johnson PS, Shifley SR, Rogers R (2001) The ecology and silviculture of oaks. CABI Publishing, New York

Jose S, Merritt S, Ramsey C (2003) Growth, nutrition, photosynthesis and transpiration responses of longleaf pine seedlings to light, water and nitrogen. For Ecol Manage 180(1):335–344

Jurado E, Westoby MhDG (1992) Seedling growth in relation to seed size among species of arid Australia. J Ecol 80:407–416

Kabeya D, Sakai S (2005) The relative importance of carbohydrate and nitrogen for the resprouting ability of *Quercus crispula* seedlings. Ann Bot 96:479–488

Kays JS, Canham CD (1991) Effects of time and frequency of cutting on hardwood root reserves and sprout growth. For Sci 37:524–539

Ke G, Werger MA (1999) Different responses to shade of evergreen and deciduous oak seedlings and the effect of acorn size. Acta Oecol 20:579–586. doi:10.1016/S1146-609X(99)00103-4

Kelly D, Sork VL (2002) Mast seeding in perennial plants: why, how, where? Annu Rev Ecol Syst 33:427–447

Ketchum JS, Rose R (2000) Interaction of initial seedling size, fertilization and vegetation control. In: 21st forest vegetation management conference, Redding, CA, 18–20 Jan 2000, pp 63–69

Kittredge D, Kelty M, Ashton P (1992) The use of tree shelters with Northern red oak natural regeneration in southern New England. NJAF 9:141–145

Kjelgren R, Rupp L (1997) Establishment in tree shelters I. Shelters reduce growth, water use and hardiness, but not drought avoidance. Hort Sci 32:1281–1283

Kleiner KW, Abrams MD, Schultz JC (1992) The impact of water and nutrient deficiencies on the growth, gas exchange and water relations of red oak and chestnut oak. Tree Physiol 11:271–287

Koenig WD, Alejano R, Carbonero MD, Fernández-Rebollos P, Knops JMH, Marañon T, Padilla-Díaz MC, Pearse IS, Pérez-Ramos IM, Vázquez-Piqué J, Pesendorfer MB (2016) Is he relationship between mast-seeding and weather in oaks related to their life-history or phylogeny? Ecology 97(10):2603–2615. doi:10.1002/ecy.1490

Kozlowski T, Davies W (1975) Control of water balance in transplanted trees. Arboriculture 1:1–10

Landhäusser SM, Lieffers VJ (2002) Leaf area renewal, root retention and carbohydrate reserves in a clonal tree species following above-ground disturbance. J Ecol 90:658–665

Landis TD (1989) Seedling nutrition and irrigation. The container tree nursery manual, vol 4. USDA Forest Service, Washington, DC, pp 1–70

Landis TD (1990) Containers: types and functions. The container tree nursery manual, vol 2. USDA Forest Service, Washington, DC, pp 1–40

Landis T (1995) Nursery planning, development and management. The container tree nursery manual, vol 1. USDA Fores, USDA Forest Service, Washington, DC, 188 pp

Landis TD, Tinus RW, Barnett JP (1998) Seedling propagation. The container tree nursery manual, vol 6. USDA Forest Service, Washington, DC, 166 pp

Landis T, Dumroese R, Haase D (2010) Seedling processing, storage and outplanting. The container tree nursery manual, vol 7. USDA Forest Service, Washington, DC, 208 pp

Larcher W (2000) Temperature stress and survival ability of Mediterranean sclerophyllous plants. Plant Biosyst 134:279–295

Larcher W (2003) Physiological plant ecology, 4th edn. Springer, Berlin, Germany, p 513

Larcher W, De Moraes J, Bauer H (1981) Adaptive response of leaf water potential, CO_2-gas exchange and water use efficiency of *Olea europaea* during drying and rewatering. Components of productivity of mediterranean-climate regions. Basic and applied aspects. In: Margaris NS, Mooney HA (eds.) Dr.W. Junk Publishers, La Haya, pp 77–84

Lauridsen EB, Olesen K, Schøler E (1992) Packaging materials for tropical tree fruits and seeds. Danida Forest Seed Centre, Technical nNote no 41

Leishman MR, Wright IJ, Moles AT, Westoby M (2000) The evolutionary ecology of seed size. In: Fenner M (ed) Seeds: the ecology regeneration in plant communities. CAB International, UK, pp 31–57

Leiva MJ, Fernández-Alés R (2005) Holm-oak (*Quercus ilex* subsp. *ballota*) acorns infestation by insects in Mediterranean dehesas and shrublands: its effect on acorn germination and seedling emergence. For Ecol Manage 212:221–229. doi:10.1016/j.foreco.2005.03.036

Levitt J (1980) Responses of plants to environmental stresses, vol II. Academic Press, New York

Li G, Zhu Y, Liu Y, Wang J, Liu J, Dumroese RK (2014) Combined effects of pre-hardening and fall fertilization on nitrogen translocation and storage in *Quercus variabilis* seedlings. Eur J For Res 133:983–992. doi:10.1007/s10342-014-0816-4

Lloret F, Casanovas C, Peñuelas J (1999) Seedling survival of Mediterranean shrubland species in relation to root: shoot ratio, seed size and water and nitrogen use. Funct Ecol 13:210–216. doi:10.1046/j.1365-2435.1999.00309.x

Löf M, Thomsen A, Madsen P (2004) Sowing and transplanting of broadleaves (*Fagus sylvatica* L., *Quercus robur* L., *Prunus avium* L. and *Crataegus monogyna* Jacq.) for afforestation of farmland. For Ecol Manage 188(1–3):113–123

Löf M, Dey D, Navarro-Cerrillo R, Jacobs D (2012) Mechanical site preparation for forest restoration. New For 43(5–6):825–848

Lombardo JA, McCarthy BC (2009) Seed germination and seedling vigor of weevil-damaged acorns of red oak. Can J For Res 39:1600–1605. doi:10.1139/X09-079

López-Soria L, Castell C (1992) Comparative genet survival after fire in woody Mediterranean species. Oecologia 53:493–499

Lynch J (1995) Root architecture and plant productivity. Plant Physiol 109:7–13

Makita N, Hirano Y, Mzoguchi T, Kominami Y, Dannoura M, Ishii H, Finér L, Kanazawa Y (2011) Very fine roots respond to soil depth: biomass allocation, morphology, and physiology in a broad-leaved temperate forest. Ecol Res 26:95–104

Marañón T, Camarero JJ, Castro J, Díaz M, Espelta JM, Hampe A, Jordano P, Valladares F, Verdú M, Zamora R (2004) Heterogeneidad ambiental y nicho de regeneración. In: Valladares F (ed) Ecología del bosque mediterráneo en un mundo cambiante. Ministerio de Medio Ambiente, Madrid, pp 69–99

Margolis H, Brand D (1990) An ecophysiological basis for understanding plantation establishment. Can J For Res 20:375–390

Mariotti B, Maltoni A, Jacobs DF, Tani A (2015) Tree shelters affect shoot and root system growth and structure in *Quercus robur* during regeneration establishment. Eur J For Res 134(4): 641–652

Martínez Romero G, Planelles R, Zazo J, Bela D, Vivar A, López-Arias M (2001) Estudio de la influencia de la fertilizacion nitrogenada y la iluminacion sobre atributos morfológicos y fisiológicos de brinzales de *Q. suber* L. cultivado en vivero. Resultados del 1er año en campo. In: SECF (ed) III Congreso Forestal Español, Granada

Mattsson A (1997) Predicting field performance using seedling quality assessment. New For 13:227–252. doi:10.1023/A:1006590409595

Mayhead GJ, Boothman IR (1997) The effect of treeshelter height on the early growth of sessile oak (*Quercus petraea* (Matt.) Liebl.). Forestry 70:1991–1996

Mayhead GJ, Jones D (1991) Carbon dioxide concentrations within tree shelters. Quaterly J For 85:228–232

Mayr S, Hacke U, Schmid P, Schwienbacher F Gruber A (2006a) Frost drought in conifers at the alpine timberline: xylem dysfunction and adaptations. Ecology 87: 3175–3185

Mayr S, Wieser G, Bauer H (2006b) Xylem temperatures during winter in conifers at the alpine timberline. Agric For Met 137:81–88

McKay HM (1992) Electrolyte leakage from fine roots of conifer seedlings: a rapid index of plant vitality following cold storage. Can J For Res 22:1371–1377

Mckay HM (1997) A review of the effect of stresses between lifting and planting on nursery stock quality and performance. New For 13(1–3):363–393

McKay HM, Jinks RL, McEvoy C (1999) The effect of desiccation and rough-handling on the survival and early growth of ash, beech, birch and oak seedlings. Ann For Sci 56:391–402

Merouani H, Branco C, Almeida MH, Pereira JS (2001) Effects of acorn storage duration and parental tree on emergence and physiological status of Cork oak (*Quercus suber* L.) seedlings. Ann For Sci 58:543–554

Mexal JG, Landis TD (1990) Target seedling concepts: height and diameter. In: Rose R, Campbell SJ, Landis TD (eds) Target seedling symposium, UDSA Forest Service, Roseburg (Oregon), pp 17–35

Millard P, Grelet G-A (2010) Nitrogen storage and remobilization by trees: ecophysiological relevance in a changing world. Tree Physiol 30:1083–1095. doi:10.1093/treephys/tpq042

Mitrakos K (1980) A theory for Mediterranean plant life. Acta Oecologica-Oecologia Plant 1:245–252

Mollá S, Villar-Salvador P, García-Fayos P, Peñuelas Rubira JL (2006) Physiological and transplanting performance of *Quercus ilex* L. (holm oak) seedlings grown in nurseries with different winter conditions. For Ecol Manage 237:218–226. doi:10.1016/j.foreco.2006.09.047

Morin X, Améglio T, Ahas R, Kurz-Besson C, Lanta V, Lebourgeois F, Miglietta F, Chuine I (2007) Variation in cold hardiness and carbohydrate concentration from dormancy induction to bud burst among provenances of three European oak species. Tree Physiol 27:817–825

Muñoz A, Bonal R, Espelta JM (2012) Responses of a scatter-hoarding rodent to seed morphology: links between seed choices and seed variability. Anim Behav 84:1435–1442

Muñoz A, Bonal R, Espelta JM (2014) Acorn–weevil interactions in a mixed-oak forest: outcomes for larval growth and plant recruitment. For Ecol Manage 322:98–105

Nardini A, Ghirardelli L, Salleo S (1998a) Vulnerability to freeze stress of seedlings of *Quercus ilex* L.: an ecological interpretation. Ann For Sci 55:553–565

Nardini A, Lo Gullo M, Salleo D (1998b) Seasonal changes of root hydraulic conductance (KRL) in four forest trees: an ecological interpretation. Plant Ecol 139:81–90

Nardini A, Salleo S, Tyree M, Vertovec M (2000) Influence of the ectomycorrhizas formed by *Tuber melanosporum* Vitt. on hydraulic conductance and water relations of *Quercus ilex* L. seedlings. Ann For Sci 57:305–312

Navarro FB, Jiménez N, Ripoll M, Fernández-Ondoño E, Gallego E, de Simón E (2006) Direct sowing of holm oak acorns: effects of acorn size and soil treatment. Ann For Sci 63:961–967

Navarro-Cerrillo RM, Fragueiro B, Ceaceros C, del Campo AD, de Prado R (2005) Establishment of *Quercus ilex* L. subsp. *ballota* [Desf.] Samp. using different weed control strategies in southern Spain. Ecol Eng 25(4):332–342

Navarro-Cerrillo RM, Del Campo AD, Cortina J (2006a) Factores que afectan al éxito de una repoblación y su relación con la calidad de la planta. In: Cortina J, Peñuelas JL, Puértolas J, Vilagrosa A, Savé R (eds) Calidad de planta forestal para la restauración en ambientes Mediterráneos. Estado actual de conocimientos, Organismo Autónomo Parques Nacionales, Ministerio de Medio Ambiente, Madrid, España, pp 1–23

Navarro-Cerrillo RM, Villar-Salvador P, del Campo AD (2006b) Morfología y establecimiento de los plantones. In: Cortina J, Peñuelas JL, Puértolas J, Vilagrosa A, Savé R (eds) Calidad de planta forestal para la restauración en ambientes Mediterráneos. Estado actual de conocimientos, Organismo Autónomo Parques Nacionales, Ministerio de Medio Ambiente, Madrid, España, pp 67–88

Navarro-Cerrillo RM, del Campo AD, Ceaceros C, Quero J, de Mena J (2014) On the importance of topography, site quality, stock quality and planting date in a semiarid plantation: feasibility of using low-density LiDAR. Ecol Eng 67:25–38

Nilsson U, Allen HL (2003) Short-and long-term effects of site preparation, fertilization and vegetation control on growth and stand development of planted loblolly pine. For Ecol Manage 175(1):367–377

North G, Nobel P (1991) Changes in hydraulic conductivity and anatomy caused by drying and rewetting roots of *Agave deserti* (Agavaceae). Am J Bot 78:906–915

Ocaña L, Domínguez-Lerena S, Carrasco I, Peñuelas JL, Herrero N (1997) Influencia del tamaño de la semilla y diferentes dosis de fertilización sobre el crecimiento y supervivencia en campo de cuatro especies forestales. In: Puertas F (ed) II Congreso Forestal Español, Gobierno de Navarra, Pamplona, pp 461–467

Oddou-Muratorio S, Klein EK, Vendramin GG, Fady B (2011) Spatial vs temporal effects on demographic and genetic structures: the roles of dispersal, masting and differential mortality on patterns of recruitment in *Fagus sylvatica*. Mol Ecol 20:1997–2010

OEPM (2014) ES1098586 U. Contenedor y Bandejas Forestal. Patente Modelo de Utilidad (N°. solicitud: 201331478; N° de publicación: 1 098 586; Fecha de publicación: 27.01.2014). En: Oficina Españolas de Patentes y Marcas. Ministerio de Industria, Energía y Turismo. Gobierno de España. Fecha de consulta: 07/03/2017. Available at: http://www.oepm.es/es/invenciones/resultados.html?field=TITU_RESU&bases=0&keyword=Contenedor+y+Bandeja+Forestal

Oliet J, Jacobs D (2007) Microclimatic conditions and plant morpho-physiological development within a tree shelter environment during establishment of *Quercus ilex* seedlings. Agric For Meteorol 144:58–72

Oliet J, Navarro-Cerrillo RM, Contreras O (2003) Evaluación de la aplicación de mejoradores y tubos en repoblaciones forestales. Consejería de Medio Ambiente de la Junta de Andalucía, 234 pp

Oliet J, Planelles R, Segura ML, Artero F, Jacobs DF (2004) Mineral nutrition and growth of containerized *Pinus halepensis* seedling under controlled-release fertilizer. Sci Hortic (Amsterdam) 103:113–129

Oliet J, Tejada M, Salifu K, Collazos A, Jacobs DF (2009) Performance and nutrient dynamics of holm oak (*Quercus ilex* L.) seedlings in relation to nursery nutrient loading and post-transplant fertility. Eur J For Res 128:253–263

Oliet J, Salazar JM, Villar R, Robredo E, Valladares F (2011) Fall fertilization of holm oak affects N and P dynamics, root growth potential, and post-planting phenology and growth. Ann For Sci 68:647–656. doi:10.1007/s13595-011-0060-8

Oliet J, Puértolas J, Planelles R, Jacobs DF (2013) Nutrient loading of forest tree seedlings to promote stress resistance and field performance: a Mediterranean perspective. New For 44:649–669. doi:10.1007/s11056-013-9382-8

Oliet J, Vázquez de Castro A, Puértolas J (2015) Establishing *Quercus ilex* under Mediterranean dry conditions: sowing recalcitrant acorns versus planting seedlings at different depths and tube shelter light transmissions. New For 46:869–883

Olson D (1974) *Quercus*. In: Schopmeyer CS (ed) Seeds of woody plants in the United States. USDA Forest Service, Agriculture handbook 450, pp 692–703

Padilla FM, Pugnaire FI (2007) Rooting depth and soil moisture control Mediterranean woody seedling survival during drought. Funct Ecol 21:489–495

Padilla FM, Miranda JD, Ortega R, Hervás M, Sánchez J, Pugnaire FI (2011) Does shelter enhance early seedling survival in dry environments? A test with eight Mediterranean species. Appl Veg Sci 14:31–39

Palacios G, Navarro-Cerrillo RM, del Campo AD, Toral M (2009) Site preparation, stock quality and planting date effect on early establishment of holm oak (*Quercus ilex* L.) seedlings. Ecol Eng 35(1): 38–46

Palacios-Rodríguez G (2015) Influencia de la fecha de plantación, la preparación del terreno y la calidad de planta en repoblaciones forestales de pino piñonero (Pinus pinea L.) y encina (Quercus ilex L.) en ámbito mediterráneo. Doctoral dissertation-University of Córdoba

Pardos M (2000) Comportamiento de la planta de alcornoque (*Quercus suber* L.) producida en envase: su evaluación mediante parámetros morfológicos y fisiológicos. Universidad Politécnica de Madrid

Pardos M, Royo A, Gil L, Pardos JA (2003) Effect of nursery location and outplanting date on field performance of *Pinus halepensis* and *Quercus ilex* seedlings. Forestry 76:67–81

Pascual G, Molinas ML, Verdaguer D (2002) Comparative anatomical analysis of the cotyledonary region in three Mediterranean Basin *Quercus* (Fagaceae). Am J Bot 89:383–392

Paterson J, DeYoe D, Millson S, Galloway R (2001) The handling and planting of seedlings. In: Wagner RG, Colombo SJ (eds) Regenerating the Canadian forest: principles and practice for Ontario. Ontario Ministry of Natural Resources and Fitzhenry & Whiteside Ltd., Markham, ON, Canada, pp 325–341

Peguero G, Espelta JM (2011) Disturbance intensity and seasonality affect the resprouting ability of the neotropical dry-forest tree *Acacia pennatula*: do resources stored below-ground matter? J Trop Ecol 27:539–546

Pemán J, Gil-Pelegrín E (2008) ¿Sembrar o plantar encinas (*Quercus ilex* subsp. *ballota*)? Implicaciones de la morfología y funcionalidad del sistema radicular. Cuad Soc Esp Cien For 28:49–54

Pemán J, Voltas J, Gil-Pelegrin E (2006) Morphological and functional variability in the root system of *Quercus ilex* L. subject to confinement: consequences for afforestation. Ann For Sci 63:425–430. doi:10.1051/forest:2006022

Pemán J, Gil-Pelegrín E, Valladares F, Peguero-Pina JJ (2010) Evaluation of unventilated tree shelters in the context of Mediterranean climate: insights from a study on *Quercus faginea* seedlings assessed with a 3D architectural plant model. Ecol Eng 36:517–526. doi:10.1016/j.ecoleng.2009.11.021

Peñuelas JL, Ocaña L (1996) Cultivo de plantas forestales en contenedor. Ministerio de Agricultura, Pesca y Alimentación-Editorial Mundi-Prensa, Madrid, p 190

Perea R, González R, San Miguel A, Gil L (2011) Moonlight and shelter cause differential seed selection and removal by rodents. Anim Behav 82:717–723

Perea R, San Miguel A, Martínez-Jauregui M, Valbuena-Carabaña M, Gil L (2012) Effect of seed quality and seed location on the removal of acorns and beechnuts. Eur J For Res 131:623–631

Pérez-Ramos IM, Gómez-Aparicio L, Villar R, García LV, Marañón T (2010) Seedling growth and morphology of three oak species along field resource gradients and seed mass variation: a seedling age-dependent response. J Veg Sci 21:419–437

Pérez-Ramos IM, Urbieta IR, Zavala MA, Marañón T (2012) Ontogenetic conflicts and rank reversals in two Mediterranean oak species: implications for coexistence. J Ecol 100:467–477

Pesendorfer MB (2014) The effect of seed size variation in *Quercus pacifica* on seedling establishment and growth. Gen Tech Rep PSW-GTR 251:407–412

Piñeiro J, Maestre FT, Bartolomé L, Valdecantos A (2013) Ecotechnology as a tool for restoring degraded drylands: a meta-analysis of field experiments. Ecol Eng 61:133–144

Pinto JR, Marshallb JD, Dumroesea RK, Anthony S, Davis AS, Cobosc DR (2011) Establishment and growth of container seedlings for reforestation: a function of stocktype and edaphic conditions. For Ecol Manage 261:1876–1884

Pinto J, Marshall J, Dumroese R, Davis A, Cobos D (2012) Photosynthetic response, carbon isotopic composition, survival, and growth of three stock types under water stress enhanced by vegetative competition. Can J For Res 42:333–344

Planelles R (2004) Efectos de la fertilización NPK en vivero sobre la calidad funcional de plantas de *Ceratonia siliqua* L. Universidad Politécnica de Madrid

Poorter H, Niinemets U, Walter A, Fiorani F, Schurr U (2010) A method to construct dose-response curves for a wide range of environmental factors and plant traits by means of a meta-analysis of phenotypic data. J Exp Bot 61:2043–2055. doi:10.1093/jxb/erp358

Potter MJ (1991) Tree shelters. Forestry Comission handbook 7, 48 pp

Pratt B, Ewers FW, Lawson MC, Jacobsen AL, Brediger MM, Davis SD (2005) Mechanisms for tolerating freeze-thaw stress of two evergreen chaparral species: *Rhus ovata* and *Malosma laurina* (Anacardiaceae). Am J Bot 92:1102–1113

Pritchard HW (1991) Water potential and embryonic axis viability in recalcitrant seeds of *Quercus rubra*. Ann Bot 67:43–49

Puértolas J, Gil L, Pardos JA (2003) Effects of nutritional status and seedling size on field performance of *Pinus halepensis* planted on former arable land in the Mediterranean basin. Forestry 76:159–168

Puértolas J, Oliet JA, Jacobs DF, Benito LF, Peñuelas JL (2010) Is light the key factor for success of tube shelters in forest restoration plantings under Mediterranean climates? For Ecol Manage 260:610–617. doi:10.1016/j.foreco.2010.05.017

Puttonen P (1986) Carbohydrate reserves in *Pinus sylvestris* seediling needles as an attribute of seedling vigor. Scand J For Res 1:181–193

Querejeta JI, Roldán A, Albaladejo J, Castillo V (2001) Soil water availability improved by site preparation in a *Pinus halepensis* afforestation under semiarid climate. For Ecol Manage 149 (1):115–128

Quero J, Villar R, Marañón T, Zamora R, Poorter L (2007) Seed—mass effects in four mediterranean *Quercus* species (Fagaceae) growing in contrasting light environments. Am J Bot 94:1795–1803

Quevedo L, Rodrigo A, Espelta JM (2007) Post-fire resprouting ability of 15 non-dominant shrub and tree species in Mediterranean areas of NE Spain. Ann For Sci 64:883–890

Quilhó T, Lopes F, Pereira H (2003) The effect of tree shelter on the stem anatomy of cork oak (*Quercus suber*) plants. IAWA J 24:385–395

Radoglou K, Raftoyannis Y (2001) Effects of desiccation and freezing on vitality and field performance of broadleaved tree species. Ann For Sci 58:59–68

Radoglou K, Raftoyannis Y, Halivopoulos G (2003) The effects of planting date and seedling quality on field performance of *Castanea sativa* Mill. and *Quercus frainetto* Ten. Forestry 76:569–578

Rambal S (1984) Water balance and pattern of root water uptake by a *Quercus coccifera* L. evergreen shurb. Oecologia 62:18–25

Ramírez-Valiente JA, Valladares F, Gil L, Aranda I (2009) Population differences in juvenile survival under increasing drought are mediated by seed size in cork oak (*Quercus suber* L.). For Ecol Manage 257:1676–1683. doi:10.1016/j.foreco.2009.01.024

Retana J, Espelta JM, Gracia M, Riba M (1999) Seedling recruitment. In: Rodà F, Retana J, Gracia CA, Bellot J (eds) Ecology of Mediterranean evergreen oak forests. Springer, Berlin, pp 89–103

Rice KJ, Gordon DR, Hardison JL, Welker JM (1993) Phenotypic variation in seedlings of a "keystone" tree species (*Quercus douglasii*): the interactive effects of acorn source and competitive environment. Oecologia 96:537–547. doi:10.1007/bf00320511

Riedacker A, Belgrand M (1983) Morphogénèse des systèmes racinaires des semis el boutures de chêne pédonculè. Plant Soil 71:131–146

Riedacker A, Dexheimer J, Rahmat T, Alaoui H (1982) Modifications expérimentales de la morphogénèse et des géotropismes dans le système racinaire de jeunes chênes. Can J Bot 60:765–778

Rieger M, Litvin P (1999) Root system hydraulic conductivity in species with contrasting root anatomy. J Expr Bot 50:201–209

Rietveld W (1989) Transplanting stress in bareroot conifer seedlings: its development and progression to establishment. North J Appl For 6:99–107

Ritchie GA (1984) Assessing seedling quality. In: Duryea ML, Landis TD (eds) Forest nursery manual: production of bareroot seedlings, Martinus Nijhoff/Dr W. Junk Publishers, The Hague/Boston/Lancaster, pp 243–259

Ritchie GA (1985) Root growth potential: principles, procedures and predictive ability. In: Duryea ML (ed) Proceedings of evaluating seedling quality: principles, procedures, and predictive abilities of major tests, 16–18 Oct 1984, Corvallis, Oregon State University, Forest Research Laboratory, pp 93–105

Ritchie GA, Dunlap JR (1980) Root growth potencial: its development and expression in forest tree seedlings. New Zeal J For Sci 10:218–248

Royo A, Gil L, Pardos J (2000) Efecto de la fecha de plantación sobre la supervivencia y el crecimiento del pino carrasco. Cuad Soc Esp Cienc For 10:57–62

Royo A, Gil L, Pardos JA (2001) Effect of water stress conditioning on morphology, physiology and field performance of *Pinus halepensis* Mill. Seedlings. New For 21:127–140

Rubio E, Vilagrosa A, Cortina J, Bellot J (2001) Modificaciones morfofisiológicas en plantones de *Pistacia lentiscus* y *Quercus rotundifolia* como consecuencia del endurecimiento hídrico en vivero. Efectos sobre supervivencia y crecimiento en campo. In: SECF-Junta de Andalucía (eds) III Congreso Forestal Español, Granada, vol II, pp 527–532

Ruthrof K, Fontaine J, Buizer M, Matusick G, McHenry M, Hardy G (2013) Linking restoration outcomes with mechanism: the role of site preparation, fertilisation and revegetation timing relative to soil density and water content. Plant Ecol 214(8):987–998

Sage RD, Koenig WD, Mclaughlin BC (2011) Fitness consequences of seed size in the valley oak *Quercus lobata* Née (Fagaceae). Ann For Sci 68:477–484. doi:10.1007/s13595-011-0062-6

Salifu KF, Jacobs DF (2006) Characterizing fertility targets and multi-element interactions in nursery culture of *Quercus rubra* seedlings. Ann For Sci 63:231–237. doi:10.1051/forest

Salifu KF, Timmer VR (2003) Optimizing nitrogen loading of *Picea mariana* seedlings during nursery culture. Can J For Res 33:1287–1294

Salifu KF, Apostol KG, Jacobs DF, Islam MA (2008) Growth, physiology, and nutrient retranslocation in nitrogen-15 fertilized *Quercus rubra* seedlings. Ann For Sci 65:101. doi:10.1051/forest:2007073

Salifu KF, Jacobs DF, Birge ZKD (2009) Nursery nitrogen loading improves field performance of bareroot oak seedlings planted on abandoned mine lands. Restor Ecol 17:339–349. doi:10.1111/j.1526-100X.2008.00373.x

Salomón R, Valbuena-Carabaña M, Gil L, González-Doncel I (2013) Clonal structure influences stem growth in *Quercus pyrenaica* Willd. coppices: bigger is less vigorous. For Ecol Manage 296:108–118

Salomón R, Valbuena-Carabaña M, Rodríguez-calcerrada J, Aubrey D, McGuire MA, Teskey R, Gil L, González-Doncel I (2015) Xylem and soil CO_2 fluxes in a *Quercus pyrenaica* Willd. coppice: root respiration increases with clonal size. Ann For Sci 72:1065–1078. doi:10.1007/s13595-015-0504-7

Salto CS, Harrand L, Oberschelp GPJ, Ewens M (2016) Crecimiento de plantines de *Prosopis alba* en diferentes sustratos, contenedores y condiciones de vivero. Bosque 37(3):527–537. doi:10.4067/S0717-92002016000300010

Sánchez-Blanco MJ, Ferrández T, Navarro A, Bañon S, Alarcón JJ (2004) Effects of irrigation and air humidity preconditioning on water relations, growth and survival of *Rosmarinus officinalis* plants during and after transplanting. J Plant Physiol 161:1133–1142

Sánchez-Humanes B, Sork VL, Espelta JM (2011) Trade-offs between vegetative growth and acorn production in *Quercus lobata* during a mast year: the relevance of crop size and hierarchical level within the canopy. Oecologia 166:101–110

Sands R (1984) Water relations of *Pinus radiata*. Aust For Res 4:67–72

Sanz-Pérez V, Castro-Díez P, Valladares F (2007) Growth versus storage: responses of Mediterranean oak seedlings to changes in nutrient and water availabilities. Ann For Sci 64:201–210

Sanz-Pérez V, Castro-Díez P, Joffre R (2009) Seasonal carbon storage and growth in Mediterranean tree seedlings under different water conditions. Tree Physiol 29:1105–1116

Saquete A, Solbes M, Escarré A, Ripoll M, de-Simón E (2006) Effects of site preparation with micro-basins on *Pinus halepensis* Mill. afforestations in a semiarid ombroclimate. Ann For Sci 63:15–23

Sardans J, Peñuelas J (2006) Plasticity of leaf morphological traits, leaf nutrient content, and water capture in the Mediterranean evergreen oak *Quercus ilex* subsp. *ballota* in response to fertilization and changes in competitive conditions. Ecoscience 13:258–270

Sardans J, Rodà F (2006) Effects of a nutrient pulse supply on nutrient status of the mediterranean trees *Quercus ilex* subsp. *ballota* and *Pinus halepensis* on different soils and under different competitive pressure. Trees 20:619–632

Saura-Mas S, Lloret F (2009) Linking post-fire regenerative strategy and leaf nutrient content in Mediterranean woody plants. Perspect Plant Ecol Evol Syst 11:219–229

Schmal JL, Jacobs DF, O'Reilly C (2011) Nitrogen budgeting and quality of exponentially fertilized *Quercus robur* seedlings in Ireland. Eur J For Res 130:557–567

Schroeder WR, Walker DS (1987) Effects of moisture content and storage temperatures on germination of *Quercus macrocarpa* acorns. J Environ Hortic 5:22–24

Seifert JR, Jacobs DF, Selig MF (2006) Influence of seasonal planting date on field performance of six temperate deciduous forest tree species. For Ecol Manage 223:371–378

Sepúlveda YL, Diez MC, Moreno FH, Osorio NW (2014) Effects of light intensity and fertilization on the growth of Andean oak seedlings at nursery. Acta Biol Colomb 19:211–220

Sharew H, Hairston-Strang A (2005) A comparison of seedling growth and light transmission among tree shelters. NJAF 22:102–110

Siegel-Issem CM, Burger J, Powers R et al (2005) Seedling root growth as a function of soil density and water content. Soil Sci Soc Am J 69:215–226

Simpson DG, Ritchie GA (1997) Does RGP predict field performance? A debate. New For 13(1–3):253–277

Simpson DG, Vyse A (1995) Planting stock performance: site and RGP effects. For Chron 71:739–742

South DB, Zwolinski JB (1996) Transplant stress index: a proposed method of quantifying planting check. New For 13:311–324

South DB, Rose R, McNabb K (2001) Nursery and site preparation interaction research in the United States. New For 22(1–2):43–58

South DB, Harris S, Barnett J, Hainds M, Gjerstad D (2005) Effect of container type and seedlings size on survival and early height growth of *Pinus palustris* seedlings in Alabama, USA. For Ecol Manage 204:385–398

Sowa S, Connor KF (2003) Recalcitrant behavior of cherrybark oak seed: an FT-IR study of desiccation sensitivity in *Quercus pagoda* Raf. acorns. Seed Technol 25:110–123

St-denis A, Messier C, Kneeshaw D (2013) Seed size, the only factor positively affecting direct seeding success in an abandoned field in Quebec, Canada. Forests 4:500–516. doi:10.3390/f4020500

Steudle E (2000) Water uptake by plant roots: an integration of views. Plant Soil 226(1):45–56

Steudle E, Meshcheryakov AB (1996) Hydraulic and osmotic properties of oak roots. J Expr Bot 47:387–401

Stewart JD, Lieffers VJ (1993) Preconditioning effects of nitrogen relative addition rate and drought stress on container-grown lodgepole pine seedlings. Can J For Res 23:1663–1671

Stjernberg EI (1997) Mechanical shock during transportation: effects on seedling performance. New For 13(1–3):401–420

Sunyer P, Muñoz A, Bonal R, Espelta JM (2013) The ecology of seed dispersal by small rodents: a role for predator and conspecific scents. Funct Ecol 27:1313–1321

Sunyer P, Espelta JM, Bonal R, Muñoz A (2014) Seeding phenology influences wood mouse seed choices: the overlooked role of timing in the foraging decisions by seed-dispersing rodents. Behav Ecol Sociobiol 68:1205–1213

Sunyer P, Boixadera E, Muñoz A, Bonal R, Espelta JM (2015) The interplay among acorn abundance and rodent behavior drives the spatial pattern of seedling recruitment in mature mediterranean oak forests. PLoS ONE. doi:10.1371/journal.pone.0129844

Suszka B, Tylkowski T (1980) Storage of acorns of the English oak (*Quercus robur* L.) over 1–5 winters. Arbor Kórnikie 25:199–288

Tamasi E, Stokes A, Lasserre B, Danjon F, Berthier S, Fourcaud T, Chiatante D (2005) Influence of wind loading on root system development and architecture in oak (Quercus robur L.) seedlings. Trees 19:374–384. doi:10.1007/s00468-004-0396-x

Taylor TS, Loewenstein EF, Chappelka AH (2006) Effect of animal browse protection and fertilizer application on the establishment of planted Nuttall oak seedlings. New For 32:133–143

Thomas FM, Ahlers U (1999) Effects of excess nitrogen on frost hardiness and freezing injury of above-ground tissue in young oaks (*Quercus petraea* and *Q. robur*). New Phytol 144:73–83

Thompson BE (1985) Seedling morphological evaluation. What you can tell by looking. In: Durvea ML (ed) Proceedings: evaluating seedling quality: principles, procedures, and predictive abilities of major tests. Forest Research Laboratory, Oregon State University, Corvallis, pp 59–71

Thompson JR, Schultz RC (1995) Root system morphology of *Quercus rubra* L. planting stock and 3-year field performance in Iowa. New For 9:225–236. doi:10.1007/BF00035489

Tian MH, Tang AJ (2010) Seed desiccation sensitivity of *Quercus fabri* and *Castanopsis fissa* (Fagaceae). Seed Sci Tech 38(1):225–230

Tibbits WN, Hodge GR (2003) Genetic parameters for cold hardiness in *Eucalyptus nitens* (Deane & Maiden). Silvae Genet 52:89–96

Tilki F, Alptekin CU (2006) Germination and seedling growth of *Quercus vulcanica*: effects of stratification, desiccation, radicle pruning, and season of sowing. New For 32:243–251. doi:10.1007/s11056-006-9001-z

Timmer VR (1997) Exponential nutrient loading: a new fertilization technique to improve seedling performance on competitive sites. New For 13:279–299. doi:10.1023/A:1006502830067

Tinus R (1996) Root growth potential as an indicator of drought stress history. Tree Physiol 16:795–799

Tognetti R, Michelozzi M, Giovannelli A (1997) Geographical variation in water relations, hydraulic architecture and terpene composition of Aleppo pine seedlings from Italian provenances. Tree Physiol 17:241–250

Trubat R (2012) Estado nutricional de especies leñosas mediterráneas y su aplicación a la restauración forestal. Tesis doctoral, Universidad de Alicante

Trubat R, Cortina J, Vilagrosa A (2008) Short-term nitrogen deprivation increases field performance in nursery seedlings of Mediterranean woody species. J Arid Environ 72:879–890

Trubat R, Cortina J, Vilagrosa A (2010) Nursery fertilization affects seedling traits but not field performance in *Quercus suber* L. J Arid Environ 74:491–497. doi:10.1016/j.jaridenv.2009.10.007

Trubat R, Cortina J, Vilagrosa A (2011) Nutrient deprivation improves field performance of woody seedlings in a degraded semi-arid shrubland. Ecol Eng 37:1164–1173. doi:10.1016/j.ecoleng.2011.02.015

Tsakaldimi M (2009) A comparison of root architecture and shoot morphology between naturally regenerated and container-grown seedlings of *Quercus ilex*. Plant Soil 324:103–113

Tsakaldimi M, Zagas T, Tsitsoni T, Ganatsas P (2005) Root morphology, stem growth and field performance of seedlings of two Mediterranean evergreen oak species raised in different container types. Plant Soil 278:85–93. doi:10.1007/s11104-005-2580-1

Tsakaldimi M, Ganatsas P, Jacobs DF (2013) Prediction of planted seedling survival of five Mediterranean species based on initial seedling morphology. New For 44(3): 327-329

Tsitsoni T, Tsakaldimi M, Gousiopoulou M (2015) Studying shoot and root architecture and growth of *Quercus ithaburensis* subsp. *macrolepis* seedlings; key factors for successful restoration of Mediterranean ecosystems. Ecol Med 41(2)

Tuley G (1985) The growth of young oak trees in shelters. Forestry 58:181–195

Uscola M, Oliet JA, Villar-Salvador P, Jacobs DF (2014) Nitrogen form and concentration interact to affect the performance of two ecologically distinct Mediterranean forest trees. Eur J For Res 133:235–246. doi:10.1007/s10342-013-0749-3

Uscola M, Salifu KF, Oliet JA, Jacobs DF (2015a) An exponential fertilization dose-response model to promote restoration of the Mediterranean oak *Quercus ilex*. New For 46:795–812. doi:10.1007/s11056-015-9493-5

Uscola M, Villar-Salvador P, Gross P, Maillard P (2015b) Fast growth involves high dependence on stored resources in seedlings of Mediterranean evergreen trees. Ann Bot 115:1001–1013. doi:10.1093/aob/mcv019

Valbuena-Carabaña M, Gil L (2017) Centenary coppicing maintains high levels of genetic diversity in a root resprouting oak (*Quercus pyrenaica* Willd.). Tree Genet Genomes13:28

Valkonen S (2008) Survival and growth of planted and seeded oak (*Quercus robur* L.) seedlings with and without shelters on field afforestation sites in Finland. For Ecol Manage 255:1085–1094

Valladares F, Martínez-Ferri E, Balaguer L, Pérez-Corona E, Manrique E (2000) Low leaf-level response to light and nutrients in Mediterranean evergreen oaks: a conservative resource-use strategy? New Phytol 148:79–91

Valladares F, Vilagrosa A, Peñuelas J, Ogaya R, Camarero JJ, Corcuera l, Sisó S, Gil-Pelegrín E (2008) Estrés hídrico: Ecofisiología y escalas de sequía. In: Valladares F (ed) Ecología del bosque mediterráneo en un mundo cambiante. Ministerio de Medio Ambiente, Egraf, Madrid. pp 165–192

Vallejo VR, Aronson J, Pausas JG, Cortina J (2006) Restoration of Mediterranean woodlands. In: van Andel J, Aronson J (eds) Restoration ecology. The new frontier, Blackwell Science, USA, pp 193–207

van den Driessche R (1982) Relationship between spacing and nitrogen fertilization of seedlings in the nursery, seedling size, and outplanting performance. Can J For Res 12:865–875

van den Driessche R (1991a) Effects of nutrients on stock performance in the forest. In: Van den Driessche R (ed) Mineral nutrition of conifer seedlings, CRC Press, USA, pp 229–260

van den Driescche R (1991) Influence of container nursery regimes on drought resistance I Survival and growth. Can J For Res 21:555–565

Vázquez de Castro A, Oliet J, Puértolas J, Jacobs DF (2014) Light transmissivity of tube shelters affects root growth and biomass allocation of *Quercus ilex* L. and *Pinus halepensis* Mill. Ann For Sci 71:91–99. doi:10.1007/s13595-013-0335-3

Verdaguer DE, García-Berthou G, Pascual P, Puigderrajols (2001) Sprouting of seedlings of three *Quercus* species (*Q. humilis* Miller, *Q. ilex* L., and *Q. suber* L.) in relation to repeated pruning and the cotyledonary node. Aust J Bot 49:67–74

Vilagrosa A (2002) Estrategias de resistencia al déficit hídrico en *Pistacia lentiscus* L. y *Quercus coccifera* L. Tesis doctoral. Departamento de Ecología, Universidad de Alicante, Alicante, España

Vilagrosa A, Seva JP, Valdecantos A, Cortina J, Alloza JA, Serrasolsas I, Diego V, Abril M, Ferran A, Bellot J, Vallejo VR (1997) Plantaciones para la restauración forestal en la Comunidad Valenciana. In: Vallejo VR (ed) La restauración de la cubierta vegetal en la Comunidad Valenciana, Valencia, pp 435–548

Vilagrosa A, Cortina J, Gil-Pelegrín E, Bellot J (2003a) Suitability of drought-preconditioning techniques in mediterranean climate. Restor Ecol 11:208–216

Vilagrosa A, Bellot J, Vallejo VR, Gil-Pelegrín E (2003b) Cavitation, stomatal conductance, and leaf dieback in seedlings of two co-occurring Mediterranean shrubs during an intense drought. Jour Exp Bot 54:2015–2024

Vilagrosa A, Cortina J, Rubio E, Trubat R, Chirino E, Gil-Pelegrín E, Vallejo VR (2005) El papel de la ecofisiología en la restauración forestal de ecosistemas mediterráneos. Investig Agrar Sist y Recur For 14:446–461

Vilagrosa A, Villar-Salvador P, Puertolas J (2006) Endurecimiento en vivero de especies mediterráneas. In: Cortina J, Peñuelas JL, Puértolas J, Vilagrosa A, Savé R (eds) Calidad de planta forestal para la restauración en ambientes Mediterráneos. Estado actual de conocimientos, Organismo Autónomo Parques Nacionales, Ministerio de Medio Ambiente, Madrid

Villar-Salvador P, Ocaña L, Peñuelas J, Carrasco I (1999) Effect of water stress conditioning on the water relations, root growth capacity, and the nitrogen and non-structural carbohydrate concentration of *Pinus halepensis* Mill. (Aleppo pine) seedlings. Ann For Sci 56:459–465

Villar-Salvador P, Peñuelas JL, Carrasco I (2000) Influencia del endurecimiento por estrés hídrico y la fertilización en algunos parámetros funcionales relacionados con la calidad de la planta de *Pinus pinea*. In: Junta de Castilla-León (ed) Actas del 1er Simposio sobre el pino piñonero, Valladolid, vol. 1, pp 211–218

Villar-Salvador P, Planelles R, Enríquez E, Peñuelas-Rubira J (2004a) Nursery cultivation regimes, plant functional attributes, and field performance relationships in the Mediterranean oak *Quercus ilex* L. For Ecol Manage 196:257–266. doi:10.1016/j.foreco.2004.02.061

Villar-Salvador P, Planelles R, Oliet J, Peñuelas JL, Jacobs DF, González M (2004b) Drought tolerance and transplanting performance of holm oak (*Quercus ilex*) seedlings after drought hardening in the nursery. Tree Physiol 24:1147–1155. doi:10.1093/treephys/24.10.1147

Villar-Salvador P, Peñuelas-Rubira JL, Carrasco I, Benito LF (2005a) Influencia de las condiciones de cultivo en vivero en las características funcionales y el desarrollo en campo de *Juniperus thurifera*. IV Congreso Forestal Español, Zaragoza

Villar-Salvador P, Puértolas J, Peñuelas JL, Planelles R (2005b) Effect of nitrogen fertilization in the nursery on the drought and frost resistance of Mediterranean forest species. Investig Agrar Sist y Recur For 14:408–418. doi:10.5424/srf/2005143-00935

Villar-Salvador P, Heredia N, Millard P (2010) Remobilization of acorn nitrogen for seedling growth in holm oak (*Quercus ilex*), cultivated with contrasting nutrient availability. Tree Physiol 30:257–263. doi:10.1093/treephys/tpp115

Villar-Salvador P, Puértolas J, Cuesta B, Peñuelas JL, Uscola M, Heredia N, Rey Benayas JM (2012) Increase in size and nitrogen concentration enhances seedling survival in Mediterranean plantations. Insights from an ecophysiological conceptual model of plant survival. New For 43:755–770. doi:10.1007/s11056-012-9328-6

Villar-Salvador P, Nicolás-Peragón JL, Heredia-Guerrero N, Uscola M (2013a) *Quercus ilex*. In: Pemán J, Navarro-Cerrillo RM, Nicolás JL, Prada A, Serrada R (eds) Producción y manejo de semillas y plantas forestales. Organismo Autónomo Parques Nacionales. Serie Forestal, Madrid, pp 226–250

Villar-Salvador P, Peñuelas JL, Nicolás JL, Benito LF, Domínguez-Lerena S (2013b) Is nitrogen fertilization in the nursery a suitable tool for enhancing the performance of Mediterranean oak plantations? New For 44:733–751. doi:10.1007/s11056-013-9374-8

Villar-Salvador P, Peñuelas JL, Jacobs DF (2013c) Nitrogen nutrition and drought hardening exert opposite effects on the stress tolerance of *Pinus pinea* L. seedlings. Tree Physiol 33:221–232

Villar-Salvador P, Uscola M, Jacobs DF (2015) The role of stored carbohydrates and nitrogen in the growth and stress tolerance of planted forest trees. New For 46:813–839. doi:10.1007/s11056-015-9499-z

Vizoso S, Gerant D, Guehl JM, Joffre R, Chalot M, Gross P, Maillard P (2008) Do elevation of CO_2 concentration and nitrogen fertilization alter storage and remobilization of carbon and nitrogen in pedunculate oak saplings? Tree Physiol 28:1729–1739. doi:10.1093/treephys/28.11.1729

Vogt AR, Cox GS (1970) Evidence for the hormonal control of stump sprouting by oak. For Sci 16:165–171

Wang J, Li G, Pinto J, Liu J, Shi W, Liu Y (2015) Both nursery and field performance determine suitable nitrogen supply of nursery-grown, exponentially fertilized Chinese pine. Silva Fenn 49 (3):1–13. doi:10.14214/sf.1295

Wang J, Yu H, Li G, Zhang F (2016) Growth and nutrient dynamics of transplanted *Quercus variabilis* seedlings as influenced by pre-hardening and fall fertilization. Silva Fenn 50:1–18

Ward JS, Gent MPN, Stephens GR (2000) Effects of planting stock quality and browse protection-type on height growth of northern red oak and eastern white pine. For Ecol Manage 127:205–216

Webb DP, von Althen F (1980) Storage of hardwood planting stock: effects of various storage regimes and packaging methods on root growth and physiological quality. New Zeal J For Sci 96:83–96

Welander NT, Ottosson B (1998) The influence of shading on growth and morphology in seedlings of *Quercus robur* L. and *Fagus sylvatica* L. For Ecol Manage 107:117–126

Welander NTT, Ottosson B (2000) The influence of low light, drought and fertilization on transpiration and growth in young seedlings of *Quercus robur* L. For Ecol Manage 127: 139–151

West GB (1999) The origin of universal scaling laws in Biology. Physica 263:104–113

Wilson BC, Jacobs DF (2006) Quality assessment of temperate zone deciduous hardwood seedlings. New For 31: C, Park A (2007) Root characteristics and growth potential of container and bare-root seedlings of red oak (*Quercus rubra* L.) in Ontario, Canada. New For 34: 163–176

Wilson ER, Vitols KC, Park A (2007) Root characteristics and growth potential of container and bare-root seedlings of red oak (*Quercus rubra L.*) in Ontario, Canada. New Forests 34(2): 163–176

Xia K, Daws MI, Hay FRR, Chen WY, Zhou ZK, Pritchard HW (2012a) A comparative study of desiccation responses of seeds of Asian Evergreen Oaks, *Quercus* subgenus *Cyclobalanopsis* and *Quercus* subgenus *Quercus*. South African J Bot 78:47–54. doi:10.1016/j.sajb.2011.05.001

Xia K, Daws MI, Stuppy W, Zhou ZK, Pritchard HW (2012b) Rates of water loss and uptake in recalcitrant fruits of *Quercus* species are determined by pericarp anatomy. PLoS ONE 7(10):1–11. doi:10.1371/journal.pone.0047368

Xia K, Tan HY, Turkington R, Hu JJ, Zhou ZK (2016) Desiccation and post-dispersal infestation of acorns of *Quercus schottkyana*, a dominant evergreen oak in SW China. Plant Ecol. doi:10.1007/s11258-016-0654-1

Xiao Z, Harris MK, Zhang Z (2007) Acorn defenses to herbivory from insects. Implications for the joint evolution of resistance, tolerance and escape. For Ecol Manage 238:302–308

Yi X, Wang Z (2016) The importance of cotyledons for early-stage oak seedlings under different nutrient levels: a multi-species study. J Plant Growth Regul 35:183–189. doi:10.1007/s00344-015-9516-7

Zavala MA, Espelta JM, Caspersen J, Retana J (2011) Interspecific differences in sapling performance with respect to light and aridity gradients in Mediterranean pine–oak forests: implications for species coexistence. Can J For Res 41:1432–1444

Zhu WZ, Cao M, Wang SG, Xiao WF, Li MH (2012) Seasonal dynamics of mobile carbon supply in *Quercus aquifolioides* at the upper elevational limit. PLoS ONE 7:e34213

Zhu Y, Dumroese RK, Pinto JR, Li GL, Liu Y (2013) Fall fertilization enhanced nitrogen storage and translocation in *Larix olgensis* seedlings. New For 44:849–861. doi:10.1007/s11056-013-9370-z

Zhu Y, Wang SG, Yu DZ, Jiang Y, Li MH (2014) Elevational patterns of endogenous hormones and their relation to resprouting ability of *Quercus aquifolioides* plants on the eastern edge of the Tibetan Plateau. Trees 28(2):359–372

Chapter 15
Competition Drives Oak Species Distribution and Functioning in Europe: Implications Under Global Change

Jaime Madrigal-González, Paloma Ruiz-Benito, Sophia Ratcliffe, Andreas Rigling, Christian Wirth, Niklaus E. Zimmermann, Roman Zweifel and Miguel A. Zavala

Abstract Oaks are a widely represented woody species across the main European forest biomes, ranging from semi-arid Mediterranean shrub lands to cool temperate and transitional boreal forests. We provide quantitative evidence of large-scale distribution and abundance patterns of oaks across Europe. In addition, we present key demographic processes, such as mortality, which underlie deterministic mechanisms behind large-scale distribution patterns; chiefly competitive exclusion and complementarity versus environmental filtering along the main energy-productivity gradient of Europe. Finally, we investigate the role of concomitant climate changes, land use legacies and management regimes as key drivers of future oak forest distribution in continental Europe. Overall, oak distribution and dominance at the population and community levels is largely determined by environmental filtering together with intra- and inter-specific functional differences as major elements driving oak species distribution and dominance at the population and community levels, but these processes are strongly modulated by global change which may result in a significant alteration of current distribution of European oak forests.

J. Madrigal-González · P. Ruiz-Benito · M. A. Zavala (✉)
Grupo de Ecología y Restauración Forestal, Departamento de Ciencias de la Vida, Universidad de Alcalá, Edificio de Ciencias, Campus Universitario, 28805 Alcalá de Henares, Madrid, Spain
e-mail: madezavala@gmail.com

J. Madrigal-González
e-mail: ecojmg@hotmail.com

P. Ruiz-Benito
e-mail: palomaruizbenito@gmail.com

S. Ratcliffe · C. Wirth
Department of Systematic Botany and Functional Biodiversity, Institute of Biology, University of Leipzig, Johannisallee 21-23, 04103 Leipzig, Germany
e-mail: Sophia.ratcliffe@uni-leipzig.de

© Springer International Publishing AG 2017
E. Gil-Pelegrín et al. (eds.), *Oaks Physiological Ecology. Exploring the Functional Diversity of Genus* Quercus *L.*, Tree Physiology 7,
https://doi.org/10.1007/978-3-319-69099-5_15

15.1 Oak Forests in Europe

Oaks represent a functionally diverse group of woody species and are widely distributed across the northern hemisphere. Although most oak species are trees, some ecotypes and species are shrubs, especially in dry ecosystems. Oaks exhibit important plastic responses to the local conditions and thus can present a variety of morphological types depending on ecological factors such as herbivory, drought or competition (Kleinschmit 1993). In general, oak forests are highly valued ecosystems from the viewpoint of human economical and cultural interests, and their distribution and physiognomy has been greatly modulated by humans since the Neolithic (Barbero et al. 1990). Spatial and temporal climate variability, as well as global climate changes during the last century, are also decisive factors for understanding contemporary oak species biogeography (Comes and Kadereit 1998; Saurer et al. 2014). Thus, not only past climatic fluctuations, but also agricultural intensification and, more recently, widespread agricultural land abandonment associated with human migration from rural to urban areas are recognized as major forces leading to recent oak encroachment, expansion or decline in different European regions (Prentice et al. 1996; Küster 1997; Thomas et al. 2002; Urbieta et al. 2008; Gimmi et al. 2010; Zweifel et al. 2009). Contemporary species distribution, however, also demands an in-depth understanding of community assembly determinants along environmental gradients. For example, significant phylogenetic overdispersion in oak-rich communities in North America suggests a major role of competitive interactions and phenotypic traits in community assembly (Cavender-Bares et al. 2004). Thus, a full understanding of future oak species distribution and dominance requires not only a solid characterization of climate-oak relationships, but also the role of intra- and interspecific interactions modulated by contemporary global changes as key drivers of oaks population dynamics.

C. Wirth
e-mail: cwirth@uni-leipzig.de

A. Rigling · R. Zweifel
Forest Dynamics, Swiss Federal Research Institute, WSL-Zürcherstr. 111, 8903 Birmensdorf, Switzerland
e-mail: andreas.rigling@wsl.ch

R. Zweifel
e-mail: roman.zweifel@natkon.ch

C. Wirth
German Centre for Integrative Biodiversity Research (iDiv) Halle-Jena-Leipzig, Deutscher Platz 5E, 04103 Leipzig, Germany

N. E. Zimmermann
Landscape Dynamics, Swiss Federal Research Institute, WSL-Zürcherstr. 111, 8903 Birmensdorf, Switzerland
e-mail: niklaus.zimmermann@wsl.ch

Two main deterministic mechanisms driving species assembly along large-scale environmental gradients have been proposed: (i) environmental filtering and (ii) competitive exclusion. The environmental filtering hypothesis supports a major role of the physical environment in shaping species' biogeographical ranges. Species are sorted along environmental gradients based on functional traits that confer adaptation to the prevailing abiotic conditions (Diaz et al. 1998). In the particular case of continental Europe, a main climatic gradient associated with increased energy and reduced precipitation southwards is recognized as a primary large-scale driver of species ranges (Whittaker et al. 2007). Nonetheless, current species distributions do not reflect a perfectly stable relationship between species traits and climatic conditions (Holt 2003). Fossil evidence from pollen and wood debris has largely improved our understanding of past and current migration dynamics of oak species following the post-glacial climatic fluctuations from glacial *refugia* in southern peninsulas (Svenning and Skov 2007) and a number of northern outlier populations (Feurdean et al. 2013). Northernmost limits of oaks coincide with the temperate-boreal transition (*c.* 60°N). At the other end of the energy gradient in Europe (*c.* 36°N), oak species distribution is determined by the presence of functional adaptations to drought such as deeper root systems (Canadell et al. 1996), wood anatomical traits (Fonti et al. 2013), sclerophyllous leaves (Acherar and Rambal 1992) and reduced tree height (Poorter et al. 2012). Thus, the abiotic filtering is expected to limit oak species distributions according to their lack of functional adaptations to environmental constraints from the best-productive conditions in temperate areas to either cold or dry-warm ends of the continental energy gradient in temperate/boreal and Mediterranean biomes, respectively.

Competitive exclusion then acts as a second major filter, according to functional trade-offs between adaptations to above- and belowground competition (Tilman 1988; Zavala et al. 2000; Sánchez-Gómez et al. 2008). Interestingly, while environmental filtering implies functional convergence in co-existing species under the same abiotic constraints (Chase and Leibold 2003), the competitive exclusion principle states that two species competing for identical limited resources cannot coexist indefinitely at constant population levels (Hardin 1960). This necessarily implies that (i) the species with the functional advantages will prevail and (ii) competition is less intense in multispecies communities with increased functional dissimilarity between the co-existing species. At this point, the concept limiting similarity arises to denote a theoretical maximum functional overlap between two species for stable coexistence to be possible. The signature of competitive exclusion in multispecies communities might nonetheless be partially blurred because a myriad of biotic and abiotic influences alter the net effect of a single species on another. Thus, competition at the community level is often described as a density-dependent phenomenon that can be summarized as an aggregation or stock variable. For example, increases in tree density, stand wood volume or stand basal area are expected to reduce tree growth or increase mortality (Gómez-Aparicio et al. 2011; Ruiz-Benito et al. 2013). At the regional scale, and even at the scale of a forest stand, the most conspicuous evidence of competitive exclusion on species assembly is the spatial segregation of species (Shigesada et al. 1979; Gotelli and McCabe 2002).

A less obvious signature of competition on species assembly is the density-dependent mortality of species within communities. Most evidence of density-dependent mortality in tree species usually reflects the deleterious effects of extreme climatic events such as drought (Jump et al. 2017). This climate-induced mortality has been therefore attributed to exacerbated abiotic stresses in crowed forest communities. However, mortality associated with intra- or interspecific competition at the productive end of the distribution range of species has rarely been documented as indirect evidence of disparate light harvesting strategies and thus differential aboveground competitive abilities. In parallel, a positive role of tree diversity should be more conspicuous in highly competitive environments where functional dissimilarity among heterospecific traits might reduce the most deleterious impacts of competition (e.g. Ruiz-Benito et al. 2017a).

Recruitment limitation is expected to play an important role in community assembly, reinforcing filtering effects across large-scale environmental gradients (Hurtt and Pacala 1995; Clark et al. 1999). In addition to filtering effects, recruitment limitation in oaks can arise from a myriad of factors ranging from masting behaviour, to seed predation, to lack of parent reproductive individuals or the absence of dispersers (e.g. Pérez Ramos et al. 2008, 2015). As a result of filtering, competitive and dispersal processes in oak dynamics resemble a stochastic process and, as a result, oak species are far from being in a stable relationship with the physical environment, and in particular with climate (e.g. García-Valdés et al. 2015; Sáenz-Romero et al. 2016). The non-equilibrium dynamics reflect the balance between colonisations and extinctions and results in spatially explicit movements of species following secondary succession dynamics in changing environments. Aspects such as landscape fragmentation and patch isolation are critical as they directly constrain colonisation dynamics through uneven seed dispersal. In the particular case of oak species, seed availability and dispersal are decisive factors in understanding recent widespread passive restoration of forests after land use cessation in Europe (Pérez-Ramos et al. 2008). Animals, such as mice or jays, are responsible for the majority of oak seed dispersal, and so the probability of an oak species reaching an empty habitat can be expressed as a function of distance to suitable sites (i.e. fragmentation) and animal mobility across the landscape (Montoya et al. 2008; Purves et al. 2007a, b). Thus oak population dynamics is contingent upon spatial habitat configuration and the presence of key dispersers, which in turn can depend on the maintenance of a given ecological network.

In this chapter we provide evidence of competition as a major driver of oak species dominance through comparison between abundance and mortality data along the energy-productivity gradient of Europe using National Forest Inventory data. We show in particular how tree mortality rates are higher on the more productive side of the species ranges along the energy gradient. Secondly, we provide quantitative evidence that increased functional diversity reduces competition and thus promotes species coexistence and productivity in oak forests across Europe. Species interactions therefore modulate environmental filtering determining both transient dynamics and the outcome of succession. While competitive exclusion is thought to dominate at the high productivity end of an environmental gradient, complementarity among disparate functional types (e.g. angiosperm vs. gymnosperm) can be critical for oak recruitment dynamics particularly in harsh environments (Lookingbill and Zavala 2000; Urbieta et al. 2008). A number of experimental studies have concentrated on the ecophysiological mechanisms (Zweifel et al. 2009) underpinning environmental filtering in oaks distribution (Abrams 1996; Bréda et al. 2006; Morin et al. 2007; Arend et al. 2011, 2013; Chaps. 1 and 2) and modelling studies have documented in detail the role of competitive exclusion as a driver of secondary succession and species distributions (Huston and DeAngelis 1994; Pacala and Rees 1998; Kohyama and Takada 2009; Kunstler et al. 2012) and yet, to our knowledge, few studies have provided empirical large-scale quantitative evidence of competition and mortality as a driver of regional tree species distribution (but see Ruiz-Benito et al. 2013, 2017b). Finally, we discuss how the fate of oak forests in Europe might be crucially depending as well on several concomitant global change processes chiefly agricultural land abandonment and climate change.

15.2 Competition Drives Oak Species Distribution

Oaks are widely distributed woody species in cool temperate and Mediterranean forest biomes of Europe. Using abundance data from National Forest Inventories of Germany and Spain marked spatial segregation of oak species can be seen along the main latitudinal PET climatic gradient (i.e. humid-cool to dry-warm conditions southwards) (Fig. 15.1). Two major functional types in oak species are representative of the large-scale climatic filtering from cool temperate to Mediterranean ecosystems: i.e. deciduous versus evergreen and sclerophyllous versus non-hard-leaved. Deciduous oak trees are distributed throughout the cool temperate biome with a (distinct) dormancy period, where dominance is largely controlled by a non-conservative leaf water economy that maximizes specific leaf area and light yielding during the growing season. Conversely, sclerophyllous leaves are a more efficient strategy to minimize water loss through investments in impermeable wax coatings and stomata distributed at the underside of leaves. These general functional types, however, do not represent rigid functional strategies since high variability in leaf traits can be found both within species (i.e. phenotypic and genotypic plasticity) and between species within functional types (Table 15.1). The most abundant oak species in cool temperate latitudes are *Quercus robur* and *Quercus petraea*. They are both highly competitive species that are present in many temperate forest typologies of Europe (Eaton et al. 2016). Southwards, the growing seasonal water shortage, associated with the moisture advection blockage in the Azores, determines a gradual species replacement. In the temperate-Mediterranean transition, marcescent-deciduous oaks such as *Quercus humilis*, *Quercus faginea*, *Quercus pubescens* and *Quercus pyrenaica* dominate until their full replacement by sclerophyllous, drought-tolerant, species such as *Quercus suber* or *Quercus ilex* in Mediterranean water-limited forests. These regional patterns are complemented by

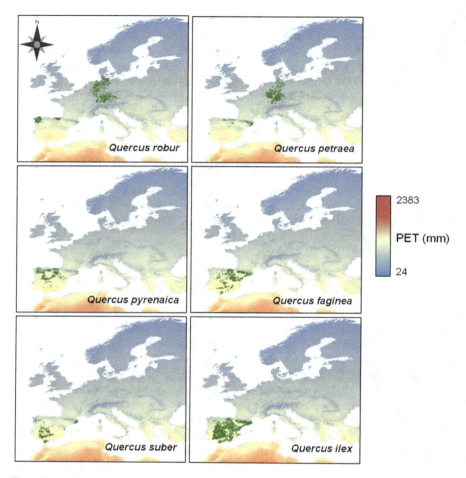

Fig. 15.1 Oak species distribution obtained from the national forest inventory of Spain (Villaescusa and Díaz 1998; Villanueva 2004) and Germany (Kändler 2009). Annual Potential Evapotranspiration (PET, mm) obtained from: Zomer et al. (2008). Global Aridity Index (Global-Aridity) and Global Potential Evapo-Transpiration (Global-PET) Geospatial Database. CGIAR Consortium for Spatial Information. Published online, available from the CGIAR-CSI GeoPortal at: http://www.csi.cgiar.org

small-scale, local patterns where topography and aspect among other factors associated with water balance determine local species segregation (i.e. sclerophyllous oaks such as *Q. suber* in equator facing slopes and marcescent oaks such as *Q. canariensis* in pole-faced slopes, Urbieta et al. 2008).

Climatic conditions are nonetheless insufficient to fully understand the species spatial arrangement along the energy-productivity gradient. Above- and below-ground competition are widely recognized as key drivers of species assembly and coexistence (Tilman 1988; Huston and DeAngelis 1994). In size-structured

Table 15.1 Traits of principal oak functional types (FT) included in the German and the Spanish forest inventories (data obtained in the TRY project)

Species	FT	MHT (m)	WD (kg/m^3)	SM (mg)	LMA (gr/m^2)	Nmass (%)
Quercus faginea	D/M	23	788.90	1747.21	109.53	1.9
Quercus ilex	SC	16	900.80	2310.53	150.80	1.4
Quercus petraea	D	36.5	716.35	2248.19	71.24	2.1
Quercus pyrenaica	D/M	24	825.00	3000.77	76.12	1.8
Quercus robur	D	37	592.22	3225.85	68.45	2.3
Quercus suber	SC	19	827.07	3837.38	93.88	1.6

3*SC* Sclerophyllous; *D/M* Deciduous marcescent; *D* Deciduous. Legend of traits: *MHT* Maximum tree height; *WD* Wood density; *SM* Seed mass; *LMA* Leaf mass per area; *Nmass* Percentage of nitrogen in leaves

communities such as forests, aboveground competition alone can greatly explain long-term species turn-over and forest dynamics (Kohyama and Takada 2009, 2012). Tree architecture and resource-use trade-offs are thus critical aspects involved in species competitive ability (Poorter et al. 2006, 2012). The spatial segregation of oak species along the PET gradient in Fig. 15.2 was strongly driven by stand basal area (SBA, as a proxy for density-dependent competition). Results of Generalized Additive Models (GAM) applied to species dominance data ($n_{species}$/N_{total}) support a significant role of SBA in driving oak species abundance (Fig. 15.2; Table 15.2). Moreover, competitive effects appear more important towards the driest end of the climate gradient, where shading conditions may aggravate stresses associated with soil water scarcity (Valladares and Pearcy 2002; Pardos et al. 2005). Adapting to shade and drought simultaneously represent a major challenge for plants to establish and survive in water-limited environments (Holmgren et al. 1997, 2012; Holmgren 2000) because, in general, tolerance to shade and drought are inversely correlated (Smith and Huston 1989; Niinemets and Valladares 2006). The two sclerophyllous species, *Quercus suber* and *Quercus ilex*, clearly segregate along the stand basal area axis in the driest side of the climatic gradient: i.e. *Quercus suber* dominates under moderately high stand basal area and *Quercus ilex* in open woodlands. This is consistent with studies showing greater *Q. ilex* regeneration rates in open forests, particularly in pine and oak mixed forests (e.g. Urbieta et al. 2011). *Quercus suber* has been found to perform better under deep shading irrespective of the water supply (Quero et al. 2006). Under such conditions, *Quercus suber* had higher specific leaf area (SLA) and carboxylation rates (A) than *Quercus ilex* and thus a more positive carbon balance. Under low stand basal area, *Quercus ilex* can also co-exist with two marcescent-deciduous trees such as *Quercus faginea* and *Quercus pyrenaica*. *Quercus faginea*, which represents a transitional taxon between sclerophyllous and deciduous (i.e. small-leathery leaves), is the least shade-tolerant oak species (Gómez-Aparicio et al. 2011) and thus tends to dominate only in open forests at intermediate positions of the climatic gradient while, *Quercus pyrenaica*, is more like a typical deciduous

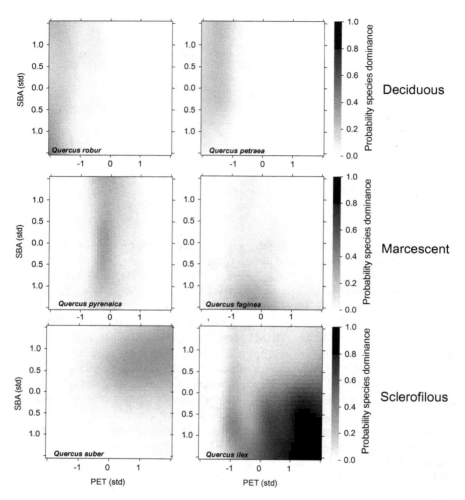

Fig. 15.2 Probability of species dominance along standardised stand basal area (SBA, std) and standardised potential evapotranspiration (PET, std) for deciduous, marcescent (deciduous-marcescent) and sclerophyllous oaks. Negative values in PET and SBA represent below-mean values and contrary positive values above-normal values

species and tends to dominate in denser transitional forests. For example Madrigal-González et al. (2014) showed indirect evidence of the density-dependent segregation of *Quercus ilex* and *Quercus pyrenaica* in shrub communities in central-western Spain. Shrubs were shown to facilitate the establishment of the sclerophyllous tree, *Quercus ilex*, only at early-open successional stages in high elevations. At mid to late successional stages in high elevations, shrub communities facilitate the deciduous oak, *Quercus pyrenaica*, and hindered the development of *Quercus ilex* saplings most likely due to increased light interception by the mature encroached shrubs. The light environment is also pivotal for the differential

Table 15.2 Statistics of generalized additive models applied to dominance data (n_i/N) of each oak species in each sampling plot (German and Spanish Forest inventories)

Species	Spline term	Chi_sq	p-value	Dev. explained (%)
Quercus suber	s(PET, SBA)	739.6	<2e−16	11.30
Quercus ilex	s(PET, SBA)	3916	<2e−16	30.60
Quercus faginea	s(PET, SBA)	523.6	<2e−16	11.90
Quercus pyrenaica	s(PET, SBA)	823.9	<2e−16	18.20
Quercus petraea	s(PET, SBA)	432	<2e−16	31.40
Quercus robur	s(PET, SBA)	367.1	<2e−16	23.20

A Chi-square statistic (Chi-sq) and associated p-value, and a deviance explained estimate (Dev. Explained) are shown

regeneration of temperate deciduous tree species along soil resource gradients in temperate forest biomes. Experimental evidence in contrasted sunlight conditions supported a better performance of *Quercus petraea* compared to *Quercus pyrenaica* seedlings due to increased SLA and lower light compensation points, which enables the maintenance of a positive carbon balance under increased shading conditions (Rodríguez-Calcerrada et al. 2008).

Trade-offs in allocation of resources between roots and shoots are decisive factors for a tree's ability to thrive under deep shade conditions. Species in dry environments tend to invest more resources in root development to enhance soil filling and water uptake (Mokany et al. 2006). However, reduced root-shoot ratios reduce a tree's capacity to rapidly reach the forest canopy and to develop an efficient light yielding capacity, and thus overcome shading limitations. Experimental evidence has shown that temperate deciduous species such as *Quercus petraea* tend to invest more resources in aboveground organs, including annual internode elongations, than marcescent-deciduous oaks such as *Quercus pyrenaica* in identical shading levels (Rodríguez-Calcerrada et al. 2008). Allometries for the most abundant oak tree species in the Iberian Peninsula strongly support these results. In particular, species segregation along tree height and crown length dimensions strongly resembles species segregation along the energy-water availability gradient in Iberia (Poorter et al. 2012). Following similar reasoning, other deciduous tree species such as *Fagus sylvatica* can even exclude *Quercus petraea* and *Quercus robur* under mild temperate conditions. Experiments conducted in protected chambers have reported a higher shade-tolerant ability for *Fagus sylvatica* compared to *Quercus robur* seedlings (Welander and Ottosson 1998). In particular, *Quercus robur* and *Fagus sylvatica* seedlings showed similar tolerance to shading in the first year. However, by the second year, greater light interception by *Fagus* caused strong negative effects on *Quercus robur*. Thus, the dominance of deciduous oaks can be strongly reduced in highly productive temperate environments under the influence of strong competitors, such as *Fagus sylvatica*. The recent advance of *Fagus sylvatica* in Western Europe to the detriment of temperate deciduous and conifer dominances has been widely documented (Küster 1997).

At the coldest end of the gradient, a question arises regarding the absence of drought tolerant species such as *Quercus ilex*, at least in open forests where aboveground competition is negligible. In such situations, only *Quercus robur* occurs up to the boreal latitudes, where all oaks are excluded. The study of cold hardiness in different oak tree species suggests a major role of soluble carbohydrates for immediate responses to freezing (Morin et al. 2007). Inter- and intraspecific variability in the concentration of soluble carbohydrates was influenced by temperature, and differed among the three species considered (i.e. *Quercus robus, Quercus humilis, Quercus ilex*) according to their position along the main latitudinal-climatic gradient of Europe. Solute concentration is directly associated with the avoidance of intra- and intercellular ice formation and the subsequent cellular dehydration (Levitt 1978). High solute concentration, chiefly low-molecular-weight carbohydrates, can stabilize membranes thus conferring cells with resistance against water loss under freezing conditions (Cavender-Bares et al. 2005).

Predicted oak species dominances along the climatic gradient (Fig. 15.2) strongly overlapped at intermediate-low stand basal area (SBA) and clearly segregated at high SBA. It is nonetheless difficult to detect the signature of competitive exclusion in natural multispecies communities. The most drastic impact of competition on a given species' distribution should reflect a full spatial segregation from other species. Nonetheless, spatial segregation often occurs at scales in which other forces such as herbivory, nutrient availability or the disturbance frequency plays a key role. Thus, complex long-term experimental designs would be needed to clearly partitioning each effect.

Assuming that species colonization/extinction dynamics are not at equilibrium, we can examine likely effects of competition on tree species distribution by comparing the relative positions of species individual mortality patterns and their occurrence. Specifically, we can compare in a multivariate space—including neighbourhood competition—the position where the probability of mortality for a given species is at its highest versus the position in which the species has the highest probability of occurring. We could explore then at which side of the species distribution range tree mortality is more likely: i.e. whether tree mortality is biased towards either the abiotic or biotic constraints of the species range along the climatic gradient. To examine this idea, we conducted a detrended correspondence analysis (DCA) using a matrix of sampling plots × species (National Forest Inventories of Germany and Spain) in which mortality of each oak species was included in the model. The highest mortality probability in oak species appeared to be biased to the side of the distribution range where they faced their main competitors (Fig. 15.3). Thus, the two sclerophyllous oaks, *Quercus ilex* and *Quercus suber*, and the transitional oak *Quercus faginea* showed maximum mortality probability oriented to *Quercus pyrenaica*, their most direct competitor in the temperate-Mediterranean bioclimatic transition. In turn, the highest mortality probability of *Quercus pyrenaica* was strongly biased towards maximum occurrence probability of *Quercus robur*. Finally, maximum mortality probability of *Quercus robur* and *Quercus petraea* was distributed along the most productive range of the climatic gradient where other highly competitive species, such as *Fagus sylvatica*, have their maximum occurrence probabilities.

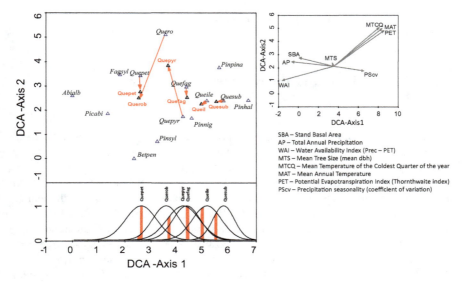

Fig. 15.3 Results of the detrended corresponded analyses including main species in Europe (Abialb *Abies alba*; Picabi *Picea abies*; Fagsyl *Fagus sylvatica*; Betpen *Betula pendula*; Quepet *Quercus petraea*; Pinsyl *Pinus sylvestris*; Querob *Quercus robur*; Quepyr *Quercus pyrenaica*; Quefag *Quercus faginea*; Pinnig *Pinus nigra*; Queile *Quercus ilex*; Quesub *Quercus suber*; Pinpina *Pinus pinaster*; Pinhal *Pinus halepensis*). In red, six more columns in the species x sampling plots matrix were included to denote presence of death individuals in each of the six *Quercus* species considered. Biplot diagram on the right shows the distribution of climatic-structural environmental variables regarding to the two main ordination axes (see legend below for variable names). Red arrows link distribution centroids of species with their corresponding mortality centroids. On the bottom of the DCA plot, Gaussian responses of each *Quercus* species are represented based on the standard deviation and the centroid (multidimensional mean) in the main ordination axis (DCA-Axis1). Red bars represent the centroid (multidimensional mean) of mortality for each oak species in the first DCA axis

Mortality due to competition has been modelled as a self-thinning process in young crowded communities (Yoda et al. 1963; Westoby 1984; Lonsdale 1990; Pretzsch 2006; West et al. 2009). As saplings grow competition gets more intense until sudden mortality reduces tree density up to a steady-state population demography (Yoda et al. 1963; Peet and Christensen 1987). However, the results presented here do not include early self-thinning phases. From then on, mortality becomes evident in understory individuals, particularly shade-intolerant species in which a negative carbon balance due to insufficient light intensity leads to carbon starvation and decline. Yet, mortality patterns associated with tree size in NFI depict an inverted-J curve along the size gradient which aggravates under increased basal area of large trees (Ruiz-Benito et al. 2013). These findings reinforce the idea that our species maximum mortality probabilities largely hold on young understory trees under intense competitive environments. Other causes of mortality such as wind throw or drought might largely affect tree demography along the species ranges. However, such oak mortality events have been described more as punctual events in time and space so their imprint in tree species distribution should demand longer time records.

15.3 Competition and the Maintenance of Oak Forest Functioning

The principle of competitive exclusion and niche theory suggest long term local mono-dominance although this may not be the case for slow growing relatively shade tolerant species such as oaks (e.g. Falster et al. 2017). Functional diversity is an important component of the functional composition of European forests, and has been shown to promote tree productivity, carbon storage and sapling abundance (Nadrowski et al. 2010; Zhang et al. 2012; Ratcliffe et al. 2016b; Van der Plas et al. 2016; Madrigal-González et al. 2016; Ruiz-Benito et al. 2017a; Chamagne et al. 2017). It has been hypothesised that the positive effects of diversity on tree productivity are due to complementarity and selection mechanisms (see e.g. Loreau and Hector 2001). Complementarity effects are hypothesised to increase ecosystem function through facilitation and niche partitioning, because functionally diverse species assemblages enhance resource acquisition and use, and nutrient retention (e.g. Loreau 2000). Oaks are angiosperms plant species that functionally contrast with gymnosperms, particularly in terms of wood density, leaf mass per area and hydraulic traits (see e.g. Carnicer et al. 2013; Ruiz-Benito et al. 2017b). Ratcliffe et al. (2016b) observed that functional identity (i.e. dominant community trait values) was particularly important for tree growth at the latitudinal extremes of Europe, and proposed that the importance of this functional identity reflects a strong trait-based differentiation due to successional transitions from gymnosperms to angiosperms in the Mediterranean forests and the reverse in boreal forests. The high importance of functional identity may also be evidence of oak replacement of pine species under climate warming and increased drought intensities (e.g. Galiano et al. 2010; Vilá-Cabrera et al. 2013; Rigling et al. 2013; Zweifel et al. 2009), together with secondary succession processes (see Ruiz-Benito et al. 2017b). These transitions agree with worldwide studies that have found that certain traits (such as those in evergreen oaks) are consistent with a larger resistance to intense droughts (see Greenwood et al. 2017). However, forest management and stand density may play a critical role on the effects of complementarity and diversity, and particularly to certain responses to extreme events such as drought (see e.g. Jump et al. 2017; Weber et al. 2008; Gimmi and Bürgi 2007). Looking at mixing effects in temperate forests, Pretzsch et al. (2013) reported overyielding in oak-beech mixtures to be dependent on site fertility according to expectations of the stress gradient hypothesis (i.e. positive interactions are expected to be more frequent under severe environmental conditions, Bertness and Callaway 1994): fertile sites exhibited under-yielding while infertile sites exhibited over-yielding. Besides stress associated with soil fertility, beech (as with Pines) can benefit from hydraulic lift abilities of oaks in temperate areas which increase soil moisture and also indirectly nutrient levels.

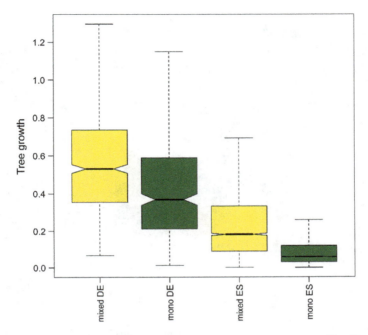

Fig. 15.4 Box-plot of tree growth in oak forests between mixed and monospecific plots in Spain (SP) and Germany (DE)

In oak forests from the forest inventories of Spain and Germany we observed larger tree growth in mixed than in monospecific forests (Fig. 15.4) although this correlation can be based on indirect effects and not necessarily on a causal diversity effect. The pattern of exploratory analysis between tree species richness and growth in both forest inventories showed a positive relationship in oak forests (Fig. 15.5) in agreement with observational studies that have generally found a positive effect of diversity on forest productivity and biomass storage (see Ratcliffe et al. 2016a). Across Europe it has been found that diversity effects on productivity show the highest importance in Mediterranean regions (see e.g. Ruiz-Benito et al. 2014) declining along the temperate biome (Ratcliffe et al. 2016a). In addition, Ruiz-Benito et al. (2017a) suggested that the effect of diversity on sapling abundance is much larger than the effect on tree growth, while no effect of diversity on mortality was found. Regeneration and mortality, however, are critical demographic processes underlying forest dynamics and succession and, therefore, the overall magnitude of the effects of diversity on forest functioning might have been underestimated. Areas with higher functional diversity could reflect the presence of individuals with contrasting functional traits, such as pine-oak forests that are typical across Europe depending on the environmental conditions (see e.g. Zavala and Zea 2004; Carnicer et al. 2014). Jucker et al. (2015) found that mixing pines and oaks at a stand increased aboveground wood production by reducing

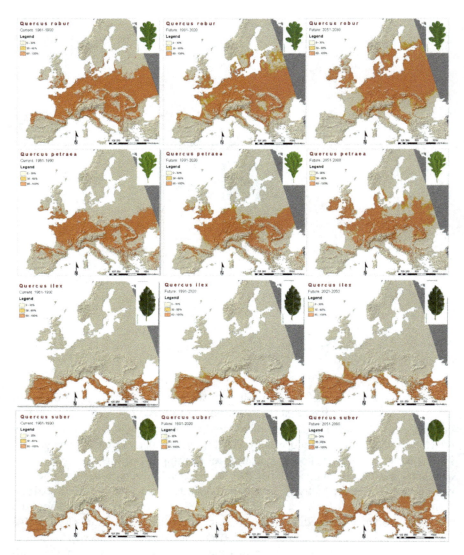

Fig. 15.5 Species niches and their respective shifts over time for four European oak species. Colour-coded are suitable habitats found in 0–30, 30–60, and 60–100% of all model combinations (regional models according to Lindner et al. 2014)

competition for light between neighbours. Pines in mixed stands received more light and had faster growth rates than in monoculture. Interestingly, the likely complementarity effects underpinning the pine-oak association fade out under extreme drought conditions (see also Zweifel et al. 2009).

The ability of species to tolerate shade is considered one of the primary mechanisms behind positive effects of species mixing on tree growth and forest functioning (Liang et al. 2016). Morin et al. (2011), using a forest succession model, showed how increased species richness promotes forest functioning, and enhanced demographic responses to mortality events, by means of aboveground complementarity only (i.e. more efficient partitioning of solar radiation). Complementarity effects can take place both in space and time (Holzwarth et al. 2016). In space, different light harvesting strategies are critical in explaining increased resource yielding in mixed forest stands (Ishii and Asano 2010; Ishii et al. 2013). In particular, specific leaf area has been shown to be directly involved with carbon yields and aboveground competition in forests worldwide (Kunstler et al. 2016). Grassland studies revealed how differences in leaf phenology determine temporal segregation of light capture and thus species coexistence (Mason et al. 2013). Tree size is directly related with access to solar radiation in forests and thus tolerance to shading of small trees plays a pivotal role in forest dynamics and tree development (Kohyama and Takada 2009, 2012). Decoupling between juvenile and adult tree leaf phenology has been reported in mixed temperate forests (Augspurger 2003; Vitasse et al. 2014). Ontogenetic differences in tree size may be thus critical in interpreting diversity-productivity relationship in forests (Nadrowski et al. 2010), especially with secondary succession (Holzwarth et al. 2016). Trees have a continuous size development that modulates growth (Stephenson et al. 2014), stand productivity (Coomes et al. 2014) and within-community interactions (Le Roux et al. 2013). A different size of individuals implies not only niche differentiation between young and adult trees (either conspecifics or heterospecifics; see Niinemets 2010) but also idiosyncratic growth responses along climatic/structural gradients (Zhang et al. 2012; Madrigal-González and Zavala 2014). This implies that individual responses to environmental factors change through ontogeny and thus, diversity-productivity relationships might shift during secondary succession. Following this reasoning, Madrigal-González et al. (2016) observed that tree size is critical for complementarity effects across Europe in support to the idea that competitive interactions for any given focal tree are contingent on its size and thus asymmetrical interactions in forests are pivotal also for the diversity-productivity relationships. In particular reversals in the strength of complementarity effects due to the size of focal trees across the European latitudinal PET gradient were observed: i.e. while complementarity effects were stronger in small trees in temperate forests, only large trees showed complementarity effects in Mediterranean forests. In contrast, small trees in Mediterranean forests showed negative growth responses to increased functional dissimilarity in the neighbourhood. This suggests that saplings of oak species such as *Quercus ilex* and *Quercus faginea* grow less when mixed with conifer species in dry forests. This is not to say that oak species are not able to recruit beneath adult pine trees (which otherwise is a common successional pathway in the Mediterranean), but to affirm that oak saplings growing faster beneath oaks might be reflecting trade-offs in water-limited forests to cope with shade and drought (e.g. Zavala et al. 2000). For large trees, the findings of Madrigal-González et al. (2016) are consistent with the larger role of diversity in Mediterranean water-limited forests at the stand scale (e.g. Ratcliffe et al. 2016a; Ruiz-Benito et al. 2017a).

15.4 Global Change and the Fate of Oak Forests in Europe

Several factors determine the response of a tree species to changing environmental conditions, such as the physiological ability of a tree species to cope with environment variability including variations in climate, the legacy effects of past forest management, as well as changes in current forest management (Gimmi and Bürgi 2007). These factors can be considered as drivers of the contemporary distribution and potential range shifts of a species. If the climatic conditions change, tree species are forced to adapt, to migrate, or they go locally extinct or at least simply persist for shorter or longer times locally with little capacity to expand from there. The adaptive and plastic response of a species largely depends on its physiological traits (Martinez-Vilalta et al. 2009; Sterck et al. 2012) and the level of competitive stress at a specific site (Purves et al. 2007a, b). There is evidence that growth but also the risk of local mortality depends on how strongly the current conditions deviate from optimal conditions for a respective species (Sáenz-Romero et al. 2016). According to the niche theory, every species has a specific range of environmental conditions within which it can maintain a non-negative population growth rate over time. This means that there are limits of climate conditions beyond which a specific tree species is not able to survive through time, either because the environmental conditions are beyond its physiological tolerance (fundamental niche is exceeded) or the local competitors are too strong for the focal species to survive (realized niche is exceeded). The more the conditions deviate from the optimum of a species' niche, the more likely we are approaching a species' niche edge, and the closer the current conditions are locate at a species' niche edge, the more likely it is that a species will respond with migration or local extinction if the conditions shift further away from species' optima.

To what extent a species is able to colonize new areas after climate change depends on the degree of climate change, the availability of suitable areas in reachable distance, and a species' capability to migrate. Seed dispersal (Snell et al. 2014) and forest habitat connectivity (Meier et al. 2012) play an important role in determining migration, the interaction with existing species (Meier et al. 2012; Nabel et al. 2013; Svenning et al. 2014) and even the origin of a species (Sáenz-Romero, GCB, 2017). Together these factors determine the maximum speed at which a tree species can migrate. Recent studies based on forest inventories show that the species may migrate considerably slower (in the range of <100 m per year) than assumed earlier (<1000 m per year) under conditions that are less strongly fragmented and managed (e.g. Powell and Zimmermann 2004). In a study focusing on Holm oak (*Quercus ilex*), one of the main reasons for the slow migration rate was found to be the performance to interact and compete with other species in the new area in a fragmented and managed landscape (Delzon et al. 2013). Independent of the exact migration rate per year, many studies on the fate of tree species in Europe state a migration speed too low to keep track with the predicted change in climate (Delzon et al. PLOS one 2013; Feurdean et al. 2013). This indicates that for a given species, a considerable proportion of the current range will fall outside of the realized niche of the species at a rate that is faster than the species can cope with by means of migration.

15.4.1 Climate Change Projections for Europe

Climate change impact studies that analyse the potential responses of tree species ideally include regional variations in climate change, the change in the frequency of extreme events, as well as uncertainties regarding the expected changes. While most species of a region are to some extent affected by the strength of mean changes in temperature and humidity, every species and every life stage of each species responds differently to changing climate variability (Lindner et al. 2014). Using simple scenarios and a limited set of climate variables and focusing on changes in climate means likely misses some of the expected dynamics. However, little is known to date on how changing extremes will affect species' population dynamic behaviour in the longer term, although we know that climate extremes (climate variability) does explain the distribution of species to some extent (Zimmermann et al. 2009). Thus, the impact of extremes on future species responses remains poorly predictable to date.

Here, we present an analysis of oak species response to climate changes using six statistical, niche-based models to simulate the species response and climate data from six regional climate models (RCMs) and included five biologically relevant climate variables in order to include relevant drivers of potential vegetation shifts (Lindner et al. 2014). Details of the selected variables, statistical models and climate data are given in Zimmermann et al. (2013). The RCMs were fed by global circulation model (GCM) output originating from IPCC AR4 projections and are in line with a projected 3.5–6.2 °C increase in global mean annual temperature by 2100 relative to pre-industrial levels (Peters et al. 2013). Summer temperatures are expected to increase in the range between 1.3 and 4.1 °C with the highest increase in Southern Europe and more pronounced away from coasts (Lindner et al. 2014). Winter temperatures are projected to increase by 1.5–4.2 °C with North-eastern Europe showing the highest increase. Summer precipitations are projected to increase in Northern Europe (0–25%) and to decrease in Central (0–25%) and Southern Europe (25–50%). Projected winter precipitations changes range from a reduction of about 35% (mostly Southern Europe) to an increase of 5–40% (mostly Central to Northern Europe). Further, the used RCMs project a strong increase in the length of drought periods in Southern and South-Central Europe and a reduction of the risk of drought stress in more northern latitudes (Lindner et al. 2014). More generally, it is expected that temperature and precipitation extremes will increase (Seneviratne et al. 2012) and that summer heat waves as experienced in 2003 and 2010 will become more frequent (Barriopedro et al. 2011). It is very likely that these extreme events will have a stronger impact on forests than the gradual change in average conditions.

15.4.2 Model Predictions for Four Oak Species in Europe

In the following, results from niche-based model projections for *Quercus ilex, Q. suber, Q robur* and *Q. petraea* are presented based on underlying information illustrated in the review by Lindner et al. (2014). Figure 15.5 shows projected potential range shifts between the calibration period 1961–90 (based on ICP Forest, Level I data, Lorenz 1995) to the periods 1991–2010 and 2051–2080 based on the above described climate change projections. The model projects the change in suitable habitats in geographic space for the four oak species in Europe. The general patterns of suitable habitats through time show a shift to the northeast for *Q. robur*. This species largely disappears in Spain and in parts of France and new habitats become suitable in the NE of Europe. A similar dynamic is projected for *Quercus petrea*, however less pronounced than for *Q. robur*.

Q. ilex is projected to disappear in the SW of Spain and to gain new suitable habitats in Western France, which is in line with the currently observed expansions on of *Q. ilex* from plantations in this region (Delzon et al. 2013). Only marginal changes occur in other areas of the current range. A shift from the SW to the NE of Spain is projected for *Q. suber*. The species will find large areas to become suitable in the West of France, in Central Italy, Macedonia, Greece, and Turkey.

All analysed oak species show more or less pronounced shifts in suitable habitats, with comparably large areas of newly suitable towards the north (east) in Europe and in higher altitudes within their current range. The projected new ranges are of similar size or will be even larger. However, the simulations do not answer the question, how fast the species are moving to these newly suitable areas, whether they are able to track the changing climate conditions without time lags, or whether they do or how fast they do disappear from areas that become unsuitable in the future. This is because the models used include competition only implicitly and have no dynamic processes included that would allow to simulate population dynamics, migration behaviour or include the effects of geographical barriers on migration (see Delzon et al. 2013; Lindner et al. 2014). A similar study where such niche-based models were combined with the dynamic model TreeMig (Lischke et al. 2006) revealed that only early successional species might be able to almost keep track of projected climate changes and that many late successional species, such as most oaks, tend to migrate very slowly and are by far less capable of tracking climate change (Meier et al. 2012).

While the simulation of realistic migration is already difficult, the simulation of local extinction is even more complex. A species usually disappears locally due to one or both of two main constraints when the climate changing to conditions that are outside of the realized niche: (a) the species still finds habitat that it can tolerate physiologically, but it does not compete well in the long run because other species are more competitive under these conditions; (b) the species cannot physiologically cope with the new climate conditions. In the first case, a species might persist for long time periods, spanning decades to centuries, and only slowly disappear once stronger competitors invade the area locally. In the second case, an extreme event might rapidly cause large scale mortality events such as those reported in Allen et al. (2010). In essence, it means that the local extinction of tree species is even more difficult to predict. The niche-based models simply project that there is no current evidence that the species can persist through time. The consequence is, that the species will be replaced by others, ant this has monetary consequences for the economies of entire Europe (Hanewinkel et al. 2013), particularly when the transition phase should lead to negative consequences for the local or regional provisioning of ecosystem services with related impacts on the human communities.

A general outcome from projections such as those presented above is that: (a) drought resistant tree species will become more wide-spread in comparison to other European species, and (b) the more drought tolerant species tend to provide lower economic benefits. Oak species will most likely gain considerable distribution range, particularly in Central Europe, due to their resilience against drought events. This conclusion from modelling studies is supported by several field experiments investigating regeneration (Arend et al. 2016), growth patterns (Weber et al. 2007, 2008; Eilmann et al. 2009), and ecophysiological responses to changing microclimate (Zweifel et al. 2007, 2009).

Acknowledgements The research leading to these results has received funding from the project FUNDIVER (MINECO, Spain; No. CGL2015-69186-C2-2-R) and the FunDivEUROPE project received funding from the European Union's Seventh Programme (FP7/2007–2013) under grant agreement No. 265171. JMG was also supported by a Postdoctoral fellowship in the Universidad de Alcalá (Spain); PRB by TALENTO Fellow funded by the Autonomous Region of Madrid and the University of Alcalá (2016-T2/AMB-1665). We thank the MAGRAMA for access to the Spanish NFI and the Johann Heinrich von Thünen-Institut for access to the German NFI. Special thanks to Gerald Kändler for his help with the German NFI.

References

Abrams MD (1996) Distribution, historical development and ecophysiological attributes of oak species in the eastern United States. Ann Sci For 53:487–512
Acherar M, Rambal S (1992) Comparative water relations of four Mediterranean oak species. Vegetatio 99–100:177–184
Allen CD, Macalady AK, Chenchouni H et al (2010) A global overview of drought and heat-induced tree mortality reveals emerging climate change risks for forests. For Ecol Manage 259:660–684
Arend M, Kuster T, Günthardt-Goerg MS, Dobbertin M (2011) Provenance-specific growth responses to drought and air warming in three European oak species (*Quercus robur*, *Q petraea* and *Q pubescens*). Tree Physiol 31:287–297
Arend M, Brem A, Kuster TM, Günthardt-Goerg MS (2013) Seasonal photosynthetic responses of European oaks to drought and elevated daytime temperature. Plant Biol 15:169–176
Arend M, Sever K, Pflug E, Gessler A, Schaub M (2016) Seasonal photosynthetic response of European beech to severe summer drought: limitation, recovery and post-drought stimulation. Agric For Meteorol 220:83–89
Augspurger CK (2003) Differences in leaf phenology between juvenile and adult trees in a temperate deciduous forest. Tree Physiol 23:517–525

Barbero M, Bonin G, Loisel R, Quézel P (1990) Changes and disturbances of forest ecosystems caused by human activities in the western part of the Mediterranean basin. Vegetatio 87: 151–173
Barriopedro D, Fischer EM, Luterbacher J, Trigo RM, García-Herrera R (2011) The hot summer of 2010: redrawing the temperature record map of Europe. Sci 332:220–224
Bertness MD, Callaway R (1994) Positive interactions in communities. Trends Ecol Evol 9: 191–193
Bréda N, Huc R, Granier A, Dreyer E (2006) Temperate forest trees and stands under severe drought: a review of ecophysiological responses, adaptation processes and long-term consequences. Ann For Sci 63:625–644
Canadell J, Jackson RB, Ehleringer JB, Mooney HA, Sala OE, Schulze ED (1996) Maximum rooting depth of vegetation types at the global scale. Oecologia 108:583–595
Carnicer J, Barbeta A, Sperlich D, Coll M, Penuelas J (2013) Contrasting trait syndromes in angiosperms and conifers are associated with different responses of tree growth to temperature on a large scale. Front Plant Sci 4(409)
Carnicer J, Coll M, Pons X, Ninyerola M, Vayreda J, Peñuelas J (2014) Large-scale recruitment limitation in Mediterranean pines: the role of *Quercus ilex* and forest successional advance as key regional drivers. Global Ecol Biogeogr 23:371–384
Cavender-Bares J, Ackerly DD, Baum DA, Bazzaz FA (2004) Phylogenetic overdispersion in Floridian oak communities. Am Nat 163:823–843
Cavender-Bares J, Cortes P, Rambal S, Joffre R, Miles B, Rocheteau A (2005) Summer and winter sensitivity of leaves and xylem to minimum freezing temperatures: a comparison of co-occurring Mediterranean oaks that differ in leaf lifespan. New Phytol 168:597–612
Chamagne J, Tanadini M, Frank D, Matula R, Paine CE, Philipson CD, Svatek M, Turnbull LA, Volařík D, Hector A (2017). Forest diversity promotes individual tree growth in central European forest stands. J Appl Ecol 54:71–79
Chase JM, Leibold MA (2003) Ecological niches: linking classical and contemporary approaches. University of Chicago Press, Chicago
Clark JS, Beckag B, Camill P, Cleveland B, HilleRisLambers J, Lichter J, Wyckoff P (1999) Interpreting recruitment limitation in forests. Am J Bot 86:1–16
Comes HP, Kadereit JW (1998) The effect of Quaternary climatic changes on plant distribution and evolution. Trends Plant Sci 3:432–438
Coomes DA, Flores O, Holdaway R, Jucker T, Lines ER, Vanderwel MC (2014) Wood production response to climate change will depend critically on forest composition and structure. Global Change Biol 20:3632–3645
Delzon S, Urli M, Samalens JC, Porté AJ (2013) Field evidence of colonisation by Holm oak, at the northern margin of its distribution range, during the Anthropocene period. PLoS ONE 8: e80443
Diaz S, Cabido M, Casanoves F (1998) Plant functional traits and environmental filters at a regional scale. J Veg Sci 9:113–122
Eaton, E, Caudullo, G, Oliveira, S, de Rigo, D (2016) *Quercus robur* and *Quercus petraea* in Europe: distribution, habitat, usage and threats In: San-Miguel-Ayanz, J, de Rigo, D, Caudullo, G, Houston Durrant, T, Mauri, A (eds) European atlas of forest tree species, Publication Office of the European Union, Luxembourg
Eilmann B, Zweifel R, Buchmann N, Fonti P, Rigling A (2009) Drought-induced adaptation of the xylem in Scots pine and pubescent oak. Tree Physiol 29:1011–1020
Falster DS, Brännström Å, Westoby M, Dieckmann U (2017) Multitrait successional forest dynamics enable diverse competitive coexistence. Proc Natl Acad Sci U S A 114: E2719–E2728
Feurdean A, Bhagwat SA, Willis KJ, Birks HJB, Lischke H, Hickler T (2013) Tree migration-rates: narrowing the gap between inferred post-glacial rates and projected rates. PLoS ONE 8:e71797
Fonti P, Heller O, Cherubini R, Rigling A, Arend M (2013) Wood anatomical responses of oak seedlings exposed to heat and drought. Plant Biol 15:210–219

Galiano L, Martínez-Vilalta J, Lloret F (2010) Drought-induced multifactor decline of Scots pine in the Pyrenees and potential vegetation change by the expansion of co-occurring oak species. Ecosystems 13:978–991

García-Valdés R, Gotelli NJ, Zavala MA, Purves DW, Araújo MB (2015) Effects of climate, species interactions, and dispersal on decadal colonization and extinction rates of Iberian tree species. Ecol Model 309:118–127

Gimmi U, Bürgi M (2007) Using oral history and forest management plans to reconstruct traditional non-timber forest uses in the Swiss Rhone valley (Valais) since the late nineteenth century. Environ Hist-UK 13:211–246

Gimmi U, Wohlgemuth T, Rigling A, Hoffmann CW, Bürgi M (2010) Land-use and climate change effects in forest compositional trajectories in a dry Central-Alpine valley. Ann For Sci 67:701

Gómez-Aparicio L, García-Valdés R, Ruiz-Benito P, Zavala MA (2011) Disentangling the relative importance of climate, size and competition on tree growth in Iberian forests: implications for forest management under global change. Global Change Biol 17:2400–2414

Gotelli NJ, McCabe DJ (2002) Species co-occurrence: a meta-analysis of jm diamond's assembly rules model. Ecology 83:2091–2096

Greenwood S, Ruiz-Benito P, Martínez-Vilalta J, Lloret F, Kitzberger T, Allen CA, Fenshman R, Laughlin D, Kattge J, Boehnish G, Kraft N, Jump AS (2017) Tree mortality across forest biomes is promoted by drought intensity, lower wood density and higher specific leaf area. Ecol Lett 20:539–553

Hanewinkel M, Cullmann DA, Schelhaas M-J, Nabuurs G-J, Zimmermann NE (2013) Climate change may cause severe loss in the economic value of European forest land. Nature Clim Change 3:203–207

Hardin G (1960) The competitive exclusion principle. Science 131:1292–1297

Holmgren M (2000) Combined effects of shade and drought on tulip poplar seedlings: trade-off in tolerance or facilitation? Oikos 90:67–78

Holmgren M, Scheffer M, Huston MA (1997) The interplay of facilitation and competition in plant communities. Ecology 78:1966–1975

Holmgren M, Gómez-Aparicio L, Quero JL, Valladares F (2012) Non-linear effects of drought under shade: reconciling physiological and ecological models in plant communities. Oecologia 169:293–305

Holt RD (2003) On the evolutionary ecology of species' ranges. Evol Ecol Res 5:159–178

Holzwarth F, Rüger N, Wirth C (2016) Taking a closer look: disentangling effects of functional diversity on ecosystem functions with a trait-based model across hierarchy and time. R Soc Open Sci 2:140541

Hurtt GC, Pacala SW (1995) The consequences of recruitment limitation: reconciling chance, history and competitive differences between plants. J Theor Biol 176:1–12

Huston MA, DeAngelis DL (1994) Competition and coexistence: the effects of resource transport and supply rates. Am Nat 144:954–977

Ishii H, Asano S (2010) The role of crown architecture, leaf phenology and photosynthetic activity in promoting complementary use of light among coexisting species in temperate forests. Ecol Res 25:715–722

Ishii H, Azuma W, Nabeshima E (2013) The need for a canopy perspective to understand the importance of phenotypic plasticity for promoting species coexistence and light-use complementarity in forest ecosystems. Ecol Res 28:191–198

Jucker T, Bouriaud O, Coomes DA (2015) Crown plasticity enables trees to optimize canopy packing in mixed-species forests. Funct Ecol 29:1078–1086

Jump AS, Ruiz-Benito P, Greenwood S, Allen CD, Kitzberger T, Fensham R, Martínez-Vilalta J, Lloret F (2017) Structural overshoot of tree growth with climate variability and the global spectrum of drought-induced forest dieback. Global Change Biol 23:3742–3757

Kändler G (2009) The design of the second German national forest inventory. In: Proceedings of the eighth annual forest inventory and analysis symposium

Kleinschmit J (1993) Intraspecific variation of growth and adaptive traits in European oak species. Ann Sci For 50:166–185

Kohyama T, Takada T (2009) The stratification theory for plant coexistence promoted by one-sided competition. J Ecol 97:463–471

Kohyama TS, Takada T (2012) One-sided competition for light promotes coexistence of forest trees that share the same adult height. J Ecol 100:1501–1511

Kunstler G, Lavergne S, Courbaud B, Thuiller W, Vieilledent G, Zimmermann NE, Coomes DA (2012) Competitive interactions between forest trees are driven by species' trait hierarchy, not phylogenetic or functional similarity: implications for forest community assembly. Ecol Lett 15:831–840

Kunstler G, Falster D, Coomes DA, Hui F, Kooyman RM, Laughlin DC, Poorter L, Vanderwel M, Vieilledent G, Wright SJ, Aiba M, Baraloto C, Caspersen J, Cornelissen JHC, Gourlet-Fleury S, Hanewinkel M, Herault B, Kattge J, Kurokawa H, Onoda Y, Peñuelas J, Poorter H, Uriarte M, Richardson S, Ruiz-Benito P, Sun IF, Ståhl G, Swenson NG, Thompson J, Westerlund B, Wirth C, Zavala MA, Zeng H, Zimmerman JK, Zimmermann NE, Westoby M (2016) Plant functional traits have globally consistent effects on competition. Nature 529: 204–207

Küster H (1997) The role of farming in the postglacial expansion of beech and hornbeam in the oak woodlands of central Europe. Holocene 7:239–242

Le Roux PC, Shaw JD, Chown SL (2013) Ontogenetic shifts in plant interactions vary with environmental severity and affect population structure. New Phytol 200:241–250

Levitt J (1978) An overview of freezing injury and survival, and its interrelationships to other stresses. In: Li PH, Sakai A (eds) Plant cold hardiness and freezing stress. Academic Press, New York

Liang J, Crowther TW, Picard N, Wiser S, Zhou M, Alberti G, Schulze E-D, McGuire AD, Bozzato F, Pretzsch H, de-Miguel S, Paquette A, Hérault B, Scherer-Lorenzen M, Barrett CB, Glick HB, Hengeveld GM, Nabuurs G-J, Pfautsch S, Viana H, Vibrans AC, Ammer C, Schall P, Verbyla D, Tchebakova N, Fischer M, Watson JV, Chen HYH, Lei X, Schelhaas M-J, Lu H, Gianelle D, Parfenova EI, Salas C, Lee E, Lee B, Kim HS, Bruelheide H, Coomes DA, Piotto D, Sunderland T, Schmid B, Gourlet-Fleury S, Sonké B, Tavani R, Zhu J, Brandl S, Vayreda J, Kitahara F, Searle EB, Neldner VJ, Ngugi MR, Baraloto C, Frizzera L, Bałazy R, Oleksyn J, Zawiła-Niedźwiecki T, Bouriaud O, Bussotti F, Finér L, Jaroszewicz B, Jucker T, Valladares F, Jagodzinski AM, Peri PL, Gonmadje C, Marthy W, O'Brien T, Martin EH, Marshall AR, Rovero F, Bitariho R, Niklaus PA, Alvarez-Loayza P, Chamuya N, Valencia R, Mortier F, Wortel V, Engone-Obiang NL, Ferreira LV, Odeke DE, Vasquez RM, Lewis SL, Reich PB (2016) Positive biodiversity-productivity relationship predominant in global forests. Sci 354

Lindner M, Fitzgerald JB, Zimmermann NE, Reyer C, Delzon S, van der Maaten E, Schelhaas M-J, Lasch P, Eggers J, van der Maaten-Theunissen M, Suckow F, Psomas A, Poulter B, Hanewinkel M (2014) Climate change and European forests: What do we know, what are the uncertainties, and what are the implications for forest management? J Env Manag 146:69–83

Lischke H, Zimmermann NE, Bolliger J, Rickebusch S, Löffler TJ (2006) TreeMig: a forest-landscape model for simulating spatio-temporal patterns from stand to landscape scale. Ecol Model 199:409–420

Lonsdale WM (1990) The self-thinning rule: dead or alive? Ecology 71:1373–1388

Lookingbill TR, Zavala MA (2000) Spatial pattern of *Quercus ilex* and *Quercus pubescens* recruitment in *Pinus halepensis* dominated woodlands. J Veg Sci 11:607–612

Loreau M (2000) Biodiversity and ecosystem functioning: recent theoretical advances. Oikos 91:3–17

Loreau M, Hector A (2001) Partitioning selection and complementarity in biodiversity experiments. Nature 412:72–76

Lorenz M (1995) International co-operative programme on assessment and monitoring of air pollution effects on forests. Water Air Soil Pollut 85:1221–1226

Madrigal-González J, Zavala MA (2014) Competition and tree age modulated last century pine growth responses to high frequency of dry years in a water limited forest ecosystem. Agric For Meteor 192–193:18–26

Madrigal-González J, García-Rodríguez JA, Zavala MA (2014) Shrub encroachment shifts the bioclimatic limit between marcescent and sclerophyllous oaks along an elevation gradient in west-central Spain. J Veg Sci 25:514–524

Madrigal-González J, Ruiz-Benito P, Ratcliffe S, Calatayud J, Kändler G, Lehtonen A, Dahlgren J, Wirth C, Zavala MA (2016) Complementarity effects on tree growth are contingent on tree size and climatic conditions across Europe. Sci Rep-UK 6:32233

Martinez-Vilalta J, Cochard H, Mencuccini M, Sterck FJ, Herrero A, Korhonen JFJ, Llorens P, Nikinmaa E, Poyatos R, Ripullone F, Sass-Klaassen U, Zweifel R (2009) Hydraulic adjustment of Scots pine across Europe. New Phytol 184:353–364

Mason NW, Pipenbaher N, Škornik S Kaligarič M (2013) Does complementarity in leaf phenology and inclination promote co-existence in a species-rich meadow? Evidence from functional groups. J Veg Sci 24:94–100

Meier ES, Lischke H, Schmatz DR, Zimmermann NE (2012) Climate, competition and connectivity affect future migration and ranges of European trees. Global Ecol Biogeogr 21:164–178

Mokany K, Raison R, Prokushkin AS (2006) Critical analysis of root: shoot ratios in terrestrial biomes. Global Change Biol 12:84–96

Montoya D, Zavala MA, Rodríguez MA, Purves DW (2008) Animal versus wind dispersal and the robustness of tree species to deforestation. Science 320:1502–1504

Morin X, Améglio T, Ahas R, Kurz-Besson C, Lanta V, Lebourgeois F, Chuine I (2007) Variation in cold hardiness and carbohydrate concentration from dormancy induction to bud burst among provenances of three European oak species. Tree Physiol 27:817–825

Morin X, Fahse L, Scherer-Lorenzen M, Bugmann H (2011) Tree species richness promotes productivity in temperate forests through strong complementarity between species. Ecol Lett 14:1211–1219

Nabel JEMS, Zurbriggen N, Lischke H (2013) Interannual climate variability and population density thresholds can have a substantial impact on simulated tree species' migration. Ecol Model 257:88–100

Nadrowski K, Wirth C, Scherer-Lorenzen M (2010) Is forest diversity driving ecosystem function and service? Curr Opin Environ Sustain 2:75–79

Niinemets Ü (2010) A review of light interception in plant stands from leaf to canopy in different plant functional types and in species with varying shade tolerance. Ecol Res 25:693–714

Niinemets Ü, Valladares F (2006) Tolerance to shade, drought, and waterlogging of temperate Northern Hemisphere trees and shrubs. Ecol Monogr 76:521–547

Pacala SW, Rees M (1998) Models suggesting field experiments to test two hypotheses explaining successional diversity. Am Nat 152:729–737

Pardos M, Jiménez MD, Aranda I, Puértolas J, Pardos JA (2005) Water relations of cork oak (*Quercus suber* L) seedlings in response to shading and moderate drought. Ann For Sci 62:377–384

Peet RK, Christensen NL (1987) Competition and tree death. Bioscience 37:586–595

Pérez-Ramos IM, Urbieta IR, Marañón T, Zavala MA, Kobe RK (2008) Seed removal in two coexisting oak species: ecological consequences of seed size, plant cover and seed-drop timing. Oikos 117:1386–1396

Pérez-Ramos IM, Padilla-Díaz CM, Koenig WD, Marañón T (2015) Environmental drivers of mast-seeding in Mediterranean oak species: does leaf habit matter? J Ecol 103:691–700

Peters GP, Andrew RM, Boden T, Canadell JG, Ciais P, Le Quere C, Marland G, Raupach MR, Wilson C (2013) The challenge to keep global warming below 2 C. Nature Clim Change 3:4–6

Poorter L, Bongers L, Bongers F (2006) Architecture of 54 moist-forest tree species: traits, trade-offs, and functional groups. Ecology 87:1289–1301

Poorter L, Lianes E, Moreno-de Las Heras M, Zavala MA (2012) Architecture of Iberian canopy tree species in relation to wood density, shade tolerance and climate. Plant Ecol 213:707–722

Powell JA, Zimmermann NE (2004) Multiscale analysis of active seed dispersal contributes to resolving Reid's paradox. Ecology 85:490–506

Prentice C, Guiot J, Huntley B, Jolly D, Cheddadi R (1996) Reconstructing biomes from palaeoecological data: a general method and its application to European pollen data at 0 and 6 ka. Clim Dynam 12:185–194

Pretzsch H (2006) Species-specific allometric scaling under self-thinning: evidence from long-term plots in forest stands. Oecologia 146:572–583

Pretzsch H, Bielak K, Block J, Bruchwald A, Dieler J, Ehrhart H-P, Kohnle U, Nagel J, Spellmann H, Zasada M, Zingg A (2013) Productivity of mixed versus pure stands of oak (*Quercus petraea* (Matt) Liebl *and Quercus robur* L) and European beech (*Fagus sylvatica* L) along an ecological gradient. Eur J For Res-Jpn 132:263–280

Purves DW, Zavala MA, Ogle K, Prieto F, Benayas JMR (2007a) Environmental heterogeneity, bird-mediated directed dispersal and oak woodland dynamics in Mediterranean Spain. Ecol Monogr 77:77–97

Purves DW, Lichstein JW, Pacala SW (2007b) Crown plasticity and competition for canopy space: a new spatially implicit model parameterized for 250 North American tree species. PLoS ONE 2:e870

Quero JL, Villar R, Marañón T, Zamora R (2006) Interactions of drought and shade effects on seedlings of four Quercus species: physiological and structural leaf responses. New Phytol 170:819–834

Ratcliffe S, Ruiz-Benito P, Kändler G, Zavala MA (2016a) Retos y oportunidades en el uso de Inventarios Forestales Nacionales para el estudio de la relación entre la diversidad y el aprovisionamiento de servicios ecosistémicos en bosques. Ecosistemas 25:60–69

Ratcliffe S, Liebergesell M, Ruiz-Benito P et al (2016b) Modes of functional biodiversity control on tree productivity across the European continent. Global Ecol Biogeogr 25:251–262

Rigling A, Bigler C, Eilmann B, Feldmeyer-Christe E, Gimmi U, Ginzler C, Graf U, Mayer P, Vacchiano G, Weber P, Wohlgemuth T, Zweifel R ,Dobbertin M (2013) Driving factors of a vegetation shift from Scots pine to pubescent oak in dry Alpine forests. Global Change Biol 19:229–240

Rodríguez-Calcerrada J, Pardos JA, Gil L, Reich PB, Aranda I (2008) Light response in seedlings of a temperate (*Quercus petraea*) and a sub-Mediterranean species (*Quercus pyrenaica*): contrasting ecological strategies as potential keys to regeneration performance in mixed marginal populations. Plant Ecol 195:273–285

Ruiz-Benito P, Lines ER, Gómez-Aparicio L, Zavala MA, Coomes DA (2013) Patterns and drivers of tree mortality in Iberian forests: climatic effects are modified by competition. PLoS ONE 8: e56843

Ruiz-Benito P, Gómez-Aparicio L, Paquette A, Messier C, Kattge J, Zavala MA (2014) Diversity increases carbon storage and tree productivity in Spanish forests. Global Ecol Biogeogr 23:311–322

Ruiz-Benito P, Ratcliffe S, Jump AS, Gómez-Aparicio L, Madrigal-González J, Wirth C, Kändler G, Lehtonen A, Dahlgren J, Kattge J, Zavala MA (2017a) Functional diversity underlies demographic responses to environmental variation in European forests. Global Ecol Biogeogr 26:128–141

Ruiz-Benito P, Ratcliffe S, Zavala MA, Martínez-Vilalta J, Vilà-Cabrera A, Lloret F, Madrigal-González J, Wirth C, Greenwood S, Kändler G, Lehtonen A, Kattge J, Dahlgren J, Jump AS (2017b) Climate- and successional-related changes in functional composition of European forests are strongly driven by tree mortality. Global Change Biol (in press)

Sáenz-Romero C, Lamy JB, Ducousso A, Musch B, Ehrenmann F, Delzon S, Cavers S, Chałupka W, Dağdaş S, Hansen JK, Lee SJ (2016) Adaptive and plastic responses of *Quercus petraea* populations to climate across Europe. Global Change Biol 23: 2831–2847

Sánchez-Gómez D, Zavala MA, Van Schalkwijk DB, Urbieta IR, Valladares F (2008) Rank reversals in tree growth along tree size, competition and climatic gradients for four forest canopy dominant species in Central Spain. Ann For Sci 65:1

Saurer M, Spahni R, Frank DC, Joos F, Leuenberger M, Loader NJ, Andreu-Hayles L (2014) Spatial variability and temporal trends in water-use efficiency of European forests. Global Change Biol 20:3700–3712

Seneviratne SI, Nicholls N, Easterling D, Goodess CM, Kanae S, Kossin J, Luo Y, Marengo J, McInnes K, Rahimi M, Reichstein M, Sorteberg A, Vera C, Zhang X (2012) Changes in climate extremes and their impacts on the natural physical environment. In: Field CB, Barros V, Stocker TF, Qin D, Dokken DJ, Ebi KL, Mastrandrea MD, Mach KJ, Plattner G-K, Allen SK, Tignor M, Midgle PM (eds) Managing the risks of extreme events and disasters to advance climate change adaptation. pp. 109-230. Cambridge University Press, Cambridge, UK, and New York, NY, USA

Shigesada N, Kawasaki K, Teramoto E (1979) Spatial segregation of interacting species. J Theor Biol 79:83–99

Smith TM, Huston MA (1989) A theory of the spatial and temporal dynamics of plant communities. Vegetatio 83:49–69

Snell RS, Huth A, Nabel JEMS, Bocedi G, Travis JMJ, Gravel D, Bugmann H, Gutiérrez AG, Hickler T, Higgins SI, Reineking B, Scherstjanoi M, Zurbriggen N, Lischke H (2014) Using dynamic vegetation models to simulate plant range shifts. Ecography 37:1184–1197

Stephenson NL, Das AJ, Condit R, Russo SE, Baker PJ, Beckman NG, Alvarez E (2014) Rate of tree carbon accumulation increases continuously with tree size. Nature 507:90–93

Sterck FJ, Martinez-Vilalta J, Mencuccini M, Cochard H, Gerrits P, Zweifel R, Herrero A, Korhonen JFJ, Llorens P, Nikinmaa E, Nole A, Poyatos R, Ripullone F, Sass-Klaassen U (2012) Understanding trait interactions and their impacts on growth in Scots pine branches across Europe. Funct Ecol 26:541–549

Svenning JC, Skov F (2007) Could the tree diversity pattern in Europe be generated by postglacial dispersal limitation? Ecol Lett 10:453–460

Svenning JC, Gravel D, Holt RD, Schurr FM, Thuiller W, Münkemüller T, Schiffers KH, Dullinger S, Edwards TC, Hickler T, Higgins SI (2014) The influence of interspecific interactions on species range expansion rates. Ecography 37:1198–1209

Thomas FM, Blank R, Hartmann G (2002) Abiotic and biotic factors and their interactions as causes of oak decline in Central Europe. For Pathol 32:277–307

Tilman D (1988) Plant strategies and the dynamics and structure of plant communities. Princeton University Press, New Jersey

Urbieta IR, Zavala MA, Marañón T (2008) Human and non-human determinants of forest composition in southern Spain: evidence of shifts towards cork oak dominance as a result of management over the past century. J Biogeogr 35:1688–1700

Urbieta IR, García LV, Zavala MA, Marañón T (2011) Mediterranean pine and oak distribution in southern Spain: is there a mismatch between regeneration and adult distribution? J Veg Sci 22:18–31

Valladares F, Pearcy RW (2002) Drought can be more critical in the shade than in the sun: a field study of carbon gain and photo-inhibition in a Californian shrub during a dry El Niño year. Plant Cell Environ 25:749–759

Van der Plas F, Manning P, Allan E, Scherer-Lorenzen M, Verheyen K, Wirth C, Zavala MA, Hector A, Ampoorter E, Baeten L, Barbaro L (2016) Jack-of-all-trades effects drive biodiversity-ecosystem multifunctionality relationships in European forests. Nat Commun 7

Vilà-Cabrera A, Martínez-Vilalta J, Galiano L, Retana J (2013) Patterns of forest decline and regeneration across Scots pine populations. Ecosystems 16:323–335

Villaescusa R, Díaz R (1998) Segundo Inventario Forestal Nacional (1986–1996) Ministerio de Medio Ambiente. ICONA, Madrid

Villanueva JA (2004) Tercer Inventario Forestal Nacional (1997–2007) Comunidad de Madrid Ministerio de Medio Ambiente, Madrid

Vitasse Y, Lenz A, Hoch G, Körner C (2014) Earlier leaf-out rather than difference in freezing resistance puts juvenile trees at greater risk of damage than adult trees. J Ecol 102:981–988

Weber P, Bugmann H, Rigling A (2007) Radial growth responses to drought of Pinus sylvestris and Quercus pubescens in an inner-Alpine dry valley. J Veg Sci 18:777–779

Weber P, Bugmann H, Fonti P, Rigling A (2008) Using a retrospective dynamic competition index to reconstruct forest succession. For Ecol Manage 294:96–106

Welander NT, Ottosson B (1998) The influence of shading on growth and morphology in seedlings of *Quercus robur* L and *Fagus sylvatica* L. For Ecol Manage 107:117–126

West GB, Enquist BJ, Brown JH (2009) A general quantitative theory of forest structure and dynamics. Proc Natl Acad Sci U S A 106:7040–7045

Westoby M (1984) The self-thinning rule. Adv Ecol Res 14:167–225

Whittaker RJ, Nogués-Bravo D, Araújo MB (2007) Geographical gradients of species richness: a test of the wáter-energy conjecture of Hawkins et al. (2003) using European data for five taxa. Global Ecol Biogeogr 16:76–89

Yoda K, Kira T, Ogawa H, Hozumi K (1963) Self-thinning in overcrowded pure stands under cultivated and natural conditions (Intraspecific competition among higher plants XI). J Inst Polytech Osaka City Univ (Japan) Ser D14:107–129

Zavala MA, Zea E (2004) Mechanisms maintaining biodiversity in Mediterranean pine-oak forests: insights from a spatial simulation model. Plant Ecol 171:197–207

Zavala MA, Espelta JM, Retana J (2000) Constraints and trade-offs in Mediterranean plant communities: The case of holm oak-aleppo pine forests. Bot Rev 66:119–149

Zhang Y, Chen HYH, Reich PB (2012) Forest productivity increases with evenness, species richness and trait variation: a global meta-analysis. J Ecol 100:742–749

Zimmermann NE, Yoccoz NG, Edwards TC, Meier ES, Thuiller W, Guisan A, Schmatz DR, Pearman PB (2009) Climatic extremes improve predictions of spatial patterns of tree species. Proc Natl Acad Sci U S A 106: 19723–19728

Zimmermann NE, Normand S, Pearman PB, Psomas A, (2013) Future ranges in European tree species, pp 15–21. In: Fitzgerald J, Lindner M (eds) Adapting to climate change in european forests—results of the MOTIVE project. Pensoft Publishers, Sofia, 108 pp

Zomer RJ, Trabucco A, Bossio DA, Verchot LV (2008) Climate change mitigation:a spatial analysis of global land suitability for clean development mechanism afforestation and reforestation. Agric Ecosystems Envir 126:67–80

Zweifel R, Rigling A, Dobbertin M (2009) Species-specific stomatal response of trees to drought —a link to vegetation dynamics. J Veg Sci 20:442–454

Zweifel R, Steppe K, Sterck FJ (2007) Stomatal regulation by microclimate and tree water relations: interpreting ecophysiological field data with a hydraulic plant model. J Exp Bot 58:2113–2131

Index

A

Abiotic, 5, 39, 149, 154, 248, 330, 331, 334–336, 342, 346, 381, 384, 395, 396, 398, 406, 407, 410, 419, 420, 422, 424, 430, 434, 435, 472, 481, 488, 515, 516, 522
Acorn, 1, 22, 50, 52, 56, 58, 61, 78, 206, 246–249, 342, 420, 438, 439, 454–456, 460, 464–467, 477, 490
Acoustic emissions, 262, 285
Adaptation, 79, 107–109, 111, 112, 115, 118, 119, 126, 130, 145, 146, 150, 156, 167, 170, 173, 208, 240, 274, 374, 407, 428, 436, 515
Adaptive differentiation, 108, 115, 118, 119, 130
AFLPs, 244
Africa, 16, 25, 27, 28, 40, 69, 72, 73, 124, 125, 139, 174, 309, 310, 314, 370, 420, 430, 431
Agrilus biguttatus, 433
Air injection, 285, 287
America, 4, 5, 14, 16, 17, 19–21, 23–26, 28–32, 39–43, 45–47, 49–55, 57, 59, 60, 76, 80, 81, 84, 86, 87, 108–110, 115, 118, 139, 145, 148, 197–199, 309, 310, 314, 370, 375, 376, 382, 384, 420, 430, 431, 433, 435, 437, 472, 479, 514
Anisohydric, 419, 425, 426
Anteraxanthin (A), 7
Anthocyanins, 114, 115, 117, 364, 369, 382, 383
Antioxidants, 364, 367, 369, 370, 373, 374, 377–379
Archaefagaecea, 46
Architecture, 6, 67, 128, 129, 261–263, 283, 424, 460, 462, 463, 472, 519
Arid, 139, 142–146, 152–154, 156, 160–162, 167–169, 173, 177, 261, 264, 271, 273, 283, 285, 289, 290, 293, 294, 296, 329, 331, 439, 481, 513
Aridity, 118, 138–140, 143, 145, 148, 168, 169, 173, 428, 456, 518
Armillaria gallica, 432
Ascorbate (Asc), 369
Asia, 4, 16, 20, 25, 27–32, 40–43, 48, 49, 61, 70, 74–76, 80, 81, 84, 86, 87, 139, 143, 149, 242, 309, 310, 370
Assimilation, 6, 7, 120, 123, 148, 149, 196, 200, 201, 204, 205, 207, 209, 285, 303, 305, 306, 309, 311, 312, 314, 319, 328, 330, 335, 350, 365, 374, 399, 432, 488–490
Atmosphere, atmospheric, 156, 165, 170, 304, 305, 329, 335, 342, 395
Autumn senescence, 364, 381–383

B

Bareroot, 462, 468–470, 472, 479
Bark, 4, 21, 241, 242, 251–254, 281, 342, 433, 434
Barrel, 4
Basimetric area
Bench dehydration, 285, 286, 288
Biogeography, 514
Biomass, 7, 108, 117, 118, 143, 199, 208, 283, 330, 335, 380, 394, 395, 397–399, 402, 403, 406–409, 412, 424, 453, 459, 460, 473, 481, 488, 490, 525
Biomass allocation, 331, 398, 402, 403, 413
Biotic, 39, 119, 149, 154, 330, 331, 339, 342, 346, 381, 406, 419, 420, 422, 424, 425, 430, 433–435, 482, 515, 522
Biscogniauxia mediterranea, 432, 433
Brevideciduous, 109, 111, 370
Broadleaf, 79, 80, 82, 205, 470, 476
Bulk modulus of elasticity, 159
Bundle, 159, 160

C

Camus, Aimée Antoinette, 13
Carbohydrates, 154, 202, 214, 262, 331, 348, 419, 425, 427, 429, 434, 438, 460, 470, 480, 481, 522
Carbon gain, 6, 147–149, 163, 195, 218, 370, 399, 419, 426, 429, 435
Carboxylation, 156, 305, 306, 308, 311, 312, 314, 519
Carotenoids, 363, 364, 366, 367, 369, 382, 383
Castanea, 18, 19, 41, 51, 62, 63, 86, 457
Castanopsis, 18, 19, 41, 45, 51, 57
Catkins, 24, 48, 54, 62, 87
Cavitation, 120, 137, 159, 162, 171, 173, 426, 429, 430, 434, 439
Cavitron, 287, 288
C balance, 328, 333, 334, 338
C budget, 329, 348, 350
Cellulose, 203, 212, 214
Cell wall, 112, 152, 153, 156, 158, 159, 170, 212, 252, 277, 315–317, 319, 341, 480
Cenophytic, 39
Cenozoic, 5, 30, 31, 39, 43, 44, 47, 50, 75, 77, 79–81, 83, 85, 87
Centrifugation, 285, 286
Cerambyx cerdo, 433
Ceratocystis fagacearum, 430
Cerris, 5, 13, 14, 16–18, 20, 26–28, 30–32, 42, 50, 53, 54, 61, 64, 65, 67, 69, 70, 72–78, 81, 83, 84, 86, 150, 206, 241, 251, 261, 263, 264, 266–268, 270, 271, 273, 283–285, 289, 290, 292, 293, 295, 310, 337, 338, 343, 375, 378, 380, 420
Chilling, 108, 112, 370–373, 436
Chlorophyll (Chl), 111, 363, 382, 466
Chlorophyll fluorescence, 6, 111, 315, 366, 480
Chloroplast, 242–244, 303–306, 313, 315–317, 335, 343, 362, 365, 369, 380
Chloroplast surface area exposed to intercellular air spaces (S_c/S), 317, 372
Chlororespiration, 365
Chrysolepis, 23, 41, 55, 80, 145, 150, 266, 310, 314
Classification history, 108, 129, 420
Classification, modern
Classification, phylogenetic, 5, 13, 14, 16, 18, 32, 42, 152, 175, 264, 336
Climate change, 7, 108, 128, 295, 350, 383, 395, 435, 436, 517, 528–530
Climate reconstruction, 265
C loss, 328, 329, 336, 344–350
CO_2, 6, 7, 155, 156, 170, 204, 205, 303–316, 319, 328–330, 333, 335–340, 344, 346–349, 365, 367, 374, 380, 395, 411, 421, 423, 427, 429, 432, 436, 439, 488, 490
CO_2 internal transport, 335, 336
Cold, 5, 6, 81, 82, 86, 108, 109, 111–116, 123–126, 137, 138, 140, 143, 147, 152, 154, 156, 196, 208, 219, 244, 289, 331, 370–374, 383, 423, 424, 436, 460, 466, 468, 470, 471, 475, 478, 479, 482, 515, 522
Cold stress, 113, 198, 208, 474, 478, 479
Common garden, 109, 114, 119, 120, 122, 124, 125, 129, 130, 152–154, 156, 157, 172, 319, 370, 371, 373
Construction costs, 124, 196, 197, 212, 214, 217, 218, 283
Container, 440, 462, 469, 470, 472–475, 479
Coppice, 3, 340
Coppicing, 3, 423, 424, 459, 460
Cork, 3, 239, 241, 242, 244, 246, 247, 249, 251, 252, 255, 396
Cork oak, 4, 72, 74, 124, 241, 242, 245–248, 253, 395, 396
Correlative model, 435, 436
cpDNA, 242–245
Cretaceous, 43, 46–49, 85
Crown transparency, 425, 429, 435
C sink, 350
Cultivation, 4, 8, 454, 455, 462, 466, 472, 473, 477, 483, 484, 487
Cuticle, 43, 79, 171
Cuticular permeance, 171
Cyclobalanopsis, 5, 13, 14, 16–20, 22, 26, 31, 32, 41, 42, 50, 53, 55, 68, 69, 74, 75, 77–81, 84, 86, 87, 264, 266–271, 273, 283, 310, 369, 375

D

Dark-acclimated Fv/Fm, 111, 113, 430, 478
Deciduous, 3, 6, 7, 21, 24, 41, 48, 70, 72, 74, 78–80, 82–87, 111, 118, 119, 122, 124, 137, 140, 143, 145–148, 151–155, 157–162, 164, 165, 169, 171, 172, 174–176, 195–203, 205–217, 219, 261, 264, 265, 271, 274, 284, 285, 289, 290, 292–294, 296, 303, 306, 309–315, 317, 318, 329, 331, 333, 336, 340, 363, 369, 370, 372, 374–378, 381, 395, 398–400, 407, 420, 431, 456, 459, 466, 469–471, 478, 517, 519–521
Decline, 3, 7, 8, 112, 113, 116, 141, 295, 318, 319, 333, 334, 409, 419–426, 428–439, 514, 523
Dehesa, 3, 396

Dendrochronology, 265
Dendroecology
Diagnosis, 40, 62
Diagnostic traits, 20
Diameter at breast heigh (DBH), 402, 403, 409–412
Dicotylophyllum, 48
Dieback, 295, 420, 422, 423, 425, 429, 438
Diffuse-porous, 6, 21, 261, 262, 264, 266–272, 281, 284, 285, 287, 289, 290, 293, 296, 336
Diplodia, 430
Dispersers, 342, 456, 516
Dissipation, 120, 362, 363, 365–367, 369, 371, 372, 376, 378, 381, 383
Dormancy, 471, 479, 482, 517
Drought, 3, 5–7, 81, 83, 107–109, 118–122, 124–128, 130, 137, 138, 142, 144, 146–150, 152, 153, 157, 159, 161, 162, 171, 172, 174, 196–200, 208, 209, 219, 244, 248, 262–264, 282, 285, 289, 293, 295, 296, 304, 307, 308, 315, 318, 319, 331, 332, 334, 336–339, 346–348, 350, 365, 366, 369, 374–381, 383, 396, 399, 402, 408, 419–430, 432–437, 439, 440, 454, 461–463, 466, 470, 472–475, 478–484, 487–490, 514–517, 519, 522–524, 526, 527, 529, 531
Drought-induced embolism, 171, 173, 174, 176, 263, 289, 316
Drought-stress, 328, 361, 374–377
Drought tolerance, 107, 120, 121, 123, 127, 130, 174, 208, 363, 426, 470, 475, 481
Dryophyllum, 43, 47, 63, 66, 80
Dry season severity, 122
Dyrana, 48

E
Earlywood, 280, 423
Ecotypic variation, 316
Electrolyte leakage, 113, 469, 470
Electron transport, 305, 311, 344, 365, 366, 370, 372
Energy, 120, 162, 163, 170, 214, 216, 217, 304, 329, 330, 334, 346, 362, 363, 365–367, 369, 371, 372, 376, 378, 382, 383, 395, 423, 428, 430, 462, 513, 515, 516, 518, 521
Environmental regulation, 3, 79, 147, 148, 216, 241, 246, 308, 329, 333, 338, 344, 373, 399, 403, 406, 413, 435, 453, 456, 478, 484, 488, 513, 515, 516, 523, 527

Eocene–Oligocene boundary, 53, 55
Eotrigonobalanus, 47, 48, 66
Epicuticular waxes, 155, 167, 168
Epidermal structures, 31, 160, 167
Epidermis, 170, 209, 210, 363
Epoxide-lutein cycle (LxL-cycle), 366, 367, 369, 376, 377, 383
Eurasia, 5, 14, 16, 20, 25, 27, 28, 31, 39–41, 63, 70, 81–87, 145, 315, 433
Europe, 3, 4, 7, 14, 29–32, 43, 46–48, 61, 62, 65, 67, 69, 70, 72–74, 76, 81–84, 86, 87, 124, 139, 141, 145, 148, 161, 197–199, 309, 310, 314, 318, 334, 370, 374–376, 379, 382, 384, 420, 430, 431, 433, 435, 436, 440, 472, 481, 513–517, 521–525, 527–531
Evergreen, 4, 6, 7, 14, 16, 17, 21, 24, 27, 32, 48, 52, 53, 61, 67, 69, 70, 72, 74, 78–80, 82, 83, 85–87, 111, 118, 119, 122–125, 137, 140, 144–149, 151–162, 164–176, 195–198, 201–203, 205–212, 214–217, 219, 261, 264, 265, 271, 274, 284, 285, 289, 290, 292–294, 296, 303, 308–315, 318, 319, 329, 331, 333, 336, 338, 340, 343, 348, 363–365, 369–372, 374–379, 383, 395, 398, 399, 408, 420, 456, 459, 466, 468, 470, 477, 479, 481, 517, 524
Evolution, 6, 17, 39, 42, 44, 87, 108, 109, 113, 115, 118, 123, 126, 128, 130, 146, 174, 239, 240, 255, 308, 342, 343, 461, 477

F
Fagaceae, 4, 5, 17–19, 30, 32, 39–41, 46, 47, 50–52, 59, 63, 69, 274, 281, 457
Fagales, 47
Fagopsiphyllum, 48
Fagus, 26, 62, 63, 65, 317, 362, 457, 471, 521–523
Fertilization, 241, 246, 247, 249, 440, 470, 471, 473–478, 482
Fitness, 112, 119, 126, 240, 251
Flora, 16, 17, 21, 44, 45, 48, 50–54, 56, 58–61, 72, 74–77, 79–82, 144, 145, 151, 152, 154, 157, 171
Foliage, 30, 31, 44, 45, 51, 58, 303, 307, 312, 316, 317, 319, 339, 341, 343, 346, 382, 398, 420, 429, 434, 440, 469
Fossil, 5, 29–31, 40, 42–46, 52, 53, 55, 57, 62, 64, 66, 70, 74–80, 85, 87, 515
Fossil record, 14, 30–32, 40, 42–45, 49, 50, 54, 77, 78, 85–87
Freeze-induced embolism, 273, 289

Freezing tolerance, 107, 112–116, 120, 130, 371, 466, 478, 479

G
Gall, 45, 310, 318, 369, 375, 376, 378, 380, 381
Gas exchange, 6, 126, 206, 315, 328, 347, 428, 429, 469, 480, 490
Gas-phase conductance, 313
Geneflow, 87, 244
Genetically-based variation, 109
Germination, 246–249, 456, 460, 461, 464–467
Global, 16, 17, 46, 81, 108, 123, 137, 141, 145, 149, 155, 161, 172, 178, 264, 272, 281, 291–293, 306–308, 328, 330, 331, 383, 420, 421, 427, 430, 435, 471, 513, 514, 517, 518, 528, 529
Glutathione (GSH), 366, 369, 373, 378
Green Leaf Volatiles, 341
Growth respiration, 330, 337

H
Hardening, 365, 377, 475, 478–482
Hardiness, 460, 470, 479, 482, 522
Heat, 162, 198, 346, 362, 366, 374, 376, 379–381
Heat waves, 140, 350, 379–381, 421, 529
Hemisphere, 1, 4, 5, 7, 23, 30, 39, 40, 43, 81, 85–87, 137, 138, 140, 141, 144–146, 198, 309, 329, 363, 370, 420, 455, 514
Herbarium, 32, 44, 114, 178
Heterobalanus, 16, 20, 27, 42, 78, 79, 86
Historical classification, 14, 16, 17, 20, 28, 151, 264, 339
Holm oak, 69, 124, 125, 241, 242, 245–247, 249, 252, 383, 396, 460, 461, 489, 490, 528
Huber value, 264, 271, 291, 292, 294, 295
Hybridization, 42, 239–248, 251, 255
Hydraulic conductance, 118, 262, 437, 462, 464, 473, 480
Hydraulic safety-efficiency, 264, 274, 291, 294
Hydraulic safety margin, 294

I
Iberian Peninsula, 2, 62, 124, 125, 141, 143, 147, 148, 173, 174, 199, 206, 243, 247, 373, 395–397, 399, 409, 433, 521
Ilex, 2, 4–7, 13, 14, 16–18, 20, 27, 29, 31, 32, 42, 48, 49, 61, 62, 64, 65, 69, 70, 73, 78, 80, 82, 83, 86, 124, 125, 143, 145–147, 149–153, 156, 160–162, 164–166, 168, 170, 173, 174, 176, 206, 211, 215, 239, 241–252, 254, 255, 262, 263, 265, 268, 272, 274–277, 282, 287, 288, 307–310, 314–319, 330, 331, 333, 334, 336–338, 340, 343, 345, 346, 363–367, 369–371, 373–376, 378, 380, 381, 383, 395, 396, 399, 403, 407, 409, 411, 420, 428, 431, 432, 434–436, 440, 456, 458–466, 469, 470, 472–482, 485–489, 517, 519–523, 527, 528, 530
Index of moisture, 120–127
Intraspecific variation, 109, 118, 319
Introgression, 6, 239–241, 243–246, 252, 255, 343
Isohydric, 425
Isoprene, 341–344, 347, 348, 381
Isoprenoids, 341, 343, 344, 346, 348

K
Köppen, Wladimir Peter, 138, 177

L
Land use, 140, 420, 422, 424, 513, 516
Latewood, 49, 423, 428, 429, 437
Latitude, 14, 19, 20, 31, 32, 113, 139, 142, 196, 440
Laurophyllous, 67, 82, 144
Leaf abscission, 113, 121, 123, 126, 127, 200, 382
Leaf age, 205, 319
Leaf anatomical traits, 303, 313, 315, 317
Leaf angle, 363
Leaf area, 117, 118, 124, 147, 148, 150, 151, 153–155, 161, 162, 164, 171, 197, 200, 202–204, 209, 218, 264, 283, 291, 292, 295, 313, 373, 397–399, 402, 410, 425, 428, 429, 455, 456
Leaf area ratio (LAR), 397, 399, 401, 402, 413, 425
Leaf dark respiration, 331
Leaf density, 121, 157, 203, 208–211, 313
Leaf economics spectrum, 7, 201, 202, 303, 306, 307, 319
Leaf emissions, 339
Leaf habit, 7, 118, 147, 153, 176, 196, 202, 206, 208, 210, 213–215, 264, 398, 400
Leaf life span, 6, 124, 127, 166, 196, 202, 204–207, 218, 308, 319, 333
Leaf mass fraction (LMF), 397, 398, 401, 402, 404, 406, 409
Leaf mass per area (LMA), 7, 149–161, 195, 197, 202, 203, 207–215, 218, 219, 303, 306–309, 311–316, 428, 519, 524
Leaf morphology, 28, 43, 70, 121, 122, 125, 126, 150, 161, 219

Index 543

Leaf nitrogen, 124, 156, 331, 397, 411, 412
Leaf per area ratio (LAR), 397–399, 401, 402, 407
Leaf phenology, 198, 261, 264, 265, 271, 273, 274, 284, 289, 290, 293, 294, 296, 310, 527
Leaf reddening, 364
Leaf shedding, 147, 148, 420, 423, 425, 429
Leaf thickness, 120, 121, 152–157, 161, 207, 209–211, 307, 313, 317
Light absorption, 304, 362, 380
Light availability, 196, 304, 317, 319, 439, 440
Light harvesting, 363, 369, 439, 516, 527
Light scattering, 51, 53, 59, 196, 217, 305, 317, 344, 364, 366, 371, 489, 521
Lignin, 120, 202, 203, 212, 214, 252, 280
Lineage, 5, 29, 30, 62, 107, 109, 110, 115, 242–244, 289, 371
Liquid-phase conductance, 313
Lithocarpoxylon, 53
Lithocarpus, 16, 18, 41, 43, 45, 46, 50, 52, 53, 55, 57, 264
Lobatae, 5, 13, 14, 16–18, 20, 25, 28, 30–32, 42, 43, 45, 46, 49, 51–55, 57–59, 61–63, 65, 67, 69, 70, 73, 81, 82, 84, 86, 145, 261, 264, 266–271, 273, 282–285, 289, 290, 292, 295, 310, 420, 465
Lobe, 76
Loudon, John Claudius, 13
Lutein (L), 1, 4, 6, 21, 25, 27, 28, 39–42, 46, 55, 59, 70, 72, 73, 76, 77, 82, 87, 145, 240, 241, 252, 253, 366, 367, 369, 372, 383, 433, 485, 486
Lutein epoxide (Lx), 7, 366, 367, 369
Lymantria dispar, 433

M

Macrobioclimate, 139
Maintenance costs, 195, 201, 212, 214, 215, 217
Maintenance respiration, 215–217, 330
Malformations, 250
Mean annual precipitation (MAP), 82, 83, 110, 172, 403, 407–410
Mean annual temperature (MAT), 81–83, 141, 398, 403, 407–410, 529
Mediterranean Basin, 1, 137, 139–141, 143, 145–147, 149, 173, 198, 201, 241, 242, 370, 376, 472, 481
Mediterranean seasonality, 74
Mediterranean-type climate, 2, 138–140, 142, 143, 145–149, 151, 152, 154, 155, 173, 460
Mehler reaction, 366

Meliosma, 48
Menitsky, Yuri Leonárdovich, 13
Mesophyll, 6, 7, 79, 156–161, 168, 209, 210, 212, 303–306, 308–318, 331, 334, 364
Mesophyll conductance (g_m), 305, 311, 314
Mesophyll surface area exposed intercellular air spaces (S_m/S), 303, 306
Mesophytic, 64, 67, 69, 72, 83, 152
Messinian, 83, 84
Metabolic acclimation, 115, 201, 329, 336, 341, 343, 362, 369, 429, 466
MicroCT, 288
Microsatellite, 245
Microsite, 439, 483
Minimum leaf water potential, 293
Miocene, 30, 31, 41, 43, 45, 50, 54–61, 63–65, 67, 69–74, 77–84, 87
Mio-Pliocene, 59, 61
Molecular systematics, 13, 16, 19, 128, 242, 367
Monoterpenes, 342–344
Montado, 3, 241
Morphogenus, 53, 66, 74
Morphological traits, 14, 19, 33, 241, 246, 468, 481
Morphotypes, 31, 49, 73, 75, 80
mtDNA, 243
Mycorrhizal, 4, 348, 430, 434, 435, 440

N

Nemoral, 150, 151, 161, 309, 318, 374, 375, 379
Neogene, 30–32, 44, 50, 53–55, 57, 59, 69, 70, 72–74, 77–80, 82, 84, 86, 87
Net assimilation rate (NAR), 312, 397, 399, 401, 407
Net CO_2 assimilation (A_N), 309, 319, 374, 432
Net primary production, 206, 328, 349
Nitrogen (N), 21, 62, 81, 87, 124, 153, 156, 157, 172, 202, 203, 207, 208, 214, 215, 273, 284, 288–290, 292, 293, 306–308, 310, 312, 317, 331, 333, 339, 397, 411, 412, 425, 427, 434, 435, 439, 440, 456, 458–460, 470, 473–478, 480, 482, 519, 521
Nixon, Kevin C., 13
Non-photochemical quenching (NPQ), 366, 367, 369, 371, 376, 377, 381, 383
Non structural carbohydrates (NSC), 419, 425–429, 434, 438, 481
Nursery, 8, 440, 466, 471–474, 477–479, 481–484
Nutrient, 117, 119, 124, 152, 202, 216, 308, 330, 331, 333, 348, 382, 399, 406, 407,

409, 413, 421, 425, 430, 435, 440, 458, 461, 462, 467–470, 472, 474, 476, 478–481, 484, 522, 524

O

Oak, 2–8, 14, 17–19, 28–30, 32, 33, 44, 46, 49, 52–55, 63, 65, 67, 69, 72–76, 79–84, 86, 87, 108–110, 116–118, 120, 123–125, 128, 129, 137, 140–147, 149, 151, 153–157, 159, 161, 162, 164, 165, 167–170, 174–176, 195, 197–200, 205–209, 211, 215, 240–242, 245–249, 251–253, 255, 263, 265, 272–276, 281–283, 285, 287–296, 303, 307, 309, 311–313, 319, 329–331, 333–340, 342–344, 346–350, 362, 370, 372–374, 376, 378–381, 383, 384, 395, 396, 420, 421, 423–440, 453–457, 459–461, 464, 466–473, 476, 479, 481, 482, 489, 490, 492, 513–531
Oligocene, 29–31, 41, 43, 50, 51, 53–55, 57, 59, 61, 63–70, 75–77, 80, 82, 84, 86, 87
Open vessel artefact, 286
Operophtera brumata, 434
Ørsted, Anders Sandøe, 13
Osmoregulation, 127
Osmotic adjustment, 127, 156, 427, 480, 481
Outplanting, 467, 468, 470, 471, 473–476, 478, 479, 481, 482, 487
Overcrowded, 419, 423
Overstory, 439, 440
Oxidative damage, 362, 367, 377, 378, 380

P

Palaeoclimate, 83
Palaeotropical, 5, 6, 70
Paleocene, 5, 43, 47–49, 62, 85
Paleogene, 30–32, 39, 40, 44, 49, 50, 53, 54, 70, 73, 82, 84, 86, 87
Palynology, 16, 44, 49, 62, 65, 73, 75, 84
Paraquercus, 47
Pathogen, 252, 281, 282, 342, 421, 424, 431, 432, 434
Pay-back time, 217
Phellem, 251–254
Phellogen, 251–253
Phenology, 118, 197–199, 246, 247, 261, 264–266, 271, 273, 274, 284, 289, 290, 292–294, 296, 310, 336, 436, 456, 527
Phenotype, 250, 365
Phenotypic, 126, 127, 241, 242, 250, 251, 420, 422, 424, 425, 430, 514, 517
Phloem, 251, 262, 426
Phosphorus, 151, 152, 348, 470, 473

Photoinhibition, 7, 364, 376–378, 423, 428, 439
Photoprotection, 120, 362–366, 370–372, 374–376, 379, 381–383
Photosynthesis, 6, 114, 120, 123, 125, 148, 155, 166, 195–199, 201, 203, 208, 217, 218, 263, 303–305, 307, 308, 312, 315, 318, 319, 328, 330, 344–347, 349, 350, 365, 370–372, 374, 382, 398, 409–411, 423, 434, 436–440, 460, 468, 470, 474, 485
Photosynthetic limitations, 303, 309, 314, 334, 428
Photosystem II (PSII), 366, 367, 371, 372, 377, 428, 469
Phylogenomics, 28
Phytophthora cinnamomi, 3, 282, 435
Phytophthora quercina, 282
Phytophthora ramorum, 431
Plant functional types, 7, 82, 202, 205–207, 329
Planting, 439, 454–456, 460–464, 467–472, 474, 479, 482–488, 490, 492
Plasticity, 109, 112, 113, 115, 126, 127, 130, 167, 198, 246, 315, 317, 319, 333, 362, 367, 369, 424, 426, 439, 460, 481, 517
Plastic variation, 319
Platypus cylindrus, 433
Platypus quercivorus, 433
Pliocene, 30, 49, 50, 55, 58–61, 67, 69, 70, 72–74, 76, 79–82, 84, 86, 87
Pollen, 5, 14, 21–32, 39, 41, 43, 44, 46–52, 54, 57–65, 73–75, 78–80, 85, 244–248, 515
Pollination, 32, 246, 247
Population, 1, 2, 29, 109, 111–116, 118, 122–128, 130, 153, 242, 244, 245, 247, 251, 286, 343, 371, 374, 379, 381, 453, 513–516, 523, 528–530
Post-zygotic barriers, 245, 246, 248
Potassium, 470
Precipitation, 5, 82, 83, 107–110, 118–126, 130, 137–142, 144–146, 148, 172, 201, 242, 272, 379, 395, 398, 403, 406–410, 413, 485, 515, 529
Pre-zygotic Barriers, 246
Prinus, 14, 16, 17, 24, 76, 77, 274, 276, 337
Protobalanus, 5, 13, 14, 16–19, 23, 30, 32, 42, 43, 49, 52–55, 57, 62, 81, 84, 86, 264, 266, 283, 310
Provenance, 374, 377, 479
Pruning, 438, 439, 462

Q

Quaternary, 40, 43, 44, 85, 144

Index

Quercinium, 52, 53, 55, 62, 69, 74, 87
Quercoidites, 48, 51, 62, 64, 73, 75, 78, 84, 85
Quercophyllum, 43, 47, 85
Quercopollenites, 59, 62, 64, 65, 73, 78
Quercoxylon Savanna, 53, 55, 62, 74, 87
Quercus, 1–8, 13, 14, 16–32, 39–63, 65–70, 72–87, 108, 109, 117, 118, 121–124, 128, 129, 137, 139, 140, 143–148, 150–153, 157, 158, 160–164, 167, 168, 170, 171, 173, 174, 176, 197, 200, 207–211, 213, 216, 217, 239–242, 247, 251–253, 255, 261–284, 287–292, 295, 303–305, 307–316, 318, 319, 329, 337, 339, 347, 362–367, 369–371, 373–377, 379–381, 383, 395–411, 413, 420, 426, 431, 433, 455, 460–466, 470–477, 479–482, 484–486, 488–490, 517, 519–523, 527, 528, 530
Quercus section Cerris, 27, 28
Quercus section Cyclobalanopsis, 5, 59
Quercus section Ilex, 27
Quercus section Lobatae, 25
Quercus section Ponticae, 23
Quercus section Protobalanus, 23
Quercus section Quercus, 59
Quercus section Virentes, 5, 24, 108, 109
Quercus subgenus Cerris, 26
Quercus subgenus Quercus, 78

R
Raffaelea quercivora, 433
Rainfal, 108, 118, 138, 142, 424
Reactive oxygen species (ROS), 114, 216, 365–367, 369, 373, 374, 376–379, 382
Reciprocal transplant, 107, 119
Recruitment, 8, 420, 439, 455–457, 464, 516, 517
Regeneration, 2, 8, 305, 329, 369, 373, 408, 421, 437, 439, 440, 453–457, 460, 463, 487, 519, 521, 525, 531
Relative growth rate (RGR), 7, 126, 394–403, 405–413
Resistance, 108, 109, 118, 122, 125, 126, 137, 154, 156, 157, 159, 161, 162, 168, 171–173, 175, 200, 207, 208, 212, 244, 275, 283, 285, 286, 288, 295, 306, 313, 315, 316, 331, 338, 342, 344, 425, 426, 430, 434, 470, 480, 482, 483, 522, 524
Respiration, 6, 7, 216–218, 305, 315, 329–332, 335, 337, 340, 349, 350, 382, 409, 411, 426, 428, 429, 466, 470
Respiration-photosynthesis ratio, 215, 332

Respiratory alternative oxidase resprouting, 216
Restoration, 178, 239, 428, 454, 460, 467, 478, 483, 484, 487, 488, 492, 516
Rhizodeposition, 328, 329, 348–350
Rhytidome, 251–253
Ring-porous, 6, 21, 26, 28, 46, 52, 261, 262, 264, 266–272, 275, 279–281, 284–290, 293, 296, 336, 423
Root, 7, 29, 118, 215, 216, 262, 288–290, 335–339, 348, 349, 398, 399, 402, 413, 423–430, 432, 434, 435, 438, 440, 453, 454, 456–464, 466, 468–476, 478, 480, 482–487, 515, 521
Root exudates
Root mass fraction (RMF), 398, 400–404, 406
Rubisco, 304, 305, 308, 312, 460

S
Schimper, Andreas Franz Wilhelm, 40, 43, 146, 150, 152
Schwarz, Otto Karl Anton, 13
Sclerophyllodrys, 16, 61
Sclerophyllous, 6, 16, 63, 67, 69, 70, 72, 73, 78, 79, 83, 86, 120, 121, 137, 144–146, 149–152, 155, 161, 196, 197, 239, 240, 316, 331, 333, 515, 517–520, 522
Scolytus intricatus, 433
Seasonally Dry Tropical Forest (SDTF), 118, 119, 125
Seed, 24, 27, 32, 342, 399, 400, 453, 455–457, 460, 464, 466, 467, 489, 516, 528
Seeding, 279, 454, 460–464, 466, 467, 490
Seedling, 7, 8, 119, 202, 248–250, 399, 412, 420, 439, 440, 453–456, 460–462, 464, 466–479, 481–485, 487, 488
Seed mass, 398, 399, 402, 519
Semideciduous, 205, 207, 213–215
Senescence, 121, 199, 318, 364, 381–383
Shelter, 454, 456, 488, 490
Shoot, 118, 147, 164, 171, 172, 249, 250, 423, 424, 426, 427, 429, 438, 454, 456, 460, 462, 464, 468, 477–480, 485, 487, 521
Sink strength
Soil, 2, 3, 6, 117–121, 140, 146–148, 152, 163, 164, 166, 170, 196, 197, 200, 201, 244, 264, 294, 318, 328, 334–336, 338–340, 344, 346, 348, 349, 380, 399, 403, 407, 409, 413, 419, 421–424, 426, 429, 432, 434, 437, 439, 440, 454, 456, 458, 460–463, 470, 472, 473, 476–478, 483–485, 490, 519, 521, 524

Sowing, 249, 439, 440, 477
Spanish National Forest Inventory, 397
Specific leaf area (SLA), 117, 119–122, 124–126, 331, 397–402, 407, 409, 411–413, 456, 517, 519, 521, 527
Stagnation, 420, 423, 459
Starch, 371, 424, 458, 470, 482
Starvation, 426, 523
Static centrifuge techniques, 288
Stem mass fraction (SMF), 402–404, 406
Stomata, 67, 155, 164, 165, 167–171, 305, 307, 313, 344, 427, 517
Stomatal conductance (g_s), 163, 209, 306, 311
Submediterranean, 147, 148
Subtropical, 4, 5, 13, 16, 27, 41, 48, 75, 79, 80, 82, 85, 87, 139, 142, 265, 329
Summer, 3, 5, 6, 70, 81, 82, 123, 124, 137–150, 152, 153, 155, 161, 168, 170, 171, 177, 196–201, 208, 209, 212, 215, 216, 263, 280, 318, 336, 338, 346, 366, 367, 374–380, 383, 408, 423, 429, 440, 459, 461, 463, 475, 479, 485, 488–490, 529
Superoxide dismutases (SODs), 369
Survival, 1, 112, 119, 148, 171, 195, 198, 208, 239, 249, 274, 329, 330, 397, 399, 423–425, 436, 439, 440, 454, 456, 458, 459, 462, 466–474, 476, 478, 482–485, 487–490, 492
Sustained dissipation, 372

T
Taproot, 454, 460–462, 473
Teeth, 24, 26, 49, 67, 76–78, 80
Temperate, 4–6, 14, 28, 29, 41, 46, 48, 80–82, 86, 87, 107–109, 111–117, 123, 124, 139, 142, 143, 145, 147, 149, 151, 153–165, 167, 169, 171–177, 198, 199, 241, 261, 264, 265, 271, 273, 281, 283–285, 290, 293, 294, 296, 309, 310, 313, 315, 318, 329, 339, 340, 348, 374, 375, 379, 384, 433, 435, 436, 469, 471–473, 479, 481, 488, 490, 513, 515, 517, 521, 522, 524, 525, 527
Temperature, 5, 80–83, 107–110, 112–114, 116, 123, 130, 138, 139, 141, 146, 149, 152, 162, 163, 196–198, 216, 242, 306, 330–339, 342, 344–347, 364, 365, 371, 372, 380, 395, 398, 403, 406–410, 413, 420, 436, 454, 457, 466, 472, 475, 478, 479, 482, 484, 485, 488–490, 522, 529
Temperature sensitivity, 336, 337
Terpenoids, 280, 341
Tertiary, 44, 77–79, 84, 144, 145

Thermophilous, 67, 73, 82, 83, 242
Thinning, 3, 403, 437–439, 455, 523
Timber, 4, 423
Tocopherol, 373
Tolerance, 3, 107, 108, 112–116, 120, 121, 123, 127, 130, 161, 166, 174, 208, 248, 263, 308, 362, 363, 371, 375, 378, 380, 383, 384, 396, 402, 426, 427, 429, 466, 470, 471, 474, 475, 478–481, 490, 519, 521, 527, 528
Trade-off, 108, 113, 116, 127, 158, 207, 261, 283, 290, 291, 295, 306, 307, 317, 411, 430, 515, 519, 521, 527
Trait, 17, 20, 40, 111, 129, 130, 150, 168, 171, 173, 195, 197, 202, 203, 207, 208, 265, 274, 307, 308, 313, 316, 399, 430, 469, 524
Transition, transitional, 48, 62, 143, 144, 148, 169, 513, 517, 520, 524, 531
Transplanting, 119, 454, 464, 467–470, 473, 476, 482–485
Tree diameter, 402, 404
Tree height, 402, 403, 515, 519, 521
Tree rings, 423
Tree size, 7, 395, 398, 402, 406, 407, 409, 410, 412, 413, 523, 527
Trelease, William, 13
Trichomes, 23, 62, 67, 79, 87, 167–170, 242, 363, 364, 380
Trigonobalanus, 18, 41, 43, 61
Tropical, 5, 6, 13, 16, 27, 41, 46, 48, 72, 75, 80–82, 85, 87, 107–109, 111–120, 122–125, 139, 141, 153–157, 160–162, 169, 171, 176–178, 261, 264, 271, 273, 283, 285, 290, 293, 296, 309, 310, 371, 384
Truffle, 4
Tuber melanosporum, 4, 464
Turgor, 120, 123, 156, 157, 337, 426, 427, 481
Turgor loss point, 121, 123, 126, 127, 157, 158, 426, 480

V
Vascular tissue, 209–211
VAZ-cycle, 366, 367, 369, 383
Vegetative activity, 149, 152, 489
Vegetative period, 151
Vegetative season, 6, 21, 148, 154, 485
Verrutricolporites, 65
Vessel, 46, 59, 261–265, 271–277, 279, 280, 282, 294–296, 430
Vessel clustering, 245, 290
Vessel lumen, 265, 280
Viability, 2, 246, 248, 464, 465

Vigor, 420, 422, 423, 436–438, 464, 466, 467
Violaxanthin (V), 65, 130, 243, 275, 296, 366, 369, 377
Volatile Organic Compounds, 6, 7, 328, 339, 348, 349, 428

W

Walter, Heinrich Karl, 25, 138, 139, 146, 177
Water content, 146, 152, 166, 338, 423, 429, 432, 433, 471, 481
Water limitation, 109, 116, 117, 119, 376, 407
Waterlogging, 7, 421–423, 426
Water potential, 121, 123, 126, 127, 137, 156, 157, 161, 166, 174–176, 263, 271, 285, 287, 288, 291, 293, 295, 334, 347, 376, 377, 425–429, 440, 473, 480, 489, 490
Water status, 332, 336, 424, 440, 462, 487, 489
Water use efficiency (WUE), 121, 124, 307, 315, 469, 474, 480
Weakened, 423, 425, 432–434, 437, 438
Winter, 5–7, 81, 107, 108, 111–114, 123–125, 137–143, 146–149, 151–154, 158, 159, 162, 168, 174, 176, 177, 196, 198, 199, 201, 208, 212, 216, 248, 280, 310, 365–367, 369–374, 381, 383, 421, 423, 475, 478, 479, 482, 529
Wood density, 120, 261, 263–265, 271, 283–285, 295, 296, 409, 519, 524
Woodland, 3, 171
Wood structure, 281

X

Xantophylls, 7
Xeromorphic, 137, 167, 168, 170, 171
Xeromorphism, 150, 151
X-ray computed microtomography, 288
Xylem [CO_2], 6, 164, 261, 295, 410, 478
Xylem fibers
Xylem hydraulic conductivity
Xylem hydraulic efficiency, 272
Xylem hydraulic resistance, 279
Xylem hydraulic safety, 261, 285
Xylem intervessel pits, 275, 279
Xylem lumen fraction, 272–274, 295
Xylem P50, 171, 263, 264, 274, 285, 288, 291, 295
Xylem parenchyma, 282, 426
Xylem structure, 265, 283
Xylem tension, 171, 429
Xylem vessel density, 289
Xylem vessel diameter, 289, 433
Xylem vessel length
Xylem vulnerability curves, 175, 262, 285–288, 290, 294, 295
Xylem water potential, 285, 427, 428
Xylem water transport, 161, 169, 248, 262, 419, 425, 428, 481

Z

Zeaxanthin (Z), 7, 365–367, 369, 372, 374, 377
Zonobiome, 139, 143, 144

CPSIA information can be obtained
at www.ICGtesting.com
Printed in the USA
LVHW021722090619
620633LV00003B/26/P